Springer Proceedings in Mathematics & Statistics

Volume 324

Springer Proceedings in Mathematics & Statistics

This book series features volumes composed of selected contributions from workshops and conferences in all areas of current research in mathematics and statistics, including operation research and optimization. In addition to an overall evaluation of the interest, scientific quality, and timeliness of each proposal at the hands of the publisher, individual contributions are all refereed to the high quality standards of leading journals in the field. Thus, this series provides the research community with well-edited, authoritative reports on developments in the most exciting areas of mathematical and statistical research today.

More information about this series at http://www.springer.com/series/10533

Bruno Tuffin · Pierre L'Ecuyer
Editors

Monte Carlo and Quasi-Monte Carlo Methods

MCQMC 2018, Rennes, France, July 1–6

Springer

Editors
Bruno Tuffin
INRIA Rennes Bretagne-Atlantique
Campus Universitaire de Beaulieu
Rennes, France

Pierre L'Ecuyer
Département d'informatique et de Recherche
Opérationnelle (DIRO)
Université de Montréal
Montreal, QC, Canada

ISSN 2194-1009 ISSN 2194-1017 (electronic)
Springer Proceedings in Mathematics & Statistics
ISBN 978-3-030-43467-0 ISBN 978-3-030-43465-6 (eBook)
https://doi.org/10.1007/978-3-030-43465-6

Mathematics Subject Classification (2010): 65C05, 62D05, 62E10, 65B99, 62-06, 65Cxx, 62Pxx, 62Lxx

This Springer imprint is published by the registered company Springer Nature Switzerland AG
The registered company address is: Gewerbestrasse 11, 6330 Cham, Switzerland

Preface

This volume represents the refereed proceedings of the Thirteenth International Conference on Monte Carlo and Quasi-Monte Carlo Methods in Scientific Computing, which was held at the University of Rennes, France, and organized by Inria, from 1–6 July, 2018. It contains a limited selection of articles based on presentations made at the conference. The program was arranged with the help of an international committee consisting of the following members:

Zdravko I. Botev (Australia, University of New South Wales)
Hector Cancela (Uruguay, Universidad de la Republica)
Frédéric Cérou (France, Inria)
Nicolas Chopin (France, ENSAE)
Ronald Cools (Belgium, KU Leuven)
Josef Dick (Australia, University of New South Wales)
Arnaud Doucet (UK, Oxford University)
Paul Dupuis (USA, Brown University)
Michael B. Giles (UK, Oxford University)
Mark Girolami (UK, Uinversity of Warwick)
Paul Glasserman (USA, Columbia University)
Peter W. Glynn (USA, Stanford University)
Michael Gnewuch (Germany, Universität Kiel)
Emmanuel Gobet (France, Ecole Polytechnique)
Takashi Goda (Japan, The University of Tokyo)
Arnaud Guyader (France, Université Pierre et Marie Curie)
Stefan Heinrich (Germany, Universität Kaiserslautern)
Fred J. Hickernell (USA, Illinois Institute of Technology)
Aicke Hinrichs (Austria, JKU Linz)
Wenzel Jakob (Switzerland, ETH and Disney)
Alexander Keller (Germany, NVIDIA)
Dirk P. Kroese (Australia, University of Queensland)
Frances Y. Kuo (Australia, University of New South Wales)
Gerhard Larcher (Austria, JKU Linz)

Christian Lécot (France, Université de Savoie)
Pierre L'Ecuyer (Canada, Université de Montréal)
Christiane Lemieux (Canada, University of Waterloo)
Faming Liang (USA, University of Florida, Gainesville)
Makoto Matsumoto (Japan, Hiroshima University)
Eric Moulines (France, Ecole Polytechnique)
Thomas Mueller-Gronbach (Germany, Universität Passau)
Harald Niederreiter (Austria, Academy of Sciences)
Erich Novak (Germany, Universität Jena)
Dirk Nuyens (Belgium, KU Leuven)
Art B. Owen (USA, Stanford University)
Gareth Peters (UK, University College London)
Friedrich Pillichshammer (Austria, JKU Linz)
Klaus Ritter (Germany, Universität Kaiserslauten)
Gerardo Rubino (France, Inria)
Wolfgang Ch. Schmid (Austria, Universität Salzburg)
Ian H. Sloan (Australia, University of New South Wales)
Raul Tempone (Saudi Arabia, KAUST)
Xiaoqun Wang (China, Tsinghua University)
Grzegorz W. Wasilkowski (USA, University of Kentucky)
Henryk Wozniakowski (USA, Columbia University).

This conference continued the tradition of biennial MCQMC conferences initiated by Harald Niederreiter. They were begun at the University of Nevada in Las Vegas, Nevada, USA, in June 1994 and followed by conferences at the University of Salzburg, Austria, in July 1996; the Claremont Colleges in Claremont, California, USA, in June 1998; Hong Kong Baptist University in Hong Kong, China, in November 2000; the National University of Singapore, Republic of Singapore, in November 2002; the Palais des Congrès in Juan-les-Pins, France, in June 2004; Ulm University, Germany, in July 2006; Université de Montréal, Canada, in July 2008; University of Warsaw, Poland, in August 2010; the University of New South Wales, Sydney, Australia, in February 2012; KU Leuven, Belgium, in April 2014; and Stanford University, USA, in August 2016. The next MCQMC conference will be held in Oxford, UK, on August 9–14, 2020.

The proceedings of these previous conferences were all published by Springer-Verlag, under the following titles:

- *Monte Carlo and Quasi-Monte Carlo Methods in Scientific Computing* (H. Niederreiter and P. J.-S. Shiue, eds.)
- *Monte Carlo and Quasi-Monte Carlo Methods 1996* (H. Niederreiter, P. Hellekalek, G. Larcher and P. Zinterhof, eds.)
- *Monte Carlo and Quasi-Monte Carlo Methods 1998* (H. Niederreiter and J. Spanier, eds.)
- *Monte Carlo and Quasi-Monte Carlo Methods 2000* (K.-T. Fang, F. J. Hickernell and H. Niederreiter, eds.)

- *Monte Carlo and Quasi-Monte Carlo Methods 2002* (H. Niederreiter, ed.)
- *Monte Carlo and Quasi-Monte Carlo Methods 2004* (H. Niederreiter and D. Talay, eds.)
- *Monte Carlo and Quasi-Monte Carlo Methods 2006* (A. Keller and S. Heinrich and H. Niederreiter, eds.)
- *Monte Carlo and Quasi-Monte Carlo Methods 2008* (P. L'Ecuyer and A. Owen, eds.)
- *Monte Carlo and Quasi-Monte Carlo Methods 2010* (L. Plaskota and H. Woźniakowski, eds.)
- *Monte Carlo and Quasi-Monte Carlo Methods 2012* (J. Dick, F. Y. Kuo, G. W. Peters and I. Sloan, eds.)
- *Monte Carlo and Quasi-Monte Carlo Methods 2014* (R. Cools and D. Nuyens, eds.)
- *Monte Carlo and Quasi-Monte Carlo Methods 2016* (A. Owen and P. W. Glynn, eds.).

The program of the conference was rich and varied with over regular 190 talks being presented and more than 230 registered participants. Highlights were the invited plenary talks given by Christophe Andrieu (Bristol, UK), Pierre Henry-Labordère (Société Générale, Paris, France), Éric Moulines (Ecole Polytechnique, France), Marvin Nakayama (NJIT, USA), Barry L. Nelson (Northwestern University, USA), Friedrich Pillichshammer (JKU Linz, Austria), Clémentine Prieur (Université Grenoble Alpes & Inria, France), and Christoph Schwab (ETH Zurich, Swizerland).

The papers in this volume were carefully screened and cover both the theory and the applications of Monte Carlo and quasi-Monte Carlo methods. We thank the anonymous reviewers for their reports and many others who contributed enormously to the excellent quality of the conference presentations and to the high standards for publication in these proceedings by careful review of the abstracts and manuscripts that were submitted.

We gratefully acknowledge generous financial support of the conference by Inria, the University of Rennes 1, Région Bretagne, and Rennes Métropole.

Finally, we want to express our gratitude to Springer-Verlag for publishing this volume.

Rennes, France
Montreal, Canada
September 2019

Pierre L'Ecuyer
Bruno Tuffin

Contents

Part I
Invited Talks

A Tutorial on Quantile Estimation via Monte Carlo

Hui Dong and Marvin K. Nakayama

Abstract Quantiles are frequently used to assess risk in a wide spectrum of application areas, such as finance, nuclear engineering, and service industries. This tutorial discusses Monte Carlo simulation methods for estimating a quantile, also known as a percentile or value-at-risk, where p of a distribution's mass lies below its p-quantile. We describe a general approach that is often followed to construct quantile estimators, and show how it applies when employing naive Monte Carlo or variance-reduction techniques. We review some large-sample properties of quantile estimators. We also describe procedures for building a confidence interval for a quantile, which provides a measure of the sampling error.

Keywords Percentile · Value-at-risk · Variance-reduction techniques · Confidence intervals

1 Introduction

Numerous application settings have adopted quantiles as a way of measuring risk. For a fixed constant $0 < p < 1$, the p-quantile of a continuous random variable is a constant ξ such that p of the distribution's mass lies below ξ. For example, the median is the 0.5-quantile. In finance, a quantile is called a *value-at-risk*, and risk managers commonly employ p-quantiles for $p \approx 1$ (e.g., $p = 0.99$ or $p = 0.999$) to help determine capital levels needed to be able to cover future large losses with high probability; e.g., see [33].

Hui Dong—This work is not related to Amazon, regardless of the affiliation.

H. Dong
Amazon.com Corporate LLC, Seattle, WA 98109, USA
e-mail: huidong@amazon.com

M. K. Nakayama (✉)
Computer Science Department, New Jersey Institute of Technology,
Newark, NJ 07102, USA
e-mail: marvin@njit.edu

B. Tuffin and P. L'Ecuyer (eds.), *Monte Carlo and Quasi-Monte Carlo Methods*,
Springer Proceedings in Mathematics & Statistics 324,
https://doi.org/10.1007/978-3-030-43465-6_1

Nuclear engineers use 0.95-quantiles in *probabilistic safety assessments* (PSAs) of nuclear power plants. PSAs are often performed with Monte Carlo, and the U.S. Nuclear Regulatory Commission (NRC) further requires that a PSA accounts for the Monte Carlo sampling error; e.g., see [50], Sect. 3.2 of [49], and Sect. 24.9 of [51]. This can be accomplished by providing a confidence interval for ξ.

Quantiles also arise as risk measures in service industries. For out-of-hospital patient care, a 0.9-quantile is commonly employed to assess response times of emergency vehicles and times to transport patients to hospitals [5]. In addition, [20] examines the 0.9-quantile of customer waiting times at a call center.

This tutorial discusses various Monte Carlo methods for estimating a quantile. Section 2 lays out the mathematical setting. In Sect. 3 we outline a general approach for quantile estimation via Monte Carlo, and illustrate it for the special case of naive Monte Carlo (NMC). We examine large-sample properties of quantile estimators in Sect. 4. Section 5 shows how the basic procedure in Sect. 3 can also be used when employing *variance-reduction techniques* (VRTs), which can produce quantile estimators with smaller sampling error than when NMC is applied. We describe different methods for constructing confidence intervals for ξ in Sect. 6.

2 Mathematical Framework

Consider the following example, which we will revisit throughout the paper to help illustrate ideas and notation. The particular stochastic model in the example turns out to be simple enough that it can actually be solved through a combination of analytical and numerical methods, making Monte Carlo simulation unnecessary. But the tractability allows us to compute exact quantiles, which are useful for our numerical studies in Sects. 5.7 and 6.4 comparing different Monte Carlo methods. Larger, more complicated versions of the model are usually analytically intractable.

Example 1 (*Stochastic activity network (SAN)*) A contractor is preparing a bid to work on a project, such as developing a software product, or constructing a building. She wants to determine a time ξ to use as the bid's promised completion date so that there is a high probability of finishing the project by ξ to avoid incurring a penalty. To try to figure out such a ξ, she builds a stochastic model of the project's duration.

The project consists of d activities, numbered $1, 2, \ldots, d$. Certain activities must be completed before others can start, e.g., building permits must be secured prior to laying the foundation. Figure 1, which has been previously studied in [13, 15, 29, 47], presents a directed graph that specifies the precedence constraints of a project with $d = 5$ activities. The nodes in the graph represent particular epochs in time, and edges denote activities. For a given node v, all activities corresponding to edges into v must be completed before starting any of the activities for edges out of v. Hence, activity 1 must finish before beginning activities 2 and 3. Also, activity 5 can commence only after activities 3 and 4 are done.

Fig. 1 A stochastic activity network with $d = 5$ activities

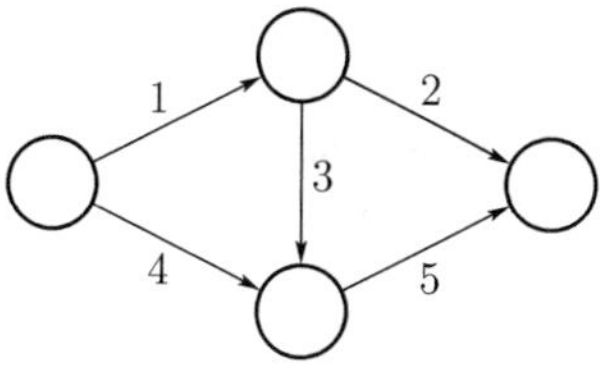

For each $j = 1, 2, \ldots, d$, activity j has a random duration X_j, which is the length of edge j and has marginal *cumulative distribution function* (CDF) G_j, where each G_j is an exponential distribution with mean 1; i.e., $G_j(x) = \mathsf{P}(X_j \le x) = 1 - e^{-x}$ for $x \ge 0$, and $G_j(x) = 0$ for $x < 0$. We further assume that $X_1, X_2, \ldots, X_d$ are mutually independent. The (random) time Y to complete the project is then the length of the longest path from the source, which is the leftmost node in Fig. 1, to the sink, the rightmost node. The graph has $r = 3$ paths from source to sink,

$$\mathscr{P}_1 = \{1, 2\}, \quad \mathscr{P}_2 = \{4, 5\}, \quad \mathscr{P}_3 = \{1, 3, 5\}; \tag{1}$$

e.g., path $\mathscr{P}_3$ consists of activities 1, 3, and 5. For each $k = 1, 2, \ldots, r$, let $T_k = \sum_{j \in \mathscr{P}_k} X_j$ be the (random) length of path $\mathscr{P}_k$. Thus,

$$Y = \max_{k=1,2,\ldots,r} T_k = \max(X_1 + X_2,\ X_4 + X_5,\ X_1 + X_3 + X_5) \tag{2}$$

represents the project's completion time, and we denote its CDF by F. □

More generally, consider a (complicated) stochastic model, and define P and E as the probability measure and expectation operator, respectively, induced by the model. Let Y be an $\Re$-valued output of the model representing its random performance or behavior, and define F as the CDF of Y, i.e.,

$$F(y) = \mathsf{P}(Y \le y) = \mathsf{E}[I(Y \le y)] \text{ for each } y \in \Re, \tag{3}$$

where $I(\cdot)$ denotes the indicator function, which takes value 1 (resp., 0) when its argument is true (resp., false). For a fixed constant $0 < p < 1$, define the p-quantile ξ of F as the generalized inverse of F; i.e.,

$$\xi = F^{-1}(p) \equiv \inf\{y : F(y) \ge p\}. \tag{4}$$

If F is continuous at ξ, then $F(\xi) = p$, but $F(\xi) \ge p$ in general.

Example 1 (*continued*) In her bid for the project, the contractor may specify the 0.95-quantile ξ as the promised completion date. Hence, according to the model, the project will complete by time ξ with probability $p = 0.95$. □

We assume that the complexity of the stochastic model prevents F from being computed, but we can simulate the model using Monte Carlo to produce an output

$Y \sim F$, where the notation $\sim$ means "is distributed as." Thus, our goal is to use Monte Carlo simulation to develop an estimator of ξ and also to provide a confidence interval for ξ as a measure of the estimator's statistical error.

A special case of our framework arises when the random variable Y has the form

$$Y = c_Y(U_1, U_2, \ldots, U_d) \sim F \tag{5}$$

for a given function $c_Y : [0, 1)^d \to \Re$, and $U_1, U_2, \ldots, U_d$ are independent and identically distributed (i.i.d.) unif[0, 1), where unif[0, 1) denotes a (continuous) uniform distribution on the interval [0, 1). We can think of c_Y as a computer code that takes $\mathbf{U} \equiv (U_1, U_2, \ldots, U_d)$ as input, transforms it into a random vector having a specified joint CDF with some (stochastic) dependence structure (independence being a special case), performs computations using the random vector, and then finally outputs Y. When Y satisfies (5), we can express its CDF F in (3) as

$$F(y) = \mathsf{P}(c_Y(\mathbf{U}) \leq y) = \mathsf{E}[I(c_Y(\mathbf{U}) \leq y)] = \int_{\mathbf{u}\in[0,1)^d} I(c_Y(\mathbf{u}) \leq y)\, \mathrm{d}\mathbf{u}$$

for any constant $y \in \Re$, which we will later exploit in Sect. 5.3 when considering a VRT known as Latin hypercube sampling. For smooth integrands, computing a d-dimensional integral when d is small (say no more than 4 or 5) can be more efficiently handled through numerical quadrature techniques [14] rather than Monte Carlo simulation. But when d is large or the integrand is not smooth, Monte Carlo may be more attractive.

As we will later see in Sect. 5.4 when considering a VRT known as importance sampling, it is sometimes more convenient to instead consider Y having the form

$$Y = c'_Y(X_1, X_2, \ldots, X_{d'}) \sim F \tag{6}$$

for a given function $c'_Y : \Re^{d'} \to \Re$, and $\mathbf{X} = (X_1, X_2, \ldots, X_{d'})$ is a random vector with known joint CDF G from which we can generate observations. The joint CDF G specifies a dependence structure (independence being a special case) for $\mathbf{X}$, and the marginal distributions of the components of $\mathbf{X}$ may differ. We can see that (5) is a special case of (6) by taking $d' = d$, and assuming that $X_1, X_2, \ldots, X_{d'}$ are i.i.d. unif[0, 1). When Y has the form in (6), the CDF F in (3) satisfies

$$F(y) = \mathsf{P}(c'_Y(\mathbf{X}) \leq y) = \mathsf{E}[I(c'_Y(\mathbf{x}) \leq y)] = \int_{\mathbf{x}\in\Re^{d'}} I(c'_Y(\mathbf{x}) \leq y)\, \mathrm{d}G(\mathbf{x}). \tag{7}$$

Let G_j be the marginal CDF of X_j. In the special case when $X_1, X_2, \ldots, X_{d'}$ are mutually independent under G and each G_j has a density g_j, we have that $\mathrm{d}G(\mathbf{x}) = \prod_{j=1}^{d'} g_j(x_j)\, \mathrm{d}x_j$ for $\mathbf{x} = (x_1, x_2, \ldots, x_{d'})$.

Example 1 (*continued*) For our SAN model in Fig. 1 with Y in (2),

$$c'_Y(X_1, X_2, \ldots, X_{d'}) = \max(X_1 + X_2,\ X_4 + X_5,\ X_1 + X_3 + X_5)$$

is the function c'_Y in (6), where $d' = d = 5$. To define the function c_Y in (5) for this model, let $U_1, U_2, \ldots, U_d$ be $d = 5$ i.i.d. unif[0, 1) random variables. For each activity $j = 1, 2, \ldots, d$, we can use the inverse transform method (e.g., Sect. II.2a of [4] or Sect. 2.2.1 of [22]) to convert $U_j \sim \text{unif}[0, 1)$ into $X_j \sim G_j$ by letting $X_j = G_j^{-1}(U_j) = -\ln(1 - U_j)$. Hence, for $(u_1, u_2, \ldots, u_d) \in [0, 1)^d$,

$$\begin{aligned} c_Y(u_1, u_2, \ldots, u_d) = \max(&G_1^{-1}(u_1) + G_2^{-1}(u_2),\ G_4^{-1}(u_4) + G_5^{-1}(u_5), \\ &G_1^{-1}(u_1) + G_3^{-1}(u_3) + G_5^{-1}(u_5)) \end{aligned} \tag{8}$$

specifies the function c_Y in (5) to generate $Y \sim F$. □

3 Quantile Point Estimation via Monte Carlo

As seen in (4), the p-quantile ξ is the (generalized) inverse of the true CDF F evaluated at p. Thus, a common (but not the only) approach for devising a point estimator for ξ follows a generic recipe.

Step 1. Use a Monte Carlo method to construct $\hat{F}_n$ as an estimator of F, where n denotes the computational budget, typically the number of times the simulation model (e.g., a computer code as in (5)) is run.
Step 2. Compute $\hat{\xi}_n = \hat{F}_n^{-1}(p)$ as an estimator of ξ.

How we accomplish Step 1 depends on the particular Monte Carlo method being applied. Different methods will yield different CDF estimators, which in turn will produce different quantile estimators in Step 2.

3.1 Naive Monte Carlo

We next illustrate how to accomplish the two steps when applying naive Monte Carlo (NMC). Alternatively called crude Monte Carlo, standard simulation, and simple random sampling, NMC simply employs Monte Carlo without applying any variance-reduction technique. Note that (3) suggests estimating $F(y)$ by averaging i.i.d. copies of $I(Y \leq y)$. To do this, generate $Y_1, Y_2, \ldots, Y_n$ as n i.i.d. copies of $Y \sim F$. We then compute the NMC estimator $\hat{F}_{\text{NMC},n}$ of the CDF F as

$$\hat{F}_{\text{NMC},n}(y) = \frac{1}{n} \sum_{i=1}^{n} I(Y_i \leq y), \tag{9}$$

completing Step 1. For each y, $\hat{F}_{\text{NMC},n}(y)$ is an unbiased estimator of $F(y)$ because

$$\mathsf{E}[\hat{F}_{\text{NMC},n}(y)] = \frac{1}{n}\sum_{i=1}^{n} \mathsf{E}[I(Y_i \le y)] = \frac{1}{n}\sum_{i=1}^{n} \mathsf{P}(Y_i \le y) = F(y) \tag{10}$$

as each $Y_i \sim F$. Then applying Step 2 yields the NMC quantile estimator

$$\hat{\xi}_{\text{NMC},n} = \hat{F}^{-1}_{\text{NMC},n}(p). \tag{11}$$

We can compute $\hat{\xi}_{\text{NMC},n}$ in (11) via *order statistics*. Let $Y_{1:n} \le Y_{2:n} \le \cdots \le Y_{n:n}$ be the sorted values of the sample $Y_1, Y_2, \ldots, Y_n$, so $Y_{i:n}$ is the ith smallest value. Then we have that $\hat{\xi}_{\text{NMC},n} = Y_{\lceil np \rceil:n}$, where $\lceil \cdot \rceil$ is the ceiling (or round-up) function. Although (10) shows that the CDF estimator is unbiased, the p-quantile estimator typically has bias; e.g., see Proposition 2 of [6].

In the special case of (5), we can obtain n i.i.d. copies of $Y \sim F$ by generating $n \times d$ i.i.d. unif[0, 1) random numbers $U_{i,j}$, $i = 1, 2, \ldots, n$, $j = 1, 2, \ldots, d$, which we arrange in an $n \times d$ grid:

$$\begin{matrix} U_{1,1} & U_{1,2} & \ldots & U_{1,d} \\ U_{2,1} & U_{2,2} & \ldots & U_{2,d} \\ \vdots & \vdots & \ddots & \vdots \\ U_{n,1} & U_{n,2} & \ldots & U_{n,d} \end{matrix}. \tag{12}$$

Now apply the function c_Y in (5) to each row to get

$$\begin{matrix} Y_1 = c_Y(U_{1,1}, & U_{1,2}, & \ldots, & U_{1,d}), \\ Y_2 = c_Y(U_{2,1}, & U_{2,2}, & \ldots, & U_{2,d}), \\ \vdots \;\; \vdots \;\;\; \vdots & \vdots & \ddots & \vdots \\ Y_n = c_Y(U_{n,1}, & U_{n,2}, & \ldots, & U_{n,d}). \end{matrix} \tag{13}$$

Because each row i of (12) has d independent unif[0, 1) random numbers, we see that $Y_i \sim F$ by (5). Moreover, the independence of the rows of (12) ensures that $Y_1, Y_2, \ldots, Y_n$ are also independent.

Example 1 (*continued*) To apply NMC to our SAN model, we employ c_Y from (8) with $d = 5$ in (13) to obtain $Y_1, Y_2, \ldots, Y_n$, which are used to compute the NMC CDF estimator $\hat{F}_{\text{NMC},n}$ in (9) and the NMC p-quantile estimator $\hat{\xi}_{\text{NMC},n}$ in (11). □

We have considered the NMC p-quantile estimator $\hat{\xi}_{\text{NMC},n}$ in (11) obtained by inverting the CDF estimator $\hat{F}_{\text{NMC},n}$ in (9), but other NMC quantile estimators have also been developed. For example, we may replace the step function $\hat{F}_{\text{NMC},n}$ with a linearly interpolated version, and [30] examines several such variants. Although these alternative quantile estimators may behave differently when the sample size n

is small, they typically share the same large-sample properties (to be discussed in Sect. 4) as (11).

3.2 A General Approach to Construct a CDF Estimator

In addition to NMC, there are other ways of accomplishing Steps 1 and 2 of Sect. 3 to obtain CDF and quantile estimators. Constructing another CDF estimator often entails deriving and exploiting an alternative representation for F. To do this, we may perform Step 1 through the following:

Step 1a. Identify a random variable $J(y)$, whose value depends on y, such that

$$\mathsf{E}[J(y)] = F(y) \text{ for each } y \in \Re. \tag{14}$$

Step 1b. Construct the estimator $\hat{F}_n(y)$ of $F(y)$ as the sample average of n identically distributed copies of $J(y)$, possibly with some adjustments.

Note that we built the NMC CDF estimator $\hat{F}_{\mathrm{NMC},n}$ in (9) by following Steps 1a and 1b, with $J(y) = I(Y \leq y)$, which satisfies (14) by (3).

Section 5 will review other Monte Carlo methods for performing Steps 1 and 2 of Sect. 3. Many (but not all) of the approaches handle Step 1 via Steps 1a and 1b. The n copies of $J(y)$ in Step 1b are often generated independently, but certain Monte Carlo methods sample them in a dependent manner; e.g., see Sect. 5.3.

4 Large-Sample Properties of Quantile Estimators

Although $\hat{\xi}_n$ is often *not* a sample average, it still typically obeys a *central limit theorem* (CLT) as the sample size n grows large. To establish this, let f be the derivative (when it exists) of F. Throughout the rest of the paper, whenever examining large-sample properties, we assume that $f(\xi) > 0$, which ensures that $F(\xi) = p$ and that $y = \xi$ is the unique root of the equation $F(y) = p$. Under various conditions that depend on the Monte Carlo method used to construct the CDF estimator $\hat{F}_n$ in Step 1 of Sect. 3, the corresponding p-quantile estimator $\hat{\xi}_n = \hat{F}_n^{-1}(p)$ satisfies a CLT

$$\sqrt{n}[\hat{\xi}_n - \xi] \Rightarrow N(0, \tau^2), \quad \text{as } n \to \infty, \tag{15}$$

where $\Rightarrow$ denotes convergence in distribution (e.g., see Sect. 25 of [9]), $N(a, b^2)$ represents a normal random variable with mean a and variance b^2, and the *asymptotic variance* τ^2 has the form

$$\tau^2 = \frac{\psi^2}{f^2(\xi)}. \tag{16}$$

The numerator ψ^2 on the right side of (16) is the asymptotic variance in the CLT for the CDF estimator at ξ:

$$\sqrt{n}[\hat{F}_n(\xi) - p] \Rightarrow N(0, \psi^2), \quad \text{as } n \to \infty, \tag{17}$$

where $p = F(\xi)$ because $f(\xi) > 0$. The CLT (17) typically holds (under appropriate conditions) because $\hat{F}_n(\xi)$ is often a sample average, e.g., as in (9) for NMC.

There are various ways to prove the CLT (15); e.g., see Sects. 2.3.3 and 2.5 of [46] for NMC. A particularly insightful approach exploits a *Bahadur representation* [8], which shows for large n, a quantile estimator $\hat{\xi}_n = \hat{F}_n^{-1}(p)$ is well approximated by a linear transformation of its corresponding CDF estimator $\hat{F}_n$ at ξ:

$$\hat{\xi}_n \approx \xi + \frac{p - \hat{F}_n(\xi)}{f(\xi)}. \tag{18}$$

To heuristically justify this, note that $\hat{F}_n(y) \approx F(y)$ for each y when n is large, so

$$\hat{F}_n(\hat{\xi}_n) - \hat{F}_n(\xi) \approx F(\hat{\xi}_n) - F(\xi) \approx f(\xi)[\hat{\xi}_n - \xi] \tag{19}$$

by a first-order Taylor approximation. Also, $\hat{\xi}_n = \hat{F}_n^{-1}(p)$ implies $\hat{F}_n(\hat{\xi}_n) \approx p$, which we put into (19) and rearrange to finally get (18).

Bahadur [8] makes rigorous the heuristic argument for the NMC setting of Sect. 3.1. Specifically, if F is twice differentiable at ξ (with $f(\xi) > 0$), then

$$\hat{\xi}_n = \xi + \frac{p - \hat{F}_n(\xi)}{f(\xi)} + R_n \tag{20}$$

for $\hat{\xi}_n = \hat{\xi}_{\mathrm{NMC},n}$ from (11), such that with probability 1,

$$R_n = O(n^{-3/4}(\log\log n)^{3/4}), \quad \text{as } n \to \infty. \tag{21}$$

(The statement that "with probability 1, $A_n = O(h(n))$ as $n \to \infty$" for some function $h(n)$ means that there exists an event Ω_0 such that $\mathsf{P}(\Omega_0) = 1$ and for each outcome $\omega \in \Omega_0$, there exists a constant $K(\omega)$ such that $|A_n(\omega)| \le K(\omega)h(n)$ for all n sufficiently large.) (The almost-sure rate at which R_n vanishes in (21) is sharper than what [8] originally proved; see Sect. 2.5 of [46] for details.) Assuming only $f(\xi) > 0$, [21] proves a weaker result,

$$\sqrt{n}R_n \Rightarrow 0 \quad \text{as } n \to \infty, \tag{22}$$

which is sufficient for most applications. Note that (21) implies (22), and we call (20) combined with (21) (resp., (22)) a *strong* (resp., *weak*) Bahadur representation. The paper [13] provides a general framework for establishing a weak Bahadur representation, which may be verified for different variance-reduction techniques.

A (strong or weak) Bahadur representation ensures that $\hat{\xi}_n$ obeys a CLT. To see why, rearrange (20) and scale by $\sqrt{n}$ to get

$$\sqrt{n}\left[\hat{\xi}_n - \xi\right] = \frac{\sqrt{n}}{f(\xi)}\left[p - \hat{F}_n(\xi)\right] + \sqrt{n}R_n. \tag{23}$$

As $\hat{F}_n(\xi)$ is typically a sample average (e.g., (9)), it satisfies the CLT in (17). The second term on the right side of (23) vanishes weakly (resp., strongly) by (22) (resp., (21)), so Slutsky's theorem (e.g., Theorem 1.5.4 of [46]) verifies the CLT in (15).

When we apply Steps 1 and 2 in Sect. 3 to obtain $\hat{\xi}_n = \hat{F}_n^{-1}(p)$, (23) clarifies the reason the asymptotic variance τ^2 in the CLT (15) for $\hat{\xi}_n$ has the ratio form $\psi^2/f^2(\xi)$ in (16). The numerator ψ^2 arises from the CLT in (17) for the CDF estimator $\hat{F}_n$ at ξ, so ψ^2 is determined by the particular Monte Carlo method used to construct $\hat{F}_n$. For NMC, the CLT (17) uses $\hat{F}_{\mathrm{NMC},n}(\xi)$ from (9), which averages i.i.d. copies of $I(Y \leq \xi)$, and the numerator in (16) is then

$$\psi^2_{\mathrm{NMC}} = \mathsf{Var}[I(Y \leq \xi)] = p(1-p), \tag{24}$$

with Var the variance operator. But the denominator $f^2(\xi)$ in (16) is the *same* for each method.

5 Variance-Reduction Techniques for Quantile Estimation

Section 3.1 showed how to construct a quantile estimator when employing NMC. We next illustrate how the general approach of quantile estimation described in Sects. 3 and 3.2 can be applied for other Monte Carlo methods using VRTs.

5.1 Control Variates

Suppose that along with the response Y, the simulation model also outputs another random variable V whose mean $\mu_V = \mathsf{E}[V]$ is known. The method of control variates (CV) exploits this additional information to produce an estimator with typically reduced variance compared to its NMC counterpart. Section V.2 of [4] and Sect. 4.1 of [22] review CV for estimating a mean, and [13, 27, 29] apply this approach to estimate the CDF F, which is inverted to obtain an estimator of the p-quantile ξ.

When Y has the form in (5), we assume that the control variate V is generated by

$$V = c_V(U_1, U_2, \ldots, U_d) \tag{25}$$

for some function $c_V : [0, 1)^d \to \Re$, where again we require that $\mu_V = \mathsf{E}[V]$ is known. Because the inputs $U_1, U_2, \ldots, U_d$ are the same in (25) and (5), V and Y are typically dependent. As will be later seen in (35), the CV method works best when V is strongly (positively or negatively) correlated with $I(Y \leq \xi)$.

Example 1 (*continued*) Figure 1 has $r = 3$ paths from source to sink in (1). Of those, the length $T_3 = X_1 + X_3 + X_5$ of path $\mathscr{P}_3$ has the largest mean. We then choose the CV as $V = I(T_3 \leq \zeta)$, where ζ is the p-quantile of the CDF $\widetilde{G}_3$ of T_3. As X_1, X_3, X_5 are i.i.d. exponential with mean 1, the CDF $\widetilde{G}_3$ is an Erlang with shape parameter 3 and scale parameter 1; i.e., $\widetilde{G}_3(x) = 1 - (1 + x + x^2)e^{-x}$ for $x \geq 0$. We can then compute $\zeta = \widetilde{G}_3^{-1}(p)$, and $\mu_V = p$. Hence,

$$c_V(U_1, U_2, \ldots, U_d) = I(G_1^{-1}(U_1) + G_3^{-1}(U_3) + G_5^{-1}(U_5) \leq \zeta)$$

is the function c_V in (25). □

To design a CDF estimator when applying CV, we can follow the approach described in Sect. 3. For any constant $\beta \in \Re$, note that

$$F(y) = \mathsf{E}[I(Y \leq y) + \beta(V - \mu_V)] \tag{26}$$

as $\mathsf{E}[V] = \mu_V$. Thus, take $J(y) = I(Y \leq y) + \beta(V - \mu_V)$ in Step 1a of Sect. 3.2, and Step 1b suggests estimating $F(y)$ by averaging copies of $I(Y \leq y) + \beta(V - \mu_V)$. Specifically, let (Y_i, V_i), $i = 1, 2, \ldots, n$, be i.i.d. copies of (Y, V), and define

$$\hat{F}'_{\text{CV},\beta,n}(y) = \frac{1}{n}\sum_{i=1}^{n}[I(Y_i \leq y) - \beta(V_i - \mu_V)] \tag{27}$$

$$= \hat{F}_{\text{NMC},n}(y) - \beta(\hat{\mu}_{V,n} - \mu_V), \tag{28}$$

where $\hat{F}_{\text{NMC},n}(y)$ is the NMC CDF estimator in (9), and $\hat{\mu}_{V,n} = (1/n)\sum_{i=1}^{n} V_i$. For each y and β, $\hat{F}'_{\text{CV},\beta,n}(y)$ is an unbiased estimator of $F(y)$ by (26) and (27).

Although the choice of β does not affect the mean of $\hat{F}'_{\text{CV},\beta,n}(y)$ by (26), it does have an impact on its variance, which by (27) equals

$$\begin{aligned}\mathsf{Var}[\hat{F}'_{\text{CV},\beta,n}(y)] &= \frac{1}{n}\mathsf{Var}\big[I(Y \leq y) - \beta(V - \mu_V)\big] \\ &= \frac{1}{n}\Big(F(y)[1 - F(y)] + \beta^2\mathsf{Var}[V] - 2\beta\mathsf{Cov}[I(Y \leq y), V]\Big),\end{aligned} \tag{29}$$

where Cov denotes the covariance operator. As (29) is a quadratic function in β, we can easily find the value $\beta = \beta_y^*$ minimizing (29) as

$$\beta_y^* = \frac{\mathsf{Cov}[I(Y \le y), V]}{\mathsf{Var}[V]} = \frac{\mathsf{E}[I(Y \le y)V] - \mathsf{E}[I(Y \le y)]\,\mathsf{E}[V]}{\mathsf{E}[(V - \mu_V)^2]}. \tag{30}$$

The values of $\mathsf{Var}[V]$ and $\mathsf{Cov}[I(Y \le y), V]$ may be unknown, so we estimate them from our data (Y_i, V_i), $i = 1, 2, \ldots, n$. We then arrive at an estimator for β_y^* in (30) as

$$\hat{\beta}_{y,n}^* = \frac{[(1/n)\sum_{i=1}^n I(Y_i \le y)V_i] - \hat{F}_{\mathrm{NMC},n}(y)\hat{\mu}_{V,n}}{(1/n)\sum_{i=1}^n (V_i - \hat{\mu}_{V,n})^2}, \tag{31}$$

which uses $\hat{F}_{\mathrm{NMC},n}(y)$ to estimate $\mathsf{E}[I(Y \le y)] = F(y)$. Replacing β in (28) with the estimator $\hat{\beta}_{y,n}^*$ of its optimal value leads to the CV estimator of $F(y)$ as

$$\hat{F}_{\mathrm{CV},n}(y) = \hat{F}_{\mathrm{NMC},n}(y) - \hat{\beta}_{y,n}^*(\hat{\mu}_{V,n} - \mu_V). \tag{32}$$

For any constant β, (26) ensures that $\hat{F}'_{\mathrm{CV},\beta,n}(y)$ in (27) is an unbiased estimator of $F(y)$ for each $y \in \Re$, but the estimator $\hat{F}_{\mathrm{CV},n}(y)$ typically no longer enjoys this property as $\hat{\beta}_{y,n}^*$ and $\hat{\mu}_{V,n}$ are dependent. We finally obtain the CV p-quantile estimator

$$\hat{\xi}_{\mathrm{CV},n} = \hat{F}_{\mathrm{CV},n}^{-1}(p). \tag{33}$$

Computing the inverse in (33) appears to be complicated by the fact that the estimator $\hat{\beta}_{y,n}^*$ in (31) of the optimal β_y^* depends on y. However, [27] derives an algebraically equivalent representation for $\hat{F}_{\mathrm{CV},n}(y)$ that avoids this complication. It turns out that we can rewrite the CV CDF estimator in (32) as

$$\hat{F}_{\mathrm{CV},n}(y) = \sum_{i=1}^n W_i I(Y_i \le y) \quad \text{with} \quad W_i = \frac{1}{n} + \frac{(\hat{\mu}_{V,n} - V_i)(\hat{\mu}_{V,n} - \mu_V)}{\sum_{\ell=1}^n (V_\ell - \hat{\mu}_{V,n})^2}, \tag{34}$$

which satisfies $\sum_{i=1}^n W_i = 1$. While it is possible for $W_i < 0$, [27] notes it is unlikely. Because of (34), we can view $\hat{F}_{\mathrm{CV},n}(y)$ as a weighted average of the $I(Y_i \le y)$.

The weights W_i reduce to a simple form when the control $V = I(\widetilde{V} \le \zeta)$, where $\widetilde{V}$ is an auxiliary random variable, and ζ is the (known) p-quantile of the CDF of $\widetilde{V}$. (This is the setting of Example 1, in which $\widetilde{V} = T_3$.) Let $(Y_i, \widetilde{V}_i)$, $i = 1, 2, \ldots, n$, be i.i.d. copies of $(Y, \widetilde{V})$, and define $V_i = I(\widetilde{V}_i \le \zeta)$. Also, let $M = \sum_{i=1}^n V_i$. Then each weight becomes $W_i = p/M$ if $V_i = 1$, and $W_i = (1-p)/(n-M)$ if $V_i = 0$.

The key point of the representation in (34) is that each W_i does not depend on the argument y at which the CDF estimator $\hat{F}_{\mathrm{CV},n}$ is evaluated, simplifying the computation of its inverse. Specifically, let $Y_{i:n}$ be the ith smallest value among $Y_1, Y_2, \ldots, Y_n$, and let $W_{i::n}$ correspond to $Y_{i:n}$. Then the CV p-quantile estimator in (33) satisfies $\hat{\xi}_{\mathrm{CV},n} = Y_{i_p:n}$, where $i_p = \min\{k : \sum_{i=1}^k W_{i::n} \ge p\}$.

When $0 < \mathsf{Var}[V] < \infty$, the CV p-quantile estimator $\hat{\xi}_{\mathrm{CV},n}$ in (33) satisfies the CLT in (15), where ψ^2 in (16) is given by

$$\psi^2_{\mathrm{CV}} = p(1-p) - \frac{(\mathsf{Cov}[I(Y \le \xi), V])^2}{\mathsf{Var}[V]} = (1-\rho^2)p(1-p), \tag{35}$$

and $\rho = \mathsf{Cov}[I(Y \le \xi), V]/\sqrt{\mathsf{Var}[I(Y \le \xi)]\mathsf{Var}[V]}$ is the (Pearson) correlation coefficient of $I(Y \le \xi)$ and V; see [13, 27]. Thus, (35) shows that the more strongly (negatively or positively) correlated the CV V and $I(Y \le \xi)$ are, the smaller the asymptotic variance of the CV p-quantile estimator is, by (16). Also, [13] establishes that $\hat{\xi}_{\mathrm{CV},n}$ satisfies a weak Bahadur representation, as in (20) and (22).

We have developed the CV method when there is a single control V. But the idea extends to multiple controls $V^{(1)}, V^{(2)}, \dots, V^{(m)}$, in which case the CDF estimator corresponds to a linear-regression estimator on the multiple CVs; see [27] for details. Also, rather than following the framework in Sect. 3 of constructing a p-quantile estimator as $\hat{\xi}_n = \hat{F}_n^{-1}(p)$, [27, 29, 44] consider an alternative CV estimator $\hat{\xi}'_{\mathrm{CV},n} \equiv \hat{\xi}_{\mathrm{NMC},n} - \beta(\hat{\zeta}_{\mathrm{NMC},n} - \zeta)$, where $\hat{\zeta}_{\mathrm{NMC},n}$ is the NMC estimator of the p-quantile ζ (assumed known) of the CDF of a random variable $\widetilde{V}$ (e.g., $\widetilde{V} = T_3$ in Example 1).

5.2 *Stratified Sampling*

Stratified sampling (SS) partitions the sample space into a finite number of subsets, known as *strata*, and allocates a fixed fraction of the overall sample size to sample from each stratum. Sect. 4.3 of [22] provides an overview of SS to estimate a mean, and [12, 13, 23] apply SS to estimate a quantile.

One way to partition the sample space for SS, as developed in [23], is as follows. Let S be an auxiliary random variable that is generated at the same time as the output Y. When Y has the form in (5), we assume that S is computed as

$$S = c_S(U_1, U_2, \dots, U_d) \tag{36}$$

for some function $c_S : [0,1)^d \to \Re$, where $U_1, U_2, \dots, U_d$ are the same uniforms used to generate Y in (5).

We next use S as a *stratification variable* to partition the sample space of (Y, S) by splitting the support of S into $t \ge 1$ disjoint subsets. Let $\mathscr{A}$ be the support of S, so $\mathsf{P}(S \in \mathscr{A}) = 1$. We then partition $\mathscr{A} = \cup_{s=1}^{t} \mathscr{A}_{\langle s\rangle}$ for some user-specified integer $t \ge 1$, where $\mathscr{A}_{\langle s\rangle} \cap \mathscr{A}_{\langle s'\rangle} = \emptyset$ for $s \ne s'$. For each $s = 1, 2, \dots, t$, let $\lambda_{\langle s\rangle} = \mathsf{P}(S \in \mathscr{A}_{\langle s\rangle})$. The law of total probability implies

$$\begin{aligned} F(y) = \mathsf{P}(Y \le y) &= \sum_{s=1}^{t} \mathsf{P}(Y \le y, S \in \mathscr{A}_{\langle s\rangle}) \\ &= \sum_{s=1}^{t} \mathsf{P}(S \in \mathscr{A}_{\langle s\rangle})\mathsf{P}(Y \le y \mid S \in \mathscr{A}_{\langle s\rangle}) = \sum_{s=1}^{t} \lambda_{\langle s\rangle} F_{\langle s\rangle}(y), \end{aligned} \tag{37}$$

where $F_{\langle s\rangle}(y) \equiv \mathsf{P}(Y \le y \mid S \in \mathscr{A}_{\langle s\rangle})$. In (37), $\lambda=(\lambda_{\langle s\rangle} : s = 1, 2, \ldots, t)$ is assumed known, but we need to estimate each $F_{\langle s\rangle}(y)$. We further assume that we have a way of sampling $Y_{\langle s\rangle} \sim F_{\langle s\rangle}$. A simple (but not necessarily the most efficient) way is through rejection sampling: generate (Y, S), and accept (resp., reject) Y as an observation from $F_{\langle s\rangle}$ if $S \in \mathscr{A}_{\langle s\rangle}$ (resp., if $S \notin \mathscr{A}_{\langle s\rangle}$).

To construct our SS estimator of F, we define $\gamma = (\gamma_{\langle s\rangle} : s = 1, 2, \ldots, t)$ as a vector of positive constants satisfying $\sum_{s=1}^{t} \gamma_{\langle s\rangle} = 1$. Then for our overall sample size n, we allocate a portion $n_{\langle s\rangle} \equiv \gamma_{\langle s\rangle} n$ to estimate $F_{\langle s\rangle}$ for stratum index s, where we assume that each $n_{\langle s\rangle}$ is integer-valued, so that $\sum_{s=1}^{t} n_{\langle s\rangle} = n$. For each $s = 1, 2, \ldots, t$, let $Y_{\langle s\rangle,i}$, $i = 1, 2, \ldots, n_{\langle s\rangle}$, be i.i.d. observations from $F_{\langle s\rangle}$, so our estimator of $F_{\langle s\rangle}$ is given by

$$\hat{F}_{\langle s\rangle,\gamma,n}(y) = \frac{1}{n_{\langle s\rangle}} \sum_{i=1}^{n_{\langle s\rangle}} I(Y_{\langle s\rangle,i} \le y). \tag{38}$$

Replacing each $F_{\langle s\rangle}(y)$ in (37) by its estimator $\hat{F}_{\langle s\rangle,\gamma,n}(y)$ gives

$$\hat{F}_{\mathrm{SS},\gamma,n}(y) = \sum_{s=1}^{t} \lambda_{\langle s\rangle} \hat{F}_{\langle s\rangle,\gamma,n}(y) \tag{39}$$

as the SS estimator of F. Inverting $\hat{F}_{\mathrm{SS},\gamma,n}$ leads to the SS p-quantile estimator

$$\hat{\xi}_{\mathrm{SS},\gamma,n} = \hat{F}^{-1}_{\mathrm{SS},\gamma,n}(p). \tag{40}$$

While (39) and (40) follow the general approach of Steps 1 and 2 of Sect. 3, the way we constructed (39) does not exactly fit into the scheme of Steps 1a and 1b of Sect. 3.2, although the estimator $\hat{F}_{\langle s\rangle,\gamma,n}(y)$ in (38) applies the same idea.

We can compute $\hat{\xi}_{\mathrm{SS},\gamma,n}$ in (40) as follows. Let $D_k = Y_{\langle s\rangle,i}$ for $k = \sum_{\ell=1}^{s-1} n_{\langle \ell\rangle} + i$, and let $W'_k = \lambda_{\langle s\rangle}/n_{\langle s\rangle}$, which satisfies $\sum_{k=1}^{n} W'_k - 1$. Next define $D_{1:n} \le D_{2:n} \le \cdots \le D_{n:n}$ as the order statistics of $D_1, D_2, \ldots, D_n$, and let $W'_{i::n}$ be the W'_k associated with $D_{i:n}$. Then we have that $\hat{\xi}_{\mathrm{SS},\gamma,n} = D_{i'_p:n}$ for $i'_p = \min\{\ell : \sum_{i=1}^{\ell} W'_{i::n} \ge p\}$.

Example 1 (*continued*) Let the stratification variable in (36) be

$$S = X_1 + X_3 + X_5 = G_1^{-1}(U_1) + G_3^{-1}(U_3) + G_5^{-1}(U_5) \equiv c_S(U_1, U_2, \ldots, U_5), \tag{41}$$

the (random) length of the path $\mathscr{P}_3$ in (1), which has largest expectation among all paths in (1). As in Sect. 5.1, the CDF $\widetilde{G}_S$ of S is then an Erlang with shape parameter 3 and scale parameter 1. One way of partitioning the support $\mathscr{A}$ of S into $t \ge 1$ intervals takes $\mathscr{A}_{\langle s\rangle} = [\widetilde{G}_S^{-1}((s-1)/t), \widetilde{G}_S^{-1}(s/t))$ for each $s = 1, 2, \ldots, t$.

As in [23] we can use a "bin tossing" approach to sample the $Y_{\langle s\rangle,i}$, $s = 1, 2, \ldots, t$, $i = 1, 2, \ldots, n_{\langle s\rangle}$. In one run, generate $U_1, U_2, \ldots, U_5$ as i.i.d. unif[0, 1) random numbers, and compute $Y = c_Y(U_1, U_2, \ldots, U_5)$ for c_Y in (8) and $S =$

$c_S(U_1, U_2, \ldots, U_5)$ for c_S in (41). If $S \in \mathscr{A}_{\langle s \rangle}$, then use Y as an observation from the stratum with index s. Keep independently sampling $(U_1, U_2, \ldots, U_5)$ and computing (Y, S) until each stratum index s has $n_{\langle s \rangle}$ observations, discarding any extras in a stratum. □

The SS p-quantile estimator $\hat{\xi}_{\text{SS},\gamma,n}$ in (10) satisfies the CLT in (15) with

$$\psi^2_{\text{SS},\gamma} = \sum_{s=1}^{t} \frac{\lambda^2_{\langle s \rangle}}{\gamma_{\langle s \rangle}} F_{\langle s \rangle}(\xi)[1 - F_{\langle s \rangle}(\xi)] \tag{42}$$

in (16); see [12, 13, 23]. Also, [13] shows that $\hat{\xi}_{\text{CV},n}$ satisfies a weak Bahadur representation, as in (20) and (22). The value of $\psi^2_{\text{SS},\gamma}$ depends on how the user specifies the sampling-allocation parameter γ. Setting $\gamma = \lambda$, known as the *proportional allocation*, ensures that $\psi^2_{\text{SS},\lambda} \leq \psi^2_{\text{NMC}}$, so the proportional allocation guarantees no greater asymptotic variance than NMC. The optimal value of γ to minimize $\psi^2_{\text{SS},\gamma}$ is $\gamma^* = (\gamma^*_{\langle s \rangle} : s = 1, 2, \ldots, t)$ with $\gamma^*_{\langle s \rangle} = \kappa_{\langle s \rangle} / (\sum_{s'=1}^{t} \kappa_{\langle s' \rangle})$, where $\kappa_{\langle s \rangle} = \lambda_{\langle s \rangle}(F_{\langle s \rangle}(\xi)[1 - F_{\langle s \rangle}(\xi)])^{1/2}$; e.g., see p. 217 of [12, 23]. Although the $\kappa_{\langle s \rangle}$ are unknown, [12] employs pilot runs to estimate them, which are then used to estimate γ^*, and then performs additional runs with the estimated γ^*.

5.3 Latin Hypercube Sampling

Latin hypercube sampling (LHS) can be thought of as an efficient way of implementing SS in high dimensions. Section 5.4 of [22] provides an overview of LHS to estimate a mean, and [6, 15, 17, 25, 31, 38] develop LHS for quantile estimation.

To motivate how we apply LHS to estimate ξ, recall that for NMC, (10) shows that $\hat{F}_{\text{NMC},n}(y)$ in (9) is an unbiased estimator of $F(y)$ for each y. While NMC uses a sample $Y_1, Y_2, \ldots, Y_n$ that are i.i.d. with CDF F, (10) still holds if we replace the sample with $Y_1^*, Y_2^*, \ldots, Y_n^*$ that are *dependent*, with each $Y_i^* \sim F$. Moreover, as

$$\text{Var}\Big[\sum_{i=1}^{n} I(Y_i^* \leq y)\Big] = \sum_{i=1}^{n} \text{Var}[I(Y_i^* \leq y)] + 2 \sum_{1 \leq i < j \leq n} \text{Cov}[I(Y_i^* \leq y), I(Y_j^* \leq y)],$$

if $I(Y_i^* \leq y)$ and $I(Y_j^* \leq y)$ are negatively correlated for each $i \neq j$, then the average of the $I(Y_i^* \leq y)$ will have lower variance than the average of the $I(Y_i \leq y)$. We next show for the setting of (5) how LHS samples the $Y_i^* \sim F$ in a dependent manner.

Recall that d is the number of uniform inputs to c_Y in (5). For each $j = 1, 2, \ldots, d$, let $\pi_j = (\pi_j(1), \pi_j(2), \ldots, \pi_j(n))$ be a uniform random permutation of $(1, 2, \ldots, n)$, where $\pi_j(i)$ denotes the number in $\{1, 2, \ldots, n\}$ to which i maps. Thus, π_j equals one of the particular $n!$ permutations with probability $1/n!$. Let $\pi_1, \pi_2, \ldots, \pi_d$, be d mutually independent permutations, and also independent of the $n \times d$ grid of i.i.d. unif[0, 1) random numbers $U_{i,j}$ in (12). Then define

$$U^*_{i,j} = \frac{U_{i,j} + \pi_j(i) - 1}{n}, \quad \text{for } i = 1, 2, \ldots, n, \; j = 1, 2, \ldots, d. \tag{43}$$

It is easy to show that each $U^*_{i,j} \sim \text{unif}[0, 1)$. Next arrange the $U^*_{i,j}$ in an $n \times d$ grid:

$$\begin{array}{cccc} U^*_{1,1} & U^*_{1,2} & \cdots & U^*_{1,d} \\ U^*_{2,1} & U^*_{2,2} & \cdots & U^*_{2,d} \\ \vdots & \vdots & \ddots & \vdots \\ U^*_{n,1} & U^*_{n,2} & \cdots & U^*_{n,d} \end{array}. \tag{44}$$

Each column j in (44) depends on π_j but not on any other permutation, making the d columns independent because $\pi_1, \pi_2, \ldots, \pi_d$ are. But the rows in (44) are *dependent* because for each column j, its entries $U^*_{i,j}$, $i = 1, 2, \ldots, n$, share the same permutation π_j. Now apply the function c_Y in (5) to each row of (44) to get

$$\begin{array}{ccccc} Y^*_1 = c_Y(U^*_{1,1}, & U^*_{1,2}, & \ldots, & U^*_{1,d}), \\ Y^*_2 = c_Y(U^*_{2,1}, & U^*_{2,2}, & \ldots, & U^*_{2,d}), \\ \vdots \qquad \vdots & \vdots & \ddots & \vdots \\ Y^*_n = c_Y(U^*_{n,1}, & U^*_{n,2}, & \ldots, & U^*_{n,d}). \end{array} \tag{45}$$

Because each row i of (44) has d i.i.d. unif[0, 1) random numbers, we see that $Y^*_i \sim F$ by (5). But $Y^*_1, Y^*_2, \ldots, Y^*_n$ are dependent because (44) has dependent rows.

Consider any column $j = 1, 2, \ldots, d$, in (44), and an interval $I_{k,n} = [(k - 1)/n, k/n)$ for any $k = 1, 2, \ldots, n$. By (43), exactly one $U^*_{i,j}$ from column j lies in $I_{k,n}$. Thus, each column j forms a stratified sample of size n of unif[0, 1) random numbers, so LHS simultaneously stratifies each input coordinate $j = 1, 2, \ldots, d$.

We form the LHS estimator of the CDF F as

$$\hat{F}_{\text{LHS},n}(y) = \frac{1}{n} \sum_{i=1}^{n} I(Y^*_i \leq y) \tag{46}$$

and the LHS p-quantile estimator as

$$\hat{\xi}_{\text{LHS},n} = \hat{F}^{-1}_{\text{LHS},n}(p). \tag{47}$$

We can compute (47) by $\hat{\xi}_{\text{LHS},n} = Y^*_{\lceil np \rceil:n}$, where $Y^*_{i:n}$ is the ith smallest value among $Y^*_1, Y^*_2, \ldots, Y^*_n$ in (45). Note that (46) and (47) fit into the framework of Sect. 3, where Step 1 is implemented through Steps 1a and 1b of Sect. 3.2, with $J(y) = I(Y \leq y)$. But in contrast to the other methods considered, Step 1b generates n *dependent* copies of $I(Y \leq y)$ as $I(Y^*_i \leq y)$, $i = 1, 2, \ldots, n$, where each $Y^*_i \sim F$.

Example 1 (*continued*) To apply LHS to our SAN model, we employ c_Y from (8) in (45) to obtain $Y^*_1, Y^*_2, \ldots, Y^*_n$, which are then used in (46) and (47) to compute the LHS CDF estimator $\hat{F}_{\text{LHS},n}$ and the LHS p-quantile estimator $\hat{\xi}_{\text{LHS},n}$. □

Under regularity conditions, [6] proves that the LHS p-quantile estimator $\hat{\xi}_{\mathrm{LHS},n}$ in (47) obeys the CLT (15), and gives the specific form of $\psi^2 = \psi^2_{\mathrm{LHS}}$ in (16). Also, [16] shows that $\hat{\xi}_{\mathrm{LHS},n}$ satisfies a weak Bahadur representation, as in (20) and (22).

5.4 Importance Sampling

Importance sampling (IS) is a variance-reduction technique that can be particularly effective when studying rare events. The basic idea is to change the distributions driving the stochastic model to cause the rare event of interest to occur more frequently, and then unbias the outputs by multiplying by a correction factor. Section V.1 and Chap. VI of [4] and Sect. 4.6 of [22] provide overviews of IS to estimate a mean or tail probability.

For IS quantile estimation [13, 23, 24, 48], it is more natural to consider Y having the form in (6) rather than (5), i.e., $Y = c'_Y(\mathbf{X})$ for random vector $\mathbf{X} \in \Re^{d'}$ with joint CDF G. Let H be another joint CDF on $\Re^{d'}$ such that G is *absolutely continuous* with respect to H. For example, if G (resp., H) has a joint density function g (resp., h), then G is absolutely continuous with respect to H if $g(\mathbf{x}) > 0$ implies $h(\mathbf{x}) > 0$. In general, let P_G and E_G (resp., P_H and E_H) be the probability measure and expectation operator when $\mathbf{X} \sim G$ (resp., $\mathbf{X} \sim H$). The absolute continuity permits us to apply a *change of measure* to express the tail distribution corresponding to (7) as

$$
\begin{aligned}
1 - F(y) &= \mathsf{P}_G(Y > y) = \mathsf{E}_G[I(c'_Y(\mathbf{X}) > y)] = \int_{\mathbf{x}\in\Re^{d'}} I(c'_Y(\mathbf{x}) > y)\,\mathrm{d}G(\mathbf{x}) \\
&= \int_{\mathbf{x}\in\Re^{d'}} I(c'_Y(\mathbf{x}) > y)\frac{\mathrm{d}G(\mathbf{x})}{\mathrm{d}H(\mathbf{x})}\,\mathrm{d}H(\mathbf{x}) = \int_{\mathbf{x}\in\Re^{d'}} I(c'_Y(\mathbf{x}) > y)L(\mathbf{x})\,\mathrm{d}H(\mathbf{x}) \\
&= \mathsf{E}_H[I(c'_Y(\mathbf{X}) > y)L(\mathbf{X})], \qquad (48)
\end{aligned}
$$

where $L(\mathbf{x}) = \mathrm{d}G(\mathbf{x})/\,\mathrm{d}H(\mathbf{x})$ is the *likelihood ratio* or Radon–Nikodym derivative of G with respect to H; see Sect. 32 of [9]. In the special case when $\mathbf{X} = (X_1, X_2, \ldots, X_{d'})$ has mutually independent components under G (resp., H) with each marginal CDF G_j (resp., H_j) of X_j having a density function g_j (resp., h_j), the likelihood ratio becomes $L(\mathbf{x}) = \prod_{j=1}^{d'} g_j(x_j)/h_j(x_j)$. By (48), we can obtain an unbiased estimator of $1 - F(y)$ by averaging i.i.d. copies of $I(c'_Y(\mathbf{X}) > y)L(\mathbf{X})$, with $\mathbf{X} \sim H$. Specifically, let $\mathbf{X}_1, \mathbf{X}_2, \ldots, \mathbf{X}_n$ be i.i.d., with each $\mathbf{X}_i \sim H$. Then we get an IS estimator of F as

$$
\hat{F}_{\mathrm{IS},n}(y) = 1 - \frac{1}{n}\sum_{i=1}^{n} I(c'_Y(\mathbf{X}_i) > y)L(\mathbf{X}_i). \qquad (49)
$$

An IS p-quantile estimator is then

$$\hat{\xi}_{\text{IS},n} = \hat{F}^{-1}_{\text{IS},n}(p). \tag{50}$$

Note that (49) and (50) follow the general approach of Steps 1 and 2 of Sect. 3, where (49) is obtained through Steps 1a and 1b of Sect. 3.2, with $J(y) = 1 - I(c'_Y(\mathbf{X}) > y)L(\mathbf{X})$, which satisfies (14) by (48).

As shown in [24], we can compute $\hat{\xi}_{\text{IS},n}$ in (50) as follows. Let $Y_i = c'_Y(\mathbf{X}_i)$, and define $Y_{i:n}$ as the ith smallest value among $Y_1, Y_2, \ldots, Y_n$. Also, let $L_{i::n} = L(\mathbf{X}_j)$ for $\mathbf{X}_j$ corresponding to $Y_{i:n}$. Then $\hat{\xi}_{\text{IS},n} = Y_{i'_p:n}$ for $i'_p = \max\{k : \sum_{i=k}^{n} L_{i::n} \le (1-p)n\}$.

The key to effective application of IS is choosing an appropriate IS distribution H for $\mathbf{X}$ so that the quantile estimator $\hat{\xi}_{\text{IS},n}$ has small variance. As seen in Sect. 4, the asymptotic variance τ^2 of the IS p-quantile estimator $\hat{\xi}_{\text{IS},n}$ is closely related to the asymptotic variance $\psi^2 = \psi^2_{\text{IS}}$ in the CLT (17) for $\hat{F}_{\text{IS},n}(\xi)$. Let Var_H denote the variance operator when $\mathbf{X} \sim H$, and as $\mathbf{X}_i$, $i = 1, 2, \ldots, n$, are i.i.d., we have that

$$\mathsf{Var}_H[\hat{F}_{\text{IS},n}(\xi)] = \frac{1}{n}\mathsf{Var}_H[I(c'_Y(\mathbf{X}) > \xi)L(\mathbf{X})] \equiv \frac{1}{n}\psi^2_{\text{IS}}. \tag{51}$$

A "good" choice for H is problem specific, and a poorly designed H can actually increase the variance (or even produce infinite variance). The papers [13, 23, 24] discuss particular ways of selecting H in various problem settings.

Example 1 (*continued*) For a SAN as in Fig. 1, [13] estimates the p-quantile when $p \approx 1$ via an IS scheme that combines ideas from [24, 34]. Recall that Fig. 1 has $r = 3$ paths from source to sink, which are given in (1). When estimating the SAN tail probability $\mathsf{P}_G(Y > y)$ for large y, we want to choose H so that the event $\{Y > y\}$ occurs more frequently. To do this, [34] specifies H as a mixture of r CDFs $H^{(1)}, H^{(2)}, \ldots, H^{(r)}$; i.e., $H(\mathbf{x}) = \sum_{k=1}^{r} \alpha^{(k)} H^{(k)}(\mathbf{x})$, where each $\alpha^{(k)}$ is a nonnegative constant such that $\sum_{k=1}^{r} \alpha^{(k)} = 1$. Each $H^{(k)}$ keeps all activity durations as independent exponentials but increases the mean of X_j for edges $j \in \mathscr{P}_k$, making $\{Y > y\}$ more likely. (More generally, one could choose $H^{(k)}$ to not only have different means for activities $j \in \mathscr{P}_k$ but further to have entirely different distributions.) Also, $H^{(k)}$ leaves unaltered the CDF of $X_{j'}$ for each $j' \notin \mathscr{P}_k$. Changing the mean of X_j corresponds to *exponentially twisting* its original CDF G_j; see Example 4.6.2 of [22] and Sect. V.1b of [4] for details on exponential twisting. The exponential twist requires specifying a twisting parameter $\theta \in \Re$, and [13] employs an approach in [24] to choose a value for $\theta = \theta^{(k)}$ for each $H^{(k)}$ in the mixture. Also, by adapting a heuristic from [34] for estimating a tail probability to instead handle a quantile, [13] determines the mixing weights $\alpha^{(k)}$, $k = 1, 2, \ldots, r$, by first obtaining an approximate upper bound for the second moment $\mathsf{E}_H[(I(c'_Y(\mathbf{X}) > \xi)L(\mathbf{X}))^2]$ in terms of the $\alpha^{(k)}$, and then choosing the $\alpha^{(k)}$ to minimize the approximate upper bound. Note that the mixture H used for IS does *not* satisfy the special case mentioned after (48), so the likelihood ratio $L(\mathbf{X}) = \mathrm{d}G(\mathbf{X})/\mathrm{d}H(\mathbf{X})$ is *not* simply the product $\prod_{j=1}^{d'} g_j(X_j)/h_j(X_j)$; see Eq. (33) of [13] for details. □

Glynn [24] develops other estimators of the CDF F using IS, leading to different IS quantile estimators. Through a simple example, he shows that of the IS p-quantile estimators he considers, $\hat{\xi}_{\text{IS},n}$ in (50) can be the most effective in reducing variance when $p \approx 1$, but another of his IS p-quantile estimators can be better when $p \approx 0$.

Under a variety of different sets of assumptions (see [1, 13, 24]), the IS p-quantile estimator $\hat{\xi}_{\text{IS},n}$ in (50) satisfies the CLT in (15), where ψ^2 in (16) equals ψ^2_{IS} in (51). Also, [13] shows that $\hat{\xi}_{\text{CV},n}$ satisfies a weak Bahadur representation, as in (20) and (22). Moreover, [48] shows another IS p-quantile estimator from [24] obeys a strong Bahadur representation.

5.5 Conditional Monte Carlo

Conditional Monte Carlo (CMC) reduces variance by analytically integrating out some of the variability; see Sect. V.4 of [4] for an overview of CMC to estimate a mean. We next explain how to employ CMC for estimating a quantile, as developed in [3, 17, 18, 40], which fits into the general framework given in Sect. 3.

Let $\mathbf{Z}$ be an $\Re^{\bar{d}}$-valued random vector that is generated along with the output Y. In the special case when Y has the form in (5), we assume that

$$\mathbf{Z} = c_{\mathbf{Z}}(U_1, U_2, \ldots, U_d) \tag{52}$$

for a given function $c_{\mathbf{Z}} : [0, 1)^d \to \Re^{\bar{d}}$. Because (52) and (5) utilize the same unif[0, 1) inputs $U_1, U_2, \ldots, U_d$, we see that $\mathbf{Z}$ and Y are dependent. In general, by using iterated expectations (e.g., p. 448 of [9]), we express the CDF F of Y as

$$F(y) = \mathsf{P}(Y \le y) = \mathsf{E}[\mathsf{P}(Y \le y \mid \mathbf{Z})] = \mathsf{E}[q(\mathbf{Z}, y)], \tag{53}$$

where the function $q : \Re^{\bar{d}+1} \to \Re$ is defined for each $\mathbf{z} \in \Re^{\bar{d}}$ as

$$q(\mathbf{z}, y) = \mathsf{P}(Y \le y \mid \mathbf{Z} = \mathbf{z}) = \mathsf{E}[I(Y \le y) \mid \mathbf{Z} = \mathbf{z}]. \tag{54}$$

We assume that $q(\mathbf{z}, y)$ can be computed, analytically or numerically, for each possible $\mathbf{z}$ and $y \in \Re$. By (53), we can obtain an unbiased estimator of $F(y)$ by averaging i.i.d. copies of $q(\mathbf{Z}, y)$. Specifically, let $\mathbf{Z}_1, \mathbf{Z}_2, \ldots, \mathbf{Z}_n$ be i.i.d. replicates of the conditioning vector $\mathbf{Z}$. We then define the CMC estimator of the CDF F by

$$\hat{F}_{\text{CMC},n}(y) = \frac{1}{n} \sum_{i=1}^{n} q(\mathbf{Z}_i, y), \tag{55}$$

which uses copies of $\mathbf{Z}$ but not of Y. We finally get the CMC p-quantile estimator

$$\hat{\xi}_{\text{CMC},n} = \hat{F}^{-1}_{\text{CMC},n}(p). \tag{56}$$

Thus, we obtained (55) and (56) by following Steps 1a, 1b, and 2 of Sects. 3 and 3.2, where in Step 1a, we take $J(y) = q(\mathbf{Z}, y)$, which satisfies (14) by (53). Computing the inverse in (56) typically requires employing an iterative root-finding method, such as the bisection method or Newton's method (e.g., Chap. 7 of [41]), incurring some computation cost.

Example 1 (*continued*) For a SAN, [47] develops a CMC approach for estimating the CDF F of Y, which we apply as follows. Let the conditioning vector $\mathbf{Z}$ be the (random) durations of the activities on the path $\mathscr{P}_3 = \{1, 3, 5\}$, so $\mathbf{Z} = (X_1, X_3, X_5) \in \Re^{\bar{d}}$ with $\bar{d} = 3$. Thus, the function $c_{\mathbf{Z}}$ in (52) is given by

$$c_{\mathbf{Z}}(U_1, U_2, \dots, U_5) = (G_1^{-1}(U_1), G_3^{-1}(U_3), G_5^{-1}(U_5)).$$

Recall that for each $k = 1, 2, 3$, we defined $T_k = \sum_{j \in \mathscr{P}_k} X_k$, the (random) length of path $\mathscr{P}_k$ in (1). Since $\{Y \leq y\} = \{T_1 \leq y, T_2 \leq y, T_3 \leq y\}$ by (2), we can compute the function $q(\mathbf{z}, y)$ in (54) for any constant $\mathbf{z} = (x_1, x_3, x_5) \in \Re^{\bar{d}}$ as

$$\begin{aligned}
q((x_1, x_3, x_5), y) &= \mathsf{P}(Y \leq y \mid X_1 = x_1,\ X_3 = x_3,\ X_5 = x_5) \\
&= \mathsf{P}(X_1 + X_2 \leq y,\ X_4 + X_5 \leq y,\ X_1 + X_3 + X_5 \leq y \mid X_1 = x_1,\ X_3 = x_3,\ X_5 = x_5) \\
&= \mathsf{P}(X_2 \leq y - x_1,\ X_4 \leq y - x_5,\ x_1 + x_3 + x_5 \leq y \mid X_1 = x_1,\ X_3 = x_3,\ X_5 = x_5) \\
&= \mathsf{P}(X_2 \leq y - x_1)\, \mathsf{P}(X_4 \leq y - x_5)\, \mathsf{P}(x_1 + x_3 + x_5 \leq y) \\
&= (1 - e^{-(y - x_1)})\, (1 - e^{-(y - x_5)})\, I(x_1 + x_3 + x_5 \leq y)
\end{aligned}$$

because $X_1, X_2, \dots, X_5$ are i.i.d. exponential with mean 1. □

Applying a variance decomposition (e.g., Problem 34.10 of [9]) yields

$$\begin{aligned}
\mathsf{Var}[I(Y \leq y)] &= \mathsf{Var}[\mathsf{E}[I(Y \leq y) \mid \mathbf{Z}]] + \mathsf{E}[\mathsf{Var}[I(Y \leq y) \mid \mathbf{Z}]] \\
&\geq \mathsf{Var}[\mathsf{E}[I(Y \leq y) \mid \mathbf{Z}]] = \mathsf{Var}[q(\mathbf{Z}, y)]
\end{aligned}$$

for each y, where the inequality uses the nonnegativity of conditional variance, and the last step holds by (54). Hence, for each y, averaging i.i.d. copies of $q(\mathbf{Z}, y)$, as is done in constructing $\hat{F}_{\mathrm{CMC},n}(y)$ in (55), leads to smaller variance than averaging i.i.d. copies of $I(Y \leq y)$, as in the estimator $\hat{F}_{\mathrm{NMC},n}(y)$ in (9). We thus conclude that CMC provides a CDF estimator with lower variance at each point than NMC.

The CMC p-quantile estimator $\hat{\xi}_{\mathrm{CMC},n}$ in (56) obeys the CLT (15) with ψ^2 in (16) as

$$\psi^2_{\mathrm{CMC}} = \mathsf{Var}[q(\mathbf{Z}, \xi)] \leq \mathsf{Var}[I(Y \leq \xi)] = p(1-p) = \psi^2_{\mathrm{NMC}} \tag{57}$$

By (57), so the CMC p-quantile estimator has no greater asymptotic variance than that of NMC; See [3, 17, 18, 40]. Also, $\hat{\xi}_{\mathrm{CMC},n}$ has a weak Bahadur representation, as in (20), (22).

While we have applied CMC by conditioning on a random vector $\mathbf{Z}$, the method can be more generally applied by instead conditioning on a sigma-field; see [3].

5.6 Other Approaches

LHS in Sect. 5.3 reduces variance by inducing negative correlation among the outputs, and [6] examines quantile estimation via other correlation-induction schemes, including *antithetic variates* (AV); see also [13]. (Randomized) quasi–Monte Carlo has been applied for quantile estimation [26, 32, 42]. Other simulation-based methods for estimating ξ do not follow the approach in Steps 1 and 2 of Sect. 3. For example, [45] considers quantile estimation as a root-finding problem, and applies stochastic approximation to solve it.

We can also combine different variance-reduction techniques to estimate a quantile. The integrated methods can sometimes (but not always) behave synergistically, outperforming each approach by itself. Some particularly effective mergers include combined IS+SS [13, 23], CMC+LHS [17], and SS+CMC+LHS [18].

5.7 Numerical Results of Point Estimators for Quantiles

We now provide numerical results comparing some of the methods discussed in Sects. 3.1 and 5.1–5.6 applied to the SAN model in Example 1. Using 10^3 independent replications, we estimated the bias, variance, and mean-square error (MSE) of quantile estimators with sample size $n = 640$, where we numerically computed (without simulation) the true values of the p-quantile ξ as approximately $\xi = 3.58049$ for $p = 0.6$ and $\xi = 6.66446$ for $p = 0.95$. For each method x, we computed the MSE improvement factor (IF) of x as the ratio of the MSEs for NMC and x.

Table 1 shows that each VRT reduces the variance and MSE compared to NMC. Each VRT also produces less bias for $p = 0.95$, but not always for $p = 0.6$, especially for IS. The IS approach (Sect. 5.4) for the SAN is designed to estimate the p-quantile when $p \approx 1$, and it leads to substantial MSE improvement for $p = 0.95$. But for $p = 0.6$, IS only slightly outperforms NMC. Also, observe that the IF of the combi-

Table 1 Bias, variance, and mean-square error of p-quantile estimators for $p = 0.6$ and 0.95, where a method's MSE improvement factor (IF) is the ratio of the MSEs of NMC and the method

Method	$p = 0.6$				$p = 0.95$			
	Bias ($\times 10^{-3}$)	Variance ($\times 10^{-3}$)	MSE ($\times 10^{-3}$)	MSE IF	Bias ($\times 10^{-2}$)	Variance ($\times 10^{-2}$)	MSE ($\times 10^{-2}$)	MSE IF
NMC	1.32	7.18	7.18	1.00	−3.00	5.15	5.24	1.00
CV	1.45	3.88	3.89	1.85	0.69	2.15	2.15	2.44
LHS	−0.87	2.78	2.78	2.58	−1.74	2.36	2.39	2.19
IS	12.39	6.43	6.58	1.09	1.46	1.01	1.03	5.09
CMC	3.39	5.26	5.27	1.36	0.03	4.01	4.01	1.31
CMC+LHS	0.84	1.32	1.32	5.42	−0.36	1.67	1.67	3.14

nation CMC+LHS is larger than the product of the IFs of CMC and LHS, illustrating that their combination can work synergistically together.

6 Confidence Intervals for a Quantile

Example 1 (*continued*) The contractor understands that her p-quantile estimator $\hat{\xi}_n$ does not exactly equal the true p-quantile ξ due to Monte Carlo's sampling noise. To account for the statistical error, she also desires a 90% confidence interval $\mathscr{C}_n$ for ξ, so she can be highly confident that the true value of ξ lies in $\mathscr{C}_n$. □

We want a confidence interval (CI) $\mathscr{C}_n$ for ξ based on a sample size n satisfying

$$\mathsf{P}(\xi \in \mathscr{C}_n) = 1 - \alpha \tag{58}$$

for a user-specified constant $0 < \alpha < 1$, where $1 - \alpha$ is the desired *confidence level*, e.g., $1 - \alpha = 0.9$ for a 90% CI. In a few limited cases, we can design a CI for which (58) or $\mathsf{P}(\xi \in \mathscr{C}_n) \geq 1 - \alpha$ holds for a fixed n. But for most Monte Carlo methods, we instead have to be satisfied with a large-sample CI $\mathscr{C}_n$ for which

$$\mathsf{P}(\xi \in \mathscr{C}_n) \to 1 - \alpha, \quad \text{as } n \to \infty. \tag{59}$$

6.1 *Small-Sample CIs*

Consider applying NMC as in Sect. 3.1 with a fixed sample size n. Let $Y_1, Y_2, \ldots, Y_n$ be an i.i.d. sample from F, which we assume is continuous at ξ, ensuring that $\mathsf{P}(Y_i \leq \xi) = p$. Then $B_{n,p} \equiv n\hat{F}_{\mathrm{NMC},n}(\xi) = \sum_{i=1}^{n} I(Y_i \leq \xi)$ has a binomial(n, p) distribution by (9). Recall that $Y_{i:n}$ is the ith smallest value in the sample, so $\{Y_{i:n} \leq \xi\} = \{B_{n,p} \geq i\}$, which is equivalent to $\{Y_{i:n} > \xi\} = \{B_{n,p} < i\}$, Thus, for any integers $1 \leq i_1 < i_2 \leq n$, we see that

$$\mathsf{P}(Y_{i_1:n} \leq \xi < Y_{i_2:n}) = \mathsf{P}(i_1 \leq B_{n,p} < i_2) = 1 - \mathsf{P}(B_{n,p} < i_1) - \mathsf{P}(B_{n,p} \geq i_2).$$

If we select i_1 and i_2 such that $\mathsf{P}(B_{n,p} < i_1) + \mathsf{P}(B_{n,p} \geq i_2) \leq \alpha$, then

$$\mathscr{C}_{\mathrm{bin},n} \equiv [Y_{i_1:n}, Y_{i_2:n}) \tag{60}$$

is a CI for ξ with confidence level at least $1 - \alpha$. For example, we may pick i_1 and i_2 so that $\mathsf{P}(B_{n,p} < i_1) \leq \alpha/2$ and $\mathsf{P}(B_{n,p} \geq i_2) \leq \alpha/2$. We call (60) the *binomial CI*, also known as a *distribution-free CI*; Sect. 2.6.1 of [46] provides more details.

This idea unfortunately breaks down when applying a Monte Carlo method other than NMC because $n\hat{F}_n(\xi)$ no longer has a binomial distribution in general. But [29] extends the binomial approach to a multinomial for the alternative CV p-quantile estimator $\hat{\xi}'_{\mathrm{CV},n}$ described in the last paragraph of Sect. 5.1.

6.2 Consistent Estimation of Asymptotic Variance

We can also build a *large-sample* CI $\mathscr{C}_n$ for ξ satisfying (59) by exploiting the CLT in (15) or the (weak) Bahadur representation in (20) and (22), which both hold for the Monte Carlo methods we considered in Sects. 3 and 5. One approach based on the CLT (15) requires a *consistent* estimator $\hat{\tau}_n^2$ of τ^2 from (16); i.e., $\hat{\tau}_n^2 \Rightarrow \tau^2$ as $n \to \infty$. Then we can obtain a CI $\mathscr{C}_n$ for which (59) holds as

$$\mathscr{C}_{\mathrm{con},n,b} = [\hat{\xi}_n \pm z_\alpha \hat{\tau}_n / \sqrt{n}], \tag{61}$$

where $z_\alpha = \Phi^{-1}(1 - \alpha/2)$ and Φ is the $N(0, 1)$ CDF; e.g., $z_\alpha = 1.645$ for $1 - \alpha = 0.9$. A way to construct a consistent estimator $\hat{\tau}_n^2$ of $\tau^2 = \psi^2/f^2(\xi)$ devises a consistent estimator $\hat{\psi}_n^2$ of the numerator ψ^2 and also one for the denominator $f^2(\xi)$.

To handle ψ^2, [13] develops consistent estimators $\hat{\psi}_n^2$ when ψ^2 equals ψ_{CV}^2 in (35) for CV, $\psi_{\mathrm{SS},\gamma}^2$ in (42) for SS, and ψ_{IS}^2 in (51) for IS, as well as for IS+SS. Also, [40] provides an estimator for ψ_{CMC}^2 in (57), and [16] handles LHS. For NMC, (24) shows that $\psi_{\mathrm{NMC}}^2 = p(1 - p)$, which does not require estimation.

Several techniques have been devised to consistently estimate $f(\xi)$ appearing in the denominator of (16). One approach exploits the fact that

$$\eta \equiv \frac{1}{f(\xi)} = \frac{\mathrm{d}}{\mathrm{d}p} F^{-1}(p) = \lim_{\delta \to 0} \frac{F^{-1}(p + \delta) - F^{-1}(p - \delta)}{2\delta} \tag{62}$$

by the chain rule of differentiation, which suggests estimating η by a *finite difference*

$$\hat{\eta}_n = \frac{\hat{F}_n^{-1}(p + \delta_n) - \hat{F}_n^{-1}(p - \delta_n)}{2\delta_n}, \tag{63}$$

for some user-specified *bandwidth* $\delta_n > 0$. For the case of NMC, [10, 11] establish the consistency of $\hat{\eta}_n$ when $\delta_n \to 0$ and $n\delta_n \to \infty$ as $n \to \infty$, and [13, 16] develop similar results when applying various variance-reduction techniques. Then in (61), we can use $\hat{\tau}_n^2 = \hat{\psi}_n^2 \hat{\eta}_n^2$ to consistently estimate τ^2. Kernel methods [19, 37, 43] have also been employed to estimate $f(\xi)$.

6.3 Batching, Sectioning, and Other Methods

An issue with the finite-difference estimator in (63) and with kernel methods is that for a given sample size n, the user must specify an appropriate bandwidth δ_n, which can be difficult to do in practice. To avoid this complication, we can instead build a CI for ξ via a method that does not try to consistently estimate the asymptotic variance τ^2 in (16).

Batching is such an approach; e.g., see p. 491 of [22]. Rather than computing one p-quantile estimator from a single sample, batching instead generates $b \geq 2$ independent samples, each called a *batch* (or *subsample*), and builds a p-quantile estimator from each batch. We then construct a CI from the sample average and sample variance of the b i.i.d. p-quantile estimators. Specifically, to keep the overall sample size as n, we generate the b independent batches to each have size $m = n/b$. In practice, setting $b = 10$ is often a reasonable choice. For example, for NMC with an overall sample $Y_1, Y_2, \ldots, Y_n$ of size n, batch $\ell = 1, 2, \ldots, b$, comprises observations $Y_{(\ell-1)m+i}$, $i = 1, 2, \ldots, m$. From each batch $\ell = 1, 2, \ldots, b$, we compute a p-quantile estimator $\hat{\xi}_{m,\ell}$, which is roughly normally distributed when the batch size $m = n/b$ is large, by the CLT in (15). As the batches are independent, we have that $\hat{\xi}_{m,\ell}$, $\ell = 1, 2, \ldots, b$, are i.i.d. From their sample average $\bar{\xi}_{n,b} = (1/b) \sum_{\ell=1}^{b} \hat{\xi}_{m,\ell}$ and sample variance $S_{n,b}^2 = (1/(b-1)) \sum_{\ell=1}^{b} [\hat{\xi}_{m,\ell} - \bar{\xi}_{n,b}]^2$, we obtain the *batching CI* as

$$\mathscr{C}_{\mathrm{bat},n,b} = [\bar{\xi}_{n,b} \pm t_{b-1,\alpha} S_{n,b}/\sqrt{b}], \tag{64}$$

where $t_{b-1,\alpha} = \Gamma_{b-1}^{-1}(1-\alpha/2)$ with Γ_{b-1} as the CDF of a Student-t random variable with $b-1$ degrees of freedom; e.g., $t_{b-1,\alpha} = 1.83$ when $b = 10$ and $1 - \alpha = 0.9$. The batching CI $\mathscr{C}_{\mathrm{bat},n,b}$ uses a Student t critical point $t_{b-1,\alpha}$ rather than z_α from a normal, as in (61), because $\mathscr{C}_{\mathrm{bat},n,b}$ has a *fixed* (small) number b of batches, and the quantile estimator $\hat{\xi}_{m,\ell}$ from each batch ℓ is approximately normally distributed. (When applying LHS as in Sect. 5.3, each batch is an LHS sample, as in (45), but of size m. We then sample the b batches independently; see [16] for details.)

While the batching CI $\mathscr{C}_{\mathrm{bat},n,b}$ in (64) is asymptotically valid in the sense that (59) holds for any fixed $b \geq 2$, it can have poor performance when the overall sample size n is not large. Specifically, for a generic CI $\mathscr{C}_n$ for ξ, define the CI's *coverage* as $\mathsf{P}(\xi \in \mathscr{C}_n)$, which may differ from the nominal confidence level $1 - \alpha$ for any fixed n even though (59) holds. The issue with the batching CI stems from quantile estimators being biased in general; e.g., see Proposition 2 of [6] for the case of NMC. While the bias typically vanishes as the sample size $n \to \infty$, the bias can be significant when n is not large. The bias of the batching point estimator $\bar{\xi}_{n,b}$ is determined by the batch size $m = n/b < n$, so $\bar{\xi}_{n,b}$ may be severely contaminated by bias. Hence, the batching CI $\mathscr{C}_{\mathrm{bat},n,b}$ is centered at the wrong point on average, which can lead to poor coverage when n is small.

Sectioning can produce a CI with better coverage than batching. Introduced in Sect. III.5a of [4] for NMC and extended by [16, 39] to apply when employing different VRTs, sectioning modifies batching to center its CI at the p-quantile estimator

$\hat{\xi}_n$ based on the entire sample size n rather than at the batching point estimator $\bar{\xi}_{n,b}$. For example, for NMC, we use $\hat{\xi}_n = \hat{\xi}_{\mathrm{NMC},n}$ from (11). We also replace $S_{n,b}$ in (64) with $S'_{n,b}$, where $S'^2_{n,b} = (1/(b-1)) \sum_{\ell=1}^{b} [\hat{\xi}_{m,\ell} - \hat{\xi}_n]^2$. The *sectioning CI* is then

$$\mathscr{C}_{\mathrm{sec},n,b} = [\hat{\xi}_n \pm t_{b-1,\alpha} S'_{n,b}/\sqrt{b}]. \tag{65}$$

Because we center $\mathscr{C}_{\mathrm{sec},n,b}$ at $\hat{\xi}_n$ instead of the typically more-biased $\bar{\xi}_{n,b}$, the sectioning CI $\mathscr{C}_{\mathrm{sec},n,b}$ can have better coverage than the batching CI $\mathscr{C}_{\mathrm{bat},n,b}$ when n is small. By exploiting a weak Bahadur representation, as in (20) and (22), we can rigorously justify replacing the batching point estimator $\bar{\xi}_{n,b}$ in (64) with the overall point estimator $\hat{\xi}_n$ and still maintain the asymptotic validity in (59).

For NMC, bootstrap CIs for ξ have been developed in in [7, 36]. Also, [35] develops bootstrap CI for ξ when applying IS.

6.4 Numerical Results of CIs for Quantiles

Table 2 provides numerical results comparing the methods discussed in Sects. 6.1–6.3 to construct nominal 90% CIs for a p-quantile ξ of the longest path Y in the SAN model in Example 1 for different values of p. We built the CIs using NMC with different overall sample sizes n. For the consistent CI in (61), we estimated $\eta = 1/f(\xi)$ in (62) via the finite difference in (63) with bandwidth $\delta_n = 1/\sqrt{n}$. For a given CI $\mathscr{C}_n$ based on an overall sample size n, we estimated its coverage $\mathsf{P}(\xi \in \mathscr{C}_n)$

Table 2 Average relative half width (ARHW) and coverage of nominal 90% CIs for the p-quantile for $p = 0.6$ and 0.95 with different sample sizes n when applying NMC. Batching and sectioning use $b = n/10$ batches

n	Method	$p = 0.6$		$p = 0.95$	
		ARHW	Coverage	ARHW	Coverage
400	Binomial	0.053	0.921	0.082	0.932
400	Consistent	0.051	0.893	0.094	0.952
400	Batching	0.054	0.869	0.069	0.666
400	Sectioning	0.055	0.907	0.075	0.888
1600	Binomial	0.026	0.910	0.038	0.914
1600	Consistent	0.025	0.896	0.039	0.916
1600	Batching	0.027	0.893	0.037	0.838
1600	Sectioning	0.028	0.904	0.038	0.904
6400	Binomial	0.013	0.904	0.018	0.905
6400	Consistent	0.013	0.897	0.018	0.899
6400	Batching	0.014	0.898	0.019	0.885
6400	Sectioning	0.014	0.900	0.019	0.903

from 10^4 independent replications. Also, we computed for each method the average relative half width (ARHW), defined as the average half-width of the CI divided by the true p-quantile ξ, computed numerically; Sect. 5.7 gives the values.

Comparing the results for $p = 0.6$ and $p = 0.95$, we see that the more extreme quantile is harder to estimate, which is typically the case. For example, for the same n, the ARHW for $p = 0.95$ is larger than for $p = 0.6$. To see why, recall that the NMC p-quantile estimator's asymptotic variance is $p(1-p)/f^2(\xi)$ by (16) and (24). Although the numerator shrinks as p approaches 1, the denominator $f^2(\xi)$ decreases much faster. Moreover, while each method's coverage for $p = 0.6$ is close to the nominal 0.9 for each n, the consistent CI and the batching CI from (64) for $p = 0.95$ exhibit coverages that substantially depart from 0.9 when n is small, with overcoverage (resp., undercoverage) for the consistent (resp., batching) CI. When n is large, both methods produce CIs with close to nominal coverage, illustrating their asymptotic validity. As explained in Sect. 6.3, the batching CI can suffer from poor coverage for small n because the batching point estimator can be significantly biased. In contrast, the binomial CI in (60) and sectioning CI from (65) have coverage close to 0.9 for all n. It is important to remember that the binomial CI does not apply in general when applying VRTs, but sectioning does.

7 Summary and Concluding Remarks

This tutorial reviewed various Monte Carlo methods for estimating a p-quantile ξ of the CDF F of a random variable Y. Because $\xi = F^{-1}(p)$, a common approach for estimating ξ first obtains an estimator $\hat{F}_n$ of F, and then inverts $\hat{F}_n$ to obtain a p-quantile estimator $\hat{\xi}_n = \hat{F}_n^{-1}(p)$. Sections 3 and 5 applied this approach to construct quantile estimators based on different Monte Carlo methods. We also discussed techniques for constructing confidence intervals for ξ. In addition to our paper, [28] further surveys simulation procedures for estimating ξ, along with another risk measure $\mathsf{E}[Y \mid Y > \xi]$, which is known as the conditional value-at-risk, expected shortfall, or conditional tail expectation, and often used in finance.

We focused on quantile estimation for the setting in which the outputs are i.i.d., but there has also been work covering the situation when outputs form a dependent sequence, as in a time series or stochastic process. For example, see [2, 52] and references therein.

Acknowledgements This work has been supported in part by the National Science Foundation under Grant No. CMMI-1537322. Any opinions, findings, and conclusions or recommendations expressed in this material are those of the authors and do not necessarily reflect the views of the National Science Foundation.

References

1. Ahn, J.Y., Shyamalkumar, N.D.: Large sample behavior of the CTE and VaR estimators under importance sampling. N. Am. Actuar. J. **15**, 393–416 (2011)
2. Alexopoulos, C., Goldsman, D., Mokashi, A.C., Wilson, J.R.: Automated estimation of extreme steady-state quantiles via the maximum transformation. ACM Trans. Model. Comput. Simul. **27**(4), 22:1–22:29 (2017)
3. Asmussen, S.: Conditional Monte Carlo for sums, with applications to insurance and finance. Ann. Actuar. Sci. **12**(2), 455–478 (2018)
4. Asmussen, S., Glynn, P.: Stochastic Simulation: Algorithms and Analysis. Springer, New York (2007)
5. Austin, P., Schull, M.: Quantile regression: a tool for out-of-hospital research. Acad. Emerg. Med. **10**(7), 789–797 (2003)
6. Avramidis, A.N., Wilson, J.R.: Correlation-induction techniques for estimating quantiles in simulation. Oper. Res. **46**, 574–591 (1998)
7. Babu, G.J.: A note on bootstrapping the variance of sample quantile. Ann. Inst. Stat. Math. **38**(3), 439–443 (1986)
8. Bahadur, R.R.: A note on quantiles in large samples. Ann. Math. Stat. **37**(3), 577–580 (1966)
9. Billingsley, P.: Probability and Measure, 3rd edn. Wiley, New York (1995)
10. Bloch, D.A., Gastwirth, J.L.: On a simple estimate of the reciprocal of the density function. Ann. Math. Stat. **39**, 1083–1085 (1968)
11. Bofinger, E.: Estimation of a density function using order statistics. Aust. J. Stat. **17**, 1–7 (1975)
12. Cannamela, C., Garnier, J., Iooss, B.: Controlled stratification for quantile estimation. Ann. Appl. Stat. **2**(4), 1554–1580 (2008)
13. Chu, F., Nakayama, M.K.: Confidence intervals for quantiles when applying variance-reduction techniques. ACM Trans. Model. Comput. Simul. **22**(2), 10:1–10:25 (2012)
14. Davis, P.J., Rabinowitz, P.: Methods of Numerical Integration, 2nd edn. Academic, San Diego (1984)
15. Dong, H., Nakayama, M.K.: Constructing confidence intervals for a quantile using batching and sectioning when applying latin hypercube sampling. In: Tolk, A., Diallo, S.D., Ryzhov, I.O., Yilmaz, L., Buckley, S., Miller, J.A. (eds.) Proceedings of the 2014 Winter Simulation Conference, pp. 640–651. Institute of Electrical and Electronics Engineers, Piscataway, New Jersey (2014)
16. Dong, H., Nakayama, M.K.: Quantile estimation with Latin hypercube sampling. Oper. Res. **65**(6), 1678–1695 (2017)
17. Dong, H., Nakayama, M.K.: Quantile estimattion using conditional Monte Carlo and latin hypercube sampling. In: Chan, W.K.V., D'Ambrogio, A., Zacharewicz, G., Mustafee, N., Wainer, G. Page, E. (eds.) Proceedings of the 2017 Winter Simulation Conference, pp. 1986–1997. Institute of Electrical and Electronics Engineers, Piscataway, NJ (2017)
18. Dong, H., Nakayama, M.K.: Quantile estimation using stratified sampling, conditional Monte Carlo, and Latin hypercube sampling (2018). In preparation
19. Falk, M.: On the estimation of the quantile density function. Stat. Probab. Lett. **4**, 69–73 (1986)
20. Garnett, O., Mandelbaum, A., Reiman, M.: Designing a call center with impatient customers. Manuf. Serv. Oper. Manag. **4**(3), 208–227 (2002)
21. Ghosh, J.K.: A new proof of the Bahadur representation of quantiles and an application. Ann. Math. Stat. **42**, 1957–1961 (1971)
22. Glasserman, P.: Monte Carlo Methods in Financial Engineering. Springer, New York (2004)
23. Glasserman, P., Heidelberger, P., Shahabuddin, P.: Variance reduction techniques for estimating value-at-risk. Manag. Sci. **46**, 1349–1364 (2000)
24. Glynn, P.W.: Importance sampling for Monte Carlo estimation of quantiles. In: S.M. Ermakov, V.B. Melas (eds.) Mathematical Methods in Stochastic Simulation and Experimental Design: Proceedings of the 2nd St. Petersburg Workshop on Simulation, pp. 180–185. Publishing House of St. Petersburg Univ., St. Petersburg, Russia (1996)

25. Grabaskas, D., Nakayama, M.K., Denning, R., Aldemir, T.: Advantages of variance reduction techniques in establishing confidence intervals for quantiles. Reliab. Eng. Syst. Saf. **149**, 187–203 (2016)
26. He, Z., Wang, X.: Convergence of randomized quasi-Monte Carlo sampling for value-at-risk and conditional value-at-risk (2017). ArXiv:1706.00540
27. Hesterberg, T.C., Nelson, B.L.: Control variates for probability and quantile estimation. Manag. Sci. **44**, 1295–1312 (1998)
28. Hong, L.J., Hu, Z., Liu, G.: Monte Carlo methods for value-at-risk and conditional value-at-risk: a review. ACM Trans. Model. Comput. Simul. **24**(4), 22:1–22:37 (2014)
29. Hsu, J.C., Nelson, B.L.: Control variates for quantile estimation. Manag. Sci. **36**, 835–851 (1990)
30. Hyndman, R.J., Fan, Y.: Sample quantiles in statistical packages. Am. Stat. **50**(4), 361–365 (1996)
31. Jin, X., Fu, M.C., Xiong, X.: Probabilistic error bounds for simulation quantile estimation. Manag. Sci. **49**, 230–246 (2003)
32. Jin, X., Zhang, A.X.: Reclaiming quasi-Monte Carlo efficiency in portfolio value-at-risk simulation through Fourier transform. Manag. Sci. **52**(6), 925–938 (2006)
33. Jorion, P.: Value at Risk: The New Benchmark for Managing Financial Risk, 3rd edn. McGraw-Hill, New York (2007)
34. Juneja, S., Karandikar, R., Shahabuddin, P.: Asymptotics and fast simulation for tail probabilities of maximum of sums of few random variables. ACM Trans. Model. Comput. Simul. **17**(2), 35 (2007)
35. Liu, J., Yang, X.: The convergence rate and asymptotic distribution of bootstrap quantile variance estimator for importance sampling. Adv. Appl. Probab. **44**, 815–841 (2012)
36. Maritz, J.S., Jarrett, R.G.: A note on estimating the variance of the sample median. J. Am. Stat. Assoc. **73**, 194–196 (1978)
37. Nakayama, M.K.: Asymptotic properties of kernel density estimators when applying importance sampling. In: Jain, S., Creasey, R., Himmelspach, J., White, K., Fu,M. (eds.) Proceedings of the 2011 Winter Simulation Conference, pp. 556–568. Institute of Electrical and Electronics Engineers, Piscataway, New Jersey (2011)
38. Nakayama, M.K.: Asymptotically valid confidence intervals for quantiles and values-at-risk when applying latin hypercube sampling. Int. J. Adv. Syst. Meas. **4**, 86–94 (2011)
39. Nakayama, M.K.: Confidence intervals using sectioning for quantiles when applying variance-reduction techniques. ACM Trans. Model. Comput. Simul. **24**(4), 19:1–19:21 (2014)
40. Nakayama, M.K.: Quantile estimation when applying conditional Monte Carlo. In: SIMULTECH 2014 Proceedings, pp. 280–285 (2014)
41. Ortega, J.M., Rheinboldt, W.C.: terative Solution of Nonlinear Equations in Several Variables. SIAM, Philadelphia (2000)
42. Papageorgiou, A., Paskov, S.H.: Deterministic simulation for risk management. J. Portf. Manag. **25**(5), 122–127 (1999)
43. Parzen, E.: Nonparametric statistical data modeling. J. Am. Stat. Assoc. **74**(365), 105–121 (1979)
44. Ressler, R.L., Lewis, P.A.W.: Variance reduction for quantile estimates in simulations via nonlinear controls. Commun. Stat.—Simul. Comput. **B19**(3), 1045–1077 (1990)
45. Robinson, D.W.: Non-parametric quantile estimation through stochastic approximation. Ph.D. thesis, Naval Postgraduate School, Monterey, California (1975)
46. Serfling, R.J.: Approximation Theorems of Mathematical Statistics. Wiley, New York (1980)
47. Sigal, C.E., Pritsker, A.A.B., Solberg, J.J.: The use of cutsets in Monte Carlo analysis of stochastic networks. Math. Comput. Simul. **21**(4), 376–384 (1979)
48. Sun, L., Hong, L.J.: Asymptotic representations for importance-sampling estimators of value-at-risk and conditional value-at-risk. Oper. Res. Lett. **38**(4), 246–251 (2010)
49. U.S. Nuclear Regulatory Commission: Final safety evaluation for WCAP-16009-P, revision 0, "realistic large break LOCA evaluation methodology using automated statistical treatment of uncertainty method (ASTRUM)" (TAC no. MB9483). Tech. rep., U.S. Nuclear Regulatory Commission, Washington, DC (2005). https://www.nrc.gov/docs/ML0509/ML050910159.pdf

50. U.S. Nuclear Regulatory Commission: Acceptance criteria for emergency core cooling systems for light-water nuclear power reactors. Title 10, Code of Federal Regulations §50.46, NRC, Washington, DC (2010)
51. U.S. Nuclear Regulatory Commission: Applying statistics. U.S. Nuclear Regulatory Commission Report NUREG-1475, Rev 1, U.S. Nuclear Regulatory Commission, Washington, DC (2011)
52. Wu, W.B.: On the Bahadur representation of sample quantiles for dependent sequences. Ann. Stat. **33**(4), 1934–1963 (2005)

Multilevel Quasi-Monte Carlo Uncertainty Quantification for Advection-Diffusion-Reaction

Lukas Herrmann and Christoph Schwab

Abstract We survey the numerical analysis of a class of deterministic, higher-order QMC integration methods in forward and inverse uncertainty quantification algorithms for advection-diffusion-reaction (ADR) equations in polygonal domains $D \subset \mathbb{R}^2$ with distributed uncertain inputs. We admit spatially heterogeneous material properties. For the parametrization of the uncertainty, we assume at hand systems of functions which are locally supported in D. Distributed uncertain inputs are written in countably parametric, deterministic form with locally supported representation systems. Parametric regularity and sparsity of solution families and of response functions in scales of weighted Kontrat'ev spaces in D are quantified using analytic continuation.

Keywords Higher order quasi-Monte Carlo · Parametric operator equations · Bayesian inverse problems · Uncertainty quantification

1 Introduction

Computational uncertainty quantification (UQ) addresses the efficient, quantitative numerical treatment of differential- and integral equation models in engineering and in the sciences. In the simplest setting, such models need to be analyzed for *parametric* input data with sequences $\boldsymbol{y} = (y_j)_{j\geq 1}$ of parameters y_j which range in a compact, metric space U. In [15] the authors proposed and analyzed the convergence rates of higher-order Quasi-Monte Carlo (HoQMC) approximations of conditional expectations which arise in Bayesian inverse problems for partial differential equa-

L. Herrmann
Johann Radon Institute for Computational and Applied Mathematics, Austrian Academy of Science, Altenbergerstrasse 69, 4040 Linz, Austria
e-mail: lukas.herrmann@ricam.oeaw.ac.at

C. Schwab (✉)
Seminar for Applied Mathematics, ETH Zürich, Rämistrasse 101, 8092 Zurich, Switzerland
e-mail: christoph.schwab@sam.math.ethz.ch

B. Tuffin and P. L'Ecuyer (eds.), *Monte Carlo and Quasi-Monte Carlo Methods*, Springer Proceedings in Mathematics & Statistics 324,
https://doi.org/10.1007/978-3-030-43465-6_2

tions (PDEs). The authors studied broad classes of parametric operator equations with *distributed uncertain parametric input data.* Typical examples are elliptic or parabolic partial differential equations with uncertain, spatially heterogeneous coefficients, but also differential- and integral equations in uncertain physical domains of definition are admissible. Upon suitable *uncertainty parametrization* and, in inverse uncertainty quantification, with a suitable Bayesian prior measure placed on the, in general, infinite-dimensional parameter space, the task of numerical evaluation of statistical estimates for quantities of interest (QoI's) becomes *numerical computation of parametric, deterministic integrals over a high-dimensional parameter space.*

The method of choice in many current inverse computational UQ is the Markov chain Monte Carlo (MCMC) method and its variants [6, 30]. Due to its Monte Carlo character, it affords a generally low convergence rate and, due to the intrinsically sequential nature of, e.g., the independence sampler, MCMC meet with difficulties for parallelization. As an alternative to the MCMC method, in [8, 43, 44] recently developed, dimension-adaptive Smolyak quadrature techniques were applied to the evaluation of the corresponding integrals. In [15, 16] a convergence theory for HoQMC integration for the numerical evaluation of the corresponding integrals was developed, based on earlier work [17] on these methods in forward UQ. In particular, it was shown in [16] that convergence rates of order $> 1/2$ in terms of the number N of approximate solves of the forward problem that are independent of the dimension can be achieved with judiciously chosen, *deterministic* HoQMC quadratures instead of Monte Carlo or MCMC sampling of the Bayesian posterior. The achievable, dimension-independent rate of HoQMC is, in principle, only limited by the sparsity of the forward problem. Moreover, the execution of the algorithm is "embarrassingly parallel", since for QMC algorithms, unlike MCMC and sequential Monte Carlo (SMC) methods, the forward problem may be solved simultaneously and in parallel. The error analysis in [16] was extended in [15] to the multilevel setting. As is well known in the context of Monte Carlo methods, *multilevel strategies* can lead to substantial gains in accuracy versus computational cost, see also the survey [26] on multilevel Monte Carlo (MLMC) methods. Multilevel discretizations for QMC integration were explored first for parametric, linear forward problems in [32, 34] and in the context of HoQMC for parametric operator equations in [15]. For the use of multilevel strategies in the context of MCMC methods for Bayesian inverse problems we refer to [21, 30] and the references there. The purpose of the present paper is to extend the convergence analysis of deterministic Bayesian inversion algorithms for forward problems given by PDEs with distributed random input data, which are based on Quasi-Monte Carlo integration from [15] and the references there, to uncertainty parametrization with basis functions which are locally supported in the physical domain D. Let us mention in passing that while we consider here conforming Finite Element (FE) discretization, other discretizations in D could equally be considered. We mention only discontinuous Galerkin Finite Element methods (FEM) which have been introduced for advection-diffusion-reaction (ADR) equations as considered here in [31]. The duality argument in weighted function spaces for these methods has been developed in [35].

The principal contributions of the present work are as follows: We prove, for a class of linear ADR problems in a polygon D with uncertain diffusion coefficients, drift coefficients and reaction coefficient, the well-posedness of the corresponding Bayesian inverse problem. We establish optimal convergence rate bounds of FE discretizations of the parametric forward problem, with judicious mesh refinement towards the corners $\mathscr{C}$ of D, allowing in particular also corner singularities in the uncertain input data; these appear typically in Karhunen-Loève eigenfunctions corresponding to principal components of covariance operators which are negative fractional powers of elliptic precision operators in D with boundary conditions on ∂D. We show that a singularity-adapted uncertainty parametrization with locally supported in D spline-wavelet functions allows for optimal (in the sense of convergence rate) parametrization of the uncertain input data. We establish that higher order Quasi-Monte Carlo rules of IPL ("interlaced, polynomial lattice rules") type from [17] admit, for the considered boundary value problems with high-dimensional, parametric inputs a dimension-independent convergence rate which is limited only by the sparsity of the parametric input data.

The structure of this paper is as follows. In Sect. 2, we present a class of linear, second order ADR problems in bounded, polygonal domains. Particular attention is paid to regularity in weighted function spaces which account for possible singularities at the corners of the physical domain; we base our presentation on the recent reference [5] where the corresponding regularity theory has been developed. In Sect. 3, an analysis of the consistency error of FE discretizations of the parametric ADR model in polygons is presented. The analysis is uniform w.r. to the uncertain input and accounts for the impact of numerical integration in the presence of local mesh refinement to obtain a fully discrete FE approximation. Parts of the somewhat technical proofs are postponed to Sect. 7. So-called forward UQ is studied in some detail in Sect. 4 including estimates of the ε-complexity of the proposed QMC-FE algorithm, which are free of the curse of dimensionality. In Sect. 5, we review elements of the general theory of well-posed Bayesian inverse problems in function spaces, as presented e.g. in [13]. The presentation and the setup is analogous to what was used in [15], but in technical details, there are important differences: unlike the development in [15], the *uncertainty parametrization* employed in the present paper will be achieved by *locally supported functions* ψ_j in the physical domain D. In particular, we shall admit biorthogonal, piecewise polynomial multiresolution analyses in D. These allow us, as we show, to resolve uncertain inputs with corner and interface singularities at optimal rates, and their local supports enable the use of HoQMC integration with so-called SPROD ("Smoothness driven PRODuct") weights. To this end, and as in [15], we require a novel, combined regularity theory of the parametric forward maps in weighted Kondrat'ev–Sobolev spaces in D. In particular, we present an error versus work analysis of the combined multilevel HoQMC Petrov–Galerkin algorithms.

2 UQ for Advection-Diffusion-Reaction Equations in Polygons

We review the notation and mathematical setting of forward and inverse UQ for a class of smooth, parametric operator equations. We develop here the error analysis for the multilevel extension of the algorithms in [23] for general linear, second order advection-diffusion-reaction problems in an open, polygonal domain $D \subset \mathbb{R}^2$, see also [22]. We assume the uncertain inputs comprising the operators' coefficients $u = ((a_{ij}(x), b_i(x), c(x))$ to belong to a Banach space X being a weighted Hölder space in the physical domain D. As in [15], uncertainty parametrization with an unconditional basis of X will result in a countably-parametric, deterministic boundary value problem. Unlike the Karhunen-Loève basis which is often used for uncertainty parametrization in UQ, we consider here the use of *representation systems whose elements have well-localized supports contained in D*; one example are spline wavelets.

Upon adopting such representations, both forward and (Bayesian) inverse problems become countably parametric, deterministic operator equations. In [43], Bayesian inverse UQ was expressed as countably-parametric, deterministic quadrature problem, with the integrand functions appearing in the Bayesian estimation problems stemming from a (Petrov–)Galerkin discretization of the parametric forward problem in the physical domain. Contrary to widely used MCMC algorithms (e.g. [21] and the references there), high-dimensional quadratures of Smolyak type are deterministic and were proposed for numerical integration against a (Bayesian) posterior in [43, 44]. In the present paper, we review this approach for forward and (Bayesian) inverse UQ for ADR in planar, polygonal domains D. We consider in detail high order FE discretizations of the ADR problem on meshes with local corner-refinement in D. We review the use of *deterministic* HoQMC integration methods from [17, 18, 20] and the references there in multilevel algorithms for Bayesian estimation in ADR models with uncertain input.

2.1 Model Advection-Diffusion-Reaction Problem in D

We present the parametric ADR model problem in a plane, polygonal domain D and recapitulate its well-posedness and regularity, following [5]. There, in particular, regularity in weighted function spaces in D and holomorphy of the data-to-solution map for this problem in these weighted spaces was established. Optimal FE convergence rates result for Lagrangean FEM in D with locally refined meshes near the singular points of the solution (being either corners of D or boundary points where the nature of the boundary condition changes) by invoking suitable approximation results from [1] and references there.

In the bounded, polygonal domain D with J corners $\mathscr{C} = \{\boldsymbol{c}_1, \ldots, \boldsymbol{c}_J\}$, for some $J \in \mathbb{N}$, we consider the *forward problem* being the mixed boundary value problem for the linear, second order divergence form differential operator

$$
\begin{aligned}
\mathscr{L}(u)q &:= -\sum_{i,j=1}^{2} \partial_i(a_{ij}\partial_j q) + \sum_{i=1}^{2} b_i \partial_i q - \sum_{i=1}^{2} \partial_i(b_{2+i}q) + cq = f \quad \text{in } D, \\
q\Big|_{\Gamma_1} &= 0, \quad \sum_{i=1}^{2}\left(\sum_{j=1}^{2} a_{ij}\partial_j q + b_{2+i}q\right) n_i\Big|_{\Gamma_2} = 0,
\end{aligned}
\tag{1}
$$

where n denotes the unit normal vector of the domain D and $\emptyset \neq \Gamma_1 \subset \partial D$ denotes the Dirichlet boundary and $\Gamma_2 = \partial D \backslash \Gamma_1$ denotes the Neumann boundary. We shall assume that $\mathscr{C} \subset \overline{\Gamma_1}$.

Define further

$$V := \{v \in H^1(D) : v|_{\Gamma_1} = 0\},$$

where $v|_{\Gamma_1} \in H^{1/2}(\Gamma_1)$ has to be understood in the sense of the trace of $v \in H^1(D)$. Here, for a subset $\Gamma_1 \subset \Gamma$ of positive arclength, $H^{1/2}(\Gamma_1)$ denotes the Sobolev-Slobodeckij space of order $1/2$ on Γ_1, being the space of all restrictions of functions from $H^{1/2}(\Gamma)$ to Γ_1.

In (1), the differential operator $\mathscr{L}$ depends on the uncertain, parametric coefficients

$$u(\boldsymbol{y}) := ((a_{ij}(\boldsymbol{y}^0))_{1\le i,j\le 2}, (b_i(\boldsymbol{y}^1))_{1\le i\le 4}, c(\boldsymbol{y}^2)), \quad \boldsymbol{y}^i \in \left[-\frac{1}{2}, \frac{1}{2}\right]^{\mathbb{N}}, i = 0, 1, 2,$$

where $a_{ij} = a_{ji}$ and where we have used the notation $\boldsymbol{y} := (\boldsymbol{y}^0, \boldsymbol{y}^1, \boldsymbol{y}^2)$ and further introduce the parameter set

$$U := \prod_{i=0,1,2} \left[-\frac{1}{2}, \frac{1}{2}\right]^{\mathbb{N}}.$$

The uncertain coefficient functions $u(\boldsymbol{y})$ may also depend on the spatial coordinate $x \in D$, and for each $\boldsymbol{y} \in U$ are assumed to belong to weighted Sobolev spaces $\mathscr{W}^{m,\infty}(D)$ of integer order $m \geq 0$ being given by

$$\mathscr{W}^{m,\infty}(D) := \{v : D \to \mathbb{C} : r_D^{|\alpha|}\partial^\alpha v \in L^\infty(D), |\alpha| \le m\}. \tag{2}$$

Specifically, for $m \in \mathbb{N}_0$, we assume that

$$
\begin{aligned}
u \in X_m :=&\{u : a_{ij} \in \mathscr{W}^{m,\infty}(D), i, j = 1, 2, r_D b_i \in \mathscr{W}^{m,\infty}(D), \\
&i = 1, \ldots, 4, r_D^2 c \in \mathscr{W}^{m,\infty}(D)\}.
\end{aligned}
\tag{3}
$$

Here, $D \ni x \mapsto r_D(x)$ denotes a "regularized" distance to the corners $\mathscr{C}$ of D, i.e., $r_D(x) \simeq \text{dist}(x, \mathscr{C})$ for $x \in D$. We equip X_m, $m \in \mathbb{N}_0$, with the norm

$$
\begin{aligned}
\|u\|_{X_m} :=& \max\{\|a_{ij}\|_{\mathscr{W}^{m,\infty}(D)}, i, j = 1, 2, \|r_D b_i\|_{\mathscr{W}^{m,\infty}(D)}, \\
&i = 1, \ldots, 4, \|r_D^2 c\|_{\mathscr{W}^{m,\infty}(D)}\}.
\end{aligned}
\tag{4}
$$

We introduce the parametric bilinear form

$$A(u(\boldsymbol{y}))(w, v) := \langle \mathscr{L}(u(\boldsymbol{y}))w, v\rangle_{V^*,V}, \quad \forall w, v \in V.$$

The variational formulation of the parametric, deterministic problem reads: given $\boldsymbol{y} \in U$, find $q(\boldsymbol{y}) \in V$ such that

$$A(u(\boldsymbol{y}))(q(\boldsymbol{y}), v) = \langle f, v\rangle_{V^*,V}, \quad \forall v \in V .$$

Here, $f \in V^*$ and $\langle\cdot, \cdot\rangle_{V^*,V}$ denotes the $V^* \times V$ duality pairing, with V^* denoting the Hilbertian (anti-) dual of V.[1] This parametric problem is well-posed if $u(\boldsymbol{y}) \in X_0$, $\boldsymbol{y} \in U$, is such that there exists a positive constant $c > 0$

$$\inf_{\boldsymbol{y}\in U} \Re(A(u(\boldsymbol{y}))(v, v)) \geq c\|v\|_V^2 , \quad \forall v \in V, \tag{5}$$

where $\Re(z)$ denotes the real part of $z \in \mathbb{C}$. We observe that (5) precludes implicitly that the ADR operator in (1) is singularly perturbed. This, in turn, obviates in the ensuing FE approximation theory in Sect. 3 the need for boundary layer resolution or anisotropic mesh refinements. As a consequence of (5) and of the Lax–Milgram lemma, for every $\boldsymbol{y} \in U$ the parametric solution $q(\boldsymbol{y}) \in V$ exists and satisfies the *uniform* a-priori *estimate*

$$\sup_{\boldsymbol{y}\in U} \|q(\boldsymbol{y})\|_V \leq c^{-1}\|f\|_{V^*}. \tag{6}$$

We introduce *weighted Sobolev spaces of Kondrat'ev type* $\mathscr{K}_a^m(D)$, $m \in \mathbb{N}_0 \cup \{-1\}$, $a \in \mathbb{R}$, as closures of $C^\infty(\overline{D}; \mathbb{C})$ with respect to the *homogeneous weighted norm* given by

$$\|v\|^2_{\mathscr{K}_a^m(D)} := \sum_{|\alpha|\leq m} \|r_D^{|\alpha|-a} \partial^\alpha v\|^2_{L^2(D)}. \tag{7}$$

We observe that (up to equivalence of norms) $V = \{v \in \mathscr{K}_1^1(D) : v|_{\Gamma_1} = 0\}$, which is a consequence of the Hardy inequality (see e.g. [40, Theorem 21.3]). In [5], the authors proved regularity shifts of $\mathscr{L}(u)$. Specifically, if $A(u)$ satisfies (5) and if $u \in X_m$, then by [5, Corollary 4.5 and Theorem 4.4] there exist constants $C > 0$, $a_0 > 0$ such that for every $a \in (-a_0, a_0)$, for every $f \in \mathscr{K}_{a-1}^{m-1}(D)$, and for every $\boldsymbol{y}^i \in \left[-\frac{1}{2}, \frac{1}{2}\right]^{\mathbb{N}}$, $i = 0, 1, 2$, there holds $q(\boldsymbol{y}) \in \mathscr{K}_{a+1}^{m+1}(D)$ and

$$\sup_{\boldsymbol{y}^i\in\left[-\frac{1}{2},\frac{1}{2}\right]^{\mathbb{N}}, i=0,1,2} \|q(\boldsymbol{y})\|_{\mathscr{K}_{a+1}^{m+1}(D)} \leq C\|f\|_{\mathscr{K}_{a-1}^{m-1}(D)}. \tag{8}$$

[1] For spaces V of real-valued functions, V^* denotes the Hilbertian dual; in the case that solutions $q(\boldsymbol{y})$ are complex-valued, e.g. for Helmholtz problems, V^* denotes the antidual of V. Even for parametric models with real valued solutions, complexification is required for analytic continuation to complex parameters [5].

Note that the dependence of the constant C on the coefficients u can also be made explicit, cp. [5].

2.2 Uncertainty Parametrization

For uncertainty parametrization, the data space X is assumed to be a separable, infinite-dimensional Banach space with norm $\|\cdot\|_X$ (separably valued data u in an otherwise non-separable space are equally admissible). We suppose that we have at hand representation systems $(\psi^i_j)_{j\geq 1} \subset L^\infty(D;\mathbb{R}^{k_i})$, $i \in \{0,1,2\}$ of locally supported functions in D which parametrize the uncertain coefficient functions $u = (a, b, c)$ for integers $k_0, k_1, k_2 \in \mathbb{N}$.

The smoothness scale $\{X_m\}_{m\geq 0}$ defined in (3) for $m \geq 1$ with $X = L^\infty(D)^8 = X_0 \supset X_1 \supset X_2 \supset \ldots$ (we recall $a_{ij} = a_{ji}$) and a smoothness order $t \geq 1$ is being given as part of the problem specification. We restrict the uncertain inputs u to sets X_t with "higher regularity" in order to obtain convergence rate estimates for the discretization of the forward problem. Note that for $u \in X_t$ and with ψ_j being Fourier or multiresolution analyses, higher values of t correspond to stronger decay of the ψ^i_j, $i \in \{0,1,2\}$.

For the numerical analysis of a FE discretization in D, we have to slightly strengthen the norm of X_m. To this end, we define the weighted spaces $W^{m,\infty}_\delta(D)$ for $\delta \in [0,1]$, $m \in \mathbb{N}$, as subspaces of $\mathcal{W}^{m,\infty}(D)$ equipped with the norm

$$\|v\|_{W^{m,\infty}_\delta(D)} := \max_{|\alpha|\leq m}\{\|r_D^{\max\{0,\delta+|\alpha|-1\}}\partial^\alpha v\|_{L^\infty(D)}\}.$$

Note that $W^{m,\infty}_1(D) = \mathcal{W}^{m,\infty}(D)$. For $\delta \in [0,1)$, we define $X_{m,\delta}$ by the norm

$$\|u\|_{X_{m,\delta}} := \max\{\|a_{ij}\|_{W^{m,\infty}_\delta(D)}, i,j=1,2, \|r_D b_i\|_{W^{m,\infty}_\delta(D)}, \\ i = 1,\ldots,4, \|r_D^2\|_{W^{m,\infty}_\delta(D)}\}. \tag{9}$$

It is easy to see that the embedding $X_{m,\delta} \subset X_m$ is continuous, $m \geq 1$, $\delta \in [0,1)$. We assume that the $\{\psi^i_j\}_{j\geq 1}$, $i \in \{0,1,2\}$, are scaled such that for some $\delta \in [0,1)$, $\tau \in \mathbb{N}$, and positive sequences $(\rho^i_{r,j})_{j\geq 1}$,

$$\max_{|\boldsymbol{\alpha}|\leq r}\left\|\sum_{j\geq 1}\rho^i_{r,j} r_D^{\max\{0,\delta+|\alpha|-1\}}|\partial^{\boldsymbol{\alpha}}_x((r_D)^i\psi^i_j)|\right\|_{L^\infty(D)} < \infty, \quad r = 0,\ldots,\tau. \tag{10}$$

Lemma 1 *Let $w \in \mathcal{W}^{m,\infty}(D;\mathbb{C}^k)$ for some $m,k \in \mathbb{N}$ and let $v : D \times \mathbb{C}^k \supset D \times w(D) \to \mathbb{C}$ be a function that is $\mathcal{W}^{m,\infty}$-regular in the first argument and analytic in the second. Then, $[x \mapsto v(x,w(x))] \in \mathcal{W}^{m,\infty}(D)$.*

Proof Let $\widetilde{v} := [\overline{w(D)} \ni z \mapsto v(x, z)]$ for arbitrary $x \in D$ be such that $\widetilde{v}$ is well-defined. By an application of the Faà di Bruno formula [12, Theorem 2.1],

$$
\begin{aligned}
&r_D^{|\boldsymbol{\alpha}|}\partial_x^{\boldsymbol{\alpha}}(\widetilde{v} \circ w) \\
&= \sum_{1 \le |\boldsymbol{\lambda}| \le n} \partial_y^{\boldsymbol{\lambda}} \widetilde{v} \sum_{\iota=1}^{n} \sum_{p_\iota(\boldsymbol{\alpha},\boldsymbol{\lambda})} \boldsymbol{\alpha}! \prod_{j=1}^{\iota} \frac{1}{\boldsymbol{\nu}(j)!(\boldsymbol{\eta}(j)!)^{|\boldsymbol{\nu}(j)|}} \prod_{i=1}^{k} (r_D^{|\boldsymbol{\nu}(j)|}\partial^{\boldsymbol{\nu}(j)} w_i)^{\boldsymbol{\nu}(j)_i},
\end{aligned}
$$

where $n = |\boldsymbol{\alpha}|$, $w = (w_1, \ldots, w_k)$, and

$$
\begin{aligned}
p_\iota(\boldsymbol{\alpha}, \boldsymbol{\lambda}) = \Bigg\{&(\boldsymbol{\nu}(1), \ldots, \boldsymbol{\nu}(\iota); \boldsymbol{\eta}(1), \ldots, \boldsymbol{\eta}(\iota)) : |\boldsymbol{\nu}(i)| > 0, \\
&\mathbf{0} \prec \boldsymbol{\eta}(1) \prec \ldots \prec \boldsymbol{\eta}(\iota), \sum_{i=1}^{\iota} \boldsymbol{\nu}(i) = \boldsymbol{\lambda}, \text{ and } \sum_{i=1}^{\iota} |\boldsymbol{\nu}(i)|\boldsymbol{\eta}(i) = \boldsymbol{\alpha}\Bigg\},
\end{aligned}
$$

where the multi-indices $\boldsymbol{\nu}$ are k-dimensional and the multi-indices $\boldsymbol{\eta}$ are d-dimensional (here $d = 2$, since the domain D is a polygon). The symbol $\prec$ for multi-indices $\boldsymbol{\eta}$ and $\widetilde{\boldsymbol{\eta}}$ is defined by $\boldsymbol{\eta} \prec \widetilde{\boldsymbol{\eta}}$ (here for $d = 2$) if either (i) $|\boldsymbol{\eta}| < |\widetilde{\boldsymbol{\eta}}|$ or (ii) $|\boldsymbol{\eta}| = |\widetilde{\boldsymbol{\eta}}|$ and $\eta_1 < \widetilde{\eta}_1$, where $|\boldsymbol{\eta}| = \sum_{j \ge 1} \eta_j$. Since $L^\infty(D)$ is an algebra and $|\boldsymbol{\alpha}| = \sum_{i=1}^{\iota} |\boldsymbol{\nu}(i)||\boldsymbol{\nu}(i)|$, $(\widetilde{v} \circ w) \in \mathscr{W}^{m,\infty}(D)$. The claim of the lemma now follows by another application of the Faà di Bruno formula. □

Remark 1 The statement of Lemma 1 also holds if we replace $\mathscr{W}^{m,\infty}(D)$ with $W_\delta^{m,\infty}(D)$, $\delta \in [0, 1)$, at all places.

Define the complex-parametric sets U^i for $i \in \{0, 1, 2\}$ of admissible data

$$
U^i := \left\{ \sum_{j \ge 1} z_j |\psi_j^i(x)| : \boldsymbol{z} \in \mathbb{C}^{\mathbb{N}}, |z_j| \le \rho_{0,j}^i, j \ge 1, x \in D \right\} \subset \mathbb{C}^{k_i},
$$

where $|\cdot|$ denotes component-wise absolute value. Let $g : D \times U^0 \times U^1 \times U^2 \to \mathbb{C}^8$ be a function such that $(z^0, z^1, z^2) \mapsto g(x, z^0, z^1, z^2)$ is holomorphic for almost every $x \in D$ and such that $[x \mapsto g(x, z^0, z^1, z^2)] \in X_m$ for some $m \ge 1$ and every $(z^0, z^1, z^2) \in U^0 \times U^1 \times U^2$. The uncertain coefficient $u = (a, b, c)$ is then parametrized by

$$
\begin{aligned}
u(x, \boldsymbol{y}) &= \left(a(x, \boldsymbol{y}^0), b(x, \boldsymbol{y}^1), c(x, \boldsymbol{y}^2)\right) \\
&= g\left(x, \sum_{j \ge 1} y_j^0 \psi_j^0, \sum_{j \ge 1} y_j^1 \psi_j^1, \sum_{j \ge 1} y_j^2 \psi_j^2\right), \quad \text{a.e. } x \in D, \boldsymbol{y}^i \in U, i = 0, 1, 2.
\end{aligned} \tag{11}
$$

Hence, $u = (a, b, c)$ is given through the coordinates of the function g via $a_{11} = g_1$, $a_{22} = g_2$, $a_{21} = a_{12} = g_3$, $b_i = g_{i+3}$, $i = 1, \ldots, 4$, $c = g_8$.

Elements in the space $X_{m,\delta}$ may have singularities in the corners, but can be approximated in the X_0-norm at optimal rates for example by biorthogonal wavelets with suitable refinements near vertices of D.

Proposition 1 *Let $\delta \in [0, 1)$ and $m \in \mathbb{N}$ be given. Assume further at hand a biorthogonal, compactly supported spline wavelet basis with sufficiently large number (depending on m) of vanishing moments and compactly supported dual basis. Then, there exists a constant $C > 0$ and, for every $L \in \mathbb{N}$, projection operators P_L into a biorthogonal wavelet basis such that*

$$\|w - P_L w\|_{X_0} \leq C N_L^{-m/2} \|w\|_{X_{m,\delta}}, \quad \forall w \in X_{m,\delta},$$

where N_L denotes the number of terms in the expansion $P_L w$.

The proof of this (in principle, well-known) proposition is given in Sect. 7.2, where also further details on biorthogonal wavelets are presented.

3 Finite Element Discretization

We introduce conforming Finite Element discretizations in the physical domain D and review an approximation property as a preparation for the analysis of the impact of numerical integration on locally refined meshes in D. Let $\mathfrak{T}$ denote a family of regular, simplicial triangulations of the polygon D. We assume that $\mathfrak{T}$ is obtained from a coarse, initial triangulation by *newest vertex bisection*, cp. [25]. In this section we will omit the parameter vector $\boldsymbol{y}$ in our notation with the understanding that all estimates depend on the parameter vector $\boldsymbol{y}$ only via dependencies on the coefficients $u = (a, b, c)$. We assume that there exists a constant $C > 0$ independent of h and $\beta \in (0, 1)$ such that for every $\mathcal{T} \in \mathfrak{T}$ and for every $K \in \mathcal{T}$:

$$\begin{aligned} &\text{(i) If } \overline{K} \cap \mathscr{C} = \emptyset, \text{ then } C^{-1} h r_D^{\beta}(x) \leq h_K \leq C h r_D^{\beta}(x) \text{ for every } x \in K. \\ &\text{(ii) If } \overline{K} \cap \mathscr{C} \neq \emptyset, \text{ then } C^{-1} h \sup_{x \in K}\{r_D^{\beta}(x)\} \leq h_K \leq C h \sup_{x \in K}\{r_D^{\beta}(x)\}, \end{aligned} \tag{12}$$

where

$$h_K := \operatorname{diam}(K),\, K \in \mathcal{T}, \quad \text{and} \quad h := \max_{K \in \mathcal{T}}\{h_K\}.$$

Such a mesh can be achieved with the algorithm proposed in [25, Sect. 4.1] with input values the global meshwidth h, the polynomial degree k, and the weight exponent $\gamma = (1 + k)(1 - \beta)$, assuming $(1 + k)(1 - \beta) < 1$. There are also *graded meshes* that satisfy (12), which were introduced in [3]. We define the spaces of Lagrangean Finite Elements of order $k \in \mathbb{N}$ by

$$V_{\mathcal{T}}^k := \{v \in V : v|_K \in \mathbb{P}_k(K),\, K \in \mathcal{T}\}, \quad \mathcal{T} \in \mathfrak{T},$$

where $\mathbb{P}_k(K)$ denotes the polynomials of total degree smaller than or equal to $k \geq 1$ on element $K \in \mathcal{T}$.

Proposition 2 *Let $k \in \mathbb{N}$ and let $0 < \delta < \beta < 1$ be such that $(1-\delta)/(1-\beta) > k$ and set $a = 1 - \delta$. There exists a constant $C > 0$ independent of the global mesh width h such that for every $w \in \mathcal{K}^{k+1}_{a+1}(D)$ there exist $w^{\mathcal{T}} \in V^k_{\mathcal{T}}$ such that*

$$\|w - w^{\mathcal{T}}\|_{H^1(D)} \leq Ch^k \|w\|_{\mathcal{K}^{k+1}_{a+1}(D)}.$$

This result is, in principle, known; e.g. [1, 37] and references there.

3.1 Numerical Integration

An essential component in the numerical analysis of the considered class of problems is the efficient numerical evaluation of the mass and stiffness matrices which contain the inhomogeneous, parametric coefficients. Owing to their origin as sample-wise realizations of random fields, these coefficients have, in general, only finite Sobolev regularity. Furthermore, for covariances in bounded domains which result from precision operators which include boundary conditions such as the Dirichlet Laplacean, these realizations can exhibit singular behaviour near corners of D. This is accommodated by the weighted Sobolev spaces $\mathcal{W}^{m,\infty}(D)$ comprising the data spaces X_m as defined above in (3). Efficient numerical quadrature for the evaluation of the stiffness and mass matrices which preserves the FE approximation properties on locally refined meshes is therefore needed. The numerical analysis of the impact of quadrature on FEM on locally refined meshes for uncertain coefficients in X_m is therefore required.

The impact of numerical integration in approximate computation of the stiffness matrix and load vector on the convergence rates of the FE solution is well understood for uniform mesh refinement, cp. for example [10, Sect. 4.1]. We extend this theory to regular, simplicial meshes with local refinement towards the singular points, and to possibly singular coefficients which belong to weighted spaces, i.e., $u = (a, b, c) \in X_{m,\delta}$, $m \in \mathbb{N}$, $\delta \in [0, 1)$, as defined in (9), (10). We provide a strategy to numerically approximate the stiffness matrix by quadrature so that the resulting additional consistency error is consistent with the FE approximation error, *uniformly with respect to the parameter sequences which characterize the uncertain inputs*. We denote by $\widetilde{A}$ on $V^k_{\mathcal{T}} \times V^k_{\mathcal{T}}$ the bilinear form, which has been obtained with numerical integration, i.e., for quadrature weights and nodes $(\omega_{K,\bar{k}}, x_{K,\bar{k}})_{K \in \mathcal{T}, \bar{k} \in \mathcal{I}} \subset (0, \infty) \times \overline{D}$

$$\widetilde{A}(w, v)$$

$$:= \sum_{K \in \mathcal{T}_\ell} \sum_{\bar{k} \in \mathcal{I}} \omega_{K,\bar{k}} \left(\sum_{i,j=1}^{2} a_{ij} \partial_j w \partial_i v + \sum_{i=1}^{2} b_i \partial_i w v + \sum_{i=1}^{2} b_{2+i} w \partial_i v + cwv \right) (x_{K,\bar{k}}),$$

for every $w, v \in V^k_{\mathscr{T}}$. Let us denote by $F_K : \widehat{K} \to K$ the affine element mappings that are given by $\xi \mapsto F_K(\xi) = B_K\xi + b_K$, $K \in \mathscr{T}$. Let $(\widehat{\omega_{\bar{k}}}, \widehat{x_{\bar{k}}})_{\bar{k}\in\mathscr{I}}$ be a set of positive weights and nodes (indexed elements of the finite set $\mathscr{I}$) for quadrature on the reference element $\widehat{K}$. Then, $\omega_{K,\bar{k}} := \det(B_K)\widehat{\omega_{\bar{k}}} > 0$ and $x_{K,\bar{k}} := F_K(\widehat{x_{\bar{k}}}) \in K$, $K \in \mathscr{T}$, $\bar{k} \in \mathscr{I}$. We define the element quadrature error for every $K \in \mathscr{T}$ and integrable ϕ such that point evaluation is well-defined by

$$E_K(\phi) = \int_K \phi \,\mathrm{d}x - \sum_{\bar{k}\in\mathscr{I}} \omega_{K,\bar{k}}\phi(x_{K,\bar{k}}).$$

The quadrature error $E_{\widehat{K}}$ on the reference element $\widehat{K}$ is defined analogously.

Under these assumptions, it can be shown as in the proof of [10, Theorem 4.1.2] (which covers the case that $b_i \equiv 0$ and $c \equiv 0$ in u in (1)) that the corresponding approximate sesquilinear form $\widetilde{A}(u)(\cdot,\cdot) : V^k_{\mathscr{T}} \times V^k_{\mathscr{T}} \to \mathbb{C}$ satisfies coercivity (5) with a positive coercivity constant $\tilde{c}$, possibly smaller than $c > 0$ in (5) but still independent of $\boldsymbol{y}^i \in [-1/2, 1/2]^{\mathbb{N}}$, $i = 0, 1, 2$.

Let us denote the FE solution with respect to the bilinear form $\widetilde{A}(u) : V^k_{\mathscr{T}} \times V^k_{\mathscr{T}} \to \mathbb{C}$ by $\widetilde{q}^{\mathscr{T}} \in V^k_{\mathscr{T}}$, i.e., the solution of the Galerkin-projected, parametric variational formulation

$$\widetilde{A}(u)(\widetilde{q}^{\mathscr{T}}, v) = \langle f, v\rangle_{V^*,V}, \quad \forall v \in V^k_{\mathscr{T}}.$$

The error incurred by employing numerical quadrature is consistent with the FE approximation rate, as demonstrated in the following theorem.

Theorem 1 *For $k \geq 1$, suppose that $E_{\widehat{K}}(\widehat{\phi}) = 0$ for every $\widehat{\phi} \in \mathbb{P}_{2k-1}$. Let $0 < \delta < \beta < 1$ satisfy $(1-\delta)/(1-\beta) > k$. There exists a constant $C > 0$ independent of h and of $u = (a, b, c) \in X_{k,\delta}$ such that*

$$\|q - \widetilde{q}^{\mathscr{T}}\|_V \leq Ch^k \left(1 + \|u\|_{X_{k,\delta}}\right) \|q\|_{\mathscr{K}^{k+1}_{a+1}(D)}.$$

The impact of numerical integration on linear functionals of the solution has been studied in the case of FEM with uniform mesh refinement for solutions belonging to a higher-order, unweighted Sobolev spaces for example in [4]. We extend the result in [4] to solutions to the parametric ADR problems in polygons in the following corollary.

Corollary 1 *Let $0 \leq k' \leq k$ be integers. Suppose that $E_{\widehat{K}}(\widehat{\phi}) = 0$ for every $\widehat{\phi} \in \mathbb{P}_{2k}$. Let $0 < \delta < \beta < 1$ satisfy $(1-\delta)/(1-\beta) > k + k'$. Then, there exists a constant $C > 0$ that does not depend on h such that for every $G \in \mathscr{K}^{k'-1}_{a-1}(D)$,*

$$|G(q) - G(\widetilde{q}^{\mathscr{T}})| \leq Ch^{k+k'} \left(1 + \|u\|_{X_{k,\delta}}\right)\left(1 + \|u\|_{X_{k+k',\delta}}\right) \|q\|_{\mathscr{K}^{k+1}_{a+1}(D)} \|G\|_{\mathscr{K}^{k'-1}_{a-1}(D)}.$$

The proofs of Theorem 1 and Corollary 1 are given in Sect. 7.1.

3.2 Finite Element Approximation of the Parametric Solution

To this end we suppose that we have a sequence of FE triangulations $\{\mathcal{T}_\ell\}_{\ell\geq 0}$ such that $\mathcal{T}_\ell$ satisfies the assumption in (12) with constants that are uniform in $\ell \geq 0$. The global mesh widths are denoted by $(h_\ell)_{\ell\geq 0}$. We denote by V_ℓ^k, $\ell \geq 0$, the respective FE spaces of polynomial order $k \geq 1$ and define

$$M_\ell := \dim(V_\ell^k), \quad \ell \geq 0.$$

Recall the bilinear form $\widetilde{A}(u(\boldsymbol{y}))$ on $V_\ell^k \times V_\ell^k$ that results from the application of numerical integration in the previous section. The Galerkin approximation $\widetilde{q}^{\mathcal{T}_\ell}(\boldsymbol{y}) \in V_\ell^k$ is the unique solution to

$$\widetilde{A}(u(\boldsymbol{y}))(\widetilde{q}^{\mathcal{T}_\ell}(\boldsymbol{y}), v) = \langle f, v\rangle_{V^*,V}, \quad \forall v \in V_\ell^k. \tag{13}$$

We recall the sparsity assumption in (10) for positive sequences $(\rho^i_{r,j})_{j\geq 1}$, $r = 0, \ldots, \tau$. This assumption and Lemma 1 imply that $a_{ij}(\boldsymbol{y}^0), r_D b_j(\boldsymbol{y}^1), r_D^2 c(\boldsymbol{y}^2) \in W_\delta^{\tau,\infty}(D)$ for every $\boldsymbol{y}^0, \boldsymbol{y}^1, \boldsymbol{y}^2 \in [-1/2, 1/2]^{\mathbb{N}}$ and admissible i, j. Specifically, Lemma 1 is applied coordinatewise to $[(x, z_0, z_1, z_2) \mapsto g(x, z_0, z_1, z_2)]$ composed with $\sum_{j\geq 1} z^i_j \psi^i_j$, $z^i \in U^i$, $i = 0, 1, 2$, see also (11). Note that $r^i_D \sum_{j\geq 1} z^i_j \psi^i_j \in W_\delta^{\tau,\infty}(D)$, $z^i \in U^i$, $i = 0, 1, 2$ by (10). Here, and throughout, we understand function spaces to be defined over the complex scalars. We assume that $f \in \mathcal{K}^{t-1}_{a-1}(D)$, $G \in \mathcal{K}^{t'-1}_{a-1}(D)$ for integers $t, t' \geq 0$ satisfying $t + t' \leq \tau$. Then, by Corollary 1 and (8),

$$\sup_{\boldsymbol{y}\in U} |G(q(\boldsymbol{y})) - G(\widetilde{q}^{\mathcal{T}_\ell}(\boldsymbol{y}))| \leq C M_\ell^{-(\min\{t,k\}+\min\{t',k\})/2} \|f\|_{\mathcal{K}^{t-1}_{a-1}(D)} \|G\|_{\mathcal{K}^{t'-1}_{a-1}(D)}, \tag{14}$$

where we applied that $M_\ell = \mathcal{O}(h_\ell^{-d})$, $\ell \geq 0$.

The parametric solution may be approximated consistently up to any order of h_ℓ by preconditioned, relaxed Richardson iteration in work $\mathcal{O}(M_\ell \log(M_\ell))$. Admissible preconditioners in the symmetric case, i.e., $b_i(\boldsymbol{y}^1) \equiv 0$, for $i = 1, \ldots, 4$ and for $\Gamma_2 = \emptyset$ are the so-called BPX preconditioner and the symmetric V-cycle, respectively. Respective condition numbers for local mesh refinement by newest vertex bisection for BPX and symmetric V-cycle have been studied for the Dirichlet Laplacean in [7]. These results are applicable, since the Dirichlet Laplacean is spectrally equivalent to $\mathcal{L}(u)$. For notational convenience, approximation of $\widetilde{q}^{\mathcal{T}_\ell}(\boldsymbol{y})$ by preconditioned, relaxed Richardson iteration, cp. [49, Proposition 2.3] will be denoted by the same symbol. Since the result after a finite number of steps of a Richardson iteration depends polynomially on the system matrix and since the preconditioner is independent of the parameter, holomorphic dependence on the parameters $\boldsymbol{y}$ is preserved.

We also consider parameter dimension truncation to obtain a finite-dimensional parameter set and denote by $s^0, s^1, s^2 \in \mathbb{N}$ the corresponding parameter dimensions. We denote the triple of those parameter dimensions by $\boldsymbol{s} := (s^0, s^1, s^2)$. Let us introduce further

$$\widetilde{q}^{s,\mathscr{T}_\ell}(\boldsymbol{y}) := \widetilde{q}^{\mathscr{T}_\ell}(\boldsymbol{y}^0_{\{1:s^0\}}, \boldsymbol{y}^1_{\{1:s^1\}}, \boldsymbol{y}^2_{\{1:s^2\}}), \quad \boldsymbol{y}^i \in \left[-\frac{1}{2}, \frac{1}{2}\right]^{\mathbb{N}}, i = 0, 1, 2,$$

where we have used the notation $(\boldsymbol{y}^i_{\{1:s^i\}})_j = y^i_j$ for $j \in \{1 : s^i\} := \{1, \dots, s^i\}$ and zero otherwise, $i = 0, 1, 2$. We define u^s and q^s analogously.

Lemma 2 *Let $u_1 = (a^1, b^1, c^1), u_2 = (a^2, b^2, c^2) \in X_0$ and $q_1, q_2 \in V$ satisfy $\mathscr{L}(u_i)q_i = f$, $i = 1, 2$. Assume that the bilinear forms $A(u_1)(\cdot,\cdot)$, $A(u_2)(\cdot,\cdot)$ are coercive with coercivity constants $c_1, c_2 > 0$ in the sense of* (5). *Then, there exists a constant $C > 0$ independent of q_1, q_2, u_1, u_2 such that*

$$\|q_1 - q_2\|_V \le \frac{C}{c_1 c_2} \|u_1 - u_2\|_{X_0} \|f\|_{V^*} .$$

Proof We observe $\|q_1 - q_2\|_V^2 c_1 \le A(u_1)(q_1 - q_2, q_1 - q_2) = A(u_2 - u_1)(q_2, q_1 - q_2)$. By the Hardy inequality (see e.g. [40, Theorem 21.3]) there exists a constant $C > 0$ such that for every $v \in V$

$$\|r_D^{-1} v\|_{L^2(D)} \le C \| |\nabla v| \|_{L^2(D)}. \tag{15}$$

As a consequence,

$$\begin{aligned} &|A(u_2 - u_1)(q_2, q_1 - q_2)| \\ &\le C \left(\sum_{i,j=1}^{2} \|a^1_{ij} - a^2_{ij}\|_{L^\infty(D)} + \sum_{j=1}^{4} \|r_D(b^1_j - b^2_j)\|_{L^\infty(D)} + \|r_D^2(c^1 - c^2)\|_{L^\infty(D)} \right) \\ &\quad \times \|q_2\|_V \|q_1 - q_2\|_V, \end{aligned}$$

where $C > 0$ is the constant from the Hardy inequality. In the previous step, we used multiplication by one, i.e., by $r_D r_D^{-1}$ for the advection terms and by $r_D^2 r_D^{-2}$ for the reaction term. The claim now follows with (6). □

It is easy to see that since g as introduced in (11) is in particular locally Lipschitz continuous. By Lemma 2, there exists a constant $C > 0$ such that

$$\sup_{\boldsymbol{y} \in U} \|q(\boldsymbol{y}) - q^s(\boldsymbol{y})\|_V \le C \left(\sup_{j > s^0} \{(\rho^0_{0,j})^{-1}\} + \sup_{j > s^1} \{(\rho^1_{0,j})^{-1}\} + \sup_{j > s^2} \{(\rho^2_{0,j})^{-1}\} \right).$$

Thus, by (14),

$$\sup_{\boldsymbol{y}\in U} |G(q(\boldsymbol{y})) - G(\widetilde{q}^{\boldsymbol{s},\mathscr{T}_\ell}(\boldsymbol{y}))| \leq C \left(M_\ell^{-(\min\{t,k\}+\min\{t',k\})/2} + \max_{i=0,1,2} \sup_{j>s^i} \{(\rho^i_{0,j})^{-1}\} \right). \tag{16}$$

4 Forward UQ

In this section we discuss the consistent approximation of the expectation of $G(q)$, where $G \in V^*$ is a linear functional. The expectation is taken with respect to the uniform product measure on U, which is denoted by $\mathrm{d}\boldsymbol{y} := \bigotimes_{i=0,1,2} \bigotimes_{j\geq 1} \mathrm{d}y^i_j$. The expectation of $G(q)$ will be denoted by

$$\mathbb{E}(G(q)) := \int_U G(q(\boldsymbol{y}))\mathrm{d}\boldsymbol{y}.$$

4.1 Higher Order QMC Integration

For a finite integration dimension $s \in \mathbb{N}$, QMC quadrature approximates integrals over the s-dimensional unit cube with equal quadrature weights, i.e., for a suitable integrand function F (possibly Banach space valued) and judiciously chosen, deterministic QMC points $\{\boldsymbol{y}^{(0)}, \ldots, \boldsymbol{y}^{(N-1)}\} \subset [0, 1]^s$

$$I_s(F) := \int_{[-\frac{1}{2},\frac{1}{2}]^s} F(\boldsymbol{y})\mathrm{d}\boldsymbol{y} \approx \frac{1}{N}\sum_{i=0}^{N-1} F\left(\boldsymbol{y}^{(i)} - \frac{\mathbf{1}}{\mathbf{2}}\right) =: Q_{s,N}(F),$$

where $(\frac{\mathbf{1}}{\mathbf{2}})_j = 1/2$, $j = 1, \ldots, s$.

Integration by QMC methods is able to achieve convergence rates that are independent of the dimension of integration and are higher than for Monte Carlo sampling; we refer to the surveys [19, 33]. In particular, *interlaced polynomial lattice rules* are able to achieve convergence rates which can even be of arbitrary, finite order, independent of the integration dimension s, provided the integrand satisfies certain conditions, cp. [17]. The analysis in this work will be for QMC by interlaced polynomial lattice rules. As in our previous works [22, 23], we will justify the application of interlaced polynomial lattice rules with *product weights*, which implies that the construction cost of the respective QMC points by the fast CBC construction is $\mathscr{O}(sN\log(N))$, where s is the dimension of integration and N the number of QMC points, cp. [17, 38, 39]. We state the main approximation result for interlaced polynomial lattice rules from [17] for product weights given in [17, Eq. (3.18)].

Theorem 2 ([17, Theorem 3.2]) *Let $s \in \mathbb{N}$ and $N = b^m$ for $m \in \mathbb{N}$ and b a prime number. Let $\boldsymbol{\beta} = (\beta_j)_{j\geq 1}$ be a sequence of positive numbers and assume that*

$\boldsymbol{\beta} \in \ell^p(\mathbb{N})$ for some $p \in (0, 1]$. Define the integer $\alpha = \lfloor 1/p \rfloor + 1 \geq 2$. Suppose the partial derivatives of the integrand $F : [-1/2, 1/2]^s \to \mathbb{R}$ satisfy the product bound

$$\forall \boldsymbol{y} \in [-1/2, 1/2]^s,\ \forall \boldsymbol{\nu} \in \{0, \ldots, \alpha\}^s : \quad |\partial_{\boldsymbol{y}}^{\boldsymbol{\nu}} F(\boldsymbol{y})| \leq c\boldsymbol{\nu}! \prod_{j=1}^{s} \beta_j^{\nu_j},$$

for some constant $c > 0$ which is independent of s and of $\boldsymbol{\nu}$.

Then, there exists an interlaced polynomial lattice rule which can be constructed with the CBC algorithm for product weights $(\gamma_{\mathfrak{u}})_{\mathfrak{u} \subset \mathbb{N}}$ that are given by $\gamma_\emptyset = 1$ and

$$\gamma_{\mathfrak{u}} = \prod_{j \in \mathfrak{u}} \left(C_{\alpha,b} b^{\alpha(\alpha-1)/2} \sum_{\nu=1}^{\alpha} 2^{\delta(\nu,\alpha)} \beta_j^{\nu} \right), \quad \mathfrak{u} \subset \mathbb{N},\ |\mathfrak{u}| < \infty, \tag{17}$$

($\delta(\nu, \alpha) = 1$ if $\nu = \alpha$ and zero otherwise) in $\mathcal{O}(\alpha s N \log(N))$ operations such that

$$\forall N \in \mathbb{N} : \quad |I_s(F) - Q_{s,N}(F)| \leq C_{\alpha,\boldsymbol{\beta},b,p} N^{-1/p},$$

where $C_{\alpha,\boldsymbol{\beta},b,p} < \infty$ is independent of s and N.

A numerical value for the Walsh constant $C_{\alpha,b}$ is as given in [17, Eq. (3.11)]. An improved bound for $C_{\alpha,b}$ is derived in [50].

4.2 Parametric Regularity

For the applicability of higher order integration methods such as QMC for UQ, the assumption on the partial derivatives with respect to the parameter $\boldsymbol{y}$ in Theorem 2 of the solution $q(\boldsymbol{y})$ or of functionals composed with $q(\boldsymbol{y})$ has to be verified. In [5] the authors proved analytic dependence of the solution on the coefficient in the complex valued setting. Hence, holomorphy is a direct consequence. By (5), (10), and Lemma 1, for every truncation dimension $\boldsymbol{s} = (s^0, s^1, s^2)$, the coefficients

$$u : \mathscr{D}_{\rho_r}^{\boldsymbol{s}} \to X_r$$

are holomorphic for $r = 0, \ldots, t$, where

$$\mathscr{D}_{\rho_r}^{\boldsymbol{s}} := \{z = (z^0, z^1, z^2) : z^i \in \mathbb{C}^{s^i}, |z_j^i| \leq \rho_{r,j}^i/2, i = 0, 1, 2\}.$$

As a composition of holomorphic mappings by [5, Corollary 5.1], the map

$$q : \mathscr{D}_{\rho_r}^{\boldsymbol{s}} \to \mathscr{K}_{a+1}^{r+1}(D), \quad r = 0, \ldots, t,$$

is holomorphic and

$$\sup_{s\in\mathbb{N}^3}\sup_{z\in\mathcal{D}^s_{\rho_r}} \|q(z)\|_{\mathcal{K}^{r+1}_{a+1}(D)} < \infty, \quad r = 0,\dots,t.$$

The following lemma is a version of [20, Lemma 3.1].

Lemma 3 ([20, Lemma 3.1]) *For a Banach space B and $\boldsymbol{\rho} = (\rho_j)_{j\geq 1} \subset (1,\infty)^{\mathbb{N}}$, $s \in \mathbb{N}$, let $F : \mathcal{D}^s_{\boldsymbol{\rho}} \to B$ be holomorphic, where $\mathcal{D}^s_{\boldsymbol{\rho}} := \{z \in \mathbb{C}^s : |z_j| \leq \rho_j, j = 1,\dots,s\}$. Then, for every $\boldsymbol{y} \in [-1,1]^{\mathbb{N}}$,*

$$\forall \boldsymbol{\nu} \in \mathbb{N}_0^{\mathbb{N}}, |\boldsymbol{\nu}| < \infty : \quad \|\partial^{\boldsymbol{\nu}}_{\boldsymbol{y}} F(\boldsymbol{y})\|_B \leq \sup_{z\in\mathcal{D}^s_{\boldsymbol{\rho}}}\{\|F(z)\|_B\} \prod_{j\geq 1} \frac{\rho_j}{(\rho_j - 1)^{\nu_j+1}}.$$

The argument used in the proof of this lemma is based on the Cauchy integral formula for holomorphic functions (see also [9, 11]).

Theorem 3 *Let the uncertain coefficient u be parametrized according to* (11) *and suppose that the assumption in* (10) *is satisfied for some $\tau \geq 1$. Let the assumptions of Corollary 1 be satisfied. There exists a constant $C > 0$ such that for every $\boldsymbol{\nu} = (\boldsymbol{\nu}^0, \boldsymbol{\nu}^1, \boldsymbol{\nu}^2)$, $\boldsymbol{\nu}^i \in \mathbb{N}_0^{\mathbb{N}}$, $|\boldsymbol{\nu}^i| < \infty$, and for every $\boldsymbol{s} = (s^0, s^1, s^2)$, $s^i \in \mathbb{N}$, $i = 0,1,2$, $\ell \geq 0$, $0 \leq t' \leq t \leq k$ such that $t + t' \leq \tau$, $\theta \in [0,1]$, and every $\boldsymbol{y} \in U$,*

$$\begin{aligned}
&|\partial^{\boldsymbol{\nu}}_{\boldsymbol{y}}(G(q(\boldsymbol{y})) - G(\widetilde{q}^{\boldsymbol{s},\mathcal{T}_\ell}(\boldsymbol{y})))| \\
&\quad \leq C\|G\|_{V^*}\|f\|_{V^*} \max_{i=0,1,2} \sup_{j>s^i}\{(\rho^i_{0,j})^{-\theta}\} \prod_{i=0,1,2}\prod_{j\geq 1}(\rho^i_{0,j}/2)^{-\nu^i_j(1-\theta)} \\
&\quad\quad + C\|G\|_{\mathcal{K}^{t'-1}_{a-1}}\|f\|_{\mathcal{K}^{t-1}_{a-1}} M_\ell^{-(t+t')/d} \prod_{i=0,1,2}\prod_{j\geq 1}(\rho^i_{t+t',j}/2)^{-\nu^i_j}.
\end{aligned}$$

Proof The estimate will follow by a twofold application of Lemma 3 and the holomorphic dependence of the solution on the parametric input. By the triangle inequality,

$$|G(q(\boldsymbol{y})) - G(\widetilde{q}^{\boldsymbol{s},\mathcal{T}_\ell}(\boldsymbol{y}))| \leq |G(q(\boldsymbol{y})) - G(q^{\boldsymbol{s}}(\boldsymbol{y}))| + |G(q^{\boldsymbol{s}}(\boldsymbol{y})) - G(\widetilde{q}^{\boldsymbol{s},\mathcal{T}_\ell}(\boldsymbol{y}))|.$$

By the assumption in (10) and [5, Corollary 5.1], the mapping $z \mapsto G(q(z)) - G(q^{\boldsymbol{s}}(z))$ is holomorphic on $\mathcal{D}^s_{(\rho_0)^{1-\theta}}$ and by Lemma 2 it holds that

$$\sup_{z\in\mathcal{D}^s_{(\rho_0)^{1-\theta}}} |G(q(z)) - G(q^{\boldsymbol{s}}(z))| \leq C\|G\|_{V^*}\|f\|_{V^*} \max_{i=0,1,2}\sup_{j>s^i}\{(\rho^i_{0,j})^{-\theta}\}.$$

Hence, by Lemma 3, where we scale the parameter vectors by a factor of $1/2$

$$\begin{aligned}
&|\partial^{\boldsymbol{\nu}}_{\boldsymbol{y}}(G(q(\boldsymbol{y})) - G(q^{\boldsymbol{s}}(\boldsymbol{y})))| \\
&\quad \leq C\|G\|_{V^*}\|f\|_{V^*} \max_{i=0,1,2}\sup_{j>s^i}\{(\rho^i_{0,j})^{-\theta}\} \prod_{i=0,1,2}\prod_{j\geq 1}(\rho^i_{0,j}/2)^{-\nu^i_j(1-\theta)}.
\end{aligned}$$

Furthermore by the assumption in (10) and [5, Corollary 5.1], the mapping $z \mapsto q^s(z)$ is holomorphic from $\mathscr{D}^s_{\rho_t}$ to $\mathscr{K}^{t+1}_{a+1}(D)$ and

$$\sup_{z\in\mathscr{D}^s_{\rho_t}} \|q(z)\|_{\mathscr{K}^{t+1}_{a+1}(D)} < \infty.$$

Thus, by (14), there exists $C > 0$ such that for all s and all ℓ holds

$$\sup_{z\in\mathscr{D}^s_{\rho_t}} |G(q^s(z)) - G(\widetilde{q}^{s,\mathscr{T}_\ell}(z))| \le C\|G\|_{\mathscr{K}^{t'-1}_{a-1}}\|f\|_{\mathscr{K}^{t-1}_{a-1}} M_\ell^{-(t+t')/d}.$$

The second part of the estimate now also follows by Lemma 3. □

4.3 Multilevel QMC Error Estimates

Multilevel integration schemes offer a reduction in the overall computational cost, subject to suitable regularity (see, e.g., [18, 28, 32]). For $s_{\ell=0,\dots,L}, N_{\ell=0,\dots,L}, L \in \mathbb{N}_0$, define the multilevel QMC quadrature

$$Q_L(G(\widetilde{q}^L)) := \sum_{\ell=0}^{L} Q_{|s|_\ell,N_\ell}(G(\widetilde{q}^\ell) - G(\widetilde{q}^{\ell-1})),$$

where we used the notation $\widetilde{q}^\ell := \widetilde{q}^{s_\ell,\mathscr{T}_\ell}$, $\ell = 0, \dots, L$, and $\widetilde{q}^{-1} := 0$. The QMC weights in (17) are obtained from (17) with input

$$\beta_{j(j',i)} := 2\max\{(\rho^i_{0,j'})^{-(1-\theta)}, (\rho^i_{\tau,j'})^{-1}\}, \tag{18}$$

where $\tau = t + t'$ and $j(j', i) := 3j' - i$, $j' \in \mathbb{N}$, $i = 0, 1, 2$, is an enumeration of $\mathbb{N}$ with elements in $\mathbb{N} \times \{0, 1, 2\}$.

Theorem 4 *Suppose that the weight sequence in* (18) *satisfies* $\boldsymbol{\beta} = (\beta_j)_{j\ge1} \in \ell^p(\mathbb{N})$ *for some* $p \in (0, 1]$. *Then, with an interlaced polynomial lattice rule of order* $\alpha = \lfloor 1/p\rfloor + 1$ *and product weights* (17) *with weight sequence* (18) *there exists a constant* $C > 0$ *such that for* $s_{\ell=0,\dots,L}, N_{\ell=0,\dots,L}, L \in \mathbb{N}_0$,

$$|\mathbb{E}(G(q)) - Q_L(G(\widetilde{q}^L))| \le C \left(M_L^{-(t'+t)/d} + \max_{i=0,1,2} \sup_{j>s^i}\{(\rho^i_{0,j})^{-1}\} \right.$$
$$\left. + \sum_{\ell=0}^{L} N_\ell^{-1/p} \left(M_{\ell-1}^{-(t'+t)/d} + \max_{i=0,1,2} \sup_{j>s^i_{\ell-1}} \{(\rho^i_{0,j})^{-\theta}\} \right) \right).$$

Proof By the triangle inequality, we obtain the deterministic error estimate

$$
\begin{aligned}
&|\mathbb{E}(G(q)) - Q_L(G(\widetilde{q}^L))| \\
&\quad \le |\mathbb{E}(G(q)) - I_{|s_L|}(G(\widetilde{q}^L))| + \sum_{\ell=0}^{L} |(I_{|s_\ell|} - Q_{|s_\ell|,N_\ell})(G(\widetilde{q}^\ell) - G(\widetilde{q}^{\ell-1}))|,
\end{aligned}
$$

where $|s_\ell| = s_\ell^0 + s_\ell^1 + s_\ell^2$. Then, Theorems 2 and 3 and (16) imply the claim. □

4.4 Error Versus Work Analysis

In this section we analyze the overall computational complexity of the multilevel QMC algorithm with product weights for function systems $(\psi_j^i)_{j\ge1}$, $i = 0, 1, 2$. For the analysis, we assume that the function systems $(\psi_j^i)_{j\ge1}$, $i = 0, 1, 2$, which appear in the uncertainty parametrization have a multilevel structure with control of the overlaps of the supports. Suppose for $i = 0, 1, 2$, there exist enumerations $j_i : \nabla^i \to \mathbb{N}$, where elements of $\lambda \in \nabla^i$ are tuples of the form $\lambda = (\ell, k)$, where $k \in \nabla_\ell^i$. The index sets are related by $\nabla^i = \bigcup_{\ell\ge0}(\{\ell\} \cup \nabla_\ell^i)$, $i = 0, 1, 2$. Also define $|\lambda| = |(\ell, k)| = \ell$ for every $\lambda \in \nabla^i$. We assume that $|\nabla_\ell^i| = \mathscr{O}(2^{d\ell})$, $|\mathrm{supp}(\psi_\lambda)| = \mathscr{O}(2^{-d\ell})$, $\lambda = (\ell, k) \in \nabla^i$, and there exists $K > 0$ such that for every $x \in D$ and every $\ell \in \mathbb{N}_0$,

$$
\left|\{\lambda \in \nabla^i : |\lambda| = \ell, \psi_\lambda^i(x) \neq 0\}\right| \le K. \tag{19}
$$

Moreover, we assume that

$$
\rho_{r,\lambda}^i \lesssim 2^{-|\lambda|(\widehat{\alpha}-r)}, \quad \lambda \in \nabla^i, r = 0, \dots, t, i = 0, 1, 2,
$$

for $\widehat{\alpha} > t + t'$. Note that $\rho_{r,j(\lambda)}^i \lesssim j^{-(\widehat{\alpha}-r)/d}$, $j \ge 1$. We equilibrate the sparsity contribution of the sequences $(\rho_{0,\lambda}^i)_{\lambda\in\nabla^i}$ and $(\rho_{t,\lambda}^i)_{\lambda\in\nabla^i}$ in the weight sequence in (18). Hence, we choose $\theta = (t + t')/\widehat{\alpha}$. Recall that we assume $f \in \mathscr{K}_{a-1}^{t-1}(D)$ and $G \in \mathscr{K}_{a-1}^{t'-1}(D)$. Furthermore, with this choice of θ we also equilibrate the errors in the multilevel QMC estimate from Theorem 4, where the truncation dimension s_ℓ is still a free parameter. The error contributions in Theorem 4 are equilibrated for the choice

$$
s_\ell^i \sim M_\ell, \quad i = 0, 1, 2. \tag{20}
$$

In conclusion, the overall error of multilevel QMC with $L \in \mathbb{N}_0$ levels satisfies for every $p > d/(\widehat{\alpha} - (t + t'))$,

$$
\mathrm{error}_L = \mathscr{O}\left(M_L^{-(t'+t)/d} + \sum_{\ell=0}^{L} N_\ell^{-1/p} M_{\ell-1}^{-(t'+t)d} \right). \tag{21}
$$

We assume that we have a procedure at hand that approximates the solution of a parameter instance up to an accuracy, which is consistent with the discretization error, in computational cost

$$\text{work}_{\text{PDE solver},\ell} = \mathscr{O}(M_\ell \log(M_\ell)), \quad \ell \geq 0.$$

Recall from Sect. 3.2 that in the self-adjoint case with homogeneous Dirichlet boundary conditions, i.e., $b_i(\boldsymbol{y}^1) = 0$ and $\Gamma_2 = \emptyset$, this can be achieved by a relaxed Richardson iteration preconditioned by BPX or symmetric V-cycle as preconditioners. The stiffness matrix has $\mathscr{O}(M_\ell)$ non-zero entries by using a nodal FE basis. The finite overlap property (19), the choice in (20), and the fact that the number of quadrature nodes does not depend on the dimension of the FE space (see assumption of Theorem 1 and Corollary 1) imply that the computation of every matrix entry has computational cost $\mathscr{O}(\log(M_\ell))$. The total computational cost of the multilevel QMC algorithm is the sum of the cost of the CBC construction, the cost of assembling the stiffness matrix and the cost of approximating the solution of the linear systems multiplied by the number of QMC points. Specifically,

$$\text{work}_L = \mathscr{O}\left(\sum_{\ell=0}^{L} M_\ell N_\ell \log(N_\ell) + N_\ell M_\ell \log(M_\ell)\right),$$

where we remind the reader that by (20) the dimension of integration on each discretization level ℓ in D is $\mathscr{O}(M_\ell)$. Since the QMC convergence rate $1/p$ satisfies the strict inequality $\chi := 1/p < (\widehat{\alpha} - t + t')/d$, also the rate $\chi(1+\varepsilon)$ is admissible for sufficiently small $\varepsilon > 0$. This way the sample numbers can be reduced to $N_\ell^{1/(1+\varepsilon)}$, which allows us to estimate $N_\ell^{1/(1+\varepsilon)} \log(N_\ell) \leq N_\ell^{1/(1+\varepsilon)} N_\ell^{\varepsilon/(1+\varepsilon)}(1+\varepsilon)/(e\varepsilon) \leq N_\ell(1+\varepsilon)/(e\varepsilon)$, where we used the elementary estimate $\log(N) \leq N^{\varepsilon'}/(e\varepsilon')$ for every $N \geq 1$, $\varepsilon' > 0$. Thus, we obtain the estimate of the work

$$\text{work}_L = \mathscr{O}\left(\sum_{\ell=0}^{L} M_\ell \log(M_\ell) N_\ell\right). \tag{22}$$

By [25, Lemma 4.9], it holds that $M_\ell = \mathscr{O}(2^{d\ell})$. The sample numbers are now obtained by optimizing the error versus the computational work, cp. [34, Sect. 3.7]. For the error and work estimates in (21) and in (22), sample numbers are derived in [22, Sect. 6]. Specifically, by [22, Eqs. (26) and (27)],

$$N_\ell := \left\lceil N_0 (M_\ell^{-1-(t+t')/d} \log(M_\ell)^{-1})^{p/(1+p)} \right\rceil, \quad \ell = 1, \ldots, L, \tag{23}$$

where

$$N_0 := \begin{cases} M_L^{(t+t')p/d} & \text{if } d < p(t+t'), \\ M_L^{(t+t')p/d} \log(M_L)^{p(p+2)/(p+1)} & \text{if } d = p(t+t'), \\ M_L^{(1+(t+t')/d)p/(p+1)} \log(M_L)^{p/(p+1)} & \text{if } d > p(t+t'). \end{cases} \tag{24}$$

The corresponding work satisfies (see for example [22, p. 396])

$$\text{work}_L = \begin{cases} \mathcal{O}(M_L^{(t+t')p/d}) & \text{if } d < p(t+t'), \\ \mathcal{O}(M_L^{(t+t')p/d} \log(M_L)^{p+2}) & \text{if } d = p(t+t'), \\ \mathcal{O}(M_L \log(M_L)) & \text{if } d > p(t+t'). \end{cases}$$

We summarize the preceding discussion in the following theorem stating the ε-complexity of the multilevel QMC algorithm.

Theorem 5 *For $p \in (d/(\widehat{\alpha} - (t+t')), 1]$, assuming $d < \widehat{\alpha} - (t+t')$, an error threshold $\varepsilon > 0$, i.e.,*

$$|\mathbb{E}(G(q)) - Q_L(G(\widetilde{q}^L))| = \mathcal{O}(\varepsilon)$$

can be achieved with

$$\text{work}_L = \begin{cases} \mathcal{O}(\varepsilon^{-p}) & \text{if } d < p(t+t'), \\ \mathcal{O}(\varepsilon^{-p} \log(\varepsilon^{-1})^{p+2}) & \text{if } d = p(t+t'), \\ \mathcal{O}(\varepsilon^{-d/(t+t')} \log(\varepsilon^{-1})) & \text{if } d > p(t+t'). \end{cases}$$

5 Bayesian Inverse UQ

The preceding considerations pertained to so-called *forward UQ* for the ADR problem (1) with uncertain input data $u = ((a_{ij}), (b_j), c)$ taking values in certain subsets of the function spaces X_m in (3). The goal of computation is the efficient evaluation of *ensemble averages*, i.e. the expected response over all *parametric inputs* u as in (11) with respect to a probability measure on the parameter domains U^i.

In Bayesian inverse UQ, we are interested in similar expectations of a QoI of the forward response of the ADR PDE, *conditional to noisy observations of functionals of the responses*. Again, a (prior) probability measure on the uncertain (and assumed non-observable) *parametric ADR PDE inputs u* in (11) is prescribed. As explained in [13, 43], in this setting Bayes' theorem provides a formula for the conditional expectation as a high-dimensional, parametric deterministic integral which, as shown in [15, 18, 42], is amenable to deterministic HoQMC integration affording convergence rates which are superior to those of, e.g., MCMC methods [21, 30].

5.1 Formulation of the Bayesian Inverse Problem

Specifically, assume at hand noisy observations of the ADR PDE response $q = (\mathscr{L}(u))^{-1}f$ subject to *additive Gaussian observation noise* $\boldsymbol{\eta}$, i.e.

$$\boldsymbol{\delta} = \boldsymbol{G}(q) + \boldsymbol{\eta}. \tag{25}$$

In (25), q denotes the response of the uncertain input u, $\boldsymbol{G} = (G_1, \ldots, G_K)$ is a vector of K (linear) observation functionals, i.e., $G_i \in V^*$, the additive noise $\boldsymbol{\eta}$ is assumed centered and normally distributed with positive covariance Γ, i.e.,$\boldsymbol{\eta} \sim \mathscr{N}(0, \Gamma)$, and the data $\boldsymbol{\delta} \in \mathbb{R}^K$ is supposed to be available. We introduce the so-called prior measure $\boldsymbol{\pi}$ on X_0 as the law of $U \ni \boldsymbol{y} \mapsto u(\boldsymbol{y}) \in X_0$ with respect to the uniform product measure $\mathrm{d}\boldsymbol{y}$ on U. The density of the posterior distribution with respect to the prior is given by [13, Theorem 14]

$$U \ni \boldsymbol{y} \mapsto \frac{1}{Z} \exp\left(-\Phi_\Gamma(q(\boldsymbol{y}); \boldsymbol{\delta})\right), \tag{26}$$

where the negative *log-likelihood* Φ_Γ is given by

$$\Phi_\Gamma(q(\boldsymbol{y}); \boldsymbol{\delta}) := \frac{1}{2}(\boldsymbol{\delta} - \boldsymbol{G}(q(\boldsymbol{y})))^\top \Gamma^{-1}(\boldsymbol{\delta} - \boldsymbol{G}(q(\boldsymbol{y}))) \quad \forall \boldsymbol{y} \in U.$$

Since (6) implies $\sup_{\boldsymbol{y} \in U} \|q(\boldsymbol{y})\|_V < \infty$, the *normalization constant* in (26) satisfies

$$Z := \int_{X_0} \exp\left(-\Phi_\Gamma(q; \boldsymbol{\delta})\right) \boldsymbol{\pi}(\mathrm{d}u) = \int_U \exp\left(-\Phi_\Gamma(q(\boldsymbol{y}); \boldsymbol{\delta})\right) \mathrm{d}\boldsymbol{y} > 0,$$

where we recall that $q = \mathscr{L}(u)^{-1}f$. The posterior measure will be denoted by $\boldsymbol{\pi}^{\delta}$ and the posterior with respect to $\widetilde{q}^{s,\mathscr{T}_\ell}$ will be denoted by $\widetilde{\boldsymbol{\pi}}^{\delta}_{s,\mathscr{T}_\ell}$. The QoIs, being assumed bounded linear functionals applied to $q \in V$ (which could be weakened [21]), admit a unique representer $\phi \in V^*$. For any QoI $\phi \in V^*$, denote the expectation with respect to the posterior of ϕ by

$$\mathbb{E}^{\boldsymbol{\pi}^{\delta}}(\phi) := \int_{X_0} \phi(q)\boldsymbol{\pi}^{\delta}(\mathrm{d}u) = \frac{1}{Z}\int_U \phi(q(\boldsymbol{y})) \exp\left(-\Phi_\Gamma(q(\boldsymbol{y}); \boldsymbol{\delta})\right) \mathrm{d}\boldsymbol{y}.$$

Here, $\Phi_\Gamma(q(\boldsymbol{y}); \boldsymbol{\delta})$ is Lipschitz continuous with respect to $\boldsymbol{\delta}$ and with respect to $q(\boldsymbol{y})$, $\boldsymbol{y} \in U$. As a consequence of (16), for every $\boldsymbol{s} \in \mathbb{N}^3$ and $\ell \geq 0$,

$$|\mathbb{E}^{\boldsymbol{\pi}^{\delta}}(\phi) - \mathbb{E}^{\widetilde{\boldsymbol{\pi}}^{\delta}_{s,\mathscr{T}_\ell}}(\phi)| \leq C \left(M_\ell^{-(\min\{t,k\}+\min\{t',k\})/2} + \max_{i=0,1,2} \sup_{j > s^i}\{(\rho^i_{0,j})^{-1}\} \right), \tag{27}$$

where we also used that $\Phi_\Gamma(q(\boldsymbol{y});\boldsymbol{\delta})$ and $\Phi_\Gamma(\widetilde{q}^{s,\mathcal{T}_\ell}(\boldsymbol{y});\boldsymbol{\delta})$ are uniformly upper bounded with respect to $\boldsymbol{y} \in U$. See also the discussion in [16, Sect. 3.3]. Here, the abstract assumptions made in [16, Sect. 3.3], stemming from [13], may be verified concretely. The estimate in (27) is not just a restatement of the results of [15, 16]. Here, a general parametric ADR forward problem on polygonal domains is considered and higher order FE convergence on locally refined triangulations $\mathcal{T}_\ell$ is achieved based on regularity in weighted spaces of Kondrat'ev type. The corresponding FE approximation results are proved in Sect. 7.

5.2 Multilevel HoQMC-FE Discretization

The expectation with respect to the posterior measure $\widetilde{\boldsymbol{\pi}}^{\delta}_{s,\mathcal{T}_\ell}$ is an integral over a $|\boldsymbol{s}|$-dimensional parameter space and may therefore be approximated by multilevel QMC. We recall the FE spaces V_ℓ^k based on the regular, simplicial triangulations $\mathcal{T}_\ell$ and suppose given a sequence of $\boldsymbol{s}_\ell$ of dimension truncations, $\ell = 0,\dots,L$, where $L \in \mathbb{N}$ is the maximal discretization level. The error analysis will be along the lines of [15, Sect. 4], see also [43, 44]. Following the notation in [15], we define for $\ell = 0,\dots,L$,

$$\mathbb{E}^{\widetilde{\boldsymbol{\pi}}^{\delta}_{s_\ell,\mathcal{T}_\ell}}(\phi) = \frac{\int_{[-1/2,1/2]^{|s|}} \phi(\widetilde{q}^{s_\ell,\mathcal{T}_\ell}(\boldsymbol{y}))\Theta_\ell(\boldsymbol{y})\mathrm{d}\boldsymbol{y}}{\int_{[-1/2,1/2]^{|s|}} \Theta_\ell(\boldsymbol{y})\mathrm{d}\boldsymbol{y}} =: \frac{Z'_\ell}{Z_\ell},$$

where $\Theta_\ell(\boldsymbol{y}) := \exp(-\Phi_\Gamma(\widetilde{q}^{s_\ell,\mathcal{T}_\ell}(\boldsymbol{y});\boldsymbol{\delta}))$. In [15, Sect. 4.2], multilevel QMC *ratio* and *splitting* estimators were proposed for the deterministic approximation of Z'_L/Z_L. They are, for sequences of numbers of QMC points $(N_\ell)_{\ell=0,\dots,L}$ and of dimension truncations $(\boldsymbol{s}_\ell)_{\ell=0,\dots,L}$, defined by

$$Q_{L,\text{ratio}} := \frac{Q_L(\phi(\widetilde{q}^L)\Theta_L)}{Q_L(\Theta_L)} \tag{28}$$

and, with the notation $|\boldsymbol{s}_\ell| = s_\ell^0 + s_\ell^1 + s_\ell^2$,

$$Q_{L,\text{split}} := \frac{Q_{|s_0|,N_0}(\phi(\widetilde{q}^0)\Theta_0)}{Q_{|s_0|,N_0}(\Theta_0)} + \sum_{\ell=1}^{L} \frac{Q_{|s_\ell|,N_\ell}(\phi(\widetilde{q}^\ell)\Theta_\ell)}{Q_{|s_\ell|,N_\ell}(\Theta_\ell)} - \frac{Q_{|s_\ell|,N_\ell}(\phi(\widetilde{q}^{\ell-1})\Theta_{\ell-1})}{Q_{|s_\ell|,N_\ell}(\Theta_{\ell-1})}. \tag{29}$$

The error analysis of these estimators requires that the integrands satisfy certain parametric regularity estimates. In Sect. 4.2, parametric regularity estimates of $q(\boldsymbol{y}) - \widetilde{q}^{s,\mathcal{T}_\ell}(\boldsymbol{y})$ were shown using analytic continuation. The integrands $\phi(\widetilde{q}^\ell)\Theta_\ell$ and Θ_ℓ depend analytically on $\widetilde{q}^\ell$ and are, as compositions and products of holomorphic mappings with compatible domains and ranges, again holomorphic.

The consistency errors of the ratio and splitting estimators are analyzed in [15, Sects. 4.3.2 and 4.3.3] in the setting of globally supported function systems. However, the proofs of [15, Theorem 4.1 and Theorem 4.2] are applicable.

Proposition 3 *Let the assumptions and the setting of steering parameters θ, p, t', t, s_ℓ, N_ℓ, M_ℓ, $\ell = 0, \ldots, L$ of Q_L in Theorem 4 be satisfied. Then,*

$$|\mathbb{E}^{\pi^\delta}(\phi) - Q_{L,\text{ratio}}| \leq C \left(M_L^{-(t'+t)/d} + \max_{i=0,1,2} \sup_{j>s^i} \{(\rho_{0,j}^i)^{-1}\} + \sum_{\ell=0}^{L} N_\ell^{-1/p} \left(M_{\ell-1}^{-(t'+t)/d} + \max_{i=0,1,2} \sup_{j>s_{\ell-1}^i} \{(\rho_{0,j}^i)^{-\theta}\} \right) \right).$$

Proof The estimate (6) and Theorem 1 imply that $\Phi_\Gamma(\widetilde{q}^{s_\ell,\mathcal{T}_\ell}(\boldsymbol{y}); \boldsymbol{\delta})$ can be upper bounded uniformly with respect to $\boldsymbol{y}$ as follows. There exists a constant $C_0 > 0$, which does not depend on L, such that $Q_L(\Theta_L) \geq C_0$. Now, the assertion follows as [15, Theorem 4.1]. As mentioned above, $\phi(\widetilde{q}^\ell)\Theta_\ell$ may be analytically extended to a suitable polydisc as in the proof of Theorem 3. The same line of argument used in the proof of [15, Theorem 4.1] may be applied here. Further details are left to the reader. □

The error estimate from Proposition 3 for Bayesian estimation can also be shown for the splitting estimator $Q_{L,\text{split}}$ along the lines of the proof of [15, Theorem 4.2].

Since the posterior density depends analytically on the response q, the QMC sample numbers for ratio and splitting estimators $Q_{L,\text{ratio}}$ and $Q_{L,\text{split}}$ are the same as those for forward UQ in (23) and (24). In particular, also the same ε-complexity estimates from Theorem 5 hold under the same assumptions on the steering parameters.

Remark 2 Forward and Bayesian inverse UQ for uncertain domains by pullbacks to a polygonal *nominal* or reference domain is a straightforward extension of the presented theory. It requires the extension of the PDE regularity theory to parametric right hand sides $f(\boldsymbol{y})$. Since this dependence is inherited by the parametric solution due to linearity, we did not explicitly consider it for the sake of a concise presentation, but refer to [2, 24, 29] for the numerical analysis of domain uncertainty quantification analysis with QMC.

6 Conclusions

We discussed forward and Bayesian inverse UQ by multilevel QMC for general ADR problems in polygons allowing the input coefficients and the response, i.e., the solution to the ADR problem, to be singular near corners of the domain. A wide class of uncertain input coefficients is admissible to our theory. The coefficients are assumed to depend holomorphically on a series expansion with uncertain, uniformly distributed parameters and a function system with local support. Locally supported representation systems are well-known to allow product weights in higher-order QMC integration [22, 23]. Here, we generalized this principle from isotropic diffusion problems to general ADR problems with not necessarily affine-parametric uncertainty in all coefficients of the ADR forward model. The presently developed setting also allows extension to UQ for random domains. Regularity of the uncertain input coefficients and of the response in scales of Kondrat'ev spaces and sparsity are utilized in the presented multilevel QMC algorithm by combining higher order QMC in the parametric domain and higher order FEM with mesh refinement in the physical domain. The overall approximation scheme is fully discrete, since also the impact of numerical integration in the FEM is analyzed here (to our knowledge for the first time). In the present setting, general ADR problems with possibly singular coefficients at the corners (as arise, e.g., from non-stationary covariance models with elliptic precision operators in the physical domain) are admissible. The analysis is shown to extend to the corresponding Bayesian inverse problems, where the higher order QMC-FE convergence rates from the forward UQ analysis are proved to be preserved, and to scale linearly with the number of parameters under a-priori, data-independent truncation of the prior.

7 Proofs

We provide proofs of several results in the main text. They were postponed to this section to increase readability of the main text.

7.1 Numerical Integration

In the following proofs, we require a nodal interpolant. As preparation, for $k \geq 3$, we introduce certain subsets of $\mathcal{T}$

$$\mathcal{T}^{k'} := \left\{ K \in \mathcal{T} \backslash \mathcal{T}^{k'-1} : \overline{K} \cap \bigcup_{K' \in \mathcal{T}^{k'-1}} \overline{K'} \neq \emptyset \right\}, \quad k' = 2, \dots, k-1,$$

where $\mathcal{T}^1 := \{K \in \mathcal{T} : \overline{K} \cap \mathcal{C} \neq \emptyset\}$ and $\mathcal{T}^k := \mathcal{T} \backslash \mathcal{T}^{k-1}$. For $k = 2$, $\mathcal{T}^2 := \mathcal{T} \backslash \mathcal{T}^1$ and $\mathcal{T}^1$ is defined as above. For $k = 1$, $\mathcal{T}^1 := \mathcal{T}$. We define a FE space such that in the elements abutting at a vertex, $\mathbb{P}_1$ FE are used and for the remaining, "interior" elements, $\mathbb{P}_k$ FE are used such that the polynomial degree of neighboring elements only differs by one, i.e.,

$$\widetilde{V}_{\mathcal{T}}^k := \{v \in V : v|_K \in \mathbb{P}_{k'}(K), K \in \mathcal{T}^{k'}, k' = 1, \dots, k\}.$$

The potential change of the polynomial degree in neighboring elements near the singular points constitutes a difficulty in defining a nodal interpolant for $k \geq 2$. Let $K_1 \in \mathscr{T}^{k'-1}$, $K_2 \in \mathscr{T}^{k'}$ be neighboring triangles such that $\overline{K_1} \cap \overline{K_2} =: e$ denotes the common edge. To avoid a discontinuity across the edge e, the usual nodal interpolant $I^{k'}_{K_2}$ may need to be corrected. For $v \in C^0(\overline{K}_2)$, the discontinuity $(I^{k'}_{K_2} v|_e - I^{k'-1}_e v)$ is equal to zero at the endpoints of the edge e. By [46, Lemma 4.55], there exists $(I^{k'}_{K_2} v|_e - I^{k'-1}_e v)_{\text{lift},k',e} \in \mathbb{P}^{k'}(K_2)$ such that $(I^{k'}_{K_2} v|_e - I^{k'-1}_e v)_{\text{lift},k',e} = (I^{k'}_{K_2} v|_e - I^{k'-1}_e v)$ on the edge e, $(I^{k'}_{K_2} v|_e - I^{k'-1}_e v)_{\text{lift},k',e} = 0$ on the remaining edges of K_2, and it holds

$$\begin{aligned} \|(I^{k'}_{K_2} v|_e - I^{k'-1}_e v)_{\text{lift},k',e}\|^2_{H^1(K_2)} &\leq C h_{K_2} \|(I^{k'}_{K_2} v)|_e - I^{k'}_e v|_e\|^2_{H^1(e)} \\ &\leq C' h^{2k'-1}_{K_2} |I^{k'}_e v|^2_{H^{k'}(e)} \leq C' h^{2k'-2}_{K_2} |I^{k'}_{K_2} v|^2_{H^{k'}(K_2)}, \end{aligned} \tag{30}$$

where we applied the approximation property in dimension $d-1=1$, cp. [10, Theorem 3.1.6], the shape regularity of $\mathscr{T}$, and the fact that k'-th order partial derivatives of $I^{k'}_{K_2} v$ are constant on K_2.

We will define an interpolant $I_{\mathscr{T}} : \mathscr{K}^{k+1}_{a+1} \to \widetilde{V}^k_{\mathscr{T}} \subset V^k_{\mathscr{T}}$ by

$$I_{\mathscr{T}} v = \begin{cases} I^1_K v & \text{if } K \in \mathscr{T}^1, \\ I^{k'}_K v - (I^{k'}_K v|_e - I^{k'-1}_e v)_{\text{lift},k',e} & \text{if } K \in \mathscr{T}^{k'}, e := \overline{K} \cap \overline{\mathscr{T}^{k'-1}} \neq \emptyset, \\ & k' = 2, \dots, k, \\ I^k_K v & \text{if } K \in \mathscr{T}^k, \overline{K} \cap \overline{\mathscr{T}^{k-1}} = \emptyset, \end{cases}$$

where $I^{k'}_K$ is the usual nodal interpolant of order $k' \in \mathbb{N}$ on the element K and we introduced the notation $\overline{\mathscr{T}^{k'}} := \bigcup_{K' \in \mathscr{T}^{k'}} \overline{K'}$, $k' = 1, \dots, k$. This first paragraph of Sect. 7.1 originates from [27, Sect. 3.2], where also a proof of Proposition 2 is given.

Proposition 4 *Suppose that for some integer $k \in \mathbb{N}$, $k' \in \mathbb{N}_0$,*

$$E^k_{\widehat{K}}(\widehat{\phi}) = 0, \quad \forall \widehat{\phi} \in \mathbb{P}_{k'+k-1}(\widehat{K}).$$

Then, there exists a constant $C > 0$ such that for every $K \in \mathscr{T}$, $a \in W^{k,\infty}(K)$, $v \in \mathbb{P}_k(K)$, $w \in \mathbb{P}_{k'}(K)$,

$$|E^k_K(avw)| \leq C h^k_K \left(\sum_{j=0}^{k} |a|_{W^{k-j,\infty}(K)} |v|_{H^j(K)} \right) \|w\|_{L^2(K)}.$$

Proof This is a version of [10, Theorem 4.1.4]. The claimed estimate follows by [10, Eqs. (4.1.47) and (4.1.46), Theorems 3.1.2 and 3.1.3]. We note that we did not assume here $v \in \mathbb{P}_{k-1}(K)$, which results in the sum over $j = 0, \dots, k$. However, if $v \in \mathbb{P}_{k-1}(K)$ for some $k \geq 1$, then $|v|_{H^k(K)} = 0$. $\square$

Lemma 4 *Let $K \in \mathcal{T}$ be such that $\boldsymbol{c}_i \in \overline{K}$, $i \in \{1, \dots, J\}$. There exists a constant $C > 0$ independent of K such that for every $v \in \mathbb{P}_1(K)$ satisfying $v(\boldsymbol{c}_i) = 0$*

$$\|r_D^{-1} v\|_{L^\infty(K)} \leq C \|r_D^{-1} v\|_{L^2(K)} \det(B_K)^{-1/2}.$$

Proof We will prove the main step on the reference element $\widehat{K}$. It is easy to see that $\|r_D^{-1} v\|_{L^\infty(K)} = \|\widehat{r}_D^{-1} \widehat{v}\|_{L^\infty(\widehat{K})}$ and $\|\widehat{r}_D^{-1} \widehat{v}\|_{L^2(\widehat{K})} = \|r_D^{-1} v\|_{L^2(K)} \det(B_K)^{-1/2}$.

Suppose that $\widehat{K} := \{\widehat{x} \in (0,1)^2 : 0 < \widehat{x}_1 + \widehat{x}_2 < 1\}$ and wlog. that $F_K^{-1}(\boldsymbol{c}_i) = 0$. The space $\{\widehat{v} \in \mathbb{P}_1(\widehat{K}) : v(0) = 0\}$ is spanned by the monomials $\{\widehat{x}_1, \widehat{x}_2\}$. By [10, Theorem 3.1.3] and the shape regularity of $\mathcal{T}$, there exist constants $C, C' > 0$ independent of K such that $\|B_K\| \leq C h_K$ and $\|B_K^{-1}\| \leq C' h_K^{-1}$, where $\|\cdot\|$ denotes the operator matrix norm induced by the Euclidean norm $\|\cdot\|_2$. Note that $\min_{i=1,2}\{\int_{\widehat{K}} \frac{|\widehat{x}_i|}{\|\widehat{x}\|_2} \mathrm{d}\widehat{x}\} =: C'' > 0$. This implies by elementary manipulations and the Cauchy–Schwarz inequality

$$\begin{aligned} \sup_{\widehat{x} \in \widehat{K}} \frac{\|\widehat{x}\|_2}{\|B_K \widehat{x}\|_2} = \|B_K^{-1}\| &\leq 2 \frac{C'}{C'' h_K} \int_{\widehat{K}} \frac{|\widehat{x}_i|}{\|\widehat{x}\|_2} \mathrm{d}\widehat{x} \leq 2 \frac{CC'}{C''} \int_{\widehat{K}} \frac{|\widehat{x}_i|}{\|B_K\| \|\widehat{x}\|_2} \mathrm{d}\widehat{x} \\ &\leq 2^{1/2} \frac{CC'}{C''} \left(\int_{\widehat{K}} \frac{|\widehat{x}_i|^2}{\|B_K \widehat{x}\|_2^2} \mathrm{d}\widehat{x} \right)^{1/2}. \end{aligned} \tag{31}$$

On $\widehat{K}$, $\widehat{r}_D(\widehat{x}) = r_D(F_K(\widehat{x})) \sim \|B_K \widehat{x}\|_2$. Let $\widehat{v} = \widehat{v}_1 \widehat{x}_1 + \widehat{v}_2 \widehat{x}_2$. Thus, by (31) there exist constants $C, C' > 0$ independent of K such that

$$\|\widehat{r}_D^{-1} \widehat{v}\|_{L^\infty(K)} \leq C \left(|\widehat{v}_1| \sup_{\widehat{x} \in \widehat{K}} \frac{|\widehat{x}_1|}{\|B_K \widehat{x}\|_2} + |\widehat{v}_2| \sup_{\widehat{x} \in \widehat{K}} \frac{|\widehat{x}_2|}{\|B_K \widehat{x}\|_2} \right) \leq C' \left(\int_{\widehat{K}} \frac{|\widehat{v}(\widehat{x})|^2}{\|B_K \widehat{x}\|_2^2} \mathrm{d}\widehat{x} \right)^{1/2}.$$

The proof of the lemma is complete, since $\widehat{r}_D(\widehat{x}) \sim \|B_K \widehat{x}\|_2$ on $\widehat{K}$. $\square$

Proposition 5 *Let $K \in \mathcal{T}$ be such that $\boldsymbol{c}_i \in \overline{K}$, for some $i \in \{1, \dots, J\}$. Let $E_K^1(\cdot)$ denote the error from a one point quadrature in the barycenter $\bar{x}$ of K. Let $\delta_1, \delta_2, \delta_3, \delta_4 \in [0, 1)$. Then there exists a constant $C > 0$ such that for every $(r_D^{\delta_3+\delta_4} a) \in L^\infty(K)$ satisfying $r_D^{\delta_1+\delta_2} \|\nabla a\|_2 \in L^\infty(K)$ such that point evaluation at $\bar{x}$ is well defined and for every $v, w \in \mathbb{P}_k(K)$ for some $k \geq 0$*

$$\begin{aligned} &|E_K^1(avw)| \\ &\quad \leq C h_K^{1-\delta_1} \|r_D^{\delta_1+\delta_2} \|\nabla a\|_2\|_{L^\infty(K)} \|v\|_{L^2(K)} \|r_D^{-\delta_2} w\|_{L^2(K)} \\ &\quad + C h_K^{1-\delta_3} \|r_D^{\delta_3+\delta_4} a\|_{L^\infty(K)} \left(|v|_{H^1(K)} \|r_D^{-\delta_4} w\|_{L^2(K)} + \|r_D^{-\delta_4} v\|_{L^2(K)} |w|_{H^1(K)} \right). \end{aligned}$$

If additionally $v, w \in \mathbb{P}_1(K)$ *satisfy that* $v(\boldsymbol{c}_i) = 0 = w(\boldsymbol{c}_i)$*, the above assumption can be relaxed to* $r_D^{\delta_3+i} a \in L^\infty(K)$ *and* $r_D^{\delta_1+1+i}\|\nabla a\|_2 \in L^\infty(K)$*,* $i = 0, 1$*, and it holds that*

$$|E_K^1(avw)| \leq Ch_K^{1-\delta_1}\|r_D^{\delta_1+1+i}\|\nabla a\|_2\|_{L^\infty(K)}\|r_D^{-i}v\|_{L^2(K)}\|r_D^{-1}w\|_{L^2(K)}$$
$$+ Ch_K^{1-\delta_3}\|r_D^{\delta_3+i}a\|_{L^\infty(K)}\left(|v|_{H^1(K)}\|r_D^{-i}w\|_{L^2(K)} + \|r_D^{-i}v\|_{L^2(K)}|w|_{H^1(K)}\right).$$

Proof We observe that

$$|E_K^1(avw)| \leq \int_K |a(x) - a(\bar{x})||v(x)w(x)|\mathrm{d}x$$
$$+ \int_K |a(\bar{x})|\left(|v(x) - v(\bar{x})||w(x)| + |v(\bar{x})||w(x) - w(\bar{x})|\right)\mathrm{d}x. \tag{32}$$

For any $f \in W^{1,\infty}(\widetilde{K})$ and any $x \in \widetilde{K}$ ($\widetilde{K}$ a compact subset of K),

$$|f(x) - f(\bar{x})| \leq \sup_{\widetilde{x}\in\gamma_{x,\bar{x}}([0,1])}\{\|\nabla f(\widetilde{x})\|_2\}\|x - \bar{x}\|_2,$$

where $\gamma_{x,\bar{x}}$ is a suitable smooth path such that $\gamma_{x,\bar{x}}(1) = x$ and $\gamma_{x,\bar{x}}(0) = \bar{x}$. We will estimate the two integrals in (32) separately. Since $\boldsymbol{c}_i \in \overline{K}$, the weight function is locally $r_D(x) \simeq \|x - \boldsymbol{c}_i\|_2$. Due to the radial monotonicity of $x \mapsto \|x - \boldsymbol{c}_i\|_2$, $\gamma_{x,\bar{x}}$ can be chosen such that

$$\inf_{\widetilde{x}\in\gamma_{x,\bar{x}}([0,1])}\{\|\widetilde{x} - \boldsymbol{c}_i\|_2\} \in \{\|x - \boldsymbol{c}_i\|_2, \|\bar{x} - \boldsymbol{c}_i\|_2\}.$$

Hence, there exists a constant $C > 0$ independent of K such that for every $x \in K$

$$\frac{|a(x) - a(\bar{x})|}{\|x - \bar{x}\|_2}\min\{r_D^{\delta_1+\delta_2}(x), r_D^{\delta_1+\delta_2}(\bar{x})\} \leq C\|r_D^{\delta_1+\delta_2}\|\nabla a\|_2\|_{L^\infty(K)}. \tag{33}$$

Since all norms on $\mathbb{P}_k(\widehat{K})$ are equivalent, there exists a constant $C > 0$ independent of K such that $\|v\|_{L^\infty(K)} = \|\widehat{v}\|_{L^\infty(\widehat{K})} \leq C\|\widehat{v}\|_{L^2(\widehat{K})} = C\|v\|_{L^2(K)}\det(B_K)^{-1/2}$. Moreover, since there exists a constant $C > 0$ independent of K such that for every $x \in K$, $r_D(x) \leq Cr_D(\bar{x})$, there exists a constant $C > 0$ independent of K such that

$$\|1/\min\{r_D^{\delta_1}, r_D^{\delta_1}(\bar{x})\}\|_{L^2(K)} \leq Ch_K^{1-\delta_1}.$$

Similarly, $\|w/\min\{r_D^{\delta_2}, r_D^{\delta_2}(\bar{x})\}\|_{L^2(K)} \leq C\|r_D^{-\delta_2}w\|_{L^2(K)}$. It also holds that $\|x - \bar{x}\|_2 \leq Ch_K$ and $\det(B_K) \sim h_K^2$. Hence, for constants $C, C', C'' > 0$ independent of K,

$$\begin{aligned}
&\int_K |a(x) - a(\bar{x})||v(x)w(x)|\mathrm{d}x \\
&\quad \leq C\|r_D^{\delta_1+\delta_2}\|\nabla a\|_2\|_{L^\infty(K)} \int_K |v(x)| \frac{|w(x)|}{r_D^{\delta_2}} \frac{\|x - \bar{x}\|_2}{r_D^{\delta_1}} \mathrm{d}x \\
&\quad \leq C' h_K \|r_D^{\delta_1+\delta_2}\|\nabla a\|_2\|_{L^\infty(K)} \|v\|_{L^\infty(K)} \|r_D^{-\delta_2} w\|_{L^2(K)} \|r_D^{-\delta_1}\|_{L^2(K)} \\
&\quad \leq C'' h_K^{1-\delta_1} \|r_D^{\delta_1+\delta_2}\|\nabla a\|_2\|_{L^\infty(K)} \|v\|_{L^2(K)} \|r_D^{-\delta_2} w\|_{L^2(K)}.
\end{aligned} \tag{34}$$

On K such that $\overline{K}$ contains a corner, there exists a constant $\widehat{C} > 0$ that does not depend on K such that $\widehat{C} r_D(x) \leq r_D(\bar{x})$ for every $x \in K$. For the first summand in the second integral in (32), we obtain similarly for constants $C, C' > 0$ independent of K,

$$\begin{aligned}
\int_K |a(\bar{x})||v(x) - v(\bar{x})||w(x)|\mathrm{d}x &\leq \widehat{C} \int_K \frac{r_D^{\delta_3+\delta_4}(\bar{x})}{r_D^{\delta_3+\delta_4}(x)} |a(\bar{x})||v(x) - v(\bar{x})||w(x)|\mathrm{d}x \\
&\leq \|r_D^{\delta_3+\delta_4} a\|_{L^\infty(K)} \int_K \frac{|v(x) - v(\bar{x})|}{\|x - \bar{x}\|_2} \frac{|w(x)|}{r_D^{\delta_4}(x)} \frac{\|x - \bar{x}\|_2}{r_D^{\delta_3}(x)} \mathrm{d}x \\
&\leq C\|r_D^{\delta_3+\delta_4} a\|_{L^\infty(K)} \|\nabla v\|_{L^\infty(K)} \int_K \frac{|w(x)|}{r_D^{\delta_4}(x)} \frac{\|x - \bar{x}\|_2}{r_D^{\delta_3}(x)} \mathrm{d}x \\
&\leq C' h_K^{1-\delta_3} \|r_D^{\delta_3+\delta_4} a\|_{L^\infty(K)} |v|_{H^1(K)} \|r_D^{-\delta_4} w\|_{L^2(K)}
\end{aligned}$$

using that there are constants $\widetilde{C}, \widetilde{C}', \widetilde{C}'' > 0$ independent of K such that $\|\partial_{x_i} v\|_{L^\infty(K)} \leq \widetilde{C} h_K^{-1} \|\partial_{\widehat{x}_i} \widehat{v}\|_{L^\infty(\widehat{K})} \leq \widetilde{C}' h_K^{-1} \|\partial_{\widehat{x}_i} \widehat{v}\|_{L^2(\widehat{K})} \leq \widetilde{C}'' \|\partial_{x_i} v\|_{L^2(K)} \det(B_K)^{-1/2}$. Also note that by shape regularity of the triangulations, $\det(B_K) \sim h_K^2$ and by using polar coordinates $\int_K \|x - \bar{x}\|_2 r_D^{-\delta_3}(x)\mathrm{d}x \sim h^{2-\delta_3}$, where we used that $\overline{K}$ contains a corner of the domain D. The second summand in the second integral in (32) is estimated analogously.

The second estimate follows since $\|r_D^{-1} v\|_{L^\infty(K)} < \infty$ and $\|r_D^{-1} w\|_{L^2(K)} < \infty$, which allows us to conclude similarly as in (34)

$$\begin{aligned}
\int_K |a(x) - a(\bar{x})||v(x)w(x)|\mathrm{d}x &\leq C\|r_D^{\delta_1+1+i}\|\nabla a\|_2\|_{L^\infty(K)} \int_K \frac{|v(x)|}{r_D^i(x)} \frac{|w(x)|}{r_D(x)} \frac{\|x - \bar{x}\|_2}{r_D^{\delta_1}(x)} \mathrm{d}x \\
&\leq C' h_K^{2-\delta_1} \|r_D^{\delta_1+2}\|\nabla a\|_2\|_{L^\infty(K)} \|r_D^{-i} v\|_{L^\infty(K)} \|r_D^{-1} w\|_{L^2(K)} \\
&\leq C'' h_K^{1-\delta_1} \|r_D^{\delta_1+2}\|\nabla a\|_2\|_{L^\infty(K)} \|r_D^{-i} v\|_{L^2(K)} \|r_D^{-1} w\|_{L^2(K)},
\end{aligned}$$

where we used that $\|r_D^{-1} v\|_{L^\infty(K)} \leq \widetilde{C} h_K^{-1} \|r_D^{-1} v\|_{L^2(K)}$ for a constant $\widetilde{C} > 0$ that neither depends on K nor on v, which follows by Lemma 4. Also the constants $C, C', C'' > 0$ neither depend on K nor on v. □

Proof of Theorem 1. The proof generalizes [10, Theorem 4.1.6] to the case of local mesh refinement and singularities of the solution and the coefficients. Throughout this proof $C, C' > 0$ denote generic constants that neither depend on elements of the triangulation $\mathscr{T}$ nor on functions on D. We recall the *first Strang lemma*, see for example [10, Theorem 4.1.1]

$$\|q - \widetilde{q}^{\mathscr{T}}\|_V \leq \frac{u_{\max}}{u_{\min}} \inf_{v^{\mathscr{T}} \in V^k_{\mathscr{T}}} \left\{ \|q - v^{\mathscr{T}}\|_V + \sup_{0 \neq w^{\mathscr{T}} \in V^k_{\mathscr{T}}} \frac{|A(u)(v^{\mathscr{T}}, w^{\mathscr{T}}) - \widetilde{A}(u)(v^{\mathscr{T}}, w^{\mathscr{T}})|}{\|w^{\mathscr{T}}\|_V} \right\},$$

where $u_{\max}$ and $u_{\min}$ are continuity and coercivity constants of $A(u)$, $\widetilde{A}(u)$. The right hand side of the first Strang lemma will be upper bounded by choosing $v^{\mathscr{T}} := I_{\mathscr{T}} q \in \widetilde{V}^k_{\mathscr{T}}$.

We will treat the second, first, and zero order terms separately and start with the second order term. We decompose $A(u) = \sum_{i,j=1}^{2} A(a_{ij}) + \sum_{j=1}^{4} A(b_j) + A(c)$ and $\widetilde{A}(u) = \sum_{i,j=1}^{2} \widetilde{A}(a_{ij}) + \sum_{j=1}^{4} \widetilde{A}(b_j) + \widetilde{A}(c)$. As in the proof of [27, Proposition 3.2.1], we discuss the error contributions elementwise. There, we distinguish several cases, $K \in \mathscr{T}^1$, $K \in \mathscr{T}^{k'}$ and $\overline{K} \cap \overline{\mathscr{T}^{k'-1}} \neq \emptyset$, $k' = 2, \ldots, k$, and $K \in \mathscr{T}^k$ and $\overline{K} \cap \overline{\mathscr{T}^{k-1}} = \emptyset$. We observe

$$\begin{aligned} &|\sum_{i,j=1}^{2} A(a_{ij})(I_{\mathscr{T}} q, w^{\mathscr{T}}) - \sum_{i,j=1}^{2} \widetilde{A}(a_{ij})(I_{\mathscr{T}} q, w^{\mathscr{T}})| \\ &\leq \sum_{K \in \mathscr{T}^1} \sum_{i,j=1}^{2} |E^1_K(a_{ij} \partial_j I^1_K q \partial_i w^{\mathscr{T}})| + \sum_{K \in \mathscr{T}^k, \overline{K} \cap \overline{\mathscr{T}^{k-1}} = \emptyset} \sum_{i,j=1}^{2} |E^k_K(a_{ij} \partial_j I^k_K q \partial_i w^{\mathscr{T}})| \\ &+ \sum_{k'=2}^{k} \sum_{K \in \mathscr{T}^{k'}, e := \overline{K} \cap \overline{\mathscr{T}^{k'-1}} \neq \emptyset} \sum_{i,j=1}^{2} |E^{k'}_K(a_{ij} \partial_j (I^{k'}_K q - (I^{k'}_K q|_e - I^{k'-1}_e q)_{\mathrm{lift},k',e}) \partial_i w^{\mathscr{T}})|. \end{aligned}$$

By (12), for $K \in \mathscr{T}^1$,

$$h_K^{1-\delta} \leq C h^{(1-\delta)/(1-\beta)} \leq C h^k. \tag{35}$$

For $K \in \mathscr{T}^1$, by Proposition 5 (with $\delta_1 = \delta_3 = \delta$, $\delta_2 = \delta_4 = 0$) and (35)

$$|E^1_K(a_{ij} \partial_j I^1_K q \partial_i w^{\mathscr{T}})| \leq C h^k \|a_{ij}\|_{W^{1,\infty}_\delta(K)} \|I^1_K q\|_{H^1(K)} \|w^{\mathscr{T}}\|_{H^1(K)}$$

and [46, Lemma 4.16] implies with the triangle inequality the existence of a constant $C > 0$ (depending only on the shape regularity of the triangulations $\{\mathscr{T}^k\}_{k \geq 0}$) such that for every $q \in H^2_\delta(K)$ holds $\|I^1_K q\|_{H^1(K)} \leq C(\|q\|_{H^1(K)} + h_K^{1-\delta} |q|_{H^2_\delta(K)})$. For $K \in \mathscr{T}^k$ such that $\overline{K} \cap \overline{\mathscr{T}^{k-1}} = \emptyset$, by Proposition 4

$$
\begin{aligned}
&|E_K^k(a_{ij}\partial_j I_K^k q \partial_i w^{\mathcal{T}})| \\
&\quad \le Ch^k \sum_{\ell=0}^{k-1} \inf_{x\in K} r_D^{\beta(k-\ell)}(x)|a_{ij}|_{W^{k-\ell,\infty}(K)} \inf_{x\in K} r_D^{\beta\ell}(x)|I_K^k q|_{H^{\ell+1}(K)}|w^{\mathcal{T}}|_{H^1(K)}.
\end{aligned} \tag{36}
$$

It follows directly from (12),

$$
h_K \le Ch^{1/(1-(\beta-\alpha))} r_D^{\alpha/(1-(\beta-\alpha))}(x) \quad \forall x \in K, \forall \alpha \in (0,\beta).
$$

We choose $\alpha := (1-\beta)(k'-2+\delta)/(1-\delta)$ and apply $(1-\delta)/(1-\beta) > k$,

$$
h_K \le Ch^{k/(k'-1)} r_D^{(\delta+k'-2)/(k'-1)}(x) \quad \forall x \in K, k' = 2,\dots,k. \tag{37}
$$

For $K \in \mathcal{T}^{k'}$ such that $e := \overline{K} \cap \overline{\mathcal{T}^{k'-1}} \neq \emptyset, k' = 2,\dots,k$, by Proposition 4 and (37)

$$
\begin{aligned}
&|E_K^{k'}(a_{ij}\partial_j(I_K^{k'} q - (I_K^{k'} q|_e - I_e^{k'-1} q)_{\mathrm{lift},k',e})\partial_i w^{\mathcal{T}})| \\
&\quad \le Ch_K^{k'} \sum_{\ell=0}^{k'-1} |a_{ij}|_{W^{k'-\ell,\infty}(K)} |I_K^{k'} q - (I_K^{k'} q|_e - I_e^{k'-1} q)_{\mathrm{lift},k',e}|_{H^{\ell+1}(K)} |w^{\mathcal{T}}|_{H^1(K)} \\
&\quad \le C'h^k \sum_{\ell=0}^{k'-1} \inf_{x\in K} r_D^{\delta+k'-1}(x)|a_{ij}|_{W^{k'-\ell,\infty}(K)} \\
&\qquad\qquad \times |I_K^{k'} q - (I_K^{k'} q|_e - I_e^{k'-1} q)_{\mathrm{lift},k',e}|_{H^{\ell+1}(K)} |w^{\mathcal{T}}|_{H^1(K)}.
\end{aligned}
$$

Note that $(1-\delta)/(1-\beta) > k$ implies that $\beta k' > \delta + k' - 1$, $k' = 1,\dots,k$. We observe with [10, Theorem 3.1.6]

$$
|I_K^k q|_{H^{\ell+1}(K)} \le C(|q|_{H^{\ell+1}(K)} + h_K^{k'-1}|q|_{H^{k'+1}(K)}), \quad \ell = 0,\dots,k'-1,
$$

and by a similar argument as in the proof of [27, Proposition 3.2.1] for $\ell = 0,\dots,k'-1$,

$$
|I_K^{k'} q - (I_K^{k'} q|_e - I_e^{k'-1} q)_{\mathrm{lift},k',e}|_{H^{\ell+1}(K)} \le C(|q|_{H^{\ell+1}(K)} + h_K^{k'-1}|q|_{H^{k'+1}(K)}).
$$

By the Cauchy–Schwarz inequality we conclude with the previous inequalities

$$
\begin{aligned}
&|\sum_{i,j=1}^{2} A(a_{ij})(I_{\mathcal{T}} q, w^{\mathcal{T}}) - \sum_{i,j=1}^{2} \widetilde{A}(a_{ij})(I_{\mathcal{T}} q, w^{\mathcal{T}})| \\
&\qquad \le Ch^k \sum_{i,j=1}^{2} \|a_{ij}\|_{W_\delta^{k,\infty}(K)} \|q\|_{\mathcal{K}_{a+1}^{k+1}(D)} \|w^{\mathcal{T}}\|_{H^1(D)}.
\end{aligned}
$$

The argument for the advection and reaction terms $\sum_{j=1}^{4} A(b_j)$, $A(c)$ is similar. Here, the additional weight r_D for the advection terms and r_D^2 for the reaction term needs to be accommodated. For the advection term, by the second part of Proposition 5 (with $\delta_1 = \delta_3 = \delta$, $i = 0$) and $K \in \mathscr{T}_1$ for $j = 1, 2$

$$|E_K^1(b_j(\partial_j I_K^1 q) w^{\mathscr{T}})| \leq C h^k \|r_D^\delta b_j\|_{W_\delta^{1,\infty}(K)} \|I_K^1 q\|_{H^1(K)} \left(\|r_D^{-1} w^{\mathscr{T}}\|_{L^2(K)} + |w^{\mathscr{T}}|_{H^1(K)}\right) \tag{38}$$

and for $j = 3, 4$

$$|E_K^1(b_j I_K^1 q \partial_j w^{\mathscr{T}})| \leq C h^k \|r_D^\delta b_j\|_{W_\delta^{1,\infty}(K)} \left(\|r_D^{-1} I_K^1 q\|_{L^2(K)} + |I_K^1 q|_{H^1(K)}\right) \|w^{\mathscr{T}}\|_{H^1(K)}.$$

For the interior elements $K \in \mathscr{T} \backslash \mathscr{T}^1$, the additional weight r_D can be accommodated by compensating it with $\|r_D^{-1} w^{\mathscr{T}}\|_{L^2(K)}$ as in (38) for $j = 1, 2$. If the partial derivative is on the trial function, i.e., $j = 3, 4$, the order of the Sobolev semi-norm as for example in (36) is reduced by one to $|I_K^k q|_{H^\ell(K)}$. Here, the weight $r_D^{\beta(k-\ell+1)}$ is assigned to $|b_j|_{W^{k-\ell,\infty}(K)}$, if $\ell \geq 1$. For $\ell = 0$, the additional weight r_D can be compensated by $\|r_D^{-1} I_K^k q\|_{L^2(K)}$. We recall the Hardy inequality from (15), i.e., there exists a constant $C > 0$ such that for every $v \in V$

$$\|r_D^{-1} v\|_{L^2(D)} \leq C \|\|\nabla v\|_2\|_{L^2(D)}.$$

Thus, $\|r_D^{-1} w^{\mathscr{T}}\|_{L^2(D)} \leq C \|w^{\mathscr{T}}\|_V$ and $\|r_D^{-1} I_K^k q\|_{L^2(D)} \leq C \|I_K^k q\|_V$. The rest of the proof for the advection terms is analogous to the diffusion terms, which were proved in detail. The argument for the reaction term uses the second part of Proposition 5 with $i = 1$. We omit the details. □

Proof of Corollary 1 Throughout this proof $C, C' > 0$ denote generic constants that depend on the shape regularity of the triangulation $\mathscr{T}$, but are independent of element sizes or of functions on D. For given uncertain input u, the solution $g \in V$ to the adjoint problem is characterized by

$$A(u)(w, g) = \langle G, w \rangle_{V^*, V} \quad \forall w \in V.$$

The respective FE approximation $g^{\mathscr{T}}$ is characterized by $A(u)(w^{\mathscr{T}}, g - g^{\mathscr{T}}) = 0$ for every $w^{\mathscr{T}} \in V_{\mathscr{T}}^k$. In the superconvergence analysis, we employ the usual duality argument as outlined in the proof of [4, Theorem 3.6]. By a version of [4, Lemma 3.1] for non-symmetric bilinear forms $A(u)(\cdot, \cdot)$,

$$G(q) - G(q^{\mathscr{T}}) = A(u)(q - q^{\mathscr{T}}, g - g^{\mathscr{T}}) + \widetilde{A}(u)(\widetilde{q}^{\mathscr{T}}, g^{\mathscr{T}}) - A(u)(\widetilde{q}^{\mathscr{T}}, g^{\mathscr{T}}).$$

As in the previous proof, we begin by estimating the diffusion terms related to a_{ij}. By a similar argument that we used to show (36) also using Proposition 5, we obtain

$$
\begin{aligned}
&|\widetilde{A}(a_{ij})(\widetilde{q}^{\mathcal{T}}, g^{\mathcal{T}}) - A(a_{ij})(\widetilde{q}^{\mathcal{T}}, g^{\mathcal{T}})| \\
&\leq Ch^{k+k'}\left(\sum_{K\in\mathcal{T}^1} \|a_{ij}\|_{W^{1,\infty}_\delta(D)}\|\widetilde{q}^{\mathcal{T}}\|_{H^1(K)}\|g^{\mathcal{T}}\|_{H^1(K)}\right.\\
&\left.+\sum_{K\in\mathcal{T}\backslash\mathcal{T}^1}\sum_{i,j=1}^{2}\sum_{\ell=0}^{k+k'-1}\inf_{x\in K} r_D^{\beta(k+k'-\ell)}(x)|a_{ij}|_{W^{k+k'-\ell,\infty}(K)}\inf_{x\in K} r_D^{\beta\ell}|\partial_j\widetilde{q}^{\mathcal{T}}\partial_i g^{\mathcal{T}}|_{H^\ell(K)}\right).
\end{aligned}
$$

Note that $(\partial_j\widetilde{q}^{\mathcal{T}})|_K, (\partial_i g^{\mathcal{T}})|_K \in \mathbb{P}_{k-1}(K)$, which implies that

$$
\partial^{\boldsymbol{\alpha}}(\partial_j\widetilde{q}^{\mathcal{T}})|_K = 0 = \partial^{\boldsymbol{\alpha}}(\partial_i g^{\mathcal{T}})|_K \quad \forall\boldsymbol{\alpha}\in\mathbb{N}_0^2, |\boldsymbol{\alpha}| > k-1.
$$

By the product rule and by the Cauchy–Schwarz inequality

$$
|\partial_j\widetilde{q}^{\mathcal{T}}\partial_i g^{\mathcal{T}}|_{H^\ell(K)} \leq C\sum_{\ell'=0}^{\ell}|\partial_j\widetilde{q}^{\mathcal{T}}|_{H^{\ell'}(K)}|\partial_i g^{\mathcal{T}}|_{H^{\ell-\ell'}(K)}.
$$

By the inverse inequality and the element-wise approximation property of the nodal interpolant, e.g. [10, Theorem 3.1.6] we observe that there exist constants $C, C' > 0$ such that for every $K \in \mathcal{T}\backslash\mathcal{T}^1$,

$$
\begin{aligned}
|\partial_j\widetilde{q}^{\mathcal{T}}|_{H^{\ell'}(K)} &\leq |q|_{H^{\ell'+1}(K)} + |\partial_j q - I_K^k\partial_j q|_{H^{\ell'}(K)} + |I_K^k\partial_j q - \partial_j\widetilde{q}^{\mathcal{T}}|_{H^{\ell'}(K)}\\
&\leq C|q|_{H^{\ell'+1}(K)} + Ch_K^{-\ell'}\|I_K^k\partial_j q - \partial_j\widetilde{q}^{\mathcal{T}}\|_{L^2(K)}\\
&\leq C|q|_{H^{\ell'+1}(K)} + Ch_K^{-\ell'}(\|I_K^k\partial_j q - \partial_j q\|_{L^2(K)} + \|\partial_j q - \partial_j\widetilde{q}^{\mathcal{T}}\|_{L^2(K)})\\
&\leq C'(|q|_{H^{\ell'+1}(K)} + h_K^{-\ell'}|q - \widetilde{q}^{\mathcal{T}}|_{H^1(K)}).
\end{aligned}
$$

Similarly, it holds that $|\partial_i g^{\mathcal{T}}|_{H^{\ell-\ell'}(K)} \leq C|g^{\mathcal{T}}|_{H^{\ell-\ell'+1}(K)}$. The previous elementwise estimates allow us to conclude with the Cauchy–Schwarz inequality

$$
\begin{aligned}
&|\widetilde{A}(a_{ij})(\widetilde{q}^{\mathcal{T}}, g^{\mathcal{T}}) - A(a_{ij})(\widetilde{q}^{\mathcal{T}}, g^{\mathcal{T}})|\\
&\quad\leq Ch^{k+k'}\|a_{ij}\|_{W^{k+k',\infty}_\delta(D)}(\|q\|_{\mathcal{K}^{k+1}_{a+1}(D)} + h^{-k}\|q - \widetilde{q}^{\mathcal{T}}\|_V)\|g\|_{\mathcal{K}^{k'+1}_{a+1}(D)}\\
&\quad\leq C'h^{k+k'}\|a_{ij}\|_{W^{k+k',\infty}_\delta(D)}(1 + \|u\|_{X_{k,\delta}})\|q\|_{\mathcal{K}^{k+1}_{a+1}(D)}\|g\|_{\mathcal{K}^{k'+1}_{a+1}(D)},
\end{aligned}
$$

where we used Theorem 1 in the second step. The argument for the advection and reaction terms $A(b_j)$, $j = 1, \ldots, 4$, and $A(c)$ is similar. See also the proof of Theorem 1. Since Proposition 2 and (8) imply with Céa's lemma

$$
|A(u)(q - q^{\mathcal{T}}, g - g^{\mathcal{T}})| \leq Ch^{k+k'}\|f\|_{\mathcal{K}^{k-1}_{a-1}(D)}\|G\|_{\mathcal{K}^{k'-1}_{a-1}(D)},
$$

the assertion follows. □

7.2 Approximation of Functions with Point Singularities

In this section we analyze approximation rates by biorthogonal spline wavelet expansions with compact supports for functions in the polygon D with point singularities. We consider regularity in weighted Hölder spaces $\mathscr{W}_\delta^{m,\infty}(D)$ and more generally in $X_{m,\delta}$ for $\delta \in [0, 1)$. We explicitly define *a-priori truncation* of infinite biorthogonal wavelet expansions of these functions, mimicking in this way FE mesh refinement in D as in [36] (see also [45]).

Let $(\psi_\lambda)_{\lambda\in\nabla}$ be a biorthogonal spline wavelet basis of $L^2(D)$ with dual wavelet system $(\widetilde{\psi}_\lambda)_{\lambda\in\nabla}$, we refer to [14, 41, 47, 48] for concrete constructions. We suppose that $(\psi_\lambda)_{\lambda\in\nabla}$ and $(\widetilde{\psi}_\lambda)_{\lambda\in\nabla}$ have the following properties.

1. (biorthogonality) $\int_D \psi_\lambda \widetilde{\psi}_{\lambda'} \mathrm{d}x = \delta_{\lambda\lambda'}$, $\lambda, \lambda' \in \nabla$,
2. (normalization) $\|\psi_\lambda\|_{L^\infty(D)} \lesssim 2^{d|\lambda|/2}$ and $\|\widetilde{\psi}_\lambda\|_{L^\infty(D)} \lesssim 2^{d|\lambda|/2}$ for every $\lambda \in \nabla$,
3. (compact support) $|\text{supp}(\psi_\lambda)| = \mathscr{O}(2^{-|\lambda|d})$ and $|\text{supp}(\widetilde{\psi}_\lambda)| = \mathscr{O}(2^{-|\lambda|d})$ for every $\lambda \in \nabla$,
4. (vanishing moments of order k) $\int_D x^{\boldsymbol{\alpha}} \psi_\lambda \mathrm{d}x = 0$ and $\int_D x^{\boldsymbol{\alpha}} \widetilde{\psi}_\lambda \mathrm{d}x = 0$ for all multi-indices $\boldsymbol{\alpha} \in \mathbb{N}_0^2$ such that $|\boldsymbol{\alpha}| \le k$ and for every $\lambda \in \nabla$.

We also suppose that $(\psi_\lambda)_{\lambda\in\nabla}$ satisfies the finite overlap property in (19). Denoting the $L^2(D)$ inner product by $(\cdot, \cdot)_{L^2(D)}$, for $L \in \mathbb{N}_0$ and $\beta \in [0, 1)$, define the index sets

$$\Lambda_{L,\beta} := \left\{\lambda \in \nabla : r_D^\beta(x_\lambda) \le 2^{L-|\lambda|}\right\},$$

where x_λ is the barycenter of $\text{supp}(\psi_\lambda)$, $\lambda \in \nabla$. Every function $w \in L^2(D)$ can be represented as $u = \sum_{\lambda\in\nabla} (w, \widetilde{\psi}_\lambda)_{L^2(D)} \psi_\lambda$ with equality in $L^2(D)$. With the finite index set $\Lambda_{L,\beta}$, we define the quasi-interpolant $P_{L,\beta}$ by

$$P_{L,\beta} w := \sum_{\lambda\in\Lambda_{L,\beta}} (w, \widetilde{\psi}_\lambda)_{L^2(D)} \psi_\lambda. \tag{39}$$

Proposition 6 *For $m \in \mathbb{N}$, suppose $m > k$ and $0 < \delta < \beta < 1$ satisfy $(1 - \delta)/(1 - \beta) > m$. Then, there exists a constant $C > 0$ such that for every $w \in W_\delta^{m,\infty}(D)$*

$$\|w - P_{L,\beta} w\|_{L^\infty(D)} \le C 2^{-\min\{k+1,m\}L} \|w\|_{W_\delta^{m,\infty}(D)}.$$

Proof Without loss of generality we assume that $k + 1 = m$. We distinguish the cases $\inf_{x\in\text{supp}(\widetilde{\psi}_\lambda)} r_D(x) = 0$ and $\inf_{x\in\text{supp}(\widetilde{\psi}_\lambda)} r_D(x) > 0$.

In the latter case $w \in W^{m,\infty}(\text{supp}(\widetilde{\psi}_\lambda))$. The Taylor sum $\sum_{|\boldsymbol{\alpha}|\le k} w_{\boldsymbol{\alpha}} x^{\boldsymbol{\alpha}}$ of w in $\text{supp}(\widetilde{\psi}_\lambda)$ satisfies that there exists a constant $C > 0$ independent of w such that for every $\lambda \in \nabla$,

$$\operatorname{ess\,sup}_{x\in\operatorname{supp}(\psi_\lambda)}\left|w(x)-\sum_{|\alpha|\le k} w_\alpha x^\alpha\right| \le C[\operatorname{diam}(\operatorname{supp}(\psi_\lambda))]^{k+1}\|w\|_{W^{k+1,\infty}(\operatorname{supp}(\psi_\lambda))}. \tag{40}$$

By the *vanishing moments property*, the $L^\infty(D)$ bounds and the support property of $\widetilde{\psi}_\lambda$, the Cauchy–Schwarz inequality, and (40),

$$|(w,\widetilde{\psi}_\lambda)_{L^2(D)}| \le C2^{-(k+1)|\lambda|}2^{-d|\lambda|/2}\|w\|_{W^{k+1,\infty}(\operatorname{supp}(\widetilde{\psi}_\lambda))}. \tag{41}$$

This estimate is suitable if $\inf_{x\in\operatorname{supp}(\psi_\lambda)} r_D(x) > 0$. If λ is such that $\inf_{x\in\operatorname{supp}(\psi_\lambda)} r_D(x) = 0$, which essentially implies that $\operatorname{supp}(\psi_\lambda)$ abuts at a corner of D, by the estimate in (33), there exists a constant $C > 0$ (independent of w and λ) such that

$$\operatorname{ess\,sup}_{x\in\operatorname{supp}(\widetilde{\psi}_\lambda)}\{r_D^\delta(x)|w(x)-w(x_\lambda)|\} \le C2^{-|\lambda|}\|r_D^\delta\|\nabla w\|_2\|_{L^\infty(\operatorname{supp}(\widetilde{\psi}_\lambda))}.$$

Thus,

$$\begin{aligned}|(w,\widetilde{\psi}_\lambda)_{L^2(D)}| &= |(w-w(x_\lambda),\widetilde{\psi}_\lambda)_{L^2(D)}|\\ &\le C2^{-|\lambda|}\|r_D^{-\delta}\|_{L^2(\operatorname{supp}(\widetilde{\psi}_\lambda))}\|r_D^\delta\|\nabla w\|_2\|_{L^\infty(\operatorname{supp}(\widetilde{\psi}_\lambda))}\|\widetilde{\psi}_\lambda\|_{L^2(D)}.\end{aligned} \tag{42}$$

We note that $\|r_D^{-\delta}\|_{L^2(\operatorname{supp}(\widetilde{\psi}_\lambda))} \le C2^{-|\lambda|(d/2-\delta)}$ for a constant $C > 0$ independent of λ. For $\lambda \in \nabla\backslash\Lambda_{L,\beta}$ and $\operatorname{supp}(\widetilde{\psi}_\lambda)\cap\mathscr{C}\neq\emptyset$, $r_D(x_\lambda)^\beta > 2^{L-|\lambda|}$ and $r_D(x_\lambda)^\beta \le C2^{-|\lambda|\beta}$ for a constant independent of λ. Since $(1-\delta)/(1-\beta) > k+1$,

$$2^{-|\lambda|}\|r_D^{-\delta}\|_{L^2(\operatorname{supp}(\widetilde{\psi}_\lambda))} \le C2^{-d|\lambda|/2}2^{-L(k+1)}. \tag{43}$$

For $\lambda \in \nabla\backslash\Lambda_{L,\beta}$ and $\operatorname{supp}(\widetilde{\psi}_\lambda)\cap\mathscr{C}=\emptyset$, $(1-\delta)/(1-\beta) > k+1$ implies that

$$2^{-|\lambda|(k+1)} \le C2^{-L(k+1)}r_D^{\delta+k}(x_\lambda). \tag{44}$$

Let $\widetilde{\Lambda}\subset\nabla\backslash\Lambda_{L,\beta}$ be an index set such that $\overline{D}\subset\bigcup_{\lambda\in\widetilde{\Lambda}}\operatorname{supp}(\psi_\lambda)$ and for every $\lambda,\lambda'\in\widetilde{\Lambda}$, $\operatorname{supp}(\psi_\lambda)\not\subset\operatorname{supp}(\psi'_\lambda)$. For $\lambda'\in\widetilde{\Lambda}$ such that $\operatorname{supp}(\psi_\lambda)\cap\mathscr{C}=\emptyset$, by (41), the bounded support overlap property (19) of $(\psi_\lambda)_{\lambda\in\nabla}$, and by (44), there exist constants $C, C' > 0$ such that

$$\begin{aligned}\|w-P_{L,\beta}w\|_{L^\infty(\operatorname{supp}(\psi_{\lambda'}))} &\le C\sum_{\ell\ge|\lambda'|}2^{-(k+1)\ell}\|w\|_{W^{k+1,\infty}(\operatorname{supp}(\widetilde{\psi}_{\lambda'}))}\\ &\le C2^{-(k+1)|\lambda'|}\sum_{\ell\ge 0}2^{-\ell}\|w\|_{W^{k+1,\infty}(\operatorname{supp}(\widetilde{\psi}_{\lambda'}))}\\ &\le C'2^{-(k+1)L}\|w\|_{W_\delta^{k+1,\infty}(\operatorname{supp}(\widetilde{\psi}_{\lambda'}))}.\end{aligned}$$

Similarly, for $\lambda'\in\widetilde{\Lambda}$ such that $\operatorname{supp}(\psi_\lambda)\cap\mathscr{C}\neq\emptyset$, by (42), the bounded support overlap property of $(\psi_\lambda)_{\lambda\in\nabla}$, and (43) there exists constants $C, C' > 0$ such that

$$
\begin{aligned}
&\|w - P_{L,\beta} w\|_{L^\infty(\mathrm{supp}(\psi_{\lambda'}))} \\
&\quad \le C \sum_{\ell \ge |\lambda'|} 2^{-\ell} \sum_{\lambda \in \nabla \setminus \Lambda_{L,\beta} : |\lambda| = \ell} \|r_D^{-\delta}\|_{L^2(\mathrm{supp}(\widetilde{\psi}_\lambda))} 2^{d\ell/2} \|w\|_{W_\delta^{1,\infty}(\mathrm{supp}(\widetilde{\psi}_{\lambda'}))} \\
&\quad \le C 2^{-(k+1)L} \sum_{\ell \ge 0} 2^{-\ell} \|w\|_{W^{1,\infty}(\mathrm{supp}_\delta(\widetilde{\psi}_{\lambda'}))}.
\end{aligned}
$$

Since $\overline{D} \subset \bigcup_{\lambda \in \widetilde{\Lambda}} \mathrm{supp}(\psi_\lambda)$, the proof of the proposition is complete. □

The following lemma may be shown as [36, Eqs. (5) and (13)].

Lemma 5 *For every $L \in \mathbb{N}$ and $\beta \in [0, 1)$, $|\Lambda_{L,\beta}| = \mathcal{O}(2^{dL})$.*

Proof of Proposition 1. We write $w = (a_{ij}, b_j, c) \in X_{m,\delta}$ for some $m \ge 1$. We suppose that the biorthogonal wavelets $(\psi_\lambda)_\nabla$ have vanishing moments of order $m - 1 = k \ge 0$. The statement of the theorem follows applying Proposition 6 to a_{ij}, $r_D b_j$, and to $r_D^2 c$ together with Lemma 5. □

Acknowledgements This work was supported in part by the Swiss National Science Foundation (SNSF) under grant SNF 159940. This work was completed when LH was a member of the Seminar for Applied Mathematics at ETH Zürich. The authors thank the editors and one reviewer for his/her careful reading and the constructive remarks on our initial submission which improved the presentation.

References

1. Adler, J.H., Nistor, V.: Graded mesh approximation in weighted Sobolev spaces and elliptic equations in 2D. Math. Comput. **84**(295), 2191–2220 (2015)
2. Aylwin, R., Jerez-Hanckes, C., Schwab, C., Zech, J.: Domain uncertainty quantification in computational electromagnetics. SIAM/ASA Uncertain. Quantif. **8**(1), 301–341 (2020)
3. Babuška, I., Kellogg, R.B., Pitkäranta, J.: Direct and inverse error estimates for finite elements with mesh refinements. Numer. Math. **33**(4), 447–471 (1979)
4. Babuška, I., Banerjee, U., Li, H.: The effect of numerical integration on the finite element approximation of linear functionals. Numer. Math. **117**(1), 65–88 (2011)
5. Băcuţă, C., Li, H., Nistor, V.: Differential operators on domains with conical points: precise uniform regularity estimates. Rev. Roumaine Math. Pures Appl. **62**(3), 383–411 (2017)
6. Bui-Thanh, T., Girolami, M.: Solving large-scale PDE-constrained Bayesian inverse problems with Riemann manifold Hamiltonian Monte Carlo. Inverse Probl. **30**(11), 114,014, 23 (2014)
7. Chen, L., Nochetto, R.H., Xu, J.: Optimal multilevel methods for graded bisection grids. Numer. Math. **120**(1), 1–34 (2012)
8. Chen, P., Villa, U., Ghattas, O.: Hessian-based adaptive sparse quadrature for infinite-dimensional Bayesian inverse problems. Comput. Methods Appl. Mech. Engrg. **327**, 147–172 (2017)
9. Chkifa, A., Cohen, A., Schwab, C.: Breaking the curse of dimensionality in sparse polynomial approximation of parametric PDEs. J. Math. Pures Appl. **103**(2), 400–428 (2015)
10. Ciarlet, P.G.: The Finite Element Method for Elliptic Problems. Studies in Mathematics and its Applications, vol. 4. North-Holland Publishing Co., Amsterdam (1978)
11. Cohen, A., DeVore, R., Schwab, Ch.: Analytic regularity and polynomial approximation of parametric and stochastic elliptic PDE's. Anal. Appl. (Singap.) **9**(1), 11–47 (2011)

12. Constantine, G.M., Savits, T.H.: A multivariate Faà di Bruno formula with applications. Trans. Am. Math. Soc. **348**(2), 503–520 (1996)
13. Dashti, M., Stuart, A.M.: The Bayesian approach to inverse problems. pp. 1–118. Springer International Publishing (2015)
14. Davydov, O., Stevenson, R.: Hierarchical Riesz bases for $H^s(\Omega)$, $1<s<\frac{5}{2}$. Constr. Approx. **22**(3), 365–394 (2005)
15. Dick, J., Gantner, R.N., LeGia, Q.T., Schwab, C.: Multilevel higher-order quasi-Monte Carlo Bayesian estimation. Math. Models Methods Appl. Sci. **27**(5), 953–995 (2017)
16. Dick, J., Gantner, R.N., LeGia, Q.T., Schwab, C.: Higher order quasi-Monte Carlo integration for Bayesian PDE inversion. Comput. Math. Appl. **77**(1), 144–172 (2019)
17. Dick, J., Kuo, F.Y., LeGia, Q.T., Nuyens, D., Schwab, Ch.: Higher order QMC Petrov-Galerkin discretization for affine parametric operator equations with random field inputs. SIAM J. Numer. Anal. **52**(6), 2676–2702 (2014)
18. Dick, J., Kuo, F.Y., LeGia, Q.T., Schwab, Ch.: Multilevel higher order QMC Petrov-Galerkin discretization for affine parametric operator equations. SIAM J. Numer. Anal. **54**(4), 2541–2568 (2016)
19. Dick, J., Kuo, F.Y., Sloan, I.H.: High-dimensional integration: the quasi-Monte Carlo way. Acta Numer. **22**, 133–288 (2013)
20. Dick, J., LeGia, Q.T., Schwab, C.: Higher order quasi-Monte Carlo integration for holomorphic, parametric operator equations. SIAM/ASA J. Uncertain. Quantif. **4**(1), 48–79 (2016)
21. Dodwell, T.J., Ketelsen, C., Scheichl, R., Teckentrup, A.L.: A hierarchical multilevel Markov chain Monte Carlo algorithm with applications to uncertainty quantification in subsurface flow. SIAM/ASA J. Uncertain. Quantif. **3**(1), 1075–1108 (2015)
22. Gantner, R.N., Herrmann, L., Schwab, C.: Multilevel QMC with product weights for affine-parametric, elliptic PDEs. In: Dick, J., Kuo, F.Y., Woźniakowski, H. (eds.) Contemporary Computational Mathematics-A Celebration of the 80th Birthday of Ian Sloan, pp. 373–405. Springer, Cham (2018)
23. Gantner, R.N., Herrmann, L., Schwab, C.: Quasi-Monte Carlo integration for affine-parametric, elliptic PDEs: local supports and product weights. SIAM J. Numer. Anal. **56**(1), 111–135 (2018)
24. Gantner, R.N., Peters, M.D.: Higher-order Quasi-Monte Carlo for Bayesian shape inversion. SIAM/ASA J. Uncertain. Quantif. **6**(2), 707–736 (2018)
25. Gaspoz, F.D., Morin, P.: Convergence rates for adaptive finite elements. IMA J. Numer. Anal. **29**(4), 917–936 (2009)
26. Giles, M.B.: Multilevel Monte Carlo methods. Acta Numer. **24**, 259–328 (2015)
27. Herrmann, L.: Quasi-Monte Carlo integration in uncertainty quantification for PDEs with log-Gaussian random field inputs. Ph.D. thesis. ETH Zürich, Diss. ETH No. 25849 (2019)
28. Herrmann, L., Schwab, C.: Multilevel Quasi-Monte Carlo integration with product weights for elliptic PDEs with lognormal coefficients. ESAIM: Math. Model. Numer. Anal. **53**(5), 1507–1552 (2019)
29. Herrmann, L., Schwab, C., Zech, J.: Uncertainty quantification for spectral fractional diffusion: Sparsity analysis of parametric solutions. SIAM/ASA J. Uncertain. Quantif. **7**(3), 913–947 (2019)
30. Hoang, V.H., Schwab, C., Stuart, A.M.: Complexity analysis of accelerated MCMC methods for Bayesian inversion. Inverse Probl. **29**(8), 085010, 37 (2013)
31. Houston, P., Schwab, C., Süli, E.: Discontinuous *hp*-finite element methods for advection-diffusion-reaction problems. SIAM J. Numer. Anal. **39**(6), 2133–2163 (2002)
32. Kuo, F.Y., Scheichl, R., Schwab, C., Sloan, I.H., Ullmann, E.: Multilevel Quasi-Monte Carlo methods for lognormal diffusion problems. Math. Comput. **86**(308), 2827–2860 (2017)
33. Kuo, F.Y., Schwab, Ch., Sloan, I.H.: Quasi-Monte Carlo methods for high-dimensional integration: the standard (weighted Hilbert space) setting and beyond. ANZIAM J. **53**(1), 1–37 (2011)
34. Kuo, F.Y., Schwab, Ch., Sloan, I.H.: Multi-level Quasi-Monte Carlo finite element methods for a class of elliptic PDEs with random coefficients. Found. Comput. Math. **15**(2), 411–449 (2015)

35. Müller, F., Schötzau, D., Schwab, C.: Symmetric interior penalty discontinuous Galerkin methods for elliptic problems in polygons. SIAM J. Numer. Anal. **55**(5), 2490–2521 (2017)
36. Nitsche, P.A.: Sparse approximation of singularity functions. Constr. Approx. **21**(1), 63–81 (2005)
37. Nochetto, R.H., Veeser, A.: Primer of adaptive finite element methods. In: Multiscale and Adaptivity: Modeling, Numerics and Applications. Lecture Notes in Mathematics, vol. 2040, pp. 125–225. Springer, Heidelberg (2012)
38. Nuyens, D., Cools, R.: Fast algorithms for component-by-component construction of rank-1 lattice rules in shift-invariant reproducing kernel Hilbert spaces. Math. Comput. **75**(254), 903–920 (electronic) (2006)
39. Nuyens, D., Cools, R.: Fast component-by-component construction of rank-1 lattice rules with a non-prime number of points. J. Complex. **22**(1), 4–28 (2006)
40. Opic, B., Kufner, A.: Hardy-type inequalities. Pitman Research Notes in Mathematics Series, vol. 219. Longman Scientific Technical, Harlow (1990)
41. Rekatsinas, N., Stevenson, R.: A quadratic finite element wavelet Riesz basis. Int. J. Wavelets Multiresolut. Inf. Process. **16**(4), 1850033, 17 (2018)
42. Scheichl, R., Stuart, A.M., Teckentrup, A.L.: Quasi-Monte Carlo and multilevel Monte Carlo methods for computing posterior expectations in elliptic inverse problems. SIAM/ASA J. Uncertain. Quantif. **5**(1), 493–518 (2017)
43. Schillings, C., Schwab, C.: Sparse, adaptive Smolyak quadratures for Bayesian inverse problems. Inverse Probl. **29**(6), 065011, 28 (2013)
44. Schillings, C., Schwab, C.: Sparsity in Bayesian inversion of parametric operator equations. Inverse Probl. **30**(6), 065007, 30 (2014)
45. Schneider, R.: Optimal convergence rates of adaptive algorithms for finite element multiscale methods. In: Boundary value problems and integral equations in nonsmooth domains (Luminy, 1993). Lecture Notes in Pure and Applied Mathematics, vol. 167, pp. 269–284. Dekker, New York (1995)
46. Schwab, C.: *p*- and *hp*-Finite Element Methods. Theory and Applications in Solid and Fluid Mechanics, Numerical Mathematics and Scientific Computation. The Clarendon Press, Oxford University Press, New York (1998)
47. Stevenson, R.: Stable three-point wavelet bases on general meshes. Numer. Math. **80**(1), 131–158 (1998)
48. Stevenson, R.: Locally supported, piecewise polynomial biorthogonal wavelets on nonuniform meshes. Constr. Approx. **19**(4), 477–508 (2003)
49. Xu, J.: Iterative methods by space decomposition and subspace correction. SIAM Rev. **34**(4), 581–613 (1992)
50. Yoshiki, T.: Bounds on Walsh coefficients by dyadic difference and a new Koksma-Hlawka type inequality for Quasi-Monte Carlo integration. Hiroshima Math. J. **47**(2), 155–179 (2017)

Selecting the Best Simulated System: Thinking Differently About an Old Problem

Barry L. Nelson

Abstract The methods known collectively as "ranking & selection" have been a theoretical and practical success story for the optimization of simulated stochastic systems: they are widely used in practice, have been implemented in commercial simulation software, and research has made them more and more statistically efficient. However, "statistically efficient" has meant minimizing the number of simulation-generated observations required to make a selection, or maximizing the strength of the inference given a budget of observations. Exploiting high-performance computing, and specifically the capability to simulate many feasible solutions in parallel, has challenged the ranking & selection paradigm. In this paper we review the challenge and suggest an entirely different approach.

Keywords Stochastic simulation · Simulation optimization · Ranking & selection · Parallel simulation

1 Introduction

A generic stochastic simulation optimization (SO) problem has the form

Maximize E[Simulated Performance]
Subject to: Resource constraints

The types of simulations that are the focus of this paper are dynamic, often nonstationary, and may be computationally expensive to execute. SO is difficult because the lack of a mathematical expression for, or even a deterministic numerical method to evaluate, E[Simulated Performance], implies that algorithms must make progress by *estimating* the performance of specific feasible solutions. This leads to the three sources of error in SO:

B. L. Nelson (✉)
Northwestern University, Evanston, IL, USA
e-mail: nelsonb@northwestern.edu

B. Tuffin and P. L'Ecuyer (eds.), *Monte Carlo and Quasi-Monte Carlo Methods*, Springer Proceedings in Mathematics & Statistics 324,
https://doi.org/10.1007/978-3-030-43465-6_3

1. The SO algorithm never simulates the optimal solution.
2. The SO algorithm does not recognize the best feasible solution it simulated.
3. The estimated performance of the sample-best solution returned by the SO algorithm is biased.

This paper addresses methods collectively known as Ranking & Selection (R&S). R&S originated with Bechhofer [2] and Gupta [7] in the 1950s for biostatistics and industrial applications, such as evaluating the efficacy of three drug treatments and a placebo. Typical problem characteristics included a small number of treatments k; normally distributed responses; relatively equal (maybe even known) variances; and a requirement to be easy to implement (e.g., since human subjects were involved). At the 1983 Winter Simulation Conference Goldsman [6] presented a tutorial on R&S and organized a session with both Bechhofer and Gupta, arguing that R&S was useful for optimizing simulated systems as well.

Since 1983 R&S has been an area of intense theoretical and practical interest in stochastic simulation. However, simulators were interested in problems with different characteristics:

- *Much* larger numbers of "treatments" (system designs) k.
- Possibly non-normal (nominal) simulation output data.
- Significantly unequal variances across system designs.
- Intentionally induced dependence across the outputs of simulated system designs due to Common Random Numbers (CRN).
- Highly sequential procedures to reduce the number of expensive simulation runs required to select the best system.

R&S has been a theoretical and practical success for simulation, including innovative theory; asymptotic regimes for non-normal data; and effective use of concepts from "statistical learning." Further, R&S is routinely applied in real problems and is included in many commercial software packages. The appeal of R&S is that it can control all three SO errors:

1. R&S is exhaustive, simulating all feasible solutions, so the optimal solution is always simulated.
2. R&S is explicitly concerned with recognizing the best solution with statistical confidence.
3. R&S may provide confidence intervals on the true performance of the selected solution.

Thus, it is desirable to turn a SO problem into a R&S problem if at all possible, and high-performance computing, and in particular parallel computing, would seem to facilitate treating problems with larger and larger numbers of feasible solutions as R&S problems. Unfortunately, nearly all the methodological developments in R&S assume single-processor computing, and define "cost" as synonymous with the number of simulated observations. The topic of this paper is how parallel computing changes (nearly) everything, and a suggestion for how to think differently.

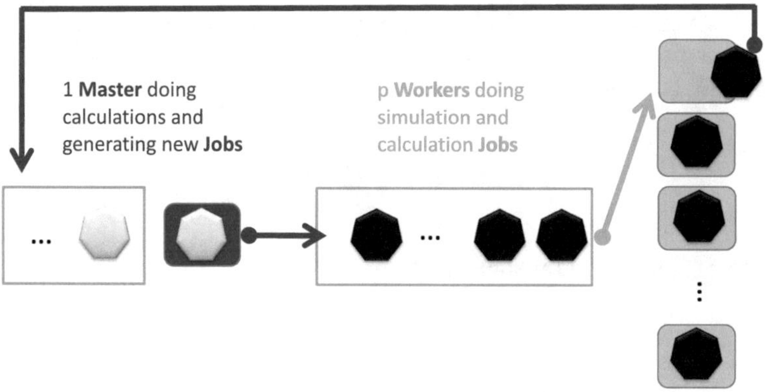

Fig. 1 Master-Worker environment

Remark 1 There is a connection between R&S and multi-arm bandit (MAB) problems that will not be explored here, other than to say that the objectives of MAB and R&S are often different (e.g., MAB minimizes regret); the MAB focus is online decision making, while R&S is always offline; and the two literatures have different standards for what constitutes a "good procedure." See for instance [10].

Remark 2 The particular parallel computing architecture within which we implement R&S matters, but we will not address those details other than to assume that there are $p+1$ processors in a "Master-Worker" environment in which the Master processor performs calculations and decides what jobs to send to the Worker processors; see Fig. 1. We define a "job" more precisely later.

Remark 3 While it is possible to treat many SO problems as R&S problems, clearly not all of them can be attacked in this way. We now consider $k = 10{,}000$ systems to be routine, but there are practical problems for which k is several orders of magnitude larger, and can even be uncountably infinite if systems are defined by continuous-valued decision variables. Further, the resource constraints may also be stochastic, requiring simulation to assess feasibility.

2 R&S Basics

For notation, let the true, but unknown, expected values (means) of the k feasible solutions (systems) be denoted by

$$\mu_1 \leq \mu_2 \leq \cdots \leq \mu_{k-1} \leq \mu_k.$$

We refer to system k, or any system tied with system k, as the best, and of course we do not actually know which system is system k. Let Y_{ij} be the jth output from system i, which has mean μ_i and variance σ_i^2. For system i we can estimate μ_i with a consistent estimator, which for the purpose of this paper is the sample mean of n_i independent and identically distributed (i.i.d.) replications:

$$\bar{Y}_i(n_i) = \frac{1}{n_i}\sum_{j=1}^{n_i} Y_{ij}.$$

The R&S procedure returns something like

$$\hat{K} = \text{argmax}_{i\in\{1,2,\ldots,k\}}\bar{Y}_i(n_i)$$

as the selected solution, where what the procedure specifies is the values of n_i.

One categorization of R&S procedures is *fixed-precision* versus *fixed-budget*. For the former, we simulate until a prespecified confidence level is achieved, ideally probability of correct selection (PCS): $\Pr\{\mu_{\hat{K}} = \mu_k\} \geq 1-\alpha$. Since attaining this goal can be computationally infeasible if, say, the best and second-best systems' means are very close, a compromise is made such as the following:

- **Indifference zone:** $\Pr\left\{\hat{K} = k \mid \mu_k - \mu_{k-1} \geq \delta\right\} \geq 1-\alpha$
- **Good selection:** $\Pr\left\{\mu_k - \mu_{\hat{K}} \leq \delta\right\} \geq 1-\alpha$
- **Top** m**:** $\Pr\left\{\hat{K} \in [k, k-1, \ldots, k-m+1]\right\} \geq 1-\alpha$
- **Subset:** Find $\widehat{S} \subseteq \{1,2,\ldots,k\}$ such that $\Pr\{k \in \widehat{S}\} \geq 1-\alpha$.

These are typically *frequentist* guarantees to be achieved as efficiently as possible. Here δ is taken as the smallest difference that is practically relevant.

A fundamental building block for many fixed-precision procedures is the standardized sums of differences:

$$\left[\frac{\sigma_k^2}{n_k}+\frac{\sigma_i^2}{n_i}\right]^{-1}\left[\bar{Y}_k(n_k)-\bar{Y}_i(n_i)\right] \stackrel{\mathscr{D}}{\approx} \mathscr{B}_{\mu_k-\mu_i}\left(\left[\frac{\sigma_k^2}{n_k}+\frac{\sigma_i^2}{n_i}\right]^{-1}\right)$$

where $\mathscr{B}_{\mu_k-\mu_i}(t)$ is Brownian motion (BM) with drift $\mu_k-\mu_i$ and the sample sizes n_k and n_i are independent of the sample means. This relationship is true in finite samples if the Y_{ij} are normally distributed (see [8]), and may be true asymptotically for appropriately standardized statistics. Much is known about BM processes crossing various boundaries (see, for instance, [12]), but for the purpose of this paper notice that employing this building block involves $k(k-1)/2$ pairwise comparisons, a number that can become a computational bottleneck when k is large.

For fixed-budget procedures, the goal is to obtain as strong an inference as possible within a fixed computation budget. This is typically formulated as minimizing some expected loss for the chosen solution, $\mathrm{E}[\mathscr{L}(\hat{K})]$, and the inference is typically *Bayesian*:

0-1 Loss: Maximize posterior PCS
Opportunity cost: Minimize posterior expected optimality gap.

The fixed-budget paradigm is to attain information in an optimal, sequential fashion; see Frazier [5]. Tools for doing so include "Expected Improvement" and the "Knowledge Gradient (KG)." For instance if our prior is

$$(\mu_1, \mu_2, \ldots, \mu_k)^\top \sim \mathrm{N}(\boldsymbol{\mu}_0, \boldsymbol{\Sigma}_0)$$

and the simulation output are normal, then we can compute the Complete Expected Improvement of solution i over the current sample best $\hat{K}$,

$$\mathrm{CEI}(i, \hat{K}) = \mathrm{E}\left[\max\{0, \mu_i - \mu_{\hat{K}}\} \mid Y_{ij}'s \text{ collected through stage } t\right]$$

from $\mathrm{N}(\boldsymbol{\mu}_t, \boldsymbol{\Sigma}_t \mid Y_{ij}'s\text{collected through stage } t)$, the posterior (normal) distribution. Notice that, implemented naively, this statistical learning approach takes only one simulated observation "optimally" at a time, and therefore does not exploit parallelization. In addition, calculation of the posterior distribution and searching for the maximum CEI or KG solution can be numerically challenging for large k.

Remark 4 R&S addresses a more diverse set of problems than selecting the system with the best mean; see [1] for a comprehensive reference.

3 R&S Computation

This section is based on [9].

Instead of thinking in terms of the statistical efficiency of a R&S procedure, here we consider the overall computation involved. All R&S procedures perform simulation *replications* and numerical *calculations*. Therefore, we define a R&S "job" j as the ordered list

$$J_j \equiv \{\underbrace{(\mathcal{Q}_j, \Delta_j, \mathcal{U}_j)}_{\text{simulate}}, \underbrace{(\mathcal{P}_j, \mathcal{C}_j)}_{\text{calculate}}\}$$

where

- $\mathcal{Q}_j \subseteq \{1, 2, \ldots, k\}$ indices of systems to be simulated;
- $\Delta_j = \{\Delta_{ij}\}$ how many replications to take from each system $i \in \mathcal{Q}_j$;
- $\mathcal{U}_j$ (optional) the assigned block of random numbers;
- $\mathcal{C}_j$ is a list of non-simulation calculations or operations to perform; and
- $\mathcal{P}_j$ is a list of jobs that must complete *before* the calculation $\mathcal{C}_j$.

We allow $(\mathcal{Q}_j, \Delta_j, \mathcal{U}_j)$ or $(\mathcal{P}_j, \mathcal{C}_j)$ to be null, or for a job to contain multiple simulate and calculate sub-jobs. The random numbers $\mathcal{U}_j$ are important to insure independence or dependence (CRN), if desired. Since we do not discuss CRN here, we suppress the specification of random numbers $\mathscr{U}_j$ from here on.

From the perspective of the jobs required, a generic R&S procedure looks something like this:

Generic R&S Procedure

1. For job $\ell = 1, 2, \ldots$ until termination, do

   a. *Simulation jobs*

$$\mathscr{J}_\ell = [\{(\text{system } 1, 1\text{rep}), (\emptyset)\}, \ldots, \{(\text{system } i, 1 \text{ rep}), (\emptyset)\}, \ldots]$$

   b. *Comparison jobs*
$$J'_\ell = \{(\emptyset), (\text{all jobs in } \mathscr{J}_\ell, \mathscr{C}_\ell)\}$$

   where $\mathscr{C}_\ell$ performs calculations on all (non-eliminated) systems.

2. Report $\hat{K}$.

This generic model enforces many of the assumptions necessary for both small-sample and asymptotic analysis by "synchronized coupling:" simulate all required replications, perform calculations on the collected output to decide what to simulate next, simulate all required replications, and so on.

Now suppose that we want to parallelize this. Recall that we initially have k systems and $p+1$ processors, 1 Master and p Workers. Perhaps the most natural way to think about adapting the Generic R&S Procedure to this setting is for the Master to maintain a round robin queue of systems from which a replication is needed, and whenever a replication result is returned from some processor the Master assigns another system to it from its queue. Based on the returned replications the Master then makes comparisons, eliminates systems, updates posterior distributions, etc.

The obvious problem with this approach is that when k is very large, the computations required of the Master may be so significant that the p Workers are starved for additional simulation assignments. But there is also a more subtle issue. Define the input sequence and output sequence as follows:

Input sequence: X_{ij} is the jth *requested* observation from system i by the Master, with execution time T_{ij}.
Output sequence: Y_{ij} is the jth *returned* observation to the Master from system i.

The validity of a R&S procedure is established based on properties of the *returned* sequence, which will not be the same as the *requested* sequence when there are $p > 1$ Workers and the execution times are random variables. As shown in [13], this can lead to statistical problems, including random sample sizes, non-i.i.d. outputs from any specific system, and a dependence induced across systems outputs by eliminations, all of which invalidate the statistical guarantees of R&S procedures. Of course $X_{ij} = Y_{ij}$ can be assured by having the Master wait for and reorder the output, insuring the statistical validity but significantly diminishing the computational efficiency.

This suggests that when we have the capability to simulate in parallel we need to refine our goals for R&S. We now formally define a R&S Procedure as the collection of jobs generated by the Master: $\mathscr{J} = \{J_j : 1 \le j \le M\}$, where M is determined by the procedure and may be either random or fixed. Both *wall-clock* ending time of the procedure and the *cost* of purchasing time on $p + 1$ processors matter:

- Let $0 < T_j < \infty$ be the wall-clock time job J_j finishes, so
$$T_e(\mathscr{J}) = \max_{j=1,2,\ldots,M} T_j$$
is the ending time of the procedure.
- $c(p, s)$ = cost to purchase p processors for s time units.
- $t(p, b)$ = maximum time we can purchase on p processors for budget \$$b$
$$t(p, b) = \max\{s : c(p, s) \le b\}.$$

We can now define revised objectives:

Fixed precision: Requires statistical guarantees while being efficient.

$$\begin{aligned} \text{minimize}_{p,\mathscr{J}} \quad & \mathrm{E}[\beta_t \underbrace{T_e(\mathscr{J})}_{\text{time}} + \beta_c \underbrace{c(p, T_e(\mathscr{J}))}_{\text{cost}}] \\ \text{s.t.} \quad & \Pr\{\underbrace{G(\hat{K}, k)}_{\text{good event}}\} \ge 1 - \alpha \end{aligned}$$

where β_t and β_c are weights or relative costs; typically one of β_t or β_c is zero and the other is one.

Fixed budget: Provides an efficiency guarantee within a budget.

$$\begin{aligned} \text{minimize}_{p,\mathscr{J}} \quad & \mathrm{E}[\underbrace{\mathscr{L}(G^c(\hat{K}, k), \mathscr{J})}_{\text{loss of bad event}}] \\ \text{s.t.} \quad & \underbrace{t(p, b)}_{\text{processor time}} \le t^\star \end{aligned}$$

where $t^\star$ is the wall-clock-time budget.

To the best of our knowledge, no one has yet formulated a parallel R&S procedure specifically to solve one of these optimization problems. Instead, the procedures shown in Table 1 either try to balance the Master-Worker load in a way that keeps the Workers busy, or they weaken the assumptions behind the Generic R&S Procedure so that it is still (at least asymptotically) valid when $X_{ij} \ne Y_{ij}$.

Remark 5 The clever approaches cited in Table 1 all try to adapt the existing R&S paradigms to the parallel environment. However, if we have, say, $k >$ 1,000,000 systems, then is it sensible to insist on locating the single best/near-best with high

Table 1 Existing parallel R&S procedures

R&S procedure	Load balancing (standard assumptions)	Comparison timing (relaxed assumptions)
Fixed-precision	Simple divide and conquer [3]	Asymptotic parallel selection [13]
	Vector-filling procedure [13]	
	Good selection procedure [16]	
Fixed-budget	Parallel OCBA [14]	
	Asynchronous OCBA/KG [11]	

probability? We should expect many bad systems, but also a lot of good ones. Guarantees like PCS also run counter to approaches in large-scale statistical inference of controlling "error rates." In fact, to control PCS requires more effort/system as k increases, while error rates such as "false discovery" can be attained with little or no "k effect."

4 Thinking Differently

The section is based on [17].

We want to disassemble the R&S paradigm and start over with the expectation of a very large number of systems k and number of parallel processors $p+1$. Our goals are (a) to provide a more scalable—but still useful and understandable—error control than PCS; and (b) avoid coupled operations and synchronization by exploiting the idea of comparisons with a standard [15]. The result is our Parallel Adaptive Survivor Selection (PASS) framework.

Again, let $Y_{i1}, Y_{i2}, \ldots$ be i.i.d. with mean μ_i and from here on we assume $\mu_k > \mu_{k-1} > \cdots > \mu_1$. For some known constant $\mu^\star$ that we refer to as the *standard*, let

$$S_i(n) = \sum_{j=1}^{n}(Y_{ij} - \mu^\star) = \sum_{j=1}^{n} Y_{ij} - n\mu^\star.$$

We will employ a non-decreasing function $c_i(\cdot)$ with the property that

$$\Pr\{S_i(n) \le -c_i(n),\ \text{some } n < \infty\} \begin{cases} \le \alpha & \mu_i \ge \mu^\star \\ = 1 & \mu_i < \mu^\star. \end{cases}$$

For normally distributed output such functions can be derived from the results in [4]. Finally, let $\mathscr{G} = \{i:\ \mu_i \ge \mu^\star\}$, the set of systems as good or better than the standard $\mu^\star$, which we assume is not empty; if it is empty then there is no false elimination. For any algorithm, let $\mathscr{E}$ be the set of systems that the algorithm decides are are not in $\mathscr{G}$ when they actually are. Then we define the expected false elimination rate for the algorithm as EFER $= \mathrm{E}[|\mathscr{E}|]/|\mathscr{G}|$.

Before tackling the case of unknown $\mu^\star$, consider the following algorithm:

Parallel Survivor Selection (PSS)

1. given a standard $\mu^\star$, an increment $\Delta n \geq 1$ and a budget
2. let $\mathsf{W} = \{1, 2, \ldots, p\}$ be the set of available Workers; $\mathsf{I} = \{1, 2, \ldots, k\}$ the set of surviving systems; and $n_i = 0$ for all $i \in \mathsf{I}$.
3. until the budget is consumed
   a. while an available Worker in W, do in parallel:
      i. remove next system $i \in \mathsf{I}$ and assign to available Worker $w \in \mathsf{W}$
      ii. $j = 1$
      iii. while $j \leq \Delta n$
         simulate Y_{i,n_i+j}
         if $S_i(n_i + j) \leq -c_i(n_i + j)$ then eliminate system i and break loop
         else $j = j + 1$
      iv. if i not eliminated then return to $\mathsf{I} = \mathsf{I} \cup \{i\}$, $n_i = n_i + \Delta n$
      v. release Worker w to available Workers W
4. return I

Notice that PSS requires no coupling and keeps the Workers constantly busy. And from the properties of $c(\cdot)$, PSS maintains EFER $\leq \alpha$ and, if run forever, will eliminate all systems with means $<\mu^\star$. Further, the EFER is still controlled at $\leq\alpha$ and elimination of systems not in $\mathscr{G}$ still occurs with probability 1, if we let Δn_i depend on the system i, and we replace $\mu^\star$ by $\mu(n) \leq \mu^\star$ where $\mu(n) \to \mu^\star$. This is the case because a system eliminated by a smaller standard would also have been eliminated by a larger standard, and a system protected from a larger standard would also be protected from a smaller one. This suggests that in the practical case in which $\mu^\star$ is *unknown* we may be able to *learn* the standard in a way that still that achieves our objectives; we call this Parallel *Adaptive* Survivor Selection.

Generically, we define the standard to be $\mu^\star = g(\mu_1, \mu_2, \ldots, \mu_k)$. Some examples of possibly interesting standards include

- Protect the best: $\mu^\star = \mu_k$, which we focus on here.
- Protect the top b: $\mu^\star = \mu_{k-b+1}$.
- Protect best and everything as good as some known value μ^+: $\mu^\star = \min\{\mu^+, \mu_k\}$.

We want to learn the standard's value in a way that still avoids synchronized coupling but does not compromise the EFER.

Consider PSS but with the adaptive standard

$$\bar{\mu} = \frac{1}{|\mathsf{I}|} \sum_{i \in I} \bar{Y}_i(n_i)$$

which is the average of the sample means of the current survivors. Thus, the adaptive standard acts like a bisection search. We call algorithm PSS with this standard bi-PASS. Under some conditions, including normally distributed output, we can show that the EFER for system k is still $\leq\alpha$ [17]. Thus, we can achieve nearly uncoupled parallelization and controlled EFER with an unknown standard. When $\mu^\star = \mu_k$ this means the chance that we eliminate the best system is $\leq\alpha$. However, since EFER is controlled marginally, α can be set even smaller than the traditional $\alpha = 0.1, 0.05, 0.01$ values with little penalty on efficiency and greater protection for system k.

5 Conclusions

When a simulation optimization problem can be treated as a R&S problem then it can be "solved" with statistical guarantees: that is, all three SO errors can be controlled. High-performance, parallel computing extends the "R&S limit" to larger problems, but introduces new statistical and computational challenges, including violation of standard assumptions and "cost" not being captured by the number of observations. The PASS framework introduced here replaces guarantees like PCS that do not scale well with k, with EFER which does, while at the same time making it easier to achieve "embarrassingly parallel" speed up by comparing each system only to an adaptive standard, rather than to each other.

Acknowledgements Portions of this work were supported by NSF Grant CMMI-1537060 and SAS Institute.

References

1. Bechhofer, R., Santner, T., Goldsman, D.: Designing Experiments for Statistical Selection, Screening, and Multiple Comparisons. Wiley, New York (1995)
2. Bechhofer, R.E.: A single-sample multiple decision procedure for ranking means of normal populations with known variances. Ann. Math. Stat. **25**(1), 16–39 (1954)
3. Chen, E.J.: Using parallel and distributed computing to increase the capability of selection procedures. In: M.E. Kuhl, N.M. Steiger, F.B. Armstrong, J.A. Joines (eds.) Proceedings of the 2005 Winter Simulation Conference, pp. 723–731. Institute of Electrical and Electronics Engineers, Inc., Piscataway, NJ (2005)
4. Fan, W., Hong, L.J., Nelson, B.L.: Indifference-zone-free selection of the best. Oper. Res. **64**(6), 1499–1514 (2016)
5. Frazier, P.I.: Decision-theoretic foundations of simulation optimization. Wiley Encyclopedia of Operations Research and Management Sciences. Wiley, New York (2010)
6. Goldsman, D.: Ranking and selection in simulation. In: Roberts, S., Banks, J., Schmeiser B. (eds.) Proceedings of the 1983 Winter Simulation Conference, pp. 387–393, http://informs-sim.org/wsc83papers/1983_0017.pdf (1983)
7. Gupta, S.S.: On some multiple decision (selection and ranking) rules. Technometrics **7**(2), 225–245 (1965)

8. Hong, L.J.: Fully sequential indifference-zone selection procedures with variance-dependent sampling. Naval Res. Logist. **53**(5), 464–476 (2006)
9. Hunter, S.R., Nelson, B.L.: Parallel ranking and selection. In: Tolk, A., Fowler, J., Shao, G., Yücesan, E. (eds.) Advances in Modeling and Simulation: Seminal Research from 50 Years of Winter Simulation Conferences, pp. 249–275. Springer International Publishing (2017). https://doi.org/10.1007/978-3-319-64182-9_12
10. Jamieson, K., Nowak, R.: Best-arm identification algorithms for multi-armed bandits in the fixed confidence setting. In: 2014 48th Annual Conference on Information Sciences and Systems (CISS), pp. 1–6. IEEE (2014)
11. Kamiński, B., Szufel, P.: On parallel policies for ranking and selection problems. J. Appl. Stat. **45**(9), 1690–1713 (2018)
12. Kim, S.H., Nelson, B.L.: A fully sequential procedure for indifference-zone selection in simulation. ACM Trans. Model. Comput. Simul. **11**(3), 251–273 (2001)
13. Luo, J., Hong, L.J., Nelson, B.L., Wu, Y.: Fully sequential procedures for large-scale ranking-and-selection problems in parallel computing environments. Oper. Res. **63**(5), 1177–1194 (2015). https://doi.org/10.1287/opre.2015.1413
14. Luo, Y.C., Chen, C.H., Yücesan, E., Lee, I.: Distributed web-based simulation optimization. In: Joines, J.A., Barton, R.R., Kang, K., Fishwick, P.A. (eds.) Proceedings of the 2000 Winter Simulation Conference, pp. 1785–1793. Institute of Electrical and Electronics Engineers, Inc., Piscataway, NJ (2000). https://doi.org/10.1109/WSC.2000.899170
15. Nelson, B.L., Goldsman, D.: Comparisons with a standard in simulation experiments. Manag. Sci. **47**(3), 449–463 (2001)
16. Ni, E.C., Ciocan, D.F., Henderson, S.G., Hunter, S.R.: Efficient ranking and selection in parallel computing environments. Oper. Res. **65**(3), 821–836 (2017)
17. Pei, L., Hunter, S., Nelson, B.L.: A new framework for parallel ranking & selection using an adaptive standard. In: Rabe, M., Juan, A.A., Mustafee, N., Skoogh, A., Jain, S., Johansson, B. (eds.) Proceedings of the 2018 Winter Simulation Conference, pp. 2201–2212. IEEE Press (2018)

Discrepancy of Digital Sequences: New Results on a Classical QMC Topic

Friedrich Pillichshammer

Abstract The theory of digital sequences is a fundamental topic in QMC theory. Digital sequences are prototypes of sequences with low discrepancy. First examples were given by Il'ya Meerovich Sobol' and by Henri Faure with their famous constructions. The unifying theory was developed later by Harald Niederreiter. Nowadays there is a magnitude of examples of digital sequences and it is classical knowledge that the star discrepancy of the initial N elements of such sequences can achieve a rate of order $(\log N)^s/N$, where s denotes the dimension. On the other hand, very little has been known about the L_p norm of the discrepancy function of digital sequences for finite p, apart from evident estimates in terms of star discrepancy. In this article we give a review of some recent results on various types of discrepancy of digital sequences. This comprises: star discrepancy and weighted star discrepancy, L_p-discrepancy, discrepancy with respect to bounded mean oscillation and exponential Orlicz norms, as well as Sobolev, Besov and Triebel–Lizorkin norms with dominating mixed smoothness.

Keywords Discrepancy · Digital sequences · Digital Kronecker sequence · Tractability · Quasi-Monte Carlo integration

Preamble

This paper is devoted to Henri Faure who celebrated his 80th birthday on July 12, 2018. Henri is well known for his pioneering work on low-discrepancy sequences. As an example we would like to mention his famous paper [31] from 1982 in which he gave one of the first explicit constructions of digital sequences in arbitrary dimension with low star discrepancy. These sequences are nowadays known as *Faure sequences*.

I met Henri for the first time at the MCQMC conference 2002 in Singapore. Later, during several visits of Henri in Linz, we started a fruitful cooperation which

Dedicated to Henri Faure on the occasion of his 80th birthday.

F. Pillichshammer (✉)
Johannes Kepler University Linz, Altenbergerstraße 69, 4040 Linz, Austria
e-mail: friedrich.pillichshammer@jku.at

B. Tuffin and P. L'Ecuyer (eds.), *Monte Carlo and Quasi-Monte Carlo Methods*, Springer Proceedings in Mathematics & Statistics 324,
https://doi.org/10.1007/978-3-030-43465-6_4

continues to this day. I would like to thank Henri for this close cooperation and for his great friendship and wish him and his family all the best for the future.

1 Introduction

We consider infinite sequences $\mathscr{S} = (\boldsymbol{x}_n)_{n\geq 0}$ of points $\boldsymbol{x}_n$ in the s-dimensional unit cube $[0, 1)^s$. For $N \in \mathbb{N}$ let $\mathscr{S}_N = (\boldsymbol{x}_n)_{n=0}^{N-1}$ be the initial segment of $\mathscr{S}$ consisting of the first N elements.

According to Weyl [94] a sequence $\mathscr{S} = (\boldsymbol{x}_n)_{n\geq 0}$ is uniformly distributed (u.d.) if for every axes-parallel box $J \subseteq [0, 1)^s$ it is true that

$$\lim_{N\to\infty} \frac{\#\{n \in \{0, \ldots, N-1\} \ : \ \boldsymbol{x}_n \in J\}}{N} = \mathrm{Volume}(J).$$

An extensive introduction to the theory of uniform distribution of sequences can be found in the book of Kuipers and Niederreiter [53].

There are several equivalent definitions of uniform distribution of a sequence and one of them is of particular importance for quasi-Monte Carlo (QMC) integration. Weyl proved that a sequence $\mathscr{S}$ is u.d. if and only if for every Riemann-integrable function $f : [0, 1]^s \to \mathbb{R}$ we have

$$\lim_{N\to\infty} \frac{1}{N} \sum_{n=0}^{N-1} f(\boldsymbol{x}_n) = \int_{[0,1]^s} f(\boldsymbol{x}) \,\mathrm{d}\boldsymbol{x}. \tag{1}$$

The average of function evaluations on the left-hand side is nowadays called a *QMC rule*,

$$Q_N(f) = \frac{1}{N} \sum_{n=0}^{N-1} f(\boldsymbol{x}_n).$$

Hence, in order to have a QMC rule converging to the true value of the integral of a function it has to be based on a u.d. sequence. A quantitative version of (1) can be stated in terms of discrepancy.

Definition 1 For a finite initial segment $\mathscr{S}_N$ of a sequence (or a finite point set) in $[0, 1)^s$ the *local discrepancy function* $\Delta_{\mathscr{S}_N} : [0, 1]^s \to \mathbb{R}$ is defined as

$$\Delta_{\mathscr{S}_N}(\boldsymbol{t}) = \frac{\#\{n \in \{0, 1, \ldots, N-1\} \ : \ \boldsymbol{x}_n \in [\boldsymbol{0}, \boldsymbol{t})\}}{N} - t_1 t_2 \cdots t_s,$$

where $\boldsymbol{t} = (t_1, t_2, \ldots, t_s)$, $[\boldsymbol{0}, \boldsymbol{t}) = [0, t_1) \times [0, t_2) \times \cdots \times [0, t_s)$, and hence $t_1 t_2 \cdots t_s = \mathrm{Volume}([\boldsymbol{0}, \boldsymbol{t}))$.

For $p \geq 1$ the L_p *discrepancy* of $\mathscr{S}_N$ is defined as the L_p norm of the local discrepancy function

$$L_{p,N}(\mathscr{S}_N) = \|\Delta_{\mathscr{S}_N}\|_{L_p([0,1]^s)} = \left(\int_{[0,1]^s} |\Delta_{\mathscr{S}_N}(\boldsymbol{t})|^p \mathrm{d}\boldsymbol{t}\right)^{1/p}$$

with the usual adaptions if $p = \infty$. In this latter case one often talks about *star discrepancy* which is denoted by $D_N^*(\mathscr{S}_N) := L_{\infty,N}(\mathscr{S}_N)$.

For an infinite sequence $\mathscr{S}$ in $[0,1)^s$ we denote the L_p discrepancy of the first N points by $L_{p,N}(\mathscr{S}) = L_{p,N}(\mathscr{S}_N)$ for $N \geq 1$.

It is well-known that a sequence $\mathscr{S}$ is u.d. if and only if $\lim_{N\to\infty} L_{p,N}(\mathscr{S}) = 0$ for some $p \geq 1$. A quantitative version of (1) is the famous *Koksma–Hlawka inequality* which states that for every function $f : [0,1]^s \to \mathbb{R}$ with bounded variation $V(f)$ in the sense of Hardy and Krause and for every finite sequence $\mathscr{S}_N$ of points in $[0,1)^s$ we have

$$\left|\int_{[0,1]^s} f(\boldsymbol{x})\,\mathrm{d}\boldsymbol{x} - \frac{1}{N}\sum_{n=0}^{N-1} f(\boldsymbol{x}_n)\right| \leq V(f) D_N^*(\mathscr{S}_N).$$

The Koksma–Hlawka inequality is the fundamental error estimate for QMC rules and the basis for QMC theory. Nowadays there exist several versions of this inequality which may also be based on the L_p discrepancy or other norms of the local discrepancy function. One often speaks about "Koksma–Hlawka type inequalities". For more information and for introductions to QMC theory we refer to [22, 24, 61, 72].

It is clear that QMC requires sequences with low discrepancy in some sense and this motivates the study of "low discrepancy sequences". On the other hand discrepancy is also an interesting topic by itself that is intensively studied (see, e.g., the books [4, 14, 24, 29, 53, 69, 72]).

In the following we collect some well-known facts about L_p discrepancy of finite and infinite sequences.

2 Known Facts About the L_p Discrepancy

We begin with results on finite sequences: for every $p \in (1,\infty]$ and $s \in \mathbb{N}$ there exists a $c_{p,s} > 0$ such that for every finite N-element sequence $\mathscr{S}_N$ in $[0,1)^s$ with $N \geq 2$ we have

$$L_{p,N}(\mathscr{S}_N) \geq c_{p,s}\frac{(\log N)^{\frac{s-1}{2}}}{N} \quad \text{and} \quad D_N^*(\mathscr{S}_N) \geq c_{\infty,s}\frac{(\log N)^{\frac{s-1}{2}+\eta_s}}{N}$$

for some $\eta_s \in (0, \frac{1}{2})$. The result on the left hand side for $p \geq 2$ is a celebrated result by Roth [81] from 1954 that was extended later by Schmidt [84] to the case $p \in (1,2)$. The general lower bound for the star discrepancy is an important result of Bilyk, Lacey and Vagharshakyan [8] from 2008. As shown by Halász [42], the L_p estimate is also true for $p = 1$ and $s = 2$, i.e., there exists a positive constant $c_{1,2}$ with the following property: for every finite sequence $\mathscr{S}_N$ in $[0,1)^2$ with $N \geq 2$ we have

$$L_{1,N}(\mathscr{S}_N) \geq c_{1,2}\frac{\sqrt{\log N}}{N}. \tag{2}$$

Schmidt showed for $s = 2$ the improved lower bound on star discrepancy

$$D_N^*(\mathscr{S}_N) \geq c_{\infty,2}\frac{\log N}{N}$$

for some $c_{\infty,2} > 0$. On the other hand, it is known that for every $s, N \in \mathbb{N}$ there exist finite sequences $\mathscr{S}_N$ in $[0, 1)^s$ such that

$$D_N^*(\mathscr{S}_N) \lesssim_s \frac{(\log N)^{s-1}}{N}.$$

First examples for such sequences are the Hammersley point sets, see, e.g., [24, Sect. 3.4.2] or [72, Sect. 3.2].

Similarly, for every $s, N \in \mathbb{N}$ and every $p \in [1, \infty)$ there exist finite sequences $\mathscr{S}_N$ in $[0, 1)^s$ such that

$$L_{p,N}(\mathscr{S}_N) \lesssim_{s,p} \frac{(\log N)^{\frac{s-1}{2}}}{N}. \tag{3}$$

Hence, for $p \in (1, \infty)$ and arbitrary $s \in \mathbb{N}$ we have matching lower and upper bounds. For both $p = 1$ and $p = \infty$ we have matching lower and upper bounds only for $s = 2$. The result in (3) was proved by Davenport [15] for $p = 2$, $s = 2$, by Roth [82] for $p = 2$ and arbitrary s and finally by Chen [11] in the general case. Other proofs were found by Frolov [40], Chen [12], Dobrovol'skiĭ [27], Skriganov [85, 86], Hickernell and Yue [45], and Dick and Pillichshammer [23]. For more details on the early history of the subject see the monograph [4]. Apart from Davenport, who gave an explicit construction in dimension $s = 2$, these results are pure existence results and explicit constructions of point sets were not known until the beginning of this millennium. First explicit constructions of point sets with optimal order of L_2 discrepancy in arbitrary dimensions have been provided in 2002 by Chen and Skriganov [13] for $p = 2$ and in 2006 by Skriganov [87] for general p. Other explicit constructions are due to Dick and Pillichshammer [25] for $p = 2$, and Dick [19] and Markhasin [68] for general p.

Before we summarize results about infinite sequences some words about the conceptual difference between the discrepancy of finite and infinite sequences are appropriate. Matoušek [69] explained this in the following way: while for finite sequences one is interested in the distribution behavior of the whole sequence $(\boldsymbol{x}_0, \boldsymbol{x}_1, \ldots, \boldsymbol{x}_{N-1})$ with a fixed number of elements N, for infinite sequences one is interested in the discrepancy of all initial segments $(\boldsymbol{x}_0)$, $(\boldsymbol{x}_0, \boldsymbol{x}_1)$, $(\boldsymbol{x}_0, \boldsymbol{x}_1, \boldsymbol{x}_2)$, ..., $(\boldsymbol{x}_0, \boldsymbol{x}_1, \boldsymbol{x}_2, \ldots, \boldsymbol{x}_{N-1})$, simultaneously for $N \in \mathbb{N}$. In this sense the discrepancy of finite sequences can be viewed as a static setting and the discrepancy of infinite sequences as a dynamic setting.

Using a method from Proĭnov [80] (see also [25]) the results about lower bounds on L_p discrepancy for finite sequences can be transferred to the following lower bounds for infinite sequences: for every $p \in (1, \infty]$ and every $s \in \mathbb{N}$ there exists a $C_{p,s} > 0$ such that for every infinite sequence $\mathscr{S}$ in $[0, 1)^s$

$$L_{p,N}(\mathscr{S}) \geq C_{p,s} \frac{(\log N)^{\frac{s}{2}}}{N} \quad \text{infinitely often} \tag{4}$$

and

$$D_N^*(\mathscr{S}) \geq C_{\infty,s} \frac{(\log N)^{\frac{s}{2}+\eta_s}}{N} \quad \text{infinitely often,} \tag{5}$$

where $\eta_s \in (0, \frac{1}{2})$ is independent of the concrete sequence. For $s = 1$ the result holds also for the case $p = 1$, i.e., for every $\mathscr{S}$ in $[0, 1)$ we have

$$L_{1,N}(\mathscr{S}) \geq c_{1,1} \frac{\sqrt{\log N}}{N} \quad \text{infinitely often,}$$

and the result on the star discrepancy can be improved to (see Schmidt [83]; see also [5, 56, 60])

$$D_N^*(\mathscr{S}) \geq c_{\infty,1} \frac{\log N}{N} \quad \text{infinitely often.} \tag{6}$$

On the other hand, for every dimension s there exist infinite sequences $\mathscr{S}$ in $[0, 1)^s$ such that

$$D_N^*(\mathscr{S}) \lesssim_s \frac{(\log N)^s}{N} \quad \text{for all } N \geq 2. \tag{7}$$

Informally one calls a sequence a *low-discrepancy sequence* if its star discrepancy satisfies the bound (7). Examples of low-discrepancy sequences are:

- Kronecker sequences $(\{n\boldsymbol{\alpha}\})_{n\geq 0}$, where $\boldsymbol{\alpha} \in \mathbb{R}^s$ and where the fractional part function $\{\cdot\}$ is applied component-wise. In dimension $s = 1$ and if $\alpha \in \mathbb{R}$ has bounded continued fraction coefficients, then the Kronecker sequence $(\{n\alpha\})_{n\geq 0}$ has star discrepancy of exact order of magnitude $\log N/N$; see [72, Chap. 3] for more information.
- Digital sequences: the prototype of a digital sequence is the van der Corput sequence in base b which was introduced by van der Corput [93] in 1935. For an integer $b \geq 2$ (the "basis") the nth element of this sequence is given by $x_n = n_0 b^{-1} + n_1 b^{-2} + n_2 b^{-3} + \cdots$ whenever n has b-adic expansion $n = n_0 + n_1 b + n_2 b^2 + \cdots$. The van der Corput sequence has star discrepancy of exact order of magnitude $\log N/N$; see the recent survey article [36] and the references therein.

  Multi-dimensional extensions of the van der Corput sequence are the Halton sequence [43], which is the component-wise concatenation of van der Corput sequences in pairwise co-prime bases, or digital (t, s)-sequences, where the basis b is the same for all coordinate directions. First examples of such sequences have been given by Sobol' [90] and by Faure [31]. Later the general unifying concept has been introduced by Niederreiter [71] in 1987. Halton sequences in pairwise co-prime bases as well as digital (t, s)-sequences have star discrepancy of order of magnitude of at most $(\log N)^s/N$; see Sect. 3.2.

Except for the one-dimensional case, there is a gap for the $\log N$ exponent in the lower and upper bound on the star discrepancy of infinite sequences (cf. Eqs. (5) and (7)) which seems to be very difficult to close. There is a grand conjecture in discrepancy theory which share many colleagues (but it must be mentioned that there are also other opinions; see, e.g., [7]):

Conjecture 1 *For every $s \in \mathbb{N}$ there exists a $c_s > 0$ with the following property: for every $\mathscr{S}$ in $[0,1)^s$ it holds true that*

$$D_N^*(\mathscr{S}) \geq c_s \frac{(\log N)^s}{N} \quad \textit{infinitely often.}$$

For the L_p discrepancy of infinite sequences with finite p the situation is different. It was widely assumed that the general lower bound of Roth–Schmidt–Proĭnov in Eq. (4) is optimal in the order of magnitude in N but until recently there was no proof of this conjecture (although it was some times quoted as a proven fact). In the meantime there exist explicit constructions of infinite sequences with optimal order of L_p discrepancy in the sense of the general lower bound (4). These constructions will be presented in Sect. 3.4.

3 Discrepancy of Digital Sequences

In the following we give the definition of digital sequences in prime bases b. For the general definition we refer to [72, Sect. 4.3]. From now on let b be a prime number and let $\mathbb{F}_b$ be the finite field of order b. We identify $\mathbb{F}_b$ with the set of integers $\{0, 1, \ldots, b-1\}$ equipped with the usual arithmetic operations modulo b.

Definition 2 (*Niederreiter* [71]) A digital sequence is constructed in the following way:

- choose s infinite matrices $C_1, \ldots, C_s \in \mathbb{F}_b^{\mathbb{N} \times \mathbb{N}}$;
- for $n \in \mathbb{N}_0$ of the form $n = n_0 + n_1 b + n_2 b^2 + \cdots$ and $j = 1, 2, \ldots, s$ compute (over $\mathbb{F}_b$) the matrix-vector products

$$C_j \begin{pmatrix} n_0 \\ n_1 \\ n_2 \\ \vdots \end{pmatrix} =: \begin{pmatrix} x_{n,j,1} \\ x_{n,j,2} \\ x_{n,j,3} \\ \vdots \end{pmatrix};$$

- put

$$x_{n,j} = \frac{x_{n,j,1}}{b} + \frac{x_{n,j,2}}{b^2} + \frac{x_{n,j,3}}{b^3} + \cdots \quad \text{and} \quad \boldsymbol{x}_n = (x_{n,1}, \ldots, x_{n,s}).$$

The resulting sequence $\mathscr{S}(C_1, \ldots, C_s) = (\boldsymbol{x}_n)_{n\geq 0}$ is called a *digital sequence over* $\mathbb{F}_b$ and $C_1, \ldots, C_s$ are called the *generating matrices* of the digital sequence.

3.1 A Metrical Result

It is known that almost all digital sequences in a fixed dimension s are low-discrepancy sequences, up to some $\log\log N$-term. The "almost all" statement is with respect to a natural probability measure on the set of all s-tuples $(C_1, \ldots, C_s)$ of $\mathbb{N} \times \mathbb{N}$ matrices over $\mathbb{F}_b$. For the definition of this probability measure we refer to [58, p. 107].

Theorem 1 (Larcher [54], Larcher & Pillichshammer [58, 59]) *Let $\varepsilon > 0$. For almost all s-tuples $(C_1, \ldots, C_s)$ with $C_j \in \mathbb{F}_b^{\mathbb{N}\times\mathbb{N}}$ the corresponding digital sequences $\mathscr{S} = \mathscr{S}(C_1, \ldots, C_s)$ satisfy*

$$D_N^*(\mathscr{S}) \lesssim_{b,s,\varepsilon} \frac{(\log N)^s (\log\log N)^{2+\varepsilon}}{N} \quad \forall N \geq 2$$

and

$$D_N^*(\mathscr{S}) \geq c_{b,s} \frac{(\log N)^s \log\log N}{N} \quad \textit{infinitely often.}$$

The upper estimate has been shown by Larcher in [54] and a proof for the lower bound can be found in [58] (see Sect. 4 for a definition).

A corresponding result for the sub-class of so-called digital Kronecker sequences can be found in [55] (upper bound) and [59] (lower bound). These results correspond to metrical discrepancy bounds for classical Kronecker sequences by Beck [3].

The question now arises whether there are s-tuples $(C_1, \ldots, C_s)$ of generating matrices such that the resulting digital sequences are low-discrepancy sequences and, if the answer is *yes*, which properties of the matrices guarantee low discrepancy. Niederreiter found out that this depends on a certain linear independence structure of the rows of the matrices $C_1, \ldots, C_s$. This leads to the concept of digital (t, s)-sequences.

3.2 Digital (t, s)-Sequences

For $C \in \mathbb{F}_b^{\mathbb{N}\times\mathbb{N}}$ and $m \in \mathbb{N}$ denote by $C(m)$ the left upper $m \times m$ submatrix of C.

For technical reasons one often assumes that the generating matrices $C_1, \ldots, C_s$ satisfy the following condition: let $C_j = (c_{k,\ell}^{(j)})_{k,\ell\in\mathbb{N}}$, then for each $\ell \in \mathbb{N}$ there exists a $K(\ell) \in \mathbb{N}$ such that $c_{k,\ell}^{(j)} = 0$ for all $k > K(\ell)$. This condition, which is condition (S6) in [72, p.72], guarantees that the components of the elements of a digital sequence have a finite digit expansion in base b. For the rest of the paper we tacitly assume that this condition is satisfied. (We remark that in order to include new important constructions to the concept of digital (t, s)-sequences, Niederreiter and Xing [73, 74] use a truncation operator to overcome the above-mentioned technicalities. Such sequences are sometimes called (t, s)-sequences *in the broad sense*.)

Definition 3 (*Niederreiter*) Given $C_1, \ldots, C_s \in \mathbb{F}_b^{\mathbb{N}\times\mathbb{N}}$. If there exists a number $t \in \mathbb{N}_0$ such that for every $m \geq t$ and for all $d_1, \ldots, d_s \geq 0$ with $d_1 + \cdots + d_s = m - t$ the

$$\left.\begin{array}{l}\text{first } d_1 \text{ rows of } C_1(m),\\ \text{first } d_2 \text{ rows of } C_2(m),\\ \ldots\\ \text{first } d_s \text{ rows of } C_s(m),\end{array}\right\} \text{ are linearly independent over } \mathbb{F}_b,$$

then the corresponding digital sequence $\mathcal{S}(C_1, \ldots, C_s)$ is called a *digital (t, s)-sequence over $\mathbb{F}_b$.*

The technical condition from the above definition guarantees that every b^m-element sub-block $(\boldsymbol{x}_{kb^m}, \boldsymbol{x}_{kb^m+1}, \ldots, \boldsymbol{x}_{(k+1)b^m-1}) =: \mathcal{S}_{m,k}$ of the digital sequence, where $m \geq t$ and $k \in \mathbb{N}_0$, is a (t, m, s)-net in base b, i.e., every so-called elementary b-adic interval of the form

$$J = \prod_{j=1}^{s} \left[\frac{a_j}{b^{d_j}}, \frac{a_j + 1}{b^{d_j}}\right) \quad \text{with Volume}(J) = b^{t-m}$$

contains the right share of elements from $\mathcal{S}_{m,k}$, which is exactly b^t. For more information we refer to [24, Chap. 4] and [72, Chap. 4]. Examples for digital (t, s)-sequences are generalized Niederreiter sequences which comprise the concepts of Sobol'-, Faure- and original Niederreiter-sequences, Niederreiter-Xing sequences, ... We refer to [24, Chap. 8] for a collection of constructions and for further references. An overview of the constructions of Niederreiter and Xing can also be found in [74, Chap. 8].

It has been shown by Niederreiter [71] that every digital (t, s)-sequence is a low-discrepancy sequence. The following result holds true:

Theorem 2 (Niederreiter [71]) *For every digital (t, s)-sequence $\mathcal{S}$ over $\mathbb{F}_b$ we have*

$$D_N^*(\mathcal{S}) \leq c_{s,b}\, b^t \,\frac{(\log N)^s}{N} + O\left(\frac{(\log N)^{s-1}}{N}\right).$$

Later several authors worked on improvements of the implied quantity $c_{s,b}$, e.g. [34, 51]. The currently smallest values for $c_{s,b}$ were provided by Faure and Kritzer [34]. More explicit versions of the estimate in Theorem 2 can be found in [37–39]. For a summary of these results one can also consult [36, Sect. 4.3].

Remark 1 Theorem 2 in combination with the lower bound in Theorem 1 shows that the set of s-tuples $(C_1, \ldots, C_s)$ of matrices that generate a digital (t, s)-sequence is a set of measure zero.

Remember that the exact order of optimal star discrepancy of infinite sequences is still unknown (except for the one-dimensional case). From this point of view it might be still possible that Niederreiter's star discrepancy bound in Theorem 2 could be improved in the order of magnitude in N. However, it has been shown recently by Levin [63] that this is not possible in general. In his proofs Levin requires the concept of d-admissibility. He calls a sequence $(\boldsymbol{x}_n)_{n\geq 0}$ in $[0,1)^s$ *d-admissible* if

$$\inf_{n>k\geq 0} \|n \ominus k\|_b \|\boldsymbol{x}_n \ominus \boldsymbol{x}_k\|_b \geq b^{-d},$$

where $\log_b \|x\|_b = \lfloor \log_b x \rfloor$ and $\ominus$ is the b-adic difference. Roughly speaking, this means that the b-adic distance between elements from the sequence whose indices are close is not too small.

Theorem 3 (Levin [63]) *Let $\mathscr{S}$ be a d-admissible (t,s)-sequence. Then*

$$D_N^*(\mathscr{S}) \geq c_{s,t,d} \frac{(\log N)^s}{N} \quad \text{infinitely often.}$$

In his paper, Levin gave a whole list of digital (t,s)-sequences that have the property of being d-admissible for certain d. This list comprises the concepts of generalized Niederreiter sequences (which includes Sobol'-, Faure- and original Niederreiter-sequences), Niederreiter-Xing sequences, ... For a survey of Levin's result we also refer to [49]. It should also be mentioned that there is one single result by Faure [32] from the year 1995 who already gave a lower bound for a particular digital $(0,2)$-sequence (in dimension 2) which is also of order $(\log N)^2/N$.

Levin's results [62, 63] are important contributions to the grand problem in discrepancy theory (cf. Conjecture 1). But they only cover the important sub-class of admissible (t,s)-sequences and allow no conclusion for arbitrary (including non-digital) sequences.

3.3 Digital $(0,1)$*-Sequences over* $\mathbb{F}_2$

In this sub-section we say a few words about the discrepancy of digital $(0,1)$-sequence over $\mathbb{F}_2$, because in this case exact results are known. Let $b=2$ and let I be the $\mathbb{N} \times \mathbb{N}$ identity matrix, that is, the matrix whose entries are 0 except for the entries on the main-diagonal which are 1. The corresponding one-dimensional digital sequence $\mathscr{S}(I)$ is the van der Corput sequence in base 2 and in fact, it is also a digital $(0,1)$-sequence over $\mathbb{F}_2$. The following is known: among all digital $(0,1)$-sequences over $\mathbb{F}_2$ the van der Corput sequence, which is the prototype of all digital constructions and whose star discrepancy is very well studied, has the worst star discrepancy; see [79, Theorem 2]. More concretely, for every $\mathbb{N} \times \mathbb{N}$ matrix C which generates a digital $(0,1)$-sequence $\mathscr{S}(C)$ over $\mathbb{F}_2$ we have

$$D_N^*(\mathscr{S}(C)) \le D_N^*(\mathscr{S}(I)) \le \begin{cases} \left(\frac{\log N}{3\log 2}+1\right)\frac{1}{N}, \\ \frac{S_2(N)}{N}, \end{cases} \tag{8}$$

where $S_2(N)$ denotes the *dyadic sum-of-digits function* of the integer N. The first bound on $D_N^*(\mathscr{S}(I))$ is a result of Béjian and Faure [6]. The factor $1/(3\log 2)$ conjoined with the $\log N$-term is known to be best possible, in fact,

$$\limsup_{N\to\infty} \frac{ND_N^*(\mathscr{S}(I))}{\log N} = \frac{1}{3\log 2}.$$

(The corresponding result for van der Corput sequences in arbitrary base can be found in [30, 33, 50].) However, also the second estimate in terms of the dyadic sum-of-digits function, which follows easily from the proof of [53, Theorem 3.5 on p. 127], is very interesting. It shows that the star discrepancy of the van der Corput sequence (and of any digital (0, 1)-sequence) is not always close to the high level of order $\log N/N$. If N has only very few dyadic digits different from zero, then the star discrepancy is very small. For example, if N is a power of two, then $S_2(N) = 1$ and therefore $D_N^*(\mathscr{S}(I)) \le 1/N$. The bound in (8) is demonstrated in Fig. 1.

While the star discrepancy of any digital (0, 1)-sequence over $\mathbb{F}_2$ is of optimal order with respect to (6) this fact is not true in general for the L_p discrepancies with finite parameter p. For example, for the van der Corput sequence we have for all $p \in [1,\infty)$

$$\limsup_{N\to\infty} \frac{NL_{p,N}(\mathscr{S}(I))}{\log N} = \frac{1}{6\log 2},$$

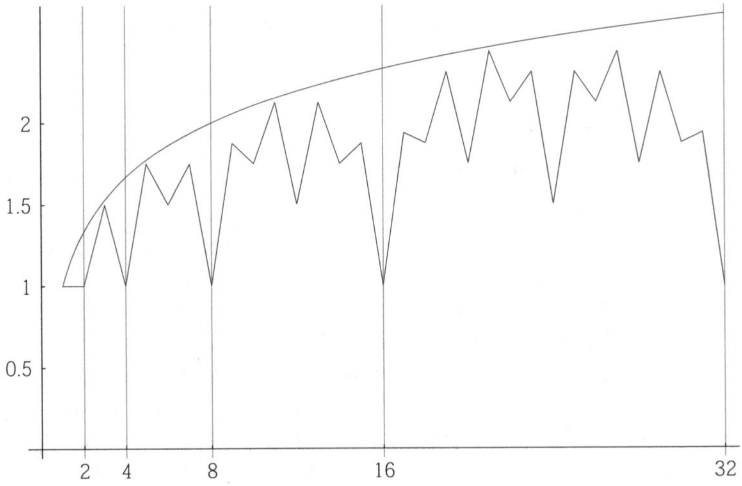

Fig. 1 $ND_N^*(\mathscr{S}(I))$ compared with $\frac{\log N}{3\log 2}+1$ (red line) for $N = 2, 3, \ldots, 32$; if N is a power of two, then $ND_N^*(\mathscr{S}(I)) = 1$

see [79]. Hence the L_p discrepancy of the van der Corput sequence is at least of order of magnitude $\log N/N$ for infinitely many N. Another example, to be found in [28], is the digital (0, 1)-sequence generated by the matrix

$$U = \begin{pmatrix} 1 \; 1 \; 1 \; \dots \\ 0 \; 1 \; 1 \; \dots \\ 0 \; 0 \; 1 \; \dots \\ \dots \end{pmatrix}$$

for which we have, with some positive real $c > 0$,

$$\limsup_{N \to \infty} \frac{N L_{2,N}(\mathcal{S}(U))}{\log N} \geq c > 0.$$

More information on the discrepancy of digital (0, 1)-sequences can be found in the survey articles [35, 36] and the references therein.

The results in dimension one show that, in general, the L_p discrepancy of digital sequences does not match the general lower bound (4) from Roth–Schmidt–Proĭnov. Hence, in order to achieve the assumed optimal order of magnitude $(\log N)^{s/2}/N$ for the L_p discrepancy with digital sequences, if at all possible, one needs more demanding properties on the generating matrices. This leads to the concept of higher order digital sequences.

3.4 Digital Sequences with Optimal Order of L_p Discrepancy

So-called *higher order digital sequences* have been introduced by Dick [17, 18] in 2007 with the aim to achieve optimal convergence rates for QMC rules applied to sufficiently smooth functions. For the definition of higher order digital sequences and for further information and references we refer to [24, Chap. 15] or to [22].

For our purposes it suffices to consider higher order digital sequences of order two. We just show how such sequences can be constructed: to this end let $d := 2s$ and let $C_1, \dots, C_d \in \mathbb{F}_2^{\mathbb{N} \times \mathbb{N}}$ be generating matrices of a digital (t, d)-sequence in dimension d, for example a generalized Niederreiter sequence. Let $\mathbf{c}_{j,k}$ denote the kth row-vector of the matrix C_j. Now define s matrices $E_1, \dots, E_s$ in the following way: the row-vectors of E_j are given by

$$\mathbf{e}_{j,2u+v} = \mathbf{c}_{2(j-1)+v,u+1} \quad \text{for } j \in \{1, 2, \dots, s\},\, u \in \mathbb{N}_0 \text{ and } v \in \{1, 2\}.$$

We illustrate the construction for $s = 1$. Then $d = 2$ and

$$C_1 = \begin{pmatrix} \mathbf{c}_{1,1} \\ \mathbf{c}_{1,2} \\ \vdots \end{pmatrix}, \ C_2 = \begin{pmatrix} \mathbf{c}_{2,1} \\ \mathbf{c}_{2,2} \\ \vdots \end{pmatrix} \Rightarrow E_1 = \begin{pmatrix} \mathbf{c}_{1,1} \\ \mathbf{c}_{2,1} \\ \mathbf{c}_{1,2} \\ \mathbf{c}_{2,2} \\ \vdots \end{pmatrix}.$$

This procedure is called *interlacing* (here the so-called “interlacing factor” is 2).

The following theorem has been shown in [20].

Theorem 4 (Dick, Hinrichs, Markhasin & Pillichshammer [20, 21]) *Assume that $E_1, \ldots, E_s \in \mathbb{F}_2^{\mathbb{N}\times\mathbb{N}}$ are constructed with the interlacing principle as given above. Then for the corresponding digital sequence $\mathcal{S} = \mathcal{S}(E_1, \ldots, E_s)$ we have*

$$L_{p,N}(\mathcal{S}) \lesssim_{p,s} 2^{2t} \frac{(\log N)^{s/2}}{N} \quad \textit{for all } N \geq 2 \textit{ and all } 1 \leq p < \infty.$$

This theorem shows, in a constructive way, that the lower bound (4) from Roth–Schmidt–Proĭnov is best possible in the order of magnitude in N for all parameters $p \in (1, \infty)$. Furthermore, the constructed digital sequences have optimal order of L_p discrepancy simultaneously for all $p \in (1, \infty)$.

For $p = 2$ there is an interesting improvement, although this improvement requires higher order digital sequences of order 5 (instead of order 2). For such sequences $\mathcal{S}$ it has been shown in [26] that

$$L_{2,N}(\mathcal{S}) \lesssim_s \frac{(\log N)^{(s-1)/2}}{N} \sqrt{S_2(N)} \quad \text{for all } N \geq 2.$$

The dyadic sum-of-digit function of N is in the worst-case of order $\log N$ and then the above L_2 discrepancy bound is of order of magnitude $(\log N)^{s/2}/N$. But if N has very few non-zero dyadic digits, for example if it is a power of 2, then the bound on the L_2 discrepancy becomes $(\log N)^{(s-1)/2}/N$ only.

The proof of Theorem 4 uses methods from harmonic analysis, in particular the estimate of the L_p norm of the discrepancy function is based on the following Littlewood–Paley type inequality: for $p \in (1, \infty)$ and $f \in L_p([0, 1]^s)$ we have

$$\|f\|_{L_p([0,1]^s)} \lesssim_{p,s} \sum_{\boldsymbol{j}\in\mathbb{N}_{-1}^s} 2^{2|\boldsymbol{j}|(1-1/\bar{p})} \left(\sum_{\boldsymbol{m}\in\mathbb{D}_{\boldsymbol{j}}} |\langle f, h_{\boldsymbol{j},\boldsymbol{m}}\rangle|^{\bar{p}} \right)^{2/\bar{p}}, \tag{9}$$

where $\bar{p} = \max(p, 2)$, $\mathbb{N}_{-1} = \mathbb{N} \cup \{-1, 0\}$, for $\boldsymbol{j} = (j_1, \ldots, j_s)$, $\mathbb{D}_{\boldsymbol{j}} = \mathbb{D}_{j_1} \times \cdots \times \mathbb{D}_{j_s}$, where $\mathbb{D}_j = \{0, 1, \ldots, 2^j - 1\}$, $|\boldsymbol{j}| = \max(j_1, 0) + \cdots + \max(j_s, 0)$, and, for $\boldsymbol{m} \in \mathbb{D}_{\boldsymbol{j}}$, $h_{\boldsymbol{j},\boldsymbol{m}}(\boldsymbol{x}) = h_{j_1,m_1}(x_1) \cdots h_{j_s,m_s}(x_s)$, where $h_{j,m}$ is the mth dyadic Haar function on level j; see [20, 66]. The L_2 inner products $\langle f, h_{\boldsymbol{j},\boldsymbol{m}}\rangle$ are the so-called Haar coefficients of f. Inequality (9) is used for the local discrepancy function of digital sequences which then requires tight estimates of the Haar coefficients of the local discrepancy function. For details we refer to [20].

With the same method one can also handle the quasi-norm of the local discrepancy function in Besov spaces and Triebel–Lizorkin spaces with dominating mixed smoothness. One reason why Besov spaces and Triebel–Lizorkin spaces are interesting in this context is that they form natural scales of function spaces including the L_p-spaces and Sobolev spaces of dominating mixed smoothness (see, e.g., [91]). The study of discrepancy in these function spaces has been initiated by Triebel [91, 92] in 2010. Further results (for finite sequences) can be found in [47, 65–68] and (for infinite sequences in dimension one) in [52]. In [21, Theorems 3.1 and 3.2] general lower bounds on the quasi-norm of the local discrepancy function in Besov spaces and Triebel–Lizorkin spaces with dominating mixed smoothness in the sense of the result of Roth–Schmidt–Proĭnov in Eq. (4) are shown. Furthermore, these lower bounds are optimal in the order of magnitude in N, since matching upper bounds are obtained for infinite order two digital sequences as constructed above. For details we refer to [21].

3.5 Intermediate Norms of the Local Discrepancy Function

While the quest for the exact order of the optimal L_p discrepancy of infinite sequences in arbitrary dimension is now solved for finite parameters $p \in (1, \infty)$ the situation for the cases $p \in \{1, \infty\}$ remains open. In this situation, Bilyk, Lacey, Parissis and Vagharshakyan [9] studied the question of what happens in intermediate spaces "close" to L_∞. Two standard examples of such spaces are:

- *Exponential Orlicz space*: for the exact definition of the corresponding norm $\|\cdot\|_{\exp(L^\beta)}$, $\beta > 0$, we refer to [9, 10, 21]. There is an equivalence which shows the relation to the L_p norm, which is stated for any $\beta > 0$,

$$\|f\|_{\exp(L^\beta)} \asymp \sup_{p>1} p^{-\frac{1}{\beta}} \|f\|_{L_p([0,1]^s)}.$$

  This equivalence suggests that the study of discrepancy with respect to the exponential Orlicz norm is related to the study of the dependence of the constant appearing in the L_p discrepancy bounds on the parameter p. The latter problem is also studied in [88].
- *BMO space* (where BMO stands for "bounded mean oscillation"): the definition of the corresponding semi-norm uses Haar functions and is given as

$$\|f\|^2_{\mathrm{BMO}^s} = \sup_{U \subseteq [0,1)^s} \frac{1}{\lambda_s(U)} \sum_{\boldsymbol{j} \in \mathbb{N}_0^s} 2^{|\boldsymbol{j}|} \sum_{\substack{\boldsymbol{m} \in \mathbb{D}_{\boldsymbol{j}} \\ \mathrm{supp}(h_{\boldsymbol{j},\boldsymbol{m}}) \subseteq U}} |\langle f, h_{\boldsymbol{j},\boldsymbol{m}} \rangle|^2,$$

where the supremum is extended over all measurable subsets U from $[0, 1)^s$. See again [9, 10, 21] and the references therein for more information.

Exponential Orlicz norm and BMO semi-norm of the local discrepancy function for finite point sets have been studied in [9] (in dimension $s = 2$) and in [10] (in the general multi-variate case). For infinite sequences we have the following results which have been shown in [21]:

Theorem 5 (Dick, Hinrichs, Markhasin & Pillichshammer [20, 21]) *Assume that $E_1, \ldots, E_s \in \mathbb{F}_2^{\mathbb{N}\times\mathbb{N}}$ are constructed with the interlacing principle as given in Sect. 3.4. Then for the corresponding digital sequence $\mathscr{S} = \mathscr{S}(E_1, \ldots, E_s)$ we have*

$$\|\Delta_{\mathscr{S}_N}\|_{\exp(L^\beta)} \lesssim_s \frac{(\log N)^{s-\frac{1}{\beta}}}{N} \quad \textit{for all } N \geq 2 \textit{ and for all } \frac{2}{s-1} \leq \beta < \infty$$

and

$$\|\Delta_{\mathscr{S}_N}\|_{\mathrm{BMO}^s} \lesssim_s \frac{(\log N)^{\frac{s}{2}}}{N} \quad \textit{for all } N \geq 2. \tag{10}$$

A matching lower bound in the case of exponential Orlicz norm on $\|\Delta_{\mathscr{S}_N}\|_{\exp(L^\beta)}$ in arbitrary dimension is currently not available and seems to be a very difficult problem, even for finite sequences (see [10, Remark after Theorem 1.3]; for matching lower and upper bounds for finite sequences in dimension $s = 2$ we refer to [9]). On the other hand, the result (10) for the BMO semi-norm is best possible in the order of magnitude in N. A general lower bound in the sense of Roth–Schmidt–Proĭnov's result (4) for the L_p discrepancy has been shown in [21, Theorem 2.1] and states that for every $s \in \mathbb{N}$ there exists a $c_s > 0$ such that for every infinite sequence $\mathscr{S}$ in $[0, 1)^s$ we have

$$\|\Delta_{\mathscr{S}_N}\|_{\mathrm{BMO}^s} \geq c_s \frac{(\log N)^{\frac{s}{2}}}{N} \quad \text{infinitely often.} \tag{11}$$

4 Discussion of the Asymptotic Discrepancy Estimates

We restrict the following discussion to the case of star discrepancy. We have seen that the star discrepancy of digital sequences, and therefore QMC rules which are based on digital sequences, can achieve error bounds of order of magnitude $(\log N)^s/N$. At first sight this seems to be an excellent result. However, the crux of these, in an asymptotic sense, optimal results, lies in the dependence on the dimension s. If we consider the function $x \mapsto (\log x)^s/x$, then one can observe, that this function is increasing up to $x = \mathrm{e}^s$ and only then it starts to decrease to 0 with the asymptotic order of almost $1/x$. This means, in order to have meaningful error bounds for QMC rules one requires finite sequences with at least e^s many elements or even larger. But e^s is already huge, even for moderate dimensions s. For example, if $s = 200$, then $\mathrm{e}^s \approx 7.2 \times 10^{86}$ which exceeds the estimated number of atoms in our universe (which is $\approx 10^{78}$).

As it appears, according to the classical theory with its excellent asymptotic results, QMC rules cannot be expected to work for high-dimensional functions. However, there is numerical evidence, that QMC rules can also be used in these cases. The work of Paskov and Traub [78] from 1995 attracted much attention in this context. They considered a real world problem from mathematical finance which resulted in the evaluation of several 360 dimensional integrals and reported on their successful use of Sobol' and Halton-sequences in order to evaluate these integrals.

Of course, it is now the aim of theory to the explain, *why* QMC rules also work for high-dimensional problems. One stream of research is to take the viewpoint of *Information Based Complexity (IBC)* in which also the dependence of the error bounds (discrepancy in our case) on the dimension s is studied. A first remarkable, and at that time very surprising result, has been established by Heinrich, Novak, Wasilkowski and Woźniakowski [44] in 2001.

Theorem 6 (Heinrich, Novak, Wasilkowski & Woźniakowski [44]) *For all* $N, s \in \mathbb{N}$ *there exist finite sequences* $\mathscr{S}_N$ *of* N *elements in* $[0, 1)^s$ *such that*

$$D_N^*(\mathscr{S}_N) \lesssim \sqrt{\frac{s}{N}},$$

where the implied constant is absolute, i.e., does neither depend on s, nor on N.

In 2007 Dick [16] extended this result to infinite sequences (in infinite dimension).

In IBC the information complexity is studied rather then direct error bounds. In the case of star discrepancy the *information complexity*, which is then also called the *inverse of star discrepancy*, is, for some error demand $\varepsilon \in (0, 1]$ and dimension s, given as

$$N^*(\varepsilon, s) = \min\{N \in \mathbb{N} : \exists\, \mathscr{S}_N \subseteq [0, 1)^s \text{ with } |\mathscr{S}_N| = N \text{ and } D_N^*(\mathscr{S}_N) \le \varepsilon\}.$$

From Theorem 6 one can deduce that

$$N^*(\varepsilon, s) \lesssim s\varepsilon^{-2}$$

and this property is called *polynomial tractability* with ε-exponent 2 and s-exponent 1. In 2004 Hinrichs [46] proved that there exists a positive c such that $N^*(\varepsilon, s) \ge cs\varepsilon^{-1}$ for all s and all small enough $\varepsilon > 0$. Combining these results we see, that *the inverse of the star discrepancy depends (exactly) linearly on the dimension s* (which is the programmatic title of the paper [44]). The exact dependence of the inverse of the star discrepancy on ε^{-1} is still unknown and seems to be a very difficult problem. In 2011 Aistleitner [1] gave a new proof of the result in Theorem 6 from which one can obtain an explicit constant in the star discrepancy estimate. He proved that there exist finite sequences $\mathscr{S}_N$ of N elements in $[0, 1)^s$ such that $D_N^*(\mathscr{S}_N) \le 10\sqrt{s/N}$ and hence $N^*(\varepsilon, s) \le 100s\varepsilon^{-2}$. Recently Gnewuch and Hebbinghaus [41] improved these implied constants to $D_N^*(\mathscr{S}_N) \le (2.5287\ldots) \times \sqrt{s/N}$ and hence $N^*(\varepsilon, s) \le (6.3943\ldots) \times s\varepsilon^{-2}$.

For a comprehensive introduction to IBC and tractability theory we refer to the three volumes [75–77] by Novak and Woźniakowski.

Unfortunately, the result in Theorem 6 is a pure existence result and until now no concrete point set is known whose star discrepancy satisfies the given upper bound. Motivated by the excellent asymptotic behavior it may be obvious to consider digital sequences also in the context of tractability. This assumption is supported by a recent metrical result for a certain subsequence of a digital Kronecker sequence. In order to explain this result we need some notation:

- Let $\mathbb{F}_b((t^{-1}))$ be the field of *formal Laurent series* over $\mathbb{F}_b$ in the variable t:

$$\mathbb{F}_b((t^{-1})) = \left\{ \sum_{i=w}^{\infty} g_i t^{-i} \ : \ w \in \mathbb{Z}, \forall i : g_i \in \mathbb{F}_b \right\}.$$

- For $g \in \mathbb{F}_b((t^{-1}))$ of the form $g = \sum_{i=w}^{\infty} g_i t^{-i}$ define the "fractional part"

$$\{g\} := \sum_{i=\max\{w,1\}}^{\infty} g_i t^{-i}.$$

- Every $n \in \mathbb{N}_0$ with b-adic expansion $n = n_0 + n_1 b + \cdots + n_r b^r$, where $n_i \in \{0, \ldots, b-1\}$, is associated in the natural way with the polynomial

$$n \cong n_0 + n_1 t + \cdots + n_r t^r \in \mathbb{F}_b[t].$$

Now a digital Kronecker sequence is defined as follows:

Definition 4 Let $\boldsymbol{f} = (f_1, \ldots, f_s) \in \mathbb{F}_b((t^{-1}))^s$. Then the sequence $\mathscr{S}(\boldsymbol{f}) = (\boldsymbol{y}_n)_{n \geq 0}$ given by

$$\boldsymbol{y}_n := \{n\boldsymbol{f}\}_{|t=b} = (\{nf_1\}_{|t=b}, \ldots, \{nf_s\}_{|t=b})$$

is called a *digital Kronecker sequence over* $\mathbb{F}_b$.

It can be shown that digital Kronecker sequences are examples of digital sequences where the generating matrices are Hankel matrices (i.e., constant ascending skew-diagonals) whose entries are the coefficients of the Laurent series expansions of $f_1, \ldots, f_s$; see, e.g., [57, 72]. Neumüller and Pillichshammer [70] studied a subsequence of digital Kronecker sequences. For $\boldsymbol{f} \in \mathbb{F}_b((t^{-1}))^s$ consider $\widetilde{\mathscr{S}}(\boldsymbol{f}) = (\boldsymbol{y}_n)_{n \geq 0}$ where

$$\boldsymbol{y}_n = \{t^n \boldsymbol{f}\}_{|t=b} = (\{t^n f_1\}_{|t=b}, \ldots, \{t^n f_s\}_{|t=b}).$$

With a certain natural probability measure on $\mathbb{F}_b((t^{-1}))^s$ the following metrical result can be shown:

Theorem 7 (Neumüller & Pillichshammer [70]) *Let $s \geq 2$. For every $\delta \in (0, 1)$ we have*

$$D_N^*(\widetilde{\mathscr{S}}(f)) \lesssim_{b,\delta} \sqrt{\frac{s \log s}{N}} \log N \quad \textit{for all } N \geq 2 \tag{12}$$

with probability at least $1 - \delta$, where the implied constant $C_{b,\delta} \asymp_b \log \delta^{-1}$.

The estimate (12) is only slightly weaker than the bound in Theorem 6. The additional $\log N$-term comes from the consideration of infinite sequences. Note that the result holds for all $N \geq 2$ simultaneously. One gets rid of this $\log N$-term when one considers only finite sequences as in Theorem 6; see [70, Theorem 3]. Furthermore, we remark that Theorem 7 corresponds to a result for classical Kronecker sequences which has been proved by Löbbe [64].

5 Weighted Discrepancy of Digital Sequences

Another way to explain the success of QMC rules for high-dimensional problems is the study of so-called weighted function classes. This study, initiated by Sloan and Woźniakowski [89] in 1998, is based on the assumption that functions depend differently on different variables and groups of variables when the dimension s is large. This different dependence should be reflected in the error analysis. For this purpose Sloan and Woźniakowski proposed the introduction of weights that model the dependence of the functions on different coordinate directions. In the context of discrepancy theory this led to the introduction of weighted L_p discrepancy. Here we restrict ourselves to the case of weighted star discrepancy:

In the following let $\boldsymbol{\gamma} = (\gamma_1, \gamma_2, \gamma_3, \ldots)$ be a sequence of positive reals, the so-called weights. Let $[s] := \{1, 2, \ldots, s\}$ and for $\mathfrak{u} \subseteq [s]$ put

$$\gamma_{\mathfrak{u}} := \prod_{j \in \mathfrak{u}} \gamma_j.$$

Definition 5 (*Sloan & Woźniakowski* [89]) For a sequence $\mathscr{S}$ in $[0, 1)^s$ the *$\boldsymbol{\gamma}$-weighted star discrepancy* is defined as

$$D_{N,\boldsymbol{\gamma}}^*(\mathscr{S}) := \sup_{\boldsymbol{\alpha} \in [0,1]^s} \max_{\emptyset \neq \mathfrak{u} \subseteq [s]} \gamma_{\mathfrak{u}} |\Delta_{\mathscr{S}_N}(\boldsymbol{\alpha}_{\mathfrak{u}}, \mathbf{1})|,$$

where for $\boldsymbol{\alpha} = (\alpha_1, \ldots, \alpha_s) \in [0, 1]^s$ and for $\mathfrak{u} \subseteq [s]$ we put $(\boldsymbol{\alpha}_{\mathfrak{u}}, \mathbf{1}) = (y_1, \ldots, y_s)$ with $y_j = \alpha_j$ if $j \in \mathfrak{u}$ and $y_j = 1$ if $j \notin \mathfrak{u}$.

Remark 2 If $\gamma_j = 1$ for all $j \geq 1$, then $D_{N,\boldsymbol{\gamma}}^*(\mathscr{S}) = D_N^*(\mathscr{S})$.

The relation between weighted discrepancy and error bounds for QMC rules is expressed by means of a *weighted Koksma–Hlawka inequality* as follows: Let

$\mathscr{W}_1^{(1,1,\ldots,1)}([0,1]^s)$ be the Sobolev space of functions defined on $[0,1]^s$ that are once differentiable in each variable, and whose derivatives have finite L_1 norm. Consider

$$\mathscr{F}_{s,1,\boldsymbol{\gamma}} = \{f \in \mathscr{W}_1^{(1,1,\ldots,1)}([0,1]^s) \ : \ \|f\|_{s,1,\boldsymbol{\gamma}} < \infty\},$$

where

$$\|f\|_{s,1,\boldsymbol{\gamma}} = |f(\mathbf{1})| + \sum_{\emptyset \neq \mathfrak{u} \subseteq [s]} \frac{1}{\gamma_{\mathfrak{u}}} \left\| \frac{\partial^{|\mathfrak{u}|}}{\partial \boldsymbol{x}_{\mathfrak{u}}} f(\boldsymbol{x}_{\mathfrak{u}}, \mathbf{1}) \right\|_{L_1}.$$

The $\boldsymbol{\gamma}$-weighted star discrepancy of a finite sequence is then exactly the worst-case error of a QMC rule in $\mathscr{F}_{s,1,\boldsymbol{\gamma}}$ that is based on this sequence, see [89] or [76, p. 65]. More precisely, we have

$$\sup_{\|f\|_{s,1,\boldsymbol{\gamma}} \le 1} \left| \int_{[0,1]^s} f(\boldsymbol{x}) \mathrm{d}\boldsymbol{x} - \frac{1}{N} \sum_{\boldsymbol{x} \in \mathscr{S}_N} f(\boldsymbol{x}) \right| = D^*_{N,\boldsymbol{\gamma}}(\mathscr{S}).$$

In IBC again the *inverse of weighted star discrepancy*

$$N^*_{\boldsymbol{\gamma}}(\varepsilon, s) := \min\{N \ : \ \exists\, \mathscr{S}_N \subseteq [0,1)^s \text{ with } |\mathscr{S}_N| = N \text{ and } D^*_{N,\boldsymbol{\gamma}}(\mathscr{S}_N) \le \varepsilon\}$$

is studied. The weighted star discrepancy is said to be *strongly polynomially tractable* (SPT), if there exist non-negative real numbers C and β such that

$$N^*_{\boldsymbol{\gamma}}(\varepsilon, s) \le C\varepsilon^{-\beta} \quad \text{for all } s \in \mathbb{N} \text{ and for all } \varepsilon \in (0,1). \tag{13}$$

The infimum β^* over all $\beta > 0$ such that (13) holds is called the ε-exponent of strong polynomial tractability. It should be mentioned, that there are several other notions of tractability which are considered in literature. Examples are polynomial tractability, weak tractability, etc. For an overview we refer to [75–77].

In [48] Hinrichs, Tezuka and the author studied tractability properties of the weighted star discrepancy of several digital sequences.

Theorem 8 (Hinrichs, Pillichshammer & Tezuka [48]) *The weighted star discrepancy of the Halton sequence (where the bases $b_1, \ldots, b_s$ are the first s prime numbers in increasing order) and of Niederreiter sequences achieve SPT with ε-exponent*

- $\beta^* = 1$, *which is optimal, if*

$$\sum_{j \ge 1} j\gamma_j < \infty, \qquad \text{e.g., if } \gamma_j = \frac{1}{j^{2+\delta}} \text{ with some } \delta > 0;$$

- $\beta^* \le 2$, *if*

$$\sup_{s \ge 1} \max_{\emptyset \neq \mathfrak{u} \subseteq [s]} \prod_{j \in \mathfrak{u}} (j\gamma_j) < \infty \qquad \text{e.g., if } \gamma_j = \frac{1}{j}.$$

This result is the currently mildest weight condition for a "constructive" proof of SPT of the weighted star discrepancy. Furthermore, it is the first "constructive" result which does not require that the weights are summable in order to achieve SPT. By a "constructive" result we mean in this context that the corresponding point set can be found or constructed by a polynomial-time algorithm in s and in ε^{-1}.

To put the result in Theorem 8 into context we recall the currently best "existence result" which has been shown by Aistleitner [2]:

Theorem 9 (Aistleitner) *If there exists a* $c > 0$ *such that*

$$\sum_{j=1}^{\infty} \exp(-c\gamma_j^{-2}) < \infty \qquad \text{e.g., if } \gamma_j = \frac{1}{\sqrt{\log j}},$$

then the weighted star discrepancy is SPT with ε*-exponent* $\beta^* \leq 2$.

Obviously the condition on the weights in Aistleitner's "existence" result is much weaker then for the "constructive" result in Theorem 8. It is now the task to find sequences whose weighted star discrepancy achieves SPT under the milder weight condition.

6 Summary

Digital (t, s)-sequences are without doubt the most powerful concept for the construction of low-discrepancy sequences in many settings. Such sequences are very much-needed as sample points for QMC integration rules. They have excellent discrepancy properties in an asymptotic sense when the dimension s is fixed and when $N \to \infty$:

- For $p \in [1, \infty)$ there are constructions of digital sequences with L_p discrepancy

$$L_p(\mathscr{S}) \lesssim_{s,p} \frac{(\log N)^{s/2}}{N} \qquad \text{for all } N \geq 2 \text{ and } p \in [1, \infty)$$

  and this estimate is best possible in the order of magnitude in N for $p \in (1, \infty)$ according to the general lower bound (4).
- The star discrepancy of digital (t, s)-sequences satisfies a bound of the form

$$D_N^*(\mathscr{S}) \lesssim_s \frac{(\log N)^s}{N} \qquad \text{for all } N \geq 2$$

  and this bound is often assumed to be best possible at all.
- For discrepancy with respect to various other norms digital sequences achieve very good and even optimal results.

On the other hand, nowadays one is also very much interested in the dependence of discrepancy on the dimension s. This is a very important topic, in particular in order to justify the use of QMC in high dimensions. First results suggest that also in this IBC context digital sequences may perform very well. But here many questions are still open and require further studies. One particularly important question is how sequences can be constructed whose discrepancy achieves some notion of tractability. Maybe digital sequences are good candidates also for this purpose.

Acknowledgements F. Pillichshammer is supported by the Austrian Science Fund (FWF) Project F5509-N26.

References

1. Aistleitner, C.: Covering numbers, dyadic chaining and discrepancy. J. Complex. **27**, 531–540 (2011)
2. Aistleitner, C.: Tractability results for the weighted star-discrepancy. J. Complex. **30**, 381–391 (2014)
3. Beck, J.: Probabilistic diophantine approximationI. Kronecker-sequences. Ann. Math. **140**, 451–502 (1994)
4. Beck, J., Chen, W.W.L.: Irregularities of Distribution. Cambridge University Press, Cambridge (1987)
5. Béjian, R.: Minoration de la discrépance d'une suite quelconque sur T. Acta Arith. **41**, 185–202 (1982). (in French)
6. Béjian, R., Faure, H.: Discrépance de la suite de van der Corput. Séminaire Delange-Pisot-Poitou (Théorie des nombres) 13, 1–14 (1977/78) (in French)
7. Bilyk, D., Lacey, M.T.: The supremum norm of the discrepancy function: recent results and connections. In: Monte Carlo and Quasi-Monte Carlo Methods. Springer Proceedings in Mathematics & Statistics, 2013. Springer, Berlin, Heidelberg (2012)
8. Bilyk, D., Lacey, M.T., Vagharshakyan, A.: On the small ball inequality in all dimensions. J. Funct. Anal. **254**, 2470–2502 (2008)
9. Bilyk, D., Lacey, M.T., Parissis, I., Vagharshakyan, A.: Exponential squared integrability of the discrepancy function in two dimensions. Mathematika **55**, 2470–2502 (2009)
10. Bilyk, D., Markhasin, L.: BMO and exponential Orlicz space estimates of the discrepancy function in arbitrary dimension. J. d'Analyse Math. **135**, 249–269 (2018)
11. Chen, W.W.L.: On irregularities of distribution. Mathematika **27**, 153–170 (1981)
12. Chen, W.W.L.: On irregularities of distribution II. Q. J. Math. **34**, 257–279 (1983)
13. Chen, W.W.L., Skriganov, M.M.: Explicit constructions in the classical mean squares problem in irregularities of point distribution. J. Reine Angew. Math. **545**, 67–95 (2002)
14. Chen, W.W.L., Srivastav, A., Travaglini, G.: A Panorama of Discrepancy Theory. Lecture Notes in Mathematics, vol. 2107. Springer, Cham (2014)
15. Davenport, H.: Note on irregularities of distribution. Mathematika **3**, 131–135 (1956)
16. Dick, J.: A note on the existence of sequences with small star discrepancy. J. Complex. **23**, 649–652 (2007)
17. Dick, J.: Explicit constructions of quasi-Monte Carlo rules for the numerical integration of high-dimensional periodic functions. SIAM J. Numer. Anal. **45**, 2141–2176 (2007)
18. Dick, J.: Walsh spaces containing smooth functions and quasi-Monte Carlo rules of arbitrary high order. SIAM J. Numer. Anal. **46**, 1519–1553 (2008)
19. Dick, J.: Discrepancy bounds for infinite-dimensional order two digital sequences over $\mathbb{F}2$. J. Number Theory **136**, 204–232 (2014)

20. Dick, J., Hinrichs, A., Markhasin, L., Pillichshammer, F.: Optimal L_p-discrepancy bounds for second order digital sequences. Isr. J. Math. **221**, 489–510 (2017)
21. Dick, J., Hinrichs, A., Markhasin, L., Pillichshammer, F.: Discrepancy of second order digital sequences in function spaces with dominating mixed smoothness. Mathematika **63**, 863–894 (2017)
22. Dick, J., Kuo, F.Y., Sloan, I.H.: High dimensional numerical integration-the Quasi-Monte Carlo way. Acta Numer. **22**, 133–288 (2013)
23. Dick, J., Pillichshammer, F.: On the mean square weighted L_2 discrepancy of randomized digital (t, m, s)-nets over $\mathbb{Z}_2$. Acta Arith. **117**, 371–403 (2005)
24. Dick, J., Pillichshammer, F.: Digital Nets and Sequences: Discrepancy Theory and Quasi-Monte Carlo Integration. Cambridge University Press, Cambridge (2010)
25. Dick, J., Pillichshammer, F.: Explicit constructions of point sets and sequences with low discrepancy. Uniform Distribution and Quasi-Monte Carlo Methods. Radon Series on Computational and Applied Mathematics, vol. 15, pp. 63–86. De Gruyter, Berlin (2014)
26. Dick, J., Pillichshammer, F.: Optimal $\mathscr{L}_2$ discrepancy bounds for higher order digital sequences over the finite field $\mathbb{F}_2$. Acta Arith. **162**, 65–99 (2014)
27. Dobrovol'skiĭ, N.M.: An effective proof of Roth's theorem on quadratic dispersion. Akademiya Nauk SSSR i Moskovskoe Matematicheskoe Obshchestvo **39**, 155–156 (1984) (English translation in Russ. Math. Surv. **39**, 117–118 (1984))
28. Drmota, M., Larcher, G., Pillichshammer, F.: Precise distribution properties of the van der Corput sequence and related sequences. Manuscr. Math. **118**, 11–41 (2005)
29. Drmota, M., Tichy, R.F.: Sequences, Discrepancies and Applications. Lecture Notes in Mathematics, vol. 1651. Springer, Berlin (1997)
30. Faure, H.: Discrépances de suites associées a un système de numération (en dimension un). Bull. Soc. Math. Fr. **109**, 143–182 (1981). (in French)
31. Faure, H.: Discrépances de suites associées a un système de numération (en dimension s). Acta Arith. **41**, 337–351 (1982). (in French)
32. Faure, H.: Discrepancy lower bound in two dimensions. Monte Carlo and Quasi-Monte Carlo Methods in Scientific Computing. Lecture Notes in Statistics, vol. 106, pp. 198–204. Springer, Berlin (1995)
33. Faure, H.: Van der Corput sequences towards $(0, 1)$-sequences in base b. J. Théorie Nr. Bordx. **19**, 125–140 (2007)
34. Faure, H., Kritzer, P.: New star discrepancy bounds for (t, m, s)-nets and (t, s)-sequences. Monatshefte Math. **172**, 55–75 (2013)
35. Faure, H., Kritzer, P.: Discrepancy bounds for low-dimensional point sets. Applied Algebra and Number Theory, pp. 58–90. Cambridge University Press, Cambridge (2014)
36. Faure, H., Kritzer, P., Pillichshammer, F.: From van der Corput to modern constructions of sequences for quasi-Monte Carlo rules. Indag. Math. **26**(5), 760–822 (2015)
37. Faure, H., Lemieux, C.: Improvements on the star discrepancy of (t, s)-sequences. Acta Arith. **61**, 61–78 (2012)
38. Faure, H., Lemieux, C.: A variant of Atanassov's method for (t, s)-sequences and $(t, \boldsymbol{e}, s)$-sequences. J. Complex. **30**, 620–633 (2014)
39. Faure, H., Lemieux, C.: A review of discrepancy bounds for (t, s)- and $(t, \boldsymbol{e}, s)$-sequences with numerical comparisons. Math. Comput. Simul. **135**, 63–71 (2017)
40. Frolov, K.K.: Upper bound of the discrepancy in metric L_p, $2 \leq p <$. Dokl. Akad. Nauk. SSSR **252**, 805–807 (1980)
41. Gnewuch, M., Hebbinghaus, N.: Discrepancy bounds for a class of negatively dependent random points including Latin hypercube samples (2018)
42. Halász, G.: On Roth's method in the theory of irregularities of point distributions. Recent Progress in Analytic Number Theory, vol. 2, pp. 79–94. Academic, London (1981)
43. Halton, J.H.: On the efficiency of certain quasi-random sequences of points in evaluating multi-dimensional integrals. Numer. Math. **2**, 84–90 (1960)
44. Heinrich, S., Novak, E., Wasilkowski, G.W., Woźniakowski, H.: The inverse of the star-discrepancy depends linearly on the dimension. Acta Arith. **96**, 279–302 (2001)

45. Hickernell, F.J., Yue, R.-X.: The mean square discrepancy of scrambled (t, s)-sequences. SIAM J. Numer. Anal. **38**, 1089–1112 (2000)
46. Hinrichs, A.: Covering numbers, Vapnik-Červonenkis classes and bounds on the star-discrepancy. J. Complex. **20**, 477–483 (2004)
47. Hinrichs, A.: Discrepancy of Hammersley points in Besov spaces of dominating mixed smoothness. Math. Nachr. **283**, 478–488 (2010)
48. Hinrichs, A., Pillichshammer, F., Tezuka, S.: Tractability properties of the weighted star discrepancy of the Halton sequence. J. Comput. Appl. Math. **350**, 39–54 (2019)
49. Kaltenböck, L., Stockinger, W.: On M. B. Levin's proofs for the exact lower discrepancy bounds of special sequences and point sets (a survey). Unif. Distrib. Theory **13**(2), 103–130 (2018)
50. Kritzer, P.: A new upper bound on the star discrepancy of (0, 1)-sequences. Integers **5**(3), A11 (electronic) (2005), 9pp
51. Kritzer, P.: Improved upper bounds on the star discrepancy of (t, m, s)-nets and (t, s)-sequences. J. Complex. **22**, 336–347 (2006)
52. Kritzinger, R.: L_p- and $S_{p,q}^r B$-discrepancy of the symmetrized van der Corput sequence and modified Hammersley point sets in arbitrary bases. J. Complex. **33**, 145–168 (2016)
53. Kuipers, L., Niederreiter, H.: Uniform Distribution of Sequences. Wiley, New York (1974); Reprint, Dover Publications, Mineola (2006)
54. Larcher, G.: On the distribution of digital sequences. Monte Carlo and Quasi-Monte Carlo Methods 1996 (Salzburg). Lecture Notes in Statistics, vol. 127, pp. 109–123. Springer, New York (1998)
55. Larcher, G.: On the distribution of an analog to classical Kronecker-sequences. J. Number Theory **52**, 198–215 (1995)
56. Larcher, G.: On the star discrepancy of sequences in the unit interval. J. Complex. **31**(3), 474–485 (2015)
57. Larcher, G., Niederreiter, H.: Kronecker-type sequences and nonarchimedian diophantine approximation. Acta Arith. **63**, 380–396 (1993)
58. Larcher, G., Pillichshammer, F.: A metrical lower bound on the star discrepancy of digital sequences. Monatshefte Math. **174**, 105–123 (2014)
59. Larcher, G., Pillichshammer, F.: Metrical lower bounds on the discrepancy of digital Kronecker-sequences. J. Number Theory **135**, 262–283 (2014)
60. Larcher, G., Puchhammer, F.: An improved bound for the star discrepancy of sequences in the unit interval. Unif. Distrib. Theory **11**(1), 1–14 (2016)
61. Leobacher, G., Pillichshammer, F.: Introduction to Quasi-Monte Carlo Integration and Applications. Compact Textbooks in Mathematics. Birkhäuser, Basel (2014)
62. Levin, M.B.: On the lower bound of the discrepancy of (t, s) sequences: I. Comptes Rendus Math. (Académie des Sciences, Paris) **354**(6), 562–565 (2016)
63. Levin, M.B.: On the lower bound of the discrepancy of (t, s) sequences: II. Online J. Anal. Comb. 12 (electronic) (2017), 74 pp
64. Löbbe, T.: Probabilistic star discrepancy bounds for lacunary point sets (2014). See arXiv:1408.2220
65. Markhasin, L.: Discrepancy of generalized Hammersley type point sets in Besov spaces with dominating mixed smoothness. Unif. Distrib. Theory **8**, 135–164 (2013)
66. Markhasin, L.: Quasi-Monte Carlo methods for integration of functions with dominating mixed smoothness in arbitrary dimension. J. Complex. **29**, 370–388 (2013)
67. Markhasin, L.: Discrepancy and integration in function spaces with dominating mixed smoothness. Diss. Math. **494**, 1–81 (2013)
68. Markhasin, L.: L_p- and $S_{p,q}^r B$-discrepancy of (order 2) digital nets. Acta Arith. **168**, 139–159 (2015)
69. Matoušek, J.: Geometric Discrepancy. An Illustrated Guide. Algorithms and Combinatorics, vol. 18, Springer, Berlin (1999)
70. Neumüller, M., Pillichshammer, F.: Metrical star discrepancy bounds for lacunary subsequences of digital Kronecker-sequences and polynomial tractability. Unif. Distrib. Theory **13**(1), 65–86 (2018)

71. Niederreiter, H.: Point sets and sequences with small discrepancy. Monatshefte Math. **104**, 273–337 (1987)
72. Niederreiter, H.: Random Number Generation and Quasi-Monte Carlo Methods. SIAM, Philadelphia (1992)
73. Niederreiter, H., Xing, C.: Quasirandom points and global function fields. Finite Fields and Applications. London Mathematical Society Lecture Note Series, vol. 233, pp. 269–296. Cambridge University Press, Cambridge (1996)
74. Niederreiter, H., Xing, C.: Rational Points on Curves over Finite Fields. London Mathematical Society Lecture Notes Series, vol. 285. Cambridge University Press, Cambridge (2001)
75. Novak, E., Woźniakowski, H.: Tractability of Multivariate Problems. Volume I: Linear Information. European Mathematical Society, Zürich (2008)
76. Novak, E., Woźniakowski, H.: Tractability of Multivariate Problems, Volume II: Standard Information for Functionals. European Mathematical Society, Zürich (2010)
77. Novak, E., Woźniakowski, H.: Tractability of Multivariate Problems, Volume III: Standard Information for Operators. European Mathematical Society, Zürich (2012)
78. Paskov, S.H., Traub, J.F.: Faster evaluation of financial derivatives. J. Portf. Manag. **22**, 113–120 (1995)
79. Pillichshammer, F.: On the discrepancy of (0, 1)-sequences. J. Number Theory **104**, 301–314 (2004)
80. Proĭnov, P.D.: On irregularities of distribution. Doklady Bolgarskoĭ Akademii Nauk. Comptes Rendus de l'Académie Bulgare des Sciences, vol. 39, pp. 31–34 (1986)
81. Roth, K.F.: On irregularities of distribution. Mathematika **1**, 73–79 (1954)
82. Roth, K.F.: On irregularities of distribution. IV. Acta Arith. **37**, 67–75 (1980)
83. Schmidt, W.M.: Irregularities of distribution. VII. Acta Arith. **21**, 45–50 (1972)
84. Schmidt, W.M.: Irregularities of distribution X. Number Theory and Algebra, pp. 311–329. Academic, New York (1977)
85. Skriganov, M.M.: Lattices in algebraic number fields and uniform distribution mod 1. Algebra i Anal. **1**, 207–228 (1989) (English translation in Leningr. Math. J. **1**, 535–558 (1990))
86. Skriganov, M.M.: Constructions of uniform distributions in terms of geometry of numbers. Algebra i Anal. **6**, 200–230 (1994) (English translation in St: Petersburg Math. J. **6**, 635–664 (1995))
87. Skriganov, M.M.: Harmonic analysis on totally disconnected groups and irregularities of point distributions. J. Reine Angew. Math. **600**, 25–49 (2006)
88. Skriganov, M.M.: The Khinchin inequality and Chen's theorem. St. Petersburg Math. J. **23**, 761–778 (2012)
89. Sloan, I.H., Woźniakowski, H.: When are quasi-Monte Carlo algorithms efficient for high-dimensional integrals? J. Complex. **14**, 1–33 (1998)
90. Sobol', I.M.: Distribution of points in a cube and approximate evaluation of integrals. Ž. Vyčisl. Mat. Mat. Fiz. (Akademija Nauk SSSR) **7**, 784–802 (1967)
91. Triebel, H.: Bases in Function Spaces, Sampling, Discrepancy, Numerical Integration. European Mathematical Society Publishing House, Zürich (2010)
92. Triebel, H.: Numerical integration and discrepancy. A new approach. Math. Nachr. **283**, 139–159 (2010)
93. van der Corput, J.G.: Verteilungsfunktionen I-II. Proc. Akad. Wet. (Amsterdam) **38**(813–821), 1058–1066 (1935)
94. Weyl, H.: Über die Gleichverteilung mod. Eins. Math. Ann. **77**, 313–352 (1916). (in German)

Part II
Regular Talks

Network Structure Change Point Detection by Posterior Predictive Discrepancy

Lingbin Bian, Tiangang Cui, Georgy Sofronov and Jonathan Keith

Abstract Detecting changes in network structure is important for research into systems as diverse as financial trading networks, social networks and brain connectivity. Here we present novel Bayesian methods for detecting network structure change points. We use the stochastic block model to quantify the likelihood of a network structure and develop a score we call *posterior predictive discrepancy* based on sliding windows to evaluate the model fitness to the data. The parameter space for this model includes unknown latent label vectors assigning network nodes to interacting communities. Monte Carlo techniques based on Gibbs sampling are used to efficiently sample the posterior distributions over this parameter space.

Keywords Bayesian inference · Networks · Sliding window · Stochastic block model · Gibbs sampling

1 Introduction

Time varying network models are used in a wide range of applications, including in neuroscience where they have been used to model functional connectivity of brains such as the modularity models in [1, 2], and to model interactions in social network communities such as Facebook or emails [3]. The detection of changes in commu-

L. Bian (✉) · T. Cui · J. Keith
Monash University, 9 Rainforest Walk, Melbourne, VIC 3800, Australia
e-mail: lingbin.bian@monash.edu

T. Cui
e-mail: tiangang.cui@monash.edu

J. Keith
e-mail: jonathan.keith@monash.edu

G. Sofronov
Macquarie University, 12 Wally's Walk, Sydney, NSW 2109, Australia
e-mail: georgy.sofronov@mq.edu.au

B. Tuffin and P. L'Ecuyer (eds.), *Monte Carlo and Quasi-Monte Carlo Methods*, Springer Proceedings in Mathematics & Statistics 324,
https://doi.org/10.1007/978-3-030-43465-6_5

nities, or more specifically changes in how nodes are allocated to communities, is important to understand functional variation in networks.

There is a wide range of literature exploring network change point analysis in time series. A recent method of network change point detection [4] used *spectral clustering* to partition a network into several connected components. The network structure deviance before and after the candidate change point was evaluated by computing the principal angles between two eigenspaces. The location of the change point was determined in such a way as to minimise a sum of singular values. Another method of network change point analysis named *dynamic connectivity regression* (DCR) [5, 6] used graphical LASSO (GLASSO) [7] to estimate a sparse precision matrix using an L_1-constraint, which forces a large number of edge weights to zero to represent missing edges. Both spectral clustering and DCR were integrated into the random permutation procedure [8] and stationary bootstrap procedure [9] to check whether detected change points were significant. Various criteria have been proposed as test scores to identify candidate change points in network connectivity, including summation of singular values of the network eigenspace (using spectral clustering as mentioned above) and the Bayesian information criterion (BIC) [6] in the context of dynamic connectivity regression. The BIC is a criterion for model selection that includes a penalty term for the number of parameters in the model, the implementation of which is illustrated in [10]. Apart from the greedy algorithm scheme in [5], a frequency-specific method described in [11] applied a multivariate cumulative sum procedure to detect change points. Some methods such as [12–14] mainly focused on large scale network estimation in time series. There are many papers using sliding window methods for observing the time varying network connectivity in time series analysis. For example, [15] tested the equality of the two covariance matrices in a high-dimensional setup within a sliding window to evaluate changes of connectivity in networks. Some other sliding window methods for network connectivity analysis can be found in [16–19]. Detection of communities in networks is also a relevant and topical area of statistics. How communities change or how the nodes in a network are assigned to specific communities is an important problem in characterization of networks. Theory and methods for community detection in networks are described in the works [20–22].

In this paper, we propose a new method to detect network structure change points using Bayesian model fitness assessment. There is a substantial literature on model fitness [23]. For example, West [24] used the cumulative Bayes factor to check for model failure, and Gelman [25] used posterior predictive assessment with a parameter dependent statistic to evaluate model fitness. In this work, we identify change points via checking model fitness to observations within a sliding time window using parameter dependent posterior predictive assessment. Specifically, we propose to use the stochastic block model [21, 26, 27] to quantify the likelihood of a network and Gibbs sampling to sample a posterior distribution derived from this model. The Gibbs sampling approach we adopt is based on the work of Nobile [28] for finite mixture models. We propose a posterior predictive discrepancy method to check model fitness using an adjacency matrix to represent a network. The proposed procedure involves drawing parameters from the posterior distribution and using them to generate a replicated adjacency matrix, then calculating a *disagreement matrix* to quantify the difference between the replicated adjacency matrix and realised adjacency matrix. The score *posterior predictive discrepancy* (PPD) or we call the *posterior predictive discrepancy index* (PPDI) is then evaluated by averaging the fraction of elements in the disagreement matrix that indicate disagreement. We apply another new sliding window to construct a new time series we call the *cumulative discrepancy energy* (CDE). We compute the CDE and use it to define the criterion for change point detection. The CDE increases when change points are contained within the window, and can thus be used to assess whether a statistically significant change point exists within a period of time.

This paper is organized as the follows. Section 2 describes the details of the data time series, and illustrates the models and methodologies we propose for network change point detection. Section 3 contains results of numerical experiments and simulations. Section 4 assesses the advantages and disadvantages of our methods and potential future extensions and improvements.

2 Methods

2.1 The Data Set and Sliding Window Processing

Graphical models is a pictorial representation of pair-wise statistical relations between random variables. Graphical models may involve directed or undirected graphs. Directed graphs are appropriate when the nature of the relationships between variables has a directional aspect, whereas undirected graphs are appropriate for representing bi-directional or non-directional relationships. The methods we developed in this paper apply to both directed and undirected networks.

Consider a collection of N nodes $V = \{v_1, \ldots, v_N\}$. Suppose we observe a collection of N time series $\mathbf{Y} \in \Re^{N\times T}$ where $\mathbf{Y} = (\mathbf{y}_1, \mathbf{y}_2, \ldots, \mathbf{y}_T)$, with one time series corresponding to each node, and observations made at times $\{1, \ldots, T\}$. Correlations between time series indicate direct or indirect interactions between the corresponding nodes; we therefore first process the time series to construct a sequence of graphs in which edges represent temporary correlations.

We apply a sliding window technique with window length W which is considered to be an even number. The window size should be as small as possible. Large window size will limit the detection performance for those change points located closely with each other, while small window size may create statistical complication in the model assessment due to the lack of data sample. Change points may occur only at times $t \in \{M+1, \ldots, T-M\}$ where M is a margin size used to avoid computational and statistical complications. We set the margin size $M = W/2$. For each time point $t \in \{M+1, \ldots, T-M\}$, we define $\mathbf{Y}_t = \{\mathbf{y}_{t-\frac{W}{2}}, \ldots, \mathbf{y}_t, \ldots, \mathbf{y}_{t+\frac{W}{2}-1}\}$ and calculate a sample correlation matrix $\mathbf{R}_t$ within the window $\mathbf{Y}_t$. We set a threshold ε such that only those node pairs (i, j) for which the correlation coefficient $r_{ij}^{(t)} > \varepsilon$ are

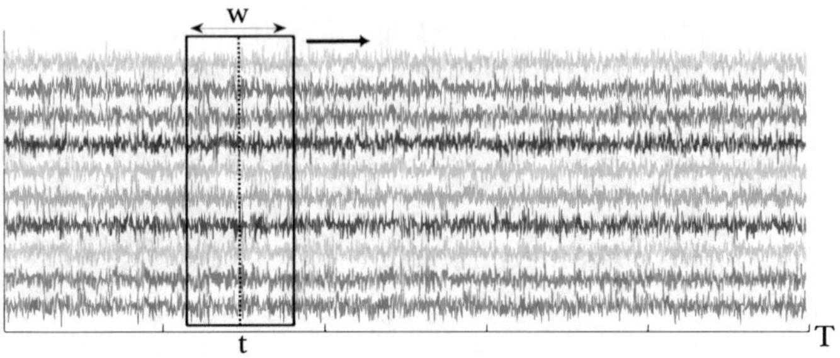

Fig. 1 Parallel time series corresponding to nodes of the network, and a sliding window of width W centred at t. The different coloured time series correspond to signal data for each node

connected by an edge in the edge set E_t representing interacting nodes at time t. It is also convenient to define an *adjacency matrix* $\mathbf{x}_t = (x_{ij}^{(t)})_{i,j=1,\ldots,N}$, where $x_{ij}^{(t)} = 1$ if there is an edge connecting nodes i and j in E_t, and $x_{ij}^{(t)} = 0$ otherwise. For each t, we then have the corresponding sample adjacency matrix $\mathbf{x}_t$ representing interacting nodes during the time window centred at time t. In what follows, we discard the signal data consisting of N time series, and instead consider the sample adjacency matrix $\mathbf{x}_t$ as the realised observation at time t. This sliding window approach is illustrated in Fig. 1.

2.2 The Stochastic Block Model

The stochastic block model is a random process generating networks on a fixed number N of nodes. A defining feature of the model is that nodes are partitioned into K communities, with interactions between nodes in the same community having a different (usually higher) probability than interactions between nodes in different communities. Taking a Bayesian perspective, we suppose that the number of communities K is a random variable drawn from a given prior distribution (for example a Poisson distribution). Determining the value of K appropriate to a given data set is a model selection problem. The stochastic block model first assigns the N nodes into the K communities, then generates edges with a probability determined by the community structure. Mathematically, we denote the *community memberships* (also called the *latent labels*) of the nodes as a random vector $\mathbf{z} = (z_1, \ldots, z_N)$ such that $z_i \in \{1, \ldots, K\}$ denotes the community containing node i. Each z_i independently follows categorical (one trial multinomial) distribution:

$$z_i \sim \text{Categorical}(1; r_1, \ldots, r_K),$$

where r_k is the probability of a node being assigned to community k and $\sum_{k=1}^{K} r_k = 1$. The multinomial probability can be expressed as

$$p(z_i|\mathbf{r}, K) = \prod_{k=1}^{K} r_k^{I_k(z_i)},$$

with the indicator function

$$I_k(z_i) = \begin{cases} 1, & \text{if } z_i = k \\ 0, & \text{if } z_i \neq k. \end{cases}$$

This implies that the N dimensional vector $\mathbf{z}$ is generated with probability

$$p(\mathbf{z}|\mathbf{r}, K) = \prod_{k=1}^{K} r_k^{m_k(\mathbf{z})},$$

where $m_k(\mathbf{z}) = \sum_{i=1}^{N} I_k(z_i)$. The vector $\mathbf{r} = (r_1, \ldots, r_K)$ is assumed to have a K-dimensional Dirichlet prior with density

$$p(\mathbf{r}|K) = N(\boldsymbol{\alpha}) \prod_{k=1}^{K} r_k^{\alpha_k - 1},$$

where the normalization factor with gamma function Γ is

$$N(\boldsymbol{\alpha}) = \frac{\Gamma(\sum_{k=1}^{K} \alpha_k)}{\prod_{k=1}^{K} \Gamma(\alpha_k)}.$$

In this work we suppose $\alpha_i = 1$ for $i = 1, \ldots, K$, so that the prior for $\mathbf{r}$ is uniform on the K-simplex.

Edges between nodes are represented using an adjacency matrix $\mathbf{x} \in \Re^{N \times N}$. Edges can be weighted or unweighted, and x_{ij} can be continuous or discrete. Here we use the *binary edge model*, in which $x_{ij} = 1$ for edges deemed present and $x_{ij} = 0$ for edges deemed absent. We define a *block* $\mathbf{x}_{kl}$ as the sub-matrix of the adjacency matrix comprised of edges connecting the nodes in community k to the nodes in community l. If the graph is undirected, there are $\frac{1}{2}K(K+1)$ blocks. If the graph is directed, there are K^2 blocks.

In the Bayesian presentation of the stochastic block model by MacDaid et al. [26], the likelihood model for edges is given by:

$$p(\mathbf{x}|\boldsymbol{\pi}, \mathbf{z}, K) = \prod_{k,l} p(\mathbf{x}_{kl}|\pi_{kl}, \mathbf{z}, K)$$

and

$$p(\mathbf{x}_{kl}|\pi_{kl}, \mathbf{z}, K) = \prod_{\{i|z_i=k\}} \prod_{\{j|z_j=l\}} p(x_{ij}|\pi_{kl}, \mathbf{z}, K)$$

where $\boldsymbol{\pi} = \{\pi_{kl}\}$ is a $K \times K$ matrix. In the binary edge model, each x_{ij} has a Bernoulli distribution, that is

$$x_{ij}|\pi_{kl}, \mathbf{z}, K \sim \text{Bernoulli}(\pi_{kl}).$$

The π_{kl} independently follow the conjugate Beta prior $\pi_{kl} \sim \text{Beta}(a, b)$. Let $n_{kl}(\mathbf{z}, \mathbf{x})$ be the number of edges in block kl (for the weighted edge model, n_{kl} becomes the sum of the edge weights). For an undirected graph, the number of edges connecting community k and community l is $n_{kl}(\mathbf{z}, \mathbf{x}) = \sum_{i,j|i\leq j, z_i=k, z_j=l} x_{ij}$. For a directed graph, $n_{kl}(\mathbf{z}, \mathbf{x}) = \sum_{i,j|z_i=k, z_j=l} x_{ij}$. We also define $w_{kl}(\mathbf{z})$ to be the maximum possible number of edges in block kl. For the off-diagonal blocks, $w_{kl}(\mathbf{z}) = m_k(\mathbf{z})m_l(\mathbf{z})$. For the diagonal blocks, if the graph is undirected, $w_{kk} = \frac{1}{2}m_k(\mathbf{z})(m_k(\mathbf{z}) + 1)$ (we consider the self-loop here), whereas if the graph is directed, $w_{kk} = m_k(\mathbf{z})^2$. With this notation, the probability associated with the edges of the block $\mathbf{x}_{kl}$ under the binary edge model is

$$p(\mathbf{x}_{kl}|\pi_{kl}, \mathbf{z}, K) = \pi_{kl}^{n_{kl}(\mathbf{z},\mathbf{x})}(1 - \pi_{kl})^{w_{kl}(\mathbf{z})-n_{kl}(\mathbf{z},\mathbf{x})}, \text{ where } 0 < \pi_{kl} < 1.$$

The corresponding conjugate prior is the Beta distribution,

$$\text{Beta}(a, b) = \frac{\pi_{kl}^{a-1}(1 - \pi_{kl})^{b-1}}{B(a, b)},$$

where $B(a, b) = \frac{\Gamma(a)\Gamma(b)}{\Gamma(a+b)}$ is the Beta function.

2.3 The Collapsed Posterior

In the change point detection applications that we consider here, a change point corresponds to a restructuring of the network, that is, a change in the clustering vector $\mathbf{z}$. We are therefore interested in the so called "collapsed" posterior distribution $p(\mathbf{z}|\mathbf{x}, K)$, the form of which we discuss in this section.

We consider K unknown and assign a Poisson random prior with the condition $K > 0$.

$$P(K) = \frac{\lambda^K}{K!}e^{-\lambda}.$$

(In practice we use $\lambda = 1$.) We then have the joint density

$$p(\mathbf{x}, \boldsymbol{\pi}, \mathbf{z}, \mathbf{r}, K) = P(K)p(\mathbf{z}, \mathbf{r}|K)p(\mathbf{x}, \boldsymbol{\pi}|\mathbf{z}).$$

The parameters $\mathbf{r}$ and $\boldsymbol{\pi}$ can be integrated out or "collapsed" to obtain the marginal density $p(\mathbf{x}, \mathbf{z}, K)$.

$$p(\mathbf{z}, K, \mathbf{x}) = P(K) \int p(\mathbf{z}, \mathbf{r}|K) d\mathbf{r} \int p(\mathbf{x}, \boldsymbol{\pi}|\mathbf{z}) d\boldsymbol{\pi},$$

then the posterior for the block-wise model can be expressed as

$$p(\mathbf{z}, K|\mathbf{x}) \propto p(\mathbf{z}, K, \mathbf{x}) = P(K) \int p(\mathbf{z}, \mathbf{r}|K) d\mathbf{r} \prod_{k,l} \int p(\mathbf{x}_{kl}, \pi_{kl}|\mathbf{z}) d\pi_{kl}.$$

The first integral

$$p(\mathbf{z}|K) = \int p(\mathbf{z}, \mathbf{r}|K) d\mathbf{r},$$

where the integral is over the K-simplex, can be calculated via the following procedure:

$$\begin{aligned} p(\mathbf{z}|K) &= \int p(\mathbf{z}, \mathbf{r}|K) d\mathbf{r} \\ &= \int p(\mathbf{r}|K) p(\mathbf{z}|\mathbf{r}, K) d\mathbf{r} \\ &= \frac{\Gamma(\sum_{k=1}^{K} \alpha_k)}{\Gamma(\sum_{k=1}^{K} (\alpha_k + m_k(\mathbf{z}))} \prod_{k=1}^{K} \frac{\Gamma(\alpha_k + m_k(\mathbf{z}))}{\Gamma(\alpha_k)}. \end{aligned}$$

Integrals of the form $\int p(\mathbf{x}_{kl}, \pi_{kl}|\mathbf{z}) d\pi_{kl}$ can be calculated as

$$\begin{aligned} p(\mathbf{x}_{kl}|\mathbf{z}) &= \int_0^1 p(\mathbf{x}_{kl}, \pi_{kl}|\mathbf{z}) d\pi_{kl} \\ &= \int_0^1 p(\pi_{kl}) p(\mathbf{x}_{kl}|\pi_{kl}, \mathbf{z}) d\pi_{kl} \\ &= \frac{B(n_{kl}(\mathbf{z}, \mathbf{x}) + a, w_{kl}(\mathbf{z}) - n_{kl}(\mathbf{z}, \mathbf{x}) + b)}{B(a, b)}. \end{aligned}$$

The derivation of the collapsing procedure is given in Appendix "Derivation of the Collapsing Procedure". Then the collapsed posterior can be expressed as

$$p(\mathbf{z}|\mathbf{x}, K) \propto \frac{1}{K!} \frac{\Gamma(\sum_{k=1}^{K} \alpha_k)}{\Gamma(\sum_{k=1}^{K} (\alpha_k + m_k))} \prod_{k=1}^{K} \frac{\Gamma(\alpha_k + m_k)}{\Gamma(\alpha_k)} \prod_{k,l} \frac{B(n_{kl} + a, w_{kl} - n_{kl} + b)}{B(a, b)}.$$

2.4 Sampling the Parameters from the Posterior

The posterior predictive method we outline below involves sampling parameters from the posterior distribution. The sampled parameters are the latent labels $\mathbf{z}$ and model parameters $\boldsymbol{\pi}$. There are several methods for estimating the latent labels and model parameters of a stochastic block model described in the literature: for example Daudin et al. [29] evaluate the model parameters by point estimation but consider the latent labels in $\mathbf{z}$ as having a distribution, making their approach similar to an EM algorithm. The method of Zhangi et al. [30] uses point estimation for both the model parameters and latent labels. Here we sample the latent labels $\mathbf{z}$ from the collapsed posterior $p(\mathbf{z}|\mathbf{x}, K)$ and then separately sample $\boldsymbol{\pi}$ from the density $p(\boldsymbol{\pi}|\mathbf{x}, \mathbf{z})$.

The estimation of K is a model selection problem [26], which we will not discuss about in this paper. It is convenient to consider the number of communities K to be fixed in the model fitness assessment in this paper (The K is supposed to be given in the numerical experiment in the later section). We use the Gibbs sampler to sample the latent labels $\mathbf{z}$ from the collapsed posterior $p(\mathbf{z}|\mathbf{x}, K)$. For each element z_i and $k \in \{1, \ldots, K\}$, we have

$$p(z_i|z_{-i}, \mathbf{x}, K) = \frac{1}{C} p(z_1, \ldots, z_{i-1}, z_i = k, z_{i+1}, \ldots, z_n|\mathbf{x}),$$

where z_{-i} represents the elements in $\mathbf{z}$ apart from z_i and the normalization term

$$C = p(z_{-i}|\mathbf{x}, K) = \sum_{k=1}^{K} p(z_1, \ldots, z_{i-1}, z_i = k, z_{i+1}, \ldots, z_n|\mathbf{x}).$$

We use the standard Gibbs sampling strategy of cycling through $z_1, \ldots, z_n$, updating each latent variable by drawing from $p(z_i|z_{-i}, \mathbf{x}, K)$.

An alternative to Gibbs sampling is to use Metropolis-Hastings moves based on the allocation sampler [31] to draw parameters $\mathbf{z}$ and K from the posterior. In this approach, a candidate vector of latent labels $\mathbf{z}^*$ is accepted with probability $\min\{1, r\}$, where

$$r = \frac{p(K, \mathbf{z}^*, \mathbf{x})p(\mathbf{z}^* \to \mathbf{z})}{p(K, \mathbf{z}, \mathbf{x})p(\mathbf{z} \to \mathbf{z}^*)}.$$

If the number of communities K is fixed, the proposal $p(\mathbf{z} \to \mathbf{z}^*)$ can be based on three kinds of moves (M1, M2, M3). If K is allowed to vary, one can use a reversible jump strategy or absorption/ejection move. The details of these approaches are illustrated in [31, 33].

To sample the model parameters $\boldsymbol{\pi}$, we first derive the posterior of the model block parameters as the following expression

$$\begin{aligned} p(\pi_{kl}|\mathbf{x}_{kl}, \mathbf{z}) &\propto p(\pi_{kl})p(\mathbf{x}_{kl}|\pi_{kl}, \mathbf{z}) \\ &\propto \pi_{kl}^{a-1}(1-\pi_{kl})^{b-1}\pi_{kl}^{n_{kl}(\mathbf{z},\mathbf{x})}(1-\pi_{kl})^{w_{kl}(\mathbf{z})-n_{kl}(\mathbf{z},\mathbf{x})} \\ &\propto \pi_{kl}^{n_{kl}(\mathbf{z},\mathbf{x})+a-1}(1-\pi_{kl})^{w_{kl}(\mathbf{z})-n_{kl}(\mathbf{z},\mathbf{x})+b-1} \end{aligned}$$

and

$$p(\boldsymbol{\pi}|\mathbf{x}, \mathbf{z}) = \prod_{k,l} p(\pi_{kl}|\mathbf{x}_{kl}, \mathbf{z}).$$

The prior and the likelihood in the above expression is the Beta-Bernoulli conjugate pair. Given the sampled $\mathbf{z}$ we can draw the sample $\boldsymbol{\pi}$ from the above posterior directly.

2.5 Posterior Predictive Discrepancy

Given inferred values of $\mathbf{z}$ and $\boldsymbol{\pi}$ under the assumptive model K, one can draw replicated data $\mathbf{x}^{rep}$ from the posterior predictive distribution $P(\mathbf{x}^{rep}|\mathbf{z}, \boldsymbol{\pi}, K)$. Note that the realised adjacency and replicated adjacency are conditionally independent,

$$P(\mathbf{x}, \mathbf{x}^{rep}|\mathbf{z}, \boldsymbol{\pi}, K) = P(\mathbf{x}^{rep}|\mathbf{z}, \boldsymbol{\pi}, K)P(\mathbf{x}|\mathbf{z}, \boldsymbol{\pi}, K).$$

Multiplying both sides of this equality by $P(\mathbf{z}, \boldsymbol{\pi}|\mathbf{x}, K)/P(\mathbf{x}|\mathbf{z}, \boldsymbol{\pi}, K)$ gives

$$P(\mathbf{x}^{rep}, \mathbf{z}, \boldsymbol{\pi}|\mathbf{x}, K) = P(\mathbf{x}^{rep}|\mathbf{z}, \boldsymbol{\pi}, K)P(\mathbf{z}, \boldsymbol{\pi}|\mathbf{x}, K).$$

Here we use replicated data in the context of posterior predictive assessment [25] to evaluate the fitness of a posited stochastic block model to a realised adjacency matrix. We generate a replicated adjacency matrix by first drawing samples $(\mathbf{z}, \boldsymbol{\pi})$ from the joint posterior $P(\mathbf{z}, \boldsymbol{\pi}|\mathbf{x}, K)$. Specifically, we sample the latent label vector $\mathbf{z}$ from $p(\mathbf{z}|\mathbf{x}, K)$ and model parameter $\boldsymbol{\pi}$ from $p(\boldsymbol{\pi}|\mathbf{x}, \mathbf{z})$ and then draw a replicated adjacency matrix from $P(\mathbf{x}^{rep}|\mathbf{z}, \boldsymbol{\pi}, K)$. We compute a discrepancy function to assess the difference between the replicated data $\mathbf{x}^{rep}$ and the realised observation $\mathbf{x}$, as a measure of model fitness.

In [25], the χ^2 function was used as the discrepancy measure, where the observation was considered as a vector. However, in the stochastic block model, the observation is an adjacency matrix and the sizes of the sub-matrices can vary. In this paper, we propose a *disagreement index* to compare binary adjacency matrices $\mathbf{x}^{rep}$ and $\mathbf{x}$. We use the exclusive OR operator to compute the disagreement matrix between the realised adjacency and replicated adjacency and calculate the fraction of non-zero elements in the disagreement matrix. This disagreement index is denoted $\gamma(\mathbf{x}^{rep}; \mathbf{x})$ and can be considered a parameter-dependent statistic. In mathematical notation, the disagreement index γ is defined as

$$\gamma(\mathbf{x}^{rep}; \mathbf{x}) = \frac{\sum_{i=1,j=1}^{N} (\mathbf{x} \bigoplus \mathbf{x}^{rep})_{ij}}{N^2},$$

where $\bigoplus$ is the exclusive OR operator. In practice we generate S replicated adjacency matrices and compute the average disagreement index, we call *posterior predictive discrepancy index* (PPDI)

$$\overline{\gamma} = \frac{\sum_{i=1}^{S} \gamma(\mathbf{x}^{rep^i}; \mathbf{x})}{S}.$$

2.6 Cumulative Discrepancy Energy via Sliding Window

Our proposed strategy to detect network change points is to assess the fitness of a stochastic block model by computing the discrepancy index $\overline{\gamma}_t$ for each $t \in \{\frac{W}{2} + 1, \ldots, T - \frac{W}{2}\}$. The key insight here is that the fitness of the model is relatively worse when there is a change point within the window used to compute $\mathbf{x}_t$. If there is a change point within the window, the data observed in the left segment and right segment are generated by different network architectures, resulting in poor model fit and a correspondingly high posterior predictive discrepancy index.

We find that the PPDI is greatest when the change point is located in the middle of the window. To identify the most plausible position of a change point, we use another window with window size W_s to smooth the results. We compute the *cumulative discrepancy energy* E_t, given by

$$E_t = \sum_{i=t-\frac{W_s}{2}}^{t+\frac{W_s}{2}-1} \overline{\gamma}_i.$$

We infer the location of change points to be local maxima of the cumulative discrepancy energy, where those maxima rise sufficiently high above the surrounding sequence. The change point detection algorithm can be summarized as the follows.

Algorithm 1 Change point detection by posterior predictive discrepancy

Input: Length of time course T, window size W, number of communities K, observations $\mathbf{Y}$.

for $t = W/2 + 1, \ldots, T - W/2$ **do**

Calculate $\mathbf{Y}_t \to \mathbf{R}_t \to \mathbf{x}_t$.

Draw the samples $\{\mathbf{z}^i, \boldsymbol{\pi}^i\}$ $(i = 1, \ldots, S)$ from the posterior $P(\mathbf{z}, \boldsymbol{\pi} | \mathbf{x}, K)$.

Simulate the replicated set $\mathbf{x}^{rep^i}$ from the predictive distribution $P(\mathbf{x}^{rep} | \mathbf{z}, \boldsymbol{\pi}, K)$.

Calculate the disagreement index $\gamma(\mathbf{x}^{rep^i}; \mathbf{x})$.

Calculate the posterior predictive discrepancy index $\overline{\gamma}_t = \frac{1}{S} \sum_{i=1}^{S} \gamma(\mathbf{x}^{rep^i}; \mathbf{x})$.

end for

for $t = \frac{W}{2} + \frac{W_s}{2} + 1, \ldots, T - \frac{W}{2} - \frac{W_s}{2}$ **do**

Calculate cumulative discrepancy energy $E_t = \sum_{I=t-\frac{W_s}{2}}^{t+\frac{W_s}{2}-1} \overline{\gamma}_I$.

end for

---

3 Simulation

3.1 Generative Model

To validate our approach, we simulate the time series consisting of three data segments from the Gaussian generative model. Within each of the resulting segment, $N = 16$ nodes are assigned to $K = 3$ communities, resulting in membership vectors $\mathbf{z}_1$, $\mathbf{z}_2$ and $\mathbf{z}_3$. Recall these are generated using the Dirichlet-Categorical conjugate pair, that is, component weights $\mathbf{r}_1$, $\mathbf{r}_2$ and $\mathbf{r}_3$ are first drawn from a uniform distribution on the K-simplex and then nodes are assigned to the communities by drawing from the corresponding categorical distributions. Time series data in $\Re^N$ are then simulated for $t = 1, \ldots, T$ by drawing from a multivariate Gaussian distribution $\mathcal{N}(\mathbf{0}, \mathbf{\Sigma})$, with

$$\Sigma_{ij} = \begin{cases} a, & \text{if } i \neq j \text{ and } i \text{ and } j \text{ are in the same communities} \\ 1, & \text{if } i = j \\ b, & \text{if } i \text{ and } j \text{ are in different communities.} \end{cases}$$

In the covariance matrix, a and b follow the uniform distribution, where $a \sim U(0.8, 1)$ and $b \sim U(0, 0.2)$. The resulting covariance matrices for the three segments we denote by $\mathbf{\Sigma}_1$, $\mathbf{\Sigma}_2$ and $\mathbf{\Sigma}_3$. The simulated data $\mathbf{Y} \in \Re^{N \times T}$ can be separated into three segments $(\mathbf{Y}_1, \mathbf{Y}_2, \mathbf{Y}_3)$.

3.2 Effect of Changing the Distance Between Change Points

We simulate the time series for a network with $N = 16$ nodes and $T = 450$ time points with different locations of true change points in four experimental settings. The sliding window size is fixed to be $W = 64$ so that the margin size is $M - 32$.

For the inference, we set the prior of π_{kl} to be $Beta(2, 2)$. During the posterior predictive procedure, according to the convergence performance of the Gibbs sampler, the Gibbs chain of the latent label vectors converges to the stationary distribution within 10 iterations. Then we draw each latent label vector every three complete Gibbs iterations. The posterior prediction replication number S determines the rate of fluctuation of the posterior predictive discrepancy index (PPDI) curve, the smaller the replication number is, the more severely the curve will vibrate. In this demonstration, we set the replication number as $S = 50$. Increasing S would lead to more accurate results, but incur additional computational cost.

The PPDI increases dramatically when the true change point begins to appear at the right of the sliding window and decreases rapidly when the true change point tend to move out the left end of the window. For the cumulative discrepancy energy (CDE), the change point is considered to be at the place where the CDE is a local maximum.

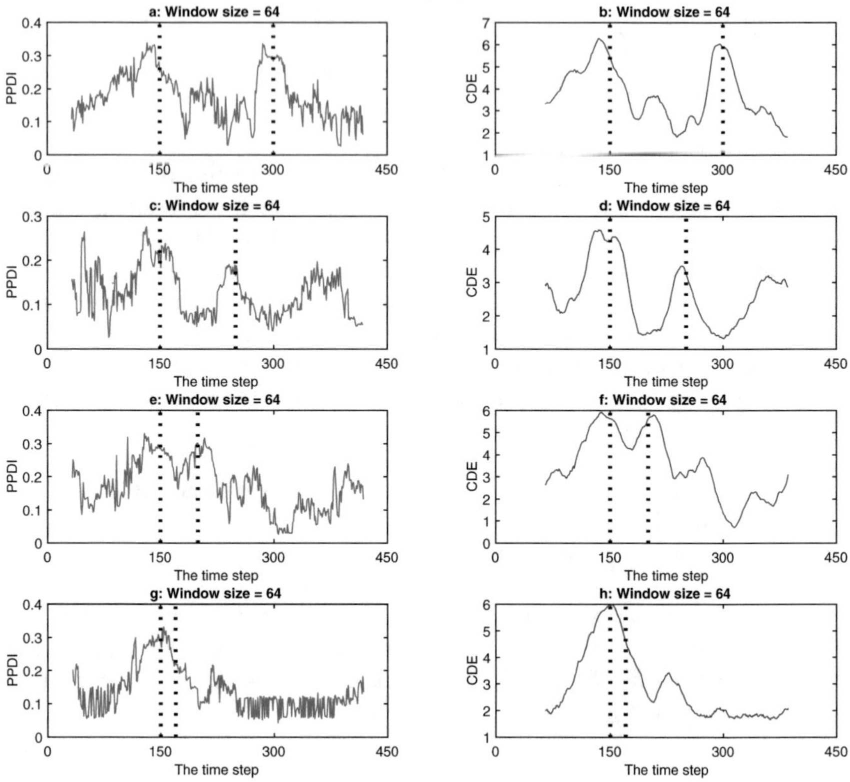

Fig. 2 The vertical lines in the figure represent the various locations of the true change points, the blue curve represents the posterior predictive discrepancy index (PPDI) and the red curve represents the cumulative discrepancy energy (CDE) with window size $W = 64$. **a** PPDI with change points at $t_1 = 150$ and $t_2 = 300$, **b** CDE with change points at $t_1 = 150$ and $t_2 = 300$; **c** PPDI with change points at $t_1 = 150$ and $t_2 = 250$, **d** CDE with change points at $t_1 = 150$ and $t_2 = 250$; **e** PPDI with change points at $t_1 = 150$ and $t_2 = 200$, **f** CDE with change points at $t_1 = 150$ and $t_2 = 200$; **g** PPDI with change points at $t_1 = 150$ and $t_2 = 170$, **h** CDE with change points at $t_1 = 150$ and $t_2 = 170$

In our first setting, true change points are placed at times $t_1 = 150$ and $t_2 = 300$ (see Fig. 2a, b). Note the minimum distance between change points is 150, which is larger than the window size. Consequently, no window can contain more than one change point. We can see from the figure that the two peaks are located around the true change points $t_1 = 150$ and $t_2 = 300$ respectively.

We repeat this experiment with the true change points at $t_1 = 150$ and $t_2 = 250$ in Fig. 2c, d so that the minimum distance between the change points is 100, which is still larger than the window size. We can see that there are two prominent peaks located around the true change points. Next, we set the true change points at $t_1 = 150$ and $t_2 = 200$ Fig. 2e, f, where the minimum distance between the change points is 50 which is slightly smaller than the window size 64. In this situation, the window may contain two change points, so that these windows cross three segments generated by different network architectures. We can still distinguish the two peaks in Fig. 2e, f because the distance of the change points is still large enough. However, in Fig. 2g, h where the change points are $t_1 = 150$ and $t_2 = 170$, we can see that there are only one peak around $t = 150$. In this case, we cannot distinguish two change points because they are closely located with each other.

3.3 *Effect of Changing the Window Size*

To investigate the effect of changing the window size, we set the true change points at $t_1 = 150$ and $t_2 = 300$ for all of the experimental settings. We apply our method with four different window sizes: $W = 24$ in Fig. 3a, b; $W = 32$ in Fig. 3c, d; $W = 48$ in Fig. 3e, f; $W = 64$ in Fig. 3g, h. Reducing the window size will increase the fluctuation of the PPDI and CDE, and renders the change point locations less distinguishable. For $W = 24$, we can see that there are multiple large peaks over the CDE time series.

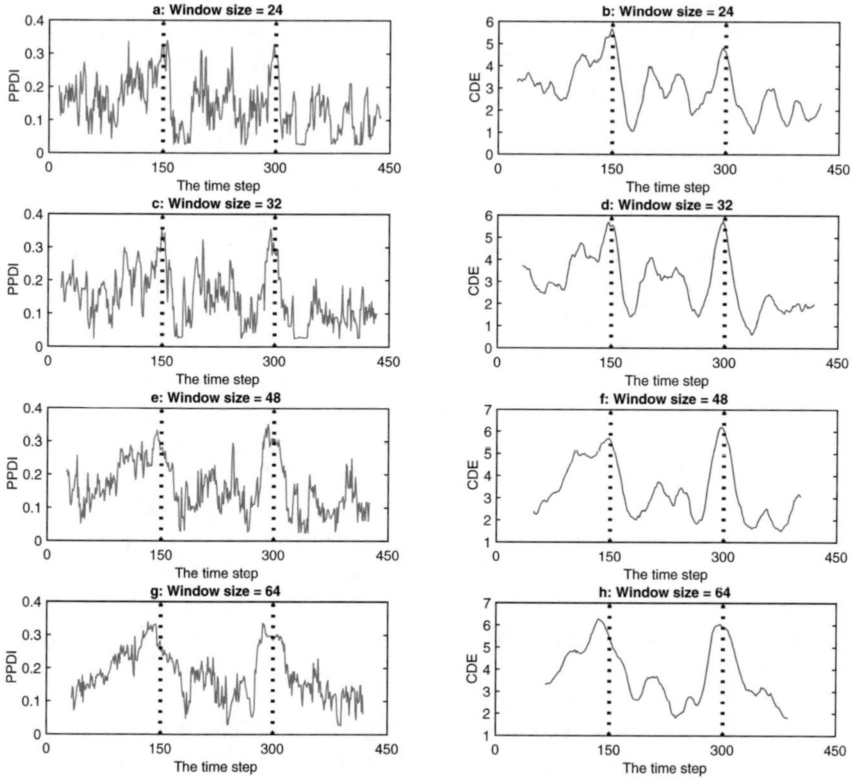

Fig. 3 The vertical lines in the figure represent the locations of the true change points, the blue curve represents the posterior predictive discrepancy index (PPDI) and the red curve represents the cumulative discrepancy energy (CDE) with window sizes 24, 32, 48 and 64

4 Discussion

The method for network structure change point detection described in this paper provides a flexible approach to modelling and estimating community structures among interacting nodes. We consider both the community latent label vector and block model parameters (block edge probabilities) as random variables to be estimated. Structural changes to the networks are reflected by changes in the latent labels and model parameters. By applying a sliding window method, we avoid partitioning the data into sub-segments recursively as in the algorithm of [4]. Compared to the method of evaluating eigen-structure of the network in [4], our approach has several advantages. Our approach is able to be used for both undirected and directed graphs. The method using stochastic block model is more flexible, because different choices of $\boldsymbol{\pi}$ can generate different connection patterns in the adjacency matrix. However, both of the methods have difficulty in detecting change points with close distances.

Ideally, the window size should be as small as possible, which can enhance the ability of detecting those change points located closely. When the window size is small, for example when $W = 24$, there may be false detections, because there are not enough samples of data in the sliding windows. In our current method, the window size cannot be made too small, which may be because we use a threshold to convert the sample correlation matrix into an adjacency matrix. This practice results in the loss of some information regarding the network architecture. If we extend the model to a weighted stochastic block model to fit the data in the future, so that the sample correlation matrix is directly considered as a weighted adjacency matrix, it may be feasible to detect change points at smaller separations and make the higher resolution of detecting the change points. For the majority of the applications in fMRI time series analysis, the time course should be around hundreds of time steps, which is because of the limitation of the sample time interval of the fMRI in the short time experiment. Therefore, the algorithm to analyse the short term network time series is important.

The computational cost of the posterior predictive discrepancy procedure in our method depends mainly on two aspects. The first includes the iterated Gibbs steps used to update the latent variables and the sampling of the model parameter. In our code, calculating $m(\mathbf{z})$ takes $O(N)$ time, calculating the probability of each element z_i to be reassigned into one of K clusters takes $O(K^2 + N^2 + KN)$ time. Therefore, iterating each latent vector $\mathbf{z}$ requires the computational cost of $O((K^2 + N^2 + KN)KN)$, sampling $\boldsymbol{\pi}$ requires $O(K^2 + N^2)$ time. The second is the number of replications needed for the predictive process. Calculating each PPDI requires $O(S)$ time. There is a natural trade off between increasing the replication number and reducing the computational cost.

In this paper, we have not considered the problem of inferring the number of communities K. In real world applications, K is unknown. Determination of K can be considered as a model selection problem, a class of problem for which many methods exist, including [34] in Bayesian statistics. For example, the allocation sampler [31] is an efficient tool for inference of K, and could potentially be integrated into our algorithm. In real word applications, some change points may not occur abruptly, but rather change gradually over time. For solving the gradual changing problem, we may potentially apply a transition matrix to the latent label vectors in the generative model between difference segments to simulate the time series with ground truth of gradual change points. We do not claim that our Gibbs sampling approach is optimal, finding alternative sampling methods is thus another possibility for improving the algorithm. One idea that is worth exploring in the future is to develop efficient sampling methods for inferring high-dimensional latent vectors in larger scale networks.

5 Conclusion

The main contribution of this paper is to demonstrate that posterior predictive discrepancy criterion can be used to detect network structure change point based on time series data. This insight is potentially applicable to a wide range of applications including analysis of fMRI data and large scale social networks.

Acknowledgements The authors are grateful to the Australian Research Council Centre of Excellence for Mathematical and Statistical Frontiers for their support of this project (CE140100049).

Appendix: Derivation of the Collapsing Procedure

We now create a new parameter vector $\boldsymbol{\eta} = \{\alpha_1 + m_1, \ldots, \alpha_K + m_K\}$. We can collapse the integral $\int p(\mathbf{z}, \mathbf{r}|K)d\mathbf{r}$ as the following procedure.

$$\begin{aligned}
\int p(\mathbf{z}, \mathbf{r}|K)d\mathbf{r} &= \int p(\mathbf{r}|K)p(\mathbf{z}|\mathbf{r}, K)d\mathbf{r} \\
&= \int N(\boldsymbol{\alpha}) \prod_{k=1}^{K} r_k^{\alpha_k - 1} \prod_{k=1}^{K} r_k^{m_k} d\mathbf{r} \\
&= \int N(\boldsymbol{\alpha}) \prod_{k=1}^{K} r_k^{\alpha_k + m_k - 1} d\mathbf{r} \\
&= \frac{N(\boldsymbol{\alpha})}{N(\boldsymbol{\eta})} \int N(\boldsymbol{\eta}) \prod_{k=1}^{K} r_k^{\alpha_k + m_k - 1} d\mathbf{r} \\
&= \frac{\Gamma(\sum_{k=1}^{K} \alpha_k)}{\Gamma(\sum_{k=1}^{K} (\alpha_k + m_k))} \prod_{k=1}^{K} \frac{\Gamma(\alpha_k + m_k)}{\Gamma(\alpha_k)}.
\end{aligned}$$

Integral of the form $\int p(\mathbf{x}_{kl}, \pi_{kl}|\mathbf{z})d\pi_{kl}$ can be calculated as

$$\begin{aligned}
\int_0^1 p(\mathbf{x}_{kl}, \pi_{kl}|\mathbf{z})d\pi_{kl} &= \int_0^1 p(\pi_{kl})p(\mathbf{x}_{kl}|\pi_{kl}, \mathbf{z})d\pi_{kl} \\
&= \int_0^1 \frac{\pi_{kl}^{a-1}(1-\pi_{kl})^{b-1}}{B(a,b)}\pi_{kl}^{n_{kl}}(1-\pi_{kl})^{w_{kl}-n_{kl}}d\pi_{kl} \\
&= \int_0^1 \frac{\pi_{kl}^{n_{kl}+a-1}(1-\pi_{kl})^{w_{kl}-n_{kl}+b-1}}{B(a,b)}d\pi_{kl} \\
&= \frac{B(n_{kl}+a, w_{kl}-n_{kl}+b)}{B(a,b)} \\
&\quad \times \int_0^1 \frac{\pi_{kl}^{n_{kl}+a-1}(1-\pi_{kl})^{w_{kl}-n_{kl}+b-1}}{B(n_{kl}+a, w_{kl}-n_{kl}+b)}d\pi_{kl} \\
&= \frac{B(n_{kl}+a, w_{kl}-n_{kl}+b)}{B(a,b)}.
\end{aligned}$$

References

1. Bassett, D.S., Porter, M.A., Wymbs, N.F., Grafton, S.T., Carlson, J.M., Mucha, P.J.: Robust detection of dynamic community structure in networks. CHAOS **23**, 013142 (2013)
2. Bassett, D.S., Wymbs, N.F., Porter, M.A., Mucha, P.J., Carlson, J.M., Grafton, S.T.: Dynamic reconfiguration of human brain networks during learning. PNAS **108**(18), 7641–7646 (2011)
3. Kawash, J., Agarwal, N., özyer, T.: Prediction and Inference from Social Networks and Social Media. Lecture Notes in Social Networks (2017)
4. Cribben, I., Yu, Y.: Estimating whole-brain dynamics by using spectral clustering. J. R. Stat. Soc., Ser. C (Appl. Stat.) **66**, 607–627 (2017)
5. Cribben, I., Haraldsdottir, R., Atlas, L.Y., Wager, T.D., Lindquist, M.A.: Dynamic connectivity regression: determining state-related changes in brain connectivity. NeuroImage **61**, 907–920 (2012)
6. Cribben, I., Wager, T.D., Lindquist, M.A.: Detecting functional connectivity change points for single-subject fMRI data. Front. Comput. Neurosci. **7**, 143 (2013)
7. Friedman, J., Hastie, T., Tibshirani, R.: Sparse inverse covariance estimation with the graphical lasso. Biostatistics **9**(3), 432–441 (2007)
8. Good, P.: Permutation Tests: A Practical Guide to Resampling Methods for Testing Hypotheses. Springer Series in Statistics (2000)
9. Politis, D.N., Romano, J.P.: The stationary bootstrap. J. Am. Stat. Assoc. **89**(428), 1303–1313 (1994)
10. Konishi, S., Kitagawa, G.: Information Criteria and Statistical Modeling. Springer Series in Statistics (2008)
11. Schröder, A.L., Ombao, H.: FreSpeD: frequency-specific change-point detection in epileptic seizure multi-channel EEG data. J. Am. Stat. Assoc. (2015)
12. Frick, K., Munk, A., Sieling, H.: Multiscale change point inference (with discussion). J. R. Stat. Society. Ser. B (Methodol.) **76**, 495–580 (2014)
13. Cho, H., Fryzlewicz, P.: Multiple-change-point detection for high dimensional time series via sparsified binary segmentation. J. R. Stat. Society. Ser. B (Methodol.) **77**, 475–507 (2015)
14. Wang, T., Samworth, R.J.: High-dimensional change point estimation via sparse projection. J. R. Stat. Society. Ser. B (Methodol.) **80**(1), 57–83 (2017)

15. Jeong, S.-O., Pae, C., Park, H.-J.: Connectivity-based change point detection for large-size functional networks NeuroImage **143**, 353–363 (2016)
16. Chang, C., Glover, G.H.: Time-frequency dynamics of resting-state brain connectivity measured with fMRI. NeuroImage **50**, 81–98 (2010)
17. Handwerker, D.A., Roopchansingh, V., Gonzalez-Castillo, J., Bandettini, P.A.: Periodic changes in fMRI connectivity. NeuroImage **63**, 1712–1719 (2012)
18. Monti, R.P., Hellyer, P., Sharp, D., Leech, R., Anagnostopoulos, C., Montana, G.: Estimating time-varying brain connectivity networks from functional MRI time series. NeuroImage **103**, 427–443 (2014)
19. Allen, E.A., Damaraju, E., Plis, S.M., Erhardt, E.B., Eichele, T., Calhoun, V.D.: Tracking whole-brain connectivity dynamics in the resting state. Cereb. Cortex **24**, 663–676 (2014)
20. Newman, M.E.J.: Modularity and community structure in networks. PNAS **103**(23), 8577–8582 (2006)
21. Wang, Y.X.R., Bickel, P.J.: Likelihood-based model selection for stochastic block models. Ann. Stat. **45**(2), 500–528 (2017)
22. Jin, J.: Fast community detection by SCORE. Ann. Stat. **43**(1), 57–89 (2015)
23. Rubin, D.B.: Bayesianly justifiable and relevant frequency calculations for the applied statistician. Ann. Stat. **12**(4), 1151–1172 (1984)
24. West, M.: Bayesian model monitoring. J. R. Stat. Society. Ser. B (Methodol.) **48**(1), 70–78 (1986)
25. Gelman, A., Meng, X.-L., Stern, H.: Posterior predictive assessment of model fitness via realised discrepancies. Stat. Sin. **6**, 733–807 (1996)
26. MacDaid, A.F., Murphy, T.B., Friel, N., Hurley, N.J.: Improved Bayesian inference for the stochastic block model with application to large networks. Comput. Stat. Data Anal. **60**, 12–31 (2012)
27. Ridder, S.D., Vandermarliere, B., Ryckebusch, J.: Detection and localization of change points in temporal networks with the aid of stochastic block models. J. Stat. Mech.: Theory Exp. (2016)
28. Nobile, A.: Bayesian analysis of finite mixture distributions. Ph.D. Dissertation (1994)
29. Daudin, J.-J., Picard, F., Robin, S.: A mixture model for random graphs. Stat. Comput. **18**, 173–183 (2008)
30. Zanghi, H., Ambroise, C., Miele, V.: Fast online graph clustering via Erdös-Rényi mixture. Pattern Recognit. **41**, 3592–3599 (2008)
31. Nobile, A., Fearnside, A.T.: Bayesian finite mixtures with an unknown number of components: The allocation sampler. Stat. Comput. **17**, 147–162 (2007)
32. Luxburg, U.V.: A tutorial on spectral clustering. Stat. Comput. **17**, 395–416 (2007)
33. Wyse, J., Friel, N.: Block clustering with collapsed latent block models. Stat. Comput. **22**, 415–428 (2012)
34. Latouche, P., Birmele, E., Ambroise, C.: Variational Bayesian inference and complexity control for stochastic block models. Stat. Model. **12**, 93–115 (2012)

Stochastic Methods for Solving High-Dimensional Partial Differential Equations

Marie Billaud-Friess, Arthur Macherey, Anthony Nouy and Clémentine Prieur

Abstract We propose algorithms for solving high-dimensional Partial Differential Equations (PDEs) that combine a probabilistic interpretation of PDEs, through Feynman–Kac representation, with sparse interpolation. Monte-Carlo methods and time-integration schemes are used to estimate pointwise evaluations of the solution of a PDE. We use a sequential control variates algorithm, where control variates are constructed based on successive approximations of the solution of the PDE. Two different algorithms are proposed, combining in different ways the sequential control variates algorithm and adaptive sparse interpolation. Numerical examples will illustrate the behavior of these algorithms.

Keywords Stochastic algorithms · High dimensional PDEs · Adaptive sparse interpolation

1 Introduction

We consider the solution of an elliptic partial differential equation

$$
\begin{aligned}
\mathscr{A}(u) &= g \quad \text{in} \quad \mathscr{D}, \\
u &= f \quad \text{on} \quad \partial\mathscr{D},
\end{aligned}
\tag{1}
$$

M. Billaud-Friess (✉) · A. Macherey · A. Nouy
Centrale Nantes, LMJL, UMR CNRS 6629, 1 rue de la Noë, 44321 Nantes, France
e-mail: marie.billaud-friess@ec-nantes.fr

A. Macherey
e-mail: arthur.macherey@ec-nantes.fr

A. Nouy
e-mail: anthony.nouy@ec-nantes.fr

A. Macherey · C. Prieur
Inria, CNRS, Grenoble INP* (*Institute of Engineering Université Grenoble Alpes), LJK, Université Grenoble Alpes, 38000 Grenoble, France
e-mail: clementine.prieur@univ-grenoble-alpes.fr

B. Tuffin and P. L'Ecuyer (eds.), *Monte Carlo and Quasi-Monte Carlo Methods*, Springer Proceedings in Mathematics & Statistics 324,
https://doi.org/10.1007/978-3-030-43465-6_6

where $u : \overline{\mathscr{D}} \to \mathbb{R}$ is a real-valued function, and $\mathscr{D}$ is an open bounded domain in $\mathbb{R}^d$. $\mathscr{A}$ is an elliptic linear differential operator and $f : \partial\mathscr{D} \to \mathbb{R}$, $g : \overline{\mathscr{D}} \to \mathbb{R}$ are respectively the boundary condition and the source term of the PDE.

We are interested in approximating the solution of (1) up to a given precision. For high dimensional PDEs ($d \gg 1$), this requires suitable approximation formats such as sparse tensors [4, 21] or low-rank tensors [1, 15, 16, 19, 20]. Also, this requires algorithms that provide approximations in a given approximation format. Approximations are typically provided by Galerkin projections using variational formulations of PDEs. Another path consists in using a probabilistic representation of the solution u through Feynman–Kac formula, and Monte-Carlo methods to provide estimations of pointwise evaluations of u (see e.g., [14]). This allows to compute approximations in a given approximation format through classical interpolation or regression [2, 3, 22]. In [11, 12], the authors consider interpolations on fixed polynomial spaces and propose a sequential control variates method for improving the performance of Monte-Carlo estimation. In this paper, we propose algorithms that combine this variance reduction method with adaptive sparse interpolation [5, 6].

The outline is as follows. In Sect. 2, we recall the theoretical and numerical aspects associated to probabilistic tools for estimating the solution of (1). We also present the sequential control variates algorithm introduced in [11, 12]. In Sect. 3 we introduce sparse polynomial interpolation methods and present a classical adaptive algorithm. In Sect. 4, we present two algorithms combining the sequential control variates algorithm from Sect. 2 and adaptive sparse polynomial interpolation. Finally, numerical results are presented in Sect. 4.

2 Probabilistic Tools for Solving PDEs

We consider the problem (1) with a linear partial differential operator defined by $\mathscr{A}(u) = -\mathscr{L}(u) + ku$, where k is a real valued function defined on $\overline{\mathscr{D}}$, and where

$$\mathscr{L}(u)(x) = \frac{1}{2}\sum_{i,j=1}^{d} (\sigma(x)\sigma(x)^T)_{ij}\partial^2_{x_i x_j} u(x) + \sum_{i=1}^{d} b_i(x)\partial_{x_i} u(x) \tag{2}$$

is the *infinitesimal generator* associated to the d-dimensional diffusion process X^x solution of the stochastic differential equation

$$dX_t^x = b(X_t^x)dt + \sigma(X_t^x)dW_t, \quad X_0^x = x \in \overline{\mathscr{D}}, \tag{3}$$

where W is a d-dimensional Brownian motion and $b := (b_1, \ldots, b_d)^T : \mathbb{R}^d \to \mathbb{R}^d$ and $\sigma : \mathbb{R}^d \to \mathbb{R}^{d\times d}$ stand for the drift and the diffusion respectively.

2.1 Pointwise Evaluations of the Solution

The following theorem recalls the Feynman–Kac formula (see [8, Theorem 2.4] or [9, Theorem 2.4] and the references therein) that provides a probabilistic representation of $u(x)$, the solution of (1) evaluated at $x \in \overline{\mathscr{D}}$.

Theorem 1 (Feynman–Kac formula) *Assume that*

$(H1)$ *$\mathscr{D}$ is an open connected bounded domain of $\mathbb{R}^d$, regular in the sense that, if $\tau^x = \inf\{s > 0 \ : \ X_s^x \notin \mathscr{D}\}$ is the first exit time of $\mathscr{D}$ for the process X^x, we have*

$$\mathbb{P}(\tau^x = 0) = 1, \quad x \in \partial\mathscr{D},$$

$(H2)$ *b, σ are Lipschitz functions,*

$(H3)$ *f is continuous on $\partial\mathscr{D}$, g and $k \geq 0$ are Hölder-continuous functions on $\overline{\mathscr{D}}$,*

$(H4)$ *(uniform ellipticity assumption) there exists $c > 0$ such that*

$$\sum_{i,j=1}^{d} \left(\sigma(x)\sigma(x)^T\right)_{ij} \xi_i\xi_j \geq c\sum_{i=1}^{d} \xi_i^2, \quad \xi \in \mathbb{R}^d, \ x \in \overline{\mathscr{D}}.$$

Then, there exists a unique solution of (1) *in $\mathscr{C}(\overline{\mathscr{D}}) \cap \mathscr{C}^2(\mathscr{D})$, which satisfies for all $x \in \overline{\mathscr{D}}$*

$$u(x) = \mathbb{E}\left[F(u, X^x)\right] \tag{4}$$

where

$$F(u, X^x) = u(X_{\tau^x}^x) \exp\left(-\int_0^{\tau^x} k(X_t^x)dt\right) + \int_0^{\tau^x} \mathscr{A}(u)(X_t^x) \exp\left(-\int_0^t k(X_s^x)ds\right) dt,$$

with $u(X_{\tau^x}^x) = f(X_{\tau^x}^x)$ and $\mathscr{A}(u)(X_t^x) = g(X_t^x)$.

Note that $F(u, X^x)$ in (4) only depends on the values of u on ∂D and $\mathscr{A}(u)$ on D, which are the given data f and g respectively. A Monte-Carlo method can then be used to estimate $u(x)$ using (4), which relies on the simulation of independent samples of an approximation of the stochastic process X^x. This process is here approximated by an Euler–Maruyama scheme. More precisely, letting $t_n = n\Delta t$, $n \in \mathbb{N}$, X^x is approximated by a piecewise constant process $X^{x,\Delta t}$, where $X_t^{x,\Delta t} = X_n^{x,\Delta t}$ for $t \in [t_n, t_{n+1}[$ and

$$\begin{aligned} X_{n+1}^{x,\Delta t} &= X_n^{x,\Delta t} + \Delta t\, b(X_n^{x,\Delta t}) + \sigma(X_n^{x,\Delta t})\, \Delta W_n, \\ X_0^{x,\Delta t} &= x. \end{aligned} \tag{5}$$

Here $\Delta W_n = W_{n+1} - W_n$ is an increment of the standard Brownian motion. For details on time-integration schemes, the reader can refer to [17]. Letting

$\{X^{x,\Delta t}(\omega_m)\}_{m=1}^M$ be independent samples of $X^{x,\Delta t}$, we obtain an estimation $u_{\Delta t,M}(x)$ of $u(x)$ defined as

$$\begin{aligned} u_{\Delta t,M}(x) &:= \frac{1}{M}\sum_{m=1}^{M} F\left(u, X^{x,\Delta t}(\omega_m)\right) \\ &= \frac{1}{M}\sum_{m=1}^{M}\Bigg[f(X^{x,\Delta t}_{\tau^{x,\Delta t}}(\omega_m)) \exp\left(-\int_0^{\tau^{x,\Delta t}} k(X_t^{x,\Delta t}(\omega_m))dt\right) \\ &+ \int_0^{\tau^{x,\Delta t}} g(X_t^{x,\Delta t}(\omega_m)) \exp\left(-\int_0^t k(X_s^{x,\Delta t}(\omega_m))ds\right) dt\Bigg] \end{aligned} \tag{6}$$

where $\tau^{x,\Delta t}$ is the first exit time of D for the process $X^{x,\Delta t}(\omega_m)$, given by

$$\tau^{x,\Delta t} = \inf\left\{t > 0 \,:\, X_t^{x,\Delta t} \notin \mathscr{D}\right\} = \min\left\{t_n > 0 \,:\, X_{t_n}^{x,\Delta t} \notin \mathscr{D}\right\}.$$

Remark 1 In practice, f has to be defined over $\mathbb{R}^d$ and not only on the boundary $\partial\mathscr{D}$. Indeed, although $X^x_{\tau^x} \in \partial\mathscr{D}$ with probability one, $X^{x,\Delta t}_{\tau^{x,\Delta t}} \in \mathbb{R}^d \setminus \overline{\mathscr{D}}$ with probability one.

The error can be decomposed in two terms

$$\begin{aligned} u(x) - u_{\Delta t,M}(x) = &\overbrace{u(x) - \mathbb{E}\left[F\left(u, X^{x,\Delta t}\right)\right]}^{\varepsilon_{\Delta t}} \\ &\underbrace{+\,\mathbb{E}\left[F\left(u, X^{x,\Delta t}\right)\right] - \frac{1}{M}\sum_{m=1}^{M} F\left(u, X^{x,\Delta t}(\omega_m)\right)}_{\varepsilon_{MC}}, \end{aligned} \tag{7}$$

where $\varepsilon_{\Delta t}$ is the time integration error and ε_{MC} is the Monte-Carlo estimation error. Before discussing the contribution of each of both terms to the error, let us introduce the following additional assumption, which ensures that $\mathscr{D}$ does not have singular points.[1]

$(H5)$ Each point of $\partial\mathscr{D}$ satisfies the *exterior cone condition* which means that, for all $x \in \partial\mathscr{D}$, there exists a finite right circular cone K, with vertex x, such that $\overline{K} \cap \overline{\mathscr{D}} = \{x\}$.

Under assumptions $(H1)$–$(H5)$, it can be proven [12, Sect. 4.1] that the time integration error $\varepsilon_{\Delta t}$ converges to zero. It can be improved to $O(\Delta t^{1/2})$ by adding differentiability assumptions on the boundary [13]. The estimation error ε_{MC} is a random variable with zero mean and standard deviation converging as $O(M^{-1/2})$. The computational complexity for computing a pointwise evaluation of $u_{\Delta t,M}(x)$ is

[1]Note that together with $(H4)$, assumption $(H5)$ implies $(H1)$ (see [12, Sect. 4.1] for details), so that the set of hypotheses $(H1)$–$(H5)$ could be reduced to $(H2)$–$(H5)$.

in $O\left(M\Delta t^{-1}\right)$ in expectation for Δt sufficiently small,[2] so that the computational complexity for achieving a precision ε (root mean squared error) behaves as $O(\varepsilon^{-4})$. This does not allow to obtain a very high accuracy in a reasonable computational time. The convergence with Δt can be improved to $O(\Delta t)$ by suitable boundary corrections [13], therefore yielding a convergence in $O(\varepsilon^{-3})$. To further improve the convergence, high-order integration schemes could be considered (see [17] for a survey). Also, variance reduction methods can be used to further improve the convergence, such as antithetic variables, importance sampling, control variates (see [14]). Multilevel Monte-Carlo [10] can be considered as a variance reduction method using several control variates (associated with processes $X^{x,\Delta t_k}$ using different time discretizations). Here, we rely on the sequential control variates algorithm proposed in [11] and analyzed in [12]. This algorithm constructs a sequence of approximations of u. At each iteration of the algorithm, the current approximation is used as a control variate for the estimation of u through Feynman–Kac formula.

2.2 A Sequential Control Variates Algorithm

Here we recall the sequential control variates algorithm introduced in [11] in a general interpolation framework. We let $V_\Lambda \subset \mathscr{C}^2(\overline{\mathscr{D}})$ be an approximation space of finite dimension $\#\Lambda$ and let $\mathscr{I}_\Lambda : \mathbb{R}^{\mathscr{D}} \to V_\Lambda$ be the interpolation operator associated with a unisolvent grid $\Gamma_\Lambda = \{x_{\boldsymbol{\nu}} : \boldsymbol{\nu} \in \Lambda\}$. We let $(l_\nu)_{\boldsymbol{\nu}\in\Lambda}$ denote the (unique) basis of V_Λ that satisfies the interpolation property $l_\nu(x_{\boldsymbol{\mu}}) = \delta_{\boldsymbol{\nu\mu}}$ for all $\boldsymbol{\nu}, \boldsymbol{\mu} \in \Lambda$. The interpolation $\mathscr{I}_\Lambda(w) = \sum_{\nu\in\Lambda} w(x_\nu) l_\nu(x)$ of function w is then the unique function in V_Λ such that

$$\mathscr{I}_\Lambda(w)(x_\nu) = w(x_\nu), \quad \nu \in \Lambda.$$

The following algorithm provides a sequence of approximations $(\tilde{u}^k)_{k\geq 1}$ of u in V_Λ, which are defined by $\tilde{u}^k = \tilde{u}^{k-1} + \tilde{e}^k$, where $\tilde{e}^k$ is an approximation of e^k, solution of

$$\begin{aligned} \mathscr{A}(e^k)(x) &= g(x) - \mathscr{A}(\tilde{u}^{k-1})(x), \; x \in \mathscr{D}, \\ e^k(x) &= f(x) - \tilde{u}^{k-1}(x), \qquad x \in \partial\mathscr{D}. \end{aligned}$$

Note that e^k admits a Feyman–Kac representation $e^k(x) = \mathbb{E}(F(e^k, X^x))$, where $F(e^k, X^x)$ depends on the residuals $g - \mathscr{A}(\tilde{u}^{k-1})$ on $\mathscr{D}$ and $f - \tilde{u}^{k-1}$ on $\partial\mathscr{D}$. The approximation $\tilde{e}^k$ is then defined as the interpolation $\mathscr{I}_\Lambda(e^k_{\Delta t,M})$ of the Monte-Carlo estimate $e^k_{\Delta t,M}(x)$ of $e^k_{\Delta t}(x) = \mathbb{E}(F(e^k, X^{x,\Delta t}))$ (using M samples of $X^{x,\Delta t}$).

[2] A realization of $X^{x,\Delta t}$ over the time interval $[0, \tau^{x,\Delta t}]$ can be computed in $O\left(\tau^{x,\Delta t}\Delta t^{-1}\right)$. Then, the complexity to evaluate $u_{\Delta t,M}(x)$ is in $O(\mathbb{E}(\tau^{x,\Delta t})M\Delta t^{-1})$ in expectation. Under $(H1)$–$(H5)$, it is stated in the proof of [12, Theorem 4.2] that $\sup_x \mathbb{E}[\tau^{x,\Delta t}] \leq C$ with C independent of Δt for Δt sufficiently small.

Algorithm 1 (Sequential control variates algorithm)

1: Set $\tilde{u}^0 = 0$, $k = 1$ and $S = 0$.
2: **while** $k \leq K$ and $S < n_s$ **do**
3: Compute $e^k_{\Delta t,M}(x_\nu)$ for $x_\nu \in \Gamma_\Lambda$.
4: Compute $\tilde{e}^k = \mathscr{I}_\Lambda(e^k_{\Delta t,M}) = \sum_{\nu\in\Lambda} e^k_{\Delta t,M}(x_\nu) l_\nu(x)$.
5: Update $\tilde{u}^k = \tilde{u}^{k-1} + \tilde{e}^k$.
6: If $\|\tilde{u}^k - \tilde{u}^{k-1}\|_2 \leq \varepsilon_{tol}\|\tilde{u}^{k-1}\|_2$ then $S = S + 1$ else $S = 0$.
7: Set $k = k + 1$.
8: **end while**

For practical reasons, Algorithm 1 is stopped using an heuristic error criterion based on stagnation. This criterion is satisfied when the desired tolerance ε_{tol} is reached for n_s successive iterations (in practice we chose $n_s = 5$).

Now let us provide some convergence results for Algorithm 1. To that goal, we introduce the time integration error at point x for a function h

$$e^{\Delta t}(h, x) = \mathbb{E}[F(h, X^{\Delta t,x})] - \mathbb{E}[F\left(h, X^x\right)]. \tag{8}$$

Then the following theorem [12, Theorem 3.1] gives a control of the error in expectation.

Theorem 2 *Assuming* $(H2)$–$(H5)$, *it holds*

$$\sup_{\nu\in\Lambda}\left|\mathbb{E}\left[\tilde{u}^{n+1}(x_\nu) - u(x_\nu)\right]\right| \leqslant C(\Delta t, \Lambda)\sup_{\nu\in\Lambda}\left|\mathbb{E}\left[\tilde{u}^n(x_\nu) - u(x_\nu)\right]\right| + C_1(\Delta t, \Lambda)$$

with $C(\Delta t, \Lambda) = \sup_{\nu\in\Lambda}\sum_{\mu\in\Lambda} |e^{\Delta t}(l_\mu, x_\nu)|$ *and* $C_1(\Delta t, \Lambda) = \sup_{\nu\in\Lambda}\left|e^{\Delta t}(u - \mathscr{I}_\Lambda(u), x_\nu)\right|$.
Moreover if $C(\Delta t, \Lambda) < 1$, *it holds*

$$\limsup_{n\to\infty}\sup_{\nu\in\Lambda}\left|\mathbb{E}\left[\tilde{u}^n(x_\nu) - u(x_\nu)\right]\right| \leqslant \frac{C_1(\Delta t, \Lambda)}{1 - C(\Delta t, \Lambda)}. \tag{9}$$

The condition $C(\Delta t, \Lambda) < 1$ implies that in practice Δt should be chosen sufficiently small [12, Theorem 4.2]. Under this condition, the error at interpolation points uniformly converges geometrically up to a threshold term depending on time integration errors for interpolation functions l_ν and the interpolation error $u - \mathscr{I}_\Lambda(u)$.

Theorem 2 provides a convergence result at interpolation points. Below, we provide a corollary to this theorem that provides a convergence result in $L^\infty(\mathscr{D})$. This result involves the Lebesgue constants in L^∞-norm associated to $\mathscr{I}_\Lambda$, defined by

$$\mathscr{L}_\Lambda = \sup_{v\in\mathscr{C}^0(\overline{\mathscr{D}})}\frac{\|\mathscr{I}_\Lambda(v)\|_\infty}{\|v\|_\infty},$$

and such that for any $v \in \mathscr{C}^0(\overline{\mathscr{D}})$,

$$\|v - \mathscr{I}_\Lambda(v)\|_\infty \leq (1 + \mathscr{L}_\Lambda) \inf_{w \in V_\Lambda} \|v - w\|_\infty. \tag{10}$$

Throughout this article, we adopt the convention that supremum exclude elements with norm 0. We recall also that the L^∞ Lebesgue constant can be expressed as $\mathscr{L}_\Lambda = \sup_{x \in \overline{\mathscr{D}}} \sum_{\nu \in \Lambda} |l_\nu(x)|$.

Corollary 1 (Convergence in L^∞) *Assuming (H2)–(H5), one has*

$$\limsup_{n \to \infty} \|\mathbb{E}\left[\tilde{u}^n - u\right]\|_\infty \leqslant \frac{C_1(\Delta t, \Lambda)}{1 - C(\Delta t, \Lambda)} \mathscr{L}_\Lambda + \|u - \mathscr{I}_\Lambda(u)\|_\infty. \tag{11}$$

Proof By triangular inequality, we have

$$\|\mathbb{E}\left[\tilde{u}^n - u\right]\|_\infty \leqslant \|\mathbb{E}\left[\tilde{u}^n - \mathscr{I}_\Lambda(u)\right]\|_\infty + \|\mathscr{I}_\Lambda(u) - u\|_\infty.$$

We can build a continuous function w such that $w(x_\nu) = \mathbb{E}\left[\tilde{u}^n(x_\nu) - u(x_\nu)\right]$ for all $\nu \in \Lambda$, and such that

$$\|w\|_\infty = \sup_{\nu \in \Lambda} |w(x_\nu)| = \sup_{\boldsymbol{\nu} \in \Lambda} \left|\mathbb{E}\left[\tilde{u}^n(x_{\boldsymbol{\nu}}) - u(x_{\boldsymbol{\nu}})\right]\right|.$$

We have then

$$\|\mathbb{E}\left[\tilde{u}^n - \mathscr{I}_\Lambda(u)\right]\|_\infty = \|\mathscr{I}_\Lambda(w)\|_\infty \leq \mathscr{L}_\Lambda \|w\|_\infty.$$

The result follows from the definition of the function w and Theorem 2. □

Remark 2 Since for bounded domains $\mathscr{D}$, we have

$$\|v\|_2 \leq |\mathscr{D}|^{1/2} \|v\|_\infty,$$

for all v in $\mathscr{C}^0(\overline{\mathscr{D}})$, where $|\mathscr{D}|$ denotes the Lebesgue measure of $\mathscr{D}$, we can deduce the convergence results in L^2 norm from those in L^∞ norm.

3 Adaptive Sparse Interpolation

We here present sparse interpolation methods following [5, 6].

3.1 Sparse Interpolation

For $1 \le i \le d$, we let $\{\varphi_k^{(i)}\}_{k\in\mathbb{N}_0}$ be a univariate polynomial basis, where $\varphi_k^{(i)}(x_i)$ is a polynomial of degree k. For a multi-index $\nu = (\nu_1, \dots, \nu_d) \in \mathbb{N}_0^d$, we introduce the multivariate polynomial

$$\varphi_{\boldsymbol{\nu}}(x) = \prod_{i=1}^{d} \varphi_{\nu_i}^{(i)}(x_i).$$

For a subset $\Lambda \subset \mathbb{N}^d$, we let $\mathscr{P}_\Lambda = \text{span}\{\varphi_{\boldsymbol{\nu}} : \nu \in \Lambda\}$. A subset Λ is said to be *downward closed* if

$$\forall \boldsymbol{\nu} \in \Lambda,\ \boldsymbol{\mu} \le \boldsymbol{\nu} \Rightarrow \boldsymbol{\mu} \in \Lambda.$$

If Λ is downward closed, then the polynomial space $\mathscr{P}_\Lambda$ does not depend on the choice of univariate polynomial bases and is such that $\mathscr{P}_\Lambda = \text{span}\{x^\nu : \nu \in \Lambda\}$, with $x^\nu = x_1^{\nu_1} \dots x_d^{\nu_d}$.

In the case where $\mathscr{D} = \mathscr{D}_1 \times \cdots \times \mathscr{D}_d$, we can choose for $\{\varphi_k^{(i)}\}_{k\in\mathbb{N}_0}$ an orthonormal basis in $L^2(\mathscr{D}_i)$ (i.e. a rescaled and shifted Legendre basis). Then $\{\varphi_\nu\}_{\nu\in\mathbb{N}_0^d}$ is an orthonormal basis of $L^2(\mathscr{D})$. To define a set of points Γ_Λ unisolvent for $\mathscr{P}_\Lambda$, we can proceed as follows. For each dimension $1 \le i \le d$, we introduce a sequence of points $\{z_k^{(i)}\}_{k\in\mathbb{N}_0}$ in $\overline{\mathscr{D}_i}$ such that for any $p \ge 0$, $\Gamma_p^{(i)} = \{z_k^{(i)}\}_{k=0}^p$ is unisolvent for $\mathscr{P}_p = \text{span}\{\varphi_k^{(i)} : 0 \le k \le p\}$, therefore defining an interpolation operator $\mathscr{I}_p^{(i)}$. Then we let

$$\Gamma_\Lambda = \{z_\nu = (z_{\nu_1}^{(1)}, \dots, z_{\nu_d}^{(d)}) : \nu \in \Lambda\} \subset \overline{\mathscr{D}}.$$

This construction is interesting for adaptive sparse algorithms since for an increasing sequence of subsets Λ_n, we obtain an increasing sequence of sets Γ_{Λ_n}, and the computation of the interpolation on $\mathscr{P}_{\Lambda_n}$ only requires the evaluation of the function on the new set of points $\Gamma_{\Lambda_n} \setminus \Gamma_{\Lambda_{n-1}}$. Also, with such a construction, we have the following property of the Lebesgue constant of $\mathscr{I}_\Lambda$ in L^∞-norm. This result is directly taken from [6, Sect. 3].

Proposition 1 *If for each dimension* $1 \le i \le d$*, the sequence of points* $\{z_k^{(i)}\}_{k\in\mathbb{N}_0}$ *is such that the interpolation operator* $\mathscr{I}_p^{(i)}$ *has a Lebesgue constant* $\mathscr{L}_p \le (p+1)^s$ *for some* $s > 0$*, then for any downward closed set* Λ*, the Lebesgue constant* $\mathscr{L}_\Lambda$ *satisfies*

$$\mathscr{L}_\Lambda \le (\#\Lambda)^{s+1}. \tag{12}$$

Leja points or magic points [18] are examples of sequences of points such that the interpolation operators $\mathscr{I}_p^{(i)}$ have Lebesgue constants not growing too fast with p. For a given Λ with $\rho_i := \max_{\nu \in \Lambda} \nu_i$, it is possible to construct univariate interpolation grids $\Gamma_{\rho_i}^{(i)}$ with better properties (e.g., Chebychev points), therefore resulting in better properties for the associated interpolation operator $\mathscr{I}_\Lambda$. However for Chebychev points, e.g., $\rho_i \leq \rho_i'$ does not ensure $\Gamma_{\rho_i}^{(i)} \subset \Gamma_{\rho_i'}^{(i)}$. Thus with such univariate grids, an increasing sequence of sets Λ_n will not be associated with an increasing sequence of sets Γ_{Λ_n}, and the evaluations of the function will not be completely recycled in adaptive algorithms. However, for some of the algorithms described in Sect. 4, this is not an issue as evaluations can not be recycled anyway.

Note that for general domains $\mathscr{D}$ which are not the product of intervals, the above constructions of grids Γ_Λ are not viable since it may yield to grids not contained in the domain $\mathscr{D}$. For such general domains, magic points obtained through greedy algorithms could be considered.

3.2 Adaptive Algorithm for Sparse Interpolation

An adaptive sparse interpolation algorithm consists in constructing a sequence of approximations $(u_n)_{n\geq 1}$ associated with an increasing sequence of downward closed subsets $(\Lambda_n)_{n\geq 1}$. According to (10), we have to construct a sequence such that the best approximation error and the Lebesgue constant are such that

$$\mathscr{L}_{\Lambda_n} \inf_{w \in \mathscr{P}_{\Lambda_n}} \|u - w\|_\infty \longrightarrow 0 \text{ as } n \to \infty$$

for obtaining a convergent algorithm. For example, if

$$\inf_{w \in \mathscr{P}_{\Lambda_n}} \|u - w\|_\infty = O((\#\Lambda_n)^{-r}) \tag{13}$$

holds[3] for some $r > 1$ and if $\mathscr{L}_{\Lambda_n} = O((\#\Lambda_n)^k)$ for $k < r$, then the error $\|u - u_n\|_\infty = O(n^{-r'})$ tends to zero with an algebraic rate of convergence $r' = r - k > 0$. Of course, the challenge is to propose a practical algorithm that constructs a good sequence of sets Λ_m.

We now present the adaptive sparse interpolation algorithm with bulk chasing procedure introduced in [5]. Let θ be a fixed bulk chasing parameter in (0, 1) and let $\mathscr{E}_\Lambda(v) = \|P_\Lambda(v)\|_2^2$, where P_Λ is the orthogonal projector over $\mathscr{P}_\Lambda$ for any subset $\Lambda \subset \mathbb{N}_0^d$.

Algorithm 2 (Adaptive interpolation algorithm)

1: Set $\Lambda_1 = \{\mathbf{0}_d\}$ and $n = 1$.
2: **while** $n \leq N$ and $\varepsilon^{n-1} > \varepsilon$ **do**

---

[3] See e.g. [7] for conditions on u ensuring such a behavior of the approximation error.

3: Compute $\mathcal{M}_{\Lambda_n}$.
4: Set $\Lambda_n^\star = \Lambda_n \cup \mathcal{M}_{\Lambda_n}$ and compute $\mathcal{I}_{\Lambda_n^\star}(u)$.
5: Select $N_n \subset \mathcal{M}_{\Lambda_n}$ the smallest such that $\mathcal{E}_{N_n}(\mathcal{I}_{\Lambda_n^\star}(u)) \geq \theta \mathcal{E}_{\mathcal{M}_{\Lambda_n}}(\mathcal{I}_{\Lambda_n^\star}(u))$
6: Update $\Lambda_{n+1} = \Lambda_n \cup N_n$.
7: Compute $u_{n+1} = \mathcal{I}_{\Lambda_{n+1}}(u)$ (this step is not necessary in practice).
8: Compute ε^n.
9: Update $n = n + 1$.
10: **end while**

At iteration n, Algorithm 2 selects a subset of multi-indices N_n in the *reduced margin* of Λ_n defined by

$$\mathcal{M}_{\Lambda_n} = \{\boldsymbol{\nu} \in \mathbb{N}^d \setminus \Lambda_n : \forall j \text{ s.t. } \nu_j > 0,\ \boldsymbol{\nu} - \boldsymbol{e}_j \in \Lambda_n\},$$

where $(\boldsymbol{e}_j)_k = \delta_{kj}$. The reduced margin is such that for any subset $S \subset \mathcal{M}_{\Lambda_n}$, $\Lambda_n \cup S$ is downward closed. This ensures that the sequence $(\Lambda_n)_{n\geq 1}$ generated by the algorithm is an increasing sequence of downward closed sets. Finally, Algorithm 2 is stopped using a criterion based on

$$\varepsilon^n = \frac{\mathcal{E}_{\mathcal{M}_n}(\mathcal{I}_{\Lambda_n^\star}(u))}{\mathcal{E}_{\Lambda_n^\star}(\mathcal{I}_{\Lambda_n^\star}(u))}.$$

4 Combining Sparse Adaptive Interpolation with Sequential Control Variates Algorithm

We present in this section two ways of combining Algorithms 1 and 2. First we introduce a perturbed version of Algorithm 2 and then an adaptive version of Algorithm 1. At the end of the section, numerical results will illustrate the behavior of the proposed algorithms.

4.1 Perturbed Version of Algorithm 2

As we do not have access to exact evaluations of the solution u of (1), Algorithm 2 can not be used for interpolating u. So we introduce a perturbed version of this algorithm, where the computation of the exact interpolant $\mathcal{I}_\Lambda(u)$ is replaced by an approximation denoted $\tilde{u}_\Lambda$, which can be computed for example with Algorithm 1 stopped for a given tolerance ε_{tol} or at step k. This brings the following algorithm.

Algorithm 3 (Perturbed adaptive sparse interpolation algorithm)
1: Set $\Lambda_1 = \{\mathbf{0}_d\}$ and $n = 1$.
2: **while** $n \leq N$ and $\tilde{\varepsilon}^{n-1} > \varepsilon$ **do**

```
    Compute 𝓜_{Λ_n}.
    Set Λ*_n = Λ_n ∪ 𝓜_{Λ_n} and compute ũ_{Λ*_n}.
    Select N_n as the smallest subset of 𝓜_{Λ_n} such that ℰ_{N_n}(ũ_{Λ*_n}) ≥ θ ℰ_{𝓜_{Λ_n}}(ũ_{Λ*_n})
    Update Λ_{n+1} = Λ_n ∪ N_n.
    Compute ũ_{Λ_{n+1}}.
    Compute ε̃^n.
    Update n = n + 1.
end while
```

4.2 Adaptive Version of Algorithm 1

As a second algorithm, we consider the sequential control variates algorithm (Algorithm 1) where at step 4, an approximation $\tilde{e}^k$ of e^k is obtained by applying the adaptive interpolation algorithm (Algorithm 3) to the function $e^k_{\Delta t,M}$, which uses Monte-Carlo estimations $e^k_{\Delta t,M}(x_\nu)$ of $e^k(x_\nu)$ at interpolation points. At each iteration, $\tilde{e}^k$ therefore belongs to a different approximation space $\mathscr{P}_{\Lambda_k}$. In the numerical section, we will call this algorithm **adaptive Algorithm** 1.

4.3 Numerical Results

In this section, we illustrate the behavior of algorithms previously introduced on different test cases. We consider the simple diffusion equation

$$\begin{aligned} -\Delta u(x) &= g(x), \quad x \in \mathscr{D}, \\ u(x) &= f(x), \quad x \in \partial\mathscr{D}, \end{aligned} \tag{14}$$

were $\mathscr{D} =]-1,1[^d$. The source terms and boundary conditions will be specified later for each test case.

The stochastic differential equation associated to (14) is the following

$$dX_t^x = \sqrt{2} dW_t, \quad X_0^x = x, \tag{15}$$

where $(W_t)_{t \geq 0}$ is a d-dimensional Brownian motion.

We use tensorized grids of magic points for the selection of interpolation points evolved in adaptive algorithms.

Small dimensional test case. We consider a first test case (TC1) in dimension $d = 5$. Here the source term and the boundary conditions in problem (14) are chosen such that the solution is given by

$$u(x) = x_1^2 + sin(x_2) + exp(x_3) + sin(x_4)(x_5 + 1), \qquad x \in \overline{\mathscr{D}}. \tag{TC1}$$

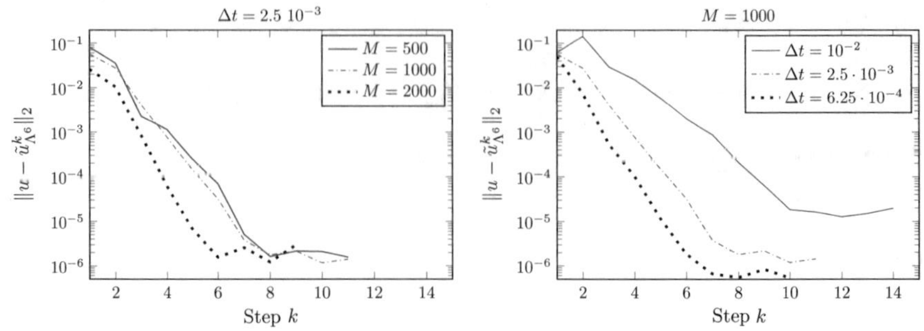

Fig. 1 (TC1) Algorithm 1 for fixed Λ: evolution of $\|u - \tilde{u}^k_{\Lambda^6}\|$ with respect to k for various M (left figure), and various Δt (right figure)

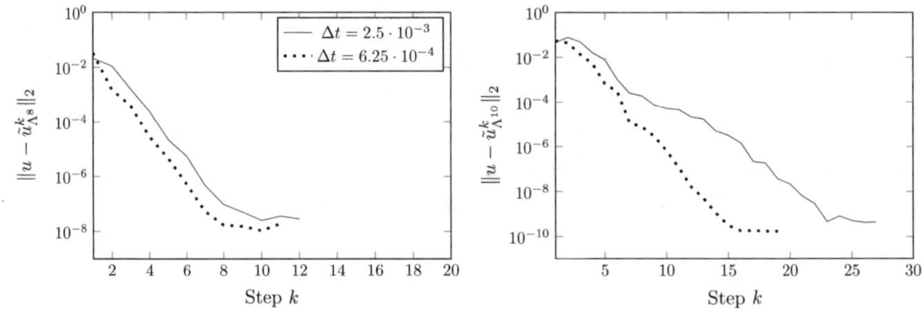

Fig. 2 (TC1) Algorithm 1 for fixed Λ^i: evolution of $\|u - \tilde{u}^k_{\Lambda^i}\|_2$ with respect to k for $i = 8$ (left figure), and $i = 10$ (right figure)

We first test the influence of Δt and M on the convergence of Algorithm 1 when Λ is fixed. In that case, Λ is selected a priori with Algorithm 2 using samples of the exact solution u for (TC1), stopped for $\varepsilon \in \{10^{-6}, 10^{-8}, 10^{-10}\}$. In what follows, the notation Λ^i stands for the set obtained for $\varepsilon = 10^{-i}$, $i \in \{6, 8, 10\}$. We represent on Fig. 1 the evolution of the absolute error in L^2-norm (similar results hold for the L^∞-norm) between the approximation and the true solution with respect to step k for $\Lambda = \Lambda^6$. As claimed in Corollary 1, we recover the geometric convergence up to a threshold value that depends on Δt. We also notice faster convergence as M increases and when Δt decreases. We fix $M = 1000$ in the next simulations.

We study the impact of the choice of Λ^i on the convergence of Algorithm 1. Again we observe on Fig. 2 that the convergence rate gets better as Δt decreases. Moreover as $\#\Lambda$ increases the threshold value decreases. This is justified by the fact that interpolation error decreases as $\#\Lambda^i$ increases (see Table 1). Nevertheless, we observe that it may also deteriorate the convergence rate if it is chosen too large together with Δt not sufficiently small. Indeed for the same number of iterations $k = 10$ and the same time-step $\Delta t = 2.5 \cdot 10^{-3}$, we have an approximate absolute error equal to 10^{-7} for Λ^8 against 10^{-4} for Λ^{10}.

Table 1 Algorithm 2 computed on the exact solution of (TC1): evolution of $\#\Lambda_n$, error criterion ε^n and interpolation errors in norms L^2 and L^∞ at each step n

Λ_n	$\#\Lambda_n$	ε^n	$\|\|u-u_n\|\|_2$	$\|\|u-u_n\|\|_\infty$
Λ^6	1	6.183372e-01	1.261601e+00	4.213566e+00
	10	2.792486e-02	1.204421e-01	3.602629e-01
	20	2.178450e-05	9.394419e-04	3.393999e-03
	26	9.632815e-07	4.270457e-06	1.585129e-05
	30	9.699704e-08	2.447475e-06	8.316435e-06
Λ^8	33	4.114730e-09	2.189518e-08	9.880306e-08
	40	1.936050e-10	6.135776e-10	1.739848e-09
Λ^{10}	41	1.008412e-11	9.535433e-11	4.781375e-10
	50	1.900248e-14	1.004230e-13	4.223288e-13
	55	7.453467e-15	2.905404e-14	1.254552e-13

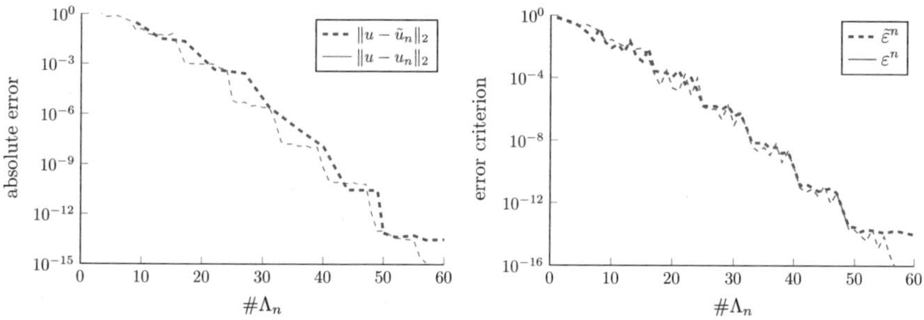

Fig. 3 (TC1) Comparison of Algorithm 2 applied to exact solution and Algorithm 3: (left) absolute error in L^2-norm (right) evolution of ε^n and $\tilde{\varepsilon}^n$ with respect to $\#\Lambda_n$

We present now the behavior of Algorithm 3. Simulations are performed with a bulk-chasing parameter $\theta = 0.5$. At each step n of Algorithm 3, we use Algorithm 1 with $(\Delta t, M) = (10^{-4}, 1000)$, stopped when a stagnation is detected. As shown on the left plot of Fig. 3, for $\#\Lambda_n = 55$ we reach approximately a precision of 10^{-14} as for Algorithm 2 performed on the exact solution (see Table 1). According to the right plot of Fig. 3, we also observe that the enrichment procedure behaves similarly for both algorithms ($\tilde{\varepsilon}^n$ and ε^n are almost the same). Here using the approximation provided by Algorithm 1 has a low impact on the behavior of Algorithm 2.

We present then results provided with the adaptive Algorithm 1. The parameters chosen for the adaptive interpolation are $\varepsilon = 5 \cdot 10^{-2}$, $\theta = 0.5$. $K = 30$ ensures the stopping of Algorithm 1. As illustrated by Fig. 4, we recover globally the same behavior as for Algorithm 1 without adaptive interpolation. Indeed as k increases, both absolute errors in L^2-norm and L^∞-norm decrease and then stagnate. Again, we notice the influence of Δt on the stagnation level. Nevertheless, the convergence rates are deteriorated and the algorithm provides less accurate approximations than Algorithm 3. This might be due to the sparse adaptive interpolation procedure, which

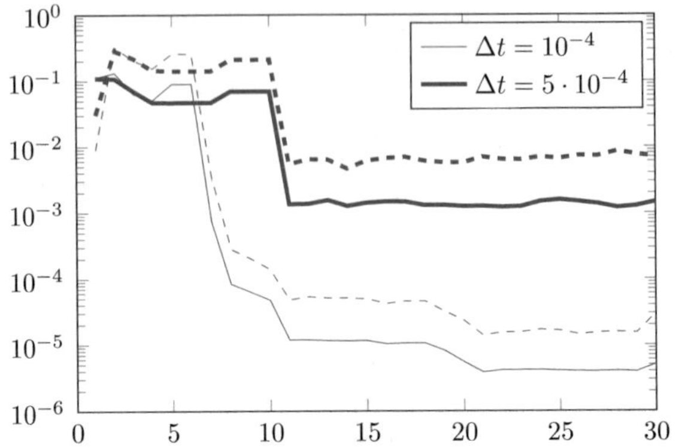

Fig. 4 (TC1) Adaptive Algorithm 1: evolution of $\|u - u^k_{\Lambda_k}\|_2$ (continuous line) and $\|u - u^k_{\Lambda_k}\|_\infty$ (dashed line) with respect to step k and Δt

Table 2 (TC1) Comparison of the algorithmic complexity to reach the precision $3 \cdot 10^{-5}$, with $(\Delta t, M) = (10^{-4}, 1000)$

	Adaptive Algorithm 1	Algorithm 3	Full-grid Algorithm 1
Th. complexity	$M(\Delta t)^{-1}(\sum_k \#\Lambda_k)$	$M(\Delta t)^{-1}(\sum_n \#\Lambda_n N_n)$	$M(\Delta t)^{-1}\#\Lambda_{max} N$
Est. complexity	$4 \cdot 10^9$ operations	$16 \cdot 10^9$ operations	$10^{12} N$ operations

uses here pointwise evaluations based on Monte-Carlo estimates, unlike Algorithm 3 which relies on pointwise evaluations resulting from Algorithm 1 stopping for a given tolerance.

Finally in Table 2, we compare the algorithmic complexity of these algorithms to reach a precision of $3 \cdot 10^{-5}$ for $(\Delta t, M) = (10^{-4}, 1000)$. For adaptive Algorithm 1, Λ_k refers to the set of multi-indices considered at step k of Algorithm 1. For Algorithm 3, N_n stands for the number of iteration required by Algorithm 1 to reach tolerance ε_{tol} at step n. Finally, Algorithm 1 is run with full-grid $\Lambda = \Lambda_{max}$ where $\Lambda_{max} = \{\nu \in \mathbb{N}^d : \nu_i \leq 10\}$ is the set of multi-indices allowing to reach the machine precision. In this case, N stands for the number of steps for this algorithm to converge.

We observe that both the adaptive version of Algorithms 1 and 3 have a similar complexity, which is better than for the full-grid version of Algorithm 1. Moreover, we observed that while adaptive version of Algorithm 1 stagnates at a precision of $3 \cdot 10^{-5}$, Algorithm 3, with the same parameters Δt and M, converges almost up to the machine precision. This is why the high-dimensional test cases will be run only with Algorithm 3.

Higher-dimensional test cases. Now, we consider two other test cases noted respectively (TC2) and (TC3) in higher dimension.

(TC2) As second test case in dimension $d = 10$, we define (14) such that its solution is the Henon–Heiles potential

Table 3 (TC2) Comparison of Algorithm 2 (first four columns) and Algorithm 3 (last four columns)

$\#\Lambda_n$	ε^n	$\|u-u_n\|_\infty$	$\|u-u_n\|_2$	$\#\Lambda_n$	$\tilde{\varepsilon}^n$	$\|u-\tilde{u}_{\Lambda_n}\|_\infty$	$\|u-\tilde{u}_{\Lambda_n}\|_2$
1	4.0523e-01	3.0151e+00	1.2094e+00	1	3.9118e-01	8.3958e-01	6.9168e-01
17	1.6243e-01	1.8876e+00	5.9579e-01	17	1.6259e-01	5.2498e-01	3.4420e-01
36	5.4494e-02	7.0219e-01	2.0016e-01	36	5.4699e-02	1.9209e-01	1.2594e-01
46	1.2767e-02	1.6715e-01	4.9736e-02	46	1.2806e-02	4.6904e-02	2.8524e-02
53	9.6987e-04	2.9343e-02	4.8820e-03	53	1.0350e-03	7.8754e-03	2.8960e-03
60	7.6753e-04	1.5475e-02	4.1979e-03	61	7.0354e-04	3.0365e-03	1.7610e-03
71	3.2532e-04	8.4575e-03	2.1450e-03	71	3.1998e-04	2.3486e-03	1.2395e-03
77	1.7434e-16	3.9968e-15	1.5784e-15	77	7.3621e-16	6.2172e-15	1.2874e-15

$$u(x) = \frac{1}{2}\sum_{i=1}^{d} x_i^2 + 0.2\sum_{i=1}^{d-1}\left(x_i x_{i+1}^2 - x_i^3\right) + 2.5\ 10^{-3}\sum_{i=1}^{d-1}\left(x_i^2 + x_{i+1}^2\right)^2, \qquad x \in \overline{\mathscr{D}}.$$

We set $(\Delta t, M) = (10^{-4}, 1000)$ and $K = 30$ for Algorithm 1.

(TC3) We also consider the problem (14) whose exact solution is a sum of non-polynomial functions, like (TC1) but now in dimension $d = 20$, given by

$$u(x) = x_1^2 + sin(x_{12}) + exp(x_5) + sin(x_{15})(x_8 + 1).$$

Here, the Monte-Carlo simulations are performed for $(\Delta t, M) = (10^{-4}, 1000)$ and $K = 30$.

Since for both test cases the exact solution is known, we propose to compare the behavior of Algorithms 3 and 2. Again, the approximations $\tilde{u}_n$, at each step n of Algorithm 3, are provided by Algorithm 1 stopped when a stagnation is detected. In both cases, the parameters for Algorithm 3 are set to $\theta = 0.5$ and $\varepsilon = 10^{-15}$.

In Tables 3 and 4, we summarize the results associated to the exact and perturbed sparse adaptive algorithms for (TC2) and (TC3) respectively. We observe that Algorithm 3 performs well in comparison to Algorithm 2, for (TC2). Indeed, we get an approximation with a precision below the prescribed value ε for both algorithms.

Similar observation holds for (TC3) in Table 4 and this despite the fact that the test case involves higher dimensional problem.

5 Conclusion

In this paper we have introduced a probabilistic approach to approximate the solution of high-dimensional elliptic PDEs. This approach relies on adaptive sparse polynomial interpolation using pointwise evaluations of the solution estimated using a Monte-Carlo method with control variates.

Table 4 (TC3) Comparison of Algorithm 2 (first four columns) and Algorithm 3 (last four columns)

$\#\Lambda_n$	ε^n	$\|u - u_n\|_\infty$	$\|u - u_n\|_2$	$\#\Lambda_n$	$\tilde{\varepsilon}^n$	$\|u - \tilde{u}_{\Lambda_n}\|_\infty$	$\|u - \tilde{u}_{\Lambda_n}\|_2$
1	7.0155e-01	3.9361e+00	1.2194e+00	1	5.5582e-01	7.2832e-01	7.0771e-01
6	1.4749e-01	2.2705e+00	5.4886e-01	6	7.4253e-02	2.7579e-01	5.1539e-01
11	2.1902e-02	2.8669e-01	1.0829e-01	11	1.4929e-02	4.4614e-02	4.1973e-02
15	7.6086e-03	1.6425e-01	4.7394e-02	15	1.2916e-02	1.5567e-02	2.5650e-02
20	2.2275e-04	2.7715e-03	7.2230e-04	20	3.4446e-04	5.6927e-04	5.3597e-04
24	1.4581e-05	1.5564e-04	7.5314e-05	24	1.6036e-05	2.5952e-05	3.0835e-05
30	1.8263e-06	8.0838e-06	2.1924e-06	30	9.0141e-07	2.8808e-06	1.9451e-06
35	3.9219e-09	8.9815e-08	2.4651e-08	35	8.1962e-09	2.1927e-08	1.5127e-08
40	1.7933e-10	2.0152e-09	6.9097e-10	40	1.6755e-10	2.8455e-10	2.6952e-10
45	5.0775e-12	2.4783e-10	4.1600e-11	45	1.4627e-11	3.3188e-11	1.7911e-11
49	1.7722e-14	4.6274e-13	8.5980e-14	49	1.7938e-14	8.6362e-14	5.0992e-14
54	3.9609e-15	2.2681e-13	3.1952e-14	54	3.2195e-15	4.8142e-14	2.6617e-14
56	4.5746e-16	8.4376e-15	3.0438e-15	56	8.2539e-16	8.4376e-15	6.3039e-15

Especially, we have proposed and compared different algorithms. First we proposed Algorithm 1 which combines the sequential algorithm proposed in [11] and sparse interpolation. For the non-adaptive version of this algorithm we recover the convergence up to a threshold as the original sequential algorithm [12]. Nevertheless it remains limited to small-dimensional test cases, since its algorithmic complexity remains high. Hence, for practical use, the adaptive Algorithm 1 should be preferred. Adaptive Algorithm 1 converges but it does not allow to reach low precision with reasonable number of Monte-Carlo samples or time-step in the Euler–Maruyama scheme. Secondly, we proposed Algorithm 3. It is a perturbed sparse adaptive interpolation algorithm relying on inexact pointwise evaluations of the function to approximate. Numerical experiments have shown that the perturbed algorithm (Algorithm 3) performs well in comparison to the ideal one (Algorithm 2) and better than the adapted Algorithm 1 with a similar algorithmic complexity. Here, since only heuristic tools have been provided to justify the convergence of this algorithm, the proof of its convergence, under assumptions on the class of functions to be approximated, should be addressed in a future work.

References

1. Bachmayr, M., Schneider, R., Uschmajew, A.: Tensor networks and hierarchical tensors for the solution of high-dimensional partial differential equations. Found. Comput. Math. **16**(6), 1423–1472 (2016)
2. Beck, C., Weinan E., Jentzen, A.: Machine learning approximation algorithms for high-dimensional fully nonlinear partial differential equations and second-order backward stochastic differential equations (2017). arXiv:1709.05963

3. Beck, C., Becker, S., Grohs, P., Jaafari, N., Jentzen, A.: Solving stochastic differential equations and Kolmogorov equations by means of deep learning (2018). arXiv:1806.00421
4. Bungartz, H.-J., Griebel, M.: Sparse grids. Acta Numer. **13**, 147–269 (2004)
5. Chkifa, A., Cohen, A., DeVore, R.: Sparse adaptive Taylor approximation algorithms for parametric and stochastic elliptic PDEs. ESAIM: Math. Model. Numer. Anal. **47**(1), 253–280 (2013)
6. Chkifa, A., Cohen, A., Schwab, C.: High-dimensional adaptive sparse polynomial interpolation and applications to parametric PDEs. Found. Comput. Math. **14**, 601–633 (2014)
7. Cohen, A., DeVore, R.: Approximation of high-dimensional parametric PDEs. Acta Numer. **24**, 1–159 (2015)
8. Comets, F., Meyre, T.: Calcul stochastique et modèles de diffusions-2ème éd. Dunod, Paris (2015)
9. Friedman, A.: Stochastic Differential Equations and Applications. Academic, New York (1975)
10. Giles, M.B.: Multilevel Monte Carlo methods. Monte Carlo and Quasi-Monte Carlo Methods 2012, pp. 83–103. Springer, Berlin, Heidelberg (2013)
11. Gobet, E., Maire, S.: A spectral Monte Carlo method for the Poisson equation. Monte Carlo Methods Appl. MCMA **10**(3–4), 275–285 (2004)
12. Gobet, E., Maire, S.: Sequential control variates for functionals of Markov processes. SIAM J. Numer. Anal. **43**(3), 1256–1275 (2005)
13. Gobet, E., Menozzi, S.: Stopped diffusion processes: boundary corrections and overshoot. Stoch. Process. Their Appl. **120**(2), 130–162 (2010)
14. Gobet, E.: Monte-Carlo Methods and Stochastic Processes: From Linear to Non-Linear. Chapman and Hall/CRC, Boca Raton (2016)
15. Grasedyck, L., Kressner, D., Tobler, C.: A literature survey of low- rank tensor approximation techniques. GAMM-Mitt. **36**(1), 53–78 (2013)
16. Hackbusch, W.: Numerical tensor calculus. Acta Numer. **23**, 651–742 (2014)
17. Kloeden, P., Platen, E.: Numerical Solution of Stochastic Differential Equations. Springer Science & Business Media, Berlin (2013)
18. Maday, Y., Nguyen, N.C., Patera, A.T., et al.: A general, multipurpose interpolation procedure: the magic points. Commun. Pure Appl. Anal. **8**(1), 383–404 (2009)
19. Nouy, A.: Low-rank methods for high-dimensional approximation and model order reduction. Model Reduction and Approximation (2017)
20. Oseledets, I.: Tensor-train decomposition. SIAM J. Sci. Comput. **33**(5), 2295–2317 (2011)
21. Shen, J., Yu, H.: Efficient spectral sparse grid methods and applications to high dimensional elliptic problems. SIAM J. Sci. Comput. **32**(6), 3228–3250 (2010)
22. Weinan, E., Jiequn, H., Jentzen, A.: Deep learning-based numerical methods for high-dimensional parabolic partial differential equations and backward stochastic differential equations. Commun. Math. Stat. **5**(4), 349–380 (2017)

Massively Parallel Construction of Radix Tree Forests for the Efficient Sampling of Discrete or Piecewise Constant Probability Distributions

Nikolaus Binder and Alexander Keller

Abstract We compare different methods for sampling from discrete or piecewise constant probability distributions and introduce a new algorithm which is especially efficient on massively parallel processors, such as GPUs. The scheme preserves the distribution properties of the input sequence, exposes constant time complexity on the average, and significantly lowers the average number of operations for certain distributions when sampling is performed in a parallel algorithm that requires synchronization. Avoiding load balancing issues of naïve approaches, a very efficient massively parallel construction algorithm for the required auxiliary data structure is proposed.

Keywords Sampling · Low discrepancy sequences · Massively parallel algorithms · GPU

1 Introduction

In many applications, samples need to be drawn according to a given discrete probability density

$$p := (p_1, p_2, \ldots, p_n),$$

where the p_i are positive and $\sum_{i=1}^{n} p_i = 1$. Defining $P_0 := 0$ and the partial sums $P_k := \sum_{i=1}^{k} p_i$ results in $0 = P_0 < P_1 < \cdots < P_n = 1$, which forms a partition of the unit interval $[0, 1)$. Then, the inverse cumulative distribution function (CDF)

$$P^{-1}(x) = i \Leftrightarrow P_{i-1} \leq x < P_i$$

N. Binder · A. Keller (✉)
NVIDIA, Berlin, Germany
e-mail: akeller@nvidia.com

N. Binder
e-mail: nbinder@nvidia.com

B. Tuffin and P. L'Ecuyer (eds.), *Monte Carlo and Quasi-Monte Carlo Methods*, Springer Proceedings in Mathematics & Statistics 324,
https://doi.org/10.1007/978-3-030-43465-6_7

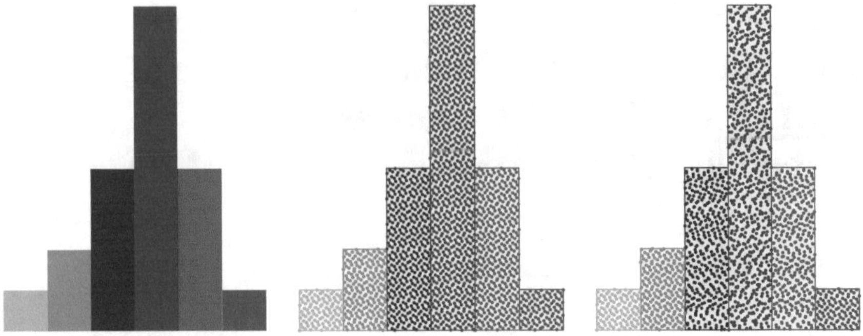

Fig. 1 Drawing samples according to a given piecewise constant probability distribution (left) using a monotonic mapping can preserve uniformity properties (middle), for example a discrepancy, of an input sample sequence, while using a non-monotonic mapping such as the Alias Method [8, 9] negatively affects uniformity (right). For illustrative purposes we distribute the samples in columns with heights proportional to the probabilities. Then, the point set is just partitioned into the columns and scaling the columns to equal heights would yield the desired density distribution. The input for both mappings is the two-dimensional Hammersley set with 1024 points. The first component is mapped to the index of the column, using the fractional part for the y position. For illustrative purposes, we use the second component to distribute the points also along the x axis inside each column. It is important to note that samples in higher columns are more often mapped to an alias and therefore become less uniform

can be used to map realizations of a uniform random variable ξ on $[0, 1)$ to $\{1, 2, \ldots, n\}$ such that

$$\text{Prob}\left(\{P_{i-1} \leq \xi < P_i\}\right) = p_i.$$

Besides identifying the most efficient method to perform such a mapping, we are interested in transforming low discrepancy sequences [7] and how such mappings affect the uniformity of the resulting warped sequence. An example for such a sampling process is shown in Fig. 1.

The remainder of the article is organized as follows: After reviewing several algorithms to sample according to a given discrete probability density by transforming uniformly distributed samples in Sect. 2, massively parallel algorithms to construct auxiliary data structures for the accelerated computation of P^{-1} are introduced in Sect. 3. The results of the scheme that preserves distribution properties, especially when transforming low discrepancy sequences, are presented in Sect. 4 and discussed in Sect. 5 before drawing the conclusions.

2 Sampling from Discrete Probability Densities

In the following we will survey existing methods to evaluate the inverse mapping P^{-1} and compare their properties with respect to computational complexity, memory requirements, memory access patterns, and sampling efficiency.

2.1 *Linear Search*

As illustrated by the example in Fig. 2, a linear search computes the inverse mapping P^{-1} by subsequently checking all intervals for inclusion of the value of the uniformly distributed variable ξ. This is simple, does not require additional memory, achieves very good performance for a small number n of values, and scans the memory in linear order. However, its average and worst case complexity of $\mathcal{O}(n)$ makes it unsuitable for large n.

2.2 *Binary Search*

Binary search lowers the average and worst case complexity of the inverse mapping P^{-1} to $\mathcal{O}(\log_2 n)$ by performing bisection. Again, no additional memory is required, but memory no longer is accessed in linear order. Figure 3 shows an example of binary search.

2.3 *Binary Trees*

The implicit decision tree traversed by binary search can be stored as an explicit binary tree structure. The leaves of the tree reference intervals, and each node stores the value q_i as well as references to the two children. The tree is then traversed by

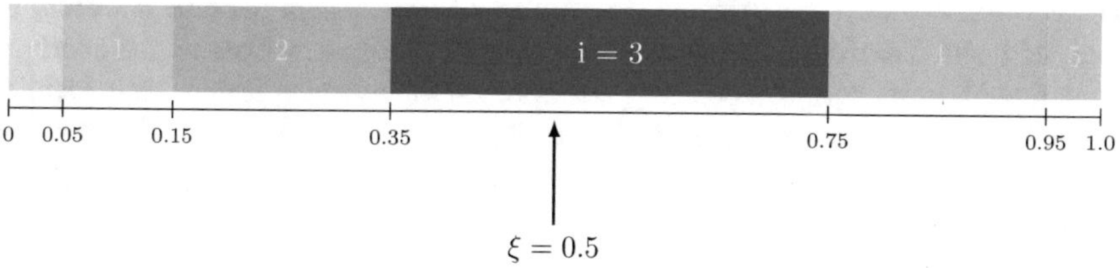

Fig. 2 Linearly searching through all intervals until the interval containing the variable ξ is found requires $\mathcal{O}(n)$ steps on the average. In this example four comparison operations are required to identify the interval $i = 3$ that includes ξ

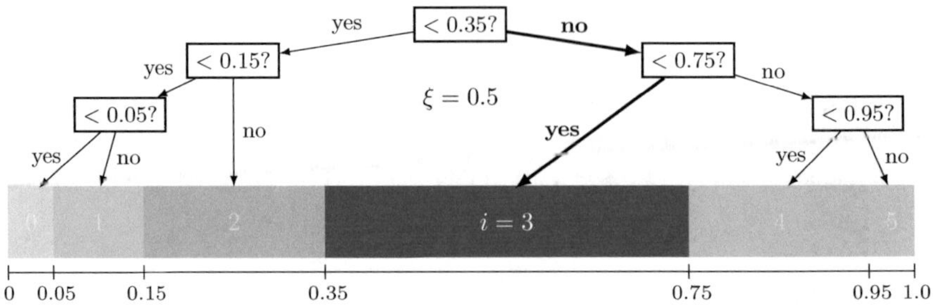

Fig. 3 Binary search bisects the list of potential intervals in each step and hence has an average case and worst case complexity of $\mathcal{O}(\log_2 n)$. In the example two comparisons are required to find that $\xi = 0.5$ is included in the interval $i = 3$

comparing ξ to q_i in each node, advancing to the left child if $\xi < q_i$, and otherwise to the right child. A common enumeration scheme uses the index i for the node that splits the unit interval in q_i, and hence can avoid explicitly storing q_i in the node data structure.

While the average case and worst case complexity of performing the inverse mapping P^{-1} with such an explicitly stored tree does not get below $\mathcal{O}(\log_2 n)$, allowing for arbitrary tree structures enables further optimization. Again, memory access is not in linear order, but due to the information required to identify the two children of each node in the tree, $\mathcal{O}(n)$ additional memory must be allocated and transferred. It is important to note that the worst case complexity may be as high as $\mathcal{O}(n)$ for degenerate trees. Such cases can be identified and avoided by an adapted re-generation of the tree.

2.4 k-ary Trees

On most hardware architectures, the smallest granularity of memory transfer is almost always larger than what would be needed to perform a single comparison and to load the index of one or the other child in a binary tree. The average case complexity for a branching factor k is $\mathcal{O}(\log_k n)$, but the number of comparisons is either increased to $\log_2 k$ (binary search in children) or even $k - 1$ (linear scan over the children) in each step, and therefore either equal or greater than for binary trees on the average. Even though more comparisons are performed on the average, it may be beneficial to use trees with a branching factor higher than two, because the additional effort in computation may be negligible as compared to the memory transfer that happens anyhow.

2.5 The Cutpoint Method

By employing additional memory, the indexed search [2], also known as the Cutpoint Method [5], can perform the inverse mapping P^{-1} in $\mathcal{O}(1)$ on the average. Therefore, the unit interval is partitioned into m cells of equal size and a *guide table* stores each first interval that overlaps a cell as illustrated in Fig. 4. Starting with this interval, linear search is employed to find the one that includes the realization of ξ. As shown in [3, Chap. 2.4], the expected number of comparisons is $1 + \frac{n}{m}$. In the worst case all but one interval are located in a single cell and since linear search has a complexity of $\mathcal{O}(n)$, the worst case complexity of the Cutpoint Method is also $\mathcal{O}(n)$. Sometimes, these worst cases can be avoided by recursively nesting another guide table in cells with many entries. In general, however, the problem persists. If nesting is performed multiple times, the structure of the nested guide tables is similar to a k-ary tree (see Sect. 2.4) with implicit split values defined by the equidistant partitioning.

Another way of improving the worst case performance is using binary search instead of linear search in each cell of the guide table. No additional data needs to be stored since the index of the first interval of the next cell can be conservatively used as the last interval of the current cell. The resulting complexity remains $\mathcal{O}(1)$ on the average, but improves to $\mathcal{O}(\log_2 n)$ in the worst case.

2.6 The Alias Method

Using the Alias Method [8, 9], the inverse P^{-1} can be sampled in $\mathcal{O}(1)$ both on the average as well as in the worst case. It avoids the worst case by cutting and reordering the intervals such that each cell contains at most two intervals, and thus neither linear search nor binary search is required. For the example used in this article, a resulting table is shown in Fig. 5. In terms of run time as well as efficiency of memory access the method is very compelling since the mapping can be performed using a single read

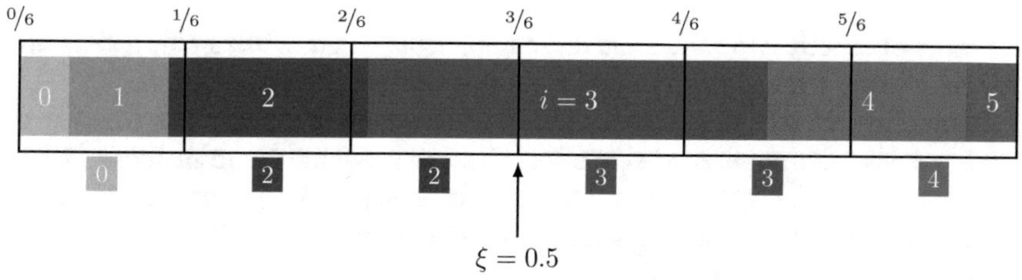

Fig. 4 The Cutpoint Method uses a guide table that uniformly partitions the unit interval and stores the first interval that overlaps each cell (shown below each cell). These indices are then used as starting points for linear search. In this example only a single lookup is required; in general the average case complexity is $\mathcal{O}(1)$

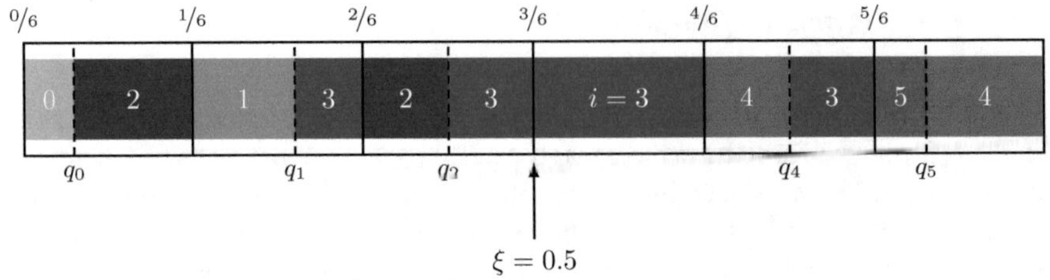

Fig. 5 Similar to the Cutpoint Method, the Alias Method uses an additional table that uniformly partitions the unit interval. However, the Alias Method cuts and redistributes the intervals so that each cell contains at most two intervals, the first one being the interval with the same index as the cell and a second one that covers the rest of the range of the cell. Then, by storing the index of the second interval in each cell and the split points q_j of each cell with index j, the mapping can *always* be evaluated with one comparison, and in our example, as before, only a single lookup is required to find the interval $i = 3$ including $\xi = 0.5$

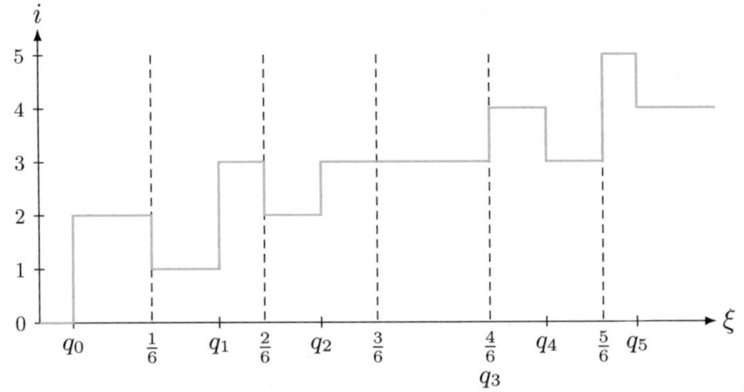

Fig. 6 The mapping of the Alias Method is not the inverse of P. While it still maps a uniform variate to one with the desired density distribution, it is not monotonic. This instance is a possible mapping $\xi \mapsto i$ for the Alias Method and the distribution shown in Fig. 1, where $i = m$ for $\xi \in \left[\frac{m}{M}, \frac{m+1}{M}\right)$ for $\xi < q_m$, and $i = \text{alias}(m)$ for $\xi \geq q_m$. Since $\text{alias}(m) \neq m + 1$ in general, the mapping cannot be monotonic

operation of exactly two values and one comparison, whereas hierarchical structures generally suffer from issues caused by multiple scattered read operations.

Reordering the intervals creates a different, non-monotonic mapping (see Fig. 6), which comes with unpleasant side effects for quasi-Monte Carlo methods [7]. As intervals are reordered, the low discrepancy of a sequence may be destroyed (see Fig. 1). This especially affects regions with high probabilities, which is apparent in the one dimensional example in Fig. 7.

In such regions, many samples are aliases of samples in low-density regions, which is intrinsic to the construction of the Alias Method. Hence the resulting set of samples cannot be guaranteed to be of low discrepancy, which may harm convergence speed.

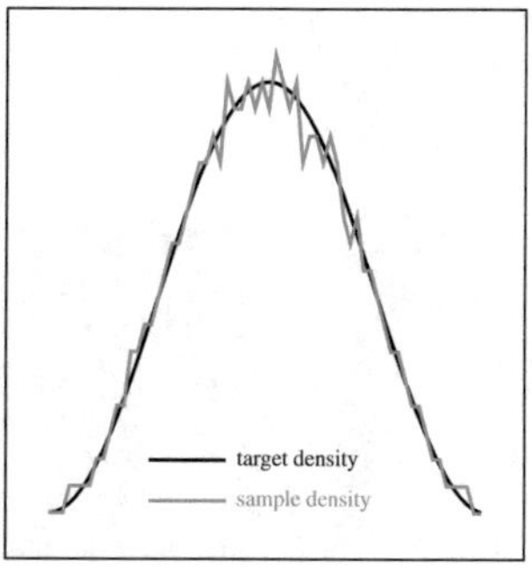

(a) Alias Method, 2^9 samples

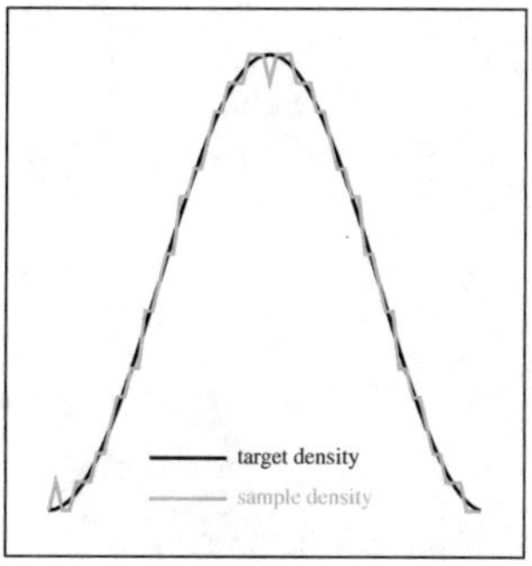

(b) monotonic mapping, 2^9 samples

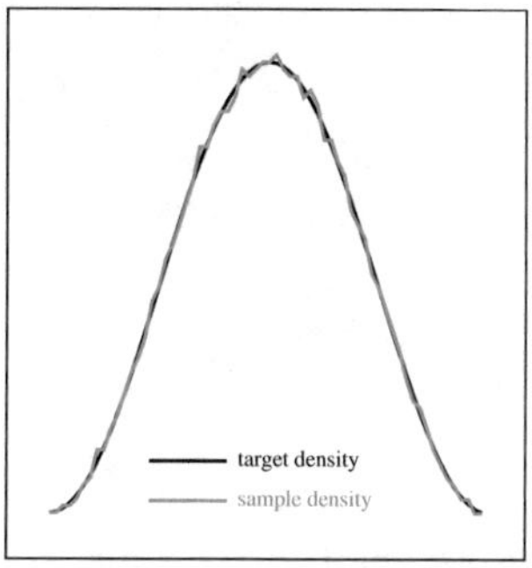

(c) Alias Method, 2^{11} samples

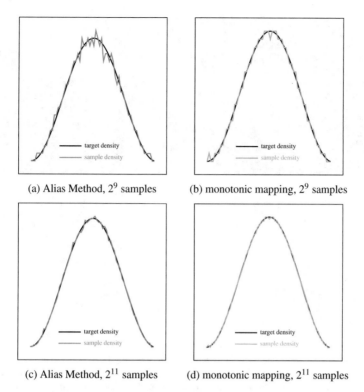

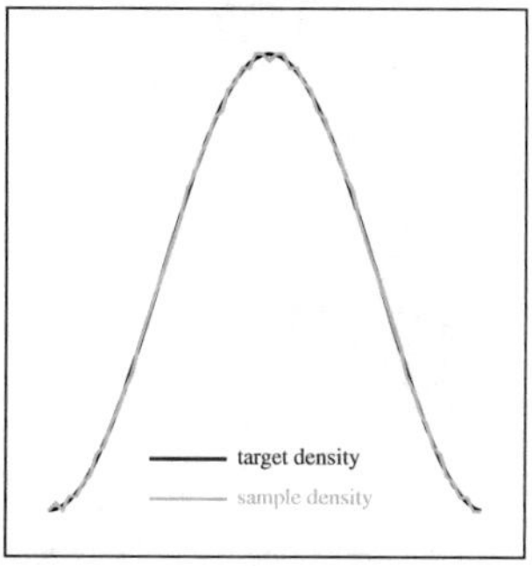

(d) monotonic mapping, 2^{11} samples

Fig. 7 Sampling proportional to a one-dimensional density with the Alias Method often converges significantly slower than sampling the inverse cumulative distribution function, especially in regions with high densities. In this example the discrete target probability distribution function is the continuous black curve sampled at 64 equidistant steps

A typical application in graphics is sampling according to a two-dimensional density map ("target density"). Figure 8 shows two regions of such a map with points sampled using the Alias Method and the inverse mapping. Points are generated by first selecting a row according to the density distribution of the rows and then selecting the column according to the distribution in this row. Already a visual comparison identifies the shortcomings of the Alias Method, which the evaluation of the quadratic deviation of the sampled density to the target density in Fig. 9 confirms.

Known algorithms for setting up the necessary data structures of the Alias Method are serial with a run time considerably higher than the prefix sum required for inversion methods, which can be efficiently calculated in parallel.

(a) target density

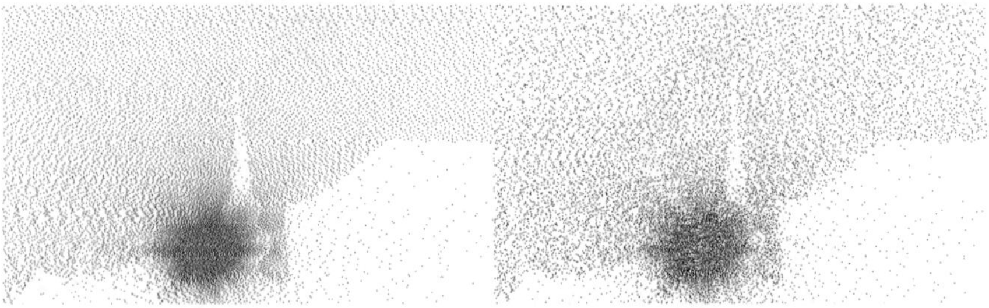

(b) monotonic mapping (c) Alias Method

Fig. 8 Sampling the two-dimensional target density in (**a**) using the two-dimensional low discrepancy Hammersley point set as a sequence of uniformly distributed points by (see the insets) (**b**) transforming them with a monotonic mapping preserves the distribution properties in a warped space, whereas (**c**) transforming with the Alias Method degrades the uniformity of the distribution. This is especially bad in regions with a high density. Image *source* https://www.openfootage.net/hdri-360-parking-lower-austria/, *license* Creative Commons Attribution 4.0 International License (https://creativecommons.org/licenses/by/4.0/)

3 An Efficient Inverse Mapping

In the following, a method is introduced that accelerates finding the inverse P^{-1} of the cumulative distribution function and preserves distribution properties, such as the star-discrepancy. Note that we aim to preserve these properties in the warped space in which regions with a high density are stretched, and regions with a low density are squashed. Unwarping the space like in Fig. 1 reveals that the sequence in warped space is just partitioned into these regions. Similar to the Cutpoint Method with binary search, the mapping can be performed in $\mathcal{O}(1)$ on the average and in $\mathcal{O}(\log_2 n)$ in the worst case. However, we improve the average case performance by explicitly storing a more optimal hierarchical structure that improves the average case of finding values that cannot be immediately identified using the guide table. While for degenerate hierarchical structures the worst case may increase to $\mathcal{O}(n)$, a fallback method constructs such a structure upon detection to guarantee logarithmic complexity.

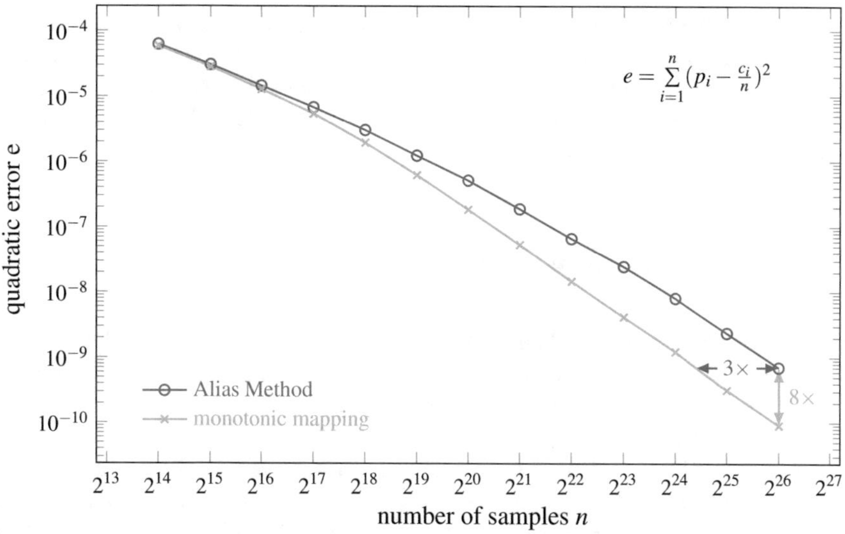

Fig. 9 As the Alias Method does not preserve the low discrepancy of the sequence, convergence speed suffers. Sampling the two-dimensional density detailed in Fig. 8 with the Alias Method, the quadratic error for 2^{26} samples is $8\times$ as high, and $3x$ as many samples are required to get the same error as sampling with the inverse mapping. n is the total number of samples, while c_i is the number of samples that realized the value i

Instead of optimizing for the minimal required memory footprint to guarantee access in constant time, we dedicate $\mathcal{O}(n)$ additional space for the hierarchical structure and propose to either assign as much memory as affordable for the guide table or to select its size proportional to the size of the discrete probability distribution p.

As explained in the previous section, using the monotonic mapping P^{-1}, and thus avoiding the Alias Method, is crucial for efficient quasi-Monte Carlo integration. At the same time, the massively parallel execution on Graphics Processing Units (GPUs) suffers more from outliers in the execution time than serial computation since computation is performed by a number of threads organized in groups ("warps"), which need to synchronize and hence only finish after the last thread in this group has terminated. Therefore, lengthy computations that would otherwise be averaged out need to be avoided.

Note that the Cutpoint Method uses the monotonic mapping P^{-1} and with binary search avoids the worst case. However, it does not yield good performance for the majority of cases in which an additional search in each cell needs to be performed. In what follows, we therefore combine the Cutpoint Method with a binary tree to especially optimize these cases.

In that context, radix trees are of special interest since they can be very efficiently built in parallel [1, 6]. Section 3.1 introduces their properties and their efficient parallel construction. Furthermore, their underlying structure that splits intervals in the middle is nearly optimal for this application.

However, as many trees of completely different sizes are required, a naïve implementation that builds these trees in parallel results in severe load balancing issues. In Sect. 3.2 we therefore introduce a method that builds the entire radix tree forest simultaneously in parallel, but instead of parallelizing over trees, parallelization is uniformly distributed over the data.

3.1 Massively Parallel Construction of Radix Trees

In a radix tree, also often called compact prefix tree, the value of each leaf node is the concatenation of the values on the path to it from the root of the tree. The number of children of each internal node is always greater than one; otherwise the values of the node and its child can be concatenated already in the node itself.

The input values referenced in the leaves of such a tree must be strictly ordered, which for arbitrary data requires an additional sorting step and an indirection from the index i' of the value in the sorted input data to the original index i. For the application of such a tree to perform the inverse mapping, i.e. to identify the interval m that includes ξ, the input values are the lower bounds of the intervals, which by construction are already sorted.

Radix trees over integers with a radix of two are of particular interest for hardware and software operating on binary numbers. Since by definition each internal node has exactly two children, there exists an enumeration scheme for these trees that determines the index of each node only given the range of input values below its children [1]: The index of each node is the lowest data index below its right, or, equivalently, the highest data index below its left child plus one. This information is not only available if the tree is built top-down by recursive bisection, but can also easily be propagated up when building the tree bottom-up. The index of the parent node, implied by this enumeration rule, is then the index of the leftmost node in the range of leaf nodes below the current root node if the node is a right child, and equals the index of the rightmost node in the range plus one if it is a left child.

Determining whether a node is a left or a right child can be done by comparing the values in the leaves at the boundaries of the range of nodes below the current root node to their neighbors outside the range. If the binary distance, determined by taking the bit-wise *exclusive or*, of the value of the leftmost leaf in the range to those of its left neighbor is smaller than the distance of the value of the rightmost leaf node in the range to those of its right neighbor, it must be a right child. Otherwise it is a left child. In case the leftmost leaf in the range is the first one, or the rightmost leaf in the range is the last one, a comparison to the neighbor is not possible, and these nodes must be a left and a right child, respectively. An example of such a decision is shown in Fig. 10.

Given this scheme, these trees can be built from bottom up completely in parallel since all information required to perform one merging step can be retrieved without synchronization. Like in the parallel construction of Radix Trees by Apetrei [1], the

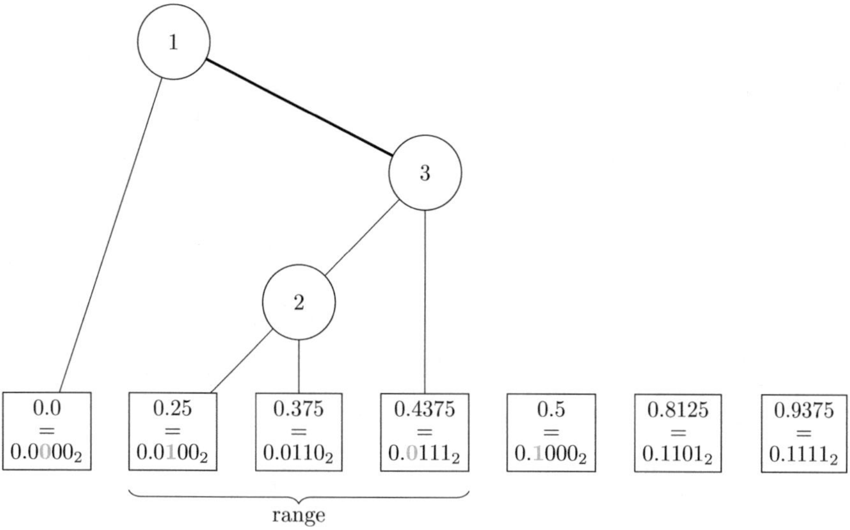

Fig. 10 In each merging step of the bottom-up construction method for the radix tree, the value of the leftmost leaf in the range of leaf nodes below the current root is compared to the value of its left neighbor, and the value of the rightmost leaf in the range is compared to the value of its right neighbor. In both comparisons, the most significant differing bit of the value in base two is determined. Since the most significant differing bit of the comparison on the left side of the current range is less significant than the most significant differing bit on the right side, the root node of the current subtree (here 3) is a right child, and the index of its parent node equals to the lowest index of the leaves in the range (here 1)

one required synchronization ensures that only one of the children merges further up, and does that after its sibling has already reported the range of nodes below it.

The implementation of the method presented in Algorithm 1 uses an atomic exchange operation (*atomicExch*) and an array initialized to -1 for synchronization. The $\oplus$ operator performs a bitwise *exclusive or* operation on the floating point representation of the input value. As the ordering of IEEE 754 floating point numbers is equal to the binary ordering, taking the *exclusive or* of two floating point numbers calculates a value in which the position of the most significant bit set to one indicates the highest level in the implicit tree induced by recursive bisection of the interval $[0, 1)$ on which the two values are not referenced by the same child. Hence, the bitwise *exclusive or* operation determines the distance of two values in such a tree.

3.2 *Massively Parallel Construction of Radix Tree Forests*

A forest of radix trees (see Fig. 11) can be built in a similar way. Again, parallelization runs over the whole data range, not over the individual trees in the forest. Therefore, perfect load balancing can be achieved. As highlighted in Algorithm 1, the key difference between building a single tree and a forest is that in each merging step it must be checked whether merging would go over partition boundaries of the forest. Avoiding such a merge operation over a boundary is as simple as setting the value of the neighbor to one, because then the distance (again computed using $\oplus$) is maximal.

Note that node indices for small (sub-) trees are always consecutive by design, which improves cache hit rates, which furthermore can be improved by interleaving the values of the cumulative distribution function and the indices of the children.

Algorithm 1 Parallel constructing a radix tree forest. Omitting the colored parts results in the construction of a radix tree, only (see [1]).

Input: $data \in [0, 1)^n$ in increasing order, number of partitions m
Output: n $nodes$, each with indices of the left and right child
$otherBounds \leftarrow (-1, ..., -1) \in \mathbb{Z}^n$
$data[-1] \leftarrow data[n] = 1$
for $i \in [0, n)$ **in parallel do**
  $nodeId \leftarrow leaf_i$
  $curCell \leftarrow \lfloor data[i] \cdot m \rfloor$
  $range \leftarrow (i, i)$
  repeat
    $valueLow \leftarrow data[range[0]]$
    $valueHigh \leftarrow data[range[1]]$
    $valueNeighborLow \leftarrow data[range[0] - 1]$
    $valueNeighborHigh \leftarrow data[range[1] + 1]$
    if $\lfloor valueNeighborLow \cdot m \rfloor < curCell$ **then**
      $valueNeighborLow \leftarrow 1$
    end if
    if $\lfloor valueNeighborHigh \cdot m \rfloor > curCell$ **then**
      $valueNeighborHigh \leftarrow 1$
    end if
    $child \leftarrow \begin{cases} 0 & valueLow \oplus valueNeighborLow > valueHigh \oplus valueNeighborHigh \\ 1 & \text{otherwise} \end{cases}$
    $parent \leftarrow \begin{cases} range[1] + 1 & child = 0 \\ range[0] & child = 1 \end{cases}$
    $nodes[parent].child[child] \leftarrow nodeId$
    $otherBound \leftarrow$ **atomicExch** $(otherBounds[parent], range[child])$
    if $otherBound \neq -1$ **then**
      $range[1 - child] \leftarrow otherBound$
      $nodeId \leftarrow parent$
    end if
  until otherBound = -1
end for

We indicate that a cell in the guide table is only overlapped by a single interval by setting the reference stored in the cell to the two's complement of the index of the interval. Then, a most significant bit set to one identifies such an interval, and one unnecessary indirection is avoided. If the size of the distribution is sufficiently

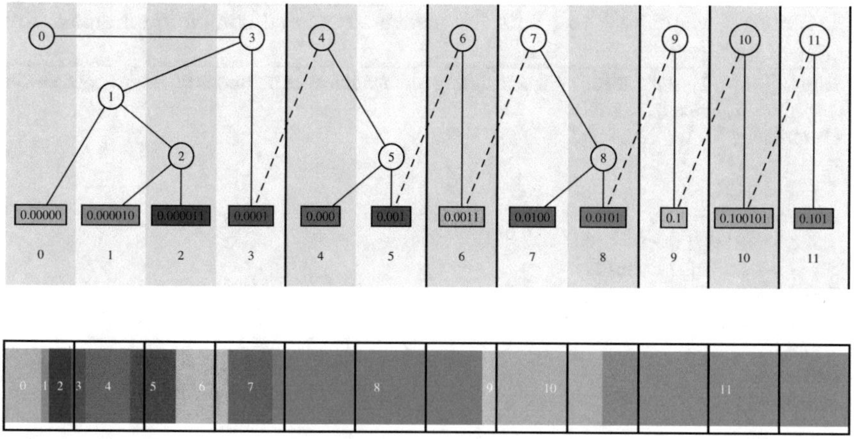

Fig. 11 A radix tree forest built with Algorithm 1. Note that all root nodes only have a right child. We manually set the reference for the left child to its left neighbor since it in practice almost always overlaps the left boundary (dashed lines). During sampling, the decision whether the left or right child must be used is purely based on the cumulative distribution function used as an input for the construction: Each node checks whether ξ is smaller than the cumulative value with the same index

small, further information could also be stored in the reference, such as a flag that there are exactly two intervals that overlap the cell. Then, only one comparison must be performed and there is no need to explicitly store a node.

Building the radix forest is slightly faster than building a radix tree over the entire distribution since merging stops earlier. However, we found the savings to be almost negligible, similarly to the effort required to set the references in the guide table.

Algorithm 2 shows the resulting combined method that maps ξ to i. First, it looks up a reference in the guide table at the index $g = \lfloor M \cdot \xi \rfloor$. If the most significant bit of the reference is one, i is the two's complement of this reference. Otherwise the reference is the root node of the tree which is traversed by iteratively comparing ξ to the value of the cumulative distribution function with the same index as the current node, advancing to its left child if ξ is smaller, and to its right child if ξ is larger or equal. Tree traversal again terminates if the most significant bit of the next node index is one. As before, i is determined by taking the two's complement (denoted by $\sim$) of the index.

4 Results

We evaluate our sampling method in two steps: First, we compare to an Alias Method in order to quantify the impact on convergence speed. Second, we compare to the Cutpoint Method with binary search which performs the identical mapping and therefore allows one to quantify execution speed.

Algorithm 2 Mapping ξ to i with the presented method combining a guide table with a radix forest.

Input: $\xi \in [0, 1)$, $data \in [0, 1)^n$ in increasing order, number of partitions m, $nodes$ created with Algorithm 1, guide table $table$.
Output: $i \in \{0, 1, ..., N-1\}$.

```
g ← ⌊ξ · m⌋
j ← table[g]
while msb(j) ≠ 1 do
  if ξ < data[j] then
    j ← nodes[j].child[0]
  else
    j ← nodes[j].child[1]
  end if
end while
return ~j
```

Table 1 Measuring the maximum and average number of memory load operations required for searching as well as the average number of load operations or idle operations of 32 simulations that need to be synchronized (Average_{32}) shows that while the maximum number of load operations is increased, for distributions with a high range the average number of load operations is typically reduced

$p_i \sim i^{20}$	Maximum	Average	Average_{32}
Cutpoint Method + binary search	8	1.25	3.66
Cutpoint Method + radix forest	16	1.23	3.46
$p_i \sim (i \mod 32 + 1)^{25}$	Maximum	Average	Average_{32}
Cutpoint Method + binary search	6	1.30	4.62
Cutpoint Method + radix forest	13	1.22	3.72
$p_i \sim (i \mod 64 + 1)^{35}$	Maximum	Average	Average_{32}
Cutpoint Method + binary search	7	1.19	4.33
Cutpoint Method + radix forest	13	1.11	2.46
"4 spikes"	Maximum	Average	Average_{32}
Cutpoint Method + binary search	4	1.60	3.98
Cutpoint Method + radix forest	5	1.67	4.93

Figure 8 illustrates how sampling is affected by the discontinuities of the Alias Method. The results indicate that sampling with the Alias Method may indeed be less efficient.

Table 1 details the performance improvement of our method as compared to the Cutpoint Method with binary search for the distributions shown in Fig. 12. As expected, the sampling performance of the new method is similar to the Cutpoint Method with binary search, which shares the same primary acceleration data structure, the guide table. For reasonable table sizes both perform almost as good as sampling with an Alias Method, however, do so without affecting the distribution quality.

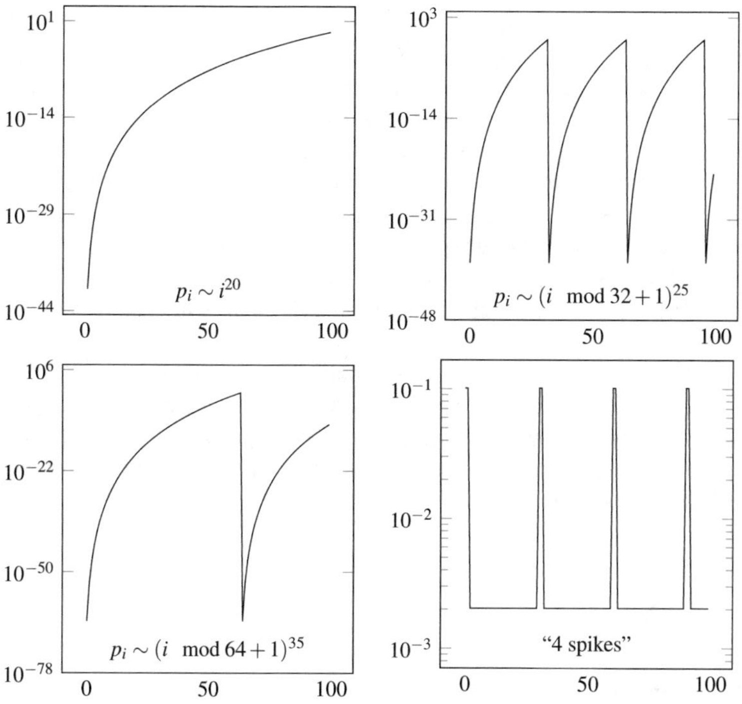

Fig. 12 Example distributions for the numerical results in Table 1

Sampling densities with a high dynamic range can be efficiently accelerated using the Cutpoint Method and its guide table. However, since then some of its cells contain many small values with largely different magnitudes, performance suffers from efficiency issues of index bisection. Radix tree forests improve on this aspect by storing an explicit tree (see the parallel Algorithm 1).

On a single processor with strictly serial execution, our method marginally improves the average search time as compared to the Cutpoint Method with binary search since the overall time is largely dominated by the time required to find large values. For every value that can be directly determined from the guide table—since it is the only one in a cell—this process is already optimal. In some cases the average search time of our new method can be slightly slower since manually assigning the index of interval overlapping from the left as the left child of the root node the tree deteriorates overall tree quality (see Table 1, example "4 spikes"). On the other hand, performing bisection to find a value inside a cell that includes many other values does not take the underlying distribution into account and is therefore suboptimal. Still, since these values are sampled with a low probability, the impact on the average sampling time is low.

The optimizations for values with a lower probability become more important in parallel simulations where the slowest simulation determines the run time for every group. The “4 spikes” example is a synthetic bad case for our method: Spikes are efficiently sampled only using the guide table, whereas all other values have uniform probabilities. Therefore binary search is optimal. The explicit tree, on the other hand, always has one sub-optimal first split to account for intervals overlapping from the left, and therefore requires one additional operation.

In practice, parallel execution of the sampling process often requires synchronization, and therefore suffers more from the outliers: Then, the slowest sampling process determines the speed of the entire group that is synchronized. On Graphics Processing Units (GPUs), such groups are typically of size 32. Under these circumstances our method performs significantly better for distributions with a high dynamic range (see Table 1, third column).

It is important to note that if the maximum execution time is of concern, binary search almost always achieves the best worst case performance. Then explicit tree structures can only improve the average time if the maximum depth does not exceed the number of comparisons required for binary search, which is the binary logarithm of the number of elements.

5 Discussion

A multi-dimensional inversion method proceeds component by component; the two-dimensional distributions in Fig. 8 have been sampled by first calculating the cumulative distribution function of the image rows and one cumulative distribution function for each row. After selecting a row its cumulative distribution function is sampled to select the column. Finally, as distributions are considered piecewise constant, a sub-pixel position, i.e. a position in the constant piece, is required. Therefore, the relative position in the pixel is calculated by rescaling the relative position in the row/column to the unit interval. Other approximations, such as piecewise linear or piecewise quadratic require an additional, simple transformation of the relative position [4].

Building multiple tables and trees simultaneously, e.g. for two-dimensional distributions, is as simple as adding yet another criterion to the extended check in Algorithm 1: If the index of the left or right neighbor goes beyond the *index boundary* of a row, it is a leftmost or a rightmost node, respectively.

Algorithm 1 constructs binary trees. Due to memory access granularity, it may be beneficial to construct 4-ary or even wider trees. A higher branching factor simply results by just collapsing two (or more) levels of the binary trees.

For reasonable table sizes, the Cutpoint Method with binary search preserves the properties of the input samples at a memory footprint comparable to the one required for the Alias Method. Radix forest trees require additional memory to reference two children for each value p_i of the distribution.

Depending on the application and space constraints, it may be beneficial to use balanced trees instead of radix trees. Balanced trees do not need to be built; their structure is implicitly defined, and for each cell we only need to determine the first and last interval that overlap it. Then, the implicit balanced tree is traversed by consecutive bisection of the index interval.

6 Conclusion

Radix tree forests trade additional memory for a faster average case search and come with a massively parallel construction algorithm with optimal load balancing independent of the probability density function p.

While the performance of evaluating the inverse cumulative distribution function P^{-1} is improved for highly nonuniform distributions p with a high dynamic range, performance is slightly worse on distributions which can already be efficiently sampled using the Cutpoint Method with binary search.

Acknowledgements The authors would like to thank Carsten Wächter and Matthias Raab for the discussion of the issues of the Alias Method when used with low discrepancy sequences that lead to the development of radix tree forests.

References

1. Apetrei, C.: Fast and simple agglomerative LBVH construction. In: Borgo, R., Tang, W. (eds.) Theory and Practice of Computer Graphics, Leeds, United Kingdom, 2014, Proceedings, pp. 41–44. The Eurographics Association (2014). https://doi.org/10.2312/cgvc.20141206
2. Chen, H.C., Asau, Y.: On generating random variates from an empirical distribution. IISE Trans. **6**(2), 163–166 (1974). https://doi.org/10.1080/05695557408974949
3. Devroye, L.: Non-uniform Random Variate Generation. Springer, New York (1986)
4. Edwards, A., Rathkopf, J., Smidt, R.: Extending the alias Monte Carlo sampling method to general distributions. Technical report, Lawrence Livermore National Lab., CA (USA) (1991)
5. Fishman, G., Moore, L.: Sampling from a discrete distribution while preserving monotonicity. Am. Stat. **38**(3), 219–223 (1984). https://doi.org/10.1080/00031305.1984.10483208
6. Karras, T.: Maximizing parallelism in the construction of BVHs, octrees, and k-d trees. In: Dachsbacher, C., Munkberg, J., Pantaleoni J. (eds.) High-Performance Graphics 2012, pp. 33–37. Eurographics Association (2012)
7. Niederreiter, H.: Random Number Generation and Quasi-Monte Carlo Methods. SIAM, Philadelphia (1992)
8. Walker, A.: New fast method for generating discrete random numbers with arbitrary frequency distributions. Electron. Lett. **10**(8), 127–128 (1974)
9. Walker, A.: An efficient method for generating discrete random variables with general distributions. ACM Trans. Math. Softw. **3**(3), 253–256 (1977). https://doi.org/10.1145/355744.355749

An Adaptive Algorithm Employing Continuous Linear Functionals

Yuhan Ding, Fred J. Hickernell and Lluís Antoni Jiménez Rugama

Abstract Automatic algorithms attempt to provide approximate solutions that differ from exact solutions by no more than a user-specified error tolerance. This paper describes an automatic, adaptive algorithm for approximating the solution to a general linear problem defined on Hilbert spaces. The algorithm employs continuous linear functionals of the input function, specifically Fourier coefficients. We assume that the Fourier coefficients of the solution decay sufficiently fast, but we do not require the decay rate to be known a priori. We also assume that the Fourier coefficients decay steadily, although not necessarily monotonically. Under these assumptions, our adaptive algorithm is shown to produce an approximate solution satisfying the desired error tolerance, without prior knowledge of the norm of the input function. Moreover, the computational cost of our algorithm is shown to be essentially no worse than that of the optimal algorithm. We provide a numerical experiment to illustrate our algorithm.

Keywords Adaptive algorithms · Cones of input functions · Fourier coefficients · Essentially optimal

Y. Ding (✉) · L. A. Jiménez Rugama
Department of Applied Mathematics, Illinois Institute of Technology, RE 208, 10 W. 32nd St., Chicago, IL 60616, USA
e-mail: yding2@hawk.iit.edu

L. A. Jiménez Rugama
e-mail: ljimene1@hawk.iit.edu

F. J. Hickernell
Center for Interdisciplinary Scientific Computation and Department of Applied Mathematics, Illinois Institute of Technology, RE 208, 10 W. 32nd St., Chicago, IL 60616, USA
e-mail: hickernell@iit.edu

B. Tuffin and P. L'Ecuyer (eds.), *Monte Carlo and Quasi-Monte Carlo Methods*, Springer Proceedings in Mathematics & Statistics 324,
https://doi.org/10.1007/978-3-030-43465-6_8

1 Introduction

Adaptive algorithms determine the design and sample size needed to solve problems to the desired accuracy based on the input function data sampled. A priori upper bounds on some norm of the input function are not needed, but some underlying assumptions about the input function are required for the adaptive algorithm to succeed. Here we consider *general linear problems* where a finite number of series coefficients of the input function are used to obtain an approximate solution. The proposed algorithm produces an approximation with guaranteed accuracy. Moreover, we demonstrate that the computational cost of our algorithm is essentially no worse than that of the best possible algorithm. Our adaptive algorithm is defined on a *cone* of input functions.

1.1 Input and Output Spaces

Let $\mathscr{F}$ be a separable Hilbert space of inputs with orthonormal basis $(u_i)_{i\in\mathbb{N}}$, and let $\mathscr{G}$ be a separable Hilbert space of outputs with orthonormal basis $(v_i)_{i\in\mathbb{N}}$. Based on Parseval's identity the norms for these two spaces may be expressed as the ℓ^2-norms of their series coefficients:

$$f = \sum_{i\in\mathbb{N}} \widehat{f}_i u_i \in \mathscr{F}, \qquad \|f\|_{\mathscr{F}} = \left\|\big(\widehat{f}_i\big)_{i\in\mathbb{N}}\right\|_2, \tag{1a}$$

$$g = \sum_{i\in\mathbb{N}} \widehat{g}_i v_i \in \mathscr{G}, \qquad \|g\|_{\mathscr{G}} = \left\|\big(\widehat{g}_i\big)_{i\in\mathbb{N}}\right\|_2. \tag{1b}$$

Let these two bases be chosen so that the linear solution operator, $S : \mathscr{F} \to \mathscr{G}$, satisfies

$$S(u_i) = \lambda_i v_i, \quad i \in \mathbb{N}, \qquad S(f) = \sum_{i\in\mathbb{N}} \lambda_i \widehat{f}_i v_i, \tag{1c}$$

$$\lambda_1 \geq \lambda_2 \geq \cdots > 0, \quad \lim_{i\to\infty} \lambda_i = 0, \qquad \|S\|_{\mathscr{F}\to\mathscr{G}} := \sup_{f\neq 0} \frac{\|S(f)\|_{\mathscr{G}}}{\|f\|_{\mathscr{F}}} = \lambda_1. \tag{1d}$$

This setting includes, for example, the recovery of functions, derivatives, indefinite integrals, and solutions of linear (partial) differential equations. We focus on cases where the exact solution requires an infinite number of series coefficients, $\widehat{f}_i$, in general, i.e., all λ_i are positive.

The existence of the $(u_i)_{i\in\mathbb{N}}$, $(v_i)_{i\in\mathbb{N}}$, and $(\lambda_i)_{i\in\mathbb{N}}$ for a given $\mathscr{F}$, $\mathscr{G}$, and S follows from the singular value decomposition. The singular value decomposition with singular values tending to zero (but never equal to zero) exists if (and only if) the solution operator is compact and not finite rank. The ease of identifying explicit expressions for these quantities depends on the particular problem of interest. Alternatively, one may start with a choice of $(u_i)_{i\in\mathbb{N}}$, $(v_i)_{i\in\mathbb{N}}$, and $(\lambda_i)_{i\in\mathbb{N}}$, which then determine the solution operator, S, and the spaces $\mathscr{F}$ and $\mathscr{G}$.

Example 1 in Sect. 1.3 illustrates this general setting. Section 5 provides a numerical example based on this example.

1.2 Solvability

Let $\mathscr{H}$ be any subset of $\mathscr{F}$, and let $\mathscr{A}(\mathscr{H})$ denote the set of deterministic algorithms that successfully approximate the solution operator $S : \mathscr{H} \to \mathscr{G}$ within the specified error tolerance for all inputs in $\mathscr{H}$:

$$\mathscr{A}(\mathscr{H}) := \{\text{algorithms } A : \mathscr{H} \times (0, \infty) \to \mathscr{G} : \\ \left\| S(f) - A(f, \varepsilon) \right\|_{\mathscr{G}} \le \varepsilon \;\forall f \in \mathscr{H},\; \varepsilon > 0 \big\}. \quad (2)$$

Algorithms in $\mathscr{A}(\mathscr{H})$ are allowed to sample adaptively any bounded, linear functionals of the input function. They must sample only a finite number of linear functionals for each input function and positive error tolerance. The definition of $\mathscr{H}$ can be used to construct algorithms in $\mathscr{A}(\mathscr{H})$, but no other a priori knowledge about the input functions is available. Following [1] we call a problem *solvable* for inputs $\mathscr{H}$ if $\mathscr{A}(\mathscr{H})$ is non-empty.

Our problem is not solvable for the whole Hilbert space, $\mathscr{F}$, as can be demonstrated by contradiction. For any potential algorithm, we show that there exists some $f \in \mathscr{F}$, that looks like 0 to the algorithm, but for which $S(f)$ is far from $S(0) = 0$. Choose any $A \in \mathscr{A}(\mathscr{F})$ and $\varepsilon > 0$, and let $L_1, \dots, L_n$ be the linear functionals that are used to compute $A(0, \varepsilon)$. Since the output space, $\mathscr{G}$, is infinite dimensional and n is finite, there exists some nonzero $f \in \mathscr{F}$ satisfying that $L_1(f) = \cdots = L_n(f) = 0$ with non-zero $S(f)$. This means that $A(\pm cf, \varepsilon) = A(0, \varepsilon)$ for any real c, and $A(\pm cf, \varepsilon)$ both have approximation error no greater than ε, i.e.,

$$\begin{aligned} \varepsilon &\ge \frac{1}{2}\left[\|S(cf) - A(cf, \varepsilon)\|_{\mathscr{G}} + \|S(-cf) - A(-cf, \varepsilon)\|_{\mathscr{G}}\right] \\ &= \frac{1}{2}\left[\|cS(f) - A(0, \varepsilon)\|_{\mathscr{G}} + \|-cS(f) - A(0, \varepsilon)\|_{\mathscr{G}}\right] \\ &\ge \|cS(f)\|_{\mathscr{G}} = |c|\,\|S(f)\|_{\mathscr{G}} \qquad \text{by the triangle inequality.} \end{aligned}$$

Since $S(f) \neq 0$, it is impossible for this inequality to hold for all real c. The presumed A does not exist, $\mathscr{A}(\mathscr{F})$ is empty, and our problem is not solvable for $\mathscr{F}$. However, our problem is solvable for well-chosen subsets of $\mathscr{F}$, as is shown in the sections below.

1.3 Computational Cost of the Algorithm and Complexity of the Problem

The computational cost of an algorithm $A \in \mathscr{A}(\mathscr{H})$ for $f \in \mathscr{H}$ and error tolerance ε is denoted $\mathrm{cost}(A, f, \varepsilon)$, and is defined as the number of linear functional values required to produce $A(f, \varepsilon)$. By overloading the notation, we define the cost of algorithms for sets of inputs, $\mathscr{H}$, as

$$\mathrm{cost}(A, \mathscr{H}, \varepsilon) := \sup\{\mathrm{cost}(A, f, \varepsilon) : f \in \mathscr{H}\} \qquad \forall \varepsilon > 0.$$

For unbounded sets, $\mathscr{H}$, this cost may be infinite. Therefore, it is meaningful to define the cost of algorithms for input functions in $\mathscr{H} \cap \mathscr{B}_\rho$, where $\mathscr{B}_\rho := \{f \in \mathscr{F} : \|f\|_{\mathscr{F}} \le \rho\}$ is the ball of radius ρ:

$$\mathrm{cost}(A, \mathscr{H}, \varepsilon, \rho) := \sup\{\mathrm{cost}(A, f, \varepsilon) : f \in \mathscr{H} \cap \mathscr{B}_\rho\} \qquad \forall \rho > 0,\ \varepsilon > 0.$$

Finally, we define the complexity of the problem as the computational cost of the best algorithm:

$$\mathrm{comp}(\mathscr{A}(\mathscr{H}), \varepsilon) := \min_{A \in \mathscr{A}(\mathscr{H})} \mathrm{cost}(A, \mathscr{H}, \varepsilon),$$
$$\mathrm{comp}(\mathscr{A}(\mathscr{H}), \varepsilon, \rho) := \min_{A \in \mathscr{A}(\mathscr{H})} \mathrm{cost}(A, \mathscr{H}, \varepsilon, \rho).$$

Note that $\mathrm{comp}(\mathscr{A}(\mathscr{H}), \varepsilon, \rho) \ge \mathrm{comp}(\mathscr{A}(\mathscr{H} \cap \mathscr{B}_\rho), \varepsilon)$. In the former case, the algorithm is unaware that the input function has norm no greater than ρ.

An optimal algorithm for $\mathscr{B}_\rho$ can be constructed in terms of interpolation with respect to the first n series coefficients of the input, namely,

$$A_n(f) := \sum_{i=1}^{n} \lambda_i \widehat{f}_i v_i, \tag{3}$$

$$\|S(f) - A_n(f)\|_{\mathscr{G}} = \left\| \left(\lambda_i \widehat{f}_i\right)_{i=n+1}^{\infty} \right\|_2 \le \lambda_{n+1} \|f\|_{\mathscr{F}}. \tag{4}$$

Define the non-adaptive algorithm as

$$\widehat{A}(f, \varepsilon) = A_{\widehat{n}}(f), \quad \text{where } \widehat{n} = \min\{n : \lambda_{n+1} \le \varepsilon/\rho\}, \qquad \widehat{A} \in \mathscr{A}(\mathscr{B}_\rho). \tag{5}$$

This algorithm is optimal among algorithms in $\mathscr{A}(\mathscr{B}_\rho)$, i.e.,

$$\mathrm{comp}(\mathscr{A}(\mathscr{B}_\rho), \varepsilon) = \mathrm{cost}(\widehat{A}, \mathscr{B}_\rho, \varepsilon) = \min\{n : \lambda_{n+1} \le \varepsilon/\rho\}.$$

To prove this, let A^* be an arbitrary algorithm in $\mathscr{A}(\mathscr{B}_\rho)$, and let $L_1, \ldots, L_{n^*}$ be the linear functionals chosen when evaluating this algorithm for the zero function with tolerance ε. Thus, $A^*(0,\varepsilon)$ is some function of $(L_1(0),\ldots,L_{n^*}(0)) = (0,\ldots,0)$. Let f be a linear combination of $u_1,\ldots,u_{n^*+1}$ with norm ρ satisfying $L_1(f) = \cdots = L_{n^*}(f) = 0$, then $A^*(\pm f,\varepsilon) = A^*(0,\varepsilon)$, and

$$\begin{aligned}
\varepsilon &\geq \max_{\pm} \left\| S(\pm f) - A^*(\pm f,\varepsilon) \right\|_{\mathscr{G}} = \max_{\pm} \left\| \pm S(f) - A^*(0,\varepsilon) \right\|_{\mathscr{G}} \\
&\geq \frac{1}{2}\left[\left\| S(f) - A^*(0,\varepsilon) \right\|_{\mathscr{G}} + \left\| -S(f) - A^*(0,\varepsilon) \right\|_{\mathscr{G}} \right] \\
&\geq \|S(f)\|_{\mathscr{G}} = \left\| \left(\lambda_i \widehat{f}_i\right)_{i=1}^{n^*+1} \right\|_2 \\
&\geq \lambda_{n^*+1} \left\| \left(\widehat{f}_i\right)_{i=1}^{n^*+1} \right\|_2 = \lambda_{n^*+1} \|f\|_{\mathscr{F}} = \lambda_{n^*+1}\rho.
\end{aligned}$$

Thus, $\lambda_{n^*+1} \leq \varepsilon/\rho$, and

$$\mathrm{cost}(A^*, \mathscr{B}_\rho, \varepsilon) \geq \mathrm{cost}(A^*, 0, \varepsilon) = n^* \geq \min\{n : \lambda_{n+1} \leq \varepsilon/\rho\} = \mathrm{cost}(\widehat{A}, \mathscr{B}_\rho, \varepsilon).$$

Hence, the algorithm $\widehat{A}$ defined in (5) is optimal for $\mathscr{A}(\mathscr{B}_\rho)$.

Example 1 Consider the case of function approximation for periodic functions defined over [0,1], and the algorithm $\widehat{A}$ defined in (5):

$$f = \sum_{k\in\mathbb{Z}} \widehat{f}(k)\widehat{u}_k = \sum_{i\in\mathbb{N}} \widehat{f}_i u_i, \qquad S(f) = \sum_{k\in\mathbb{Z}} \widehat{f}(k)\widehat{\lambda}_k\widehat{v}_k = \sum_{i\in\mathbb{N}} \widehat{f}_i\lambda_i v_i,$$

$$\widehat{v}_k(x) := \begin{cases} 1, & k=0, \\ \sqrt{2}\sin(2\pi kx), & k>0, \\ \sqrt{2}\cos(2\pi kx), & k<0, \end{cases} \qquad v_i = \begin{cases} \widehat{v}_{-i/2}, & i \text{ even}, \\ \widehat{v}_{(i-1)/2}, & i \text{ odd}, \end{cases}$$

$$\widehat{\lambda}_k := \begin{cases} 1, & k=0, \\ \dfrac{1}{|k|^r}, & k\neq 0, \end{cases} \qquad \lambda_i = \widehat{\lambda}_{\lfloor i/2\rfloor} = \frac{1}{\max(1,\lfloor i/2\rfloor)^r},$$

$$\widehat{u}_k := \widehat{\lambda}_k\widehat{v}_k, \qquad u_i = \lambda_i v_i = \begin{cases} \widehat{u}_{\,i/2}, & i \text{ even}, \\ \widehat{u}_{(i-1)/2}, & i \text{ odd}, \end{cases}$$

$$\|f\|_{\mathscr{F}}^2 = \left(\int_0^1 f(x)\,\mathrm{dx}\right)^2 + \frac{\|f^{(r)}\|_2^2}{(2\pi)^{2r}}, \quad r\in\mathbb{N}, \qquad \widehat{f}_i = \begin{cases} \widehat{f}(-i/2), & i \text{ even}, \\ \widehat{f}((i-1)/2), & i \text{ odd}, \end{cases}$$

$$\|g\|_{\mathscr{G}} = \|g\|_2,$$

$$\begin{aligned}
\mathrm{comp}(\mathscr{A}(\mathscr{B}_\rho),\varepsilon) = \mathrm{cost}(\widehat{A},\mathscr{B}_\rho,\varepsilon) &= \min\{n : \lambda_{n+1} \leq \varepsilon/\rho\} \\
&= \min\left\{ n : \frac{1}{\lfloor (n+1)/2\rfloor^r} \leq \frac{\varepsilon}{\rho} \right\} = 2\left\lceil \left(\frac{\rho}{\varepsilon}\right)^{1/r} \right\rceil - 1.
\end{aligned}$$

Here, r is positive. If r is an integer, then $\mathscr{F}$ consists of functions that have absolutely continuous, periodic derivatives of up to order $r-1$. A larger r implies a stronger $\mathscr{F}$-norm, a more exclusive $\mathscr{B}_\rho$, and a smaller $\mathrm{cost}(\widehat{A},\mathscr{B}_\rho,\varepsilon)$.

Our goal is to construct algorithms in $\mathscr{A}(\mathscr{H})$ for some $\mathscr{H}$ and also to determine whether the computational cost of these algorithms is reasonable. We define $\mathrm{cost}(A, \mathscr{H}, \varepsilon, \rho)$ to be *essentially no worse* than $\mathrm{cost}(A^*, \mathscr{H}^*, \varepsilon, \rho)$ if there exists $\omega > 0$ such that

$$\mathrm{cost}(A, \mathscr{H}, \varepsilon, \rho) \le \mathrm{cost}(A^*, \mathscr{H}^*, \omega\varepsilon, \rho) \qquad \forall \varepsilon, \rho > 0. \tag{6}$$

We extend this definition analogously if $\mathrm{cost}(A, \mathscr{H}, \varepsilon, \rho)$ is replaced by cost $(A, \mathscr{H}, \varepsilon)$ and/or $\mathrm{cost}(A^*, \mathscr{H}^*, \omega\varepsilon, \rho)$ is replaced by $\mathrm{cost}(A^*, \mathscr{H}^*, \omega\varepsilon)$. If these inequalities are not satisfied, we say that the cost of A is *essentially worse* than the cost of A^*. If the costs of two algorithms are essentially no worse than each other, then we call their costs essentially the same. An algorithm whose cost is essentially no worse than the best possible algorithm, is called *essentially optimal.*

Our condition for essentially no worse cost in (6) is not the same as

$$\mathrm{cost}(A, \mathscr{H}, \varepsilon, \rho) \le \omega\, \mathrm{cost}(A^*, \mathscr{H}^*, \varepsilon, \rho) \qquad \forall \varepsilon, \rho > 0. \tag{7}$$

If the cost grows polynomially in ε^{-1}, then conditions (6) and (7) are basically the same. If for some positive p and p^*,

$$\mathrm{cost}(A, \mathscr{H}, \varepsilon, \rho) \le C(1 + \varepsilon^{-p}\rho^{p}) \quad \text{and} \quad C^*(1 + \varepsilon^{-p^*}\rho^{p^*}) \le \mathrm{cost}(A^*, \mathscr{H}^*, \varepsilon, \rho),$$

then $\mathrm{cost}(A, \mathscr{H}, \varepsilon, \rho)$ is essentially no worse than $\mathrm{cost}(A^*, \mathscr{H}^*, \varepsilon, \rho)$ iff $p \le p^*$ under either condition (6) or (7). However, if the cost grows only logarithmically in ε^{-p}, then condition (6) makes more sense than (7). Specifically, if

$$\mathrm{cost}(A, \mathscr{H}, \varepsilon, \rho) \le C + \log\bigl(1 + \varepsilon^{-p}\rho^{p}\bigr) \quad \text{and}$$
$$C^* + \log\bigl(1 + \varepsilon^{-p^*}\rho^{p^*}\bigr) \le \mathrm{cost}(A^*, \mathscr{H}^*, \varepsilon, \rho),$$

then condition (6) requires $p \le p^*$ for $\mathrm{cost}(A, \mathscr{H}, \varepsilon, \rho)$ to be essentially no worse than $\mathrm{cost}(A^*, \mathscr{H}^*, \varepsilon, \rho)$, whereas (7) allows this to be true even for $p > p^*$. We grant that if the cost grows faster than polynomially in ε^{-1}, then (7) may be the preferred condition.

To illustrate the comparison of costs, consider a non-increasing sequence of positive numbers, $\boldsymbol{\lambda}^* = (\lambda_1^*, \lambda_2^*, \ldots)$, which converges to 0, where $\lambda_i^* \ge \lambda_i$ for all $i \in \mathbb{N}$. Also consider an unbounded, strictly increasing sequence of non-negative integers $\boldsymbol{n} = (n_0, n_1, \ldots)$. Define an algorithm A^* analogously to $\widehat{A}$ defined in (5):

$$A^*(f, \varepsilon) = A_{n^*}(f), \text{ where } n^* = n_{j^*},\ j^* = \min\{j : \lambda_{n_j+1}^* \le \varepsilon/\rho\}, \quad A^* \in \mathscr{A}(\mathscr{B}_\rho).$$

By definition, the cost of algorithm A^* is no smaller than that of $\widehat{A}$. Algorithm A^* may or may not have essentially worse cost than $\widehat{A}$ depending on the choice of $\boldsymbol{\lambda}^*$ and $\boldsymbol{n}$. The table below shows some examples. Each different case of A^* is labeled as having a cost that is either essentially no worse or essentially worse than that of $\widehat{A}$.

	$\lambda_i = \frac{C}{i^p}$	$\mathrm{cost}(\widehat{A}, \mathscr{B}_\rho, \varepsilon) \geq \left(\frac{C\rho}{\varepsilon}\right)^{1/p} - 1$ $\mathrm{cost}(\widehat{A}, \mathscr{B}_\rho, \varepsilon) < \left(\frac{C\rho}{\varepsilon}\right)^{1/p}$
no worse	$\lambda_i^* = \frac{C^*}{i^p},\ n_j = 2^j$	$\mathrm{cost}(A^*, \mathscr{B}_\rho, \varepsilon) \leq 2\left(\frac{C^*\rho}{\varepsilon}\right)^{1/p}$
worse	$\lambda_i^* = \frac{C^*}{i^q},\ q < p,\ n_j = j$	$\mathrm{cost}(A^*, \mathscr{B}_\rho, \varepsilon) \geq \left(\frac{C^*\rho}{\varepsilon}\right)^{1/q} - 1$
	$\lambda_i = \frac{C}{p^i},\ p > 1$	$\mathrm{cost}(\widehat{A}, \mathscr{B}_\rho, \varepsilon) \geq \frac{\log(C\rho/\varepsilon)}{\log(p)} - 1$ $\mathrm{cost}(\widehat{A}, \mathscr{B}_\rho, \varepsilon) < \frac{\log(C\rho/\varepsilon)}{\log(p)}$
no worse	$\lambda_i^* = \frac{C^*}{p^i},\ n_j = 2j$	$\mathrm{cost}(A^*, \mathscr{B}_\rho, \varepsilon) < \frac{\log(C^*\rho/\varepsilon)}{\log(p)} + 1$
worse	$\lambda_i^* = \frac{C^*}{p^i},\ n_j = 2^j$	$\mathrm{cost}(A^*, \mathscr{B}_\rho, \varepsilon) > 1.999\frac{\log(C^*\rho/\varepsilon)}{\log(p)}$ for some ε
worse	$\lambda_i^* = \frac{C^*}{i^q},\ q < p,\ n_j = j$	$\mathrm{cost}(A^*, \mathscr{B}_\rho, \varepsilon) \geq \frac{\log(C^*\rho/\varepsilon)}{\log(q)} - 1$

1.4 The Case for Adaptive Algorithms

For bounded sets of input functions, such as balls, non-adaptive algorithms like $\widehat{A}$ make sense. However, an a priori upper bound on $\|f\|_{\mathscr{F}}$ is typically unavailable in practice, so it is unknown which $\mathscr{B}_\rho$ contain the input function f. Our aim is to consider unbounded sets of f for which the error of the interpolatory algorithm $A_n(f)$, defined in (3), can be bounded without an a priori upper bound on $\|f\|_{\mathscr{F}}$.

Popular adaptive algorithms encountered typically employ heuristic error bounds. While any algorithm can be fooled, we would like to have precise necessary conditions for being fooled, or equivalently, sufficient conditions for the algorithm to succeed. Our adaptive algorithm has such conditions and follows in the vein of adaptive algorithms developed in [2–6].

Our rigorous, data-driven error bound assumes that the series coefficients of the input function, f, decay steadily—but not necessarily monotonically. The cone of nice input functions, $\mathscr{C}$, is defined in Sect. 2. For such inputs, we construct an adaptive algorithm, $\widetilde{A} \in \mathscr{A}(\mathscr{C})$, in Sect. 3, where $\widetilde{A}(f, \varepsilon) = A_{\widetilde{n}}(f)$ for some $\widetilde{n}$ depending on the input data and the definition of $\mathscr{C}$. The number of series coefficients sampled, $\widetilde{n}$, is adaptively determined so that $\widetilde{A}(f, \varepsilon)$ satisfies the error condition in (2). The computational cost of $\widetilde{A}$ is given in Theorem 1. Section 4 shows that our new algorithm is essentially optimal (see Theorem 3). Section 5 provides a numerical example. We end with concluding remarks.

2 Assuming a Steady Decay of the Series Coefficients of the Solution

Recall from (4) that the error of the fixed sample size interpolatory algorithm A_n is $\|S(f) - A_n(f)\|_{\mathscr{G}} = \left\|\left(\lambda_i \widehat{f_i}\right)_{i=n+1}^{\infty}\right\|_2$. The error depends on the series coefficients not yet observed, so at first it seems impossible to bound the error in terms of observed series coefficients.

However, we can observe a finite number of partial sums,

$$\sigma_j(f) := \left\|\left(\lambda_i \widehat{f_i}\right)_{i=n_{j-1}+1}^{n_j}\right\|_2, \qquad j \in \mathbb{N}, \tag{8}$$

where $\boldsymbol{n} = (n_0, n_1, \ldots)$ is an unbounded, strictly increasing sequence of non-negative integers. We define the cone of nice input functions to consist of those functions for which the $\sigma_j(f)$ decay at a given rate with respect to one another:

$$\begin{aligned} \mathscr{C} = \mathscr{C}_{\boldsymbol{n},a,b} &= \left\{f \in \mathscr{F} : \sigma_{j+r}(f) \le ab^r \sigma_j(f) \ \forall j, r \in \mathbb{N}\right\} \\ &= \left\{f \in \mathscr{F} : \sigma_j(f) \le \min_{1 \le r < j} ab^r \sigma_{j-r}(f) \ \forall j \in \mathbb{N}\right\}. \end{aligned} \tag{9}$$

Here, a and b are positive numbers that define the inclusivity of the cone $\mathscr{C}$ and satisfy

$$b < 1 < a.$$

The constant a is an inflation factor, and the constant b defines the general rate of decay of the $\sigma_j(f)$ for $f \in \mathscr{C}$. Because ab^r may be greater than one, we do not require the series coefficients of the solution, $S(f)$, to decay monotonically. However, we expect their partial sums to decay steadily.

From the expression for the error in (4) and the definition of the cone in (9), one can now derive a data-driven error bound for $j \in \mathbb{N}$:

$$\begin{aligned} \left\|S(f) - A_{n_j}(f)\right\|_{\mathscr{G}} &= \left\|\left(\lambda_i \widehat{f_i}\right)_{i=n_j+1}^{\infty}\right\|_2 = \left\{\sum_{r=1}^{\infty} \sum_{i=n_{j+r-1}+1}^{n_{j+r}} \left|\lambda_i \widehat{f_i}\right|^2\right\}^{1/2} \\ &= \left\|\left(\sigma_{j+r}(f)\right)_{r=1}^{\infty}\right\|_2 \\ &\le \left\|\left(ab^r \sigma_j(f)\right)_{r=1}^{\infty}\right\|_2 = \frac{ab}{\sqrt{1-b^2}} \sigma_j(f). \end{aligned} \tag{10}$$

This upper bound depends only on the function data and the parameters defining $\mathscr{C}$. The error vanishes as $j \to \infty$ because $\sigma_j(f) \le ab^{j-1}\sigma_1(f) \to 0$ as $j \to \infty$. Moreover, the error of $A_{n_j}(f)$ is asymptotically no worse than $\sigma_j(f)$. Our adaptive algorithm in Sect. 3 increases j until the right hand side is smaller than the error tolerance.

Consider the choice $n_j = 2^j n_0$, where the number of terms in the sums, $\sigma_j(f)$, are doubled at each step. If the series coefficients of the solution decay like $\lambda_i |\widehat{f}_i| = \Theta(i^{-p})$ for some $p > 1/2$, then it is reasonable to expect that the $\sigma_j(f)$ are bounded above and below as

$$C_{\text{lo}}(n_0 2^j)^{1/2-p} \le \sigma_j(f) \le C_{\text{up}}(n_0 2^j)^{1/2-p}, \quad j \in \mathbb{N}, \tag{11}$$

for some constants C_{lo} and C_{up}, unless the series coefficients drop precipitously in magnitude for some $n_{j-1} < i \le n_j$, and then jump back up for larger i. When (11) holds, it follows that

$$\frac{\sigma_{j+r}(f)}{\sigma_j(f)} \le \frac{C_{\text{up}}(n_0 2^{j+r})^{1/2-p}}{C_{\text{lo}}(n_0 2^j)^{1/2-p}} = \frac{C_{\text{up}} 2^{r(1/2-p)}}{C_{\text{lo}}} \qquad j \in \mathbb{N}.$$

Thus, choosing $a \ge C_{\text{up}}/C_{\text{lo}}$ and $b \ge 2^{1/2-p}$ ensures that reasonable inputs f lie inside the cone $\mathscr{C}$.

Although the definition of the cone in (9) constrains the decay rate of $\lambda_i |\widehat{f}_i|$ it is a rather weak constraint. Specifying a and b only implies a lower bound—but not an upper bound—on the p for which $\lambda_i |\widehat{f}_i|$ can be $\Theta(i^{-p})$.

Figure 1 shows three functions and their coefficients for the case of Example 1 when $r = 1$ and $n_j = 2^j$. These functions are $f(x) = \mathrm{e}^{-3x}\sin(3\pi x^2)$, $f_{\text{big}} = 100 f$ and f_{fuzzy}. The function f_{fuzzy} is obtained by taking $\sigma_8(f_{\text{fuzzy}}) = 250\sigma_8(f)$ and $\sigma_j(f_{\text{fuzzy}}) = \sigma_j(f)$ for $j \ne 8$. Both f and f_{big} lie in the same cones, $\mathscr{C}_{\boldsymbol{n},a,b}$, for all $\boldsymbol{n}$, a, and b, and both appear similarly nice to the eye. Therefore, we expect adaptive algorithms that are successful for f to also be successful for f_{big}. However, f_{big} does not lie in some of the balls, $\mathscr{B}_\rho$, that f lies in. On the other hand, the high frequency noise in f_{fuzzy} suggests that only more robust and costly adaptive algorithms would succeed for such an input. This corresponds to the fact that f_{fuzzy} does not lie in all cones, $\mathscr{C}_{\boldsymbol{n},a,b}$, that f lies in.

3 Adaptive Algorithm

Now we introduce our adaptive algorithm, $\widetilde{A} \in \mathscr{A}(\mathscr{C})$, which yields an approximate solution to the problem $S : \mathscr{C} \to \mathscr{G}$ that meets the absolute error tolerance ε.

Algorithm 1 *Given a, b, the sequence $\boldsymbol{n}$, the cone $\mathscr{C}$, the input function $f \in \mathscr{C}$, and the absolute error tolerance ε, set $j = 1$.*

Step 1. *Compute $\sigma_j(f)$ as defined in* (8).
Step 2. *Check whether j is large enough to satisfy the error tolerance, i.e.,*

$$\sigma_j(f) \le \frac{\varepsilon\sqrt{1-b^2}}{ab}.$$

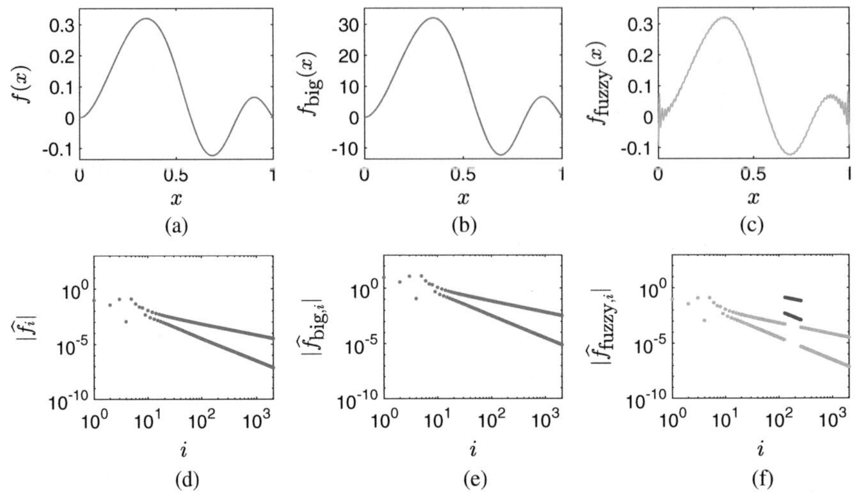

Fig. 1 **a** The function $f(x) = \mathrm{e}^{-3x}\sin(3\pi x^2)$; **b** The function $f_{\text{big}} = 100f$; **c** The function f_{fuzzy} which results from modifying some of f coefficients, in dark green; **d** The first Fourier coefficients of f; **e** The first Fourier coefficients of f_{big}; **f** The first Fourier coefficients of f_{fuzzy}

If this is true, then return $\widetilde{A}(f, \varepsilon) = A_{n_j}(f)$*, where* A_n *is defined in* (3)*, and terminate the algorithm.*

Step 3. *Otherwise, increase* j *by* 1 *and return to Step* 1.

Theorem 1 *The algorithm,* $\widetilde{A}$*, defined in Algorithm 1 lies in* $\mathscr{A}(\mathscr{C})$ *and has computational cost* $cost(\widetilde{A}, f, \varepsilon) = n_{j^*}$*, where* j^* *is defined implicitly by the inequalities*

$$j^* = \min\left\{ j \in \mathbb{N} : \sigma_j(f) \leq \frac{\varepsilon\sqrt{1-b^2}}{ab} \right\}. \tag{12}$$

Moreover, $cost(\widetilde{A}, \mathscr{C}, \varepsilon, \rho) = n_{j^\dagger}$*, where* $j^\dagger$ *satisfies the following upper bound:*

$$j^\dagger \leq \min\left\{ j \in \mathbb{N} : F(j) \geq \frac{\rho^2 a^4}{\varepsilon^2(1-b^2)} \right\}, \tag{13}$$

and F *is the strictly increasing function defined as*

$$F(j) := \sum_{k=0}^{j-1} \frac{b^{2(k-j)}}{\lambda_{n_k}^2}, \qquad j \in \mathbb{N}. \tag{14}$$

Proof This algorithm terminates for some $j = j^*$ because $\sigma_j(f) \le ab^{j-1}\sigma_1(f) \to 0$ as $j \to \infty$. The value of j^* follows directly from this termination criterion in Step 2. It then follows that the error bound on $A_{n_{j^*}}(f)$ in (10) is no greater than the error tolerance ε. So, $\widetilde{A} \in \mathscr{A}(\mathscr{C})$.

For the remainder of the proof consider ρ and ε to be fixed. To derive an upper bound on $n_{j^\dagger} = \text{cost}(\widetilde{A}, \mathscr{C}, \varepsilon, \rho)$ we first note some properties of $\sigma_j(f)$ for all $f \in \mathscr{C}$:

$$\begin{aligned}\lambda_{n_j} \left\| \left(\widehat{f_i}\right)_{i=n_{j-1}+1}^{n_j} \right\|_2 \le \left\| \left(\lambda_i \widehat{f_i}\right)_{i=n_{j-1}+1}^{n_j} \right\|_2 = \sigma_j(f) \\ \le \lambda_{n_{j-1}+1} \left\| \left(\widehat{f_i}\right)_{i=n_{j-1}+1}^{n_j} \right\|_2. \end{aligned} \tag{15}$$

A rough upper bound on $j^\dagger$ may be obtained by noting that for any $f \in \mathscr{C} \cap \mathscr{B}_\rho$ and for any $j < j^* \le j^\dagger$, it follows from (12) and (15) that

$$\rho \ge \|f\|_{\mathscr{F}} \ge \left\| \left(\widehat{f_i}\right)_{i=n_{j-1}+1}^{n_j} \right\|_2 \ge \frac{\sigma_j(f)}{\lambda_{n_{j-1}+1}} > \frac{\varepsilon\sqrt{1-b^2}}{ab\lambda_{n_{j-1}+1}} \ge \frac{\varepsilon\sqrt{1-b^2}}{ab\lambda_{n_{j-1}}}$$

Thus, one upper bound on $j^\dagger$ is the smallest j violating the above inequality:

$$j^\dagger \le \min\left\{ j \in \mathbb{N} : \lambda_{n_{j-1}}^{-1} \ge \frac{\rho ab}{\varepsilon\sqrt{1-b^2}} \right\}. \tag{16}$$

The tighter upper bound in Theorem 1 may be obtained by a more careful argument in a similar vein. For any $f \in \mathscr{C} \cap \mathscr{B}_\rho$ satisfying $n_{j^*} = \text{cost}(\widetilde{A}, f, \varepsilon) = \text{cost}(\widetilde{A}, \mathscr{C}, \varepsilon, \rho) = n_{j^\dagger}$ and for any $j < j^* = j^\dagger$,

$$\begin{aligned}\rho^2 \ge \|f\|_{\mathscr{F}}^2 &= \left\| \left(\widehat{f_i}\right)_{i=1}^{\infty} \right\|_2^2 \ge \sum_{k=1}^{j} \left\| \left(\widehat{f_i}\right)_{i=n_{k-1}+1}^{n_k} \right\|_2^2 \\ &> \sum_{k=1}^{j} \frac{\sigma_k^2(f)}{\lambda_{n_{k-1}+1}^2} \qquad \text{by (15)} \\ &\ge \sum_{k=1}^{j-1} \frac{b^{2(k-j)}\sigma_j^2(f)}{a^2\lambda_{n_{k-1}+1}^2} + \frac{\sigma_j^2(f)}{\lambda_{n_{j-1}+1}^2} \qquad \text{by (9)} \\ &= \sigma_j^2(f) \left[\sum_{k=1}^{j-1} \frac{b^{2(k-j)}}{a^2\lambda_{n_{k-1}+1}^2} + \frac{1}{\lambda_{n_{j-1}+1}^2} \right] \\ &> \frac{\sigma_j^2(f)}{a^2} \sum_{k=1}^{j} \frac{b^{2(k-j)}}{\lambda_{n_{k-1}+1}^2} \qquad \text{since } a > 1 \\ &= \frac{\sigma_j^2(f)}{a^2} \sum_{k=0}^{j-1} \frac{b^{2(k+1-j)}}{\lambda_{n_k+1}^2}\end{aligned}$$

$$\geq \frac{b^2\sigma_j^2(f)}{a^2} \sum_{k=0}^{j-1} \frac{b^{2(k-j)}}{\lambda_{n_k}^2} \qquad \text{since } \lambda_{n_k} \geq \lambda_{n_k+1}$$

$$= \frac{b^2\sigma_j^2(f)}{a^2} F(j) \qquad \text{by (14).}$$

So, for any $j < j^* = j^\dagger$ it follows from the termination criterion in (12) that

$$F(j) < \frac{\rho^2 a^4}{\varepsilon^2(1-b^2)}.$$

Note from (14) that F is an increasing function because as j increases, the sum defining F includes more positive terms and $b^{2(k-j)}$ also increases. Thus, any j that violates the above inequality, must satisfy $j \geq j^\dagger$. This establishes (13). □

We note in passing that for our adaptive algorithm,

$$\min\{\text{cost}(\widetilde{A}, f, \varepsilon) : f \in \mathscr{C} \setminus \mathscr{B}_\rho\} \begin{cases} = n_1, & n_0 > 0, \\ \leq n_2, & n_0 = 0, \end{cases} \qquad \forall \rho > 0,\ \varepsilon > 0.$$

This result may be obtained by considering functions where only $\widehat{f_1}$ is nonzero. For $n_0 > 0$, $\sigma_1(f) = 0$, and for $n_0 = 0$, $\sigma_2(f) = 0$.

The upper bound on $\text{cost}(\widetilde{A}, \mathscr{C}, \rho, \varepsilon)$ in Theorem 1 is a non-decreasing function of ρ/ε, which depends on the behavior of the sequence $(\lambda_{n_j})_{j=0}^\infty$. This in turn depends both on the increasing sequence $\boldsymbol{n}$ and on the non-increasing sequence $(\lambda_i)_{i=1}^\infty$. Considering the definition of F, one can imagine that in some cases the first term in the sum dominates, while in other cases the last term in the sum dominates, all depending on how b^{k-j}/λ_{n_k} behaves with k and j. These simplifications lead to two simpler, but coarser upper bounds on the cost of $\widetilde{A}$.

Corollary 1 *For the algorithm, $\widetilde{A}$, defined in Algorithm 1, we have cost $(\widetilde{A}, \mathscr{C}, \varepsilon, \rho) \leq n_{j^\dagger}$, where $j^\dagger$ satisfies the following upper bound:*

$$j^\dagger \leq \left\lceil \log\left(\frac{\rho a^2 \lambda_{n_0}}{\varepsilon\sqrt{1-b^2}}\right) \Big/ \log\left(\frac{1}{b}\right) \right\rceil. \tag{17}$$

Moreover, if the $\lambda_{n_{j-1}}$ decay as quickly as

$$\lambda_{n_{j-1}} \leq \alpha\beta^j, \quad j \in \mathbb{N}, \qquad \textit{for some } \alpha > 0,\ 0 < \beta < 1. \tag{18}$$

then $j^\dagger$ also satisfies the following upper bound:

$$j^\dagger \leq \left\lceil \log\left(\frac{\rho a^2 \alpha b}{\varepsilon\sqrt{1-b^2}}\right) \Big/ \log\left(\frac{1}{\beta}\right) \right\rceil. \tag{19}$$

Proof Ignoring all but the first term in the definition of F in (14) implies that

$$j^\dagger \le \min\left\{j \in \mathbb{N} : b^{-2j} \ge \frac{\rho^2 a^4 \lambda_{n_0}^2}{\varepsilon^2(1-b^2)}\right\}$$
$$= \min\left\{j \in \mathbb{N} : j \ge \log\left(\frac{\rho a^2 \lambda_{n_0}}{\varepsilon(\sqrt{1-b^2}}\right) \Big/ \log\left(\frac{1}{b}\right)\right\}.$$

This implies (17).

Ignoring all but the last term of the sum leads to the simpler upper bound similar to (16):

$$j^\dagger \le \min\left\{j \in \mathbb{N} : \lambda_{n_{j-1}}^{-1} \ge \frac{\rho a^2 b}{\varepsilon\sqrt{1-b^2}}\right\}.$$

If the $\lambda_{n_{j-1}}$ decay as assumed in (18) then

$$j^\dagger \le \min\left\{j \in \mathbb{N} : \alpha\beta^{-j} \le \frac{\rho a^2 b}{\varepsilon\sqrt{1-b^2}}\right\},$$

which implies (19). □

This corollary highlights two limiting factors on the computational cost of our adaptive algorithm, $\widetilde{A}$. When j is large enough to make $\lambda_{n_{j-1}} \|f\|_{\mathscr{F}}/\varepsilon$ small enough, $\widetilde{A}(f,\varepsilon)$ stops. This is statement (19) and its precursor, (16). Alternatively, the assumption that the $\sigma_j(f)$ are steadily decreasing, as specified in the definition of $\mathscr{C}$ in (9), means that $\widetilde{A}(f,\varepsilon)$ also must stop by the time j becomes large enough with respect to $\lambda_{n_0} \|f\|_{\mathscr{F}}/\varepsilon$.

Assumption (18) is not very restrictive. It holds if the λ_i decay algebraically and the n_j increase geometrically. It also holds if the λ_i decay geometrically and the n_j increase arithmetically.

The adaptive algorithm $\widetilde{A}$, which does not know an upper bound on $\|f\|_{\mathscr{G}}$ a priori, may cost more than the non-adaptive algorithm $\widehat{A}$, which assumes an upper bound on $\|f\|_{\mathscr{F}}$, but under reasonable assumptions, the extra cost is small.

Corollary 2 *Suppose that the sequence* $\boldsymbol{n}$ *is chosen to satisfy*

$$\lambda_{n_{j+1}} \ge c_\lambda \lambda_{n_j}, \qquad j \in \mathbb{N}, \tag{20}$$

for some positive c_λ. *Then* $cost(\widetilde{A}, \mathscr{C}, \varepsilon, \rho)$ *is essentially no worse than* $cost(\widehat{A}, \mathscr{B}_\rho, \varepsilon)$ *in the sense of* (6).

Proof Combining the upper bound on $n_{j^\dagger} = \mathrm{cost}(\widetilde{A}, \mathscr{C}, \varepsilon, \rho)$ in (16) plus (20) above, it follows that

$$\lambda_{n_{j^\dagger}} \ge c_\lambda^2 \lambda_{n_{j^\dagger-2}} > \frac{\varepsilon c_\lambda^2 \sqrt{1-b^2}}{\rho a b} = \frac{\omega\varepsilon}{\rho} \ge \lambda_{\widehat{n}+1}, \qquad \omega := \frac{c_\lambda^2\sqrt{1-b^2}}{ab},$$

where $\widehat{n} = \mathrm{cost}(\widehat{A}, \mathscr{B}_\rho, \omega\varepsilon)$. Since the λ_i are non-increasing and $\lambda_{n_{j^\dagger}} > \lambda_{\widehat{n}+1}$, it follows that $n_{j^\dagger} < \widehat{n} + 1$, and so $n_{j^\dagger} \le \widehat{n}$. Thus,

$$\mathrm{cost}(\widetilde{A}, \mathscr{C}, \varepsilon, \rho) = n_{j^\dagger} \le \widehat{n} = \mathrm{cost}(\widehat{A}, \mathscr{B}_\rho, \omega\varepsilon).$$

□

4 Essential Optimality of the Adaptive Algorithm

From Corollary 2 it is known that $\mathrm{cost}(\widetilde{A}, \mathscr{C}, \varepsilon, \rho)$ is essentially no worse than $\mathrm{cost}(\widehat{A}, \mathscr{B}_\rho, \varepsilon) = \mathrm{comp}(\mathscr{A}(\mathscr{B}_\rho), \varepsilon)$. We would like to show that $\widetilde{A} \in \mathscr{A}(\mathscr{C})$ is essentially optimal, i.e., $\mathrm{cost}(\widetilde{A}, \mathscr{C}, \varepsilon, \rho)$ is essentially no worse than $\mathrm{comp}(\mathscr{A}(\mathscr{C}), \varepsilon, \rho)$. However, $\mathrm{comp}(\mathscr{A}(\mathscr{C}), \varepsilon, \rho)$ may be smaller than $\mathrm{comp}(\mathscr{A}(\mathscr{B}_\rho), \varepsilon)$ because $\mathscr{C} \cap \mathscr{B}_\rho$ is a strict subset of $\mathscr{B}_\rho$. This presents a challenge.

A lower bound on $\mathrm{comp}(\mathscr{A}(\mathscr{C}), \varepsilon, \rho)$ is established by constructing fooling functions in $\mathscr{C}$ with norms no greater than ρ. To obtain a result that can be compared with the cost of our algorithm, we assume that

$$R = \sup_{j \in \mathbb{N}} \frac{\lambda_{n_{j-1}}}{\lambda_{n_j}} < \infty. \tag{21}$$

That is, the subsequence $\lambda_{n_1}, \lambda_{n_2}, \ldots$ cannot decay too quickly.

The following theorem establishes a lower bound on the complexity of our problem for input functions in $\mathscr{C}$. The theorem after that shows that the cost of our algorithm as given in Theorem 1 is essentially no worse than this lower bound.

Theorem 2 *Under assumption* (21), *a lower bound on the complexity of the linear problem defined in* (1) *is*

$$\mathit{comp}(\mathscr{A}(\mathscr{C}), \varepsilon, \rho) \ge n_{j^\ddagger} - 1,$$

where

$$j^\ddagger = \min\left\{ j \in \mathbb{N} : F(j+2) \ge \frac{\rho^2}{\varepsilon^2 b^2} \left[\frac{(a+1)^2 R^2}{(a-1)^2} + 1 \right]^{-1} \right\},$$

and F is the function defined in (14).

Proof Consider a fixed ρ and ε. Choose any positive integer j such that $n_j \ge \mathrm{comp}(\mathscr{A}(\mathscr{C}), \varepsilon, \rho) + 2$. The proof proceeds by carefully constructing three test input functions, f and $f_\pm$, lying in $\mathscr{C} \cap \mathscr{B}_\rho$, which yield the same approximate solution but different true solutions. This leads to a lower bound on n_j, which can be translated into a lower bound on $\mathrm{comp}(\mathscr{A}(\mathscr{C}), \varepsilon, \rho)$.

The first test function $f \in \mathscr{C}$ is defined in terms of its series coefficients as follows:

$$\widehat{f}_i := \begin{cases} \dfrac{cb^{k-j}}{\lambda_{n_k}}, & i = n_k,\ k = 1, \ldots, j, \\ 0, & \text{otherwise}, \end{cases}$$

$$c^2 := \rho^2 \left[\left(1 + \frac{(a-1)^2}{(a+1)^2 R^2} \right) \sum_{k=0}^{j} \frac{b^{2(k-j)}}{\lambda_{n_k}^2} \right]^{-1}.$$

It can be verified that the test function lies both in $\mathscr{B}_\rho$ and in $\mathscr{C}$:

$$\|f\|_{\mathscr{F}}^2 = c^2 \sum_{k=1}^{j} \frac{b^{2(k-j)}}{\lambda_{n_k}^2} \le \rho^2,$$

$$\sigma_k(f) = \begin{cases} cb^{k-j}, & k = 1, \ldots, j, \\ 0, & \text{otherwise}, \end{cases}$$

$$\sigma_{k+r}(f) = \begin{cases} b^r \sigma_k(f) \le ab^r \sigma_k(f), & k + r \le j,\ r \ge 1, \\ 0 \le ab^r \sigma_k(f), & k + r > j,\ r \ge 1. \end{cases}$$

Now suppose that $A^* \in \mathscr{A}(\mathscr{C})$ is an optimal algorithm, i.e., $\text{cost}(A^*, \mathscr{C}, \varepsilon, \rho) = \text{comp}(\mathscr{A}(\mathscr{C}), \varepsilon, \rho)$ for all $\varepsilon, \rho > 0$. For our particular input f defined above, suppose that $A^*(f, \varepsilon)$ samples $L_1(f), \ldots, L_n(f)$ where

$$n + 2 \le \text{comp}(\mathscr{A}(\mathscr{C}), \varepsilon, \rho) + 2 \le n_j.$$

Let u be a linear combination of $u_1, \cdots, u_{n_j}$, expressed as

$$u = \sum_{k=0}^{j} \frac{b^{k-j} u^{(k)}}{\lambda_{n_k}},$$

where $u^{(0)}$ is a linear combination of $u_1, \ldots, u_{n_0}$, and each $u^{(k)}$ is a linear combination of $u_{n_{k-1}+1}, \ldots, u_{n_k}$, for $k = 1, \ldots, j$. We constrain u to satisfy:

$$L_1(u) = \cdots = L_n(u) = 0, \qquad \langle u, f \rangle_{\mathscr{F}} = 0, \qquad \max_{0 \le k \le j} \left\| u^{(k)} \right\|_{\mathscr{F}} = 1.$$

Since u is a linear combination of $n_j \ge n + 2$ basis functions, these $n + 2$ constraints can be satisfied.

Let the other two test functions be constructed in terms of u as

$$
\begin{aligned}
f_{\pm} &:= f \pm \eta u, \qquad \eta := \frac{(a-1)c}{(a+1)R}, \qquad (22)\\
\|f_{\pm}\|_{\mathscr{F}}^2 &\le \|f\|_{\mathscr{F}}^2 + \|\eta u\|_{\mathscr{F}}^2 \\
&\le \sum_{k=1}^{j} \frac{b^{2(k-j)}}{\lambda_{n_k}^2}\left(c^2 + \eta^2 \|u^{(k)}\|_{\mathscr{F}}^2\right) + \eta^2 \|u^{(0)}\|_{\mathscr{F}}^2 \frac{b^{-2j}}{\lambda_{n_0}^2} \\
&\le \left(c^2+\eta^2\right)\sum_{k=0}^{j} \frac{b^{2(k-j)}}{\lambda_{n_k}^2} \\
&= \left(1 + \frac{(a-1)^2}{(a+1)^2R^2}\right) c^2 \sum_{k=0}^{j} \frac{b^{2(k-j)}}{\lambda_{n_k}^2} \\
&\le \rho^2 \quad \text{by the definition of } c \text{ above.}
\end{aligned}
$$

So, $f_{\pm} \in \mathscr{B}_\rho$. By design, $A^*(f_{\pm}, \varepsilon) = A^*(f, \varepsilon)$, which will be used below.

Now, we must check that $f_{\pm} \in \mathscr{C}$. From the definition in (8) it follows that for $k = 1, \ldots, j$ and $r \ge 1$,

$$
\sigma_k(f_{\pm}) \begin{cases} \le \sigma_k(f) + \sigma_k(\eta u) \le cb^{k-j} + \eta \lambda_{n_{k-1}+1} \dfrac{b^{k-j}\left\|u^{(k)}\right\|_{\mathscr{F}}}{\lambda_{n_k}} \le b^{k-j}(c+\eta R) \\ \ge \sigma_k(f) - \sigma_k(\eta u) \ge cb^{k-j} - \eta \lambda_{n_{k-1}+1} \dfrac{b^{k-j}\left\|u^{(k)}\right\|_{\mathscr{F}}}{\lambda_{n_k}} \ge b^{k-j}(c-\eta R), \end{cases}
$$

Therefore,

$$
\sigma_{k+r}(f_{\pm}) \le b^{k+r-j}(c+\eta R) = ab^r b^{k-j} \frac{2c}{a+1} = ab^r b^{k-j}(c - \eta R) \le ab^r \sigma_k(f_{\pm}),
$$

which establishes that $f_{\pm} \in \mathscr{C}$.

Although two test functions $f_{\pm}$ yield the same approximate solution, they have different true solutions. In particular,

$$
\begin{aligned}
\varepsilon &\ge \max\left\{\left\|S(f_+) - A^*(f_+, \varepsilon)\right\|_{\mathscr{G}}, \left\|S(f_-) - A^*(f_-, \varepsilon)\right\|_{\mathscr{G}}\right\} \\
&\ge \frac{1}{2}\left[\left\|S(f_+) - A^*(f, \varepsilon)\right\|_{\mathscr{G}} + \left\|S(f_-) - A^*(f, \varepsilon)\right\|_{\mathscr{G}}\right] \\
&\qquad \text{since } A^*(f_{\pm}, \varepsilon) = A^*(f, \varepsilon) \\
&\ge \frac{1}{2}\left\|S(f_+) - S(f_-)\right\|_{\mathscr{G}} \quad \text{by the triangle inequality} \\
&\ge \frac{1}{2}\left\|S(f_+ - f_-)\right\|_{\mathscr{G}} \quad \text{since } S \text{ is linear} \\
&= \eta \left\|S(u)\right\|_{\mathscr{G}}.
\end{aligned}
$$

Thus, we have

$$\begin{aligned}
\varepsilon^2 \geq \eta^2 \|S(u)\|_{\mathscr{G}}^2 &= \eta^2 \sum_{k=0}^{j} \left\| S(u^{(k)}) \right\|_{\mathscr{G}}^2 \frac{b^{2(k-j)}}{\lambda_{n_k}^2} \\
&\geq \eta^2 \sum_{k=0}^{j} \left\| u^{(k)} \right\|_{\mathscr{F}}^2 b^{2(k-j)} \qquad \text{since } \left\| S(u^{(k)}) \right\|_{\mathscr{G}} \geq \lambda_{n_k} \left\| u^{(k)} \right\|_{\mathscr{F}} \\
&\geq \eta^2 b^{2(k^*-j)} \qquad \text{where } k^* = \underset{0 \leq k \leq j}{\operatorname{argmax}} \left\| u^{(k)} \right\|_{\mathscr{F}} \\
&\geq \eta^2 = \frac{(a-1)^2 c^2}{(a+1)^2 R^2} \qquad \text{since } b < 1 \\
&= \frac{(a-1)^2 \rho^2}{(a+1)^2 R^2} \left[\left(1 + \frac{(a-1)^2}{(a+1)^2 R^2} \right) \sum_{k=0}^{j} \frac{b^{2(k-j)}}{\lambda_{n_k}^2} \right]^{-1} \qquad \text{by (22)} \\
&= \rho^2 \left[\left(\frac{(a+1)^2 R^2}{(a-1)^2} + 1 \right) \sum_{k=0}^{j} \frac{b^{2(k-j)}}{\lambda_{n_k}^2} \right]^{-1} \\
&= \rho^2 \left[\left(\frac{(a+1)^2 R^2}{(a-1)^2} + 1 \right) b^2 F(j+1) \right]^{-1} \qquad \text{by (14).}
\end{aligned}$$

This inequality is equivalent to

$$F(j+1) \geq \frac{\rho^2}{\varepsilon^2 b^2} \left[\frac{(a+1)^2 R^2}{(a-1)^2} + 1 \right]^{-1}.$$

This lower bound must be satisfied by j to be consistent with the assumption $\operatorname{comp}(\mathscr{A}(\mathscr{C}), \varepsilon, \rho) \leq n_j - 2$. Thus, any j violating this inequality satisfies $\operatorname{comp}(\mathscr{A}(\mathscr{C}), \varepsilon, \rho) \geq n_j - 1$. Since F is strictly increasing, the largest j violating this inequality must satisfy

$$F(j+1) < \frac{\rho^2}{\varepsilon^2 b^2} \left[\frac{(a+1)^2 R^2}{(a-1)^2} + 1 \right]^{-1} \leq F(j+2),$$

which is the definition of $j^{\ddagger}$ above in the statement of this theorem. This proves the lower bound on $\operatorname{comp}(\mathscr{A}(\mathscr{C}), \varepsilon, \rho)$. □

The next step is to show that the cost of our algorithm is essentially no worse than that of the optimal algorithm.

Theorem 3 *Under assumption* (21) *cost*$(\widetilde{A}, \mathscr{C}, \varepsilon, \rho)$ *is essentially no worse than comp*$(\mathscr{A}(\mathscr{C}), \varepsilon, \rho)$.

Proof We need an inequality relating $F(j)$, which appears in the definition of $j^{\dagger}$ in Theorem 1 and $F(j+3)$, which is related to the definition of $j^{\ddagger}$ in Theorem 2. According to (21) and the definition of F in (14), for any $\ell \in \mathbb{N}$,

$$R^{2\ell}F(j) \geq R^{2\ell}\frac{b^{-2}}{\lambda^2_{n_{j-1}}} \geq \frac{b^{-2}}{\lambda^2_{n_{j+\ell-1}}},$$

and so

$$\begin{aligned} b^{-2\ell}[1+b^2R^2\cdots+(b^2R^2)^\ell]F(j) &\geq F(j)b^{-2\ell} + \frac{b^{-2\ell}}{\lambda^2_{n_j}} + \cdots + \frac{b^{-2}}{\lambda^2_{n_{j+\ell-1}}} \\ &= \sum_{j=0}^{j-1}\frac{b^{2(k-j-\ell)}}{\lambda^2_{n_k}} + \frac{b^{-2\ell}}{\lambda^2_{n_j}} + \cdots + \frac{b^{-2}}{\lambda^2_{n_{j+\ell-1}}} \\ &= F(j+\ell). \end{aligned}$$

The $\ell = 3$ case of this inequality implies the following lower bound on $F(j)$ in terms of $F(j+3)$:

$$F(j) \geq \frac{b^6F(j+3)}{1+b^2R^2+b^4R^4+b^6R^6} = \frac{\omega^2a^4b^2}{1-b^2}\left[\frac{(a+1)^2R^2}{(a-1)^2}+1\right]F(j+3),$$

where

$$\omega = \sqrt{\frac{(1-b^2)b^4}{a^4(1+b^2R^2+b^4R^4+b^6R^6)}\left[\frac{(a+1)^2R^2}{(a-1)^2}+1\right]^{-1}}, \tag{23}$$

Note that ω does not depend on ρ or ε but only on the definition of $\mathscr{C}$.

For any positive ρ and ε, let $j^\dagger = j^\dagger(\varepsilon,\rho)$ be defined as in Theorem 1 and let $j^\ddagger = j^\ddagger(\varepsilon,\rho)$ be defined as in Theorem 2. Then by those Theorems, $j^\dagger(\varepsilon,\rho)$ can be bounded above in terms of $j^\ddagger(\omega\varepsilon,\rho)$ for ω defined in (23):

$$\begin{aligned} j^\dagger(\varepsilon,\rho) &\leq \min\left\{j\in\mathbb{N} : F(j) \geq \frac{\rho^2a^4}{\varepsilon^2(1-b^2)}\right\} \\ &\leq \min\left\{j\in\mathbb{N} : \frac{\omega^2a^4b^2}{1-b^2}\left[\frac{(a+1)^2R^2}{(a-1)^2}+1\right]F(j+3) \geq \frac{\rho^2a^4}{\varepsilon^2(1-b^2)}\right\} \\ &= \min\left\{j\in\mathbb{N} : F(j+3) \geq \frac{\rho^2}{\omega^2\varepsilon^2b^2}\left[\frac{(a+1)^2R^2}{(a-1)^2}+1\right]^{-1}\right\} \\ &= j^\ddagger(\omega\varepsilon,\rho) - 1. \end{aligned}$$

By the argument above, it follows that $j^\dagger(\varepsilon,\rho) \leq j^\ddagger(\omega\varepsilon,\rho) - 1$, and again by Theorems 1 and 2 it follows that

$$\mathrm{cost}(\widetilde{A},\mathscr{C},\varepsilon,\rho) = n_{j^\dagger}(\varepsilon,\rho) \leq n_{j^\ddagger(\omega\varepsilon,\rho)-1} \leq n_{j^\ddagger(\omega\varepsilon,\rho)} - 1 \leq \mathrm{comp}(\mathscr{A}(\mathscr{C}),\omega\varepsilon,\rho).$$

Therefore, our algorithm is essentially no more costly than the optimal algorithm.

□

5 Numerical Example

Consider the case of approximating the partial derivative with respect to x_1 of periodic functions defined on the d-dimensional unit cube:

$$\begin{aligned}
f &= \sum_{\boldsymbol{k}\in\mathbb{Z}^d} \widehat{f}(\boldsymbol{k})\widehat{u}_{\boldsymbol{k}} = \sum_{i\in\mathbb{N}} \widehat{f}_i u_i, \\
\widehat{u}_{\boldsymbol{k}}(\boldsymbol{x}) &:= \prod_{j=1}^{d} \frac{2^{(1-\delta_{k_j,0})/2}\cos(2\pi k_j x_j + \mathbb{1}_{(-\infty,0)}(k_j)\pi/2)}{[\max(1,\gamma_j k_j)]^4}, \\
S(f) &:= \frac{\partial f}{\partial x_1} = \sum_{\boldsymbol{k}\in\mathbb{Z}^d} \widehat{f}(\boldsymbol{k})\lambda(\boldsymbol{k})\widehat{v}_{\boldsymbol{k}}(\boldsymbol{x}) = \sum_{i\in\mathbb{N}} \widehat{f}_i\lambda_i v_i, \\
\widehat{v}_{\boldsymbol{k}}(\boldsymbol{x}) &:= -\operatorname{sign}(k_1)\sin(2\pi k_1 x_1 + \mathbb{1}_{(-\infty,0)}(k_1)\pi/2) \\
&\qquad \times \prod_{j=2}^{d} \cos(2\pi k_j x_j + \mathbb{1}_{(-\infty,0)}(k_j)\pi/2), \\
\lambda(\boldsymbol{k}) &:= 2\pi|k_1| \frac{\prod_{j=1}^{d} 2^{(1-\delta_{k_j,0})/2}}{\prod_{j=1}^{d}[\max(1,\gamma_j k_j)]^4}, \\
\boldsymbol{\gamma} &:= (1, 1/2, 1/4, \ldots, 2^{-d+1}).
\end{aligned}$$

Note that $\lambda_1 \geq \lambda_2 \geq \cdots$ is an ordering of the $\lambda(\boldsymbol{k})$. That ordering then determines the $\widehat{f}_i$, u_i, and v_i in terms of the $\widehat{f}(\boldsymbol{k})$, $\widehat{u}(\boldsymbol{k})$, and $\widehat{v}(\boldsymbol{k})$, respectively.

We construct a function by choosing its Fourier coefficients $\widehat{f}(\boldsymbol{k}) \overset{\text{IID}}{\sim} \mathcal{N}(0,1)$ for $d = 3$, $\boldsymbol{k} \in \{-30, -29, \ldots, 30\}^3$, and $\widehat{f}(\boldsymbol{k}) = 0$ otherwise. This corresponds to $61^3 \approx 2 \times 10^5$ nonzero Fourier coefficients. Let $a = 2$ and $b = 1/2$ and choose $\boldsymbol{n} = (0, 16, 32, 64, \ldots)$. To compute $\sigma_j(f)$, $j \in \mathbb{N}$ by (8), we need to sort $\big(\lambda(\boldsymbol{k})\big)_{\boldsymbol{k}\in\mathbb{Z}^d}$ in descending order, $\lambda_1, \lambda_2, \ldots$. Given ε, we can then find the number of series coefficients needed to satisfy the the error criterion, i.e., $n_{j^\dagger}$ where

$$j^\dagger = \min\left\{ j \in \mathbb{N} : \frac{ab\sigma_j(f)}{\sqrt{1-b^2}} \leq \varepsilon. \right\}$$

Figure 2 shows the input function, the solution, the approximate solution, and the error of the approximate solution for $\varepsilon = 0.1$. For this example, $n_{j^\dagger} = 8192$ is sufficient to satisfy the error tolerance, as is clear from Fig. 2d. Figure 3 shows the sample size, $n_{j^\dagger}$ needed for ten different error tolerances from 0.1 to 10. Because the possible sample sizes are powers of 2, some tolerances require the same sample size.

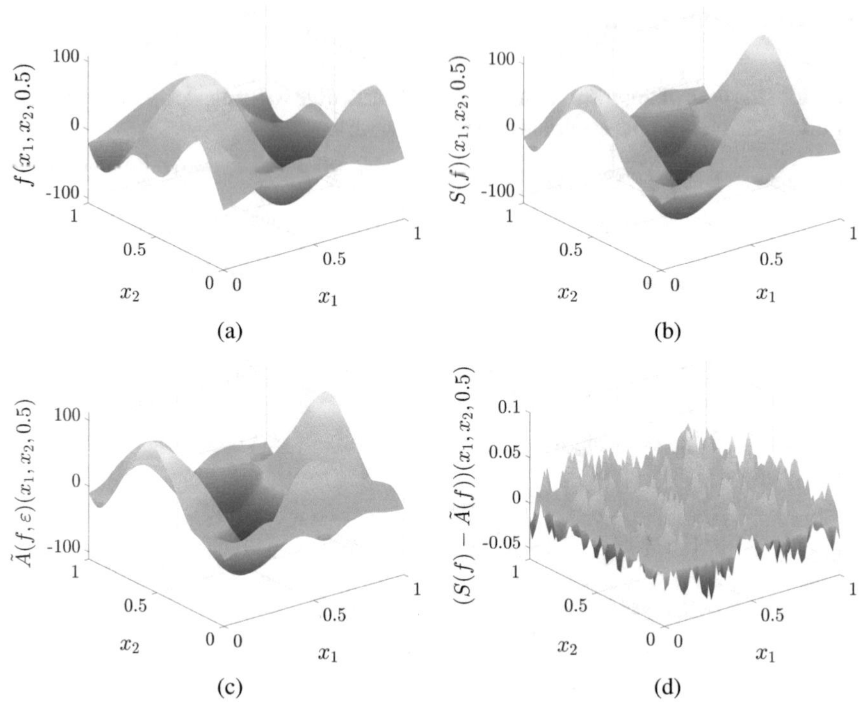

Fig. 2 For $\varepsilon = 0.1$: **a** The input function, f; **b** The true first partial derivative of f; **c** The approximate first partial derivative of f; **d** The approximation error

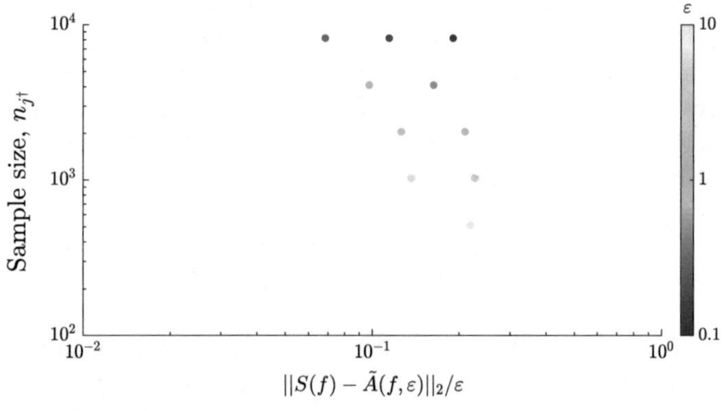

Fig. 3 Sample size $n_{j^\dagger}$, error tolerance ε, and ratio of true error to error tolerance

6 Discussion and Conclusion

Many practical adaptive algorithms lack theory, and many theoretically justified algorithms are non-adaptive. We have demonstrated for a general setting how to construct a theoretically justified, essentially optimal algorithm. The decay of the singular values determines the computational complexity of the problem and the computational cost of our algorithm.

The key idea of our algorithm is to derive an adaptive error bound by assuming the steady decay of the Fourier series coefficients of the solution. The set of such functions constitutes a cone. We do not need to know the decay rate of these coefficients a priori. The cost of our algorithm also serves as a goal for an algorithm that uses function values, which are more commonly available than Fourier series coefficients. An important next step is to identify an essentially optimal algorithm based on function values. Another research direction is to extend this setting to Banach spaces of inputs and/or outputs.

Acknowledgements The authors would like to thank the organizers for a wonderful MCQMC 2018. We also thank the referees for their many helpful suggestions. This work is supported in part by National Science Foundation grants DMS-1522687 and DMS-1638521 (SAMSI). Yuhan Ding acknowledges the support of the Department of Mathematics, Misercordia University.

References

1. Kunsch, R.J., Novak, E., Rudolf, D.: Solvable integration problems and optimal sample size selection. J. Complex. **53**, 40–67 (2019)
2. Hickernell, F.J., Jiang, L., Liu, Y., Owen, A.B.: Guaranteed conservative fixed width confidence intervals via Monte Carlo sampling. In: Dick, J., Kuo, F.Y., Peters, G.W., Sloan, I.H. (eds.) Monte Carlo and Quasi-Monte Carlo Methods 2012. Springer Proceedings in Mathematics and Statistics, vol. 65, pp. 105–128, Springer, Berlin (2013). https://doi.org/10.1007/978-3-642-41095-6
3. Clancy, N., Ding, Y., Hamilton, C., Hickernell, F.J., Zhang, Y.: The cost of deterministic, adaptive, automatic algorithms: cones, not balls. J. Complex. **30**, 21–45 (2014). https://doi.org/10.1016/j.jco.2013.09.002
4. Hickernell, F.J., Jiménez Rugama, L.A.: Reliable adaptive cubature using digital sequences. In: Cools and Nuyens, D. (eds.): Monte Carlo and Quasi-Monte Carlo methods: MCQMC, Leuven, Belgium, April 2014, Springer Proceedings in Mathematics and Statistics, vol. 163. Springer, Berlin, pp. 367–383 (2016). arXiv:1410.8615 [math.NA]
5. Jiménez Rugama, L.A., Hickernell, F.J.: Adaptive multidimensional integration based on rank-1 lattices. In: Cools and Nuyens, D. (eds.): Monte Carlo and Quasi-Monte Carlo methods: MCQMC, Leuven, Belgium, April 2014, Springer Proceedings in Mathematics and Statistics, vol. 163. Springer, Berlin, pp. 407–422 (2016). arXiv:1411.1966
6. Hickernell, F.J., Jiménez Rugama, L.A., Li, D.: Adaptive quasi-Monte Carlo methods for cubature. In: Dick, J., Kuo, F.Y., Woźniakowski, H. (eds.) Contemporary Computational Mathematics—a Celebration of the 80th Birthday of Ian Sloan, pp. 597–619. Springer, Berlin (2018). https://doi.org/10.1007/978-3-319-72456-0

Constructing QMC Finite Element Methods for Elliptic PDEs with Random Coefficients by a Reduced CBC Construction

Adrian Ebert, Peter Kritzer and Dirk Nuyens

Abstract In the analysis of using quasi-Monte Carlo (QMC) methods to approximate expectations of a linear functional of the solution of an elliptic PDE with random diffusion coefficient the sensitivity w.r.t. the parameters is often stated in terms of product-and-order-dependent (POD) weights. The (offline) fast component-by-component (CBC) construction of an N-point QMC method making use of these POD weights leads to a cost of $\mathcal{O}(sN\log(N) + s^2N)$ with s the parameter truncation dimension. When s is large this cost is prohibitive. As an alternative Herrmann and Schwab [9] introduced an analysis resulting in product weights to reduce the construction cost to $\mathcal{O}(sN\log(N))$. We here show how the reduced CBC method [5] can be used for POD weights to reduce the cost to $\mathcal{O}(\sum_{j=1}^{\min\{s,s^*\}}(m - w_j + j)\, b^{m-w_j})$, where $N = b^m$ with prime b, $w_1 \leq \cdots \leq w_s$ are nonnegative integers and s^* can be chosen much smaller than s depending on the regularity of the random field expansion as such making it possible to use the POD weights directly. We show a total error estimate for using randomly shifted lattice rules constructed by the reduced CBC method.

P. Kritzer supported by the Austrian Science Fund (FWF) Project F5506-N26, part of the Special Research Program "Quasi-Monte Carlo Methods: Theory and Applications".

A. Ebert (✉) · D. Nuyens
KU Leuven, Celestijnenlaan 200A, 3001 Leuven, Belgium
e-mail: adrian.ebert@cs.kuleuven.be; adrian.ebert@ricam.oeaw.ac.at

D. Nuyens
e-mail: dirk.nuyens@cs.kuleuven.be

P. Kritzer
Johann Radon Institute for Computational and Applied Mathematics (RICAM), Altenbergerstr. 69, 4040 Linz, Austria
e-mail: peter.kritzer@oeaw.ac.at

B. Tuffin and P. L'Ecuyer (eds.), *Monte Carlo and Quasi-Monte Carlo Methods*, Springer Proceedings in Mathematics & Statistics 324,
https://doi.org/10.1007/978-3-030-43465-6_9

Keywords Quasi-Monte Carlo methods · Infinite-dimensional integration · Elliptic partial differential equations with random coefficients · Component-by-component constructions

1 Introduction and Problem Setting

We consider the parametric elliptic Dirichlet problem given by

$$-\nabla \cdot (a(\boldsymbol{x}, \boldsymbol{y})\, \nabla u(\boldsymbol{x}, \boldsymbol{y})) = f(\boldsymbol{x}) \quad \text{for } \boldsymbol{x} \in D \subset \mathbb{R}^d, \quad u(\boldsymbol{x}, \boldsymbol{y}) = 0 \quad \text{for } \boldsymbol{x} \text{ on } \partial D, \qquad (1)$$

for $D \subset \mathbb{R}^d$ a bounded, convex Lipschitz polyhedron domain with boundary ∂D and fixed spatial dimension $d \in \{1, 2, 3\}$. The function f lies in $L^2(D)$, the parametric variable $\boldsymbol{y} = (y_j)_{j \geq 1}$ belongs to a domain U, and the differential operators are understood to be with respect to the physical variable $\boldsymbol{x} \in D$. Here we study the "uniform case", i.e., we assume that $\boldsymbol{y}$ is uniformly distributed on $U := \left[-\frac{1}{2}, \frac{1}{2}\right]^{\mathbb{N}}$ with uniform probability measure $\mu(\mathrm{d}\boldsymbol{y}) = \bigotimes_{j \geq 1} \mathrm{d}y_j = \mathrm{d}\boldsymbol{y}$. The parametric diffusion coefficient $a(\boldsymbol{x}, \boldsymbol{y})$ is assumed to depend linearly on the parameters y_j in the following way,

$$a(\boldsymbol{x}, \boldsymbol{y}) = a_0(\boldsymbol{x}) + \sum_{j \geq 1} y_j\, \psi_j(\boldsymbol{x}), \quad \boldsymbol{x} \in D, \quad \boldsymbol{y} \in U \qquad (2)$$

for a given system $\{\psi_j\}_{j \geq 1}$ of functions in $L^2(D)$. For the variational formulation of (1), we consider the Sobolev space $V = H_0^1(D)$ of functions v which vanish on the boundary ∂D with norm

$$\|v\|_V := \left(\int_D \sum_{j=1}^{d} |\partial_{x_j} v(\boldsymbol{x})|^2 \mathrm{d}\boldsymbol{x} \right)^{\frac{1}{2}} = \|\nabla v\|_{L^2(D)}.$$

The corresponding dual space of bounded linear functionals on V with respect to the pivot space $L^2(D)$ is further denoted by $V^* = H^{-1}(D)$. Then, for given $f \in V^*$ and $\boldsymbol{y} \in U$, the weak (or variational) formulation of (1) is to find $u(\cdot, \boldsymbol{y}) \in V$ such that

$$A(\boldsymbol{y}; u(\cdot, \boldsymbol{y}), v) = \langle f, v \rangle_{V^* \times V} = \int_D f(\boldsymbol{x}) v(\boldsymbol{x})\, \mathrm{d}\boldsymbol{x} \quad \text{for all} \quad v \in V, \qquad (3)$$

with parametric bilinear form $A : U \times V \times V \to \mathbb{R}$ given by

$$A(\boldsymbol{y}; w, v) := \int_D a(\boldsymbol{x}, \boldsymbol{y})\, \nabla w(\boldsymbol{x}) \cdot \nabla v(\boldsymbol{x})\, \mathrm{d}\boldsymbol{x} \quad \text{for all} \quad w, v \in V, \qquad (4)$$

and duality pairing $\langle \cdot, \cdot \rangle_{V^* \times V}$ between V^* and V. We will often identify elements $\varphi \in V$ with dual elements $L_\varphi \in V^*$. Indeed, for $\varphi \in V$ and $v \in V$, a bounded linear functional is given via $L_\varphi(v) := \int_D \varphi(\boldsymbol{x}) v(\boldsymbol{x}) \mathrm{d}\boldsymbol{x} = \langle \varphi, v \rangle_{L^2(D)}$ and by the Riesz representation theorem there exists a unique representer $\widetilde{\varphi} \in V$ such that

$L_\varphi(v) = \langle \widetilde{\varphi}, v \rangle_{L^2(D)}$ for all $v \in V$. Hence, the definition of the canonical duality pairing yields that $\langle L_\varphi, v \rangle_{V^* \times V} = L_\varphi(v) = \langle \varphi, v \rangle_{L^2(D)}$.

Our quantity of interest is the expected value, with respect to the probability measure $\mu(\mathrm{d}\boldsymbol{y})$, of a given bounded linear functional $G \in V^*$ applied to the solution $u(\cdot, \boldsymbol{y})$ of the PDE. We stress that the considered functionals G do not depend on the parametric variable $\boldsymbol{y}$. Therefore we seek to approximate this expectation by numerically integrating G applied to a finite element approximation $u_h^s(\cdot, \boldsymbol{y})$ of the solution $u^s(\cdot, \boldsymbol{y}) \in H_0^1(D) = V$ of (3) with truncated diffusion coefficient $a(\boldsymbol{x}, (\boldsymbol{y}_{\{1:s\}}; 0))$ where $\{1:s\} := \{1, \ldots, s\}$ and we write $(\boldsymbol{y}_{\{1:s\}}; 0) = (\tilde{y}_j)_{j \geq 1}$ with $\tilde{y}_j = y_j$ for $j \in \{1:s\}$ and $\tilde{y}_j = 0$ otherwise; that is,

$$\mathbb{E}[G(u)] := \int_U G(u(\cdot, \boldsymbol{y}))\, \mu(\mathrm{d}\boldsymbol{y}) = \int_U G(u(\cdot, \boldsymbol{y}))\, \mathrm{d}\boldsymbol{y} \approx Q_N(G(u_h^s)), \tag{5}$$

with $Q_N(\cdot)$ a linear quadrature rule using N function evaluations. The infinite-dimensional integral $\mathbb{E}[G(u)]$ in (5) is defined as

$$\mathbb{E}[G(u)] = \int_U G(u(\cdot, \boldsymbol{y}))\, \mathrm{d}\boldsymbol{y} := \lim_{s \to \infty} \int_{\left[-\frac{1}{2}, \frac{1}{2}\right]^s} G(u(\cdot, (y_1, \ldots, y_s, 0, 0, \ldots)))\, \mathrm{d}y_1 \cdots \mathrm{d}y_s$$

such that our integrands of interest are of the form $F(\boldsymbol{y}) = G(u(\cdot, \boldsymbol{y}))$ with $\boldsymbol{y} \in U$. The QMC finite element method will sample the random coefficients $\boldsymbol{y} = (\boldsymbol{y}_{\{1:s\}}; 0)$ by a QMC point set, solve the deterministic problem via an FEM approximation for each of the parameter samples and calculate the quantity of interest for each of these solutions, and finally take the average to approximate the expectation, see, e.g., [11]. In this article, we will employ (randomized) QMC methods of the form

$$Q_N(F) = \frac{1}{N} \sum_{k=1}^{N} F(\boldsymbol{t}_k),$$

i.e., equal-weight quadrature rules with (randomly shifted) deterministic points $\boldsymbol{t}_1, \ldots, \boldsymbol{t}_N \in \left[-\frac{1}{2}, \frac{1}{2}\right]^s$. The elliptic PDE studied is a standard problem considered in the numerical analysis of computational methods in uncertainty quantification, see, e.g., [1, 2, 6, 8–12].

1.1 Existence of Solutions of the Variational Problem

To assure that a unique solution to the weak problem (3) exists, we need certain conditions on the diffusion coefficient a. We assume $a_0 \in L^\infty(D)$ and $\operatorname{ess\,inf}_{\boldsymbol{x} \in D} a_0(\boldsymbol{x}) > 0$, which is equivalent to the existence of two constants $0 < a_{0,\min} \leq a_{0,\max} < \infty$ such that a.e. on D we have

$$a_{0,\min} \leq a_0(\boldsymbol{x}) \leq a_{0,\max}, \tag{6}$$

and that there exists a $\overline{\kappa} \in (0, 1)$ such that

$$\left\| \sum_{j\geq 1} \frac{|\psi_j|}{2a_0} \right\|_{L^\infty(D)} \leq \overline{\kappa} < 1. \tag{7}$$

Via (7), we obtain that $|\sum_{j\geq 1} y_j \psi_j(\boldsymbol{x})| \leq \overline{\kappa}\, a_0(\boldsymbol{x})$ and hence, using (6), almost everywhere on D and for any $\boldsymbol{y} \in U$

$$0 < (1 - \overline{\kappa})\, a_{0,\min} \leq a_0(\boldsymbol{x}) + \sum_{j\geq 1} y_j \psi_j(\boldsymbol{x}) = a(\boldsymbol{x}, \boldsymbol{y}) \leq (1 + \overline{\kappa})\, a_{0,\max}. \tag{8}$$

These estimates yield the continuity and coercivity of $A(\boldsymbol{y}, \cdot, \cdot)$ defined in (4) on $V \times V$, uniformly for all $\boldsymbol{y} \in U$. The Lax–Milgram theorem then ensures the existence of a unique solution $u(\cdot, \boldsymbol{y})$ of the weak problem in (3).

1.2 Parametric Regularity

Having established the existence of unique weak parametric solutions $u(\cdot, \boldsymbol{y})$, we investigate their regularity in terms of the behavior of their mixed first-order derivatives. Our analysis combines multiple techniques which can be found in the literature, see, e.g., [1, 2, 8–10]. In particular we want to point out that our POD form bounds can take advantage of wavelet like expansions of the random field, a technique introduced in [1] and used to the advantage of QMC constructions by [9] to deliver product weights to save on the construction compared to POD weights. Although we end up again with POD weights, we will save on the construction cost by making use of a special construction method, called the reduced CBC construction, which we will introduce in Sect. 2.4. Let $\boldsymbol{\nu} = (\nu_j)_{j\geq 1}$ with $\nu_j \in \mathbb{N}_0 := \{0, 1, 2, \ldots\}$ be a sequence of positive integers which we will refer to as a multi-index. We define the order $|\boldsymbol{\nu}|$ and the support $\operatorname{supp}(\boldsymbol{\nu})$ as

$$|\boldsymbol{\nu}| := \sum_{j\geq 1} \nu_j \quad \text{and} \quad \operatorname{supp}(\boldsymbol{\nu}) := \{j \geq 1 : \nu_j > 0\}$$

and introduce the sets $\mathscr{F}$ and $\mathscr{F}_1$ of finitely supported multi-indices as

$$\mathscr{F} := \{\boldsymbol{\nu} \in \mathbb{N}_0^{\mathbb{N}} : |\operatorname{supp}(\boldsymbol{\nu})| < \infty\} \quad \text{and} \quad \mathscr{F}_1 := \{\boldsymbol{\nu} \in \{0, 1\}^{\mathbb{N}} : |\operatorname{supp}(\boldsymbol{\nu})| < \infty\},$$

where $\mathscr{F}_1 \subseteq \mathscr{F}$ is the restriction containing only $\boldsymbol{\nu}$ with $\nu_j \in \{0, 1\}$. Then, for $\boldsymbol{\nu} \in \mathscr{F}$ denote the $\boldsymbol{\nu}$-th partial derivative with respect to the parametric variables $\boldsymbol{y} \in U$ by

$$\partial^{\boldsymbol{\nu}} = \frac{\partial^{|\boldsymbol{\nu}|}}{\partial y_1^{\nu_1} \partial y_2^{\nu_2} \cdots},$$

and for a sequence $\boldsymbol{b} = (b_j)_{j \geq 1} \subset \mathbb{R}_+^{\mathbb{N}}$, set $\boldsymbol{b}^{\boldsymbol{\nu}} := \prod_{j \geq 1} b_j^{\nu_j}$. We further write $\boldsymbol{\omega} \leq \boldsymbol{\nu}$ if $\omega_j \leq \nu_j$ for all $j \geq 1$ and denote by $\boldsymbol{e}_i \in \mathscr{F}_1$ the multi-index with components $e_j = \delta_{i,j}$. For a fixed $\boldsymbol{y} \in U$, we introduce the energy norm $\| \cdot \|_{a_y}^2$ in the space V via

$$\|v\|_{a_y}^2 := \int_D a(\boldsymbol{x}, \boldsymbol{y}) \, |\nabla v(\boldsymbol{x})|^2 \, \mathrm{d}\boldsymbol{x}$$

for which it holds true by (8) that

$$(1 - \overline{\kappa}) \, a_{0,\min} \|v\|_V^2 \leq \|v\|_{a_y}^2 \quad \text{for all} \quad v \in V. \tag{9}$$

Consequently, we have that $(1 - \overline{\kappa}) \, a_{0,\min} \|u(\cdot, \boldsymbol{y})\|_V^2 \leq \|u(\cdot, \boldsymbol{y})\|_{a_y}^2$ and hence the definition of the dual norm $\| \cdot \|_{V^*}$ yields the following initial estimate from (3) and (4),

$$\begin{aligned} \|u(\cdot, \boldsymbol{y})\|_{a_y}^2 &= \int_D a(\boldsymbol{x}, \boldsymbol{y}) \, |\nabla u(\boldsymbol{x}, \boldsymbol{y})|^2 \, \mathrm{d}\boldsymbol{x} = \int_D f(\boldsymbol{x}) u(\boldsymbol{x}, \boldsymbol{y}) \, \mathrm{d}\boldsymbol{x} \\ &= \langle f, u(\cdot, \boldsymbol{y}) \rangle_{V^* \times V} \leq \|f\|_{V^*} \|u(\cdot, \boldsymbol{y})\|_V \leq \frac{\|f\|_{V^*} \|u(\cdot, \boldsymbol{y})\|_{a_y}}{\sqrt{(1 - \overline{\kappa}) a_{0,\min}}} \end{aligned}$$

which gives in turn

$$\|u(\cdot, \boldsymbol{y})\|_{a_y}^2 \leq \frac{\|f\|_{V^*}^2}{(1 - \overline{\kappa}) \, a_{0,\min}}. \tag{10}$$

In order to exploit the decay of the norm sequence $(\|\psi_j\|_{L^\infty(D)})_{j \geq 1}$ of the basis functions, we extend condition (7) as follows. To characterize the smoothness of the random field, we assume that there exist a sequence of reals $\boldsymbol{b} = (b_j)_{j \geq 1}$ with $0 < b_j \leq 1$ for all j, such that, for positive constants κ and $\widetilde{\kappa}(\boldsymbol{\nu})$, with $\widetilde{\kappa}(\boldsymbol{\nu}) \leq \kappa$ for all $\boldsymbol{\nu} \in \mathscr{F}_1$, given as below, we have

$$\kappa := \left\| \sum_{j \geq 1} \frac{|\psi_j| / b_j}{2 a_0} \right\|_{L^\infty(D)} < 1, \quad \widetilde{\kappa}(\boldsymbol{\nu}) := \left\| \sum_{j \in \mathrm{supp}(\boldsymbol{\nu})} \frac{|\psi_j| / b_j}{2 a_0} \right\|_{L^\infty(D)} < 1. \tag{11}$$

We remark that condition (7) is included in this assumption by letting $b_j = 1$ for all $j \geq 1$ and that $0 < \overline{\kappa} \leq \kappa < 1$. Using the above estimations we can derive the following theorem for the mixed first-order partial derivatives.

Theorem 1 *Let $\boldsymbol{b} = (b_j)_{j \geq 1} \subset (0, 1]$ be a sequence of reals satisfying* (11). *Let $\boldsymbol{\nu} \in \mathscr{F}_1$ be a multi-index of finite support and let $k \in \{0, 1, \ldots, |\boldsymbol{\nu}|\}$. Then, for every $f \in V^*$ and every $\boldsymbol{y} \in U$,*

$$\sum_{\substack{\boldsymbol{\omega}\le\boldsymbol{\nu}\\|\boldsymbol{\omega}|=k}} \boldsymbol{b}^{-2\boldsymbol{\omega}} \|\partial^{\boldsymbol{\omega}} u(\cdot,\boldsymbol{y})\|_V^2 \le \left(\left(\frac{2\widetilde{\kappa}(\boldsymbol{\nu})}{1-\overline{\kappa}}\right)^k \frac{\|f\|_{V^*}}{(1-\overline{\kappa})\,a_{0,\min}}\right)^2,$$

with $\widetilde{\kappa}(\boldsymbol{\nu})$ *as in* (11). *Moreover, for* $k = |\boldsymbol{\nu}|$ *we obtain*

$$\|\partial^{\boldsymbol{\nu}} u(\cdot,\boldsymbol{y})\|_V \le \boldsymbol{b}^{\boldsymbol{\nu}} \left(\frac{2\widetilde{\kappa}(\boldsymbol{\nu})}{1-\overline{\kappa}}\right)^{|\boldsymbol{\nu}|} \frac{\|f\|_{V^*}}{(1-\overline{\kappa})\,a_{0,\min}}.$$

Proof For the special case $\boldsymbol{\nu} = \boldsymbol{0}$, the claim follows by combining (9) and (10). For $\boldsymbol{\nu} \in \mathscr{F}_1$ with $|\boldsymbol{\nu}| > 0$, as is known from, e.g., [2] and [11, Appendix], the linearity of $a(\boldsymbol{x},\boldsymbol{y})$ gives rise to the following identity for any $\boldsymbol{y} \in U$:

$$\|\partial^{\boldsymbol{\nu}} u(\cdot,\boldsymbol{y})\|_{a_{\boldsymbol{y}}}^2 = -\sum_{j\in\mathrm{supp}(\boldsymbol{\nu})} \int_D \psi_j(\boldsymbol{x})\, \nabla\partial^{\boldsymbol{\nu}-\boldsymbol{e}_j} u(\boldsymbol{x},\boldsymbol{y}) \cdot \nabla\partial^{\boldsymbol{\nu}} u(\boldsymbol{x},\boldsymbol{y})\,\mathrm{d}\boldsymbol{x}. \tag{12}$$

For sequences of $L^2(D)$-integrable functions $\boldsymbol{f} = (f_{\boldsymbol{\omega},j})_{\boldsymbol{\omega}\in\mathscr{F},j\ge1}$ with $f_{\boldsymbol{\omega},j} : D \to \mathbb{R}$, we define the inner product $\langle \boldsymbol{f},\boldsymbol{g}\rangle_{\boldsymbol{\nu},k}$ as follows,

$$\langle \boldsymbol{f},\boldsymbol{g}\rangle_{\boldsymbol{\nu},k} := \sum_{\substack{\boldsymbol{\omega}\le\boldsymbol{\nu}\\|\boldsymbol{\omega}|=k}} \int_D \sum_{j\in\mathrm{supp}(\boldsymbol{\omega})} f_{\boldsymbol{\omega},j}(\boldsymbol{x})\, g_{\boldsymbol{\omega},j}(\boldsymbol{x})\,\mathrm{d}\boldsymbol{x}.$$

We can then apply the Cauchy–Schwarz inequality to $\boldsymbol{f} = (f_{\boldsymbol{\omega},j})$ and $\boldsymbol{g} = (g_{\boldsymbol{\omega},j})$ with $f_{\boldsymbol{\omega},j} = \boldsymbol{b}^{-\boldsymbol{e}_j/2}|\psi_j|^{\frac12}\,\boldsymbol{b}^{-(\boldsymbol{\omega}-\boldsymbol{e}_j)}\nabla\partial^{\boldsymbol{\omega}-\boldsymbol{e}_j}u(\cdot,\boldsymbol{y})$ and $g_{\boldsymbol{\omega},j} = \boldsymbol{b}^{-\boldsymbol{e}_j/2}|\psi_j|^{\frac12}\,\boldsymbol{b}^{-\boldsymbol{\omega}}\nabla\partial^{\boldsymbol{\omega}}u(\cdot,\boldsymbol{y})$ to obtain, with the help of (12),

$$\begin{aligned}
&\sum_{\substack{\boldsymbol{\omega}\le\boldsymbol{\nu}\\|\boldsymbol{\omega}|=k}} \boldsymbol{b}^{-2\boldsymbol{\omega}} \|\partial^{\boldsymbol{\omega}} u(\cdot,\boldsymbol{y})\|_{a_{\boldsymbol{y}}}^2 \\
&\quad = -\sum_{\substack{\boldsymbol{\omega}\le\boldsymbol{\nu}\\|\boldsymbol{\omega}|=k}} \int_D \sum_{j\in\mathrm{supp}(\boldsymbol{\omega})} \boldsymbol{b}^{-\boldsymbol{e}_j}\boldsymbol{b}^{-(\boldsymbol{\omega}-\boldsymbol{e}_j)}\boldsymbol{b}^{-\boldsymbol{\omega}}\psi_j(\boldsymbol{x})\,\nabla\partial^{\boldsymbol{\omega}-\boldsymbol{e}_j}u(\boldsymbol{x},\boldsymbol{y})\cdot\nabla\partial^{\boldsymbol{\omega}}u(\boldsymbol{x},\boldsymbol{y})\,\mathrm{d}\boldsymbol{x} \\
&\quad \le \left(\int_D \sum_{\substack{\boldsymbol{\omega}\le\boldsymbol{\nu}\\|\boldsymbol{\omega}|=k}} \sum_{j\in\mathrm{supp}(\boldsymbol{\omega})} \boldsymbol{b}^{-\boldsymbol{e}_j}|\psi_j(\boldsymbol{x})|\,\left|\boldsymbol{b}^{-(\boldsymbol{\omega}-\boldsymbol{e}_j)}\nabla\partial^{\boldsymbol{\omega}-\boldsymbol{e}_j}u(\boldsymbol{x},\boldsymbol{y})\right|^2\mathrm{d}\boldsymbol{x}\right)^{\frac12} \\
&\qquad \times \left(\int_D \sum_{\substack{\boldsymbol{\omega}\le\boldsymbol{\nu}\\|\boldsymbol{\omega}|=k}} \sum_{j\in\mathrm{supp}(\boldsymbol{\omega})} \boldsymbol{b}^{-\boldsymbol{e}_j}|\psi_j(\boldsymbol{x})|\,\left|\boldsymbol{b}^{-\boldsymbol{\omega}}\nabla\partial^{\boldsymbol{\omega}}u(\boldsymbol{x},\boldsymbol{y})\right|^2\mathrm{d}\boldsymbol{x}\right)^{\frac12}.
\end{aligned}$$

The first of the two factors above is then bounded as follows,

$$\int_D \sum_{\substack{\boldsymbol{\omega}\le\boldsymbol{\nu}\\|\boldsymbol{\omega}|=k}} \sum_{j\in\mathrm{supp}(\boldsymbol{\omega})} \boldsymbol{b}^{-\boldsymbol{e}_j}|\psi_j(\boldsymbol{x})|\, \left|\boldsymbol{b}^{-(\boldsymbol{\omega}-\boldsymbol{e}_j)}\nabla\partial^{\boldsymbol{\omega}-\boldsymbol{e}_j}u(\boldsymbol{x},\boldsymbol{y})\right|^2 \mathrm{d}\boldsymbol{x}$$

$$= \int_D \sum_{\substack{\boldsymbol{\omega}\le\boldsymbol{\nu}\\|\boldsymbol{\omega}|=k-1}} \left(\sum_{\substack{j\in\mathrm{supp}(\boldsymbol{\nu})\\ \boldsymbol{\omega}+\boldsymbol{e}_j\le\boldsymbol{\nu}}} \boldsymbol{b}^{-\boldsymbol{e}_j}|\psi_j(\boldsymbol{x})| \right) \left|\boldsymbol{b}^{-\boldsymbol{\omega}}\nabla\partial^{\boldsymbol{\omega}}u(\boldsymbol{x},\boldsymbol{y})\right|^2 \mathrm{d}\boldsymbol{x}$$

$$\le \left\| \sum_{j\in\mathrm{supp}(\boldsymbol{\nu})} \frac{|\psi_j|/b_j}{a(\cdot,\boldsymbol{y})} \right\|_{L^\infty(D)} \sum_{\substack{\boldsymbol{\omega}\le\boldsymbol{\nu}\\|\boldsymbol{\omega}|=k-1}} \boldsymbol{b}^{-2\boldsymbol{\omega}} \int_D a(\boldsymbol{x},\boldsymbol{y}) \left|\nabla\partial^{\boldsymbol{\omega}}u(\boldsymbol{x},\boldsymbol{y})\right|^2 \mathrm{d}\boldsymbol{x}$$

$$= \left\| \sum_{j\in\mathrm{supp}(\boldsymbol{\nu})} \frac{|\psi_j|/b_j}{a(\cdot,\boldsymbol{y})} \right\|_{L^\infty(D)} \sum_{\substack{\boldsymbol{\omega}\le\boldsymbol{\nu}\\|\boldsymbol{\omega}|=k-1}} \boldsymbol{b}^{-2\boldsymbol{\omega}} \|\partial^{\boldsymbol{\omega}}u(\cdot,\boldsymbol{y})\|_{a_{\boldsymbol{y}}}^2,$$

while the other factor can be bounded trivially. Furthermore, using (8), we have for any $\boldsymbol{y}\in U$

$$\left\| \sum_{j\in\mathrm{supp}(\boldsymbol{\nu})} \frac{|\psi_j|/b_j}{a(\cdot,\boldsymbol{y})} \right\|_{L^\infty(D)} \le \frac{1}{1-\overline{\kappa}} \left\| \sum_{j\in\mathrm{supp}(\boldsymbol{\nu})} \frac{|\psi_j|/b_j}{a_0} \right\|_{L^\infty(D)} := \frac{2\widetilde{\kappa}(\boldsymbol{\nu})}{1-\overline{\kappa}},$$

so that, combining these three estimates, we obtain

$$\sum_{\substack{\boldsymbol{\omega}\le\boldsymbol{\nu}\\|\boldsymbol{\omega}|=k}} \boldsymbol{b}^{-2\boldsymbol{\omega}} \|\partial^{\boldsymbol{\omega}}u(\cdot,\boldsymbol{y})\|_{a_{\boldsymbol{y}}}^2$$

$$\le \frac{2\widetilde{\kappa}(\boldsymbol{\nu})}{1-\overline{\kappa}} \left(\sum_{\substack{\boldsymbol{\omega}\le\boldsymbol{\nu}\\|\boldsymbol{\omega}|=k-1}} \boldsymbol{b}^{-2\boldsymbol{\omega}} \|\partial^{\boldsymbol{\omega}}u(\cdot,\boldsymbol{y})\|_{a_{\boldsymbol{y}}}^2 \right)^{\frac{1}{2}} \left(\sum_{\substack{\boldsymbol{\omega}\le\boldsymbol{\nu}\\|\boldsymbol{\omega}|=k}} \boldsymbol{b}^{-2\boldsymbol{\omega}} \|\partial^{\boldsymbol{\omega}}u(\cdot,\boldsymbol{y})\|_{a_{\boldsymbol{y}}}^2 \right)^{\frac{1}{2}}.$$

Therefore, we finally obtain that

$$\sum_{\substack{\boldsymbol{\omega}\le\boldsymbol{\nu}\\|\boldsymbol{\omega}|=k}} \boldsymbol{b}^{-2\boldsymbol{\omega}} \|\partial^{\boldsymbol{\omega}}u(\cdot,\boldsymbol{y})\|_{a_{\boldsymbol{y}}}^2 \le \left(\frac{2\widetilde{\kappa}(\boldsymbol{\nu})}{1-\overline{\kappa}}\right)^2 \sum_{\substack{\boldsymbol{\omega}\le\boldsymbol{\nu}\\|\boldsymbol{\omega}|=k-1}} \boldsymbol{b}^{-2\boldsymbol{\omega}} \|\partial^{\boldsymbol{\omega}}u(\cdot,\boldsymbol{y})\|_{a_{\boldsymbol{y}}}^2$$

which inductively gives

$$\sum_{\substack{\boldsymbol{\omega}\le\boldsymbol{\nu}\\|\boldsymbol{\omega}|=k}} \boldsymbol{b}^{-2\boldsymbol{\omega}} \|\partial^{\boldsymbol{\omega}}u(\cdot,\boldsymbol{y})\|_{a_{\boldsymbol{y}}}^2 \le \left(\frac{2\widetilde{\kappa}(\boldsymbol{\nu})}{1-\overline{\kappa}}\right)^{2k} \|u(\cdot,\boldsymbol{y})\|_{a_{\boldsymbol{y}}}^2 \le \left(\frac{2\widetilde{\kappa}(\boldsymbol{\nu})}{1-\overline{\kappa}}\right)^{2k} \frac{\|f\|_{V^*}^2}{(1-\overline{\kappa})\,a_{0,\min}},$$

where the last inequality follows from the initial estimate (10). The estimate (9) then gives

$$\sum_{\substack{\boldsymbol{\omega}\leq\boldsymbol{\nu}\\|\boldsymbol{\omega}|=k}} \boldsymbol{b}^{-2\boldsymbol{\omega}}\|\partial^{\boldsymbol{\omega}}u(\cdot,\boldsymbol{y})\|_V^2 \leq \frac{1}{(1-\overline{\kappa})\,a_{0,\min}} \sum_{\substack{\boldsymbol{\omega}\leq\boldsymbol{\nu}\\|\boldsymbol{\omega}|=k}} \boldsymbol{b}^{-2\boldsymbol{\omega}}\|\partial^{\boldsymbol{\omega}}u(\cdot,\boldsymbol{y})\|_{a_{\boldsymbol{y}}}^2$$

$$\leq \left(\frac{2\widetilde{\kappa}(\boldsymbol{\nu})}{1-\overline{\kappa}}\right)^{2k} \frac{\|f\|_{V^*}^2}{(1-\overline{\kappa})^2\,a_{0,\min}^2},$$

which yields the first claim. The second claim follows since the sum over the $\boldsymbol{\omega}\leq\boldsymbol{\nu}$ with $|\boldsymbol{\omega}|=|\boldsymbol{\nu}|$ and $\boldsymbol{\nu}\in\mathscr{F}_1$ consists only of the term corresponding to $\boldsymbol{\omega}=\boldsymbol{\nu}$. □

Corollary 1 *Under the assumptions of Theorem 1, there exists a number* $\kappa(k)$ *for each* $k\in\mathbb{N}$*, given by*

$$\kappa(k) := \sup_{\substack{\boldsymbol{\nu}\in\mathscr{F}_1\\|\boldsymbol{\nu}|=k}} \widetilde{\kappa}(\boldsymbol{\nu}),$$

such that $\widetilde{\kappa}(\boldsymbol{\nu})\leq\kappa(k)\leq\kappa<1$ *for all* $\boldsymbol{\nu}\in\mathscr{F}_1$ *with* $|\boldsymbol{\nu}|=k$*. Then for* $\boldsymbol{\nu}\in\mathscr{F}_1$*, every* $f\in V^*$*, and every* $\boldsymbol{y}\in U$*, the solution* $u(\cdot,\boldsymbol{y})$ *satisfies*

$$\|\partial^{\boldsymbol{\nu}}u(\cdot,\boldsymbol{y})\|_V \leq \boldsymbol{b}^{\boldsymbol{\nu}}\left(\frac{2\kappa(|\boldsymbol{\nu}|)}{1-\overline{\kappa}}\right)^{|\boldsymbol{\nu}|}\frac{\|f\|_{V^*}}{(1-\overline{\kappa})\,a_{0,\min}}. \tag{13}$$

Note that since $0<\overline{\kappa}\leq\kappa<1$, the results of Theorem 1 and Corollary 1 remain also valid for $\overline{\kappa}$ replaced by κ.

The obtained bounds on the mixed first-order derivatives turn out to be of product-and-order-dependent (so-called POD) form; that is, they are of the general form

$$\|\partial^{\boldsymbol{\nu}}u(\cdot,\boldsymbol{y})\|_V \leq C\,\boldsymbol{b}^{\boldsymbol{\nu}}\,\Gamma(|\boldsymbol{\nu}|)\,\|f\|_{V^*} \tag{14}$$

with a map $\Gamma:\mathbb{N}_0\to\mathbb{R}$, a sequence of reals $\boldsymbol{b}=(b_j)_{j\geq1}\in\mathbb{R}^{\mathbb{N}}$ and some constant $C\in\mathbb{R}_+$. This finding motivates us to consider this special type of bounds in the following error analysis.

2 Quasi-Monte Carlo Finite Element Error

We analyze the error $\mathbb{E}[G(u)]-Q_N(G(u_h^s))$ obtained by applying QMC rules to the finite element approximation u_h^s to approximate the expected value

$$\mathbb{E}[G(u)] = \int_U G(u(\cdot,\boldsymbol{y}))\,\mathrm{d}\boldsymbol{y}.$$

To this end, we introduce the finite element approximation $u_h^s(\boldsymbol{x}, \boldsymbol{y}) := u_h(\boldsymbol{x}, (\boldsymbol{y}_{\{1:s\}}; 0))$ of a solution of (3) with truncated diffusion coefficient $a(\boldsymbol{x}, (\boldsymbol{y}_{\{1:s\}}; 0))$, where u_h is a finite element approximation as defined in (16) and $(\boldsymbol{y}_{\{1:s\}}; 0) = (y_1, \ldots, y_s, 0, 0, \ldots)$. The overall absolute QMC finite element error is then bounded as follows

$$\begin{aligned} &|\mathbb{E}[G(u)] - Q_N(G(u_h^s))| \\ &= |\mathbb{E}[G(u)] - \mathbb{E}[G(u^s)] + \mathbb{E}[G(u^s)] - \mathbb{E}[G(u_h^s)] + \mathbb{E}[G(u_h^s)] - Q_N(G(u_h^s))| \\ &\le |\mathbb{E}[G(u - u^s)]| + |\mathbb{E}[G(u^s - u_h^s)]| + |\mathbb{E}[G(u_h^s)] - Q_N(G(u_h^s))|. \end{aligned} \tag{15}$$

The first term on the right hand side of (15) will be referred to as (dimension) truncation error, the second term is the finite element discretization error and the last term is the QMC quadrature error for the integrand u_h^s. In the following sections we will analyze these different error terms separately.

2.1 Finite Element Approximation

Here, we consider the approximation of the solution $u(\cdot, \boldsymbol{y})$ of (3) by a finite element approximation $u_h(\cdot, \boldsymbol{y})$ and assess the finite element discretization error. To this end, denote by $\{V_h\}_{h>0}$ a family of subspaces $V_h \subset V$ with finite dimensions M_h of order h^{-d}. The spaces V_h contain continuous, piecewise linear finite elements defined on a sequence $\{\mathscr{T}_h\}_{h>0}$ of shape regular triangulations of D. We define the parametric finite element (FE) approximation as follows: for $f \in V^*$ and given $\boldsymbol{y} \in U$, find $u_h(\cdot, \boldsymbol{y}) \in V_h$ such that

$$A(\boldsymbol{y}; u_h(\cdot, \boldsymbol{y}), v_h) = \langle f, v_h \rangle_{V^* \times V} = \int_D f(\boldsymbol{x}) v_h(\boldsymbol{x})\, \mathrm{d}\boldsymbol{x} \quad \text{for all} \quad v_h \in V_h. \tag{16}$$

To establish convergence of the finite element approximations, we need some further conditions on $a(\boldsymbol{x}, \boldsymbol{y})$. To this end, we define the space $W^{1,\infty}(D) \subseteq L^\infty(D)$ endowed with the norm $\|v\|_{W^{1,\infty}(D)} = \max\{\|v\|_{L^\infty(D)}, \|\nabla v\|_{L^\infty(D)}\}$ and require that

$$a_0 \in W^{1,\infty}(D) \quad \text{and} \quad \sum_{j \ge 1} \|\psi_j\|_{W^{1,\infty}(D)} < \infty. \tag{17}$$

Under these conditions and using that $f \in L^2(D)$, it was proven in [12, Theorem 7.1] that for any $\boldsymbol{y} \in U$ the approximations $u_h(\cdot, \boldsymbol{y})$ satisfy, as $h \to 0$,

$$\|u(\cdot, \boldsymbol{y}) - u_h(\cdot, \boldsymbol{y})\|_V \le C_1\, h\, \|f\|_{L^2}.$$

In addition, if (the representer of) the bounded linear functional $G \in V^*$ lies in $L^2(D)$ we have by [12, Theorem 7.2] and [11, Eq. (3.11)] that for any $\boldsymbol{y} \in U$, as $h \to 0$,

$$
\begin{aligned}
|G(u(\cdot, \boldsymbol{y})) - G(u_h(\cdot, \boldsymbol{y}))| &\leq C_2 h^2 \, \|f\|_{L^2} \, \|G\|_{L^2}, \\
|\mathbb{E}[G(u(\cdot, \boldsymbol{y}) - u_h(\cdot, \boldsymbol{y}))]| &\leq C_3 h^2 \, \|f\|_{L^2} \, \|G\|_{L^2},
\end{aligned} \tag{18}
$$

where the constants $C_1, C_2, C_3 > 0$ are independent of h and $\boldsymbol{y}$. Since the above statements hold true for any $\boldsymbol{y} \subset U$, they remain also valid for $u^s(\boldsymbol{x}, \boldsymbol{y}) := u(\boldsymbol{x}, (\boldsymbol{y}_{\{1:s\}}; 0))$ and $u_h^s(\boldsymbol{x}, \boldsymbol{y}) := u_h(\boldsymbol{x}, (\boldsymbol{y}_{\{1:s\}}; 0))$ with associated constants C_1, C_2, C_3 independent of the value of s.

2.2 Dimension Truncation

For every $s \in \mathbb{N}$ and $\boldsymbol{y} \in U$, we formally define the solution of the parametric weak problem (3) corresponding to the diffusion coefficient $a(\boldsymbol{x}, (\boldsymbol{y}_{\{1:s\}}; 0))$ with sum truncated to s terms as

$$
u^s(\cdot, \boldsymbol{y}) := u(\cdot, (\boldsymbol{y}_{\{1:s\}}; 0)). \tag{19}
$$

In [8, Proposition 5.1] it was shown that for the solution u^s the following error estimates are satisfied.

Theorem 2 *Let* $\overline{\kappa} \in (0, 1)$ *be such that* (7) *is satisfied and assume furthermore that there exists a sequence of reals* $\boldsymbol{b} = (b_j)_{j \geq 1}$ *with* $0 < b_j \leq 1$ *for all* j *and a constant* $\kappa \in [\overline{\kappa}, 1)$ *as defined in* (11). *Then, for every* $\boldsymbol{y} \in U$ *and each* $s \in \mathbb{N}$

$$
\|u(\cdot, \boldsymbol{y}) - u^s(\cdot, \boldsymbol{y})\|_V \leq \frac{a_{0,\max} \, \|f\|_{V^*}}{(a_{0,\min}(1 - \overline{\kappa}))^2} \sup_{j \geq s+1} b_j.
$$

Moreover, if it holds for κ *that* $\frac{\kappa \, a_{0,\max}}{(1-\overline{\kappa}) \, a_{0,\min}} \sup_{j \geq s+1} b_j < 1$, *then for every* $G \in V^*$ *we have*

$$
\begin{aligned}
&\left| \mathbb{E}[G(u)] - \int_{[-\frac{1}{2}, \frac{1}{2}]^s} G(u^s(\cdot, (\boldsymbol{y}_{\{1:s\}}; 0))) \, \mathrm{d}\boldsymbol{y}_{\{1:s\}} \right| \\
&\quad \leq \frac{\|G\|_{V^*} \, \|f\|_{V^*}}{(1 - \overline{\kappa}) \, a_{0,\min} - a_{0,\max} \, \kappa \sup_{j \geq s+1} b_j} \left(\frac{a_{0,\max}}{(1 - \overline{\kappa}) \, a_{0,\min}} \kappa \sup_{j \geq s+1} b_j \right)^2.
\end{aligned} \tag{20}
$$

In the following subsection, we will discuss how to approximate the finite-dimensional integral of solutions of the form (19) by means of QMC methods.

2.3 Quasi-Monte Carlo Integration

For a real-valued function $F : \left[-\frac{1}{2}, \frac{1}{2}\right]^s \to \mathbb{R}$ defined over the s-dimensional unit cube centered at the origin, we consider the approximation of the integral $I_s(F)$ by N-point QMC rules $Q_N(F)$, i.e.,

$$I_s(F) := \int_{\left[-\frac{1}{2}, \frac{1}{2}\right]^s} F(\boldsymbol{y})\,\mathrm{d}\boldsymbol{y} \approx \frac{1}{N}\sum_{k=1}^{N} F(\boldsymbol{t}_k) =: Q_N(F),$$

with quadrature points $\boldsymbol{t}_1, \ldots, \boldsymbol{t}_N \in \left[-\frac{1}{2}, \frac{1}{2}\right]^s$. As a quality criterion of such a rule, we define the worst-case error for QMC integration in some Banach space $\mathscr{H}$ as

$$e^{\mathrm{wor}}(\boldsymbol{t}_1, \ldots, \boldsymbol{t}_N) := \sup_{\substack{F \in \mathscr{H} \\ \|F\|_{\mathscr{H}} \le 1}} |I_s(F) - Q_N(F)|.$$

In this article, we consider randomly shifted rank-1 lattice rules as randomized QMC rules, with underlying points of the form

$$\widetilde{\boldsymbol{t}}_k(\boldsymbol{\Delta}) = \{(k\boldsymbol{z})/N + \boldsymbol{\Delta}\} - (1/2, \ldots, 1/2), \quad k = 1, \ldots, N,$$

with generating vector $\boldsymbol{z} \in \mathbb{Z}^s$, uniform random shift $\boldsymbol{\Delta} \in [0,1]^s$ and component-wise applied fractional part, denoted by $\{\boldsymbol{x}\}$. For simplicity, we denote the worst-case error using a shifted lattice rule with generating vector $\boldsymbol{z}$ and shift $\boldsymbol{\Delta}$ by $e_{N,s}(\boldsymbol{z}, \boldsymbol{\Delta})$.

For randomly shifted QMC rules, the probabilistic error bound

$$\sqrt{\mathbb{E}_{\boldsymbol{\Delta}}\left[|I_s(F) - Q_N(F)|^2\right]} \le \widehat{e}_{N,s}(\boldsymbol{z})\,\|F\|_{\mathscr{H}},$$

holds for all $F \in \mathscr{H}$, with shift-averaged worst-case error

$$\widehat{e}_{N,s}(\boldsymbol{z}) := \left(\int_{[0,1]^s} e_{N,s}^2(\boldsymbol{z}, \boldsymbol{\Delta})\,\mathrm{d}\boldsymbol{\Delta}\right)^{1/2}.$$

As function space $\mathscr{H}$ for our integrands F, we consider the weighted, unanchored Sobolev space $\mathscr{W}_{s,\boldsymbol{\gamma}}$, which is a Hilbert space of functions defined over $\left[-\frac{1}{2}, \frac{1}{2}\right]^s$ with square integrable mixed first derivatives and general non-negative weights $\boldsymbol{\gamma} = (\gamma_{\mathfrak{u}})_{\mathfrak{u} \subseteq \{1:s\}}$. More precisely, the norm for $F \in \mathscr{W}_{s,\boldsymbol{\gamma}}$ is given by

$$\|F\|_{\mathscr{W}_{s,\boldsymbol{\gamma}}} := \left(\sum_{\mathfrak{u} \subseteq \{1:s\}} \gamma_{\mathfrak{u}}^{-1} \int_{\left[-\frac{1}{2}, \frac{1}{2}\right]^{|\mathfrak{u}|}} \left(\int_{\left[-\frac{1}{2}, \frac{1}{2}\right]^{s-|\mathfrak{u}|}} \frac{\partial^{|\mathfrak{u}|} F}{\partial \boldsymbol{y}_{\mathfrak{u}}}(\boldsymbol{y}_{\mathfrak{u}}; \boldsymbol{y}_{-\mathfrak{u}})\,\mathrm{d}\boldsymbol{y}_{-\mathfrak{u}}\right)^2 \mathrm{d}\boldsymbol{y}_{\mathfrak{u}}\right)^{1/2}, \tag{21}$$

where $\{1:s\} := \{1,\ldots,s\}$, $\frac{\partial^{|\mathfrak{u}|}F}{\partial \boldsymbol{y}_{\mathfrak{u}}}$ denotes the mixed first derivative with respect to the variables $\boldsymbol{y}_{\mathfrak{u}} = (y_j)_{j\in\mathfrak{u}}$ and we set $\boldsymbol{y}_{-\mathfrak{u}} = (y_j)_{j\in\{1:s\}\setminus\mathfrak{u}}$.

For the efficient construction of good lattice rule generating vectors, we consider the so-called reduced component-by-component (CBC) construction introduced in [5]. For $b \in \mathbb{N}$ and $m \in \mathbb{N}_0$, we define the group of units of integers modulo b^m via

$$\mathbb{Z}_{b^m}^{\times} := \left\{z \in \mathbb{Z}_{b^m} : \gcd(z, b^m) = 1\right\},$$

and note that $\mathbb{Z}_{b^0}^{\times} = \mathbb{Z}_1^{\times} = \{0\}$ since $\gcd(0,1) = 1$. Henceforth, let b be prime and recall that then, for $m \geq 1$, $|\mathbb{Z}_{b^m}^{\times}| = \varphi(b^m) = b^{m-1}\varphi(b)$ and $|\mathbb{Z}_b^{\times}| = \varphi(b) = (b-1)$, where φ is Euler's totient function. Let $\boldsymbol{w} := (w_j)_{j\geq 1}$ be a non-decreasing sequence of integers in $\mathbb{N}_0$, the elements of which we will refer to as reduction indices. In the reduced CBC algorithm the components $\widetilde{z}_j$ of the generating vector $\widetilde{\boldsymbol{z}}$ of the lattice rule will be taken as multiples of b^{w_j}.

In [5], the reduced CBC construction was introduced to construct rank-1 lattice rules for 1-periodic functions in a weighted Korobov space $\mathcal{H}(K_{s,\alpha,\boldsymbol{\gamma}})$ of smoothness α (see, e.g., [14]). We denote the worst-case error in $\mathcal{H}(K_{s,\alpha,\boldsymbol{\gamma}})$ using a rank-1 lattice rule with generating vector $\boldsymbol{z}$ by $e_{N,s}(\boldsymbol{z})$. Following [5], the reduced CBC construction is then given in Algorithm 1.

Algorithm 1 Reduced component-by-component construction

Input: Prime power $N = b^m$ with $m \in \mathbb{N}_0$ and integer reduction indices $0 \leq w_1 \leq \cdots \leq w_s$.

For j from 1 to s and as long as $w_j < m$ do:

- Select $z_j \in \mathbb{Z}_{b^{m-w_j}}^{\times}$ such that

$$z_j := \underset{z\in\mathbb{Z}_{b^{m-w_j}}^{\times}}{\operatorname{argmin}}\ e_{N,j}^2(b^{w_1}z_1,\ldots,b^{w_{j-1}}z_{j-1},b^{w_j}z).$$

Set all remaining $z_j := 0$ (for j with $w_j \geq m$).

Return: Generating vector $\widetilde{\boldsymbol{z}} := (b^{w_1}z_1,\ldots,b^{w_s}z_s)$ for $N = b^m$.

The following theorem, proven in [5], states that the algorithm yields generating vectors with a small integration error for general weights $\gamma_{\mathfrak{u}}$ in the Korobov space.

Theorem 3 *For a prime power $N = b^m$ let $\widetilde{\boldsymbol{z}} = (b^{w_1}z_1,\ldots,b^{w_s}z_s)$ be constructed according to Algorithm 1 with integer reduction indices $0 \leq w_1 \leq \cdots \leq w_s$. Then for every $d \in \{1:s\}$ and every $\lambda \in (1/\alpha, 1]$ it holds for the worst-case error in the Korobov space $\mathcal{H}(K_{s,\alpha,\boldsymbol{\gamma}})$ with $\alpha > 1$ that*

$$e_{N,d}^2(b^{w_1}z_1,\ldots,b^{w_d}z_d) \leq \left(\sum_{\emptyset\neq\mathfrak{u}\subseteq\{1:d\}} \gamma_{\mathfrak{u}}^{\lambda}\,(2\zeta(\alpha\lambda))^{|\mathfrak{u}|}\, b^{\min\{m,\max_{j\in\mathfrak{u}} w_j\}}\right)^{\frac{1}{\lambda}} \left(\frac{2}{N}\right)^{\frac{1}{\lambda}},$$

where $\zeta(x) := \sum_{h=1}^{\infty}\frac{1}{h^x}$, $x > 1$, denotes the Riemann zeta function.

This theorem can be extended to the weighted unanchored Sobolev space $\mathscr{W}_{s,\boldsymbol{\gamma}}$ using randomly shifted lattice rules as follows.

Theorem 4 *For a prime power $N = b^m$, $m \in \mathbb{N}_0$, and for $F \in \mathscr{W}_{s,\boldsymbol{\gamma}}$ belonging to the weighted unanchored Sobolev space defined over $\left[-\frac{1}{2}, \frac{1}{2}\right]^s$ with weights $\boldsymbol{\gamma} = (\gamma_{\mathfrak{u}})_{\mathfrak{u}\subseteq\{1:s\}}$, a randomly shifted lattice rule can be constructed by the reduced CBC algorithm, see Algorithm 1, such that for all $\lambda \in (1/2, 1]$,*

$$\sqrt{\mathbb{E}_{\boldsymbol{\Delta}}\left[|I_s(F) - Q_N(F)|^2\right]}$$
$$\leq \left(\sum_{\emptyset\neq\mathfrak{u}\subseteq\{1:s\}} \gamma_{\mathfrak{u}}^{\lambda}\, \varrho^{|\mathfrak{u}|}(\lambda)\, b^{\min\{m, \max_{j\in\mathfrak{u}} w_j\}} \right)^{1/(2\lambda)} \left(\frac{2}{N}\right)^{1/(2\lambda)} \|F\|_{\mathscr{W}_{s,\boldsymbol{\gamma}}},$$

with integer reduction indices $0 \leq w_1 \leq \cdots \leq w_s$ and $\varrho(\lambda) = 2\zeta(2\lambda)(2\pi^2)^{-\lambda}$.

Proof Using Theorem 3 and the connection that the shift-averaged kernel of the Sobolev space equals the kernel of the Korobov space $\mathscr{H}(K_{s,\alpha,\widetilde{\boldsymbol{\gamma}}})$ with $\alpha = 2$ and weights $\widetilde{\gamma}_{\mathfrak{u}} = \gamma_{\mathfrak{u}}/(2\pi^2)^{|\mathfrak{u}|}$, see, e.g., [7, 13], the result follows directly from

$$\sqrt{\mathbb{E}_{\boldsymbol{\Delta}}\left[|I(F) - Q_N(F)|^2\right]} \leq \sqrt{\mathbb{E}_{\boldsymbol{\Delta}}\left[e^2_{N,s}(\boldsymbol{z}, \boldsymbol{\Delta})\, \|F\|^2_{\mathscr{W}_{s,\boldsymbol{\gamma}}}\right]} = \widehat{e}_{N,s}(\boldsymbol{z})\, \|F\|_{\mathscr{W}_{s,\boldsymbol{\gamma}}}. \qquad \square$$

Theorem 4 implies that, under appropriate conditions on the weights $\gamma_{\mathfrak{u}}$ and the w_j's, see [5], the constructed randomly shifted lattice rules achieve an error convergence rate close to the optimal rate $\mathscr{O}(N^{-1})$ in the weighted Sobolev space. It follows that we can construct the lattice rule in the weighted Korobov space using the connection mentioned in the proof of the previous theorem.

2.4 Implementation of the Reduced CBC Algorithm

Similar to other variants of the CBC construction, we present a fast version of the reduced CBC method for POD weights in Algorithm 2 for which Theorems 3 and 4 still hold. The full derivation of Algorithm 2 is given in Sect. 5, here we only introduce the necessary notation. The squared worst-case error for POD weights $\boldsymbol{\gamma} = (\gamma_{\mathfrak{u}})_{\mathfrak{u}\subseteq\{1:s\}}$ with $\gamma_{\mathfrak{u}} = \Gamma(|\mathfrak{u}|)\prod_{j\in\mathfrak{u}}\gamma_j$ and $\gamma_\emptyset = 1$ in the weighted Korobov space $\mathscr{H}(K_{s,\alpha,\boldsymbol{\gamma}})$ with $\alpha > 1$ can be written as

$$e^2_{N,s}(\boldsymbol{z}) = \frac{1}{N}\sum_{k=0}^{N-1}\sum_{\ell=1}^{s}\sum_{\substack{\mathfrak{u}\subseteq\{1:s\}\\|\mathfrak{u}|=\ell}} \Gamma(\ell)\prod_{j\in\mathfrak{u}}\gamma_j\, \omega\!\left(\left\{\frac{kz_j}{N}\right\}\right),$$

where $\omega(x) = \sum_{0\neq h\in\mathbb{Z}} e^{2\pi i hx}/|h|^\alpha$, see, e.g., [7, 13], and for $n \in \mathbb{N}$ we define Ω_n as

$$\Omega_n := \left[\omega\left(\frac{kz \bmod n}{n}\right)\right]_{\substack{z\in\mathbb{Z}_n^\times \\ k\in\mathbb{Z}_n}} \in \mathbb{R}^{\varphi(n)\times n}.$$

We assume that the values of the function ω can be computed at unit cost. For integers $0 \le w' \le w'' \le m$ and given base b we define the "fold and sum" operator, which divides a length $b^{m-w'}$ vector into blocks of equal length $b^{m-w''}$ and sums them up, i.e.,

$$P^m_{w'',w'} : \mathbb{R}^{b^{m-w'}} \to \mathbb{R}^{b^{m-w''}} : P^m_{w'',w'}\,\boldsymbol{v} = [\,\underbrace{I_{b^{m-w''}} | \cdots | I_{b^{m-w''}}}_{b^{w''-w'} \text{ times}}\,]\,\boldsymbol{v}, \tag{22}$$

where $I_{b^{m-w''}}$ is the identity matrix of size $b^{m-w''} \times b^{m-w''}$. The computational cost of applying $P^m_{w'',w'}$ is the length of the input vector $\mathcal{O}(b^{m-w'})$. It should be clear that $P^m_{w''',w''}\,P^m_{w'',w'}\,\boldsymbol{v} = P^m_{w''',w'}\,\boldsymbol{v}$ for $0 \le w' \le w'' \le w''' \le m$. In step 4 of Algorithm 2 the notation $.*$ denotes the element-wise product of two vectors and $\Omega_{b^{m-w_j}}(z_j, :)$ means to take the row corresponding to $z = z_j$ from the matrix. Furthermore, Algorithm 2 includes an optional step in which the reduction indices are adjusted in case $w_1 > 0$, the auxiliary variable $w_0 = 0$ is introduced to satisfy the recurrence relation.

Algorithm 2 Fast reduced CBC construction for POD weights

Input: Prime power $N = b^m$ with $m \in \mathbb{N}_0$, integer reduction indices $0 \le w_1 \le \cdots \le w_s$, and weights $\Gamma(\ell)$, $\ell \in \mathbb{N}_0$ with $\Gamma(0) = 1$, and γ_j, $j \in \mathbb{N}$ such that $\gamma_{\mathfrak{u}} = \Gamma(|\mathfrak{u}|)\prod_{j\in\mathfrak{u}}\gamma_j$.

Optional: Adjust $m := \max\{0, m - w_1\}$ and for j from s down to 1 adjust $w_j := w_j - w_1$.

Set $\boldsymbol{q}_{0,0} := \mathbf{1}_{b^m}$ and $\boldsymbol{q}_{0,1} := \mathbf{0}_{b^m}$, set $w_0 := 0$.

For j from 1 to s and as long as $w_j < m$ do:

1. Set $\overline{\boldsymbol{q}}_j := \sum_{\ell=1}^{j} \frac{\Gamma(\ell)}{\Gamma(\ell-1)}\left[P^m_{w_j,w_{j-1}}\boldsymbol{q}_{j-1,\ell-1}\right] \in \mathbb{R}^{b^{m-w_j}}$ (with $\boldsymbol{q}_{j-1,\ell-1} \in \mathbb{R}^{b^{m-w_{j-1}}}$).
2. Calculate $\boldsymbol{T}_j := \Omega_{b^{m-w_j}}\,\overline{\boldsymbol{q}}_j \in \mathbb{R}^{\varphi(b^{m-w_j})}$ by exploiting the block-circulant structure of the matrix $\Omega_{b^{m-w_j}}$ using FFTs.
3. Set $z_j := \operatorname{argmin}_{z\in\mathbb{Z}^\times_{b^{m-w_j}}} \boldsymbol{T}_j(z)$, with $\boldsymbol{T}_j(z)$ the component corresponding to z.
4. Set $\boldsymbol{q}_{j,0} := \mathbf{1}_{b^{m-w_j}}$ and $\boldsymbol{q}_{j,j+1} := \mathbf{0}_{b^{m-w_j}}$ and for ℓ from j down to 1 set
$$\boldsymbol{q}_{j,\ell} := \left[P^m_{w_j,w_{j-1}}\boldsymbol{q}_{j-1,\ell}\right] + \frac{\Gamma(\ell)}{\Gamma(\ell-1)}\gamma_j\,\Omega_{b^{m-w_j}}(z_j,:) .* \left[P^m_{w_j,w_{j-1}}\boldsymbol{q}_{j-1,\ell-1}\right] \in \mathbb{R}^{b^{m-w_j}}.$$
5. Optional: Calculate squared worst-case error by $e_j^2 := \frac{1}{b^m}\sum_{k\in\mathbb{Z}_{b^{m-w_j}}}\sum_{\ell=1}^{j}\boldsymbol{q}_{j,\ell}(k)$.

Set all remaining $z_j := 0$ (for j with $w_j \ge m$).

Return: Generating vector $\widetilde{\boldsymbol{z}} := (b^{w_1}z_1, \ldots, b^{w_s}z_s)$ for $N = b^m$.
(Note: the w_j's and m might have been adjusted to make $w_1 = 0$.)

The standard fast CBC algorithm for POD weights has a computational complexity of $\mathcal{O}(s\,N\log(N) + s^2N)$, see, e.g., [7, 13]. The cost of our new algorithm can be substantially lower as is stated in the following theorem. We stress that the presented algorithm is the first realization of the reduced CBC construction for POD weights. Due to its lower computational cost, our new algorithm improves upon the one stated in [5] which only considers product weights, but the same technique can be used there since POD weights are more general and include product weights.

Theorem 5 *Given a sequence of integer reduction indices* $0 \le w_1 \le w_2 \le \cdots$, *the reduced CBC algorithm for a prime power* $N = b^m$ *points in* s *dimensions as specified in Algorithm 2 can construct a lattice rule whose worst-case error satisfies the error bound in Theorem 4 with an arithmetic cost of*

$$\mathcal{O}\left(\sum_{j=1}^{\min\{s,s^*\}} (m - w_j + j)\, b^{m-w_j}\right),$$

where s^* *is defined to be the largest integer such that* $w_{s^*} < m$. *The memory cost is* $\mathcal{O}(\sum_{j=1}^{\min\{s,s^*\}} b^{m-w_j})$. *In case of product weights* $\mathcal{O}(\sum_{j=1}^{\min\{s,s^*\}} (m - w_j)\, b^{m-w_j})$ *operations are required for the construction with memory* $\mathcal{O}(b^{m-w_1})$.

Proof We refer to Algorithm 2. Step 1 can be calculated in $\mathcal{O}(j\, b^{m-w_{j-1}})$ operations (and we may assume $w_0 = w_1$ since the case $w_1 > 0$ can be reduced to the case $w_1 = 0$). The matrix-vector multiplication in step 2 can be done by exploiting the block-circulant structure to obtain a fast matrix-vector product by FFTs at a cost of $\mathcal{O}((m - w_j)\, b^{m-w_j})$, see, e.g., [3, 4]. We ignore the possible saving by pre-computation of FFTs on the first columns of the blocks in the matrices $\Omega_{b^{m-w_j}}$ as this has cost $\mathcal{O}((m - w_1)\, b^{m-w_1})$ and therefore is already included in the cost of step 2. Finally, the vectors $\boldsymbol{q}_{j,\ell}$ for $\ell = 1, \ldots, j$ in step 4 can be calculated in $\mathcal{O}(j\, b^{m-w_{j-1}})$. To obtain the total complexity we remark that the applications of the "fold and sum" operator, marked by the square brackets could be performed in iteration $j - 1$ such that the cost of steps 1 and 4 in iteration j are only $\mathcal{O}(j\, b^{m-w_j})$ instead of $\mathcal{O}(j\, b^{m-w_{j-1}})$. The cost of the additional fold and sum to prepare for iteration j in iteration $j - 1$, which can be performed after step 4, is then equal to the cost of step 4 in that iteration. Since we can assume $w_0 = w_1$ we obtain the claimed construction cost. Note that the algorithm is written in such a way that the vectors $\boldsymbol{q}_{j,\ell-1}$ can be reused for storing the vectors $\boldsymbol{q}_{j,\ell}$ (which might be smaller). Similarly for the vectors $\overline{\boldsymbol{q}}_j$. Therefore the memory cost is $\mathcal{O}(\sum_{j=1}^{\min\{s,s^*\}} b^{m-w_j})$. The result for product weights can be obtained similarly, see, e.g., [13]. □

3 QMC Finite Element Error Analysis

We now combine the results of the previous subsections to analyze the overall QMC finite element error. We consider the root mean square error (RMSE) given by

$$e_{N,s,h}^{\mathrm{RMSE}}(G(u)) := \sqrt{\mathbb{E}_{\boldsymbol{\Delta}}\left[|\mathbb{E}[G(u)] - Q_N(G(u_h^s))|^2\right]}.$$

The error $\mathbb{E}[G(u)] - Q_N(G(u_h^s))$ can be written as

$$\mathbb{E}[G(u)] \quad Q_N(G(u_h^s)) = \mathbb{E}[G(u)] - I_s(G(u_h^s)) + I_s(G(u_h^s)) - Q_N(G(u_h^s))$$

such that due to the fact that $\mathbb{E}_{\boldsymbol{\Delta}}(Q_N(f)) = I_s(f)$ for any integrand f we obtain

$$\begin{aligned}\mathbb{E}_{\boldsymbol{\Delta}}\left[(\mathbb{E}[G(u)] - Q_N(G(u_h^s)))^2\right] &= (\mathbb{E}[G(u)] - I_s(G(u_h^s)))^2 + \mathbb{E}_{\boldsymbol{\Delta}}\left[(I_s - Q_N)^2(G(u_h^s))\right] \\ &\quad + 2(\mathbb{E}[G(u)] - I_s(G(u_h^s)))\,\mathbb{E}_{\boldsymbol{\Delta}}\left[(I_s - Q_N)(G(u_h^s))\right] \\ &= (\mathbb{E}[G(u)] - I_s(G(u_h^s)))^2 + \mathbb{E}_{\boldsymbol{\Delta}}\left[(I_s - Q_N)^2(G(u_h^s))\right].\end{aligned}$$

Then, noting that $\mathbb{E}[G(u)] - I_s(G(u_h^s)) = \mathbb{E}[G(u)] - I_s(G(u^s)) + I_s(G(u^s)) - I_s(G(u_h^s$

$$\begin{aligned}(\mathbb{E}[G(u)] - I_s(G(u_h^s)))^2 &= (\mathbb{E}[G(u)] - I_s(G(u^s)))^2 + (I_s(G(u^s)) - I_s(G(u_h^s)))^2 \\ &\quad + 2(\mathbb{E}[G(u)] - I_s(G(u^s)))(I_s(G(u^s)) - I_s(G(u_h^s)))\end{aligned}$$

and since for general $x, y \in \mathbb{R}$ it holds that $2xy \leq x^2 + y^2$, we obtain furthermore

$$(\mathbb{E}[G(u)] - I_s(G(u_h^s)))^2 \leq 2(\mathbb{E}[G(u)] - I_s(G(u^s)))^2 + 2(I_s(G(u^s)) - I_s(G(u_h^s)))^2.$$

From the previous subsections we can then use (20) for the truncation part, (18), which holds for general $\boldsymbol{y} \in U$ and thus also for $\boldsymbol{y}_{\{1:s\}}$, for the finite element error, and Theorem 4 for the QMC integration error to obtain the following error bound for the mean square error $\mathbb{E}_{\boldsymbol{\Delta}}[|\mathbb{E}[G(u)] - Q_N(G(u_h^s))|^2] =: e_{N,s,h}^{\mathrm{MSE}}(G(u))$,

$$\begin{aligned}e_{N,s,h}^{\mathrm{MSE}}(G(u)) \leq K_1 \|f\|_{V^*}^2 \|G\|_{V^*}^2 &\left(\frac{1}{(1-\overline{\kappa})\, a_{0,\min} - a_{0,\max}\, \kappa \sup_{j\geq s+1} b_j}\right)^2 \\ \times &\left(\frac{a_{0,\max}}{(1-\overline{\kappa})\, a_{0,\min}} \kappa \sup_{j\geq s+1} b_j\right)^4 + K_2 \|f\|_{L^2}^2 \|G\|_{L^2}^2\, h^4 \\ + &\left(\sum_{\emptyset\neq \mathrm{u}\subseteq\{1:s\}} \gamma_{\mathrm{u}}^{\lambda}\, \varrho^{|\mathrm{u}|}(\lambda)\, b^{\min\{m, \max_{j\in\mathrm{u}} w_j\}}\right)^{1/\lambda} \left(\frac{2}{N}\right)^{1/\lambda} \|G(u_h^s)\|_{\mathscr{W}_{s,\gamma}}^2\end{aligned} \tag{23}$$

for some constants $K_1, K_2 \in \mathbb{R}_+$ and provided that $\frac{a_{0,\max}}{(1-\overline{\kappa})\, a_{0,\min}} \kappa \sup_{j\geq s+1} b_j < 1$.

3.1 Derivative Bounds of POD Form

In the following we assume that we have general bounds on the mixed partial derivatives $\partial^{\boldsymbol{\nu}} u(\cdot, \boldsymbol{y})$ which are of POD form; that is,

$$\|\partial^{\boldsymbol{\nu}} u(\cdot, \boldsymbol{y})\|_V \le C \widetilde{\boldsymbol{b}}^{\boldsymbol{\nu}} \Gamma(|\boldsymbol{\nu}|) \|f\|_{V^*} \tag{24}$$

with a map $\Gamma : \mathbb{N}_0 \to \mathbb{R}$, a sequence of reals $\widetilde{\boldsymbol{b}} = (\widetilde{b}_j)_{j\ge1} \in \mathbb{R}^{\mathbb{N}}$ and some constant $C \in \mathbb{R}_+$. Such bounds can be found in the literature and we provided a new derivation in Theorem 1 also leading to POD weights.

For bounding the norm $\|G(u_h^s)\|_{\mathscr{W}_{s,\boldsymbol{\gamma}}}$, we can then use (24) and the definition in (21) to proceed as outlined in [11], to obtain the estimate

$$\|G(u_h^s)\|_{\mathscr{W}_{s,\boldsymbol{\gamma}}} \le C \|f\|_{V^*} \|G\|_{V^*} \left(\sum_{\mathrm{u} \subseteq \{1:s\}} \frac{\Gamma(|\mathrm{u}|)^2 \prod_{j\in\mathrm{u}} \widetilde{b}_j^2}{\gamma_{\mathrm{u}}} \right)^{1/2}. \tag{25}$$

Denoting $\boldsymbol{w} := (w_j)_{j\ge1}$ and using (25), the contribution of the quadrature error to the mean square error $e_{N,h,s}^{\mathrm{MSE}}(G(u))$ can be upper bounded by

$$\begin{aligned} &\left(\sum_{\emptyset \ne \mathrm{u} \subseteq \{1:s\}} \gamma_{\mathrm{u}}^{\lambda} \varrho^{|\mathrm{u}|}(\lambda)\, b^{\min\{m, \max_{j\in\mathrm{u}} w_j\}} \right)^{1/\lambda} \left(\frac{2}{N}\right)^{1/\lambda} \|G(u_h^s)\|_{\mathscr{W}_{s,\boldsymbol{\gamma}}}^2 \\ &\quad \le C \|f\|_{V^*} \|G\|_{V^*}\, C_{\boldsymbol{\gamma},\boldsymbol{w},\lambda} \left(\frac{2}{N}\right)^{1/\lambda}, \end{aligned} \tag{26}$$

where we define

$$C_{\boldsymbol{\gamma},\boldsymbol{w},\lambda} := \left(\sum_{\emptyset \ne \mathrm{u} \subseteq \{1:s\}} \gamma_{\mathrm{u}}^{\lambda} \varrho^{|\mathrm{u}|}(\lambda)\, b^{\min\{m, \max_{j\in\mathrm{u}} w_j\}} \right)^{1/\lambda} \left(\sum_{\mathrm{u} \subseteq \{1:s\}} \frac{\Gamma(|\mathrm{u}|)^2 \prod_{j\in\mathrm{u}} \widetilde{b}_j^2}{\gamma_{\mathrm{u}}} \right).$$

The term $C_{\boldsymbol{\gamma},\boldsymbol{w},\lambda}$ can be bounded as

$$C_{\boldsymbol{\gamma},\boldsymbol{w},\lambda} \le \left(\sum_{\mathrm{u} \subseteq \{1:s\}} \gamma_{\mathrm{u}}^{\lambda} \varrho^{|\mathrm{u}|}(\lambda)\, b^{\sum_{j\in\mathrm{u}} w_j - \sum_{\ell=1}^{|\mathrm{u}|-1} w_\ell} \right)^{1/\lambda} \left(\sum_{\mathrm{u} \subseteq \{1:s\}} \frac{\Gamma(|\mathrm{u}|)^2 \prod_{j\in\mathrm{u}} \widetilde{b}_j^2}{\gamma_{\mathrm{u}}} \right).$$

Due to [12, Lemma 6.2] the latter term is minimized by choosing the weights γ_{u} as

$$\gamma_{\mathrm{u}} := \left(\frac{\Gamma(|\mathrm{u}|)^2 \prod_{j\in\mathrm{u}} \widetilde{b}_j^2 \prod_{\ell=1}^{|\mathrm{u}|-1} b^{w_\ell}}{\prod_{j\in\mathrm{u}} \rho(\lambda)\, b^{w_j}} \right)^{1/(1+\lambda)}. \tag{27}$$

In an effort to further estimate $C_{\boldsymbol{\gamma},\boldsymbol{w},\lambda}$, we introduce the auxiliary quantity A_λ as

$$A_\lambda := \sum_{\mathfrak{u}\subseteq\{1:s\}} \gamma_{\mathfrak{u}}^{\lambda}\, \varrho^{|\mathfrak{u}|}(\lambda)\, b^{\sum_{j\in\mathfrak{u}} w_j - \sum_{\ell=1}^{|\mathfrak{u}|-1} w_\ell}$$

$$= \sum_{\mathfrak{u}\subseteq\{1:s\}} \left[\left(\frac{\Gamma(|\mathfrak{u}|)^{2\lambda}}{\prod_{\ell=1}^{|\mathfrak{u}|-1} b^{w_\ell}} \right) \left(\prod_{j\in\mathfrak{u}} \rho(\lambda)\, \widetilde{b}_j^{2\lambda}\, b^{w_j} \right) \right]^{\frac{1}{1+\lambda}}$$

and easily see that also

$$\sum_{\mathfrak{u}\subseteq\{1:s\}} \gamma_{\mathfrak{u}}^{-1} \left(\Gamma(|\mathfrak{u}|)^2 \prod_{j\in\mathfrak{u}} \widetilde{b}_j^2 \right) = A_\lambda,$$

which implies that $C_{\boldsymbol{\gamma},\boldsymbol{w},\lambda} \le A_\lambda^{1+1/\lambda}$. We demonstrate how the term A_λ can be estimated for the derivative bounds obtained in Sect. 1.2.

In view of Theorem 1, assume in the following that

$$\Gamma(|\mathfrak{u}|) = \kappa^{|\mathfrak{u}|}, \quad \widetilde{b}_j = \frac{2\,b_j}{1-\kappa}, \quad \sum_{j=1}^{\infty} (b_j b^{w_j})^p < \infty \quad \text{for} \quad p \in (0,1). \tag{28}$$

Note that we could also choose $\Gamma(|\mathfrak{u}|) = \kappa(|\mathfrak{u}|)^{|\mathfrak{u}|}$ above, in which case the subsequent estimate of A_λ can be done analogously, but to make the argument less technical, we consider the slightly coarser variant $\Gamma(|\mathfrak{u}|) = \kappa^{|\mathfrak{u}|}$ here. In this case,

$$A_\lambda = \sum_{\mathfrak{u}\subseteq\{1:s\}} \left[\kappa^{|\mathfrak{u}|}\right]^{\frac{2\lambda}{1+\lambda}} \left(\prod_{\ell=1}^{|\mathfrak{u}|-1} b^{\frac{-w_\ell}{2\lambda}} \right)^{\frac{2\lambda}{1+\lambda}} \prod_{j\in\mathfrak{u}} \left(\left(\frac{2\,b_j}{1-\kappa} \right)^{2\lambda} b^{w_j}\, \rho(\lambda) \right)^{\frac{1}{1+\lambda}}.$$

Note that, as $\lambda \le 1$, it holds that $b^{\frac{-w_\ell}{2\lambda}} \le b^{\frac{-w_\ell}{2}}$ and hence

$$A_\lambda \le \sum_{\mathfrak{u}\subseteq\{1:s\}} \left(\kappa^{|\mathfrak{u}|} \prod_{\ell=1}^{|\mathfrak{u}|-1} b^{\frac{-w_\ell}{2}} \right)^{\frac{2\lambda}{1+\lambda}} \prod_{j\in\mathfrak{u}} \left(\left(\frac{2\,b_j}{1-\kappa} \right)^{2\lambda} b^{w_j}\, \rho(\lambda) \right)^{\frac{1}{1+\lambda}}.$$

We now proceed similarly to the proof of Theorem 6.4 in [12]. Let $(\alpha_j)_{j\ge1}$ be a sequence of positive reals, to be specified below, which satisfies $\Sigma := \sum_{j=1}^{\infty} \alpha_j < \infty$. Dividing and multiplying by $\prod_{j\in\mathfrak{u}} \alpha_j^{(2\lambda)/(1+\lambda)}$, and applying Hölder's inequality with conjugate components $p = (1+\lambda)/(2\lambda)$ and $p^* = (1+\lambda)/(1-\lambda)$, gives

$$A_\lambda \le \sum_{\mathfrak{u}\subseteq\{1:s\}} \left(\kappa^{|\mathfrak{u}|}\prod_{\ell=1}^{|\mathfrak{u}|-1} b^{\frac{-w_\ell}{2}}\right)^{\frac{2\lambda}{1+\lambda}} \left(\prod_{j\in\mathfrak{u}} \alpha_j^{\frac{2\lambda}{1+\lambda}}\right) \prod_{j\in\mathfrak{u}} \left(\left(\frac{2\,b_j}{1-\kappa}\right)^{2\lambda} \frac{b^{w_j}\,\rho(\lambda)}{\alpha_j^{2\lambda}}\right)^{\frac{1}{1+\lambda}}$$

$$\le \left(\sum_{\mathfrak{u}\subseteq\{1:s\}} \kappa^{|\mathfrak{u}|}\left(\prod_{\ell=1}^{|\mathfrak{u}|-1} b^{\frac{-w_\ell}{2}}\right)\prod_{j\in\mathfrak{u}}\alpha_j\right)^{\frac{2\lambda}{1+\lambda}}$$

$$\times \left(\sum_{\mathfrak{u}\subseteq\{1:s\}} \prod_{j\in\mathfrak{u}} \left(\left(\frac{2\,b_j}{1-\kappa}\right)^{2\lambda} \frac{b^{w_j}\,\rho(\lambda)}{\alpha_j^{2\lambda}}\right)^{\frac{1}{1-\lambda}}\right)^{\frac{1-\lambda}{1+\lambda}} = B^{\frac{2\lambda}{1+\lambda}}\cdot \widetilde{B}^{\frac{1-\lambda}{1+\lambda}},$$

where we define

$$B := \sum_{\mathfrak{u}\subseteq\{1:s\}} \kappa^{|\mathfrak{u}|}\left(\prod_{\ell=1}^{|\mathfrak{u}|-1} b^{\frac{-w_\ell}{2}}\right)\prod_{j\in\mathfrak{u}}\alpha_j,\quad \widetilde{B} := \sum_{\mathfrak{u}\subseteq\{1:s\}} \prod_{j\in\mathfrak{u}} \left(\left(\frac{2\,b_j}{1-\kappa}\right)^{2\lambda} \frac{b^{w_j}\,\rho(\lambda)}{\alpha_j^{2\lambda}}\right)^{\frac{1}{1-\lambda}}.$$

For the first factor we estimate

$$B \le \sum_{\mathfrak{u}:\,|\mathfrak{u}|<\infty} \kappa^{|\mathfrak{u}|}\prod_{\ell=1}^{|\mathfrak{u}|-1} b^{\frac{-w_\ell}{2}}\prod_{j\in\mathfrak{u}}\alpha_j = \sum_{k=1}^{\infty}\left(\kappa^k\prod_{\ell=1}^{k-1} b^{\frac{-w_\ell}{2}}\right)\sum_{\substack{\mathfrak{u}:\,|\mathfrak{u}|<\infty\\ |\mathfrak{u}|=k}}\prod_{j\in\mathfrak{u}}\alpha_j$$

$$\le \sum_{k=1}^{\infty}\left(\kappa^k\prod_{\ell=1}^{k-1} b^{\frac{-w_\ell}{2}}\right)\frac{1}{k!}\sum_{\boldsymbol{u}\in\mathbb{N}^k}\prod_{i=1}^{k}\alpha_{u_i} = \sum_{k=1}^{\infty}\left(\kappa^k\prod_{\ell=1}^{k-1} b^{\frac{-w_\ell}{2}}\right)\frac{1}{k!}\Sigma^k.$$

By the ratio test, the latter expression is finite if we choose $(\alpha_j)_{j\ge1}$ such that $L := \sup_{k\in\mathbb{N}} \kappa\, b^{\frac{-w_k}{2}}(k+1)^{-1} = \kappa b^{\frac{-w_1}{2}}/2 < 1/\Sigma$. Hence we assume that $(\alpha_j)_{j\ge1}$ is chosen such that indeed $L < 1/\Sigma$. Note that L is small if κ is small, which means that Σ can be allowed to be large in this case. Consider now the term $\widetilde{B}$ for which

$$\widetilde{B} \le \sum_{\mathfrak{u}:\,|\mathfrak{u}|<\infty}\prod_{j\in\mathfrak{u}}\left(\left(\frac{2\,b_j}{1-\kappa}\right)^{2\lambda} b^{w_j}\,\rho(\lambda)/\alpha_j^{2\lambda}\right)^{\frac{1}{1-\lambda}}$$

$$\le \exp\left(\sum_{j=1}^{\infty}\left(\left(\frac{2\,b_j}{1-\kappa}\right)^{2\lambda} b^{w_j}\,\rho(\lambda)/\alpha_j^{2\lambda}\right)^{\frac{1}{1-\lambda}}\right)$$

$$\le \exp\left(\sum_{j=1}^{\infty}\left(\frac{1}{1-\kappa}\right)^{\frac{2\lambda}{1-\lambda}}(\rho(\lambda))^{\frac{1}{1-\lambda}}4^\lambda\left(b_j b^{w_j}\frac{1}{\alpha_j}\right)^{\frac{2\lambda}{1-\lambda}}\right)$$

$$= \exp\left((1-\kappa)^{\frac{-2\lambda}{1-\lambda}}(\rho(\lambda))^{\frac{1}{1-\lambda}}4^\lambda\sum_{j=1}^{\infty}\left(b_j b^{w_j}\alpha_j^{-1}\right)^{\frac{2\lambda}{1-\lambda}}\right).$$

Based on the previous two estimates, in order to assure that B, $\widetilde{B}$ and thus A_λ are finite, we require that

$$L < 1/\Sigma = 1/\sum_{j=1}^{\infty} \alpha_j \quad \text{and} \quad \sum_{j=1}^{\infty} \left(b_j b^{w_j} \alpha_j^{-1}\right)^{\frac{2\lambda}{1-\lambda}} < \infty. \tag{29}$$

To this end, we choose $\alpha_j := \frac{(b_j b^{w_j})^p}{\theta}$, where $\frac{\theta}{L} > \sum_{j=1}^{\infty} \left(b_j b^{w_j}\right)^p$. Then,

$$A_\lambda \leq \left(\sum_{k=1}^{\infty} \left(\kappa^k \prod_{\ell=1}^{k-1} b^{\frac{-w_\ell}{2}} \right) \frac{1}{k!} \Sigma^k \right)^{\frac{2\lambda}{1+\lambda}}$$
$$\times \exp\left(\frac{1-\lambda}{1+\lambda} \left(\frac{1}{1-\kappa} \right)^{\frac{2\lambda}{1-\lambda}} (\rho(\lambda))^{\frac{1}{1-\lambda}} 4^\lambda \sum_{j=1}^{\infty} \left(b_j b^{w_j} \frac{1}{\alpha_j} \right)^{\frac{2\lambda}{1-\lambda}} \right) \tag{30}$$

as long as we choose λ such that

$$\sum_{j=1}^{\infty} \left(b_j b^{w_j} \alpha_j^{-1}\right)^{2\lambda/(1-\lambda)} < \infty. \tag{31}$$

We denote the upper bound in (30) by $\overline{A}(\lambda)$. Similarly to what is done in [12, Proof of Theorem 6.4], we see that Condition (31) is satisfied if $\lambda \geq \frac{p}{2-p}$. Again, similarly to [12, Proof of Theorem 6.4] we see that the latter can be achieved by choosing

$$\lambda_p = \begin{cases} 1/(2-2\delta) \quad \text{for some } \delta \in (0, 1/2) & \text{if } p \in (0, 2/3], \\ p/(2-p) & \text{if } p \in (2/3, 1). \end{cases} \tag{32}$$

Hence by choosing λ equal to λ_p, we get an efficient bound on $C_{\boldsymbol{\gamma},\mathbf{w},\lambda_p} = A_{\lambda_p}^{1+1/\lambda_p}$, as long as the w_j are chosen to guarantee convergence of $\sum_{j=1}^{\infty} \left(b_j b^{w_j}\right)^p$.

4 Combined Error Bound

The derivation in the previous section leads to the following result, where the notation $a \lesssim b$ indicates that $a \leq Cb$ for some independent positive constant C.

Theorem 6 *Given the PDE in* (1) *for which we characterized the regularity of the random field by a sequence of* b_j *with sparsity* $p \in (0, 1)$ *and determined a sequence of* w_j *such that* $\sum_{j=1}^{\infty} (b_j\, b^{w_j})^p < \infty$, *we can construct the generating vector for an* N*-point randomized lattice rule using the reduced CBC algorithm (Algorithm 2), at the cost of* $\mathcal{O}(\sum_{j=1}^{\min\{s,s^*\}} (m - w_j + j)\, b^{m-w_j})$ *operations, such that, assuming that*

(11), (17) *and* $\frac{\kappa\, a_{0,\max}}{(1-\overline{\kappa})\, a_{0,\min}} \sup_{j \geq s+1} b_j < 1$ *hold, we obtain an upper bound*

$$e^{MSE}_{N,s,h}(G(u)) \lesssim \left(\sup_{j \geq s+1} b_j\right)^2 + h^4 + \left(\frac{2}{N}\right)^{1/\lambda_p}, \tag{33}$$

where the implied constant is independent of s, h *and* N.

Observe that if the w_j increase sufficiently fast, the construction cost of Algorithm 2 does not depend anymore on the increasing dimensionality. Further note that the first term on the right-hand side of (33) is small if $\sup_{j \geq s+1} b_j$ is small, and, since we assumed that b_j must tend to zero by assumption (28), we can shrink the first summand by choosing s sufficiently large. By choosing h sufficiently small, and N sufficiently large, we can also make the other two summands in the overall error bound small.

Note that $\sup_{j \geq s+1} b_j \leq \sum_{j \geq s+1} b_j$, and that (28) yields $\sum_{j=1}^{\infty} b_j^p < \infty$, which implies that one can use the machinery developed in [12] to obtain a cost analysis similar to [12, Theorem 8.1]. Note, in particular, that it is sufficient to choose N of order $\mathcal{O}(\varepsilon^{-\lambda_p/2})$, independently of s, to meet an error threshold of ε.

5 Derivation of the Fast Reduced CBC Algorithm

Finally in this last section the derivation of the fast reduced CBC algorithm for POD weights in Algorithm 2 is given. For prime b and $m \in \mathbb{N}$ let $N = b^m$. Consider a generating vector $\widetilde{z} = (b^{w_1} z_1, \ldots, b^{w_d} z_d)$ with $z_j \in \mathbb{Z}^{\times}_{b^{m-w_j}}$ and integer $0 \leq w_j \leq m$ for each $j = 1, \ldots, d$. Furthermore, for an integer $0 \leq w' \leq m$, the squared worst-case error can be written as

$$\begin{aligned}
e^2_{b^m,d}(\widetilde{z}) &= \frac{1}{b^m} \sum_{k \in \mathbb{Z}_{b^m}} \sum_{\ell=1}^{d} \sum_{\substack{\mathfrak{u} \subseteq \{1:d\} \\ |\mathfrak{u}|=\ell}} \Gamma(\ell) \prod_{j \in \mathfrak{u}} \gamma_j\, \omega\!\left(\frac{k\, b^{w_j} z_j \bmod b^m}{b^m}\right) \\
&= \frac{1}{b^m} \sum_{k \in \mathbb{Z}_{b^m}} \sum_{\ell=1}^{d} \sum_{\substack{\mathfrak{u} \subseteq \{1:d\} \\ |\mathfrak{u}|=\ell}} \Gamma(\ell) \prod_{j \in \mathfrak{u}} \gamma_j\, \omega\!\left(\frac{k\, z_j \bmod b^{m-w_j}}{b^{m-w_j}}\right) \\
&= \frac{1}{b^m} \sum_{k' \in \mathbb{Z}_{b^{m-w'}}} \sum_{\ell=1}^{d} \underbrace{\sum_{t \in \mathbb{Z}_{b^{w'}}} \sum_{\substack{\mathfrak{u} \subseteq \{1:d\} \\ |\mathfrak{u}|=\ell}} \Gamma(\ell) \prod_{j \in \mathfrak{u}} \gamma_j\, \omega\!\left(\frac{(k' + t\, b^{m-w'})\, z_j \bmod b^{m-w_j}}{b^{m-w_j}}\right)}_{=: q_{d,\ell,w'}(k') \text{ for } k' \in \mathbb{Z}_{b^{m-w'}}} \\
&= \frac{1}{b^m} \sum_{k' \in \mathbb{Z}_{b^{m-w'}}} \sum_{\ell=1}^{d} q_{d,\ell,w'}(k').
\end{aligned}$$

We note that this holds for any integer $0 \le w' \le m$ and, in particular, for $w = 0$ this is the vector being used in the normal fast CBC algorithm. We now write the error in terms of the previous error, as is standard for CBC algorithms, by splitting the expression into subsets $\mathfrak{u} \subseteq \{1 : d\}$ for which $d \notin \mathfrak{u}$ and $d \in \mathfrak{u}$, to obtain

$$e^2_{b^m,d}(\widetilde{z}) = e^2_{b^m,d-1}(\widetilde{z}_1, \dots, \widetilde{z}_{d-1}) + \frac{1}{b^m} \sum_{k \in \mathbb{Z}_{b^m}} \sum_{\ell=0}^{d-1} \frac{\Gamma(\ell+1)}{\Gamma(\ell)} \sum_{\substack{\mathfrak{u} \subseteq \{1:d-1\} \\ |\mathfrak{u}|=\ell}} \Gamma(\ell) \times$$
$$\times \prod_{j \in \mathfrak{u}} \gamma_j \, \omega\left(\frac{k\, z_j \bmod b^{m-w_j}}{b^{m-w_j}}\right) \gamma_d \, \omega\left(\frac{k\, z_d \bmod b^{m-w_d}}{b^{m-w_d}}\right).$$

Since the choice of $z_d \in \mathbb{Z}^\times_{b^{m-w_d}}$ is modulo b^{m-w_d}, we can make a judicious choice for splitting up $k = k' + t\, b^{m-w_d}$ for which the effect of dimension d (for a choice of z_d) is then constant for all $t \in \mathbb{Z}_{b^{w_d}}$. We obtain

$$e^2_{b^m,d}(\widetilde{z}) = e^2_{b^m,d-1} + \frac{1}{b^m} \sum_{k' \in \mathbb{Z}_{b^{m-w_d}}} \sum_{\ell=0}^{d-1} \frac{\Gamma(\ell+1)}{\Gamma(\ell)} q_{d-1,\ell,w_d}(k')\, \gamma_d\, \omega\left(\frac{k'\, z_d \bmod b^{m-w_d}}{b^{m-w_d}}\right). \quad (34)$$

Then we observe that for all $0 \le w_{d-1} \le w_d \le m$, with $k' \in \mathbb{Z}_{b^{m-w_d}}$, writing $t = t' + t''\, b^{w_d - w_{d-1}} \in \mathbb{Z}_{b^{w_d}}$ with $t' \in \mathbb{Z}_{b^{w_d - w_{d-1}}}$ and $t'' \in \mathbb{Z}_{b^{w_{d-1}}}$, leads to

$$\begin{aligned} q_{d-1,\ell,w_d}(k') &= \sum_{t' \in \mathbb{Z}_{b^{w_d - w_{d-1}}}} \sum_{t'' \in \mathbb{Z}_{b^{w_{d-1}}}} \sum_{\substack{\mathfrak{u} \subseteq \{1:d-1\} \\ |\mathfrak{u}|=\ell}} \Gamma(\ell) \times \\ &\quad \times \prod_{j \in \mathfrak{u}} \gamma_j\, \omega\left(\frac{(k' + (t' + t''\, b^{w_d - w_{d-1}})\, b^{m-w_d})\, z_j \bmod b^{m-w_j}}{b^{m-w_j}}\right) \\ &= \sum_{t' \in \mathbb{Z}_{b^{w_d - w_{d-1}}}} q_{d-1,\ell,w_{d-1}}(k' + t'\, b^{m-w_d}), \end{aligned}$$

where $k'' = k' + t'\, b^{m-w_d} \in \mathbb{Z}_{b^{m-w_{d-1}}}$ as required for $q_{d-1,\ell,w_{d-1}}(k'')$. Note that this is the property of the "fold and sum" operator as introduced in (22) and mentioned there. Using matrix-vector notation, we rewrite the expression in (34) for all $z_d \in \mathbb{Z}^\times_{b^{m-w_d}}$ as

$$\boldsymbol{e}^2_{b^m,d} = \boldsymbol{e}^2_{b^m,d-1} + \frac{\gamma_d}{b^m} \Omega_{b^{m-w_d}} \left(\sum_{\ell=0}^{d-1} \frac{\Gamma(\ell+1)}{\Gamma(\ell)} \left[P^m_{w_d, w_{d-1}}\, \boldsymbol{q}_{d-1,\ell,w_{d-1}} \right] \right),$$

where $\boldsymbol{e}^2_{b^m,d} \in \mathbb{R}^{\varphi(b^{m-w_d})}$ is the vector with components $e^2_{b^m,d}(b^{w_1} z_1, \dots, b^{w_d} z_d)$ for all $z_d \in \mathbb{Z}^\times_{b^{m-w_d}}$. After z_d has been selected we can calculate (for $\ell = 1, \dots, d$)

$$\boldsymbol{q}_{d,\ell,w_d} = \left[P^m_{w_d,w_{d-1}}\, \boldsymbol{q}_{d-1,\ell,w_{d-1}}\right] + \frac{\Gamma(\ell)}{\Gamma(\ell-1)} \gamma_d\, \Omega_{b^{m-w_d}}(z_d, :) .* \left[P^m_{w_d,w_{d-1}}\, \boldsymbol{q}_{d-1,\ell-1,w_{d-1}}\right].$$

In Algorithm 2 the vectors $\boldsymbol{q}_{j,\ell,w_j}$ are denoted by just $\boldsymbol{q}_{j,\ell}$.

References

1. Bachmayr, M., Cohen, A., Migliorati, G.: Sparse polynomial approximation of parametric elliptic PDEs. Part I: affine coefficients. ESAIM: Math. Model. Numer. Anal. **51**, 321–339 (2017)
2. Cohen, A., De Vore, R., Schwab, Ch.: Convergence rates of best N-term Galerkin approximations for a class of elliptic sPDEs. Found. Comput. Math. **10**, 615–646 (2010)
3. Cools, R., Kuo, F.Y., Nuyens, D.: Constructing embedded lattice rules for multivariate integration. SIAM J. Sci. Comput. **28**, 2162–2188 (2006)
4. Cools, R., Nuyens, D.: Fast component-by-component construction of rank-1 lattice rules with a non-prime number of points. J. Complex. **22**, 4–28 (2006)
5. Dick, J., Kritzer, P., Leobacher, G., Pillichshammer, F.: A reduced fast component-by-component construction of lattice points for integration in weighted spaces with fast decreasing weights. J. Comput. Appl. Math. **276**, 1–15 (2015)
6. Dick, J., Kuo, F.Y., Le Gia, Q.T., Nuyens, D., Schwab, Ch.: Higher order QMC Galerkin discretization for parametric operator equations. SIAM J. Numer. Anal. **52**, 2676–2702 (2014)
7. Dick, J., Kuo, F.Y., Sloan, I.H.: High-dimensional integration: the quasi-Monte Carlo way. Acta Numer. **22**, 133–288 (2013)
8. Gantner, R.N., Herrmann, L., Schwab, Ch.: Quasi-Monte Carlo integration for affine-parametric, elliptic PDEs: local supports and product weights. SIAM J. Numer. Anal. **56**, 111–135 (2018)
9. Herrmann, L., Schwab, Ch.: QMC integration for lognormal-parametric, elliptic PDEs: local supports and product weights. Numerische Mathematik **141**, 63–102 (2019)
10. Kazashi, Y.: Quasi-Monte Carlo integration with product weights for elliptic PDEs with log-normal coefficients. IMA J. Numer. Anal. **39**(3), 1563–1593 (2019)
11. Kuo, F.Y., Nuyens, D.: Application of quasi-Monte Carlo methods to elliptic PDEs with random diffusion coefficients: a survey of analysis and implementation. Found. Comput. Math. **16**, 1631–1696 (2016)
12. Kuo, F.Y., Schwab, Ch., Sloan, I.H.: Quasi-Monte Carlo finite element methods for a class of elliptic partial differential equations with random coefficients. SIAM J. Numer. Anal. **50**, 3351–3374 (2012)
13. Nuyens, D.: The construction of good lattice rules and polynomial lattice rules. In: Kritzer, P., Niederreiter, H., Pillichshammer, F., Winterhof, A. (eds.) Uniform Distribution and Quasi-Monte Carlo Methods: Discrepancy, Integration and Applications, pp. 223–255. De Gruyter, Berlin (2014)
14. Sloan, I.H., Woźniakowski, H.: Tractability of multivariate integration for weighted Korobov classes. J. Complex. **17**, 697–721 (2001)

Sudoku Latin Square Sampling for Markov Chain Simulation

Rami El Haddad, Joseph El Maalouf, Christian Lécot and Pierre L'Ecuyer

Abstract We are interested in Monte Carlo simulations of discrete-time Markov chains on discrete and totally ordered state spaces. To improve simulation efficiency, we use a technique previously introduced in the context of quasi-Monte Carlo simulation of an array of N Markov chains. This method simulates the N copies of the chain simultaneously, reorders the chains at each step by increasing order of their states, and samples the next state by using N two-dimensional points in the unit square. The first coordinate of each point is used to match a chain, and the second coordinate is used to sample the next state by inversion from its cumulative distribution function conditional on the current state. We study the case where the N points are obtained at each step from Sudoku Latin square sampling, which means that (1) if the unit square is uniformly divided into N identical subsquares, exactly one point lies in each subsquare, (2) for each axis, the N projections of the points are distributed with exactly one projection in each of the N subintervals of length $1/N$ that partition the unit interval, and (3) in both cases, each individual point has the uniform distribution in the subsquare and interval to which it belongs. We prove that the variance of the Sudoku Latin square sampling estimator is of order $\mathcal{O}(N^{-3/2})$. The same conver-

R. El Haddad (✉)
Laboratoire de Mathématiques et Applications, U.R. Mathématiques et modélisation, Faculté des sciences, Université Saint-Joseph, B.P. 7-5208, Mar Mikhael Beyrouth 1104 2020, Lebanon
e-mail: rami.haddad@usj.edu.lb

J. El Maalouf
École Supérieure d'Ingénieurs de Beyrouth, Université Saint-Joseph, B.P. 1514, Riad El Solh, Beyrouth 1107 2050, Lebanon
e-mail: joseph.maalouf@usj.edu.lb

C. Lécot
Université Grenoble Alpes, Université Savoie Mont Blanc, CNRS, LAMA, 73000 Chambéry, France
e-mail: christian.lecot@univ-smb.fr

P. L'Ecuyer
DIRO, Université de Montréal, C.P. 6128, Succ. Centre-Ville, Montréal H3C 3J7, Canada
e-mail: lecuyer@iro.umontreal.ca

B. Tuffin and P. L'Ecuyer (eds.), *Monte Carlo and Quasi-Monte Carlo Methods*, Springer Proceedings in Mathematics & Statistics 324,
https://doi.org/10.1007/978-3-030-43465-6_10

gence rate is obtained when property (2) is removed, which gives simple stratified sampling. However, in our numerical experiments, we observe empirically a much smaller variance and better efficiency with the Sudoku Latin square sampling than with simple stratified sampling alone.

Keywords Monte Carlo simulation · Markov chains · Stratifed sampling · Sudoku Latin square sampling

1 Introduction

We consider a discrete-time Markov chain $\{X_n,\ n \geq 0\}$ over a countable state space $\mathscr{X}$, where $X_n \in \mathscr{X}$ is the state at step n, and we are interested in estimating by simulation the expected cost at step n, $\mathbb{E}[c(X_n)]$, for one or several cost functions $c : \mathscr{X} \to [0, \infty)$ (we assume non-negative cost functions for simplicity). If the state space is finite with small cardinality and sometimes when the chain has a very special structure, it is possible to compute the exact distribution of X_n and the exact expected cost at step n, for any n. Otherwise, one can use standard Monte Carlo (MC): simulate the chain until step n, repeat N times independently, and average the N realizations of $c(X_n)$. The main drawback of this general approach is its slow convergence: The variance of the Monte Carlo estimator of $\mathbb{E}[c(X_n)]$ typically converges as $\mathscr{O}(N^{-1})$ for any n.

A (deterministic) *quasi-Monte Carlo* (QMC) method for Markov chains has been proposed in [9] for the case where the chain has a totally ordered state space. The method simulates an array of N copies of the chain in parallel. At each step n, it reorders the chains by increasing order of their states, and it uses two-dimensional quasi-random points to move them ahead by one step. Convergence (in the deterministic sense) of the average cost to the expectation when $N \to \infty$ was established, and the QMC approach outperformed plain MC in numerical experiments. However, QMC error bounds are typically too loose and inconvenient for practical error assessment. A *randomized quasi-Monte Carlo* (RQMC) approach named *Array-RQMC*, which resembles the previous QMC scheme, was proposed and analyzed in [11, 12], in the setting of a Markov chain model with general state space. The method was shown to provide an unbiased estimator of $\mathbb{E}[c(X_n)]$ for any n, and variance bounds for this estimator were proved under certain conditions. In particular, it was proved that for a Markov chain with a one-dimensional state space, if stratified sampling as in [2, 8] is used at each step to advance the array of chains by one step, and under some technical conditions, the variance converges as $\mathscr{O}(N^{-3/2})$, which beats Monte Carlo. In numerical experiments with Markov chains having one-dimensional and higher-dimensional states, the empirical variance was typically much smaller than the Monte Carlo variance, and was observed to decrease often at better rates than $\mathscr{O}(N^{-3/2})$, sometimes even faster than $\mathscr{O}(N^{-2})$: see [12–14], for example. However, no proof of these faster rates is available so far, and the $\mathscr{O}(N^{-3/2})$ rate has been proved only for ordinary stratification of the unit square in identical subsquares, for a one-dimensional state. A related convergence-rate result worth mentioning was

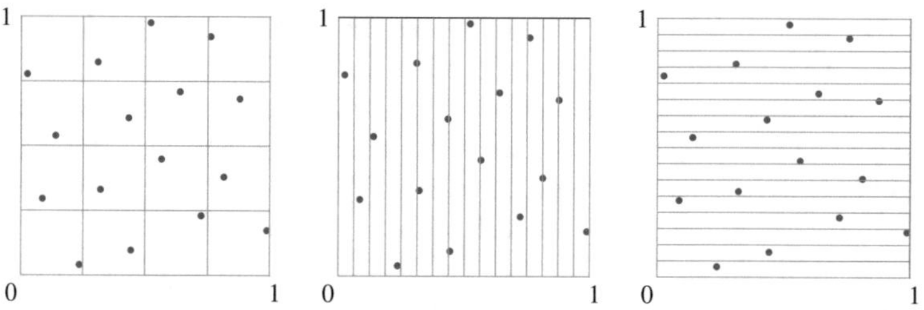

Fig. 1 Example of a Sudoku Latin square with 16 points

obtained in [7], in the context of particle filters. The authors proved that if the RQMC point set used at each step is a (t, m, s)-net with a nested uniform scramble [18] and if the states are sorted using a Hilbert-curve when their dimension is larger than 1, then the variance of the Array-RQMC estimator converges as $o(N^{-1})$, which is faster than Monte Carlo.

The aim of this paper is to increase our theoretical understanding of the method by expanding the class of sampling methods for which an $\mathscr{O}(N^{-3/2})$ convergence rate is proved. We revisit the simple stratified sampling (SSS) setting and we consider a *Sudoku Latin square sampling* (SLSS) setting, which combines two-dimensional stratified sampling with Latin hypercube sampling [15, 20]. Our theoretical results are consolidated by three numerical experiments in which we observe a significantly lower variance with SLSS than with simple stratification.

SLSS turns out to be a special case of the *U-sampling* method of [21] for sampling N points in the unit hypercube. The U-sampling first generates a random *orthogonal array-based Latin hypercube design*, which is a selection of N small cubic boxes of side size $1/N$ that form an orthogonal array of strength t [16, 17] and a Latin hypercube at the same time. Then it samples one point uniformly inside each selected small box, independently across the boxes. For the special case where $t = 2$, this type of design (the selection of the boxes) gives a *Sudoku Latin square* [19] for each two-dimensional projection of the points. Thus, each two-dimensional projection satisfies the properties (1)–(3) mentioned in the abstract. An example of a Sudoku Latin square is given in Fig. 1. Since our SLSS is in two dimensions, there is a single two-dimensional projection and it must form a Sudoku Latin square. Sudoku Latin squares are also studied in [22], although these authors are only considering discrete designs and space filling constructions, and not in sampling random points uniformly in the unit hypercube.

A different sampling method that generalizes the SLSS to more than two dimensions was studied in [5]. In d dimensions, that method generates $N = p^d$ points in a way that (1) there is always one point per subcube when we partition the d-dimensional unit cube into N identical subcubes, (2) there is one value in each subinterval when we project all the points over a single coordinate to obtain N values in the unit interval, and we partition the unit interval into N subintervals of length $1/N$, and (3) each point taken individually has the uniform distribution in the subcube to which it belongs. Variance bounds have been obtained when the integral of a function over the unit cube is estimated by the average of the function values at the N points, under certain assumptions on the integrand. Different types of variance results, in terms of the ANOVA decomposition of the integrand, were proved in [21] for the same type of integration problem with U-sampling. SLSS is the two-dimensional special case of each of these two methods.

Our paper is the first to study the use of SLSS in the context of simulating an array of Markov chains. SLSS gives stronger constructions than simple stratified points over the unit square. Our aim is to investigate if, and how much, this strengthening has an impact on the variance of expected cost estimators.

The remainder is organized as follows. In Sect. 2, we define our setting for Monte Carlo simulation of discrete-state Markov chains and explain how we proceed with plain (standard) Monte Carlo, with SSS, and with SLSS. With SSS, the way we map the points to chain states at each step follows [6, 9] and differs from what was done in [12]. In Sect. 3, we analyze the variance of these schemes. We prove that the variance of the simulation estimator of an expected state-dependent cost at any given step n is $\mathscr{O}(N^{-3/2})$ for both SSS and SLSS. This beats the known rate of $\mathscr{O}(N^{-1})$ for standard Monte Carlo. Results of computational experiments and comparison between standard Monte Carlo, SSS, and SLSS are given in Sect. 4. The empirical convergence rates of the variance for SSS and SLSS are close to those established in Sect. 3, but the variance with SLSS is significantly smaller than with SSS. In Sect. 5, we give the technical proofs of some results and we conclude in Sect. 6.

2 Monte Carlo Simulations of Markov Chains

Let $\{X_n,\ n \geq 0\}$ be a stationary discrete-time Markov chain over a countable and ordered state space $\mathscr{X}$. Without loss of generality one can assume that $\mathscr{X} = \mathbb{N}$ or $\mathbb{Z}$. Let $P(i, j) = \mathbb{P}(X_{n+1} = j | X_n = i)$ denote the transition probabilities and $\boldsymbol{P} = (P(i, j) : (i, j) \in \mathscr{X}^2)$ the transition probability matrix. We denote by $\mu_n(i) = \mathbb{P}(X_n = i)$ the state probabilities at step n and $\mu_n = (\mu_n(i) : i \in \mathscr{X})$ the probability vector for step n. We assume that the initial probability vector μ_0 is given (often, it is degenerate over a single state).

For $i, j \in \mathscr{X}$ we set

$$q_j(i) := \sum_{h \leq j} P(i, h). \tag{1}$$

We define the conditional cumulative distribution function $F_i(j) := \mathbb{P}(X_{n+1} \leq j | X_n = i) = q_j(i)$. If I denotes the unit interval $(0, 1]$ we have a disjoint union $I = \bigcup_{j \in \mathscr{X}} I_{i,j}$, where $I_{i,j} := (q_{j-1}(i), q_j(i)]$. So that for any $i \in \mathscr{X}$ and $u \in I$, there exists a unique $j \in \mathscr{X}$ such that $u \in I_{i,j}$: we denote it by $F_i^{-1}(u)$. If $X_n = i$, then $F_i^{-1}(u)$ is the next state.

Let δ_i be the Dirac measure at i, defined by

$$\delta_i(j) = \begin{cases} 1 & \text{if } j = i, \\ 0 & \text{otherwise.} \end{cases}$$

For any integer n, the distribution μ_n is approximated by

$$\widehat{\mu}_n := \frac{1}{N} \sum_{k=1}^{N} \delta_{i_k^n},$$

where N is a fixed integer and $i_1^n, \ldots, i_N^n$ are calculated iteratively.

First, a set $(i_k^0 : 1 \le k \le N)$ of N states is sampled from μ_0: several techniques are proposed in [4]. In many applications, the initial state of the chain is fixed, and then $\widehat{\mu}_0 = \mu_0$.

We describe the transition from step n to $n+1$ for three Monte Carlo methods. We introduce $\widetilde{\mu}_{n+1} := \widehat{\mu}_n \boldsymbol{P}$ as an intermediate distribution (which is not used in effective calculations). This $\widetilde{\mu}_{n+1}$ is an approximation of μ_{n+1}, but it is generally not an equally-weighted sum of Dirac measures, like $\widehat{\mu}_n$, so that an additional step is needed. We formulate this step as a quadrature: the MC methods correspond to quadrature algorithms, possibly combined with variance reduction techniques. To that end, let us consider an arbitrary sequence $s = (s(i) : i \in \mathscr{X})$ (a column vector); we assume that s is non-negative, just to avoid worrying about convergence of series. Then

$$\widetilde{\mu}_{n+1} s = \widehat{\mu}_n \boldsymbol{P} s = \frac{1}{N} \sum_{k=1}^{N} \sum_{j \in \mathscr{X}} P(i_k^n, j) s(j).$$

Let 1_k be the indicator function of the interval $I_k := ((k-1)/N, k/N]$ and $1_{i,j}$ denote the indicator function of the interval $I_{i,j}$. If we associate to s the function C_s^n defined by

$$C_s^n(u) := \sum_{k=1}^{N} \sum_{j \in \mathscr{X}} 1_k(u_1) 1_{i_k^n, j}(u_2) s(j), \quad u = (u_1, u_2) \in I^2, \tag{2}$$

then we have

$$\widetilde{\mu}_{n+1} s = \int_{I^2} C_s^n(u) du.$$

We obtain $\widehat{\mu}_{n+1}$ by approximating the integral with Monte Carlo estimation. In the following, if m is an integer, we denote $[1, m] := \{1, 2, \ldots, m\}$. The notation $U \sim \mathscr{U}(\mathscr{E})$ means that U is a random variable uniformly distributed over the set $\mathscr{E}$.

2.1 Standard Monte Carlo

The transition from step n to step $n+1$ acts as follows: if the state of the chain is i, i.e. $X_n = i$, then a random number U with $U \sim \mathcal{U}(I)$ is generated and the new state of the chain is $F_i^{-1}(U)$, i.e. $X_{n+1} = F_{X_n}^{-1}(U)$. The operation is repeated N times independently, in order to advance N copies of the chain. With our notations, this may be written as follows. Let $\{U_k : 1 \le k \le N\}$ be independent random variables with $U_k \sim \mathcal{U}(I)$, then

$$i_k^{n+1} = F_{i_k^n}^{-1}(U_k), \quad 1 \le k \le N.$$

That is, if, for any non-negative sequence s,

$$\widehat{X}_s^{n+1} := \frac{1}{N} \sum_{k=1}^{N} C_s^n \left(\frac{k-1}{N}, U_k \right), \tag{3}$$

then

$$\widehat{\mu}_{n+1} s = \widehat{X}_s^{n+1}. \tag{4}$$

2.2 Simple Stratified Sampling

We suppose that $N = p^2$, for some integer $p > 0$. The transition from n to $n+1$ has two steps: renumbering of the states and numerical integration.

(S1) The states are relabeled so that $i_1^n \le \cdots \le i_N^n$. The technique was used in the QMC context and ensures theoretical and numerical convergence of the scheme (see [9]) .

(S2) Consider a partition of I^2 into N squares: $I_\ell = H_{\ell_1} \times H'_{\ell_2}$, where, for $\ell \in [1, p]^2$: $H_{\ell_1} := ((\ell_1 - 1)/p, \ell_1/p]$ and $H_{\ell_2} := ((\ell_2 - 1)/p, \ell_2/p]$. Let $\{V_\ell : \ell \in [1, p]^2\}$ be independent random variables, where $V_\ell = (V_{\ell,1}, V_{\ell,2}) \sim \mathcal{U}(I_\ell)$.

For an arbitrary non-negative sequence s, let

$$\widehat{Y}_s^{n+1} := \frac{1}{N} \sum_{\ell \in [1,p]^2} C_s^n(V_\ell), \tag{5}$$

then

$$\widehat{\mu}_{n+1} s = \widehat{Y}_s^{n+1}. \tag{6}$$

If $u \in I$, let

$$\kappa(u) := \lceil Nu \rceil, \tag{7}$$

where $\lceil x \rceil$ is the least integer greater than or equal to x. Hence Eq. (6) means that the next states are calculated as follows:

$$i^{n+1}_{(\ell_1-1)p+\ell_2} = F^{-1}_{i^n_{\kappa(V_{\ell,1})}}(V_{\ell,2}), \quad \ell \in [1, p]^2$$

(the numbering of the states i^{n+1}_k is arbitrary). The first projection $V_{\ell,1}$ of V_ℓ is used for selecting the state at step n and the second projection $V_{\ell,2}$ is used for performing the transition to step $n+1$. Note that with this scheme, the mapping between the N points and the N states is not necessarily one-to-one: it is possible to pick the same state more than once and leave out some of the states. This differs from the SSS scheme used in [12, 14].

2.3 Sudoku Latin Square Sampling

As before, we assume $N = p^2$, and the transition from n to $n+1$ has two steps: renumbering of the states and numerical integration.

(S1) The states are relabeled so that $i^n_1 \le \cdots \le i^n_N$.

(S2) We consider the same partition of I^2 as before: I_ℓ for $\ell \in [1, p]^2$. Let $\{W_\ell : \ell \in [1, p]^2\}$ be random variables, where $W_\ell = (W_{\ell,1}, W_{\ell,2})$, with

$$W_{\ell,1} = \frac{\ell_1 - 1}{p} + \frac{\sigma_1(\ell_2) - 1 + \xi^1_\ell}{p^2} \quad W_{\ell,2} = \frac{\ell_2 - 1}{p} + \frac{\sigma_2(\ell_1) - 1 + \xi^2_\ell}{p^2}.$$

Here σ_1 and σ_2 are random permutations of $[1, p]$ and $\xi^1_\ell \sim \mathcal{U}(I)$ and $\xi^2_\ell \sim \mathcal{U}(I)$. All these random variables being independent. The set of values of the random variable W_ℓ is included in I_ℓ and has the properties:

(P1) for any $\ell \in [1, p]^2$, there is a unique point of this set in each square I_ℓ,

(P2) for any $k \in [1, N]$, there is a unique point of this set in each rectangle $I \times I_k$ or $I_k \times I$.

In addition $W_\ell \sim \mathcal{U}(I_\ell)$. For an arbitrary non-negative sequence s, let

$$\widehat{Z}^{n+1}_s := \frac{1}{N} \sum_{\ell \in [1,p]^2} C^n_s(W_\ell). \tag{8}$$

Then

$$\widehat{\mu}_{n+1} s = \widehat{Z}^{n+1}_s. \tag{9}$$

Due to property (P2), the mapping (see (7))

$$\ell := (\ell_1, \ell_2) \in [1, p]^2 \to \kappa(W_{\ell,1}) \in [1, N]$$

is one-to-one: each state of step n is considered exactly once for a transition (this is not the case with SSS). Equation (9) means that the next states are calculated as follows:

$$i^{n+1}_{(\ell_1-1)p+\ell_2} = F^{-1}_{i^n_{\kappa(W_{\ell,1})}}(W_{\ell,2}), \quad \ell \in [1, p]^2$$

(as before, the numbering of the states i^{n+1}_k is arbitrary). The first projection $W_{\ell,1}$ of W_ℓ is used for selecting the state at step n and the second projection $W_{\ell,2}$ is used for performing the transition to step $n+1$.

3 Convergence Analysis

In this section we prove, for each method, that the estimator of the expected cost at each step is unbiased and we establish that the variance of the estimator used is $\mathscr{O}(N^{-1})$ for standard MC and $\mathscr{O}(N^{-3/2})$ for SSS and SLSS, where N is the number of simulation paths. In the following, λ is the Lebesgue measure and λ_2 the two-dimensional Lebesgue measure; we put $|\mathscr{E}|$ for the number of elements of a set $\mathscr{E}$. We use the sequence s_h, for $h \in \mathscr{X}$:

$$s_h(i) := \begin{cases} 1 & \text{if } i \le h, \\ 0 & \text{otherwise.} \end{cases}$$

The total variation of a sequence $s = (s(i) : i \in \mathscr{X})$ is defined by

$$TV(s) := \sum_{i \in \mathscr{X}} |s(i+1) - s(i)|. \tag{10}$$

We use below the total variation of q_h, for $h \in \mathscr{X}$. We recall that we have from (1):

$$q_h(i) = \mathbb{P}(X_{n+1} \le h | X_n = i),$$

and from (10):

$$TV(q_h) = \sum_{i \in \mathscr{X}} |\mathbb{P}(X_{n+1} \le h | X_n = i+1) - \mathbb{P}(X_{n+1} \le h | X_n = i)|.$$

In the following, we assume that

$$M := \sup_{h \in \mathscr{X}} TV(q_h) < +\infty.$$

There are situations for which q_h is monotone and situations for which $M < 1$ (or both), but this is not always true. See [13, 14] for examples and further discussion. Our M corresponds to Λ_j in [12].

3.1 Standard Monte Carlo

Lemma 1 *Let s be a non-negative sequence. The standard Monte Carlo estimator of $\widetilde{\mu}_{n+1}s$:*

$$\widehat{X}_s^{n+1} := \frac{1}{N}\sum_{k=1}^{N} C_s^n\left(\frac{k-1}{N}, U_k\right)$$

has the following properties.

1. *$\widehat{X}_s^{n+1}$ is unbiased.*
2. *If $s = s_h$, for $h \in \mathscr{X}$, then*

$$\mathrm{Var}(\widehat{X}_{s_h}^{n+1}) \leq \frac{1}{4N}.$$

Proof 1. We have

$$\mathbb{E}\left[C_s^n\left(\frac{k-1}{N}, U_k\right)\right] = \sum_{j \in \mathscr{X}} P(i_k^n, j)s(j),$$

so that $\mathbb{E}[\widehat{X}_s^{n+1}] = \widetilde{\mu}_{n+1}s$.

2. The variable

$$C_{s_h}^n\left(\frac{k-1}{N}, U_k\right) = \sum_{j \in \mathscr{X}, j \leq h} 1_{i_k^n, j}(U_k)$$

is a Bernoulli random variable, with variance $\leq 1/4$. Hence the result.

□

We then obtain an error bound by using the same techniques as in [12]. We assume that, for any non-negative sequence s, the standard Monte Carlo estimator $\widehat{\mu}_0 s$ of $\mu_0 s$ is unbiased and that, for any $h \in \mathscr{X}$,

$$\mathrm{Var}(\widehat{\mu}_0 s_h) \leq \frac{x_0}{N},$$

for some $x_0 \geq 0$ (as noticed before, in many applications, $\widehat{\mu}_0 = \mu_0$).

Proposition 1 *For the standard Monte Carlo method, it holds:*

1. *for any non-negative sequence s*

$$\mathbb{E}\left[\widehat{\mu}_n s\right] = \mu_n s,$$

2. *for any $h \in \mathscr{X}$,*

$$\operatorname{Var}(\widehat{\mu}_n s_h) \leq \frac{x_n}{N},$$

where $x_{n+1} = M^2 x_n + 1/4$ $(n \geq 0)$.

Proof 1. We have

$$\mu_{n+1}s - \widehat{\mu}_{n+1}s = \mu_{n+1}s - \widetilde{\mu}_{n+1}s + \widetilde{\mu}_{n+1}s - \widehat{\mu}_{n+1}s = \mu_n \boldsymbol{P}s - \widehat{\mu}_n \boldsymbol{P}s + \widetilde{\mu}_{n+1}s - \widehat{X}_s^{n+1},$$

so, by using Lemma 1, the result follows by induction.

2. The variables $\mu_{n+1}s_h - \widetilde{\mu}_{n+1}s_h$ and $\widetilde{\mu}_{n+1}s_h - \widehat{\mu}_{n+1}s_h$ are uncorrelated and $\widetilde{\mu}_{n+1}s_h - \widehat{\mu}_{n+1}s_h = \widetilde{\mu}_{n+1}s_h - \widehat{X}_{s_h}^{n+1}$ is a centered variable, consequently

$$\operatorname{Var}(\widehat{\mu}_{n+1}s_h) = \mathbb{E}\left[(\mu_{n+1}s_h - \widetilde{\mu}_{n+1}s_h)^2\right] + \mathbb{E}\left[(\widetilde{\mu}_{n+1}s_h - \widehat{\mu}_{n+1}s_h)^2\right]. \quad (11)$$

For any $i \in \mathscr{X}$, we have $\boldsymbol{P}s_h(i) = F_i(h)$, hence

$$\begin{aligned}\mu_{n+1}s_h - \widetilde{\mu}_{n+1}s_h &= \mu_n \boldsymbol{P}s_h - \widehat{\mu}_n \boldsymbol{P}s_h = \sum_{i\in\mathscr{X}} \mu_n(i)F_i(h) - \sum_{i\in\mathscr{X}} \widehat{\mu}_n(i)F_i(h)\\ &= -\sum_{i\in\mathscr{X}} \mu_n s_i(F_{i+1}(h) - F_i(h)) + \sum_{i\in\mathscr{X}} \widehat{\mu}_n s_i(F_{i+1}(h) - F_i(h))\\ &= \sum_{i\in\mathscr{X}} (\widehat{\mu}_n s_i - \mu_n s_i)(F_{i+1}(h) - F_i(h)).\end{aligned}$$

On the one hand, we write

$$\begin{aligned}\mathbb{E}\left[(\mu_{n+1}s_h - \widetilde{\mu}_{n+1}s_h)^2\right] &= \mathbb{E}\left[\left(\sum_{i\in\mathscr{X}} (\widehat{\mu}_n s_i - \mu_n s_i)(q_h(i+1) - q_h(i))\right)^2\right]\\ &= \sum_{(i,j)\in\mathscr{X}^2} \mathbb{E}\left[(\widehat{\mu}_n s_i - \mu_n s_i)(q_h(i+1) - q_h(i))(\widehat{\mu}_n s_j - \mu_n s_j)(q_h(j+1) - q_h(j))\right]\\ &\leq \sum_{(i,j)\in\mathscr{X}^2} \sqrt{\operatorname{Var}(\widehat{\mu}_n s_i)\operatorname{Var}(\widehat{\mu}_n s_j)}|q_h(i+1) - q_h(i)| \times |q_h(j+1) - q_h(j)|.\end{aligned}$$

On the other hand, Lemma 1 gives

$$\mathbb{E}\left[(\widetilde{\mu}_{n+1}s_h - \widehat{\mu}_{n+1}s_h)^2\right] = \mathbb{E}\left[(\widetilde{\mu}_{n+1}s_h - \widehat{X}^{n+1}_{s_h})^2\right] \le \frac{1}{4N}.$$

So, by using (11), the result follows by induction.

□

The bounds for $\mathrm{Var}(\widehat{Y}^{n+1}_{s_h})$ (SSS) and $\mathrm{Var}(\widehat{Z}^{n+1}_{s_h})$ (SLSS) are not so easily obtained, and the proofs of Lemmas 2 and 3 are given in Sect. 5.

3.2 Simple Stratified Sampling

Lemma 2 *Let s be a non-negative sequence. The SSS estimator of $\widetilde{\mu}_{n+1}s$:*

$$\widehat{Y}^{n+1}_s := \frac{1}{N} \sum_{\ell\in[1,p]^2} C^n_s(V_\ell)$$

has the following properties.

1. *$\widehat{Y}^{n+1}_s$ is unbiased.*
2. *If $s = s_h$, for $h \in \mathscr{X}$, then*

$$\mathrm{Var}(\widehat{Y}^{n+1}_{s_h}) \le \frac{1}{4N^{3/2}}(TV(q_h) + 2).$$

A similar result (with the same $N^{-3/2}$ order) was established in [12], but the SSS method studied there differs from the one used here, so we provide a different proof (see Sect. 5). Intermediate results from this proof (Eqs. (12) and (13)) will be re-used afterwards for the analysis of SLSS.

The proof of the next result is similar to the proof of Proposition 1. We assume that, for any non-negative sequence s, the SSS estimator $\widehat{\mu}_0 s$ of $\mu_0 s$ is unbiased and that, for any $h \in \mathscr{X}$,

$$\mathrm{Var}(\widehat{\mu}_0 s_h) \le \frac{y_0}{N^{3/2}}$$

for some $y_0 \ge 0$.

Proposition 2 *For the SSS method, it holds:*

1. *for any non-negative sequence s*

$$\mathbb{E}\left[\widehat{\mu}_n s\right] = \mu_n s,$$

2. *for any* $h \in \mathscr{X}$,

$$\mathrm{Var}(\widehat{\mu}_n s_h) \leq \frac{y_n}{N^{3/2}},$$

where $y_{n+1} = M^2 y_n + (M+2)/4 \ (n \geq 0)$.

3.3 Sudoku Latin Square Sampling

Lemma 3 *Let* s *be a non-negative sequence. The SLSS estimator of* $\widetilde{\mu}_{n+1}s$*:*

$$\widehat{Z}_s^{n+1} := \frac{1}{N} \sum_{\ell \in [1,p]^2} C_s^n(W_\ell)$$

has the following properties.

1. $\widehat{Z}_s^{n+1}$ *is unbiased.*
2. *If* $s = s_h$, *for* $h \in \mathscr{X}$, *and if* q_h *is a piecewise monotonic sequence, with* r *pieces, then*

$$\mathrm{Var}(\widehat{Z}_{s_h}^{n+1}) \leq \frac{1}{N^{3/2}} \left(\left(\frac{13}{4} + r \right) (TV(q_h) + 2) + 2(TV(q_h) + 2)^2 \right).$$

The constant involved in the $\mathscr{O}(N^{-3/2})$ bound of $\mathrm{Var}(\widehat{Z}_{s_h}^{n+1})$ (SLSS) is larger than the corresponding constant for $\mathrm{Var}(\widehat{Y}_{s_h}^{n+1})$ (SSS), since $r \geq 1$ in Lemma 3; this would suggest degraded performance. But in the examples of Sect. 4 we see that it is not necessarily the case.

The proof of the next result is similar to the proof of Proposition 1. We assume that, for any non-negative sequence s, the SLSS estimator $\widehat{\mu}_0 s$ of $\mu_0 s$ is unbiased and that, for any $h \in \mathscr{X}$,

$$\mathrm{Var}(\widehat{\mu}_0 s_h) \leq \frac{z_0}{N^{3/2}},$$

for some $z_0 \geq 0$.

Proposition 3 *For the SLSS method, it holds:*

1. *for any non-negative sequence* s

$$\mathbb{E}\,[\widehat{\mu}_n s] = \mu_n s,$$

2. *for any* $h \in \mathscr{X}$,

$$\mathrm{Var}(\widehat{\mu}_n s_h) \leq \frac{z_n}{N^{3/2}},$$

where $z_{n+1} = M^2 z_n + (13/4 + r)(M+2) + 2(M+2)^2 \qquad n \geq 0.$

Remark 1 The bounds in Propositions 1, 2, and 3 increase exponentially with n when $M > 1$, and remains bounded if $M < 1$, which is not uncommon; see [14].

Remark 2 The variance of each estimator is bounded for a test sequence of the form s_h, for $h \in \mathscr{X}$. We obtain a bound for a nonnegative cost function c by the same reasoning as in Proposition 1 (see [12]):

$$\mu_n c - \widehat{\mu}_n c = \sum_{i \in \mathscr{X}} (\widehat{\mu}_n s_i - \mu_n s_i)(c(i+1) - c(i)),$$

hence

$$\mathbb{E}\left[(\mu_n c - \widehat{\mu}_n c)^2\right] \le \sum_{(i,j) \in \mathscr{X}^2} \sqrt{\mathrm{Var}(\widehat{\mu}_n s_i)\mathrm{Var}(\widehat{\mu}_n s_j)} |c(i+1) - c(i)| \times |c(j+1) - c(j)|,$$

and then

$$\mathrm{Var}(\widehat{\mu}_n c) \le TV(c)^2 \times \sup_{h \in \mathscr{X}} \mathrm{Var}(\widehat{\mu}_n s_h).$$

4 Numerical Examples

In this section, we compare standard Monte Carlo to the variance reduction strategies analyzed previously: SSS and SLSS, for three examples. For each example, each strategy, and each N considered, we compute the unbiased estimator $\hat{\mu}_n s$ of $\mu_n s$ for the selected n, replicate this 100 times, and compute the empirical variance Var of the 100 realizations of $\hat{\mu}_n s$. We then plot $\log_{10}$ Var as a function of $\log_{10} N$. Assuming that Var $\approx KN^{-\alpha}$ for some positive constants K and α, we estimate the variance rate α by linear regression. We also compute the (empirical) *efficiency* of each simulation estimator, defined as the inverse of the product of Var by the CPU time [10], and we plot $\log_{10}$ efficiency as a function of $\log_{10} N$. Note that for standard Monte Carlo, the efficiency does not depend on N. For SSS and SLSS, it takes into account the additional work to compute the estimators.

4.1 A Geo/Geo/1 Queue

We consider a discrete-time Geo/Geo/1 queue (see [1]): the queue is empty at the initial time. During each unit of time, the customer in service (if there is one) completes it with probability 0.5, and one new customer arrives with probability 0.6. We estimate the mean number of customers in the queue at time $n = 12$. Figure 2 (top) shows $\log_{10}$ Var as a function of $\log_{10} N$ on the left and $\log_{10}$ efficiency as a function of $\log_{10} N$ on the right, for $N = 10^2, 50^2, 100^2, \ldots, 1\,000^2$. We find from the plots that SSS and SLSS give not only smaller variances than standard MC (for the same

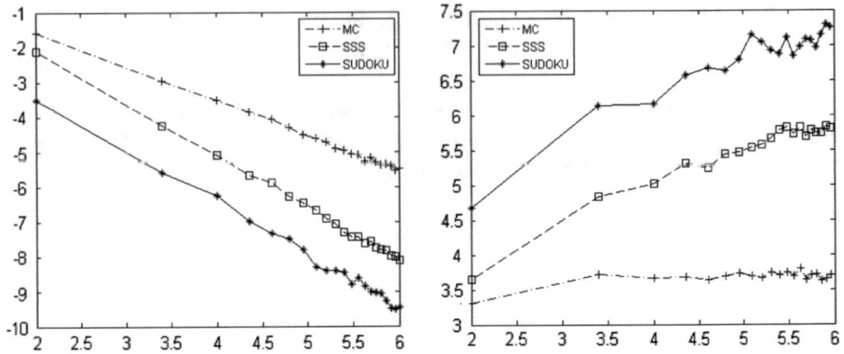

1. Geo/Geo/1 queue: sample variance (left) and efficiency (right) of 100 copies of the calculation of $E[X_{12}]$ as a function of N ($N = 10^2, 50^2, 100^2, \ldots, 1\,000^2$) (log-log scale)

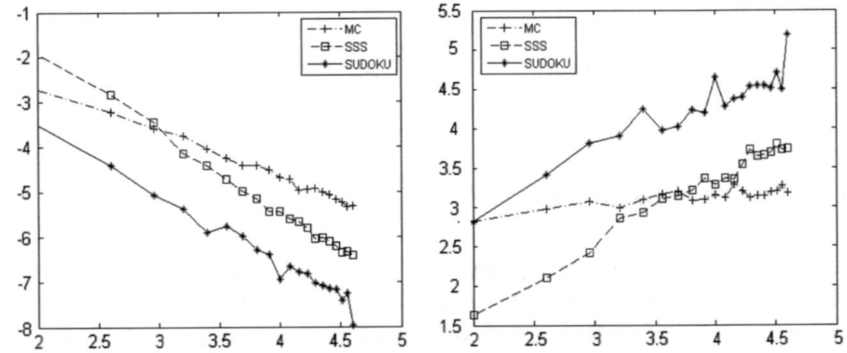

2. Gambler in a casino: sample variance (left) and efficiency (right) of 100 copies of the calculation of $\mathbb{P}(X_{1440} > 1\,500)$ as a function of N ($N = 10^2, 20^2, 30^2, \ldots, 200^2$) (log-log scale)

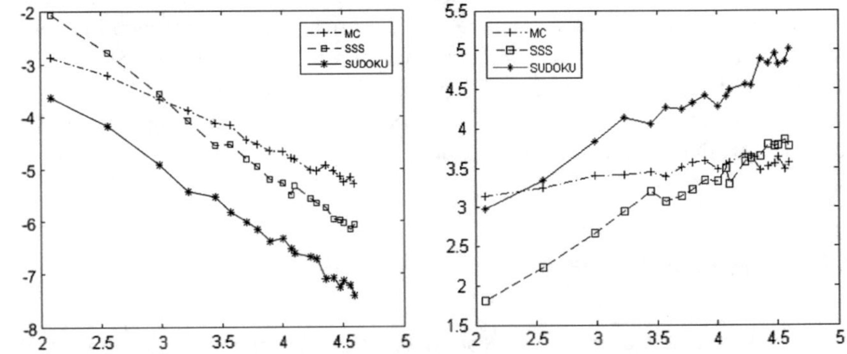

3. Diffusion: sample variance (left) and efficiency (right) of 100 copies of the calculation of $\int_{-1/2}^{1/2} c(x,T)dx$ as a function of N ($N = 11^2, 19^2, 31^2, \ldots, 199^2$) (log-log scale)

Fig. 2 Comparison of standard Monte Carlo (MC) to SSS and SLSS (Sudoku) for three examples

Table 1 Calculation of order α of the sample variance: comparison of standard Monte Carlo (MC), SSS, and SLSS for three examples and estimators

	Calculation	MC	SSS	SLSS
Geo/Geo/1 queue	$E[X_{12}]$	0.99	1.50	1.51
Gambler in a casino	$\mathbb{P}(X_{1\,440} > 1\,500)$	1.02	1.76	1.53
Diffusion	$\int_{-1/2}^{1/2} c(x, T)dx$	0.98	1.61	1.49

N), but also better efficiencies, and that SLSS outperforms SSS. The regression estimates of α are given in the first row of Table 1. They match the upper bounds of $\mathcal{O}(N^{-3/2})$ established in Sect. 3.

4.2 A Gambler in a Casino

A gambler is going to a casino for four hours. He plans to play the same game every ten seconds (so he will play 1 440 times). At this game, for each Euro that he bids, he gets 0 with probability 0.9 and $m \in \{1, 2, \ldots, 10\}$ with probability 0.01 each. His policy is the following: if he has more than 100 Euros, he plays 2 Euros, but if he has 100 Euros or less, he plays only 1. To make sure that he can play during the four hours, he brings 2 780 Euros with him. The model is a Markov chain on state space $E = [0, 28\,700]$. We estimate the probability that the gambler has more than 1 500 Euros at the end of the game. Here we use $N = 10^2, 20^2, 30^2, \ldots, 200^2$. The results are reported in the middle rows of Fig. 2 and Table 1. We find that SSS and SLSS produce both smaller variances and better efficiencies than standard MC for large enough N, and that SLSS outperforms SSS. The regression estimate of α for SLSS corresponds to the bound established in Sect. 3, but for SSS it is better. However, the variance itself is smaller for SLSS than for SSS, and the better rate of SSS might not hold beyond the observed range (it is likely caused by a few poor values of Var for the smallest values of N).

4.3 Diffusion

The 1-D diffusion equation

$$\frac{\partial c}{\partial t}(x, t) = D\frac{\partial^2 c}{\partial t^2}(x, t), \ x \in \mathbb{R}, \ t > 0 \ \text{ and } \ c(x, 0) = c_0(x), \ x \in \mathbb{R}$$

(where $c_0 \geq 0$ and $\int_{\mathbb{R}} c_0(x)dx = 1$) may be discretized with a time step Δt and a spatial step Δx and the solution is approximated using a random walk: $P(i, i-1) = P(i, i+1) = D\Delta t/\Delta x^2$, $P(i, i) = 1 - 2D\Delta t/\Delta x^2$ (we refer to [3] in a QMC context). Here we specify $D = 1$ and c_0 is the indicator function of the interval

$[-1/2, 1/2]$; we want to approximate $\int_{-1/2}^{1/2} c(x, T)dx$. We choose $T = 1$, with $\Delta t = 6.25\ 10^{-4}$ and $\Delta x = 5.0\ 10^{-2}$. Here we take $N = 11^2, 19^2, 31^2, \ldots, 199^2$. In a previous version of the paper, we had $N = 10^2, 20^2, 30^2, \ldots, 200^2$, but this gave oscillations in SLSS outcomes, with better results when $p = \sqrt{N}$ was a multiple of 20, because of interactions with other discretization parameters. To avoid this, we changed our choices of p. So we use prime numbers near the previous ones for p. The bottom part of Fig. 2 and the last row of Table 1 give the results, which are very similar to those of the previous example. Again, SLSS outperforms the other methods.

5 The Proofs

5.1 Proof of Lemma 2

Proof 1. We have

$$\mathbb{E}\left[C_s^n(V_\ell)\right] = N \sum_{k=1}^{N} \sum_{j \in \mathscr{X}} \left(\int_{I_\ell} 1_k(u_1) 1_{i_k^n, j}(u_2) du \right) s(j).$$

Hence

$$\mathbb{E}[\widehat{Y}_s^{n+1}] = \frac{1}{N} \sum_{k=1}^{N} \sum_{j \in \mathscr{X}} P(i_k^n, j) s(j) = \widetilde{\mu}_{n+1} s.$$

2. The function $C_{s_h}^n$ is the indicator function of the set

$$J_h^n := \bigcup_{k=1}^{N} \left(I_k \times \bigcup_{j \in \mathscr{X}, j \le h} I_{i_k^n, j} \right) = \bigcup_{k=1}^{N} I_k \times (0, q_h(i_k^n)]. \quad (12)$$

The variable $C_{s_h}^n(V_\ell)$ is a Bernoulli random variable, with expectation $f_{h,\ell}^n = N\lambda_2(J_h^n \cap I_\ell)$. Here $f_{h,\ell}^n = 1$ if $I_\ell \subset J_h^n$ and $f_{h,\ell}^n = 0$ if $I_\ell \cap J_h^n = \emptyset$. Consequently, $\mathrm{Var}(C_{s_h}^n(V_\ell)) = f_{h,\ell}^n(1 - f_{h,\ell}^n) \le 1/4$ and $\mathrm{Var}(C_{s_h}^n(V_\ell)) = 0$ if $I_\ell \subset J_h^n$ or if $I_\ell \cap J_h^n = \emptyset$, so that

$$\mathrm{Var}(\widehat{Y}_{s_h}^{n+1}) \le \frac{1}{4N^2} \left| \{\ell \in [1, p]^2 : I_\ell \not\subset J_h^n \text{ and } I_\ell \cap J_h^n \neq \emptyset\} \right|.$$

a. If $I_\ell \not\subset J_h^n$, then there exists (u_1, u_2) which belongs to I_ℓ and not to J_h^n; since this u_1 is in some I_k, we have:

$$\exists k \in [1, N], \exists (u_1, u_2) \in I_\ell : u_1 \in I_k \subset \left(\frac{\ell_1 - 1}{p}, \frac{\ell_1}{p}\right] \text{ and } u_2 \notin (0, q_h(i_k^n)],$$

so that

$$\exists k \in \{p(\ell_1 - 1) + 1, p(\ell_1 - 1) + 2, \dots, p\ell_1\}, \exists u_2 \in \left(\frac{\ell_2 - 1}{p}, \frac{\ell_2}{p}\right] : u_2 > q_h(i_k^n),$$

consequently

$$\ell_2 > p \min_{p(\ell_1-1)<k\leq p\ell_1} q_h(i_k^n).$$

b. Analogously, if $I_\ell \cap J_h^n \neq \emptyset$, then there exists (u_1, u_2) which belongs to I_ℓ and also to J_h^n; and we eventually obtain:

$$\ell_2 < p \max_{p(\ell_1-1)<k\leq p\ell_1} q_h(i_k^n) + 1.$$

We then have the following bounds

$$\begin{aligned}
&\left|\{\ell \in [1, p]^2 : I_\ell \not\subset J_h^n \text{ and } I_\ell \cap J_h^n \neq \emptyset\}\right| \\
&\quad \leq p \left(\sum_{\ell_1=1}^{p} \left(\max_{p(\ell_1-1)<k\leq p\ell_1} q_h(i_k^n) - \min_{p(\ell_1-1)<k\leq p\ell_1} q_h(i_k^n) \right) + 2 \right) \\
&\quad \leq N^{1/2} \left(\sum_{i\in\mathscr{X}} |q_h(i+1) - q_h(i)| + 2 \right),
\end{aligned}$$

because the states are relabeled so that $i_1^n \leq \cdots \leq i_N^n$. Consequently,

$$\left|\{\ell \in [1, p]^2 : I_\ell \not\subset J_h^n \text{ and } I_\ell \cap J_h^n \neq \emptyset\}\right| \leq N^{1/2}(TV(q_h) + 2), \tag{13}$$

and the result follows.

□

5.2 Proof of Lemma 3

Proof 1. Since $W_\ell \sim \mathscr{U}(I_\ell)$, the demonstration is the same as in Lemma 2.
2. In the following, we have many summations with $\ell, \ell', m, m' \in [1, p]^2$. In order to lighten the notations, we omit this set. We have

$$\mathrm{Var}(\widehat{Z}_{s_h}^{n+1}) = V_0(\widehat{Z}_{s_h}^{n+1}) + \frac{1}{N^2} \sum_{(\ell,\ell'):\ell\neq\ell'} \mathrm{Cov}\left(C_{s_h}^n(W_\ell), C_{s_h}^n(W_{\ell'})\right),$$

where

$$V_0(\widehat{Z}^{n+1}_{s_h}) := \frac{1}{N^2}\sum_{\ell} \mathrm{Var}\left(C^n_{s_h}(W_\ell)\right).$$

a. The function $C^n_{s_h}$ is the indicator function of the set J^n_h defined by (12). Since $W_\ell \sim \mathscr{U}(I_\ell)$, we have, as in Lemma 2:

$$V_0(\widehat{Z}^{n+1}_{s_h}) \leq \frac{1}{4N^2}\left|\{\ell \in [1, p]^2;\ I_\ell \not\subset J^n_h \text{ and } I_\ell \cap J^n_h \neq \emptyset\}\right|.$$

From the bound (13), we deduce

$$V_0(\widehat{Z}^{n+1}_{s_h}) \leq \frac{1}{4N^{3/2}}(TV(q_h) + 2).$$

b. We split $\mathrm{Var}(\widehat{Z}^{n+1}_{s_h}) = V_0(\widehat{Z}^{n+1}_{s_h}) + V_1(\widehat{Z}^{n+1}_{s_h}) + V_2(\widehat{Z}^{n+1}_{s_h}) + V_3(\widehat{Z}^{n+1}_{s_h})$, with

$$V_1(\widehat{Z}^{n+1}_{s_h}) := \frac{1}{N^2}\sum_{(\ell,\ell'):\ell_1\neq\ell'_1,\ell_2=\ell'_2} \mathrm{Cov}\left(C^n_{s_h}(W_\ell), C^n_{s_h}(W_{\ell'})\right),$$

$$V_2(\widehat{Z}^{n+1}_{s_h}) := \frac{1}{N^2}\sum_{(\ell,\ell'):\ell_1=\ell'_1,\ell_2\neq\ell'_2} \mathrm{Cov}\left(C^n_{s_h}(W_\ell), C^n_{s_h}(W_{\ell'})\right),$$

$$V_3(\widehat{Z}^{n+1}_{s_h}) := \frac{1}{N^2}\sum_{(\ell,\ell'):\ell_1\neq\ell'_1,\ell_2\neq\ell'_2} \mathrm{Cov}\left(C^n_{s_h}(W_\ell), C^n_{s_h}(W_{\ell'})\right).$$

We introduce the N^2 squares $I_{\ell,m} = H_{\ell_1,m_1} \times H_{\ell_2,m_2}$, where, for $(\ell, m) \in [1, p]^4$:

$$H_{\ell_1,m_1} := ((\ell_1 - 1)/p + (m_1 - 1)/N, (\ell_1 - 1)/p + m_1/N],$$
$$H_{\ell_2,m_2} := ((\ell_2 - 1)/p + (m_2 - 1)/N, (\ell_2 - 1)/p + m_2/N].$$

We have

$$V_1(\widehat{Z}^{n+1}_{s_h}) = \sum_{(\ell,\ell'):\ell_1\neq\ell'_1,\ell_2=\ell'_2}\left(\frac{N}{p-1}\sum_{(m,m'):m_1=m'_1,m_2\neq m'_2} \lambda_2(I_{\ell,m}\cap J^n_h)\lambda_2(I_{\ell',m'}\cap J^n_h) - \lambda_2(I_\ell\cap J^n_h)\lambda_2(I_{\ell'}\cap J^n_h)\right),$$

$$V_2(\widehat{Z}^{n+1}_{s_h}) = \sum_{(\ell,\ell'):\ell_1=\ell'_1,\ell_2\neq\ell'_2}\left(\frac{N}{p-1}\sum_{(m,m'):m_1\neq m'_1,m_2=m'_2} \lambda_2(I_{\ell,m}\cap J^n_h)\lambda_2(I_{\ell',m'}\cap J^n_h) - \lambda_2(I_\ell\cap J^n_h)\lambda_2(I_{\ell'}\cap J^n_h)\right),$$

$$V_3(\widehat{Z}^{n+1}_{s_h}) = \sum_{(\ell,\ell'):\ell_1\neq\ell'_1,\ell_2\neq\ell'_2}\left(\frac{N}{(p-1)^2}\sum_{(m,m'):m_1\neq m'_1,m_2\neq m'_2} \lambda_2(I_{\ell,m}\cap J^n_h)\lambda_2(I_{\ell',m'}\cap J^n_h)\right.$$

$$- \lambda_2(I_\ell \cap J_h^n)\lambda_2(I_{\ell'} \cap J_h^n)\Big).$$

i. We have

$$V_1(\widehat{Z}_{s_h}^{n+1}) = \sum_{\ell: I_\ell \not\subset J_h^n, I_\ell \cap J_h^n \neq \emptyset} \; \sum_{\ell': \ell'_1 \neq \ell_1, \ell'_2 = \ell_2} V_1(\ell, \ell'),$$

where

$$\begin{aligned} V_1(\ell, \ell') :=& \frac{N}{p-1} \sum_{(m,m'): m_1 = m'_1, m_2 \neq m'_2} \lambda_2(I_{\ell,m} \cap J_h^n)\lambda_2(I_{\ell',m'} \cap J_h^n) \\ & - \lambda_2(I_\ell \cap J_h^n)\lambda_2(I_{\ell'} \cap J_h^n). \end{aligned}$$

We split $V_1(\ell, \ell') = \hat{V}_1(\ell, \ell') + \check{V}_1(\ell, \ell')$, with

$$\begin{aligned} \hat{V}_1(\ell, \ell') :=& \frac{N}{p-1} \sum_{(m,m'): m_1 = m'_1, m_2 \neq m'_2} \lambda_2(I_{\ell,m} \cap J_h^n)\lambda_2(I_{\ell',m'} \cap J_h^n) \\ & - \frac{N}{p} \sum_{(m,m'): m_1 = m'_1} \lambda_2(I_{\ell,m} \cap J_h^n)\lambda_2(I_{\ell',m'} \cap J_h^n), \\ \check{V}_1(\ell, \ell') :=& p \\ & \times \sum_{(m,m'): m_1 = m'_1} \lambda_2(I_{\ell,m} \cap J_h^n)\lambda_2(I_{\ell',m'} \cap J_h^n) - \lambda_2(I_\ell \cap J_h^n)\lambda_2(I_{\ell'} \cap J_h^n). \end{aligned}$$

On one side

$$\begin{aligned} \hat{V}_1(\ell, \ell') =& N \sum_m \lambda_2(I_{\ell,m} \cap J_h^n) \\ & \times \left(\frac{1}{p(p-1)} \sum_{m': m'_1 = m_1, m'_2 \neq m_2} \lambda_2(I_{\ell',m'} \cap J_h^n) - \frac{1}{p}\lambda_2(I_{\ell',m} \cap J_h^n) \right). \end{aligned}$$

Since both terms inside the parentheses are bounded by $1/(pN^2)$, we have $|\hat{V}_1(\ell, \ell')| \leq 1/(pN^2)$ and so

$$\left| \sum_{\ell': \ell'_1 \neq \ell_1, \ell'_2 = \ell_2} \hat{V}_1(\ell, \ell') \right| \leq \frac{p-1}{pN^2}.$$

On the other side

$$\check{V}_1(\ell, \ell') = \sum_{m_1 \in [1,p]} \lambda_2((H_{\ell_1,m_1} \times H_{\ell_2}) \cap J_h^n) \sum_{m'_1 \in [1,p]} \check{V}_1(\ell, \ell', m_1, m'_1),$$

where $\check{V}_1(\ell,\ell',m_1,m_1') := \lambda_2((H_{\ell_1',m_1}\times H_{\ell_2})\cap J_h^n) - \lambda_2((H_{\ell_1',m_1'}\times H_{\ell_2})\cap J_h^n)$. We have

$$\check{V}_1(\ell,\ell',m_1,m_1') = \frac{1}{N}\left(\lambda\left(H_{\ell_2}\cap(0,q_h(i^n_{p(\ell_1'-1)+m_1})]\right)\right.\\ \left.-\lambda\left(H_{\ell_2}\cap(0,q_h(i^n_{p(\ell_1'-1)+m_1'})]\right)\right).$$

As we have

$$|\check{V}_1(\ell,\ell',m_1,m_1')| \le \frac{1}{N}\\ \times\lambda\left(H_{\ell_2}\cap\left[\min_{p(\ell_1'-1)<k\le p\ell_1'} q_h(i_k^n), \max_{p(\ell_1'-1)<k\le p\ell_1'} q_h(i_k^n)\right]\right),$$

we deduce

$$|\check{V}_1(\ell,\ell')| \le \frac{1}{Np}\lambda\left(H_{\ell_2}\cap\left[\min_{p(\ell_1'-1)<k\le p\ell_1'} q_h(i_k^n), \max_{p(\ell_1'-1)<k\le p\ell_1'} q_h(i_k^n)\right]\right).$$

Consequently

$$\left|\sum_{\ell':\ell_1'\neq\ell_1,\ell_2'=\ell_2}\check{V}_1(\ell,\ell')\right| \le \frac{1}{Np}\\ \times\sum_{\ell_1'\in[1,p]:\ell_1'\neq\ell_1}\lambda\left(H_{\ell_2}\cap\left[\min_{p(\ell_1'-1)<k\le p\ell_1'} q_h(i_k^n), \max_{p(\ell_1'-1)<k\le p\ell_1'} q_h(i_k^n)\right]\right).$$

Since q_h is a piecewise monotonic sequence, and because the states are relabeled so that $i_1^n \le \cdots \le i_N^n$, the intervals

$$\left(\min_{p(\ell_1'-1)<k\le p\ell_1'} q_h(i_k^n), \max_{p(\ell_1'-1)<k\le p\ell_1'} q_h(i_k^n)\right), \quad \ell_1'\in[1,p]$$

are pairwise disjoint on each of the r pieces where q_h is monotonic, and we obtain

$$\left|\sum_{\ell':\ell_1'\neq\ell_1,\ell_2'=\ell_2}\check{V}_1(\ell,\ell')\right| \le \frac{r}{Np}\lambda(H_{\ell_2}) \le \frac{r}{N^2}.$$

And so, using the bound (13):

$$|V_1(\widehat{Z}^{n+1}_{s_h})| \le \frac{(r+1)p-1}{pN^{3/2}}(TV(q_h)+2).$$

ii. We have

$$V_2(\widehat{Z}^{n+1}_{s_h}) = \sum_{\ell: I_\ell \not\subset J^n_h, I_\ell \cap J^n_h \neq \emptyset} \sum_{\ell': \ell'_1 = \ell_1, \ell'_2 \neq \ell_2} V_2(\ell, \ell'),$$

where

$$V_2(\ell, \ell') := \frac{N}{p-1} \times \sum_{(m,m'): m_1 \neq m'_1, m_2 = m'_2} \lambda_2(I_{\ell,m} \cap J^n_h)\lambda_2(I_{\ell',m'} \cap J^n_h) - \lambda_2(I_\ell \cap J^n_h)\lambda_2(I_{\ell'} \cap J^n_h).$$

We split $V_2(\ell, \ell') = \hat{V}_2(\ell, \ell') + \check{V}_2(\ell, \ell')$, with

$$\begin{aligned}
\hat{V}_2(\ell, \ell') :=& \frac{N}{p-1} \sum_{(m,m'): m_1 \neq m'_1, m_2 = m'_2} \lambda_2(I_{\ell,m} \cap J^n_h)\lambda_2(I_{\ell',m'} \cap J^n_h) \\
&- \frac{N}{p} \sum_{(m,m'): m_2 = m'_2} \lambda_2(I_{\ell,m} \cap J^n_h)\lambda_2(I_{\ell',m'} \cap J^n_h), \\
\check{V}_2(\ell, \ell') :=& p \sum_{(m,m'): m_2 = m'_2} \lambda_2(I_{\ell,m} \cap J^n_h)\lambda_2(I_{\ell',m'} \cap J^n_h) - \lambda_2(I_\ell \cap J^n_h)\lambda_2(I_{\ell'} \cap J^n_h).
\end{aligned}$$

On one side

$$\begin{aligned}
\hat{V}_2(\ell, \ell') =& N \sum_m \lambda_2(I_{\ell,m} \cap J^n_h) \\
& \times \left(\frac{1}{p(p-1)} \sum_{m': m'_1 \neq m_1, m'_2 = m_2} \lambda_2(I_{\ell',m'} \cap J^n_h) - \frac{1}{p}\lambda_2(I_{\ell',m} \cap J^n_h) \right).
\end{aligned}$$

Since both terms inside the parentheses are bounded by $1/(pN^2)$, we have $|\hat{V}_2(\ell, \ell')| \leq 1/(pN^2)$ and so

$$\left| \sum_{\ell': \ell'_1 = \ell_1, \ell'_2 \neq \ell_2} \hat{V}_2(\ell, \ell') \right| \leq \frac{p-1}{pN^2}.$$

On the other side

$$\check{V}_2(\ell, \ell') = \sum_{m_2 \in [1,p]} \lambda_2((H_{\ell_1} \times H_{\ell_2,m_2}) \cap J^n_h) \sum_{m'_2 \in [1,p]} \check{V}_2(\ell, \ell', m_2, m'_2),$$

where $\check{V}_2(\ell, \ell', m_2, m'_2) := \lambda_2((H_{\ell_1} \times H_{\ell'_2,m_2}) \cap J^n_h) - \lambda_2((H_{\ell_1} \times H_{\ell'_2,m'_2}) \cap J^n_h)$; we have

$$\check{V}_2(\ell, \ell', m_2, m_2') = \frac{1}{N} \sum_{m_1' \in [1,p]} \left(\lambda\left(H_{\ell_2', m_2} \cap (0, q_h(i^n_{p(\ell_1-1)+m_1'})]\right)\right.$$
$$\left. -\lambda\left(H_{\ell_2', m_2'} \cap (0, q_h(i^n_{p(\ell_1-1)+m_1'})]\right)\right).$$

Note that the difference in the parentheses is equal to 0 if $\ell_2' \neq \ell_2'(\ell_1, m_1') := \lceil n q_h(i^n_{p(\ell_1-1)+m_1'})\rceil + 1$. Consequently

$$\left|\sum_{\ell':\ell_1'=\ell_1,\ell_2'\neq\ell_2} \check{V}_2(\ell, \ell')\right| \leq \frac{1}{N} \sum_{m_2 \in [1,p]} \lambda_2((H_{\ell_1} \times H_{\ell_2,m_2}) \cap J_h^n)$$
$$\times \sum_{m'} \left| \lambda\left(H_{\ell_2'(\ell_1,m_1'),m_2} \cap (0, q_h(i^n_{p(\ell_1-1)+m_1'})]\right)\right.$$
$$\left. -\lambda\left(H_{\ell_2'(\ell_1,m_1'),m_2'} \cap (0, q_h(i^n_{p(\ell_1-1)+m_1'})]\right)\right|$$
$$\leq \frac{1}{N} \sum_{m_2 \in [1,p]} \lambda_2((H_{\ell_1} \times H_{\ell_2,m_2}) \cap J_h^n) \leq \frac{1}{N}\lambda_2(I_\ell) = \frac{1}{N^2}.$$

And so, using the bound (13):

$$|V_2(\widehat{Z}^{n+1}_{s_h})| \leq \frac{2p-1}{pN^{3/2}}(TV(q_h) + 2).$$

iii. We have

$$V_3(\widehat{Z}^{n+1}_{s_h}) = \sum_{\ell:I_\ell \not\subset J_h^n, I_\ell \cap J_h^n \neq \emptyset} \sum_{\substack{\ell':\ell_1'\neq\ell_1,\ell_2'\neq\ell_2 \\ I_{\ell'} \not\subset J_h^n, I_{\ell'} \cap J_h^n \neq \emptyset}} V_3(\ell, \ell'),$$

where

$$V_3(\ell, \ell') := \frac{N}{(p-1)^2}$$
$$\times \sum_{(m,m'):m_1\neq m_1',m_2\neq m_2'} \lambda_2(I_{\ell,m} \cap J_h^n)\lambda_2(I_{\ell',m'} \cap J_h^n) - \lambda_2(I_\ell \cap J_h^n)\lambda_2(I_{\ell'} \cap J_h^n).$$

We split $V_3(\ell, \ell') = V_3^a(\ell, \ell') - V_3^b(\ell, \ell') - V_3^c(\ell, \ell') - V_3^d(\ell, \ell')$, with

$$V_3^a(\ell, \ell') := \left(\frac{N}{(p-1)^2} - 1\right) \sum_{(m,m'):m_1\neq m_1',m_2\neq m_2'} \lambda_2(I_{\ell,m} \cap J_h^n)\lambda_2(I_{\ell',m'} \cap J_h^n),$$
$$V_3^b(\ell, \ell') := \sum_{(m,m'):m_1\neq m_1',m_2= m_2'} \lambda_2(I_{\ell,m} \cap J_h^n)\lambda_2(I_{\ell',m'} \cap J_h^n),$$
$$V_3^c(\ell, \ell') := \sum_{(m,m'):m_1= m_1',m_2\neq m_2'} \lambda_2(I_{\ell,m} \cap J_h^n)\lambda_2(I_{\ell',m'} \cap J_h^n),$$

$$V_3^d(\ell, \ell') := \sum_m \lambda_2(I_{\ell,m} \cap J_h^n)\lambda_2(I_{\ell',m} \cap J_h^n).$$

Since

$$V_3^a(\ell, \ell') \leq \frac{2p-1}{N^3}, \; V_3^b(\ell, \ell') \leq \frac{p-1}{N^3}, \; V_3^c(\ell, \ell') \leq \frac{p-1}{N^3}, \; V_3^d(\ell, \ell') \leq \frac{1}{N^3},$$

using the bound (13), we obtain:

$$|V_3(\widehat{Z}_{s_h}^{n+1})| \leq \frac{2p-1}{N^2}(TV(q_h)+2)^2.$$

Hence the final result.

□

6 Conclusion

In this article, we analyze the convergence of Monte Carlo methods, possibly combined with variance reduction techniques, for the simulation of Markov chains on one-dimensional discrete state spaces. We prove a bound of the variance of estimators used in one step of the simulation. We show that the convergence order (relative to the number of simulation paths) corresponds to the experimental order as calculated in various numerical experiments.

Albeit the theoretical convergence rates of SSS and Sudoku Latin square sampling are the same, for the numerical examples we see that the variance reduction of the Sudoku Latin square sampling approach is superior to that of the SSS. The difference is smaller for pure integration problems (see [5]). A drawback of the SSS approach for simulation is that it is not guaranteed that each state is considered exactly once for a transition.

The study aims to fill a gap between the theoretical results on variance reduction techniques used in Monte Carlo simulations and the actual improvements observed in computations. The Sudoku Latin square sampling is suited to situations where the state space is one-dimensional, with a natural order. Our interest has been in physical problems, where the states are related to particles. A numerical constraint is that N must be a square number. The method should be extended in many directions, such as continuous state spaces and multi-dimensional problems. This will be the subject of forthcoming research.

Acknowledgements The fourth author was supported by an Inria International Chair, and IVADO Research Grant, and an NSERC Discovery Grant.

References

1. Alfa, A.S.: Applied Discrete-Time Queues, 2nd edn. Springer, Berlin (2016)
2. Cheng, R.C.H., Davenport, T.: The problem of dimensionality in stratified sampling. Manag. Sci. **35**, 1278–1296 (1989)
3. Coulibaly, I., Lécot, C.: Simulation of diffusion using quasi-random methods. Math. Comput. Simul. **47**, 153–163 (1998)
4. Devroye, L.: Non-Uniform Random Variate Generation. Springer, Berlin (1986)
5. El Haddad, R., Fakhereddine, R., Lécot, C., Venkiteswaran, G.: Extended Latin hypercube sampling for integration and simulation. In: Dick, J., Kuo, F.Y., Peters, G.W., Sloan I.H. (eds.) Monte Carlo and Quasi-Monte Carlo Methods 2012, pp. 317–330. Springer, Berlin (2013)
6. Fakhereddine, R., El Haddad, R., Lécot, C., El Maalouf, J.: Stratified Monte Carlo simulation of Markov chains. Math. Comput. Simul. **135**, 51–62 (2017)
7. Gerber, M., Chopin, N.: Sequential quasi-Monte Carlo. J. R. Stat. Soc.: Ser. B **77**, 509–579 (2015)
8. Haber, S.: A modified Monte Carlo quadrature. Math. Comput. **20**, 361–368 (1966)
9. Lécot, C., Tuffin, B.: Quasi-Monte Carlo methods for estimating transient measures of discrete time Markov chains. In: Niederreiter, H. (ed) Monte Carlo and Quasi-Monte Carlo Methods 2002, pp. 329–344. Springer, Berlin (2004)
10. L'Ecuyer, P (1994) Efficiency improvement and variance reduction. In: Tew, J.D., Manivannan, S., Sadowski D.A., Seila, A.F. (eds.) Proceedings of the 1994 Winter Simulation Conference, pp. 122–132. IEEE Press (1994)
11. L'Ecuyer, P., Lécot, C., Tuffin, B.: Randomized quasi-Monte Carlo simulation of Markov chains with an ordered state space. In: Niederreiter, H., Talay, D. (eds.) Monte Carlo and Quasi-Monte Carlo Methods 2004, pp. 331–342. Springer, Berlin (2006)
12. L'Ecuyer, P., Lécot, C., Tuffin, B.: A randomized quasi-Monte Carlo simulation method for Markov chains. Oper. Res. **56**(4), 958–975 (2008)
13. L'Ecuyer, P., Lécot, C., L'Archevêque-Gaudet, A.: On array-RQMC for Markov chains: mapping alternatives and convergence rates. In: LÉcuyer, P., Owen, A. (eds.) Monte Carlo and Quasi-Monte Carlo Methods 2008, pp. 485–500. Springer, Berlin (2009)
14. L'Ecuyer, P., Munger, D., Lécot, C., Tuffin, B.: Sorting methods and convergence rates for Array-RQMC: some empirical comparisons. Math. Comput. Simul. **143**, 191–201 (2018)
15. McKay, M.D., Beckman, R.J., Conover, W.J.: A comparison of three methods for selecting values of input variables in the analysis of output from a computer code. Technometrics **21**, 239–245 (1979)
16. Owen, A.B.: Orthogonal arrays for computer experiments, integration and visualization. Stat. Sin. **2**, 439–452 (1992)
17. Owen, A.B.: Lattice sampling revisited: Monte Carlo variance of means over randomized orthogonal arrays. Ann. Stat. **22**, 930–945 (1994)
18. Owen, A.B.: Monte Carlo variance of scrambled net quadrature. SIAM J. Numer. Anal. **34**, 1884–1910 (1997)
19. Pedersen, R.M., Vis, T.L.: Sets of mutually orthogonal Sudoku Latin squares. Coll. Math. J. **40**, 174–180 (2009)
20. Stein, M.: Large sample properties of simulations using Latin hypercube sampling. Technometrics **29**, 143–151 (1987)
21. Tang, B.: Orthogonal array-based Latin hypercubes. J. Am. Stat. Assoc. **88**, 1392–1397 (1993)
22. Xu, X., Haaland, B., Qian, P.Z.G.: Sudoku-based space-filling designs. Biometrika **98**, 711–720 (2011)

Avoiding the Sign Problem in Lattice Field Theory

Tobias Hartung, Karl Jansen, Hernan Leövey and Julia Volmer

Abstract In lattice field theory, the interactions of elementary particles can be computed via high-dimensional integrals. Markov-chain Monte Carlo (MCMC) methods based on importance sampling are normally efficient to solve most of these integrals. But these methods give large errors for oscillatory integrands, exhibiting the so-called sign problem. We developed new quadrature rules using the symmetry of the considered systems to avoid the sign problem in physical one-dimensional models for the resulting high-dimensional integrals. This article gives a short introduction to integrals used in lattice QCD where the interactions of gluon and quark elementary particles are investigated, explains the alternative integration methods we developed and shows results of applying them to models with one physical dimension. The new quadrature rules avoid the sign problem and can therefore be used to perform simulations at until now not reachable regions in parameter space, where the MCMC errors are too big for affordable sample sizes. However, it is still a challenge to develop these techniques further for applications with physical higher-dimensional systems.

Keywords Sign problem · Polynomially exact integration · 1-dimensional QCD · Chemical potential · Lattice systems

---

K. Jansen · J. Volmer (✉)
DESY Zeuthen, Platanenallee 6, 15738 Zeuthen, Germany
e-mail: julia.volmer@desy.de

K. Jansen
e-mail: karl.jansen@desy.de

T. Hartung
Department of Mathematics, Kings College London, Strand, London WC2R 2LS, United Kingdom
e-mail: tobias.hartung@kcl.ac.uk

H. Leövey
Structured Energy Management, Axpo Trading, Parkstrasse 23, 5400 Baden, Germany
e-mail: HernanEugenio.Leoevey@axpo.com

B. Tuffin and P. L'Ecuyer (eds.), *Monte Carlo and Quasi-Monte Carlo Methods*, Springer Proceedings in Mathematics & Statistics 324,
https://doi.org/10.1007/978-3-030-43465-6_11

1 Introduction

Monte Carlo (MC) methods are in general very efficient to solve high-dimensional integrals. They use the law of large numbers to approximate an integral with quadrature rules that use random sampling points. But MC methods are highly *inefficient* for oscillatory integrand functions, e.g. the function shown in Fig. 1a. An *exact* integration of oscillatory functions would, of course, result in the cancellation of large negative and positive contributions to the integral—in the example in Fig. 1a this would give an integral of zero. However, the random choice of sampling points in MC methods, shown as black points in Fig. 1a, does lead only to approximate cancellation when the number of points is relatively big and hence, it is very difficult to obtain accurate results with affordable sample sizes. This non-perfect cancellation of negative and positive parts in the integration method, usually resulting in large quadrature rule errors, is called the *sign problem*. The sign problem is for example the reason why important physical interactions in the early universe cannot be simulated which could explain why there is more matter than anti-matter in our universe today. To acquaint better knowledge of these fundamental phenomena, it is essential to develop alternative quadrature rules to MC that avoid the sign problem.

In physical applications, the function to-be-integrated describes some characteristic in a given physical model. We investigated methods that use some symmetry of the physical model to result in the exact cancellation of positive and negative parts in the quadrature rule. If the model behind the function in Fig. 1b has a reflection symmetry, few MC sampling points—in black—can be chosen and together with their reflected—white—points they form a set of sampling points that results in an exact quadrature rule. In this specific example even one MC point with its reflection point would give an exact result, for more complicated functions more sampling points are needed.

This article first gives a short introduction to the high-dimensional integrals that have to be solved in particle physics, more precisely in lattice QCD. Readers that are mostly interested in the integration methods can easily skip this part. The main part of this article presents the methods we developed and tested to avoid the sign problem for high-dimensional integration in physical one-dimensional systems.

We found that symmetrically chosen quadrature rules can avoid the sign problem and can efficiently be applied also to high-dimensional integrals. These rules can help to perform simulations in important, not-yet reachable regions in parameter space, at least in physical one-dimensional systems so far. To apply them to higher physical dimensions, in particular to physical four-dimensional systems in high energy physics as lattice QCD, they clearly need to be developed further.

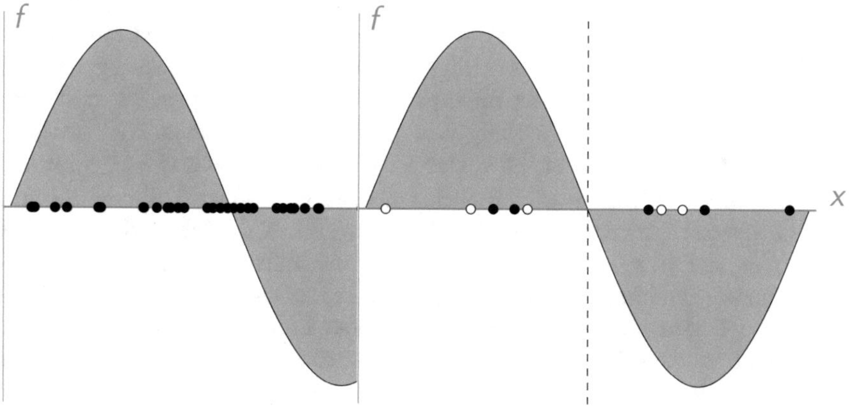

(a) MC sampling points (b) Symmetric sampling points

Fig. 1 MC integration of an oscillatory function results in large errors, known as the sign problem. This problem is due to the non-cancellation of positive and negative contributions to the quadrature rule (**a**). Choosing sampling points by using the symmetry of the underlying model results in an exact quadrature rule (**b**)

2 Integration in Lattice QCD

In theoretical physics the interaction between elementary particles such as the electron, is described by *quantum field theories* (QFT), see e.g. [21]. The mathematical formalism in QFT defines particles as classical fields that are functions in three space dimensions and one time dimension, $P(x, y, z, t)$. Operators, $O[P]$, are functionals of these fields and describe the interactions between them. An expectation value A of this interaction or operator $O[P]$, also called amplitude, is computed via the path integral,

$$A = \frac{\int O[P]B[P]\,\mathrm{d}P}{\int B[P]\,\mathrm{d}P}. \tag{1}$$

$\int \mathrm{d}P$ is the infinite-dimensional integration over all possible states of the field P in time and space. The path integral becomes a well defined expression, if a Euclidean metric is used and the fields are defined on a finite dimensional, discrete lattice.[1] In (1), $B[P]$ is called the Boltzmann-weight and provides a probability which weights the particle (field) interactions. The denominator in (1) insures the proper normalization of A. The expectation value A is interesting because physical observables can be derived from it and their numerical values can be compared with experimental results or can give new results that are not yet possible to reach with experiments.

[1]For an alternative definition using the ζ-regularization see [16, 17].

In *lattice field theory*, space-time and the involved functionals $O[P]$ and $B[P]$ are discretized in Euclidean space, such that (1) can be computed numerically. Often, the Boltzmann-weight is a highly peaked function suggesting that this computation can be done using importance sampling techniques. In most computations, this importance sampling is done by a Markov chain MC (MCMC) algorithm using a Markov chain that leaves the distribution density $\frac{B[P]}{\int B[P]\mathrm{d}P}$ invariant. To compute A numerically, four-dimensional space-time is discretized on a four-dimensional lattice with four directions $\mu \in \{1, 2, 3, 4\}$, lattice sites $\boldsymbol{n} \in \Lambda = \{(n_1, n_2, n_3, n_4)|n_1, n_2, n_3, n_4 \in \{1, \ldots, d\}\}$ and discretized fields P. This results in an $4d$-dimensional integration over the Haar measure of the compact group $\mathscr{S}\mathscr{U}(3)$. For real applications, d can be very large, reaching orders of magnitude of several thousands nowadays. Thus, we are left with an extremely high dimensional integration problem. Moreover, for some physically very important questions MCMC methods cannot be applied successfully. This concerns, for example, the very early universe or the matter anti-matter asymmetry which leads to our sheer existence. Thus, a number of interesting questions remain completely unanswered and it is exactly here where new high dimensional integration methods could be extremely helpful.

Still, the MCM methods have led to very successful computations already. By performing numerical computations on massively parallel super computers a very impressive result of such a lattice MCMC can be obtained: namely, the mass spectrum of the lightest composite particles made out of quarks and gluons that agrees completely with the experimental values, see Fig. 2. To get similar precise results for

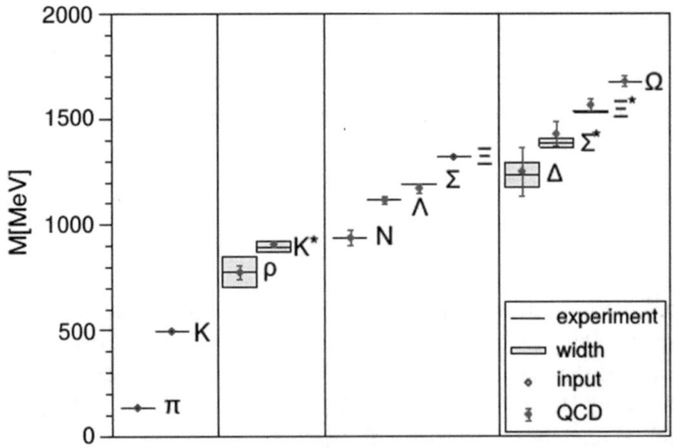

Fig. 2 The via lattice QCD computed masses of different composite particles (dots with vertical error bars) agree with the experimentally measured values (horizontal lines with error boxes) [11]. The masses of π, K and $\sum$ (dots without error bars) were input values to the computation

other, more error-prone observables, research is going on to develop new methods to make this high-dimensional integration faster and the results more precise.

A more detailed introduction to lattice QCD is for example given in the textbooks [10, 13, 22].

3 Quadrature Rules for One-Dimensional Lattices

In lattice QCD, the amplitude of interactions between quarks and gluons in physical four-dimensional space-time can be computed via a high-dimensional integral using the Haar-measure over the compact group $\mathscr{SU}(3)$, see Sect. 2. This integral is typically solved numerically using MCMC methods. If the integrand is an oscillatory function, this method results in the sign problem that gives large errors and avoids physical insights in important processes. We developed alternative methods that avoid the sign problem and at the same time are efficient for high-dimensional integration over compact groups. Due to various complications with physical four-dimensional lattice QCD, we developed and tested the methods for physical one-dimensional models that involve low-dimensional and high-dimensional integration over compact groups. As suggested in Sect. 1, we developed quadrature rules using the symmetry of the models.

This section is structured from low-dimensional to high-dimensional integration: First, it introduces symmetric quadrature rules for one-dimensional integration over compact groups to avoid the sign problem here. Then, it presents the recursive numerical integration (RNI), a method to reduce high-dimensional integrals to nested one-dimensional integrals. Finally, it shows how to combine both methods to avoid the sign problem for high-dimensional integration over compact groups. For all three presented methods, the section shows results of applying them to simple physical, one-dimensional models. More detailed explanations of the methods and applications can be found in [24].

3.1 *Avoiding the Sign Problem in Physical One-Dimensional Systems*

The sign problem can already arise in a one-dimensional integration, solving

$$I(f) = \int_G f(U)\,\mathrm{d}U \tag{2}$$

with MC methods over the Haar-measure of $G \in \{\mathscr{U}(N), \mathscr{SU}(N)\}$. Finding an alternative suitable quadrature rule $Q(f)$ ad-hoc to approximate this integral is not straightforward. The articles [8, 9] suggest that using symmetrically distributed sampling points can be beneficial for avoiding the sign problem, possibly resulting in an exact cancellation of positive and negative contributions to the integral, as stated in Sect. 1. The article of Genz [15] gives efficient quadrature rules for integrations over spheres, choosing the sampling points symmetrically on the spheres. We searched for measure preserving homeomorphisms to apply the symmetric quadrature rules on spheres to the integration over compact groups. This section describes the two steps to create the symmetric quadrature rules $Q(f)$ for (2):

Sym 1. Rewrite the integral $I(f)$ over the compact group G into an integral over spheres. We restricted ourselves to $G \in \{\mathscr{U}(1), \mathscr{U}(2), \mathscr{U}(3), \mathscr{SU}(2), \mathscr{SU}(3)\}$.

Sym 2. Approximate each integral over one spheres by a symmetric quadrature rule as proposed in Genz [15], and combine them to a product rule $Q(f)$.

Finally, this section shows results of applying $Q(f)$ to the one-dimensional QCD model with a sign problem. A more detailed explanation of the method can be found in [2, 5].

By finding measure preserving homeomorphisms between the compact groups and products of spheres we created polynomially exact quadrature rules for compact groups. The application of these rules to the one-dimensional QCD model gave results on machine precision where the standard MC method shows a sign problem. Therefore the symmetric quadrature rules avoid the sign problem and give rise to solve integrals in beforehand non-reachable parameter regions.

3.1.1 Sym 1. Rewriting the Integral

The symmetric quadrature rules of Genz [15] are designed for the integration over k-dimensional spheres S^k. To use them for the integration over the compact groups $\mathscr{U}(N)$ and $\mathscr{SU}(N)$ with $N \in \{2, 3\}$ in (2), the compact groups have to be associated with spheres. The facts that $\mathscr{U}(N)$ is isomorphic to the semidirect product of $\mathscr{SU}(N)$ acting on $\mathscr{U}(1)$ $\big(U(N) \cong SU(N) \rtimes U(1)\big)$, that $\mathscr{U}(1)$ is isomorphic to S^1 $\big(\mathscr{U}(1) \cong S^1\big)$ and that $\mathscr{SU}(N)$ is a principal $\mathscr{SU}(N-1)$ bundle over S^{2N-1} result in

$$\mathscr{SU}(N) \simeq S^3 \times S^5 \times \cdots \times S^{2N-1}, \tag{3}$$

$$\mathscr{U}(N) \simeq S^1 \times S^3 \times \cdots \times S^{2N-1}. \tag{4}$$

Then, the integral over the Haar-measure of G in (2) can be rewritten as the integral over products of spheres,

$$\int_G \mathrm{d}U\, f(U) = \int_{S^{2N-1}} \Bigg(\int_{S^{2N-3}} \Bigg(\cdots \int_{S^{n+2}} \Bigg(\int_{S^n} f\big(\Phi(\boldsymbol{x}_{S^{2N-1}}, \boldsymbol{x}_{S^{2N-3}}, \ldots, \boldsymbol{x}_{S^{n+2}}, \boldsymbol{x}_{S^n})\big) \mathrm{d}\boldsymbol{x}_{S^n} \Bigg) \mathrm{d}\boldsymbol{x}_{S^{n+2}} \cdots \Bigg) \mathrm{d}\boldsymbol{x}_{S^{2N-3}} \Bigg) \mathrm{d}\boldsymbol{x}_{S^{2N-1}}, \tag{5}$$

with $n = 1$ for $\mathscr{U}(N)$ and $n = 3$ for $\mathscr{SU}(N)$ [1]. Here, $\boldsymbol{x}_{S^k}$ is an element on the k-sphere and $\Phi : \bigtimes_j S^{2j-1} \to G$ with $G \in \{\mathscr{U}(N), \mathscr{SU}(N)\}$ is a measure preserving homeomorphism. We found the homeomorphisms $\Phi_G \equiv \Phi$ for the compact groups $G \in \{\mathscr{U}(1), \mathscr{U}(2), \mathscr{U}(3), \mathscr{SU}(2), \mathscr{SU}(3)\}$:

- For $\mathscr{SU}(2)$, Φ is an isomorphism, given by

$$\begin{aligned} \Phi_{\mathscr{SU}(2)} &: S^3 \to \mathscr{SU}(2), \\ \boldsymbol{x} &\mapsto \begin{pmatrix} x_1 + ix_2 & -(x_3 + ix_4)^* \\ x_3 + ix_4 & (x_1 + ix_2)^* \end{pmatrix}. \end{aligned} \tag{6}$$

- For $\mathscr{SU}(3)$, spherical coordinates of S^5 are needed,

$$\begin{aligned} \Psi &: [0, 2\pi)^3 \times [0, \frac{\pi}{2}) \to S^5, \\ (\alpha_1, \alpha_2, \alpha_3, \phi_1, \phi_2) &\mapsto \begin{pmatrix} \cos\alpha_1 \sin\phi_1 \\ \sin\alpha_1 \sin\phi_1 \\ \sin\alpha_2 \cos\phi_2 \sin\phi_2 \\ \cos\alpha_2 \cos\phi_2 \sin\phi_2 \\ \sin\alpha_3 \cos\phi_1 \cos\phi_2 \\ \cos\alpha_3 \cos\phi_1 \cos\phi_2 \end{pmatrix}. \end{aligned} \tag{7}$$

  Then, Φ is given by

$$\begin{aligned} \Phi_{\mathscr{SU}(3)} &: S^5_1 \times S^3 \to \mathscr{SU}(3), \\ (\boldsymbol{x}, \boldsymbol{y}) &\mapsto A(\Psi^{-1}(\boldsymbol{x})) \cdot B(\boldsymbol{y}), \end{aligned} \tag{8}$$

  with the matrices

$$A(\Psi^{-1}(\boldsymbol{x})) = \begin{pmatrix} e^{i\alpha_1} \cos\phi_1 & 0 & e^{i\alpha_1} \sin\phi_1 \\ -e^{i\alpha_2} \sin\phi_1 \sin\phi_2 & e^{-i(\alpha_1+\alpha_3)} \cos\phi_2 & e^{i\alpha_2} \cos\phi_1 \sin\phi_2 \\ -e^{i\alpha_3} \sin\phi_1 \cos\phi_2 & -e^{-i(\alpha_1+\alpha_2)} \sin\phi_2 & e^{i\alpha_3} \cos\phi_1 \cos\phi_2 \end{pmatrix}, \tag{9}$$

$$B(\boldsymbol{y}) = \begin{pmatrix} x_1 + ix_2 & -(x_3 + ix_4)^* & 0 \\ x_3 + ix_4 & (x_1 + ix_2)^* & 0 \\ 0 & 0 & 1 \end{pmatrix}. \tag{10}$$

  $\Psi^{-1}(\boldsymbol{x})$ is the inverse transformation of (7) from Euclidean to spherical coordinates. S^5_1 denotes S^5 without its poles, $\phi_1 = 0$ or $\phi_2 = 0$, because at these points the inverse transformation is not unique. The therefore excluded set is a null set, thus $\Phi_{\mathscr{SU}(3)}$ can still be used in (6).
- For $\mathscr{U}(1)$, Φ is an isomorphism,

$$\begin{aligned} \Phi_{\mathscr{U}(1)} &: S^1 \to \mathscr{U}(1), \\ \alpha &\mapsto e^{i\alpha}, \end{aligned} \tag{11}$$

with $\alpha \in [0, 2\pi)$.

- For $\mathcal{U}(2)$, Φ is an isomorphism,

$$\Phi_{\mathcal{U}(2)} : S^3 \times S^1 \to \mathcal{U}(2),$$
$$(\boldsymbol{x}, \alpha) \mapsto \Phi_{\mathcal{SU}(2)}(\boldsymbol{x}) \cdot \mathrm{diag}(\mathrm{e}^{i\alpha}, 1). \tag{12}$$

- For $\mathcal{U}(3)$, Φ is given by

$$\Phi_{\mathcal{SU}(3)} : S_1^5 \times S^3 \times S^1 \to \mathcal{U}(3),$$
$$(\boldsymbol{x}, \boldsymbol{y}, \alpha) \mapsto \Phi_{\mathcal{SU}(3)}(\boldsymbol{x}, \boldsymbol{y}) \cdot \mathrm{diag}(\mathrm{e}^{i\alpha}, 1, 1). \tag{13}$$

3.1.2 Sym 2. Quadrature Rule for Spheres

With the measure preserving homeomorphism Φ in Sect. 3.1.1, the integral (2) can be written as an integral over a product of spheres as in (6). To approximate the full integral numerically, one can use a product quadrature rule with quadratures $Q_{S^k}(g)$ that are specifically designed for integrations over spheres. The full integral can be computed efficiently if the number of involved spheres is small. As pointed out in the last subsection, in practice we are interested to build product rules for at most $S_1^5 \times S^3 \times S^1$. The quadratures over each sphere can be built in many ways. Since we are aiming for resulting quadratures that exhibit some symmetry characteristics to hopefully overcome the sign problem, it seems that quadrature rules given in [15] exhibit all required properties, i.e. high accuracy due to polynomial exactness over spheres, numerical stability of the resulting weights, and being fully symmetric. The quadratures over each sphere take the form

$$Q_{S^k}(g) = \sum_{\gamma=1}^{N_{\mathrm{sym}}} w_\gamma \, g(\boldsymbol{t}_\gamma). \tag{14}$$

The sampling points $\boldsymbol{t} \in S^k$ are chosen symmetrically on the k-sphere and are weighted via $w \in \mathbb{R}$. The specific definitions of $\boldsymbol{t}$, w and N_{sym} for different k are given in [15]. (Note that in this reference, the notation U_k is equivalent to the here used S^{k-1}.) It is possible to randomize these quadrature rules, such that an error estimate for each quadrature rule can be computed via independent replication [15].

The final quadrature rule $Q(f)$ of the full integral in (6) is a combination of different single-sphere quadrature rules given in (14). Due to the symmetric choice of the sampling points on spheres, the rule $Q(f)$ is in the following called *symmetrized quadrature rule*. A more detailed description of $Q_{S^k}(g)$ and $Q(f)$ is given in [24], Sect. 6.1.

3.1.3 Application to One-Dimensional QCD

We applied these constructed quadrature rules to physical one-dimensional QCD problems [7], which is a simplified model of strong interactions in elementary particle physics. This model is a good test model because it can be solved analytically, giving a well defined measure for the uncertainties computed by different numerical integration methods. This model has one integration variable $U \in G$ and three real input parameters: a mass m, a chemical potential μ and a length scale d. A small mass ($m \ll d\mu$) introduces a sign problem which makes it very hard for standard methods as MC to compute amplitudes as in (1) numerically.

We computed the chiral condensate in this model, given by

$$\chi = \frac{\int_G \partial_m B[U]\,\mathrm{d}U}{\int_G B[U]\,\mathrm{d}U}, \tag{15}$$

with the Boltzmann-weight

$$B[U] = \det\left(c_1(m) + c_2(d,\mu)U^\dagger + c_3(d,\mu)U\right), \tag{16}$$

expressed via the parameters

$$c_1(m) = \prod_{j=1}^{L} \tilde{m}_j, \qquad \tilde{m}_1 = m,$$

$$\tilde{m}_j = m + \frac{1}{4\tilde{m}_{j-1}} \quad \forall j \in \{2,3,\dots,d-1\},$$

$$\tilde{m}_d = m + \frac{1}{4\tilde{m}_{d-1}} + \sum_{j=1}^{d-1} \frac{(-1)^{j+1}2^{-2j}}{\tilde{m}_j \prod_{k=1}^{j-1} \tilde{m}_k^2}, \tag{17}$$

$$c_2(d,\mu) = 2^{-d}\,\mathrm{e}^{-d\mu}, \tag{18}$$

$$c_3(d,\mu) = (-1)^d 2^{-d}\,\mathrm{e}^{d\mu}. \tag{19}$$

For brevity, the dependencies of these parameters are in the following only written when needed.

In all numerical calculations, we first computed both numerator and denominator of (15) separately and then divided them. We computed the numerator by symbolically differentiating $B[U]$ and computing the integral over the result numerically.

We compared the results for χ using the symmetrized quadrature rules that are described in Sect. 3.1.2, with a standard integration method, ordinary MC sampling. The latter quadrature rule is given by

$$Q(f) = \frac{1}{N_{\mathrm{MC}}} \sum_{\gamma=1}^{N_{\mathrm{MC}}} f(V_\gamma), \tag{20}$$

where the V are matrices that are chosen randomly from a uniform distribution. We chose N_{MC} to be as large as the number of used symmetric sampling points.

Because the analytic results of χ can be calculated straightforwardly, we computed the error estimates of the numerical solutions—MC and symmetrized quadrature rules—directly via the relative deviation from the analytic value,

$$\Delta\chi = \frac{|\chi_{\text{numerical}} - \chi_{\text{analytic}}|}{|\chi_{\text{analytic}}|} \tag{21}$$

and derived the standard deviation of this error by repeatedly using on the one hand the MC quadrature rules with different random matrices V's and on the other hand the randomized symmetrized quadrature rules as indicated in Sect. 3.1.2.

The results for $\Delta\chi$ of both MC and symmetrized quadrature rule can be roughly split into a small m ($m < 10^{-1}$), a large m ($m > 10^{0.5}$) and a transition region, shown in Fig. 3 for constant $\mu = 1$ and $d = 8$, extended 1024-bit machine precision and different compact groups. For both quadrature rules we used the sampling sizes $N \equiv N_{\text{sym}} = N_{\text{MC}} = 8$ for $\mathcal{SU}(2)$, $N = 96$ for $\mathcal{SU}(3)$, $N = 4$ for $\mathcal{U}(1)$, $N = 32$ for $\mathcal{U}(2)$ and $N = 384$ for $\mathcal{U}(3)$.

First, we describe the MC results: In the small m region, $\Delta\chi$ for all groups are large—equal or larger than one. It can be shown that in this region the numerator of χ is such small that the MC evaluation cannot resolve these values for affordable sample sizes, resulting in large errors [24]. This is the manifestation of the sign

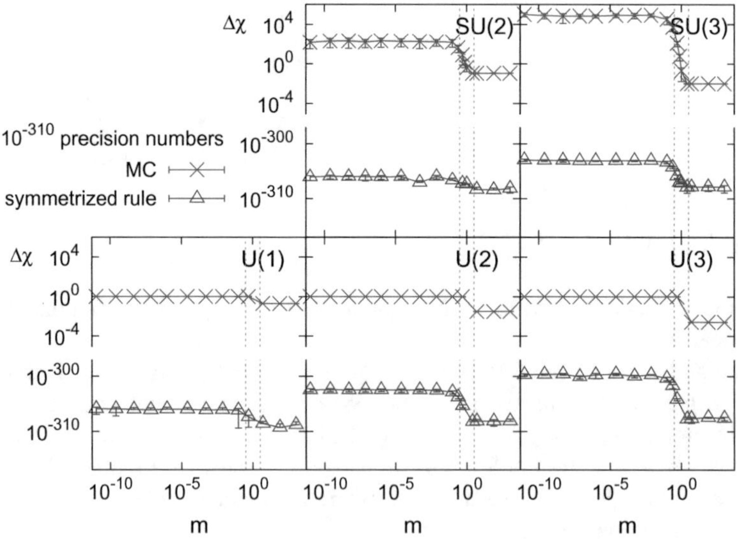

Fig. 3 The sign problem arises for MC results with small m constants, giving errors of the order of one. On the contrast, the symmetrized quadrature rules avoid the sign problem in this region completely, giving errors approximately at machine precision for all shown groups

problem, making it almost impossible to compute reasonable values of χ with MC in the small m region. On the other side, for large m all groups have a smaller MC error estimate than in the small m region. Here the numerator of χ tends to be larger and especially the denominator becomes very large, both resulting in a slightly better error estimate for the MC results.

Opposed to MC results, the symmetrized quadrature rules give error estimates approximately at machine precision up to very small m values, see Fig. 3. These numerical results show that the symmetrized quadrature rules give significant results in the sign problem region in practice, where MC simulations have error estimates of order one.

3.2 *Reducing High-Dimensional Integrals to Nested One-dimensional Integrals*

The previous section shows efficient quadrature rules for physical one-dimensional integration to avoid the sign problem. Most physical models have more than one integration variable. In general, it is not straightforward to find an efficient quadrature rule, and usually restricted Monte Carlo methods are applied to high-dimensional integrals. As a first alternative, we investigated the recursive numerical integration(RNI) method. This method reduces the d-dimensional integral

$$I(f) = \int_{D^d} f[\varphi]\,\mathrm{d}\varphi \tag{22}$$

with $\mathrm{d}\varphi = \prod_{i=1}^{d} \mathrm{d}\varphi_i$ and $D = [0, 2\pi)$ into many recursive one-dimensional integrals, and can be applied for several physical models of interest.

This is done by utilizing the typical structure of the integrand $f[\varphi]$. This section focuses on the RNI method and how to find an efficient quadrature rule for a high-dimensional integral. It does not discuss the sign problem which is investigated further in Sect. 3.3. More specifically, this section describes the two steps to create an efficient quadrature rules $Q(f)$ for the integral $I(f)$ in (22):

RNI 1. Use the structure of the integrand of the high-dimensional integral to rewrite it into recursive one-dimensional integrals.

RNI 2. Choose an efficient quadrature rule to compute each one-dimensional integral numerically. Recursively doing this results in the full quadrature rule $Q(f)$.

Finally, this section shows results of applying the method to a physical model called the topological oscillator. A more detailed explanation of the method and the results can be found in [3, 4].

3.2.1 RNI 1. Using the Structure of the Integrand

Many models in one physical dimensional have integrands with the structure

$$f[\varphi] = \prod_{i=1}^{d} f_i(\varphi_{i+1}, \varphi_i), \tag{23}$$

with periodic boundary conditions $\varphi_{d+1} = \varphi_1$. These models have only next-neighbor couplings.

The integral of (23) can be rewritten using recursive integration as described in [14, 19]: Because of next-neighbor couplings, each variable φ_i appears only twice in $f[\varphi]$, in f_i and f_{i-1}, and therefore the integral can be written as d nested one-variable integrals I_i,

$$I(f) = \int_D \cdots \int_D \prod_{i=1}^{d} f_i(\varphi_i, \varphi_{i+1})\, \mathrm{d}\varphi_d \cdots \mathrm{d}\varphi_1 \tag{24}$$

$$= \underbrace{\int_D \Bigg(\underbrace{\cdots \Bigg(\int_D f_{d-2}(\varphi_{d-2}, \varphi_{d-1}) \cdot \underbrace{\left(\int_D f_{d-1}(\varphi_{d-1}, \varphi_d) \cdot f_d(\varphi_d, \varphi_{d+1})\, \mathrm{d}\varphi_d \right)}_{I_d} \mathrm{d}\varphi_{d-1} \Bigg) \cdots}_{I_{d-1}} \Bigg) \mathrm{d}\varphi_1}_{I_1}.$$

This full integral can be computed recursively: I_d integrates out φ_d first, then I_{d-1} integrates out φ_{d-1} and so on until finally $I_1 = I(f)$ integrates out φ_1.

To avoid under- and overflow of the single quadrature rule results, we actually used quadrature rules to approximate $I_i^* = \frac{1}{c_i} I_i$ with $c_i > 0$ chosen adaptively. Then, the final integral is computed via $I = \left(\prod_{i=1}^{d} c_i\right) I^*$. For brevity, the method is described in the following without this trick.

Each integral is approximated by using an N_{quad}-point quadrature rule. The first integrand in (24) (last from the right) depends on three variables φ_{d-1}, φ_d and φ_{d+1}. The variable φ_d is integrated out, therefore the quadrature rule $Q_d(f_{d-1} \cdot f_d) \equiv Q_d$ of I_d depends on two variables,

$$Q_d(\varphi_{d-1}, \varphi_{d+1}) = \sum_{\gamma=1}^{N_{\text{quad}}} w_\gamma\ f_{d-1}(\varphi_{d-1}, t_\gamma)\ f_d(t_\gamma, \varphi_{d+1}), \tag{25}$$

with sampling points t and weights w. The next integral I_{d-1} is approximated by the quadrature rule

$$Q_{d-1}(\varphi_{d-2}, \varphi_{d+1}) = \sum_{\gamma=1}^{N_{\text{quad}}} w_\gamma\ f_{d-2}(\varphi_{d-2}, t_\gamma)\ Q_d(t_\gamma, \varphi_{d+1}), \tag{26}$$

and includes the quadrature rule Q_d given in (25). The quadrature rules Q_{d-2}, …, Q_1 are created analogically to (26). Using the same sampling points w_γ and weights t_γ, $\gamma \in \{1, \ldots, N_{\text{quad}}\}$ in all quadrature rules Q_i results in the full quadrature rule for (24),

$$Q = Q_1 = \sum_{\gamma=1}^{N_{\text{quad}}} w_\gamma Q_2(t_\gamma, t_\gamma) = \operatorname{tr}\left[\prod_{i=1}^{d} \left(M_i \cdot \operatorname{diag}(w_1, \ldots, w_{N_{\text{quad}}})\right)\right], \tag{27}$$

with M_i beeing an $N_{\text{quad}} \times N_{\text{quad}}$ matrix with entries $(M_i)_{\alpha\beta} = f_i(t_\alpha, t_\beta)$.

3.2.2 RNI 2. Choosing an Efficient Quadrature Rule

We used the Gaussian-Legendre N_{quad}-point quadrature rule, see [23] to define the sampling points t and weights w. For this rule, the error scales asymptotically (for large N_{quad}) as $\sigma \sim \mathcal{O}\left(\frac{1}{(2N_{\text{quad}})!}\right)$. (For Legendre polynomials the correct asymptotic error scaling is $\frac{(N_{\text{quad}}!)^4}{((2N_{\text{quad}})!)^3}$ [20] which is slightly improved over $\frac{1}{(2N_{\text{quad}})!}$.) The Stirling formula ($N_{\text{quad}}! \approx \sqrt{2\pi N_{\text{quad}}}\left(\frac{N_{\text{quad}}}{e}\right)^{N_{\text{quad}}}$ asymptotically) approximates the factorial to give

$$\sigma \sim \mathcal{O}\left(\exp(-2N_{\text{quad}} \ln N_{\text{quad}}) \frac{1}{\sqrt{N_{\text{quad}}}}\right) \tag{28}$$

asymptotically. This is a huge improvement over the MC error scaling $1/\sqrt{N_{\text{MC}}}$.

3.2.3 Application to the Topological Oscillator

We applied the RNI method to the topological oscillator [6], also called quantum rotor, which is a simple, physically one-dimensional model that has non-trivial characteristics which are also present in more complex models. It has d variables $\varphi_i \in [0, 2\pi)$, a length scale T and a coupling constant c. We investigated the topological charge susceptibility of this model,

$$\chi_{\text{top}} = \frac{\int O[\varphi] B[\varphi]\, d\varphi}{\int B[\varphi]\, d\varphi}, \tag{29}$$

with Boltzmann-weight

$$B[\varphi] = \exp\left(-c \sum_{i=1}^{d} (1 - \cos(\varphi_{i+1} - \varphi_i))\right), \tag{30}$$

and a squared topological charge

$$O[\varphi] = \frac{1}{T}\left(\frac{1}{2\pi}\sum_{i=1}^{d}(\varphi_{i+1} - \varphi_i) \mod 2\pi\right)^2. \tag{31}$$

With RNI, we computed both numerator and denominator of χ_{top} separately, both differing in the factorization (23) of their integrands. Straightforwardly, the denominator integrand consists out of local exponential factors. The numerator consists out of summands with varying factorization schemes, each of these summands is computed separately with RNI and they are presented in more detail in [24], Sect. 5.2. We estimated the error of χ_{top} by choosing a large number of samples N^g_{quad} in (25), (26) and similar ones for which we assumed that $\chi_{\text{top}}(N^g_{\text{quad}})$ has converged to the actual value and computed the difference of $\chi_{\text{top}}(N_{\text{quad}})$ for $N_{\text{quad}} < N^g_{\text{quad}}$ to this value,

$$\Delta\chi_{\text{top}}(N_{\text{quad}}) = |\chi_{\text{top}}(N_{\text{quad}}) - \chi_{\text{top}}(N^g_{\text{quad}})|. \tag{32}$$

We tested beforehand that this truncation error behaves exponentially for large N_{quad} in practice, as expected from (28), [4].

We compared the results of the RNI method with results using the Cluster algorithm [25], which we found is an optimal MCMC method for the application to the topological oscillator [4]. Due to the exponential error scaling of the Gauss-Legendre rule, the new method advances MCMC for large enough N_{quad}. We found that the RNI method is also advantageous for lower N_{quad}-values: our simulations showed that the RNI method needs orders of magnitude less runtime than the Cluster algorithm to result in a specified error estimate on an observable, compare Fig. 4 for $c = 2.5, T = 20, d = 200$. The Cluster algorithm measurements resulted in an error estimate that decreases proportional to $t^{-1/2}$ for runtime t, consistent with the typical MC error scaling [12]. We used between 10^2 and 10^6 sampling points here. The RNI method, using between 10 and 300 sampling points with $N^g_{\text{quad}} = 400$, resulted in orders of magnitude smaller errors. The exponential error scaling in (28) is not visible here, the asymptotic regime of the method is not yet reached with the used numbers of sampling points.

All in all, the RNI method results in orders of magnitude smaller errors than the Cluster algorithm for a fixed runtime or equivalently, the RNI method needs orders of magnitude less runtime than the Cluster algorithm to arrive at a fixed error estimate, even for a number of sampling points where the RNI error does not yet scale exponentially.

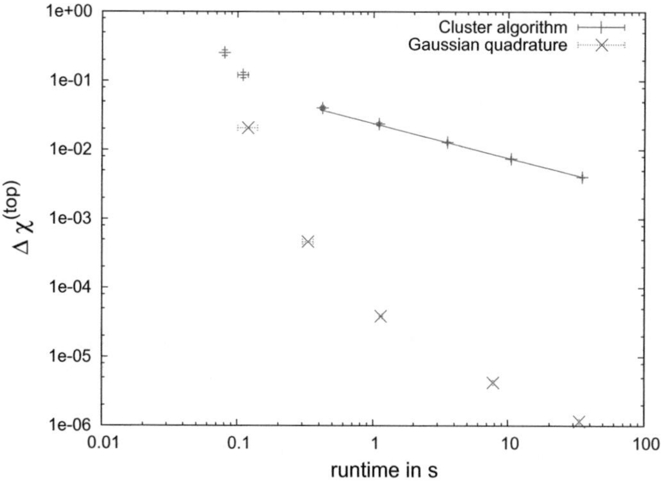

Fig. 4 The runtime to arrive at a given error estimate is orders of magnitudes smaller when using the RNI method with Gauss-Legendre points than using the Cluster MCMC algorithm

3.3 Avoiding the Sign Problem in High-Dimensional Integrals

Section 3.1 shows that the sign problem can be avoided for one-dimensional integrals using symmetric quadrature rules. But what about the sign problem for high-dimensional integrals? A quadrature rule for high-dimensional integrals over compact groups,

$$I(f) = \int_{G^d} f[U]\,\mathrm{d}U, \tag{33}$$

with $\mathrm{d}U = \prod_{i=1}^{d} \mathrm{d}U_i$ is needed that also avoids the sign problem. We combined both already presented methods, the symmetric quadrature rules in Sect. 3.1 and the RNI in Sect. 3.2 to find an efficient quadrature rule $Q(f)$ for $I(f)$ in (33). An alternative attempt to generalize the symmetrized quadrature rules to high-dimensional integrals is discussed in [18].

3.3.1 Combining Recursive Numerical Integration and Symmetric Quadrature Rules

RNI can be used to transform the high-dimensional integral $I(f)$ in (33) into one-dimensional integrals. These one-dimensional integrals can be approximated recursively, using the symmetric quadrature rules. In the following, these steps are described in more detail:

RNI 1. Find the structure, i.e. all f_i, of the integrand

$$f[U] = \prod_{i=1}^{d} f_i(U_{i+1}, U_i), \tag{34}$$

to be able to write the full integral as nested one-dimensional integrals, similar to (24).

RNI 2. Apply symmetric quadrature rules to each one-dimensional integration over U_i. Here is an example how to do this for the innermost integral I_d, integrating over U_d:

Sym 1. Rewrite the integral over U_d into an integral over the products of spheres as done in (6).

Sym 2. Approximate each iterated integral $I_d(g)$ by a product rule of quadratures over spheres parametrising the group U_d to be integrated. Note that the group U_d is parametrised at most as the product of S^1, S^3 and S^5.

3.3.2 Application to Topological Oscillator with Sign Problem

We applied this combined method again to the topological oscillator discussed in Sect. 3.2.3. This time we transformed the variables φ_i to new variables $U_j = e^{i\varphi_j} \in \mathscr{U}(1)$. Additionally, we added a sign problem to the model by using an additional factor $\prod_{j=1}^{d} U_j^{-\theta}$ in the Boltzmann-weight,

$$B[U] = \exp\left(-c\sum_{i=1}^{d} \Re(1 - U_{i+1}U_i^*)\right) \cdot \prod_{j=1}^{d} U_j^{-\theta}, \tag{35}$$

with a new parameter $\theta \in \mathbb{R}$. If this parameter is larger than zero, the sign problem arises and is most severe for $\theta = \pi$.

In this model we computed the plaquette,

$$plaquette = \frac{\int O[U]B[U]\,\mathrm{d}U}{\int B[U]\,\mathrm{d}U}, \tag{36}$$

with

$$O[U] = \frac{1}{d}\Re\left(\sum_{i=1}^{d} U_{i+1}U_i^*\right). \tag{37}$$

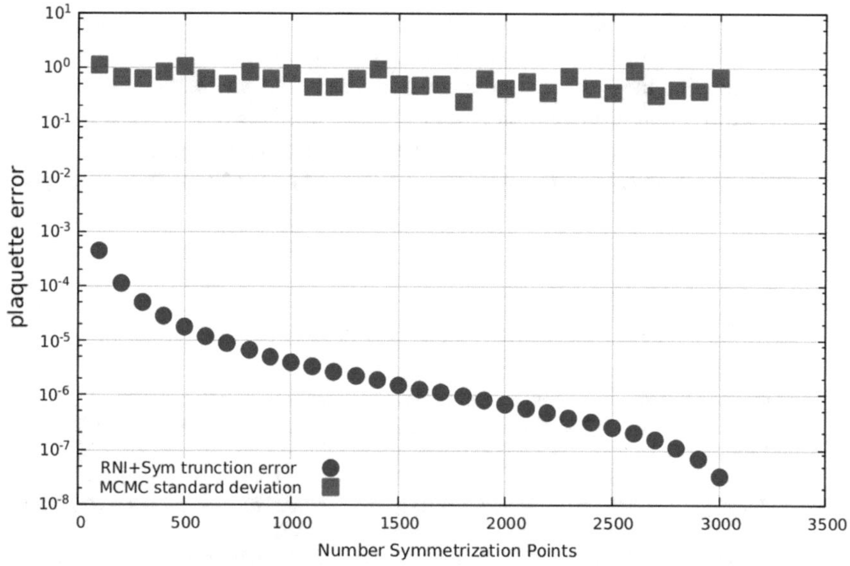

Fig. 5 The combined method avoids the sign problem that exists when using the MC method

For the combined method, we computed both numerator and denominator of (36) separately and divided the values. We used a truncation error, similar to the one given in (32). We compared the method with a standard MC method as used in Sect. 3.1.3. The MC error is computed via the standard deviation.

For $\theta = \pi$ we found that the combined method avoids the sign problem that is visible with the MC computation, compare Fig. 5. It gives orders of magnitude smaller errors that shrink the more symmetrization points are used. Therefore the combination of RNI and symmetric quadrature rules is suitable to avoid the sign problem for high-dimensional integration.

4 Conclusion

In this contribution we have demonstrated that through symmetric quadrature rules exact symmetrization and recursive numerical integration techniques problems in high energy physics can be solved which constitute a major, if not unsurmountable obstacle for standard Markov chain Monte Carlo methods. The examples we have considered here involve only a time lattice and are hence 0+1-dimensional in space-time, where as real physical problem include spatials dimensions of up to 3. We are presently investigating whether the methods we have presented here can be extended to higher, i.e. including also spacial, dimensions. While for the recursive numerical integration technique we have first results which are promising, for the full symmetrization method we were so far not successful.

Also combining the symmetrized quadrature rules with MC methods did not lead to a practically feasible method in higher dimensions. However, we are following a path to combine Quasi Monte Carlo, recursive numerical integration and a full symmetrization to overcome this problem and hope to report about these attempts in the future.

References

1. Ammon, A., Hartung, T., Jansen, K., Leövey, H., Volmer, J.: New polynomially exact integration rules on U(N) and SU(N) (2016). https://inspirehep.net/record/1490030/files/arXiv:1610.01931.pdf
2. Ammon, A., Hartung, T., Jansen, K., Leövey, H., Volmer, J.: Overcoming the sign problem in one-dimensional QCD by new integration rules with polynomial exactness. Phys. Rev. **D94**(11), 114508 (2016). https://doi.org/10.1103/PhysRevD.94.114508
3. Ammon, A., Genz, A., Hartung, T., Jansen, K., Leövey, H., Volmer, J.: Applying recursive numerical integration techniques for solving high dimensional integrals. PoS LATTICE **2016**, 335 (2016)
4. Ammon, A., Genz, A., Hartung, T., Jansen, K., Leövey, H., Volmer, J.: On the efficient numerical solution of lattice systems with low-order couplings. Comput. Phys. Commun. **198**, 71–81 (2016). https://doi.org/10.1016/j.cpc.2015.09.004
5. Ammon, A., Hartung, T., Jansen, K., Leövey, H., Volmer, J.: New polynomially exact integration rules on $U(N)$ and $SU(N)$. PoS LATTICE **2016**, 334 (2016)
6. Bietenholz, W., Gerber, U., Pepe, M., Wiese, U.J.: Topological lattice actions. JHEP **1012**, 020 (2010). https://doi.org/10.1007/JHEP12(2010)020
7. Bilic, N., Demeterfi, K.: One-dimensional QCD with finite chemical potential. Phys. Lett. B **212**, 83–87 (1988). https://doi.org/10.1016/0370-2693(88)91240-3
8. Bloch, J., Bruckmann, F., Wettig, T.: Subset method for one-dimensional QCD. JHEP **10**, 140 (2013). https://doi.org/10.1007/JHEP10(2013)140
9. Bloch, J., Bruckmann, F., Wettig, T.: Sign problem and subsets in one-dimensional QCD. PoS LATTICE **2013**, 194 (2014)
10. DeGrand, T., Detar, C.E.: Lattice Methods for Quantum Chromodynamics, p. 345. World Scientific, New Jersey (2006)
11. Durr, S., et al.: Ab-Initio determination of light hadron masses. Science **322**, 1224–1227 (2008). https://doi.org/10.1126/science.1163233
12. Fristedt, B., Gray, L.: A Modern Approach to Probability Theory. Probability and Its Applications. Birkhäuser, Basel (1997)
13. Gattringer, C., Lang, C.B.: Quantum chromodynamics on the lattice. Lect. Notes Phys. **788**, 1–343 (2010). https://doi.org/10.1007/978-3-642-01850-3
14. Genz, A., Kahaner, D.K.: The numerical evaluation of certain multivariate normal integrals. J. Comput. Appl. Math. **16**(2), 255–258 (1986). https://doi.org/10.1016/0377-0427(86)90100-7. http://www.sciencedirect.com/science/article/pii/0377042786901007
15. Genz, A.: Fully symmetric interpolatory rules for multiple integrals over hyper-spherical surfaces. J. Comput. Appl. Math. **157**, 187–195 (2003). https://doi.org/10.1016/S0377-0427(03)00413-8
16. Hartung, T., Jansen, K.: Integrating gauge fields in the ζ-formulation of Feynman's path integral. arXiv:1902.09926 [math-ph]
17. Hartung, T., Jansen, K.: Quantum computing of zeta-regularized vacuum expectation values. arXiv:1808.06784 [quant-ph]

18. Hartung, T., Jansen, K., Leövey, H., Volmer, J.: Improving monte carlo integration by symmetrization. In: Böttcher, A., Potts, D., Stollmann, P., Wenzel, D. (eds.) The Diversity and Beauty of Applied Operator Theory, pp. 291–317. Springer International Publishing, Cham (2018)
19. Hayter, A.: Recursive integration methodologies with statistical applications. J. Stat. Plan. Inference **136**(7), 2284–2296 (2006). https://doi.org/10.1016/j.jspi.2005.08.024. http://www.sciencedirect.com/science/article/pii/S0378375805002223. In Memory of Dr. Shanti Swarup Gupta
20. Kahaner, D., Moler, C., Nash, S., Forsythe, G.: Numerical Methods and Software. Prentice-Hall Series in Computational Mathematics. Prentice Hall, Upper Saddle River (1989)
21. Peskin, M.E., Schroeder, D.V.: An Introduction to quantum field theory
22. Rothe, H.J.: Lattice gauge theories: an Introduction. World Sci. Lect. Notes Phys. **43**, 1 (1992); World Sci. Lect. Notes Phys. **59**, 1 (1997); World Sci. Lect. Notes Phys. **74**, 1 (2005); World Sci. Lect. Notes Phys. **82**, 1 (2012)
23. Stoer, J., Bartels, R., Gautschi, W., Bulirsch, R., Witzgall, C.: Introduction to Numerical Analysis. Texts in Applied Mathematics. Springer, New York (2013)
24. Volmer, J.L.: New attempts for error reduction in lattice field theory calculations. Ph.D. thesis, Humboldt-Universität zu Berlin, Mathematisch-Naturwissenschaftliche Fakultät (2018). https://doi.org/10.18452/19350
25. Wolff, U.: Collective Monte Carlo updating for spin systems. Phys. Rev. Lett. **62**, 361 (1989). https://doi.org/10.1103/PhysRevLett.62.361

On Hybrid Point Sets Stemming from Halton-Type Hammersley Point Sets and Polynomial Lattice Point Sets

Roswitha Hofer

Abstract In this paper we consider finite hybrid point sets that are the digital analogs to finite hybrid point sets introduced by Kritzer. Kritzer considered hybrid point sets that are a combination of lattice point sets and Hammersley point sets constructed using the ring of integers and the field of rational numbers. In this paper we consider finite hybrid point sets whose components stem from Halton-type Hammersley point sets and lattice point sets which are constructed using the arithmetic of the ring of polynomials and the field of rational functions over a finite field. We present existence results for such finite hybrid point sets with low discrepancy.

Keywords Hybrid point sets · Polynomial lattice point sets · Halton-type sequences · Discrepancy

1 Introduction and Preliminaries

This work is motivated by applications of the theory of uniform distribution modulo one to numerical integration that is based on the Koksma–Hlawka inequality. This inequality states an upper bound for the integration error for a probably very high dimensional function $f : [0, 1]^s \to \mathbb{R}$ when using a simple, equally weighted quadrature rule with N nodes $z_0, z_1, \ldots, z_{N-1}$. More exactly,

$$\left| \int_{[0,1]^s} f(z)dz - \frac{1}{N} \sum_{n=0}^{N-1} f(z_n) \right| \leq V(f) D_N^*(z_n).$$

Here $V(f)$ denotes the variation of f in the sense of Hardy and Krause and $D_N^*(z_n)$ denotes the *star discrepancy* of the node set $z_0, z_1, \ldots, z_{N-1}$ which is defined next.

The star discrepancy D_N^* of a point set $\mathscr{P} = (z_n)_{n=0,1,\ldots,N-1}$ in $[0, 1)^s$ is given by

R. Hofer (✉)
Institute of Financial Mathematics and Applied Number Theory,Johannes Kepler University Linz,Altenbergerstr.69,4040 Linz, Austria
e-mail: roswitha.hofer@jku.at

B. Tuffin and P. L'Ecuyer (eds.), *Monte Carlo and Quasi-Monte Carlo Methods*, Springer Proceedings in Mathematics & Statistics 324,
https://doi.org/10.1007/978-3-030-43465-6_12

$$D_N^*(\mathscr{P}) = D_N^*(z_n) = \sup_J \left| \frac{A(J,N)}{N} - \lambda_s(J) \right|$$

where the supremum is extended over all half-open subintervals J of $[0, 1)^s$ with the left corner in the origin, λ_s denotes the s-dimensional Lebesgue measure, and the counting function $A(J, N)$ stands for

$$\#\{0 \le n < N : z_n \in J\}.$$

We define $\mathrm{Log}(x) := \max(1, \log(x))$ for real numbers $x > 0$. Furthermore we use the Landau symbol $h(N) = O(H(N))$ to express $|h(N)| \le CH(N)$ for all $N \in \mathbb{N}$ with some positive constant C independent of N and a function $H : \mathbb{N} \to \mathbb{R}^+$. If the implied constant C depends on some parameters, then these parameters will appear as a subscript in the Landau symbol. A symbol O without a subscript indicates, if nothing else is written, an absolute implied constant.

So far the best known upper bounds for the star discrepancy of concrete examples of point sets $(z_n)_{0 \le n < N}$ are of the form

$$ND_N^*(z_n) = O(\mathrm{Log}^{s-1} N)$$

where the implied constant might depend on some parameters but is independent of N. Examples of such *low-discrepancy point sets* are Hammersley point sets and (t, m, s)-nets. A slightly weaker discrepancy bound, i.e. $ND_N^*(z_n) = O((\mathrm{Log}^{s-1} N)\, \mathrm{LogLog}\, N)$, holds for good lattice point sets and good polynomial lattice point sets [1, 22].

Numerical integration based on low-discrepancy point sets, is well established as *quasi-Monte Carlo* (qMC) method. The stochastic counterparts of quasi-Monte Carlo methods, namely *Monte Carlo* (MC) methods, work with sequences of pseudorandom numbers. For more details on qMC and MC integration and low-discrepancy point sets we refer to [2, 30].

The potency of qMC methods and MC methods for multidimensional numerical integration depends on the nature and the dimensionality of the integrand. As a general rule of thumb, qMC methods are more effective in low dimensions and Monte Carlo methods work reasonably well in arbitrarily high dimensions. This has led to the idea, first suggested and applied by Spanier [36], of melding the advantages of qMC methods and MC methods by using so-called *hybrid sequences*. The principle here is to sample a relatively small number of dominating variables of the integrand by low-discrepancy sequences and the remaining variables by pseudorandom sequences. Application of hybrid sequences to challenging computational problems can be found in the literature (see e.g. [3, 33, 35, 36]).

In view of the Koksma–Hlawka inequality the analysis of numerical integration methods based on hybrid sequences requires the study of their discrepancy. There are probabilistic results on the discrepancy of hybrid sequences, e.g., in [5, 32, 34]. Niederreiter [25] was the first one who established nontrivial deterministic discrepancy bounds for hybrid sequences, where the qMC components are Halton sequences or Kronecker sequences. Those results where improved, extended, and unified in a series of papers [6, 26–29, 31]. In these and in several other papers, see e.g. [4, 9, 11, 14–18, 20, 23], also hybrid sequences and hybrid point sets made by combining different qMC sequences were treated. The motivation is here to combine the advantages of different qMC point sets and sequences. The challenge is to handle the different structures of the qMC point sets and sequences when studying the discrepancy of such hybrid point sets and hybrid sequences.

In this paper we mention results of Kritzer [20] on hybrid point sets where the components stem from Hammersley point sets on the one hand, and lattice point sets in the sense of Hlawka [10] and Korobov [19] on the other hand.

For the definition of Hammersley point sets we need the radical inverse function $\varphi_b : \mathbb{N}_0 \to [0, 1)$ where b is a natural number greater or equal to 2. To compute $\varphi_b(n)$ represent n in base b, that is $n = n_0 + n_1 b + n_2 b^2 + \cdots$ with $n_i \in \{0, 1, \ldots, b-1\}$, and set

$$\varphi_b(n) = \sum_{i=0}^{\infty} \frac{n_i}{b^{i+1}}.$$

For an s-dimensional Halton sequence $(\boldsymbol{x}_n)_{n\geq 0}$ [7] we choose s pairwise coprime bases $b_1, \ldots, b_s \geq 2$ and set

$$\boldsymbol{x}_n := (\varphi_{b_1}(n), \ldots, \varphi_{b_s}(n)).$$

Now for an $(s+1)$-dimensional Hammersley point set we choose in addition a natural number N and define the point set $(\boldsymbol{y}_n)_{0\leq n<N}$ by

$$\boldsymbol{y}_n := (n/N, \varphi_{b_1}(n), \ldots, \varphi_{b_s}(n)).$$

For a t-dimensional lattice point set $(\boldsymbol{y}_n)_{0\leq n<N}$ choose first a positive integer N and t integers $g_1, \ldots, g_t$. Then set

$$\boldsymbol{y}_n := (\{ng_1/N\}, \ldots, \{ng_t/N\}), \; 0 \leq n < N.$$

If $(g_1, \ldots, g_t)$ are of the specific form $(1, g, \ldots, g^{t-1})$ then we speak of a lattice point set of Korobov type.

Kritzer ensured existence of lattice point sets and lattice point sets of Korobov type as well, such that they can be combined with Hammersley point sets, and the obtained hybrid point sets satisfy low-discrepancy bounds.

[20, Theorem 1] *Let $s, t \in \mathbb{N}$. Let $p_1, \ldots, p_s$ be distinct prime numbers and let N be a prime number that is different from $p_1, \ldots, p_s$. Let $(\boldsymbol{x}_n)_{n\geq 0}$ be the Halton sequence in bases $p_1, \ldots, p_s$. Then there exist generating $g_1, \ldots, g_t \in \{1, \ldots, N-1\}$ such that the point set*

$$\mathscr{S}_N := (n/N, \boldsymbol{x}_n, \boldsymbol{y}_n)_{0\leq n<N}$$

in $[0,1]^{1+s+t}$ with $\boldsymbol{y}_n := (\{ng_1/N\}, \ldots, \{ng_t/N\})$, *satisfies*

$$N D_N^*(\mathscr{S}_N) = O(\mathrm{Log}^{s+t+1} N)$$

with an implied constant independent of N.

[20, Theorem 3] *Let* $s, t \in \mathbb{N}$. *Let* $p_1, \ldots, p_s$ *be distinct prime numbers and let* N *be a prime number that is different from* $p_1, \ldots, p_s$. *Let* $(\boldsymbol{x}_n)_{n \geq 0}$ *be the Halton sequence in bases* $p_1, \ldots, p_s$. *Then there exists a generating* $g \in \{1, \ldots, N-1\}$ *such that the point set*

$$\mathscr{S}_N := (n/N, \boldsymbol{x}_n, \boldsymbol{y}_n)_{0 \leq n < N}$$

in $[0,1]^{1+s+t}$ *with* $\boldsymbol{y}_n := (\{ng/N\}, \ldots, \{ng^t/N\})$, *satisfies*

$$N D_N^*(\mathscr{S}_N) = O(\mathrm{Log}^{s+t+1} N)$$

with an implied constant independent of N.

Kritzer used a slightly different lattice point set of Korobov type by setting $(g_1, \ldots, g_t) = (g, \ldots, g^t)$ instead of $(1, g, \ldots, g^{t-1})$. Note that $g_1 = 1$ won't mix well with the first component, that is n/N.

In the next section we will define the analogs to Hammersley point sets and lattice point sets that are using the arithmetics in the ring of polynomials and the field of rational functions over a finite field instead of the arithmetic in the ring of integers and the field of rational numbers, before we state two theorems that represent analogs to the two theorems of Kritzer.

2 Halton-Type Hammersley Point Sets, Polynomial Lattice Point Sets, and Results on the Star Discrepancy of Their Hybrid Point Sets

Let p be a prime number. Let $\mathbb{F}_p$ be the finite field with p elements. Let $\mathbb{F}_p[X]$ be the ring of polynomials over $\mathbb{F}_p$, $\mathbb{F}_p(X)$ the field of rational functions over $\mathbb{F}_p$, and $\mathbb{F}_p((X^{-1}))$ the field of formal Laurent series over $\mathbb{F}_p$.

Let $s \in \mathbb{N}$, and let $b_1(X), \ldots, b_s(X)$ be distinct monic pairwise coprime non-constant polynomials over $\mathbb{F}_p$ with degrees $e_1, \ldots, e_s$. We define the Halton type sequence $(\boldsymbol{x}_n)_{n \geq 0}$ in bases $(b_1(X), \ldots, b_s(X))$ by

$$\boldsymbol{x}_n := (\varphi_{b_1(X)}(n(X)), \ldots, \varphi_{b_s(X)}(n(X))),$$

that is based on a construction principle of sequences which are introduced in [12]. Here $\varphi_{b(X)}(n(X))$ is the radical inverse function in the ring $\mathbb{F}_p[X]$ defined as follows.

Expand n in base p of the form $n = n_0 + n_1 p + n_2 p^2 + \cdots$ with $n_i \in \{0, 1, \ldots, p-1\}$ and associate the polynomial $n(X) = n_0 X^0 + n_1 X + n_2 X^2 + \cdots$ where we do not distinguish between the set $\mathbb{F}_p$ and the set $\{0, 1, \ldots, p-1\}$. Now expand $n(X)$ in base $b(X)$ with $\deg(b(X)) = e \geq 1$ as

$$n(X) = \rho_0(X) b^0(X) + \rho_1(X) b^1(X) + \rho_2(X) b^2(X) + \cdots$$

with $\deg(\rho_j(X)) < e$ for $j \in \mathbb{N}_0$. Finally, define a bijection

$$\sigma : \{\rho(X) \in \mathbb{F}_p[X] : \deg(\rho(X)) < e\} \to \{0, 1, \ldots, p^e - 1\}$$

and set

$$\varphi_{b(X)}(n(X)) := \sum_{j=0}^{\infty} \frac{\sigma(\rho_j(X))}{p^{e(j+1)}}.$$

To avoid technical difficulties we restrict to bijections σ that are mapping 0 to 0.

Let $m \in \mathbb{N}$. Using the Halton type sequence in bases $b_1(X), \ldots, b_s(X)$ we can define an $(s+1)$-dimensional Halton-type Hammersley point set of $N = p^m$ points by using the nth point of the form

$$\left(\frac{n}{N}, \boldsymbol{x}_n\right)$$

where $n = 0, 1, \ldots, N-1$.

For the definition of polynomial lattice point sets we identify $\mathbb{F}_p$ again with the set $\{0, 1, \ldots, p-1\}$.

Let $t, m \in \mathbb{N}$. Let $p(X) \in \mathbb{F}_p[X]$ be irreducible, monic, and with degree m. Furthermore, let $\boldsymbol{q}(X) = (q_1(X), \ldots, q_t(X)) \in \mathbb{F}_p^t[X]$. The ith component $y_n^{(i)}$ of the nth point $\boldsymbol{y}_n$ is computed as follows. Expand $\{n(X) q_i(X)/p(X)\}$ in its formal Laurent series

$$\left\{\frac{n(X) q_i(X)}{p(X)}\right\} = \sum_{j=1}^{\infty} u_j X^{-j}$$

and evaluate it by exchanging X with p and summing up to the index m. Hence

$$y_n^{(i)} = \sum_{j=1}^{m} u_j p^{-j}.$$

We can also compute the ith component $y_n^{(i)}$ of the nth point $\boldsymbol{y}_n$ by using the base p representation of $n = \sum_{j=0}^{\infty} n_j p^j$ and a generating matrix $C_i \in \mathbb{F}_p^{m \times m}$. Let $\sum_{j=1}^{\infty} a_j X^{-j}$ be the formal Laurent series of $\left\{\frac{q_i(X)}{p(X)}\right\}$. Define

$$C_i := \begin{pmatrix} a_1 & a_2 & \cdots & a_m \\ a_2 & a_3 & \cdots & a_{m+1} \\ \vdots & \vdots & \cdots & \vdots \\ a_m & a_{m+1} & \cdots & a_{2m-1} \end{pmatrix}.$$

Compute $C_i \cdot (n_0, n_1, \ldots, n_{m-1})^T = (u_1, u_2, \ldots, u_m)^T \in \mathbb{F}_p^m$ and set

$$y_n^{(i)} = \sum_{j=1}^{m} u_j p^{-j}.$$

Finally letting n take values in the set $\{0, 1, \ldots, p^m - 1\}$ we obtain the polynomial lattice point set $\mathscr{P}(\boldsymbol{q}(X), p(X)) = \{\boldsymbol{y}_0, \boldsymbol{y}_1, \ldots, \boldsymbol{y}_{p^m-1}\} \subset [0, 1]^t$.

If we choose $\boldsymbol{q}(X)$ of the specific form $(g(X), g^2(X), \ldots, g^t(X))$ with $g(X) \in \mathbb{F}_p[X]$, we will speak of a polynomial lattice point set of Korobov type abbreviated by $\mathscr{K}(t, g(X), p(X))$.

In the following two theorems we ensure existence of polynomial lattice point sets as well as polynomial lattice point sets of Korobov type, such that they can be combined with a Halton-type Hammersley point set and result in hybrid point sets satisfying low-discrepancy bounds. Theorem 1 represents an analog to [20, Theorem 1] and Theorem 2 is the pendant to [20, Theorem 3].

Theorem 1 *Let $s, t \in \mathbb{N}$ and p be a prime number, let $b_1(X), \ldots, b_s(X)$ be monic pairwise coprime nonconstant polynomials in $\mathbb{F}_p[X]$ and $(\boldsymbol{x}_n)_{n \geq 0}$ be a Halton type sequence in bases $(b_1(X), \ldots, b_s(X))$. Furthermore, let $p(X)$ be a monic, irreducible polynomial in $\mathbb{F}_p[X]$ of degree m coprime with all base polynomials of the Halton-type sequence, and set $N = p^m$. Then there exists a t-tuple of polynomials $\boldsymbol{q}(X) \in \mathbb{F}_p^t[X]$ with degrees less than m such that the star discrepancy D_N^* of the point set $(n/p^m, \boldsymbol{x}_n, \boldsymbol{y}_n)_{0 \leq n < p^m} \in [0, 1]^{s+t+1}$ satisfies*

$$N D_N^*(n/p^m, \boldsymbol{x}_n, \boldsymbol{y}_n) = O_{b_1(X), \ldots, b_s(X), p, t}(\mathrm{Log}^{s+t+1} N).$$

Here $\boldsymbol{y}_n$ is the nth point of the polynomial lattice point set $\mathscr{P}(\boldsymbol{q}(X), p(X))$.

Theorem 2 *Let $s, t \in \mathbb{N}$ and p be a prime number, let $b_1(X), \ldots, b_s(X)$ be monic pairwise coprime nonconstant polynomials in $\mathbb{F}_p[X]$ and $(\boldsymbol{x}_n)_{n \geq 0}$ be a Halton type sequence in bases $(b_1(X), \ldots, b_s(X))$. Furthermore, let $p(X)$ be a monic, irreducible polynomial in $\mathbb{F}_p[X]$ of degree m, coprime with all base polynomials of the Halton-type sequence, and set $N = p^m$. Then there exists a polynomial $g(X)$ over $\mathbb{F}_p$ with degree less than m such that the star discrepancy D_N^* of the point set $(n/p^m, \boldsymbol{x}_n, \boldsymbol{y}_n)_{0 \leq n < p^m} \in [0, 1]^{s+t+1}$ satisfies*

$$N D_N^*(n/p^m, \boldsymbol{x}_n, \boldsymbol{y}_n) = O_{b_1(X), \ldots, b_s(X), p, t}(\mathrm{Log}^{s+t+1} N).$$

Here $\mathbf{y}_n$ is the nth point of the polynomial lattice point set of Korobov type $\mathcal{K}(t, g(X), p(X))$.

The rest of the paper is organized as follows. Section 3 collects auxiliary results needed for the proofs of Theorems 1 and 2, which are formulated in Sects. 4 and 5. Finally Sect. 6 suggests two problems for future research.

3 Auxiliary Results

From the construction of the Halton-type sequence we immediately obtain the following lemma.

Lemma 1 *Let $(\mathbf{x}_n)_{n\geq 0}$ be a Halton-type sequence in pairwise coprime bases $b_1(X), \ldots, b_s(X)$ with degrees $e_1, \ldots, e_s$, and let*

$$I := \prod_{i=1}^{s} \left[\frac{a_i}{p^{e_i l_i}}, \frac{a_i+1}{p^{e_i l_i}} \right)$$

with $l_i \geq 0$, $0 \leq a_i < p^{e_i l_i}$ for $i = 1, \ldots, s$. Then $\mathbf{x}_n \in I$ if and only if

$$n(X) \equiv R(X) \quad \left(\operatorname{mod} \prod_{i=1}^{s} b_i^{l_i}(X) \right)$$

where $R(X)$ depends on the a_i and $\deg(R(X)) < \sum_{i=1}^{s} e_i l_i$. Furthermore there is a one-to-one correspondence between all possible choices for $a_1, \ldots, a_s$ and $R(X)$.

Let $e(x) := \exp(2\pi\sqrt{-1}x)$ for $x \in \mathbb{R}$. We define the $\mathbf{k}$th Walsh function $\mathrm{wal}_{\mathbf{k}}$ in base p on $[0, 1)^t$ as follows. Let $\Phi_0 : \{0, 1, \ldots, p-1\} \to \{z \in \mathbb{C} : |z| = 1\}$, $a \mapsto e(a/p)$. Note that for a given $b \in \{0, 1, \ldots, p-1\}$ we have that $\sum_{a=0}^{p-1} (\Phi_0(a))^b$ equals p if $b = 0$ and 0 else.

The kth Walsh function wal_k, for $k \geq 0$, to the base p is defined by

$$\mathrm{wal}_k(x) := \prod_{j=0}^{\infty} (\Phi_0(x_j))^{k_j}$$

where $x = \sum_{j=0}^{\infty} x_j p^{-j-1}$ is the unique base p expansion of $x \in [0, 1)$ with infinitely many $x_j \neq p-1$ and $k = \sum_{j=0}^{\infty} k_j p^j$ is the base p expansion of $k \in \mathbb{N}_0$. For vectors $\mathbf{k} = (k_1, \ldots, k_t) \in \mathbb{N}_0^t$ and $\mathbf{x} = (x_1, \ldots, x_t) \in [0, 1)^t$ the Walsh function $\mathrm{wal}_{\mathbf{k}}$ on $[0, 1)^t$ denotes

$$\mathrm{wal}_{\mathbf{k}}(\mathbf{x}) := \prod_{i=1}^{t} \mathrm{wal}_{k_i}(x_i).$$

Lemma 2 ([8, Theorem 1]) *Let $\mathscr{P} = \{\boldsymbol{y}_0, \boldsymbol{y}_1, \dots, \boldsymbol{y}_{N-1}\}$ be a finite point set in $[0, 1)^t$ with $\boldsymbol{y}_n$ of the form $\boldsymbol{y}_n = \{\boldsymbol{w}_n/M\}$, $\boldsymbol{w}_n \in \mathbb{Z}^t$. Suppose that $M = p^m$, where m is a positive integer. Then the following estimate holds:*

$$D_N^*(\boldsymbol{y}_n) \leq \underbrace{1 - (1 - 1/M)^t}_{\leq t/M} + \sum_{\boldsymbol{k} \in \Delta_m^*} \rho_{\text{wal}}(\boldsymbol{k}) |S_N(\text{wal}_{\boldsymbol{k}})|,$$

where

$$S_N(\text{wal}_{\boldsymbol{k}}) := \frac{1}{N} \sum_{n=0}^{N-1} \text{wal}_{\boldsymbol{k}}(\boldsymbol{y}_n),$$

$$\Delta_m := \{\boldsymbol{k} \in \mathbb{Z}^t : 0 \leq k_i < p^m, \ \textit{for } i = 1, \dots, t\},$$

$\Delta_m^* = \Delta_m \setminus \{\boldsymbol{0}\}$, and

$$\rho_{\text{wal}}(\boldsymbol{k}) := \prod_{i=1}^{t} \rho_{\text{wal}}(k_i)$$

with

$$\rho_{\text{wal}}(k) = \begin{cases} 1 & \textit{if } k = 0, \\ \frac{1}{p^{g+1} \sin \pi k_g / p} & \textit{if } p^g \leq k < p^{g+1}, \ g \geq 0 \end{cases},$$

where k_g is the gth digit of k in the base p expansion of k.

Lemma 3 ([2, Lemma 10.22]) *Let $t, \ m \in \mathbb{N}$. For any prime number p, we have*

$$\sum_{\boldsymbol{k} \in \Delta_m} \rho_{\text{wal}}(\boldsymbol{k}) = \left(1 + m \frac{p^2 - 1}{3p}\right)^t.$$

For the statement of the next auxiliary result we define the following magnitudes:

$$G_{p,m} = \{f(X) \in \mathbb{F}_p[X] : \deg(f(X)) < m\} \quad \text{and} \quad G_{p,m}^* = G_{p,m} \setminus \{0\}.$$

Furthermore for the rational function $p(X)/q(X)$ with $p(X)$ and $q(X)$ in $\mathbb{F}_p[X] \setminus \{0\}$ we define the degree evaluation ν by

$$\nu(p(X)/q(X)) := \deg(p(X)) - \deg(q(X))$$

and we set $\nu(0) = -\infty$.

Lemma 4 *Let $p(X)$ be a monic irreducible polynomial in $\mathbb{F}_p[X]$ with degree m. Let $u \in \mathbb{N}_0$ such that $u \leq m$. Then*

$$\#\{a(X) \in G_{p,m}^* : \nu(a(X)/p(X)) < -u\} \leq p^{m-u} - 1.$$

Proof The restriction $\nu(a(X)/p(X)) < -u$ means $\deg(a(X)) < \deg(p(X)) - u = m - u$ and the result follows. □

With a number $k \in \{0, 1, \ldots, p^m - 1\}$ we associate the polynomial $k(X) = \sum_{i=0}^{m-1} k_i X^i$ where the coefficients are determined by the base p representation of $k = \sum_{i=0}^{m-1} k_i p^i$. For a tuple $\boldsymbol{k} \in \Delta_m$ we associate a polynomial with each component and write $\boldsymbol{k}(X)$ for the t-tuple of polynomials.

Lemma 5 ([13, Lemma 1]) *Let $e \in \mathbb{N}_0$, $B(X), R(X) \in \mathbb{F}_p[X]$ with $\deg(R(X)) < \deg(B(X)) = e$, and let $B(X)$ be monic. Furthermore, let $u \in \mathbb{N}$ and $K \in \mathbb{N}_0$. Let $n = Kp^{u+e}, Kp^{u+e} + 1, \ldots, (K+1)p^{u+e} - 1$. We regard all associated polynomials $n(X)$ that satisfy $n(X) \equiv R(X) \pmod{B(X)}$. Then they are of the form*

$$n(X) = k(X)B(X) + R(X)$$

with $k(X)$ out of the set

$$k(X) = r(X) + X^u C(X)$$

with a fixed $C(X) \in \mathbb{F}_p[X]$ and $r(X)$ ranges over all polynomials of degree less than u.

Lemma 6 ([24, Theorem 2.6]) *For $1 \leq i \leq k$ let w_i be a point set of N_i elements in $[0, 1]^s$. Let w be the superposition of $w_1, \ldots, w_k$, that is a point set of $N = N_1 + \cdots + N_k$ points. Then*

$$N D_N^*(w) \leq \sum_{i=1}^{k} N_i D_{N_i}^*(w_i).$$

4 Proof of Theorem 1

In this section we investigate the distribution of the point set $(z_n)_{0 \leq n < p^m} \in [0, 1)^{1+s+t}$ with $m, s, t \in \mathbb{N}$ and

$$z_n := (n/p^m, \boldsymbol{x}_n, \boldsymbol{y}_n)$$

where $(\boldsymbol{y}_n)_{0 \leq n < p^m}$ is a polynomial lattice point set $\mathscr{P}(\boldsymbol{q}(X), p(X))$ in $[0, 1)^t$ with $p(X)$ monic, irreducible, and with degree m, and where $(\boldsymbol{x}_n)_{n \geq 0}$ is a Halton-type sequence in bases $(b_1(X), \ldots, b_s(X))$, all monic, pairwise coprime, coprime with $p(X)$, and with positive degrees $e_1, \ldots, e_s$.

We set $N = p^m$. Using a well-known result in discrepancy theory (see, e.g., [30, Lemma 3.7]), we have

$$N D_N^*(z_n) \leq \max_{1 \leq \tilde{N} \leq N} \tilde{N} D_{\tilde{N}}^*((\boldsymbol{x}_n, \boldsymbol{y}_n)) + 1.$$

Let $\tilde{N} \in \{1, \ldots, N\}$ be fixed. We expand $\tilde{N}$ in base p, $\tilde{N} = N_0 + N_1 p + \cdots + N_r p^r$ with $N_i \in \{0, 1, \ldots, p-1\}$ and $r \leq m$. For $u = 0, \ldots, r$ and $v = 1, \ldots, N_u$ we define the point set

$$w_{u,v} := \{(\boldsymbol{x}_n, \boldsymbol{y}_n) : n \in \mathbb{N}_0,\ (v-1)p^u + \cdots + N_r p^r \leq n < vp^u + \cdots + N_r p^r\}.$$

Then $|w_{u,v}| = p^u$ and

$$\{0, 1, \ldots, \tilde{N} - 1\}$$

is obtained by the disjoint union

$$\bigcup_{u=0}^{r} \bigcup_{v=1}^{N_u} \{n \in \mathbb{N}_0 : (v-1)p^u + \cdots + N_r p^r \leq n < vp^u + \cdots + N_r p^r\}$$

of at most $pm = p \log_p N$ sets.

We apply Lemma 6, which results in one $\operatorname{Log} N$ factor in Theorem 1 with a constant depending on p. Next we concentrate on

$$p^u D^*_{p^u}(w_{u,v}).$$

We define

$$f_i := \left\lceil \frac{u}{e_i} \right\rceil$$

for $1 \leq i \leq s$.

The first aim in the proof is to compute or estimate the counting function $A(J, p^u)$ relative to the pointset $w_{u,v}$, where $J \subseteq [0, 1)^{s+t}$ is an interval of the form

$$J = \prod_{i=1}^{s} [0, v_i p^{-e_i f_i}) \times \prod_{j=1}^{t} [0, \beta_j) \tag{1}$$

with $v_1, \ldots, v_s \in \mathbb{Z}$, $1 \leq v_i \leq p^{e_i f_i}$ for $1 \leq i \leq s$, and $0 < \beta_j \leq 1$ for $1 \leq j \leq t$.

The crucial step is to exploit special properties of the Halton type sequence. By Lemma 1, for any integer $n \geq 0$ we have

$$(\varphi_{b_1(X)}(n(X)), \ldots, \varphi_{b_s(X)}(n(X))) \in \prod_{i=1}^{s} [0, v_i p^{-e_i f_i}) \text{ if and only if } n(X) \in \bigcup_{k=1}^{M} \mathscr{R}_k,$$

where

$$1 \leq M \leq p^{e_1} \cdots p^{e_s} f_1 \cdots f_s = O_{p, b_1(X), \ldots, b_s(X)}(\operatorname{Log}^s N).$$

Each $\mathscr{R}_k$ is a residue class in $\mathbb{F}_p[X]$, and $\mathscr{R}_1, \ldots, \mathscr{R}_M$ are (pairwise) disjoint. The moduli $B_k(X)$ of the residue classes $\mathscr{R}_k$ are of the form $b_1(X)^{j_1} \cdots b_s(X)^{j_s}$ with integers $1 \leq j_i \leq f_i$ for $1 \leq i \leq s$ and the residues $R_k(X)$ satisfy $\deg(R_k(X)) < \deg(B_k(X))$ for $1 \leq k \leq M$. The sets $\mathscr{R}_1, \ldots, \mathscr{R}_M$ depend only on $b_1(X), \ldots, b_s(X), v_1, \ldots, v_s, f_1, \ldots, f_s$ and are thus independent of n. Furthermore, one can easily prove for the Lebesgue measure of $\prod_{i=1}^{s}[0, v_i p^{-e_i f_i})$ that

$$\begin{aligned}
\lambda_s\left(\prod_{i=1}^{s}[0, v_i p^{-e_i f_i})\right) &= \prod_{i=1}^{s} v_i p^{-e_i f_i} \\
&= \lim_{N\to\infty} \#\{0 \leq n < N : (\varphi_{b_1(X)}(n(X)), \ldots, \varphi_{b_s(X)}(n(X))) \in \prod_{i=1}^{s}[0, v_i p^{-e_i f_i})\} \\
&= \lim_{N\to\infty} \#\{0 \leq n < N : n(X) \in \bigcup_{k=1}^{M} \mathscr{R}_k\} \\
&= \sum_{k=1}^{M} \lim_{N\to\infty} \#\{0 \leq n < N : n(X) \equiv R_k(X) \pmod{B_k(X)}\} \\
&= \sum_{k=1}^{M} \frac{1}{p^{\deg(B_k(X))}},
\end{aligned}$$

by applying the uniform distribution of the Halton-type sequence and the disjointness of $\mathscr{R}_1, \ldots, \mathscr{R}_M$.

Now we split up the counting function $A(J, p^u)$ into M parts as follows: $A(J, p^u) = \sum_{k=1}^{M} S_k$, where

$$\begin{aligned}
S_k = \#\Big\{&(v-1)p^u + \cdots + N_r p^r \leq n < v p^u + \cdots + N_r p^r : n(X) \equiv R_k(X) \pmod{B_k(X)} \\
&\text{and } \boldsymbol{y}_n \in \prod_{j=1}^{t}[0, \beta_j)\Big\}
\end{aligned}$$

for $1 \leq k \leq M$. Then

$$|A(J, p^u) - p^u \lambda_{s+t}(J)| \leq \sum_{k=1}^{M} \underbrace{\left| S_k - p^u \frac{1}{p^{\deg(B_k(X))}} \prod_{j=1}^{t} \beta_j \right|}_{=:\delta_k}.$$

This summation over k then results in s Log N factors in Theorem 1 with a constant depending on $p, b_1(X), \ldots, b_s(X)$.

We fix k with $1 \leq k \leq M$ for the moment. Note that if $p^u < p^{\deg(B_k(X))}$, then $S_k = 0$ or 1, and so in this case $\delta_k \leq 1$. Assume now that $p^u \geq p^{\deg(B_k(X))}$. We define the set $\mathscr{L}_k$ as

$$\left\{(v-1)p^u + \cdots + N_r p^r \leq n < v p^u + \cdots + N_r p^r : n(X) \equiv R_k(X) \pmod{B_k(X)}\right\}.$$

Then by Lemma 5, we know $|\mathcal{L}_k| = p^{u-\deg(B_k(X))} =: L_k$. We define the point set

$$\mathcal{P}_k = \{\boldsymbol{y}_n : n \in \mathcal{L}_k\}.$$

Then

$$\delta_k \leq L_k D^*_{L_k}(\mathcal{P}_k).$$

We summarize

$$|A(J, p^u) - p^u \lambda_{s+t}(J)| \leq O(M) + \sum_{\substack{k=1 \\ \deg(B_k(X)) \leq u}}^{M} L_k D^*_{L_k}(\mathcal{P}_k).$$

An arbitrary interval $I \subseteq [0, 1)^{s+t}$ of the form

$$I = \prod_{i=1}^{s} [0, \alpha_i) \times \prod_{j=1}^{t} [0, \beta_j) \tag{2}$$

with $0 < \alpha_i \leq 1$ for $1 \leq i \leq s$ and $0 < \beta_j \leq 1$ for $1 \leq j \leq t$ can be approximated from below and above by an interval J of the form (1), by taking the nearest fraction to the left and to the right, respectively, of α_i of the form $v_i p^{-e_i f_i}$ with $v_i \in \mathbb{Z}$. We easily get

$$\left|A(I, p^u) - p^u \lambda_{s+t}(I)\right| \leq \underbrace{p^u \sum_{i=1}^{s} p^{-e_i f_i}}_{\leq s} + |A(J, N) - N \lambda_{s+t}(J)|.$$

We ensure the existence of a "good" generating vector $\boldsymbol{q}(X)$ of polynomials by an averaging argument. So the core of the proof is the study of the average

$$\frac{1}{|(G^*_{p,m})|^t} \sum_{\boldsymbol{q}(X) \in (G^*_{p,m})^t} L_k D^*_{L_k}(\mathcal{P}_k), \tag{3}$$

after exchanging the order of summation.

Note that $B_k(X)$ is monic and coprime with $p(X)$, and $\deg(B_k(X)) \leq u$. In the following we set $d := u - \deg(B_k(X))$ and we will omit the index k.

First we compute the subset $\mathcal{P}$ of the polynomial lattice point set $\mathcal{P}(\boldsymbol{q}(X), p(X))$ using the congruence $n(X) \equiv R(X) \pmod{B(X)}$ and bearing in mind Lemma 5.

Let $l \in \{0, 1, \ldots, p^d - 1\}$. Note that

$$\left\{\frac{\Big(\big(l(X)+X^dC(X)\big)B(X)+R(X)\Big)q_i(X)}{p(X)}\right\}$$
$$=\left\{\frac{l(X)B(X)q_i(X)}{p(X)}\right\}+\left\{\frac{(X^dC(X)B(X)+R(X))q_i(X)}{p(X)}\right\}.$$

Let $\sum_{j=1}^{\infty} r_j^{(i)} X^{-j}$ be the formal Laurent series of $\left\{\frac{(X^dC(X)B(X)+R(X))q_i(X)}{p(X)}\right\}$ and $\sum_{j=1}^{\infty} a_j^{(i)} X^{-j}$ be the Laurent series of $\left\{\frac{B(X)q_i(X)}{p(X)}\right\}$. We define

$$C_{i,d}=\begin{pmatrix} a_1^{(i)} & \cdots & a_d^{(i)} \\ \vdots & \ddots & \vdots \\ a_d^{(i)} & \cdots & a_{2d-1}^{(i)} \\ \vdots & \ddots & \vdots \\ a_m^{(i)} & \cdots & a_{m+d-1}^{(i)} \end{pmatrix},$$

then compute

$$C_{i,d}\cdot\underbrace{\begin{pmatrix} l_0 \\ l_1 \\ \vdots \\ l_{d-1}\end{pmatrix}}_{=:\boldsymbol{l}}+\begin{pmatrix} r_1^{(i)} \\ \vdots \\ r_d^{(i)} \\ \vdots \\ r_m^{(i)}\end{pmatrix}=\begin{pmatrix} y_{l,1}^{(i)} \\ \vdots \\ y_{l,d}^{(i)} \\ \vdots \\ y_{l,m}^{(i)}\end{pmatrix}\in\mathbb{F}_p^m,$$

and set

$$y_{n_l}^{(i)}=\sum_{j-1}^{m} y_{l,j}^{(i)} p^{-j}.$$

Finally, letting l take the values between 0 and p^d-1 we obtain the subset $\mathscr{P}$.

We apply Lemma 2 to $LD_L^*(\mathscr{P})$ and obtain

$$\frac{1}{|(G_{p,m}^*)|^t}\sum_{\boldsymbol{q}(X)\in(G_{p,m}^*)^t} LD_L^*(\mathscr{P})\leq$$
$$\frac{tp^d}{p^m}+\frac{1}{|(G_{p,m}^*)|^t}\sum_{\boldsymbol{q}(X)\in(G_{p,m}^*)^t}\sum_{\boldsymbol{k}\in\Delta_m^*}\rho_{\mathrm{wal}}(\boldsymbol{k})\left|\sum_{l=0}^{p^d-1}\mathrm{wal}_{\boldsymbol{k}}(\boldsymbol{y}_{n_l})\right|.$$

We concentrate on

$$\left|\sum_{l=0}^{p^d-1} \mathrm{wal}_{\boldsymbol{k}}(\boldsymbol{y}_{n_l})\right| = \left|\sum_{l=0}^{p^d-1} e\left(\sum_{i=1}^{t}\sum_{j=1}^{m} y_{l,j}^{(i)} k_{j-1}^{(i)}\right)\right|,$$

where we expanded $k_i = \sum_{j=0}^{m-1} k_j^{(i)} p^j$ in base p. We abbreviate the jth row of $C_{i,d}$ to the row vector $\boldsymbol{c}_j^{(i)}$ and remember that

$$y_{l,j}^{(i)} = r_j^{(i)} + \boldsymbol{c}_j^{(i)} \cdot \boldsymbol{l} \pmod{p}.$$

Hence

$$\left|\sum_{l=0}^{p^d-1} \mathrm{wal}_{\boldsymbol{k}}(\boldsymbol{y}_{n_l})\right| = \left|\sum_{l=0}^{p^d-1} e\left(\left(\sum_{i=1}^{t}\sum_{j=1}^{m} k_{j-1}^{(i)} \boldsymbol{c}_j^{(i)}\right) \cdot \boldsymbol{l}\right)\right|$$
$$= \begin{cases} p^d & \text{if } C_{1,d}^T \mathrm{k}_1 + \cdots + C_{t,d}^T \mathrm{k}_t = \mathbf{0} \in \mathbb{F}_p^d \\ 0 & \text{else.} \end{cases}$$

Here k_i denotes the m-dimensional column vector $(k_0^{(i)}, \ldots, k_{m-1}^{(i)})^T$ built up by the base p digits of the ith component of $\boldsymbol{k}$.

Note that

$$C_{1,d}^T \mathrm{k}_1 + \cdots + C_{t,d}^T \mathrm{k}_t = \mathbf{0} \in \mathbb{F}_p^d,$$

can be reformulated as

$$\sum_{i=1}^{t} \begin{pmatrix} a_1^{(i)} & \cdots & a_m^{(i)} \\ \vdots & \ddots & \vdots \\ a_d^{(i)} & \cdots & a_{m+d-1}^{(i)} \end{pmatrix} \cdot \begin{pmatrix} k_0^{(i)} \\ \vdots \\ k_{m-1}^{(i)} \end{pmatrix} = \mathbf{0} \in \mathbb{F}_p^d.$$

Following the argumentation of [2, Proof of Lemma 10.6] we end up with

$$\frac{\boldsymbol{k}(X)B(X)\boldsymbol{q}(X)}{p(X)} = g + H$$

with $g \in \mathbb{F}_p[X]$ and $H \in \mathbb{F}_p((X^{-1}))$ of the form $\sum_{j=d+1}^{\infty} h_j X^{-j}$ which is equivalent to

$$\nu\left(\left\{\frac{\boldsymbol{k}(X)B(X)\boldsymbol{q}(X)}{p(X)}\right\}\right) < -d.$$

We define

$$\mathscr{D}'_{\boldsymbol{q},p,B} = \{\boldsymbol{k} \in \{0, 1, \ldots, p^m - 1\}^t : \nu\left(\left\{\frac{\boldsymbol{k}(X) \cdot B(X) \cdot \boldsymbol{q}(X)}{p(X)}\right\}\right) < -d\} \setminus \{\mathbf{0}\}.$$

and its subset

$$\mathscr{D}'_{\boldsymbol{q},p} = \{\boldsymbol{k} \in \{0,1,\dots,p^m-1\}^t : \boldsymbol{k}(X)\cdot\boldsymbol{q}(X) \equiv 0 \pmod{p(X)}\} \setminus \{\boldsymbol{0}\}.$$

Using the above considerations we obtain

$$\frac{1}{|(G^*_{p,m})|^t} \sum_{\boldsymbol{q}(X)\in(G^*_{p,m})^t} \sum_{\boldsymbol{k}\in\Delta^*_m} \rho_{\mathrm{wal}}(\boldsymbol{k}) \left| \sum_{l=0}^{p^d-1} \mathrm{wal}_{\boldsymbol{k}}(\boldsymbol{y}_{n_l}) \right| = \sum_{\boldsymbol{k}\in\Delta^*_m} \rho_{\mathrm{wal}}(\boldsymbol{k}) \frac{1}{|(G^*_{p,m})|^t} \sum_{\substack{\boldsymbol{q}(X)\in(G^*_{p,m})^t \\ \boldsymbol{k}\in\mathscr{D}'_{\boldsymbol{q},p,B}}} p^d.$$

Altogether we have to compute for $\boldsymbol{k} \in \Delta^*_m$ the number

$$\#\{\boldsymbol{q}(X) \in (G^*_{p,m})^t : \boldsymbol{k} \in \mathscr{D}'_{\boldsymbol{q},p,B}\} = \underbrace{\#\{\boldsymbol{q}(X) \in (G^*_{p,m})^t : \boldsymbol{k} \in \mathscr{D}'_{\boldsymbol{q},p}\}}_{=:K_1} + \underbrace{\#\{\boldsymbol{q}(X) \in (G^*_{p,m})^t : \boldsymbol{k} \in \mathscr{D}'_{\boldsymbol{q},p,B} \setminus \mathscr{D}'_{\boldsymbol{q},p}\}}_{=:K_2}.$$

The easy part is to compute K_1 which equals $(p^m-1)^{t-1}$ (confer, e.g., [2, Proof of Theorem 10.21]).

We now concentrate on K_2. Let τ be maximal such that $k_\tau \neq 0$. We denote by $\boldsymbol{b}^{(i)}$ the projection of $\boldsymbol{b}$ onto the first i components of $\boldsymbol{b}$. We define the following two conditions on $\boldsymbol{q}$ depending on τ,

$$\nu(\{(\boldsymbol{k}^{(\tau-1)}(X)B(X)\boldsymbol{q}^{(\tau-1)}(X) + k_\tau(X)B(X)q_\tau(X))/p(X)\}) < -d \tag{4}$$

and

$$\boldsymbol{k}^{(\tau-1)}(X)\cdot\boldsymbol{q}^{(\tau-1)}(X) \not\equiv -k_\tau(X)q_\tau(X) \pmod{p(X)} \tag{5}$$

Then

$$\frac{K_2}{(p^m-1)^{t-\tau}} = \#\{\boldsymbol{q}^{(\tau)}(X) \in (G^*_{p,m})^\tau : (4) \text{ and } (5)\} \le \sum_{\boldsymbol{q}^{(\tau-1)}(X)\in(G^*_{p,m})^{\tau-1}} \#\{q_\tau(X) \in G_{p,m} : (4) \text{ and } (5)\}.$$

Since $p(X)$ is irreducible there is exactly one $a(X) \in G_{p,m}$ such that

$$\boldsymbol{k}^{(\tau-1)}(X)\cdot\boldsymbol{q}^{(\tau-1)}(X) \equiv -k_\tau(X)a(X) \pmod{p(X)}.$$

Thus

$$\frac{K_2}{(p^m-1)^{t-\tau}} \le \sum_{\boldsymbol{q}^{(\tau-1)}(X)\in (G^*_{p,m})^{\tau-1}} \#\{q_\tau(X)\in G_{p,m}\setminus\{a(X)\} : (4)\}.$$

Now as $q_\tau(X)$ runs through $G_{p,m}\setminus\{a(X)\}$,

$$\boldsymbol{k}^{(\tau-1)}(X)\cdot \boldsymbol{q}^{(\tau-1)}(X)+k_\tau(X)q_\tau(X) \pmod{p(X)}$$

runs through all polynomials in $G^*_{p,m}$. As $B(X)$ and $p(X)$ were assumed coprime we have that

$$B(X)\boldsymbol{k}^{(\tau-1)}(X)\cdot \boldsymbol{q}^{(\tau-1)}(X)+B(X)k_\tau(X)q_\tau(X) \pmod{p(X)}$$

runs through all polynomials in $G^*_{p,m}$.

Hence

$$\frac{K_2}{(p^m-1)^{t-\tau}} \le \sum_{\boldsymbol{q}^{(\tau-1)}\in (G^*_{p,m})^{\tau-1}} \#\{b(X)\in G^*_{p,m} : \nu(b(X)/p(X)) < -d\}.$$

Altogether the core estimate provides Lemma 4 which states

$$\#\{b(X)\in G^*_{p,m} : \nu(b(X)/p(X)) < -d\} \le p^{m-d}-1.$$

Thus

$$K_2 \le (p^m-1)^{t-1}(p^{m-d}-1).$$

So we can summarize

$$K_1+K_2 \le (p^m-1)^{t-1}p^{m-d}.$$

Finally, application of Lemma 3 yields

$$\frac{1}{|G^*_{p,m}|^t}\sum_{q\in (G^*_{p,m})^t} LD^*_L(\mathscr{P}) \le t+\frac{p^m}{p^m-1}\left(1+m\frac{p^2-1}{3p}\right)^t = O_{p,t}(\mathrm{Log}^t N).$$

5 Proof of Theorem 2

The proof follows the same steps as the proof of Theorem 1, until we have to compute the average

$$\frac{1}{|(G^*_{p,m})|}\sum_{g(X)\in (G^*_{p,m})} L_k D^*_{L_k}(\mathscr{P}_k).$$

We show again that it is of the form $O_{p,t}(\mathrm{Log}^t N)$.

Using the same argumentation as in the proof of Theorem 1 we obtain

$$\#\{g(X) \in G^*_{p,m} : \boldsymbol{k} \in \mathscr{D}'_{\boldsymbol{q},p,B}\} = \underbrace{\#\{g(X) \in G^*_{p,m} : \boldsymbol{k} \in \mathscr{D}'_{\boldsymbol{q},p}\}}_{=:K_1} + \underbrace{\#\{g(X) \in G^*_{p,m} : \boldsymbol{k} \in \mathscr{D}'_{\boldsymbol{q},p,B} \setminus \mathscr{D}'_{\boldsymbol{q},p}\}}_{=:K_2}$$

where $\boldsymbol{q}(X) = (g(X), g^2(X), \ldots, g^t(X))$.

The easy part is again to estimate K_1, which satisfies $K_1 \leq t$, since

$$k_1(X)Y + k_2(X)Y^2 + \cdots + k_t(X)Y^t \equiv 0 \pmod{p(X)}$$

has at most t solutions for Y modulo $p(X)$.

In the following we show that $K_2 \leq t(p^{m-d} - 1)$. We know that for each $a(X) \in G^*_{p,m}$ the congruence

$$k_1(X)Y + k_2(X)Y^2 + \cdots + k_t(X)Y^t \equiv a(X) \pmod{p(X)}$$

has at most t solutions for Y modulo $p(X)$. As $B(X)$ and $p(X)$ are coprime

$$B(X)k_1(X)Y + B(X)k_2(X)Y^2 + \cdots + B(X)k_t(X)Y^t \equiv b(X) \pmod{p(X)}$$

has at most t solutions for each $b(X) \in G^*_{p,m}$. By Lemma 4 only $p^{m-d} - 1$ values of $b(X)$ have to be considered. Hence we have $K_2 \leq t(p^{m-d} - 1)$.

Then the result follows exactly by the same arguments as in the proof of Theorem 1.

Remark 1 Note that, mixing a polynomial point set $\mathscr{P}(1, p(X))$ with the first component $(n/p^m)_{n=0,1,\ldots,p^m-1}$ won't result in good discrepancy bounds. Also bounding K_2 in the proof of Theorem 2 won't work if we take generating vectors of the form $(1, g(X), \ldots, g^{t-1}(X))$ instead of $(g(X), g^2(X), \ldots, g^t(X))$. This is the reason why we defined Korobov polynomial lattice point sets based on generating vectors $(g(X), g^2(X), \ldots, g^t(X))$.

6 Open Problems

As already noted in the introductory Sect. 1 the best known discrepancy bound for lattice point sets and polynomial lattice point sets are of the form $ND^*_N = O((\mathrm{Log}^{s-1} N)\,\mathrm{LogLog}\, N)$. An interesting problem is to improve one $\mathrm{Log} N$ term in the discrepancy bounds for the hybrid sequences proved in this paper as well as in [20] to $\mathrm{LogLog} N$. Kritzer, Leobacher, and Pillichshammer [21] introduced a component-by-component algorithm for the construction of hybrid point sets based on Hammersley and lattice point sets. It is an open problem to do the same for hybrid point sets based on Halton-type Hammersley points sets and polynomial lattice point sets.

Acknowledgements The author is supported by the Austrian Science Fund (FWF): Project F5505-N26, which is a part of the Special Research Program "Quasi-Monte Carlo Methods: Theory and Applications". Furthermore, the author appreciates several valuable comments by the anonymous referees.

References

1. Bykovskii, V.A.: The discrepancy of Korobov lattice points. Izv. Math. **76**(3), 446–465 (2012)
2. Dick, J., Pillichshammer, F.: Digital Nets and Sequences: Discrepancy Theory and Quasi-Monte Carlo Integration. Cambridge University Press, Cambridge (2010)
3. Del Chicca, L., Larcher, G.: Hybrid Monte Carlo methods in credit risk management. Monte Carlo Methods Appl. **20**(4), 245–260 (2014)
4. Drmota, M., Hofer, R., Larcher, G.: On the discrepancy of Halton-Kronecker sequences. In: Elsholtz, C., Grabner, P. (eds.) Number Theory Diophantine Problems, Uniform Distribution and Applications, pp. 219–226. Springer, Berlin (2017)
5. Gnewuch, M.: On probabilistic results for the discrepancy of a hybrid-Monte Carlo sequence. J. Complex. **25**(4), 312–317 (2009)
6. Gómez-Pérez, D., Hofer, R., Niederreiter, H.: A general discrepancy bound for hybrid sequences involving Halton sequences. Unif. Distrib. Theory **8**(1), 31–45 (2013)
7. Halton, J.H.: On the efficiency of certain quasi-random sequences of points in evaluating multi-dimensional integrals. Numer. Math. **2**, 84–90 (1960)
8. Hellekalek, P.: General discrepancy estimates: the Walsh function system. Acta Arith. LXVI **I**(3), 209–218 (1994)
9. Hellekalek, P., Kritzer, P.: On the diaphony of some finite hybrid point sets. Acta Arith. **156**(3), 257–282 (2012)
10. Hlawka, E.: Zur angenäherten Berechnung mehrfacher Integrale. Monatshefte für Mathematik **66**, 140–151 (1962)
11. Hofer, R.: On the distribution of Niederreiter-Halton sequences. J. Number Theory **129**(2), 451–463 (2009)
12. Hofer, R.: A construction of low-discrepancy sequences involving finite-row digital (t, s)-sequences. Monatshefte für Mathematik **171**(1), 77–89 (2013)
13. Hofer, R.: Kronecker-Halton sequences in $F_p((X^{-1}))$. Finite Fields Their Appl. **50**, 154–177 (2018)
14. Hofer, R., Kritzer, P.: On hybrid sequences built from Niederreiter-Halton sequences and Kronecker sequences. Bull. Aust. Math. Soc. **84**(2), 238–254 (2011)
15. Hofer, R., Kritzer, P., Larcher, G., Pillichshammer, F.: Distribution properties of generalized van der Corput-Halton sequences and their subsequences. Int. J. Number Theory **5**(4), 719–746 (2009)
16. Hofer, R., Larcher, G.: On existence and discrepancy of certain digital Niederreiter-Halton sequences. Acta Arith. **141**(4), 369–394 (2010)
17. Hofer, R., Larcher, G.: Metrical results on the discrepancy of Halton-Kronecker sequences. Math. Z. **271**(1–2), 1–11 (2012)
18. Hofer, R., Puchhammer, F.: On the discrepancy of two-dimensional perturbed Halton-Kronecker sequences and lacunary trigonometric products. Acta Arith. **180**(4), 365–392 (2017)
19. Korobov, N.M.: Approximate evaluation of repeated integrals. Dokl. Akad. Nauk. SSSR **124**, 1207–1210 (1959)

20. Kritzer, P.: On an example of finite hybrid quasi-Monte Carlo point sets. Monatshefte für Mathematik **168**(3–4), 443–459 (2012)
21. Kritzer, P., Leobacher, G., Pillichshammer, F.: Component-by-component construction of hybrid point sets based on Hammersley and lattice point sets. In: Dick, J., Kuo, F.Y., Peters, G.W., Sloan, I.H. (eds.) Monte Carlo and Quasi-Monte Carlo Methods 2012, pp. 501–515. Springer, Heidelberg (2013)
22. Kritzer, P., Pillichshammer, F.: Low discrepancy polynomial lattice point sets. J. Number Theory **132**, 2510–2534 (2012)
23. Kritzer, P., Pillichshammer, F.: On the existence of low-diaphony sequences made of digital sequences and lattice points. Math. Nachr. **286**(2–3), 224–235 (2013)
24. Kuipers, L., Niederreiter, H.: Uniform Distribution of Sequences. Wiley, New York (1974); Reprint: Dover Publications, Mineola (2006)
25. Niederreiter, H.: On the discrepancy of some hybrid sequences. Acta Arith. **138**(4), 373–398 (2009)
26. Niederreiter, H.: A discrepancy bound for hybrid sequences involving digital explicit inversive pseudorandom numbers. Unif. Distrib. Theory **5**(1), 53–63 (2010)
27. Niederreiter, H.: Further discrepancy bounds and an Erdős-Turán-Koksma inequality for hybrid sequences. Monatshefte für Mathematik **161**(2), 193–222 (2010)
28. Niederreiter, H.: Discrepancy bounds for hybrid sequences involving matrix-method pseudorandom vectors. Publ. Math. Debr. **79**(3–4), 589–603 (2011)
29. Niederreiter, H.: Improved discrepancy bounds for hybrid sequences involving Halton sequences. Acta Arith. **155**(1), 71–84 (2012)
30. Niederreiter, H.: Random Number Generation and Quasi-Monte Carlo Methods. CBMS-NSF Regional Conference Series in Applied Mathematics, vol. 63. SIAM, Philadelphia (1992)
31. Niederreiter, H., Winterhof, A.: Discrepancy bounds for hybrid sequences involving digital explicit inversive pseudorandom numbers. Unif. Distrib. Theory **6**(1), 33–56 (2011)
32. Ökten, G.: A probabilistic result on the discrepancy of a hybrid-Monte Carlo sequence and applications. Monte Carlo Methods Appl. **2**(4), 255–270 (1996)
33. Ökten, G.: Applications of a hybrid-Monte Carlo sequence to option pricing. In: Niederreiter, H., Spanier, J. (eds.) Monte Carlo and Quasi-Monte Carlo Methods 1998, pp. 391–406. Springer, Berlin (2000)
34. Ökten, G., Gnewuch, M.: A correction of a proof in "A probabilistic result on the discrepancy of a hybrid-Monte Carlo sequence and applications". Monte Carlo Methods Appl. **15**(2), 169–172 (2009)
35. Ökten, G., Tuffin, B., Burago, V.: A central limit theorem and improved error bounds for a hybrid-Monte Carlo sequence with applications in computational finance. J. Complex. **22**(4), 435–458 (2006)
36. Spanier, J.: Quasi-Monte Carlo methods for particle transport problems. In: Niederreiter, H., Shiue, P.J.-S. (eds.) Monte Carlo and Quasi-Monte Carlo Methods in Scientific Computing. Lecture Notes in Statistics, vol. 106, pp. 121–148. Springer, Berlin (1995)

Robust Estimation of the Mean with Bounded Relative Standard Deviation

Mark Huber

Abstract Many randomized approximation algorithms operate by giving a procedure for simulating a random variable X which has mean μ equal to the target answer, and a relative standard deviation bounded above by a known constant c. Examples of this type of algorithm includes methods for approximating the number of satisfying assignments to 2-SAT or DNF, the volume of a convex body, and the partition function of a Gibbs distribution. Because the answer is usually exponentially large in the problem input size, it is typical to require an estimate $\hat{\mu}$ satisfy $\mathbb{P}(|\hat{\mu}/\mu - 1| > \varepsilon) \leq \delta$, where ε and δ are user specified nonnegative parameters. The current best algorithm uses $\lceil 2c^2\varepsilon^{-2}(1+\varepsilon)^2 \ln(2/\delta) \rceil$ samples to achieve such an estimate. By modifying the algorithm in order to balance the tails, it is possible to improve this result to $\lceil 2(c^2\varepsilon^{-2}+1)/(1-\varepsilon^2)\ln(2/\delta) \rceil$ samples. Aside from the theoretical improvement, we also consider how to best implement this algorithm in practice. Numerical experiments show the behavior of the estimator on distributions where the relative standard deviation is unknown or infinite.

Keywords Monte Carlo · Robust estimation · M-estimator

1 Introduction

Suppose we are interested in approximating a target value μ. Then many randomized approximation algorithms work by constructing a random variable X such that $\mathbb{E}[X] = \mu$ and $\text{Var}(X)/\mu^2 \leq c^2$ for a known constant c. The randomized algorithm then simulates $X_1, X_2, \ldots, X_n$ as independent identically distributed (iid) draws from X. Finally, the values are input into a function to give an estimate $\hat{\mu}$ for μ.

Examples of this type of algorithm include when μ is the number of solutions to a logic formula in Disjunctive Normal Form (DNF) [9], the volume of a convex body [4], and the partition function of a Gibbs distribution [6]. For all of these

M. Huber (✉)
Claremont McKenna College, 850 Columbia Avenue, Claremont, CA 91711, USA
e-mail: mhuber@cmc.edu

B. Tuffin and P. L'Ecuyer (eds.), *Monte Carlo and Quasi-Monte Carlo Methods*, Springer Proceedings in Mathematics & Statistics 324,
https://doi.org/10.1007/978-3-030-43465-6_13

problems, the random variable X used is nonnegative with probability 1, and so we shall only consider this case for the rest of the paper.

For these types of problems, generating the samples from a high dimensional distribution is the most computationally intensive part of the algorithm. Hence we will measure the running time of the algorithm by the number of samples from X that must be generated.

The answer μ for these problems typically grows exponentially quickly in the size of the problem input size. Therefore, it is usual to desire an approximation $\hat{\mu}$ that is accurate when measured by relative error. We use $\varepsilon > 0$ as our bound on the relative error, and $\delta > 0$ as the bound on the probability that the relative error restriction is violated.

Definition 1 Say $\hat{\mu}$ is an (ε, δ)-*randomized approximation scheme* ((ε, δ)-ras) if

$$\mathbb{P}\left(\left|\frac{\hat{\mu}}{\mu} - 1\right| > \varepsilon\right) \leq \delta.$$

This is equivalent to considering a loss function L that is $L(\hat{\mu}, \mu) = \mathbb{1}(|(\hat{\mu}/\mu) - 1| > \varepsilon)$, and requiring that the expected loss be no more than δ.

Note that this question is slightly different than the problem most statistical estimates are designed to handle. For instance, the classic work of [8] is trying to minimize the asymptotic variance of the estimator, not determine how often the relative error is at most ε.

The first approach to this problem is often the standard sample average

$$S_n = \frac{X_1 + \cdots + X_n}{n}.$$

For this estimator, only knowing that $\text{Var}(X) \leq c^2\mu^2$, the best we can say about the probability that the relative error is large comes from Chebyshev's inequality [10].

$$\mathbb{P}(|S_n - \mu| > \varepsilon\mu) \leq \frac{\text{Var}(X)}{n\varepsilon^2\mu^2} \leq c^2\varepsilon^{-2}n^{-1}.$$

In particular, setting $n = c^2\varepsilon^{-2}\delta^{-1}$ gives an (ε, δ)-ras. There are simple examples where Chebyshev's inequality is tight.

While this bound works for all distributions, for most distributions in practice the probability of error will go down exponentially (and not just polynomially) in n. We desire an estimate that matches this speed of convergence.

For example, suppose that X is a normal random variable with mean μ and variance $c^2\mu^2$ (write $X \sim \mathsf{N}(\mu, c^2\mu^2)$. Then the sample average is normally distributed as well. To be precise, $S_n \sim \mathsf{N}(\mu, c^2\mu^2/n)$, and it is straightforward to show that

$$\mathbb{P}(|S_n - \mu| > \varepsilon\mu) = \mathbb{P}(|Z| \geq \varepsilon\sqrt{n}/c),$$

where Z is a standard normal random variable. This gives

$$n = 2c^2\varepsilon^{-2}[\ln(2/\delta) - o(\delta)]. \tag{1}$$

samples being necessary and sufficient to achieve an (ε, δ)-ras.

We did this calculation for normally distributed samples, but in fact this gives a lower bound on the number of samples needed. For any $X_1, \ldots, X_n$ with mean $\mu > 0$ and variance $c^2 > 0$, and an estimate $\hat{\theta} = \hat{\theta}(X_1, \ldots, X_n)$,

$$\mathbb{P}(|\hat{\theta} - \mu| > \varepsilon\mu) \geq (1/2)\mathbb{P}(|Z| \geq \varepsilon\sqrt{n}/c).$$

See, for instance, Proposition 6.1 of [1]. Hence the minimum number of samples required for all instances is

$$n \geq 2c^2\varepsilon^{-2}[\ln(1/\delta) - o(\delta)].$$

The method presented in [5] comes close to this lower bound, requiring

$$n = \lceil 2c^2\varepsilon^{-2}(1+\varepsilon)^2\ln(2/\delta)\rceil$$

samples. So it is larger than optimal by a factor of $(1+\varepsilon)^2$.

Our main result is to modify the estimator slightly. This has two beneficial effects.

1. The nuisance factor is reduced from first order in ε to second order. To be precise, the new nuisance factor is $(1 + \varepsilon^2/c^2)/(1 - \varepsilon^2)$.
2. It is possible to solve exactly for the value of the M-estimator (using square and cube roots) rather than through numerical approximation if so desired.

Theorem 1 *Suppose that* $\mathbb{P}(X \geq 0) = 1$, $\mathbb{E}[X] = \mu$, *and the standard deviation of* X *is at most* $c\mu$, *where* $c > 0$. *Then there exists an* (ε, δ)*-ras* $\hat{\mu}_1$ *where at most*

$$\left\lceil 2(c^2\varepsilon^{-2} + 1)(1 - \varepsilon^2)^{-1}\ln(2/\delta)\right\rceil$$

draws from X *are used.*

Ignoring the ceiling function, the new method uses a number of samples bounded by the old method times a factor of

$$\frac{1 + \varepsilon^2/c^2}{(1+\varepsilon)^2(1-\varepsilon^2)}.$$

For instance, when $\varepsilon = 0.1$ and $c = 2$, this factor is $0.8556\ldots$, and so we obtain an improvement of over 14% in the running time. At first, this might seem slight, but remember that the best improvement we can hope to make based on normal random variables is $1/(1+\varepsilon)^2 = 0.8264$, or a bit less than 18%. Therefore, this does not quite obtain the maximum improvement of $1/(1+2\varepsilon+\varepsilon^2)$, but does obtain a factor of $(1+O(\varepsilon^2))/(1+2\varepsilon+\varepsilon^2)$.

The remainder of this paper is organized as follows. In Sect. 2, we describe what M-estimators are, and show how to down weight samples that are far away from the mean. Section 3 shows how to find the estimator both approximately and exactly. In Sect. 4 we consider using these estimators on some small examples and see how they behave numerically.

2 Ψ-Estimators

The median is a robust centrality estimate, but it is easier to build a random variable whose mean is the target value. Huber first showed in [8] how to build an estimate that has the robust nature of the median estimate while still converging to the mean.

Consider a set of real numbers $x_1, \ldots, x_n$. The sample average of the $\{x_i\}$ is the point m where the sum of the distances to points to the right of m equals the sum of the distances to points to the left of m.

This idea can be generalized as follows. First, begin with a function $\psi : \mathbb{R} \times \mathbb{R}^n \to \mathbb{R}$. Second, using ψ, form the function

$$\Psi(m) = \sum_{i=1}^{n} \psi(x_i, m).$$

For convenience, we suppress the dependence of Ψ on $(x_1, \ldots, x_n)$ in the notation.

Consider the set of zeros of Ψ. These zeros form the set of *Ψ-type M-estimators* for the center of the points $(x_1, \ldots, x_n)$.

For example, suppose our ψ is $f : \mathbb{R}^2 \to \mathbb{R}$ defined as

$$f(x, m) = \begin{cases} (x/m) - 1 & m \neq 0 \\ 0 & x = m = 0. \end{cases}$$

Then consider

$$\sum_{i=1}^{m} f(x_i, m) = 0.$$

For x and m with the same units, the right hand side is unitless. If the x_i are nonnegative $\{x_i\}$ and at least one x_i is positive,

$$\sum_{i=1}^{n} f(x_i, m) = 0 \Leftrightarrow m = \frac{\sum_{i=1}^{n} x_i}{n}.$$

That is, the unique M-estimator using f as our ψ function is the sample average.

Now consider

$$g(x, m) = \mathbb{1}(f(x, m) > 0) - \mathbb{1}(f(x, m) < 0).$$

Then we wish to find m such that

$$\sum_{i=1}^{n} g(x_i, m) = 0.$$

Summing $g(x_i, m)$ adds 1 when x_i greater than m, and subtracts 1 when x_i is m. Then the sum of the $g(x_i, m)$ is zero exactly when there are an equal number of x_i that are above and below m. When the number of distinct x_i values is odd, then m has a unique solution equal to the sample median.

The sample median has the advantage of being robust to large changes in the x_i values, but this will converge to the median of the distribution (under mild conditions) rather than the mean. The sample average actually converges to the mean when applied to iid draws from X, and this is the value we care about finding. However, the sample average can be badly thrown off by a single large outlier. Our goal is to create a Ψ function that combines the good qualities of each while avoiding the bad qualities.

Both f and g can be expressed using weighted differences. Let $d_f : \mathbb{R} \to \mathbb{R}$ and $d_g : \mathbb{R} \to \mathbb{R}$ be defined as

$$d_f(u) = u$$
$$d_g(u) = \mathbb{1}(u > 0) - \mathbb{1}(u < 0).$$

Let $u(x_i, m) = x_i/m - 1$ for $m \neq 0$, and 0 for $x_i = m = 0$. Then $f(x_i, m) = d_f(u)$ and $g(x_i, m) = d_g(u)$.

Catoni [1] improved upon this Ψ-estimator by using a function that approximated d_f for $|u| < 1$, and approximated d_u for $|u| > 1$, and could be analyzed using the Chernoff bound approach [3]. Catoni and Guillini then created an easier to use version of the Ψ-estimator in [2].

The following is a modification of the Catoni and Guillini Ψ-estimator. Unlike [2], the weighted difference here does not have square roots in the constants, which makes them slightly easier to work with from a computational standpoint.

$$d_h(u) = \left(\frac{5}{6}\right) \mathbb{1}(u > 1) + \left(u - \frac{u^3}{6}\right) \mathbb{1}(u \in [-1, 1]) - \left(\frac{5}{6}\right) \mathbb{1}(u < -1).$$

Then define

$$h(x_i, m) = d_h(u(x_i, m)).$$

The function d_h behaves like the mean weight for values near 0, but like the median weights for values far away from 0. See Fig. 1. In order to link this function to the

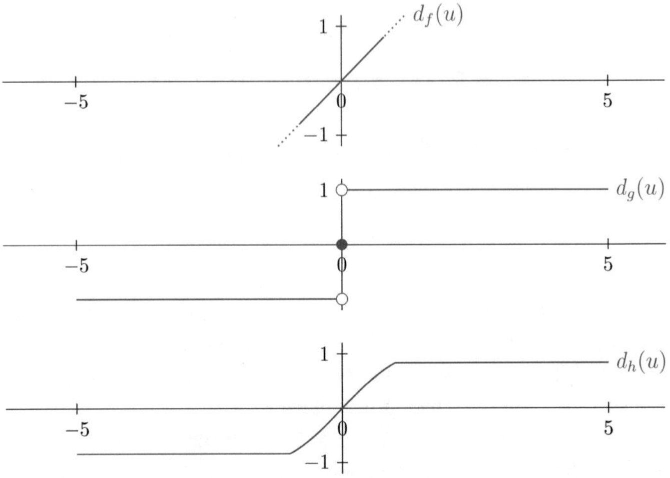

Fig. 1 The functions d_f, d_g, and d_h

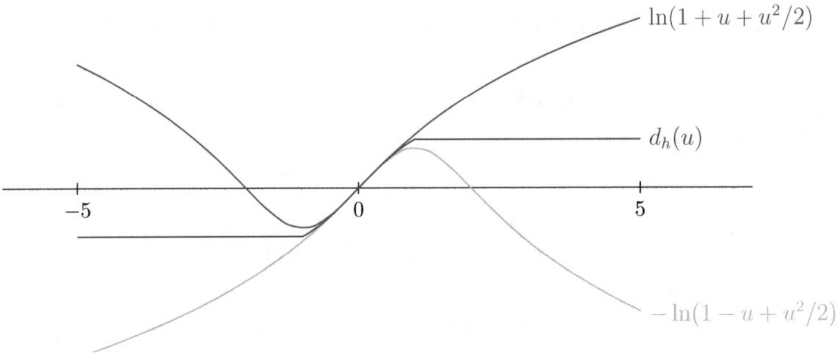

Fig. 2 Lower and upper bounds for d_h

first and second moments of the random variable, the following links d_h to weighted differences introduced in [1]. See Fig. 2.

Lemma 1 *Let*

$$d_L(u) = -\ln(1 - u + u^2/2)$$
$$d_U(u) = \ln(1 + u + u^2/2).$$

Then for all $u \in \mathbb{R}$,

$$d_L(u) \le d_h(u) \le d_U(u)$$

Proof Note $d_U'(u) = (1+u)/(1+u+u^2/2)$. This derivative is positive over $[1, \infty)$ so d_U is increasing in this region. Since $d_U(1) = \ln(2.5) > 5/6 = d_h(1)$, $d_U(u) \geq d_h(u)$ over $[1, \infty)$.

Similarly, $d_U'(u) < 0$ for $u \in (-\infty, -1]$, and so it is decreasing in this region, and $d_U(-1) = \ln(1/2) \geq -5/6 = d_h(u)$ so $d_U(u) \geq d_h(u)$ in $(-\infty, 1]$. Finally, inside $[-1, 1]$, $d_h(u) = u - u^3/6$. So a minimum of $d_U(u) - d_h(u)$ occurs either at -1, 1, or a critical point where $d_U'(u) - (1 - u^2/2) = 0$. The unique critical point in $[-1, 1]$ is at $u = 0$, and $d_U(0) = d_h(0)$. This means $d_U(u) - d_h(u) \geq 0$ for all $u \in [-1, 1]$.

The lower bound follows from $d_L(u) = -d_U(-u) \leq -d_h(-u) = d_h(u)$. □

It helps to introduce a scale factor λ that allows us to extend the effective range where $d(u) \approx u$. Let

$$h_\lambda(x_i, m) = \lambda^{-1} d_h(\lambda(x_i/m - 1)).$$

Then $\lambda > 0$ is a parameter that can be chosen by the user ahead of running the algorithm based upon ε and c.

The following was shown as Lemma 13 of [5].

Lemma 2 *For given* $\varepsilon, \delta > 0$, *let* $\varepsilon' = \varepsilon/(1+\varepsilon)$ *and* $\lambda = \varepsilon'/c^2$. *Set*

$$n = \left\lceil 2c^2\varepsilon^{-2}\ln(2/\delta)(1+\varepsilon)^2 \right\rceil.$$

For $X_1, \ldots, X_n$ *iid* X, *form the function*

$$\Psi_\lambda(m) = \frac{1}{n}\sum_{i=1}^{n} \lambda^{-1} d_h(\lambda u(X_i, m)).$$

Let $\hat{\mu}$ *be any value of* m *such that* $|\Psi_\lambda(m)| \leq (\varepsilon')^2/2$. *Then* $\hat{\mu}$ *is an* (ε, δ)*-ras for* μ.

To improve upon this result and obtain Theorem 1, we must be more careful about our choice of λ.

To meet the requirement of an (ε, δ)-ras, we must show that $\Psi_\lambda(m)$ has all its zeros in the interval $((1-\varepsilon)\mu, (1+\varepsilon)\mu)$ with probability at least $1 - \delta$. Given that Ψ_λ is continuous and decreasing, it suffices to show that $\Psi_\lambda((1-\varepsilon)\mu) > 0$ and $\Psi_\lambda((1+\varepsilon)\mu) < 0$.

Lemma 3 *Let* $X_1, \ldots, X_n$ *be iid* X. *Then for* $\lambda > 0$,

$$\mathbb{P}(\Psi_\lambda((1+\varepsilon)\mu) \geq 0) \leq \left[1 + \mathbb{E}[u_\varepsilon] + \frac{1}{2}\mathbb{E}[u_\varepsilon^2]\right]^n,$$

$$\mathbb{P}(\Psi_\lambda((1-\varepsilon)\mu) \leq 0) \leq \left[1 - \mathbb{E}[u_{-\varepsilon}] + \frac{1}{2}\mathbb{E}[u_{-\varepsilon}^2]\right]^n.$$

where

$$u_\varepsilon = \frac{X}{(1+\varepsilon)\mu} - 1.$$

Proof Note that

$$\begin{aligned}\mathbb{P}(\Psi_\lambda((1+\varepsilon)\mu) \geq 0) &= \mathbb{P}(\exp(\lambda\Psi_\lambda((1+\varepsilon)\mu) \geq 1) \\ &\leq \mathbb{E}[\exp(\lambda\Psi_\lambda((1+\varepsilon)\mu))]\end{aligned}$$

by Markov's inequality. Note

$$\exp(\lambda\Psi_\lambda(m)) = \prod_{i=1}^{n} \exp(\lambda d_h(X_i/m - 1)).$$

Each term in the product is independent, therefore the mean of the product is the product of the means (which are identical.)

Let $u = \lambda(X/m - 1)$. Then

$$\mathbb{P}(\Psi_\lambda((1+\varepsilon)\mu) \geq 0) \leq [\mathbb{E}(\exp(\lambda d_h(u)))]^n$$

From the previous lemma,

$$\begin{aligned}\exp(d_h(u)) &\leq \exp(\ln(1 + u + u^2/2)) \\ &= 1 + u + u^2/2,\end{aligned}$$

Hence

$$\mathbb{E}(\exp(d_h(u))) = 1 + \mathbb{E}(u) + \mathbb{E}(u^2/2).$$

Putting $m = (1+\varepsilon)\mu$ into this expression then gives the first inequality.

For the second inequality, the steps are nearly identical, but we begin by multiplying by $-\lambda$. This completes the proof. □

In particular, if we choose n so that $\mathbb{P}(\Psi_\lambda((1+\varepsilon)\mu) > 0) \leq \delta/2$, and $\mathbb{P}(\Psi_\lambda((1-\varepsilon)\mu) < 0) \leq \delta/2$, then by the union bound the probability that Ψ has a root in $[(1-\varepsilon)\mu, (1+\varepsilon)\mu]$ is at least $\delta/2 + \delta/2 = \delta$.

As is well known, for $g > 0$,

$$(1-g)^n \leq \exp(-gn),$$

and so if $n \geq (1/g)\ln(2/\delta)$,

$$(1-g)^n \leq \delta/2.$$

We refer to g as the *gap*. Since the number of samples needed is inversely proportional to the gap, we wish the gap to be as large as possible. We can lower bound the gap given by the previous lemma in terms of λ and m.

Lemma 4 *For $u = \lambda(X/m - 1)$, and $a(m) = 1 - (m/\mu)$,*

$$1 + \mathbb{E}[u] + \mathbb{E}[u^2/2] \leq 1 + \frac{\lambda\mu}{m}a(m) + \frac{1}{2}\left(\frac{\lambda\mu}{m}\right)^2 \left[c^2 + a(m)^2\right]. \tag{2}$$

Similarly,

$$1 + \mathbb{E}[u] + \mathbb{E}[u^2/2] \leq 1 - \frac{\lambda\mu}{m}a(m) + \frac{1}{2}\left(\frac{\lambda\mu}{m}\right)^2 \left[c^2 + a(m)^2\right]. \tag{3}$$

Proof From linearity of expectations,

$$\mathbb{E}[u] = \lambda\left(\frac{\mu}{m} - 1\right) = \frac{\lambda\mu}{m}a(m).$$

The second moment is the sum of the variance plus the square of the first moment. Using $\text{Var}(c_1 X + c_2) = c_1^2 \text{Var}(X)$, we get

$$\begin{aligned}\mathbb{E}[u^2] &= \text{Var}(u) + \mathbb{E}[u]^2 \\ &= \left(\frac{\lambda}{m}\right)^2 \text{Var}(X) + \left[\frac{\lambda\mu}{m}a(m)\right]^2.\end{aligned}$$

Since $\text{Var}(X) \leq c^2\mu^2$,

$$\mathbb{E}[u^2] \leq \left(\frac{\lambda\mu}{m}\right)^2 [c^2 + a(m)^2],$$

giving the first result. The proof of the second statement is similar. □

Lemma 5 *Let $p_\ell = \mathbb{P}(\Psi_\lambda((1-\varepsilon)\mu) \leq 0)$ and $p_r = \mathbb{P}(\Psi_\lambda((1+\varepsilon)\mu) \geq 0)$. Then*

$$p_\ell \leq \left[1 - \frac{\lambda\varepsilon}{1-\varepsilon} + \frac{1}{2}\left(\frac{\lambda}{1-\varepsilon}\right)^2 (c^2 + \varepsilon^2)\right]^n,$$

$$p_r \leq \left[1 - \frac{\lambda\varepsilon}{1+\varepsilon} + \frac{1}{2}\left(\frac{\lambda}{1+\varepsilon}\right)^2 (c^2 + \varepsilon^2)\right]^n.$$

Proof Note

$$a((1-\varepsilon)\mu) = 1 - (1-\varepsilon)\mu/\mu = \varepsilon.$$

Combine Lemmas 3, 4, and 6 with m equal to $(1-\varepsilon)\mu$ to get the first inequality. Then set $m = (1+\varepsilon)\mu$ and use $a((1+\varepsilon)\mu) = -\varepsilon$ to get the second. □

Our goal is to use as few samples as possible, which means simultaneously minimizing the quantities inside the brackets in Lemma 5. Both of the upper bounds are upward facing parabolas, and they have a unique minimum value.

Lemma 6 *The minimum value of* $f(\lambda) = 1 + a_1\lambda + (1/2)a_2\lambda^2$ *is*

$$1 - \frac{a_1^2}{2a_2},$$

at $\lambda^* = -a_1/a_2$.

Proof Complete the square. □

However, because the coefficients in the quadratic upper bounds are different, it is not possible to simultaneously minimize these bounds with the same choice of λ.

What can be said is that these bounds are quadratic with positive coefficient on λ^2, and so the best we can do is to choose a λ such that the upper bounds are equal to one another. This gives us our choice of λ.

Lemma 7 *Let*

$$\lambda = \frac{\varepsilon}{c^2 + \varepsilon^2}(1 - \varepsilon^2).$$

Then

$$\max\{p_\ell, p_r\} \le \left[1 - \frac{1}{2} \cdot \frac{\varepsilon^2}{c^2 + \varepsilon^2}(1 - \varepsilon^2)\right]^n.$$

Proof Follows directly from Lemma 5. □

This makes the inverse gap

$$\varepsilon^{-2}(c^2 + \varepsilon^2)(1 - \varepsilon^2)^{-1} = (c^2\varepsilon^{-2} + 1)(1 - \varepsilon^{-2})^{-1}$$

and immediately gives Theorem 1.

3 Computation

Set $\lambda = \varepsilon(1 - \varepsilon^2)(c^2 + \varepsilon^2)^{-1}$. Consider how to locate any root of Ψ_λ for a given set of $X_1, \ldots, X_n$. As before, we assume that the X_i are nonnegative and not all identically zero. The function $\Psi_\lambda(m)$ is continuous and decreasing, although not necessarily strictly decreasing. Therefore, it might have a set of zeros that form a closed interval.

3.1 An Approximate Method

Suppose that the points X_i are sorted into their order statistics,

$$X_{(1)} \leq X_{(2)} \leq \cdots X_{(n)}.$$

Then $\Psi_\lambda(X_{(1)}) > 0$ and $\Psi_\lambda(X_{(n)}) < 0$, so there exists some i such that $\Psi_\lambda(X_{(i)}) \geq 0$ and $\Psi_\lambda(X_{i+1}) \leq 0$. Since any particular value of Ψ_λ requires $O(n)$ time to compute, this index i can be found using binary search in $O(n \ln(n))$ time.

At this point we can switch from a discrete binary search over $\{1, \ldots, n\}$ to a continuous binary search over the interval $[X_i, X_{i+1}]$. This allows us to quickly find a root to any desired degree of accuracy.

This is the method that would most likely be used in practice.

3.2 An Exact Method

Although the approximation procedure is what would be used in practice because of its speed, there does exist a polynomial time exact method for this problem. For $i < j$, and a value m, suppose that the points of $\{X_i\}$ that fall into the interval $[m - \lambda, m + \lambda]$ are exactly $\{X_{(i)}, X_{(i+1)}, \ldots, X_{(j)}\}$. Then say that $m \in m(i, j)$.

That is, define the set $m(i, j)$ for $i < j$ as follows.

$$m(i, j) = \{m : X_{(k)} \in [m - \lambda, m + \lambda] \Leftrightarrow k \in \{i, i + 1, \ldots, j\}\}.$$

See Fig. 3.

Note that

$$(\forall m \in [X_{(1)}, X_{(n)}])(\exists i < j)(m \in m(i, j)).$$

There are at most n choose 2 such $i < j$ where $m(i, j)$ is nonempty. In fact, since as we slide m from $X_{(1)}$ to $X_{(n)}$, each point can enter or leave the interval $[m - \lambda, m + \lambda]$ exactly once. Therefore there are at most $2n$ pairs (i, j) where $m(i, j)$ is nonempty.

That means that to find a root of Ψ_λ we need merely check if there is a root $m \in m(i, j)$ for all $i < j$ such that $m(i, j) \neq \emptyset$.

For each $m \in m(i, j)$, the contribution of $X_{(1)}, \ldots, X_{(i-1)}$ to $\Psi_\lambda(m)$ is $-(i - 1)$, and the contribution of $X_{(j+1)}, \ldots, X_{(n)}$ is $n - j$. Hence $m \in m(i, j)$ is a zero of Ψ_λ if an only if $m = 1/r$ and

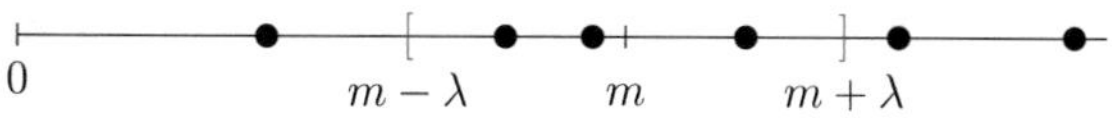

Fig. 3 Since the interval $[m - \lambda, m + \lambda]$ includes $X_{(2)}$, $X_{(3)}$, and $X_{(4)}$, $m \in m(2, 4)$

$$n - j - (i - 1) + \sum_{k=i}^{j} (rX_{(k)} - 1) - (rX_{(k)} - 1)^3/6 = 0.$$

This last equation is a cubic equation in r, and so the value of r that satisfies it (assuming such exists) can be determined exactly using the cubic formula.

4 Numerical Experiments

The M-estimator presented here can be thought of as a principled interpolation between the sample mean and the sample median. Because for u small $d(u) \approx u$, as $\lambda \to 0$, the estimator converges to the sample mean.

At the other extreme, as $\lambda \to \infty$, all of the $d(\lambda(X_i/m - 1)$ will evaluate to either 0, 1, or -1. Hence the estimator converges to the sample median.

When $\lambda = \varepsilon(1 - \varepsilon^2)/(c^2 + \varepsilon^2)$, we can precisely bound the chance of the relative error being greater than ε. However, the estimator can be used for any value of λ.

For instance, Table 1 records the result of using the estimator for 100 exponential random variables with mean 2 and median $2\ln(2) = 1.386\ldots$.

When λ is small, the result is nearly identical to the sample mean. As λ increases, the result moves towards the median value.

Unlike the sample average, however, this M-estimator will always converge to a value, even when the mean does not exist. Consider the following draws from the absolute value of a Cauchy distribution. The mean of these random variables is infinite so the sample average will not converge. The median of this distribution is 1 (Table 2).

Even for values such as $\lambda = 1$, the result is fairly close to the median. For any $\lambda > 0$, the M-estimator will not go to infinity as the sample average does, but instead to converge to a fixed value as the number of samples goes to infinity.

4.1 Timings

To test the time required to create the new estimates, the algorithm presented here together with the algorithm from [5] were implemented in R. Table 3 shows the

Table 1 Behavior of the sample average, sample mean, and M-estimator for 100 exponential random variables with mean 2 here. Repeated five times to show variation

Mean	Median	$\lambda = 0.1$	$\lambda = 1$	$\lambda = 5$
2.34	1.86	2.33	1.93	1.87
1.89	1.35	1.88	1.53	1.40
2.29	1.78	2.28	1.83	1.77
2.02	1.37	2.01	1.48	1.35
2.17	1.37	2.16	1.70	1.39

Table 2 Behavior of the sample mean, sample median, and M-estimator for 100 draws from the absolute value of a Cauchy distribution. Repeated five times to show variation

Mean	Median	$\lambda = 0.1$	$\lambda = 1$	$\lambda = 5$
2.30	0.70	1.95	0.88	0.69
2.70	1.10	2.56	1.27	1.11
10.29	1.01	2.96	1.24	1.02
3.59	1.09	2.58	1.32	1.13
8.07	1.25	4.83	1.68	1.34

Table 3 Behavior of the sample mean, sample median, and M-estimator for 100 draws from the absolute value of a Cauchy distribution. Repeated five times to show variation

Epsilon	Delta	CG	New	Relative change
0.10	1e-06	551.94	461.48	−0.1638946
0.05	1e-06	2013.24	1822.40	−0.0947925

results of running both algorithms using an Euler-Maruyama simulation of a simple SDE as a test case.

As can be seen from the data, the relative change is near -2ε, and the relative change gets closer to -2ε the smaller ε becomes.

The complete code used to generate the data in these tables can be found in Sect. 6 of [7] at https://arxiv.org/pdf/1908.05386.pdf.

5 Conclusion

The modified Catoni M-estimator presented here gives a means of interpolating between the sample mean and the sample average. The estimator is designed for the output of Monte Carlo simulations where the distribution is usually unknown, but often it is possible to compute a bound on the relative standard deviation. It is fast to calculate in practice and can be computed exactly in terms of square and cube roots in polynomial time. The estimator has a parameter λ which controls how close the estimate is to the sample mean or sample median.

Given a known upper bound c on the relative standard deviation of the output, λ can be chosen as $\varepsilon(1 - \varepsilon^2)/(c^2 + \varepsilon^2)$ to yield an (ε, δ)-randomized approximation scheme that uses a number of samples (to first order) equal to that if the data was normally distributed. Even if c is unknown (or infinite), the estimator will still converge to a fixed measure of centrality for any choice of λ.

Acknowledgements This work supported by National Science Foundation grant DMS-1418495.

References

1. Catoni, O.: Challenging the empirical mean and empirical variance: a deviation study. Ann. Inst. H. Poincaré Probab. Statist. **48**, 1148–1185 (2012)
2. Catoni, O., Giulini, I.: Dimension free PAC-Bayesian bounds for matrices, vectors, and linear least squares regression with a random design. arXiv:1712.02747 (2017)
3. Chernoff, H.: A measure of asymptotic efficiency for tests of a hypothesis based on the sum of observations. Ann. Math. Stat. **23**, 493–509 (1952)
4. Dyer, M., Frieze, A., Kannan, R.: A random polynomial-time algorithm for approximating the volume of convex bodies. J. Assoc. Comput. Mach. **38**(1), 1–17 (1991)
5. Huber, M.: An optimal (ε, δ)-approximation scheme for the mean of random variables with bounded relative variance. Random Structures Algorithms To appear
6. Huber, M.: Approximation algorithms for the normalizing constant of Gibbs distributions. Ann. Appl. Probab. **51**(1), 92–105. arXiv:1206.2689 (2015)
7. Huber, M.: Robust estimation of the mean with bounded relative standard deviation. arXiv:1908.05386 (2019)
8. Huber, P.J.: Robust estimation of a location parameter. Ann. Math. Stat. **35**(1), 73–101 (1964). https://doi.org/10.1214/aoms/1177703732
9. Karp, R.M., Luby, M.: Monte-carlo algorithms for enumerating and reliability problems. In: Proceedings of the FOCS, pp. 56–64 (1983)
10. Tchebichef, P.: Des valeurs moyennes. Journal de Mathématique Pures et Appliquées **2**(12), 177–184 (1867)

Infinite Swapping Algorithm for Training Restricted Boltzmann Machines

Henrik Hult, Pierre Nyquist and Carl Ringqvist

Abstract Given the important role latent variable models play, for example in statistical learning, there is currently a growing need for efficient Monte Carlo methods for conducting inference on the latent variables given data. Recently, Desjardins et al. (JMLR Workshop and Conference Proceedings: AISTATS 2010, pp. 145–152, 2010 [3]) explored the use of the parallel tempering algorithm for training restricted Boltzmann machines, showing considerable improvement over the previous state-of-the-art. In this paper we continue their efforts by comparing previous methods, including parallel tempering, with the infinite swapping algorithm, an MCMC method first conceived when attempting to optimise performance of parallel tempering (Dupuis et al. in J. Chem. Phys. 137, 2012 [7]), for the training task. We implement a Gibbs-sampling version of infinite swapping and evaluate its performance on a number of test cases, concluding that the algorithm enjoys better mixing properties than both persistent contrastive divergence and parallel tempering for complex energy landscapes associated with restricted Boltzmann machines.

Keywords Infinite swapping · Restricted Boltzmann machines · Statistical learning · Latent variable models · Gibbs sampling

H. Hult · P. Nyquist (✉) · C. Ringqvist
KTH Royal Institute of Technology, Lindstedtsvagen 25,
100 44 Stockholm, Sweden
e-mail: pierren@kth.se

H. Hult
e-mail: hult@kth.se

C. Ringqvist
e-mail: carrin@kth.se

B. Tuffin and P. L'Ecuyer (eds.), *Monte Carlo and Quasi-Monte Carlo Methods*,
Springer Proceedings in Mathematics & Statistics 324,
https://doi.org/10.1007/978-3-030-43465-6_14

1 Introduction

Consider a latent variable model with probability density of the form

$$p(\mathbf{v}) = \frac{1}{Z(\theta)} \sum_{\mathbf{h}} \exp\{-E_\theta(\mathbf{v}, \mathbf{h})\}, \tag{1}$$

where $\mathbf{v}$ represent visible units and $\mathbf{h}$ hidden units, $Z(\theta)$ is an unknown normalising constant, and θ an unknown parameter. The function E_θ is referred to as the *energy* function. Training such models by maximum likelihood through, e.g., Stochastic Gradient Descent (SGD) using a training set of independent samples of visible units often requires Markov chain Monte Carlo (MCMC) methods. When the energy landscape is complex, convergence to the desired stationary distribution may be slow because the Markov chain tends to get stuck near local minima. In the context of training in machine learning, this phenomenon might result in significant gradient estimation error.

A simple, yet interesting model of the form (1) is the Restricted Boltzmann Machine (see Sect. 2), which is prominent in statistical learning and used in various deep architectures. It is particularly successful in collaborative filtering, for instance in assigning ratings of movies to users, see [27]. Training of restricted Boltzmann machines have been gradually improved from contrastive divergence [15, 16], to persistent contrastive divergence [31] and most recently parallel tempering [3]. In a sense this paper continues the effort initiated in [3] by proposing the infinite swapping (INS) algorithm, designed to overcome rare-event sampling issues, for training restricted Boltzmann machines. Moreover, we investigate via an empirical study the impact the choice of training algorithm has on classification. This partially answers a question posed in [3], namely what impact the use of parallel tempering, and more generally extended ensemble Monte Carlo methods, may have on classification tasks.

Parallel tempering (PT) [9, 12, 30] has become a standard tool for molecular dynamics simulations, see for example [9, 13, 18, 21, 25, 29] and the references therein. The idea is that for models where there is a parameter acting like a temperature—the canonical case is a Gibbs measure and the associated (inverse) temperature, similar to (1)—one runs multiple Markov chains, each with a different "temperature", and couple them via swaps of the particle locations at random times according to a given intensity, see Sect. 3 for further details.

The infinite swapping algorithm was introduced in [8] as an improvement of parallel tempering, with documented success in a variety of chemical and biological physics settings. Consequently, it serves as a natural candidate for potentially improving training of machine learning models. It can be viewed as the limit of PT when the swap rate is sent to infinity; in [8] the corresponding sampling scheme is shown to be optimal from a large deviations perspective and in [5] a more in-depth analysis of PT and INS is carried out in the setting of continuous-time jump Markov processes. Recently [22] studied the ergodicity properties of INS at low temperature, deriving Eyring-Kramers formulae for the spectral gap and the log-Sobolev constant, showing superiority of infinite swapping over overdamped Langevin dynamics.

In addition to the theoretical results of [5, 8, 22] on the properties of PT and INS, recent empirical studies show superior performance of INS compared to PT and other Monte Carlo methods for a range of common performance measures [4, 7, 23]. So far the main application area for INS has been chemical and biological physics and adjacent areas, see for example [6, 20, 24, 33]. However, as extended ensemble methods, such as PT, are becoming increasingly popular for MCMC simulations in a wide range of areas, it is natural to consider INS in the same settings. Statistical learning and latent variable models is one such example where metastability is often a hindrance in the training phase.

In this paper we propose to use the INS algorithm in the training phase of a restricted Boltzmann machine to improve mixing of the underlying Markov chain and to facilitate accurate estimation of gradients. The contributions of this paper include

- an implementation of a Gibbs sampling version of the infinite swapping algorithm for latent variable models,
- details on the infinite swapping algorithm for training restricted Boltzmann machines,
- empirical comparison of the performance of training restricted Boltzmann machines using infinite swapping, parallel tempering, and persistent contrastive divergence.

So far infinite swapping has mainly been considered for Langevin and Glauber dynamics [5, 8, 20] in continuous time. The paper [8] also contains discrete-time large deviations results and discusses the corresponding sampling schemes. For the application to restricted Boltzmann machines, the large size of the state space under consideration, although discrete, renders Glauber dynamics unsuitable because of the need to compute the full transition matrix. To the best of our knowledge this paper is the first to implement a Gibbs sampling version of INS that circumvents this computational issue. Although the present study is limited to restricted Boltzmann machines, it is plausible that this Gibbs sampling version of the infinite swapping algorithm can be further generalised to more complex latent variable models. In [2] the authors introduce Hamiltonian Monte Carlo in the setting of variational auto-encoders, another prominent latent variable model, to obtain unbiased estimators of gradients with low variance.

The remainder of the paper is organised as follows: In Sect. 2 the restricted Boltzmann machine is introduced. The parallel tempering and infinite swapping algorithms are presented in Sect. 3. An empirical study of INS performance is provided in Sect. 4 and the conclusions are summarised in Sect. 5.

2 Restricted Boltzmann Machines

The Restricted Boltzmann Machine (RBM) [11, 16, 28, 32] is a probability distribution over an N-dimensional boolean space $\{0, 1\}^N$. Let $\mathbf{v} \in \{0, 1\}^N$, $\mathbf{h} \in \{0, 1\}^M$ be row vectors and set $\mathbf{x} = (\mathbf{v}, \mathbf{h}) \in \{0, 1\}^{N+M}$. Let $\mathbf{W}$ be a real-valued parameter matrix of dimension $N \times M$, and let $\mathbf{b} \in \mathbb{R}^N$, $\mathbf{c} \in \mathbb{R}^M$ be real-valued bias parameter row vectors. The RBM probability function is defined as

$$p(\mathbf{v}) = Z^{-1} \sum_{\mathbf{h}} e^{-E(\mathbf{v},\mathbf{h})}, \tag{2}$$

$$E(\mathbf{v}, \mathbf{h}) = -\mathbf{v}\mathbf{W}\mathbf{h}^T - \mathbf{v}\mathbf{b}^T - \mathbf{h}\mathbf{c}^T, \tag{3}$$

An observed data point enters the RBM model as a vector of visible units $\mathbf{v} \in \{0, 1\}^N$, while $\mathbf{h} \in \{0, 1\}^M$ denotes accompanying hidden units, i.e., the latent variables. The combined vector $\mathbf{x} = (\mathbf{v}, \mathbf{h}) \in \{0, 1\}^{N+M}$ is referred to as a *particle*.

The latent structure facilitates simulation from the joint distribution through block Gibbs sampling, as block conditional probabilities $p(\mathbf{v}|\mathbf{h})$ and $p(\mathbf{h}|\mathbf{v})$ are available in explicit form and are easy to sample from. Indeed, letting $\mathbf{e}^{(n)}$ denote the nth coordinate of any vector $\mathbf{e}$, it holds that

$$p(\mathbf{v}|\mathbf{h}) \propto \prod_{n=1}^{N} \exp\{\mathbf{v}^{(n)}(\mathbf{W}\mathbf{h}^T + \mathbf{b}^T)^{(n)}\}.$$

Let sigm be the sigmoid function $\mathrm{sigm}(x) = (1 + e^{-x})^{-1}$. The probability function factorises and straightforward algebra gives

$$p(\mathbf{v}^{(n)} = 1|\mathbf{h}) = \mathrm{sigm}[(\mathbf{W}\mathbf{h}^T + \mathbf{b}^T)^{(n)}], \tag{4}$$

$$p(\mathbf{h}^{(m)} = 1|\mathbf{v}) = \mathrm{sigm}[(\mathbf{v}\mathbf{W} + \mathbf{c})^{(m)}]. \tag{5}$$

Hence, sampling from the conditional distributions $p(\mathbf{v}|\mathbf{h})$, $p(\mathbf{h}|\mathbf{v})$ amounts to sampling independent Bernoulli variables, with probabilities extracted from the sigmoid forms (4) and (5). Samples from the joint distribution $p(\mathbf{v}, \mathbf{h}) \propto \exp\{-E(\mathbf{v}, \mathbf{h})\}$ can thus be obtained by running a block Gibbs Markov chain [17].

Inference for the parameters in a RBM, for a particular data point $\mathbf{v}$, is often conducted via maximum likelihood, by minimising the negative log-likelihood, $-\log p(\mathbf{v})$, with respect to the parameters $\mathbf{W}, \mathbf{b}, \mathbf{c}$. This is usually achieved with gradient descent methods, for which an estimate of the gradient $\nabla p(\mathbf{v})$ is needed.

A calculation of the gradient coordinate corresponding to the partial derivative w.r.t the parameter $w_{n,m}$ at row n and column m in the matrix $\mathbf{W}$ yields (see [10] for details)

$$\nabla_{w_{n,m}} (-\log p(\mathbf{v})) = \cdots = -\mathbf{v}^{(n)} p(\mathbf{h}^{(m)} = 1|\mathbf{v}) + \mathrm{E}[\mathbf{v}^{(n)} p(\mathbf{h}^{(m)} = 1|\mathbf{v})]. \tag{6}$$

The first and second term on the right-hand side of (6) are often referred to as the *positive phase* and the *negative phase*, respectively. The negative phase is the problematic term as it amounts to taking expectation under the joint distribution of $(\mathbf{v}, \mathbf{h})$. However estimates of this part of the gradient can be obtained via the Gibbs sampling procedure.

When the number of data points $|\mathbf{D}| = d$ is large, the standard technique for minimising the average negative log-likelihood is (mini-batch) stochastic gradient descent (SGD). In the SGD method, a subset $\mathbf{D}' \subset \mathbf{D}$ of size $|\mathbf{D}'| = d' < d$ is chosen at random and the gradient coordinate w.r.t. $w_{n,m}$ at current parameter state is estimated by

$$\frac{1}{d'}\sum_{\mathbf{v}\in\mathbf{D}'} -\mathbf{v}^{(n)} p(\mathbf{h}^{(m)} = 1|\mathbf{v}) + \mathrm{E}[\mathbf{v}^{(n)} p(\mathbf{h}^{(m)} = 1|\mathbf{v})]. \tag{7}$$

Here the expectation is taken w.r.t to $p(\mathbf{v})$. Similar to standard gradient descent, the estimate of the gradient is used to update the parameter vector in step $n+1$ through

$$(\mathbf{W}, \mathbf{b}, \mathbf{c})_{n+1} = (\mathbf{W}, \mathbf{b}, \mathbf{c})_n - \eta(\widetilde{\nabla\mathbf{W}}, \widetilde{\nabla\mathbf{b}}, \widetilde{\nabla\mathbf{c}})$$

where η is a scalar learning rate, and where $\widetilde{\nabla\mathbf{x}}$ denotes the estimated gradient coordinates for the matrix/vector $\mathbf{x}$.

For RBMs, where the negative phase is estimated using Gibbs sampling, SGD requires simulations to be run for each training step. Usually, a Gibbs chain of size d' is run for a fixed number of κ steps before an average is formed with the end samples. Starting the Gibbs chain anew *at sampled data points* in each gradient step is referred to as the contrastive divergence (CD-κ) training method. Since long burn-in periods might be expected with this approach, the Gibbs chain for a certain training step is typically started at the last samples of the previous training step. This method is referred to as persistent contrastive divergence (PCD-κ). Current state-of-the-art method for training RBMs is arguably a combination of PCD-1 and PT.

3 Parallel Tempering and Infinite Swapping for Gibbs Samplers

Consider the setting of Sect. 2, that is a RBM trained with SGD with batch size d'. Parallel tempering amounts to multiple Gibbs chains of size d' being run at different *temperatures* and particles being exchanged according to a Metropolis–Hastings rule. In the case of two temperatures, in addition to the original model (2) with temperature $\tau_1 = 1$, an additional RBM with a higher temperature $\tau_2 > 1$ is introduced:

$$p_{\tau_2}(\mathbf{v}) = Z_{\tau_2}^{-1} \sum_{\mathbf{h}} e^{-\frac{1}{\tau_2}E(\mathbf{v},\mathbf{h})}.$$

It is easy to check that all calculations above for $\tau_1 = 1$ carry over in a straightforward manner to a RBM with $\tau_2 > 1$. The PT method proceeds by running two Gibbs chains C_1, C_2, at respective temperature, each of size d'. Let $\mathbf{x}_{1,1}, \ldots, \mathbf{x}_{1,d'}, \mathbf{x}_{2,1}, \ldots, \mathbf{x}_{2,d'}$ denote the particles in C_1, C_2 respectively. After κ Gibbs steps, particles $\mathbf{x}_{1,i}, \mathbf{x}_{2,i}$ $i = 1, \ldots, d'$ are swapped with probability

$$1 \wedge \frac{\exp\{-\frac{1}{\tau_1} E(\mathbf{x}_{2,i}) - \frac{1}{\tau_2} E(\mathbf{x}_{1,i}\}}{\exp\{-\frac{1}{\tau_1} E(\mathbf{x}_{1,i}) - \frac{1}{\tau_2} E(\mathbf{x}_{2,i})\}}.$$

Swaps of this kind are attempted according to the so-called *swap rate* (the jump intensity) of the algorithm. The process then starts anew with running C_1, C_2, at their respective temperatures for another κ Gibbs steps. The resulting process is ergodic with the product measure $p_{\tau_1} \otimes p_{\tau_2}$ as stationary distribution. Thus, the chain C_1 converges in distribution to p_{τ_1}, the distribution of interest.

There are several ways of extending PT to additional temperatures. Here, we follow the common approach of only attempting swaps between neighbouring particles, see [19]. For K chains with respective temperatures $\tau_1 < \cdots < \tau_K$, all swaps of the form $[\mathbf{x}_{k,i}, \mathbf{x}_{k+1,i}] \to [\mathbf{x}_{k+1,i}, \mathbf{x}_{k,i}]$ are attempted after every κ Gibbs step, starting at $k = 1$ and working upward to $k = K - 1$. The swapping probabilities used are thus

$$1 \wedge \frac{\exp\{-\frac{1}{\tau_k} E(\mathbf{x}_{k+1,i}) - \frac{1}{\tau_{k+1}} E(\mathbf{x}_{k,i})\}}{\exp\{-\frac{1}{\tau_k} E(\mathbf{x}_{k,i}) - \frac{1}{\tau_{k+1}} E(\mathbf{x}_{k+1,i})\}}.$$

The swapping mechanism limits the degree of dependency between samples and forces quicker mixing of samples, thus speeding up the convergence of C_1.

Compared to parallel tempering, the infinite swapping algorithm proposes a different mechanism for exchanging information between the tempered Gibbs chains. In PT, the particle exchange probabilities are given but the proposed changes (only neighbouring particles, etc.) can be chosen. In INS the full mechanism for exchange is used; in a sense, *all* possible swaps are attempted.

Consider K chains $C_1, \ldots, C_K$ with temperatures $\tau_1 < \tau_2 < \cdots < \tau_K$ and assume each chain contains only one particle (extending to the case of d' particles is straightforward). Denote by $\mathbf{x}_k = (\mathbf{v}_k, \mathbf{h}_k)$ the particle in chain k and let $\mathbf{X} = (\mathbf{x}_1, \ldots, \mathbf{x}_K)$ denote the vector of particles. Let σ_j, $j = 1, \ldots, K!$ denote a permutation of the temperature indices $[1, \ldots, K]$, for some ordering of the $K!$ permutations. Write the RBM probability function of temperature τ_k as $p_k(\mathbf{x}) = Z_k^{-1} \exp\{-E(\mathbf{x})/\tau_k\}$. Define the *symmetrised distribution* $\bar{p}$ and the joint probability distribution p_{σ_j} across the chains for permutation σ_j as

$$\bar{p}(\mathbf{X}) = \frac{1}{K!} \sum_{j=1}^{K!} p_{\sigma_j}(\mathbf{X}), \quad p_{\sigma_j}(\mathbf{X}) = \prod_{k=1}^{K} p_{\sigma_j^k}(\mathbf{x}_k),$$

where σ_j^k denotes the kth component of the permutation σ_j. The INS algorithm consists of first running the Gibbs chains independently for κ Gibbs steps; each Gibbs-step amounts to first sampling a point $\mathbf{h}$ from $p(\mathbf{h}|\mathbf{v})$ and then a point $\mathbf{v}$ from $p(\mathbf{v}|\mathbf{h})$. In the next step *temperatures* are swapped between the chains according to permutation σ_j with probability

$$\rho_{\sigma_j}(\mathbf{X}) = \frac{p_{\sigma_j}(\mathbf{X})}{\sum_{j=1}^{K!} p_{\sigma_j}(\mathbf{X})} = \frac{p_{\sigma_j}(\mathbf{X})}{K!\bar{p}(\mathbf{X})}.$$

The procedure is then repeated, resulting in a Markov chain sample generation scheme. The collective Markov chain $(C_1, \ldots, C_K)$ with the INS temperature swapping mechanism is referred to as the *INS-Gibbs Markov chain*. The isolated κ Gibbs steps together with one swapping operation will be referred to as an *INS-Gibbs step* of the INS-Gibbs Markov chain. Note that for large K, since the number of permutations is $K!$, the computational cost of INS can be too high for practical purposes. One can then use so-called *partial INS* (PINS) [8], in which temperatures are arranged into subgroups, swaps are attempted within one such subgroup at a time and with a handoff-rule for changing between subgroups. This reduces the computational cost of INS significantly. For example, in [4] $K = 30$ temperatures are used for a Lennard-Jones model of 55-atoms argon cluster. Similarly, in [8] a collection of $K = 45$ temperatures are used for a Lennard-Jones cluster of 38 atoms. In these complex potential energy landscape the partial infinite swapping approach is appreciably more effective than conventional tempering approaches; see [7] for an extensive numerical study.

To obtain an estimate of $\mathrm{E}_{p_1}[f(\mathbf{x})]$ for any real-valued function f, a weighted average of the particles of each chain is formed:

$$\sum_{k=1}^{K} f(\mathbf{x}_k) \left(\sum_{\sigma;\sigma^k=1} \rho_\sigma(\mathbf{X}) \right), \tag{8}$$

where $\{\sigma : \sigma^k = 1\}$ is the subset of permutations that have 1 as their kth component (that is, that assign the kth chain temperature τ_1). Proofs that the INS-Gibbs Markov kernel has the symmetrised distribution $\bar{p}$ as invariant distribution, and that the estimate (8) has the desired expected value $\mathrm{E}_{p_1}[f(\mathbf{x})]$ if samples are generated from $\bar{p}$, are included in the Appendix as Propositions 1 and 2, respectively. The proofs are similar to existing results and are included for completeness.

Because the sample space is finite and the Markov chains are irreducible, Proposition 1 ensures that the empirical measures of the full chain converges to the symmetrised distribution. Consequently, if the INS-Gibbs Markov chain is run long enough, its empirical measure will approximate the symmetrised distribution.

Algorithm 1 is an outline of the INS algorithm for obtaining an estimate of $E_{p_1}[f(\mathbf{x})]$ in the setting of Gibbs-sampling. For training a RBM, the last step of the algorithm corresponds to using the gradient negative phase estimate as f and

Algorithm 1 INS-Gibbs Algorithm

1. Set number of chains K, temperature values $[\tau_1, \ldots, \tau_K]$, number of Gibbs steps κ between swap attempts, number of swap attempts q and initial data points.
2. Start chains with initial data points.
3. **for** i in $1 : q$
   a. Run each chain for κ Gibbs steps.
   b. Draw permutation σ with probability ρ_σ
   c. Permute the temperatures of the chains according to permutation σ
4. Form an estimate as

$$y = \sum_{k=1}^{K} f(\mathbf{x_k}) \left(\sum_{\sigma;\sigma^k=1} \rho_\sigma(\mathbf{X}) \right)$$

the tempered RBM joint distributions for forming the weights. While obtaining an estimate as described in the algorithm, one can use the so-called particle/temperature-associations as a diagnostic of non-convergence. These are quantities than can be computed while running the algorithm and used to indicate whether it is possible for the empirical measure to have converged; see [4, 5] for details and an analysis of this diagnostic.

4 Numerical Experiments

In order to evaluate the performance of INS for training RBMs a series of numerical experiments are conducted. Two types of data sets, described in Sect. 4.3, are considered. For the smaller data sets the exact likelihood and exact gradient can both be computed, which enables comparison of the training algorithms. For larger data sets neither the exact likelihood nor the exact gradient is tractable. Instead a classification Boltzmann machine will be used to evaluate the training algorithms; the quantity used for comparison is referred to as *prediction accuracy*.

4.1 Prediction Accuracy

To compare training algorithms using a classification Boltzmann machine, each data point in the data set is concatenated with a vector $\mathbf{c} \in C$ representing its class, where C denotes the subspace of $\{0, 1\}^{dim(\mathbf{c})}$ such that exactly one coordinate is nonzero, and $dim(\mathbf{c})$ is the number of classes. Such a vector is called a *one-hot* vector in the machine learning literature. For a RBM defined on an extended visible state space $\tilde{\mathbf{v}} = (\mathbf{v}, \mathbf{c})$, the conditional probability for class type $\mathbf{c}$ given $\mathbf{v}$ can be computed explicitly: it holds that

$$p(\mathbf{c}|\mathbf{v}) = \frac{\hat{p}(\mathbf{c}, \mathbf{v})/Z}{\hat{p}(\mathbf{v})/Z} = \frac{\hat{p}(\mathbf{c}, \mathbf{v})}{\hat{p}(\mathbf{v})} = \frac{\hat{p}(\mathbf{c}, \mathbf{v})}{\sum_{\mathbf{c}\in C} \hat{p}(\mathbf{c}, \mathbf{v})},$$

and the terms on the right-hand side can be computed with the *marginalisation trick* (see the Appendix for a description). The above expression can be used for classification, and for calculating classification error through comparing with the actual class type for data points. The efficiency of the algorithms can be evaluated by measuring the classification capabilities of each parameter state during training on the extended data sets. More specifically, for each parameter state the *prediction accuracy*,

$$A = \sum_{(\mathbf{v}_i, \mathbf{c}_i)\in \mathbf{D}'} \log \left(\frac{\hat{p}(\mathbf{c}_i, \mathbf{v}_i)}{\sum_{\mathbf{c}\in C} \hat{p}(\mathbf{c}, \mathbf{v}_i)} \right),$$

is calculated, where $\mathbf{D}'$ is a randomised subset of data, for each state during training.

4.2 Parameters and Settings

This section describes the parameters involved in the training of RBMs and the choices made for this work. The main objective of the numerical experiments is a comparison with [3], where performance results for PT in the RBM setting are presented. Parameters have therefore been selected accordingly, and no attempts have been made to find optimal parameter settings for the sampling and training tasks under consideration. Where possible, external recommendations have been taken into account, as have empirical observations from experiments on the impact of parameter changes.

4.2.1 Learning Rate

The learning rate η can be chosen as a fixed constant or as a function of, e.g., the number of updates in SGD. A large learning rate increases the speed of training but may result in inaccurate optima or degenerate behaviour, while a small learning rate allows for greater accuracy at the cost of training speed. Numerical experiments indicate that a learning rate of 0.1 yields satisfactory performance, and this value is chosen throughout. This is also consistent with experiments in [3].

4.2.2 Initialization

Throughout the experiments, the initial weights and biases are drawn randomly from normal distributions of small variance (order of magnitude 0.01) and zero expected value, in line with the recommendations in [14]. Empirically, different draws do not yield different results, nor smaller changes in the variance parameter.

4.2.3 Training Steps

The number of training steps are set to 10000 in all the numerical experiments, as all the relevant effects seems to appear within this range. Furthermore, in the examples of lower dimension, empirical observations suggest the likelihood is maximal after this number of steps. If the training rate is decreased or if the data dimension is significantly increased, more training steps would likely be needed.

4.2.4 Batch Size

We set the batch size to be 10 throughout. Empirical studies show that this batch size is sufficient for non-degenerate behaviour of the likelihood. A larger batch size yields a more precise gradient estimate, but causes longer computational time. Therefore, for computational efficiency, a small batch size is preferred.

4.2.5 Temperatures

Both PT and INS run on temperatures $\{1, 2, 3, 4, 5\}$. Empirical studies show no significant difference when adding more temperatures for the examples under consideration. However, for certain parameter setups, fewer temperatures can result in degenerate behaviour similar to what is observed for a single chain, see Sect. 4.4. At five temperatures, no notable difference in running time between INS and PT is present, allowing for a fair comparison of the algorithms. For a large number of temperatures, the PT algorithm is considerably faster and a fair comparison would then be with PINS rather than INS. Moreover, mixing properties are satisfactory for the temperatures selection. This is also in line with demonstrations in [3], where 5 temperatures are considered (of roughly the same magnitude, however the exact temperature values are not disclosed).

4.2.6 Number of Swaps

For PT, one swap is attempted for every step in the Markov chain. This choice makes the experiments consistent with [3]. Moreover, it is experimentally observed that an increase in the number of swap attempts per training step does not seem to have any significant effect on the results.

4.3 Data Sets

We use two types of data sets for empirical evaluation of the INS algorithm for training a RBM. The first is a collection of toy data sets similar to that used in [3]; by changing the size of the toy data sets we can move from cases where the gradient can be computed exactly to those where this is not possible for either gradients or likelihoods. The second type of data set we consider is the well-known MNIST data (described in detail in a following subsection).

4.3.1 Toy Data

The toy data sets are generated according to a generalisation of the "Toy Data" generating mechanism in [3]. The procedure involves choosing the number of modes μ, the distance between modes δ, the number of samples per modes ν and a permutation probability π; we require the number of modes to be a power of 2. The data is generated as follows:

1. Compute $i = \log_2(\mu)$, the number of binaries needed to encode the modes.
2. Create a list of the binary encoding of each mode.
3. Create a list of expanded mode encodings by expanding each encoding in 2. with δ copies.
4. Generate ν copies of each expanded mode encoding from 3.
5. Flip every binary variable in 4. independently with probability π, to obtain the *explanatory data*.
6. Add a one-hot vector representing mode type to get the joint *explanatory and response data*.

The following example is for $(\mu, \delta, \nu, \pi) = (4, 2, 2, 0.2)$:

1. Let $i = \log_2(4) = 2$
2. $(1, 1),\ (1, 0),\ (0, 1),\ (0, 0)$
3. $(1, 1, 1, 1),\ (1, 1, 0, 0),\ (0, 0, 1, 1),\ (0, 0, 0, 0)$
4. $(1, 1, 1, 1),\ (1, 1, 1, 1),\ (1, 1, 0, 0),\ (1, 1, 0, 0),$
   $(0, 0, 1, 1),\ (0, 0, 1, 1),\ (0, 0, 0, 0),\ (0, 0, 0, 0)$
5. $(1, 1, 1, 0),\ (1, 1, 1, 1),\ (1, 1, 0, 1),\ (1, 1, 1, 0),$
   $(0, 0, 1, 0),\ (0, 0, 1, 1),\ (0, 0, 0, 0),\ (1, 0, 0, 0)$
6. $(1, 1, 1, 0, 1, 0, 0, 0),\ (1, 1, 1, 1, 1, 0, 0, 0),$
   $(1, 1, 0, 1, 0, 1, 0, 0),\ (1, 1, 1, 0, 0, 1, 0, 0),$
   $(0, 0, 1, 0, 0, 0, 1, 0),\ (0, 0, 1, 1, 0, 0, 1, 0),$
   $(0, 0, 0, 0, 0, 0, 0, 1),\ (1, 0, 0, 0, 0, 0, 0, 1)$

Fig. 1 Two examples from the MNIST data set

In [3], π is varied over a toy data set without class type attached, with $\mu = 4$, $\delta = 8$, and $\nu = 2500$. This example is treated in Sect. 4.4.

4.3.2 MNIST

The MNIST data set is used as a benchmark for training and evaluating machine learning algorithms.[1] It consists of 55000 pictures of handwritten images of numbers $0, \ldots, 9$. One data point is a 28×28 matrix populated with grayscale pixel numbers between 0 and 1; Fig. 1 shows two examples from the data set. In this work we round each pixel to 0 or 1 in order for the data to fit the binary RBM as described. Attached to each image is also a one-hot vector of dimension 10 representing number type.

4.4 Evaluating the INS Algorithm for Small Toy Data Sets

Consider a small toy data set for which the likelihood can be computed exactly and can be trained with exact SGD (exact gradient computed for a data subsample in each step). We compare four algorithms:

- exact SGD
- SGD PCD-1
- SGD PCD-1 with Parallel Tempering
- SGD PCD-1 with INS-Gibbs

The exact likelihood and the prediction accuracy is computed for each parameter state during each training algorithm. In addition, for every step in the respective Markov chain for PT and INS, the Euclidean distance between the true gradient and the gradient estimate is computed at two different fixed parameter states (early and late in the training). This empirical evaluation provides insight into the effectiveness of the INS-Gibbs algorithm. Moreover, it allows us to compare the non-standard performance measure prediction accuracy with the likelihood.

The first toy data set was generated using the following parameters: $(\mu, \delta, \nu, \pi) = (4, 4, 2500, 0.2)$. That is, the number of visible units is 12 (including 4 dimensions for class type one-hot vectors), and the number of data points is 10000 (see Sect. 4.3). For

[1]The data set, and more information about it, is available on Yann LeCun's webpage: http://yann.lecun.com/exdb/mnist/.

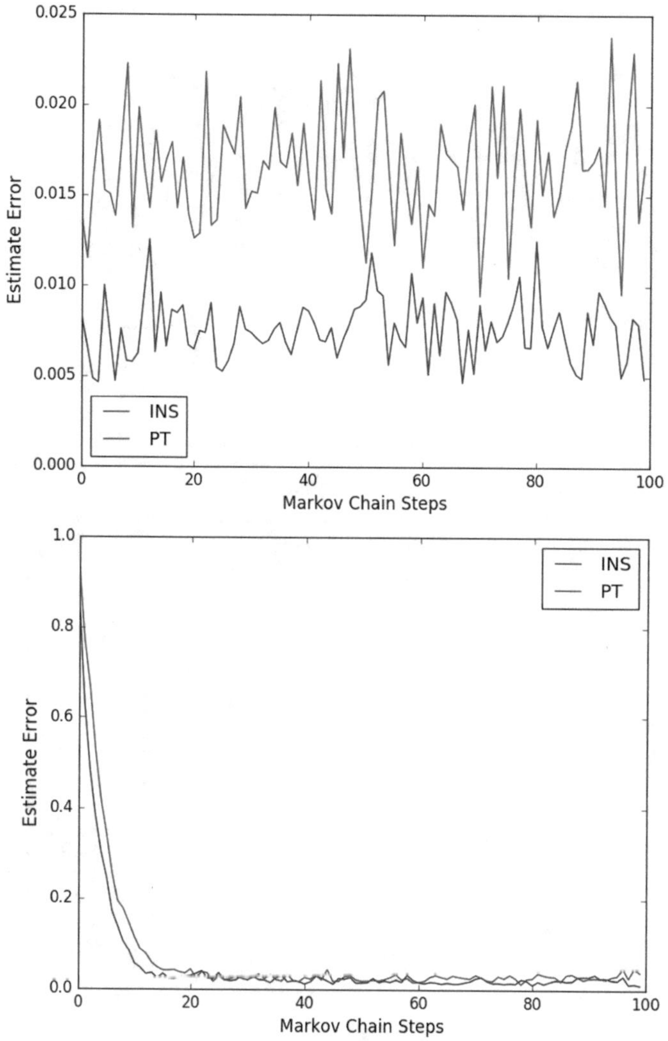

Fig. 2 Evaluation of different training algorithms on a small toy dataset. **Upper**: Euclidean distance from the gradient estimate to the true gradient, for the initial parameter state, as function over Markov chain steps. **Lower**: Euclidean distance from the gradient estimate to the true gradient, for the final parameter state, as function over Markov chain steps

the RBM model, the number of hidden units was set to $M = 4$, the starting points of the Markov chains were drawn uniformly and the training procedures were repeated 20 times for each of the algorithms.[2] The result of the experiment is illustrated in Figs. 2 and 3.

[2] All stochastic behaviour (except the data generation) were run on updated seeds in every iteration.

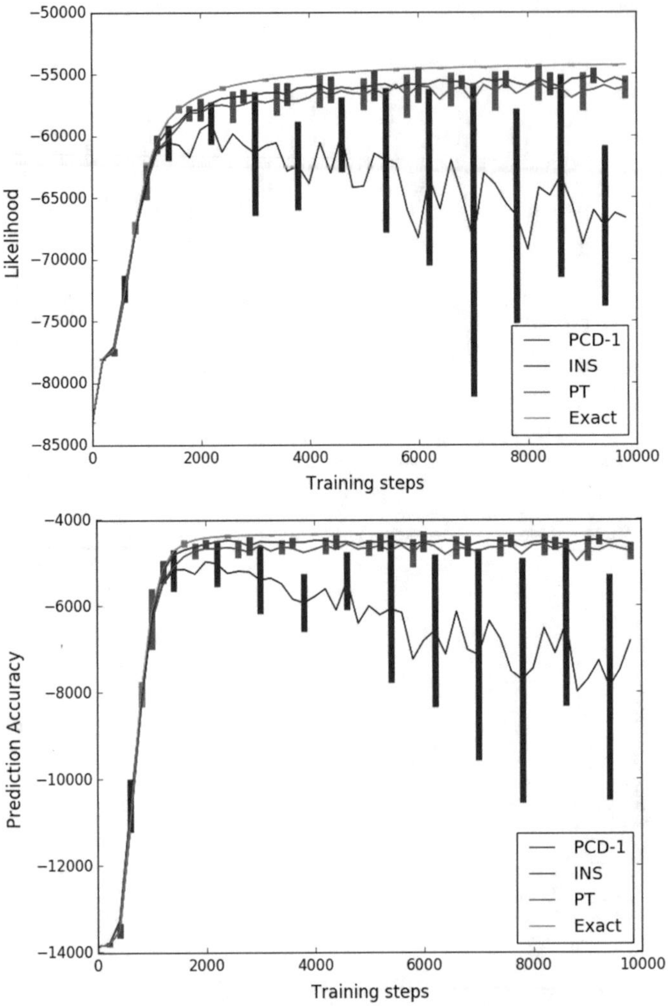

Fig. 3 Evaluation of different training algorithms on a small toy dataset. **Upper**: Average likelihood trajectory and variation (parameter state likelihood variance estimate). **Lower**: Average prediction accuracy trajectory and variation (parameter state prediction accuracy variance estimate)

Next, the experiments of [3] were recreated by using $(\mu, \delta, \nu, \pi) = (4, 8, 2500, 0.2)$. The number of hidden units was set to $M = 6$, all other parameters as in the previous experiment. The results are illustrated in Figs. 4 and 5.

Figures 1, 2, 3 and 4 show the INS algorithm generally outperforming PCD-1 and being slightly superior to PT, consistent with previous studies in different contexts. Early in the training phase gradient estimation is relatively easy and deviations from the true gradient (as a function over Markov steps) is small. However, INS seems to produce smaller estimation error and variance, resulting in a more stable behaviour

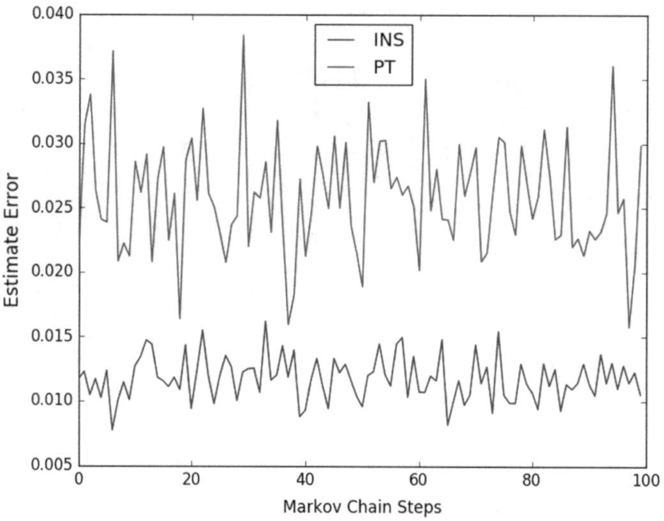

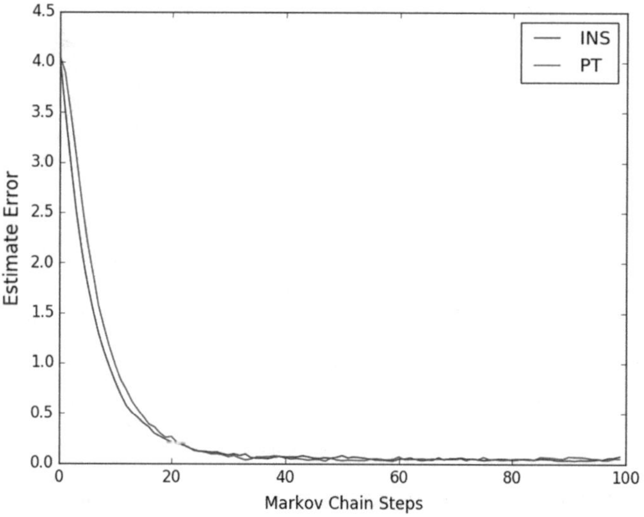

Fig. 4 Evaluation of different training algorithms on a data set generated according to [3]. See Figs. 1 and 2 for subgraph description

closer to the true gradient. Estimating the gradient becomes harder later in the training phase and both algorithms needs to be run for a considerable number of steps in order to converge. Again, we note that INS outperforms the other algorithms, converging slightly faster towards 0 distance to the true gradient than PT. The likelihood and prediction accuracy graphs paint a similar picture: both show INS performing better, in terms of average behaviour as well as estimated variance.

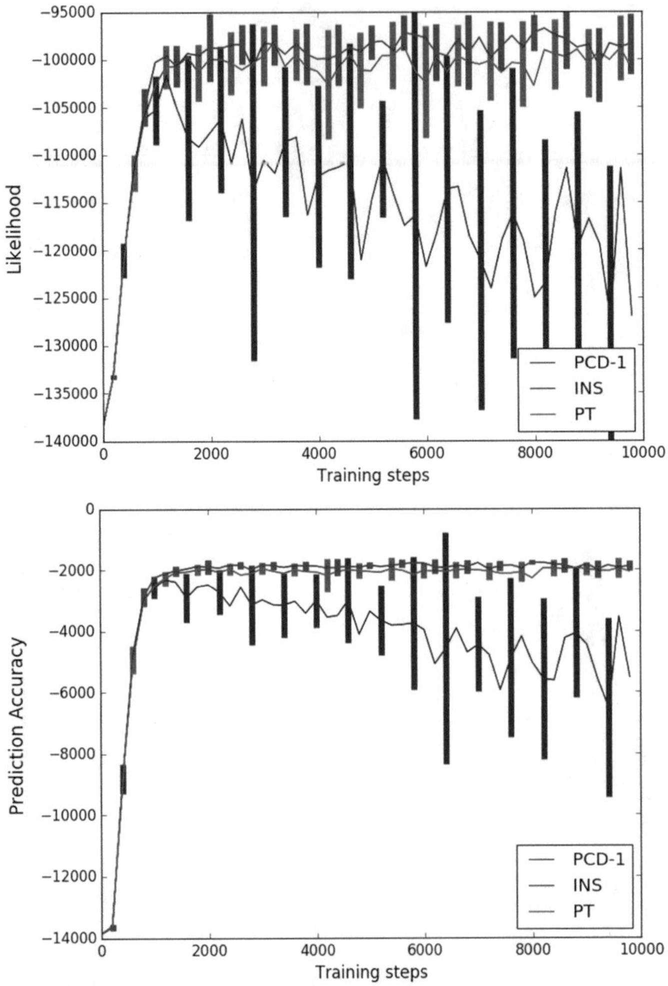

Fig. 5 Evaluation of different training algorithms on a data set generated according to [3]. See Figs. 1 and 2 for subgraph description

Remark 1 The distinction between algorithms regarding the gradient estimate quality can be expected to increase with decreasing π, as the modes in the data become more separated; this should increase the importance of good mixing. However, for the prediction accuracy to be able to distinguish between algorithms, the classification problem must be sufficiently difficult, motivating the choice of $\pi = 0.2$. Indeed, if π is too small, mixing becomes harder but classification simpler and the algorithms will all exhibit similar prediction accuracy.

4.5 *Evaluating the INS Algorithm for a Larger Toy Data Set and MNIST*

The performance of the training algorithm will now be evaluated on the MNIST data and a larger toy data set. Neither the exact training gradients nor the exact likelihood computations are now available. However, we can still use the prediction accuracy to evaluate the training algorithms. For the toy data generation the parameters were set to $(\mu, \delta, \nu, \pi) = (128, 150, 1000, 0.2)$, i.e the number of visible units is 1050, and the number of data points is 128000. Moreover, the number of hidden units were set to $M = 600$ for the toy data set and $M = 500$ for MNIST, remaining parameters were set as in the previous experiments. The outcome is illustrated in Fig. 6.

From Fig. 6, for the MNIST data INS enjoys the smallest variance initially but as the number of steps increase the differences between the algorithms have all but disappeared. Similarly for the toy data set; interesting to note for the toy data set is that PCD-1 does not display degenerate behaviour as in [3].

Remark 2 The results for MNIST, in particular when compared to those for likelihood estimation in [3], may be explained by the classification task being too simple, the modes being too few and too distinguished. In the large toy data set however, modes are greater in number and more similar. Here, we again observe a small difference in performance, both in terms of average behavior and variance. A reason might be that energy landscapes in higher dimensions tend to be less equipped with poor (in terms of training) local minima, putting less demand on gradient estimate quality [26].

5 Conclusions

We have presented the INS algorithm in a Gibbs-sampling setting for training RBMs, and conducted an empirical study of the performance of INS compared to persistent contrastive divergence and parallel tempering. The INS algorithm performs at least as well as all other training algorithms, for the cases investigated; the difference is most notable for smaller data sets. One possible explanation is that the gradient becomes hard to estimate late in training as the energy landscape becomes increasingly complex. A complex energy landscape prevents mixing, resulting different performances between the algorithms, due to their different mixing capabilities. As PT was developed to improve mixing over the PCD-1 method, and INS has been shown to be superior to PT in several models, the results of the empirical study are in line with expectations. The PCD-1 method even exhibits a degenerate behavior due to its poor mixing, as was also previously observed in [3].

For MNIST and the larger toy data set the modes of the distribution are further apart, which should prevent mixing to a stronger degree than for the other data sets. Therefore, it is at first surprising that the different algorithms perform more similarly here than for the toy data sets in Sect. 4.4. One possible explanation is

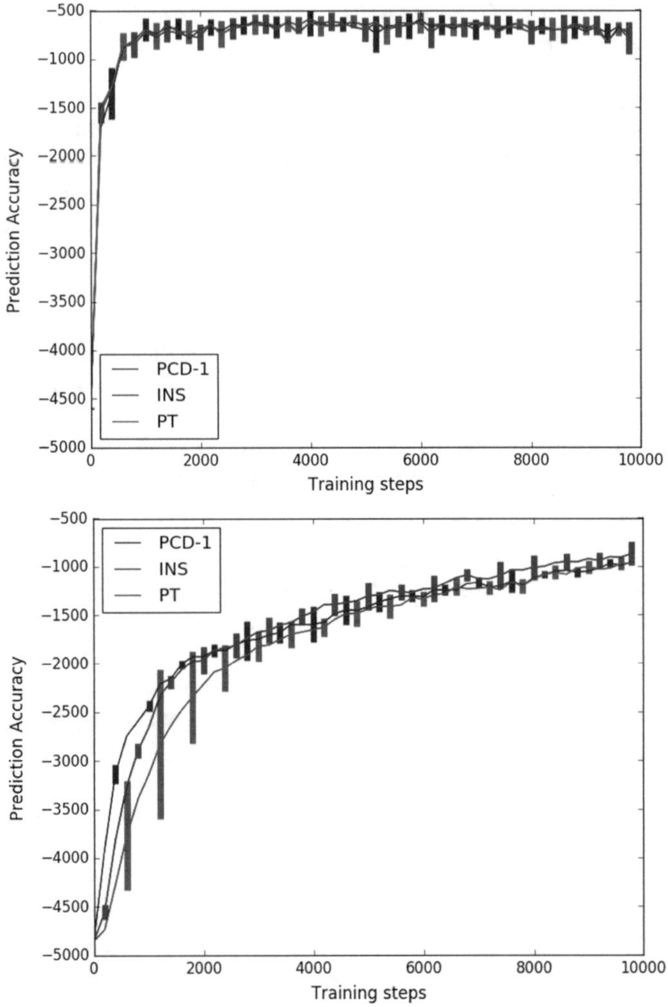

Fig. 6 Average prediction accuracy trajectory and variation for each training algorithm on the MNIST data set (upper) and the large toy dataset (lower)

provided by [26]: even though the energy landscape is complex, the collection of local minima that one is likely to end up in tends to promote good performance. However this line of reasoning does not take into account *how* SGD moves around in the energy landscape but instead looks at a static picture and "counts" the number of critical points of different indices. Recent works suggest that this view is too simplistic and that dynamics should be considered as well, see for example [1].

Another potential explanation for the observations for MNIST and the larger toy data set is that the performance measure, the prediction accuracy defined in Sect. 4.1, does not reflect the algorithms ability to mix at a fine enough level. This is combined with the fact that the classification task might be too simple, rendering the prediction accuracy incapable to distinguish between the different algorithms for training; see the remarks in Sects. 4.4 and 4.5. Indeed, the empirical study in [3] suggest significant improvements of PT over PCD-1 also for MNIST when likelihood is used to measure performance, whereas this is not observed when considering the prediction accuracy for the Boltzmann classifiers. Still, also for the toy data set in Sect. 4.5 and using the prediction accuracy, although the PCD-1 method seems of best quality early in training, INS again has slightly better classification capability later in training compared to the other methods.

Future work includes extending INS to variational auto-encoders, with an aim similar to [2], together with more extensive empirical studies, including both other data sets and comparing different performance measures (prediction accuracy, likelihood). These studies will also consider the impact of different hyperparameters in PT and INS (number of temperatures, choice of temperatures, swap rate for PT etc.), and performance when equal computational time is allotted to the different algorithms.

Appendix

The Marginalisation Trick

Let $\hat{p}(\mathbf{v}, \mathbf{h})$ denote the unnormalised joint probability function for $(\mathbf{v}, \mathbf{h})$ and $\hat{p}(\mathbf{v})$ denote the unnormalised probability function for $\mathbf{v}$ (suppressing parameter dependence):

$$\hat{p}(\mathbf{v}, \mathbf{h}) = \exp\{-E(\mathbf{v}, \mathbf{h})\}, \quad \hat{p}(\mathbf{v}) = \sum_{\mathbf{h}} \hat{p}(\mathbf{v}, \mathbf{h}).$$

For the unnormalised joint distribution, by Baye's rule it holds that, for any $\mathbf{h}$,

$$\hat{p}(\mathbf{v}) = \frac{\hat{p}(\mathbf{v}, \mathbf{h})}{\hat{p}(\mathbf{v}, \mathbf{h})/\hat{p}(\mathbf{v})} = \frac{\hat{p}(\mathbf{v}, \mathbf{h})}{p(\mathbf{h}|\mathbf{v})}.$$

The left-hand side is independent of $\mathbf{h}$ and can thus be computed by choosing $\mathbf{h}$ arbitrarily, and inserting it in the computable operation on the right hand side. Taking $\mathbf{h} \equiv 1$ yields

$$\sum_{\mathbf{h}} e^{-E(\mathbf{v},\mathbf{h})} = \frac{e^{-E(\mathbf{v},\mathbf{1})}}{p(\mathbf{1}|\mathbf{v})} = \frac{e^{-E(\mathbf{v},\mathbf{1})}}{\prod_{m=1}^{M} \mathrm{sigm}[(\mathbf{v}\mathbf{W} + \mathbf{c})^{(m)}]}.$$

In practice, $\mathbf{h}$ must be chosen with care in order to avoid numerical division by zero. For the numerical experiments in this paper

$$\mathbf{h} = \max_{\mathbf{h}} p(\mathbf{h}|\mathbf{v}) = (\mathrm{round}(p(\mathbf{h}^{(1)} = 1|\mathbf{v})), \ldots, \mathrm{round}(p(\mathbf{h}^{(M)} = 1|\mathbf{v}))),$$

is used where round denotes the rounding operator.

Propositions

Proposition 1 *The INS-Gibbs Markov kernel has the symmetrised distribution $\bar{p}$ as invariant distribution.*

Proof Let $G_{\kappa,\sigma_j}(\mathbf{X}|\mathbf{X}')$ be the probability distribution for $\mathbf{X}$ after κ Gibbs steps when starting in $\mathbf{X}'$ and temperatures are assigned according to σ_j. Given values $\mathbf{X}'$, the following probability distribution holds for sample values $\mathbf{X}$ obtained after one full INS-Gibbs step:

$$\sum_{j=1}^{K!} \rho_{\sigma_j}(\mathbf{X}') G_{\kappa,\sigma_j}(\mathbf{X}|\mathbf{X}').$$

Integration w.r.t the symmetrised distribution yields

$$\begin{aligned}
&\sum_{\mathbf{X}'} \sum_{j=1}^{K!} \rho_{\sigma_j}(\mathbf{X}') G_{\kappa,\sigma_j}(\mathbf{X}|\mathbf{X}') \bar{p}(\mathbf{X}') \\
&= \frac{1}{K!} \sum_{j=1}^{K!} \sum_{\mathbf{X}'} p_{\sigma_j}(\mathbf{X}') G_{\kappa,\sigma_j}(\mathbf{X}|\mathbf{X}') \\
&= \frac{1}{K!} \sum_{j=1}^{K!} p_{\sigma_j}(\mathbf{X}) = \bar{p}(\mathbf{X}).
\end{aligned}$$

In the first step the definition of ρ_{σ_j} is used and in the second last step the fact that the Gibbs kernel G_{κ,σ_j} has the joint distribution p_{σ_j} as its invariant distribution. □

Proposition 2 *Let* $\mathrm{E}_{\bar{p}}$ *and* E_{p_1} *denote expectation with respect to* $\bar{p}$ *and* p_1*, respectively. Then,*

$$\mathrm{E}_{\bar{p}}\left[\sum_{k=1}^{K} f(\mathbf{x_k}) \left(\sum_{\sigma;\sigma^k=1} \rho_\sigma(\mathbf{X})\right)\right] = \mathrm{E}_{p_1}[f(\mathbf{x_1})].$$

Proof For any $k = 1, \ldots, K$, it holds that

$$
\begin{aligned}
\mathrm{E}_{\bar{p}}\left[f(\mathbf{x_k})\left(\sum_{\sigma;\sigma^k=1}\rho_\sigma(\mathbf{X})\right)\right] &= \mathrm{E}_{\bar{p}}\left[f(\mathbf{x_k})\left(\sum_{\sigma;\sigma^k=1}\rho_\sigma(\mathbf{X})\right)\right] \\
&= \sum_{\mathbf{X}} f(\mathbf{x_k})\left(\sum_{\sigma;\sigma^k=1}\rho_\sigma(\mathbf{X})\right)\bar{p}(\mathbf{X}) \\
&= \sum_{\mathbf{X}} f(\mathbf{x_k})\left(\sum_{\sigma;\sigma^k=1}\frac{p_\sigma(\mathbf{X})}{K!}\right) \\
&= \frac{1}{K!}\sum_{\mathbf{X}} f(\mathbf{x_k})\left(\sum_{\sigma;\sigma^k=1}\prod_{i=1}^{K} p_{\sigma^i}(\mathbf{x}_i)\right) \\
&= \frac{1}{K!}\sum_{\mathbf{X}} f(\mathbf{x}_k)\left(\sum_{\sigma;\sigma^k=1} p_{\sigma^k}(\mathbf{x}_k)\prod_{i\neq k}^{K} p_{\sigma^i}(\mathbf{x}_i)\right) \\
&= \frac{1}{K!}\sum_{\mathbf{x}_k} f(\mathbf{x}_k)p_1(\mathbf{x}_k)\sum_{\mathbf{x}_i, i\neq k}\left(\sum_{\sigma;\sigma^k=1}\prod_{i\neq k}^{K} p_{\sigma^i}(\mathbf{x}_i)\right) \\
&= \frac{1}{K!}\sum_{\mathbf{x}_k} f(\mathbf{x}_k)p_1(\mathbf{x}_k)(K-1)! \\
&= \frac{1}{K}\mathrm{E}_{p_1}[f(\mathbf{x}_1)].
\end{aligned}
$$

Summing over $k = 1, \ldots, K$ proves the claim. □

References

1. Arora, S., Cohen, N., Golowich, N., Hu, W.: A convergence analysis of gradient descent for deep linear neural networks. arXiv:1810.0228 (2018)
2. Caterini, A.L., Doucet, A., Sejdinovic, D.: Hamiltonian variational auto-encoder. In: 32nd Conference on Neural Information Processing Systems (NeurIPS) (2018)
3. Desjardins, G., Courville, A., Bengio, Y., Vincent, P., Dellaleau, O.: Parallel tempering for training restricted Boltzmann machines. In: JMLR Workshop and Conference Proceedings: AISTATS 2010, vol. 9, pp. 145–152 (2010)
4. Doll, J.D., Dupuis, P.: On performance measures for infinite swapping Monte Carlo methods. J. Chem. Phys. **143** (2015)
5. Doll, J.D., Dupuis, P., Nyquist, P.: A large deviations analysis of certain qualitative properties of parallel tempering and infinite swapping algorithms. Appl. Math. Optim. **78**(1), 103–144 (2018)
6. Doll, J.D., Dupuis, P., Nyquist, P.: Thermodynamic integration methods, infinite swapping and the calculation of generalized averages. J. Chem. Phys. **146**, 134111 (2017)

7. Doll, J.D., Plattner, N., Freeman, D.L., Liu, Y., Dupuis, P.: Rare-event sampling: occupation-based performance measures for parallel tempering and infinite swapping Monte Carlo methods. J. Chem. Phys. **137** (2012)
8. Dupuis, P., Liu, Y., Plattner, N., Doll, J.D.: On the infinite swapping limit for parallel tempering. Multiscale Model. Simul. **10**(3), 986–1022 (2012)
9. Earl, D.J., Deem, M.W.: Parallel tempering: theory, applications, and new perspectives. Phys. Chem. Chem. Phys. **7**, 3910–3916 (2005)
10. Fischer, A., Igel, C.: An introduction to restricted Boltzmann machines. In: Alvarez, L., Mejail, M., Gomez, L., Jacobo, J. (eds.) Progress in Pattern Recognition, Image Analysis, Computer Vision, and Applications. Lecture Notes in Computer Science, vol. 7441. Springer, Berlin (2012)
11. Freund, Y., Haussler, D.: Unsupervised learning of distributions on binary vectors using two layer networks. Technical Report, University of California, Santa Cruz (1994)
12. Geyer, C.J.: Markov chain Monte Carlo maximum likelihood. In: Interface Foundation of North America. Retrieved from the University of Minnesota Digital Conservancy. http://hdl.handle.net/11299/58440 (1991)
13. Geyer, C.J., Thompson, E.A.: Annealing Markov chain Monte Carlo with applications to ancestral inference. J. Am. Stat. Assoc. **90**(431), 909–920 (1995)
14. Hinton, G.E.: A practical guide to training restricted Boltzmann machines. In: Montavon, G., Orr, G.B., Müller, K.R. (eds.) Neural Networks: Tricks of the Trade. Lecture Notes in Computer Science, vol. 7700. Springer, Berlin (2012)
15. Hinton, G.E.: Products of experts. In: Proceedings of the Ninth International Conference on Artificial Neural Networks (ICANN), vol. 1, pp. 1–6 (1999)
16. Hinton, G.E.: Training products of experts by minizing constrastive divergence. Neural Comput. **14**, 1771–1800 (2002)
17. Jensen, C.S., Kong, A., Kjaerulff, U.: Blocking-Gibbs sampling in very large probabilistic expert systems. Int. J. Hum.-Comput. Stud. 647–666 (1995)
18. Kofke, D.A.: On the acceptance probability of replica-exchange Monte Carlo trials. J. Chem. Phys. **117**(15), 6911–6914 (2002)
19. Liu, J.S.: Monte Carlo strategies in scientific computing. Springer Series in Statistics. Springer, New York (2008)
20. Lu, J., Vanden-Eijnden, E.: Infinite swapping replica exchange molecular dynamics leads to a simple simulation patch using mixture potentials. J. Chem. Phys. **138** (2013)
21. Marinari, E., Parisi, G., Ruiz-Lorenzo, J.J.: Numerical simulations of spin glass systems. In: Spin Glasses and Random Fields, vol. 12 (1997)
22. Menz, G., Schlichting, A., Tang, W.: Ergodicity of the infinite swapping algorithm at low temperature. arXiv:1811.10174 (2018)
23. Plattner, N., Doll, J.D., Dupuis, P., Wang, H., Liu, Y., Gubernatis, J.E.: An infinite swapping approach to the rare-event sampling problem. J. Chem. Phys. **135** (2011)
24. Plattner, N., Doll, J.D., Meuwly, M.: Overcoming the rare event sampling problem in biological systems with infinite swapping. J. Chem. Theory Comput. **9**(9), 4215–4224 (2013)
25. Rao, F., Caflisch, A.: Replica exchange molecular dynamics simulations of reversible folding. J. Chem. Phys. **119**, 4035 (2003)
26. Sagun, L., Ugur Guney, V., Lecun, Y.: Explorations on high dimensional landscapes, ICLR (2015)
27. Salakhutdinov, R., Mnih, A., Hinton, G.E.: Restricted Boltzmann machines for collaborative filtering, ICML (2007)
28. Smolensky, P.: Information processing in dynamical systems: foundations of harmony theory. In: Rumelhart, D.E., McClelland, J.L. (eds.) Parallel Distributed Processing (Chapter 6), vol. 1, pp. 194–281. MIT Press, Cambridge (1986)
29. Sugita, Y., Okamoto, Y.: Replica-exchange molecular dynamics method for protein folding. Chem. Phys. Lett. **314**, 141–151 (1999)
30. Swendsen, R.H., Wang, J.S.: Replica Monte Carlo simulation of spin glasses. Phys. Rev. Lett. **57**, 2607–2609 (1986)

31. Tieleman, T.: Training restricted Boltzmann machines using approximations to the likelihood gradient. In: Proceedings of the 25th International Conference on Machine Learning, pp. 1064–1071. Helsinki, Finland (2008)
32. Welling, M., Rosen-Zvi, M., Hinton, G.E.: Exponential family harmoniums with an application to information retrieval. In: NIPS 17, vol. 17. MIT Press, Cambridge (2005)
33. Yu, T.Q., Lu, J., Abrams, C.F., Vanden-Eijnden, E.: A multiscale implementation of infinite-swap replica exchange molecular dynamics. Proc. Natl. Acad. Sci. USA **113**(42), 11744–11749 (2016)

Sensitivity Ranks by Monte Carlo

Ian Iscoe and Alexander Kreinin

Abstract Application of the Monte Carlo method to the estimation of sensitivity ranks is considered. We demonstrate that the convergence rate in this problem is exponential, $\exp(-\alpha N)$, where N is the number of scenarios and $\alpha > 0$ is a constant. This result stands in contrast to the usual rate of convergence, $N^{-1/2}$, of the Monte Carlo method for estimating the mean of a random variable. This result justifies a numerical strategy of sensitivity estimation of portfolios depending on a large number of risk factors.

Keywords Sensitivity ranks · Exponential rate of convergence · Large deviations

1 Introduction

Sensitivity estimation is an important problem in the risk management of financial portfolios. This problem almost always leads to time-consuming Monte Carlo simulation (MC) which is impractical without the aid of variance reduction techniques. In particular, the calculation of sensitivities of CVA (Credit Value Adjustment) for a portfolio, is a particularly onerous task.

Estimation of the ranks of sensitivities has been found empirically to require a relatively small number of scenarios. (Some examples will be given in the last section of this paper.) In the present paper we consider a simple probabilistic model that helps to understand this phenomenon. The approach is relevant to CVA sensitivities.

The sensitivity problem can be described as follows. Suppose the value of a portfolio depends on some stochastic, risk factor, X, which is generally a discrete-time path and is vector-valued at each time point along the path. Suppose also, that the chosen model for X contains some parameters, collected into a vector $\pi \in \mathbb{R}^M$.

I. Iscoe · A. Kreinin (✉)
IBM, Quantitative Research, 185 Spadina Ave., Toronto M5T2C6, Canada
e-mail: akreinin@sscinc.com

I. Iscoe
e-mail: ian.iscoe@ca.ibm.com

B. Tuffin and P. L'Ecuyer (eds.), *Monte Carlo and Quasi-Monte Carlo Methods*, Springer Proceedings in Mathematics & Statistics 324,
https://doi.org/10.1007/978-3-030-43465-6_15

The parameters may represent volatilities, mean reversion rates or, especially, initial values of the individual, scalar risk factors. The value of a portfolio can be expressed as a risk-neutral expectation, $\mathbb{E}[f(X;\pi)]$, of an appropriate pricing function which makes the presence of the parameters explicit. (It is not just for convenience to have the parameters be absorbed into f rather than the risk-neutral probability measure; it is crucial for our approach.) If one considers a set of K vectors of parameters, $\{\pi_1, ..., \pi_K\}$ perturbing the base value of π, then we obtain the K sensitivities,

$$\mathbb{E}[f(X;\pi_k)] - \mathbb{E}[f(X;\pi)], \ k = 1, ..., K.$$

The ranks of these sensitivities, say largest to smallest, then coincide with the ranks of the values, $\mathbb{E}[f(X;\pi_k)]$, $k = 1, ..., K$, $(K \leq M)$.

In the present paper, we use this definition in a more traditional partial case: the vector of parameters π_k differs from π by the kth component only and the increment of this component is small. If one estimates the values $\mathbb{E}[f(X;\pi_k)]$, by a MC simulation, it is then of interest to know the likelihood that the estimated ranks coincide with the theoretical ranks. This is the problem that we address; namely, estimating the rate of convergence of the simulated ranks to the theoretical ranks, as the sample-size tends to infinity.

This problem had been studied in the literature on ranking and selection algorithms [2, 3, 8]. In [2], it is shown that using ordinal optimization the probability of correct selection converges at an exponential rate for a large class of systems. This result was obtained under the independence assumption of the noise of the measurements. In [3] the problem of allocating total sampling budget amongst several populations in an asymptotically optimal manner is studied. It is demonstrated that the probability of false selection is minimized.

Similar results were obtained in [5, 6] on parallel simulation of Markov processes resulting, in an associated coupled process. Suppose, from the processes we simulate in parallel, one wishes to choose the process with best performance, maximizing some expectation of an objective function. With finite simulation runs, there is a probability that the process with the best sample performance is not the one with the best expected performance. It is shown in [5, 6] that the probability of missing the processes with the best expected performance tends to 0 at an exponential rate under the assumption that the processes we wish to compare are associated.[1]

In the present paper, we extend the result on exponential convergence to the case when the observations are not necessarily independent and no additional constraints are imposed on the dependence structure. This more general assumption better describes the sensitivity estimation schema. In the next section, we will abstract the problem slightly and then present a sequence of results, in increasing generality. The most general result,[2] Theorem 2, for an arbitrary dimension K is obtained using the large-deviation techniques [4]. These techniques are less restrictive and allow

[1]Two processes are called associated if their distributions are associated. The latter means that all increasing functions of these variables are positively correlated.

[2]This result is new, to the best of our knowledge.

us to obtain the exponential rate of convergence of the probability of the correct rank estimation without additional assumptions on the dependence structure and the structure of the support of the distributions.

Numerical illustrations will be given in the final section. In closing the present section, we recall that in MC simulation, the use of quasi-random scenarios often achieves a significantly faster rate of convergence of mathematical expectations, compared with the use of pseudo-random scenarios. Although we only establish rigorous theoretical results for the latter, it will be observed in the numerical illustrations, that quasi-random scenarios have the same advantage over pseudo-random scenarios for ranking, although the advantage is not as dramatic.

2 Theoretical Results

We begin with a description of the general setting in which we will work. Let $\mathbf{Z} = (Z_1, Z_2, \ldots, Z_K)$ be a random vector such that the mean values of the coordinates are finite and distinct, and ordered[3]:

$$\mathbb{E}[Z_1] < \mathbb{E}[Z_2] < \cdots < \mathbb{E}[Z_K].$$

In this case we shall write that rank $(\mathbb{E}[Z_k]) = k$.

Suppose one has decided to estimate the order of the mean values, $\mu_k = \mathbb{E}[Z_k]$, $(k = 1, 2, \ldots, K)$, using Monte Carlo simulation.

Denote by $(\zeta_{1,n}, \zeta_{2,n}, \ldots, \zeta_{K,n})$ the nth random sample of the vector $\mathbf{Z}$, $n = 1, 2, \ldots, N$. Assume that estimation of μ_k is based on the statistics

$$\hat{\mu}_k(N) = \frac{1}{N} \sum_{n=1}^{N} \zeta_{k,n}, \quad k = 1, 2, \ldots, K.$$

Our objective is to estimate the probability

$$p_K(N) = \mathbb{P}\left(\bigcap_{k=1}^{K} \left(\text{rank}\left(\hat{\mu}_k(N)\right) = k\right) \right) \tag{1}$$

as a function of K and N. We represent

$$Z_k = \mu_k + \varepsilon_k, \quad k = 1, \ldots, K$$

and $\zeta_{k,n}$ as

$$\zeta_{k,n} = \mu_k + \varepsilon_{k,n},$$

[3]The connection to the financial setting in the Introduction, is that $\mathbf{Z}$ is a re-ordering of the $\{f(X; \pi_k)$: $1 \leq k \leq K\}$, according to the re-ordering of their expectations, in increasing order.

where $(\varepsilon_k : 1 \le k \le K)$, $(\varepsilon_{k,n} : 1 \le k \le K,\ 1 \le n \le N)$ are centred random variables.

We shall start with the Gaussian case for which very explicit results can be obtained, and then move to the general (non-Gaussian) case. For the Gaussian case, we will even start with the very simple subcase, $K = 2$, and assume that the uncorrelated random variables ε_k have the common normal distribution $\mathcal{N}(0, \sigma_0^2)$.

2.1 Gaussian Case: K = 2, Uncorrelated ε_1 and ε_2

Here we assume in addition that the $\varepsilon_{k,n}$ are mutually independent. Then we have

$$\begin{aligned} p_2(N) &= \mathbb{P}\left(\hat{\mu}_1(N) < \hat{\mu}_2(N)\right) = \mathbb{P}\left(\frac{1}{N}\sum_{n=1}^{N}\varepsilon_{1,n} + \mu_1 < \frac{1}{N}\sum_{n=1}^{N}\varepsilon_{2,n} + \mu_2\right) \\ &= \mathbb{P}\left(\frac{1}{N}\sum_{n=1}^{N}(\varepsilon_{1,n} - \varepsilon_{2,n}) < \mu_2 - \mu_1\right). \end{aligned} \tag{2}$$

The random variable

$$\frac{1}{N}\sum_{n=1}^{N}(\varepsilon_{1,n} - \varepsilon_{2,n}) \sim \mathcal{N}\left(0, \frac{2}{N}\sigma_0^2\right),$$

has a centred normal distribution, with variance, $\frac{2}{N}\sigma_0^2$. Therefore, from (2) we derive

$$p_2(N) = \Phi\left(\frac{\mu_2 - \mu_1}{\sigma_0}\sqrt{\frac{N}{2}}\right) \sim 1 - \frac{e^{-\alpha^2 N/2}}{\sqrt{2\pi N}\alpha}, \tag{3}$$

where

$$\alpha = \frac{\mu_2 - \mu_1}{\sigma_0\sqrt{2}},$$

and as usual, the cumulative distribution function of a standard normal random variable is denoted by $\Phi(\cdot)$.

The asymptotic relation (3) shows that the rate of convergence $p_2(N) \to 1$ is exponential under the assumptions of normality and independence of ε_k. However, the rate is slowed for moderate N if μ_2 is very close to μ_1. This is intuitively (qualitatively) clear and is made quantitative by the result (3). (The critical value, for fixed N, is at $\mu_2 - \mu_1 \propto 1/\sqrt{N}$.) This property persists in all of the cases which we consider.

2.2 General Gaussian Case

Denote by $\bar{\Phi}_2(x, y; \rho)$, the tail of the standard bivariate normal distribution,

$$\bar{\Phi}_2(x, y; \rho) = \mathbb{P}(\xi_1 > x, \xi_2 > y), \qquad \xi_1 \sim \mathcal{N}(0, 1), \quad \xi_2 \sim \mathcal{N}(0, 1)$$

where ρ is the correlation coefficient of the standard normal random variables ξ_1 and ξ_2. Define the constants

$$\alpha_k := \frac{\mu_{k+1} - \mu_k}{\sigma_0\sqrt{2(1-\rho_{k,k+1})}}, \quad k = 1, 2, \ldots, K-1$$

where $\rho_{k,k+1}$ denotes the correlation between ε_k and ε_{k+1}, which is assumed not to equal 1. Let $\alpha_* = \min\limits_{1 \le k < K} \alpha_k$ and denote the multiplicity of α_*, by $m_* = \#\{k : \alpha_k = \alpha_*\}$.

Theorem 1 *Suppose that the K-dimensional vector ε has a normal distribution $\varepsilon \sim \mathcal{N}\left(0, \sigma_0^2 C\right)$, where $C = \|\rho_{ij}\|$, $(1 \le i, j \le K)$ is a nonsingular correlation matrix, $K > 2$ and $\sigma_0 > 0$. Then,*

$$1 - \sum_{k=1}^{K-1} \bar{\Phi}\left(\alpha_k\sqrt{N}\right) \le p_K(N) \le 1 - \sum_{k=1}^{K-1} \bar{\Phi}\left(\alpha_k\sqrt{N}\right) + \sum_{1 \le i < j < K} \bar{\Phi}_2\left(\alpha_i\sqrt{N}, \alpha_j\sqrt{N}; \tilde{\rho}_{i,j}\right)$$

where

$$\tilde{\rho}_{i,j} := \frac{\rho_{i,j} - \rho_{i+1,j} - \rho_{i,j+1} + \rho_{i+1,j+1}}{2\sqrt{(1-\rho_{i,i+1})(1-\rho_{j,j+1})}}, \quad 1 \le i, j \le K-1.$$

Also,

$$p_K(N) \sim 1 - \frac{m_*}{\sqrt{2\pi N}\alpha_*} \exp\left(-\frac{1}{2}\alpha_*^2 N\right), \quad \text{as } N \to \infty. \tag{4}$$

Proof Denote

$$S_k \equiv S_{k,N} := \sum_{n=1}^{N} \left(\varepsilon_{k,n} - \varepsilon_{k+1,n}\right), \quad k = 1, 2, \ldots, K-1.$$

Let us express the probability, $p_K(N)$, in terms of the random variables, S_k:

$$p_K(N) = \mathbb{P}\left(\bigcap_{k=1}^{K-1}\left\{\frac{S_{k-1}}{N} < \mu_{k+1} - \mu_k\right\}\right).$$

The covariance matrix of the random vector $\mathbf{S} = (S_1, S_2, \ldots, S_{K-1})$ is given by

$$\mathrm{Var}(S_k) = 2N\sigma_0^2 \cdot (1 - \rho_{k,k+1}), \tag{5}$$

$$\mathrm{cov}(S_i, S_j) = N\sigma_0^2 \cdot \left(\rho_{i,j} - \rho_{i+1,j} - \rho_{i,j+1} + \rho_{i+1,j+1}\right). \tag{6}$$

From (5) and (6) it follows that for $1 \le i, j \le K - 1$

$$\tilde{\rho}_{i,j} := \mathrm{Corr}(S_i, S_j) = \frac{\rho_{i,j} - \rho_{i+1,j} - \rho_{i,j+1} + \rho_{i+1,j+1}}{2\sqrt{(1 - \rho_{i,i+1})(1 - \rho_{j,j+1})}}. \tag{7}$$

Let us now define the standard normal random variables

$$\eta_k \equiv \eta_{k,N} := \frac{S_{k,N}}{\sqrt{\mathrm{Var}(S_{k,N})}} = \frac{S_{k,N}}{\sigma_0\sqrt{2(1 - \rho_{k,k+1})}\sqrt{N}}.$$

We have $\mathrm{Corr}(\eta_i, \eta_j) = \tilde{\rho}_{i,j}$, $1 \le i, j \le K - 1$.

The probability $p_K(N)$ can be written as

$$p_K(N) = \mathbb{P}\Big(\eta_1 < \alpha_1\sqrt{N}, \ldots, \eta_{K-1} < \alpha_{K-1}\sqrt{N}\Big).$$

Then we have

$$\begin{aligned} 1 - p_K(N) &= \mathbb{P}\left(\bigcup_{k=1}^{K-1} \{\eta_k \ge \alpha_k\sqrt{N}\}\right) \\ &\le \sum_{k=1}^{K-1} \mathbb{P}\Big(\eta_k \ge \alpha_k\sqrt{N}\Big) \qquad (8) \\ &= \sum_{k=1}^{K-1} \bar{\Phi}\Big(\alpha_k\sqrt{N}\Big). \qquad (9) \end{aligned}$$

Denote

$$\bar{p}_k(N) = \mathbb{P}\left(\eta_k \ge \alpha_k\sqrt{N}\right) \text{ and } \bar{p}_{i,j}(N) = \mathbb{P}\left(\eta_i \ge \alpha_i\sqrt{N}, \eta_j \ge \alpha_j\sqrt{N}\right).$$

Using the well known Bonferroni inequality, we derive

$$\begin{aligned} 1 - p_K(N) &\ge \sum_{k=1}^{K-1} \bar{p}_k(N) - \sum_{1 \le i < j < K} \bar{p}_{i,j}(N) \qquad (10) \\ &= \sum_{k=1}^{K-1} \bar{\Phi}\Big(\alpha_k\sqrt{N}\Big) - \sum_{1 \le i < j < K} \bar{\Phi}_2\Big(\alpha_i\sqrt{N}, \alpha_j\sqrt{N}; \tilde{\rho}_{i,j}\Big). \qquad (11) \end{aligned}$$

Now, we have

$$\sum_{k=1}^{K-1} \bar{\Phi}\left(\alpha_k\sqrt{N}\right) \sim m_*\bar{\Phi}\left(\alpha_*\sqrt{N}\right)$$
$$\sim \frac{m_*}{\sqrt{2\pi\alpha_*\sqrt{N}}}\exp\left(-\frac{1}{2}\alpha_*^2 N\right), \quad \text{as } N \to \infty. \tag{12}$$

The inequalities (9) and (11) are equivalent to those in the first part of this theorem. We will show that the right-hand side of (9) is the dominant term on the right-hand side of (11), asymptotically as $N \to \infty$. The asymptotic result (4), in this theorem, follows immediately from the asymptotic result (12).

If either $\alpha_i > \alpha_*$ or $\alpha_j > \alpha_*$ then

$$\bar{\Phi}_2(\alpha_i\sqrt{N}, \alpha_j\sqrt{N}; \tilde{\rho}_{i,j}) = \mathbb{P}\left(\eta_i \geq \alpha_i\sqrt{N},\ \eta_j \geq \alpha_j\sqrt{N}\right) \leq \mathbb{P}\left(\eta \geq \alpha\sqrt{N}\right),$$

where $\eta \sim \mathcal{N}(0, 1)$ and $\alpha := \max(\alpha_i, \alpha_j) > \alpha_*$. For such i and j, we then have

$$\bar{\Phi}_2(\alpha_i\sqrt{N}, \alpha_j\sqrt{N}; \tilde{\rho}_{i,j}) \leq \bar{\Phi}(\alpha\sqrt{N}) = o\left(\bar{\Phi}(\alpha_*\sqrt{N})\right), \quad \text{as } N \to \infty. \tag{13}$$

There are $(m_* - 1)m_*/2$ terms of the form $\bar{\Phi}_2(\alpha_*\sqrt{N}, \alpha_*\sqrt{N}; \tilde{\rho}_{i,j})$ in (11), each of which can be estimated using the following general asymptotic result (see Lemma 1 in Sect. 4)

$$\bar{\Phi}_2(a, a; \rho) \sim \frac{(1+\rho)^2}{2\pi\sqrt{1-\rho^2}} \cdot \frac{e^{-a^2/(1+\rho)}}{a^2}, \quad \text{as } a \to \infty, \tag{14}$$

which yields

$$\bar{\Phi}_2(\alpha_*\sqrt{N}, \alpha_*\sqrt{N}; \tilde{\rho}_{i,j}) \sim \frac{(1+\tilde{\rho}_{i,j})^2}{2\pi\sqrt{1-\tilde{\rho}_{i,j}^2}} \cdot \frac{e^{-\alpha_*^2 N/(1+\tilde{\rho}_{i,j})}}{\alpha_*^2 N}$$
$$= o\left(\bar{\Phi}(\alpha_*\sqrt{N})\right), \quad \text{as } N \to \infty. \tag{15}$$

Summing over the pairs of indices (i, j), $1 \leq i < j \leq K-1$, for each of which either the estimate (13) or the estimate (15) holds, then yields that the second sum in (11) is dominated by the first sum, asymptotically as $N \to \infty$. □

2.3 General Case

For the general case, we apply large-deviation techniques. (The reference [4] can be consulted for cited results.) Set

$$\beta_k := \mu_{k+1} - \mu_k, \quad k = 1, 2, ..., K - 1. \tag{16}$$

With $S_{k,N}$ as in the proof of Theorem 1, we obtain, similarly to (8) and (10):

$$1 - p_K(N) \le \sum_{k=1}^{K-1} \mathbb{P}\big(S_{k,N} \ge \beta_k N\big) \tag{17}$$

$$1 - p_K(N) \ge \sum_{k=1}^{K-1} \mathbb{P}\big(S_{k,N} \ge \beta_k N\big) - \sum_{1 \le i < j < K} \mathbb{P}\big(S_{i,N} \ge \beta_i N,\ S_{j,N} \ge \beta_j N\big). \tag{18}$$

For $\theta \in \mathbb{R}$, set

$$M_k(\theta) := \mathbb{E}\big[e^{\theta(\varepsilon_k - \varepsilon_{k+1})}\big], \quad k = 1, 2, ..., K - 1. \tag{19}$$

We assume that each M_k is finite for sufficiently small $|\theta|$. Then, the inequality (17) already provides us with the exponential rate of convergence by a simple application of Chebyshev's inequality (the same argument used to obtain the upper bound in the classical, large-deviation result of H. Cramér—see the proof of Theorem 2.2.3 in [4]):

$$\mathbb{P}(S_{k,N} \ge \beta_k N) \le \mathbb{E}[e^{\theta S_{k,N}}]e^{-\beta_k \theta N} \ = \ \exp(-[\beta_k\theta - \log M_k(\theta)])\ \forall \theta \ge 0;$$

therefore

$$\mathbb{P}(S_{k,N} \ge \beta_k N) \le \inf_{\theta \ge 0} \exp(-[\beta_k\theta - \log M_k(\theta)]) \ \equiv \ \exp(-I_k(\beta_k)N) \tag{20}$$

where

$$I_k(\beta) := \sup_{\theta}[\beta\theta - \log M_k(\theta)] = \sup_{\theta \ge 0}[\beta\theta - \log M_k(\theta)], \tag{21}$$

the latter equality being Lemma 2.2.5(b) in [4]. Combining (17) and (20), yields

$$1 - p_K(N) \le \sum_{k=1}^{K-1} \exp(-I_k(\beta_k)N). \tag{22}$$

Let $\gamma_* = \min_{1 \le k < K} I_k(\beta_k)$ and denote $m_* = \#\{k : I_k(\beta_k) = \gamma_*\}$, the multiplicity of γ_*. Then the right-hand side of (22) is asymptotically equivalent to $m_* \exp(-\gamma_* N)$. It only remains to show that $\gamma_* > 0$; then (22) provides the convergence in probability of ranks, at a rate that is at least exponentially fast. We defer the proof that $\gamma_* > 0$ to Sect. 4

Theorem 2 *Let ε be such that the moment generating functions, in (19), are finite in a neighbourhood of $\theta = 0$. Define the functions I_k, $1 \le k \le K - 1$ as in (21). With β_k as in (16), set $\gamma_* := \min(I(\beta_k) : 1 \le k \le K - 1)$. Then $\gamma_* > 0$ and*

$$1 - p_K(N) \le \sum_{k=1}^{K-1} \exp(-I_k(\beta_k)N) \sim m_* \exp(-\gamma_* N), \ \ as\ N \to \infty, \tag{23}$$

where $m_* := \#\{k : I_k(\beta_k) = \gamma_*\}$.

3 Numerical Examples

The fast convergence of sensitivity ranks is illustrated in this section by two examples. In the first example, we consider CVA sensitivity ranks of a medium size financial portfolio. In the second example, we consider a simple equity derivative and estimate its value sensitivities.

Let $\mathbf{R} = (R_1, \ldots, R_m)$, $\mathbf{R} \in \mathbb{R}^m$, be a vector of parameters. Suppose that the value function of interest, $f(\mathbf{R})$, is a smooth real function of the parameters $\mathbf{R}$. The sensitivity of the function $f(\mathbf{R})$ to the increment of the kth component is

$$\Delta_k f = f(R_1, \ldots, R_{k-1}, R_k + \epsilon_*, R_{k+1}, \ldots R_M) - f(R_1, \ldots, R_{k-1}, R_k, R_{k+1}, \ldots R_M),$$

where $\epsilon_* = 10^{-4}$ (i.e.1 basis point).

In the first example, we illustrate the exponentially fast rate of convergence of the ranks of means, with some calculations of CVA under various increments of the initial interest rate term structure. (CVA is an expectation of a weighted sum of discounted exposures to a counterparty, at some fixed, future times; the weights are the probabilities of the counterparty defaulting in the interval just prior to the times of exposure calculations. See Sect. 21.6 in [1] for details.) The Hull-White 1-factor (extended Vasicek) model was used as a short-rate model of the term structure. For the HW model, the mean reversion rate is 0.1 and the instantaneous volatility is 0.05.

The timeline is 8 years long with all cashflow dates and exposures at yearly time intervals. The interest rate curve (IR) and discount factors are given in Table 1. A collection of interest rate increments was considered for the CVA sensitivities: 7 single-node ϵ_*-increments; the affected nodes on the term structure were the ones from years 1 to 7.

Quasi Monte Carlo (QMC) scenarios were used to calculate the Expected Discounted Exposures (EDE) which make up the CVA. The choice of 2047 sce-

Table 1 IR term structure: T is time, r is the continuously compounded rate, d is the discount factor, $d = \exp(-rT)$

T	1	2	3	4	5	6	7	8
r	0.0284	0.0301	0.0317	0.0333	0.0351	0.0369	0.0386	0.0400
d	0.9719	0.9415	0.9092	0.8750	0.8387	0.8010	0.7631	0.7259

narios was selected as a practical benchmark. The numbers of scenarios, $n = 31, 63, 127, 2047, 4095$ (all of the form $2^m - 1$ for $m = 5, 6, 7, 11, 12$; the reason for subtracting 1 is to preserve the low discrepancy of each scenario set). Some comparisons with pseudo-random scenarios are also included.

3.1 Example 1

The first example is of a portfolio of 40 instruments: 20 Swaps and 20 Caps/Flrs with some instruments at the money and others out of the money. All instruments' effective date is current time, 0, with maturities of 7 or 8 years. More details concerning Notionals, Swap fixed rates or Cap/Flr strike rates, and instrument type, including payer/receiver type, is given in Table 2.

The results of the CVA rankings, after the tweaks, are given in Table 3. Evidently, convergence has already occurred with only 31 scenarios. However, the CVA values themselves and the CVA sensitivities take longer to stabilize, as is shown in the accompanying Tables 4 and 5. We omit the calculation of percentage errors for the sensitivities because, for the very small (insignificant) ones, the large 'percentage error' is a misleading metric for quality of approximation. What is important, is that insignificance is seen to persist throughout the small sample sizes. Rather, we report the ranking of the absolute values of the sensitivities, in Table 6.

Notice that there are a couple of inversions of the rankings for small sample sizes. However, the inversions are of consecutive or closely neighbouring ranks. Therefore, when considering the largest absolute sensitivities—say the top third or even half—that group remains invariant across different sizes. This observation leads to an efficient viewpoint for sensitivity calculations familiar to the specialists in the area of statistical selection [8]:

> One can identify the most significant sensitivities as a group, in a stable manner, using only a very small number of scenarios; then re-estimate that smaller group of sensitivities more accurately, using a larger number of scenarios.

For sake of comparison, we include the analogous final results for the sensitivities, based on pseudo random samples, in Table 7. Although there is still rapid convergence of the ranks—but not the values themselves—the rate is somewhat slower than with quasi random samples, as can be expected. In Tables 3, 4, 5 and 6, n denotes the number of QMC scenarios.

Table 2 Instrument details for Example 1: 'Type' is the instrument type: 'P' for Payer Swap, 'R' for Receiver Swap, 'C' for Cap, 'F' for Flr; N is the Notional (a negative value indicates a short position); T is the maturity date; K is the fixed rate for a Swap or the strike rate for a Cap/Flr

Swaps							
Type	N	T	K	Type	N	T	K
P	12	8	0.05	P	18	8	0.04
R	18	7	0.05	R	16	7	0.04
R	20	8	0.05	R	19	8	0.04
P	25	7	0.05	P	24	7	0.04
P	22	7	0.06	P	22	7	0.04
P	17	8	0.06	P	18	8	0.04
R	15	7	0.06	R	17	7	0.04
R	20	8	0.07	R	21	8	0.04
P	25	7	0.06	P	25	7	0.04
P	22	7	0.07	P	23	7	0.05
Caps/Flrs							
Type	N	T	K	Type	N	T	K
C	12	8	0.05	C	18	8	0.04
F	−18	7	0.05	F	−16	7	0.04
F	20	8	0.05	F	19	8	0.04
C	25	7	0.05	C	24	7	0.04
C	−22	7	0.06	C	22	7	0.04
C	17	8	0.06	C	18	8	0.04
F	15	7	0.06	F	17	7	0.04
F	20	8	0.07	F	21	8	0.04
C	25	7	0.06	C	−25	7	0.04
C	22	7	0.07	C	23	7	0.05

Table 3 CVA ranking results for Example 1: In the first column we have the IR node index whose rate was increased by 1bp

Node	n				
	4095	2047	127	63	31
1	7	7	7	7	7
2	2	2	2	2	2
3	4	4	4	4	4
4	3	3	3	3	3
5	5	5	5	5	5
6	1	1	1	1	1
7	6	6	6	6	6

Table 4 CVA results for Example 1: Under 'Node', π indicates the base vector of parameters, the other cells describe the index of the IR node incremented by 1bp

Node	n				
	4095	2047	127	63	31
π	1.7576067511080	1.755083637	1.708833929	1.695534710	1.651886330
1	1.7576004087327	1.755077294	1.708827608	1.695528422	1.651879872
2	1.7576108100729	1.755087694	1.708837973	1.695538649	1.651890515
3	1.7576073730937	1.755084261	1.708834534	1.695535437	1.651887079
4	1.7576074233940	1.755084300	1.708834580	1.695535437	1.651887262
5	1.7576057280000	1.755083011	1.708832874	1.695533750	1.651885753
6	1.7576219982234	1.755098088	1.708849191	1.695550016	1.651901875
7	1.7576051209165	1.755082006	1.708832356	1.695532835	1.651883435

Table 5 CVA sensitivity results for Example 1: The first column describes index the node that was incremented by 1bp

Node	n				
	4095	2047	127	63	31
1	−0.0000063421	−0.0000063424	−0.0000063217	−0.0000062877	−0.0000064579
2	0.0000040573	0.0000040590	0.0000040438	0.0000039396	0.0000041854
3	0.0000006248	0.0000006220	0.0000006050	0.0000007272	0.0000007495
4	0.0000006633	0.0000006723	0.0000006507	0.0000007274	0.0000009321
5	−0.0000006250	−0.0000010231	−0.0000010554	−0.0000009601	−0.0000005761
6	0.0000144513	0.0000152471	0.0000152619	0.0000153062	0.0000155454
7	−0.0000016307	−0.0000016302	−0.0000015738	−0.0000018748	−0.0000028945

Table 6 Absolute CVA sensitivity ranking results for Example 1: The first column describes the IR node index that was increased by 1bp

Node	n				
	4095	2047	127	63	31
1	2	2	2	2	2
2	3	3	3	3	3
3	7	7	7	7	6
4	5	6	6	6	5
5	6	5	5	5	7
6	1	1	1	1	1
7	4	4	4	4	4

Table 7 Absolute CVA sensitivities and their ranks using pseudo MC, for Example 1: Under 'Node', we have index of the IR node incremented by 1bp; the second and the third columns describe their ranks for $n = 127$ and $n = 63$, where n denotes the number of scenarios

Node	n			
	Rank		Absolute sensitivity	
	127	63	127	63
1	2	2	0.0000078252	0.0000059181
2	3	3	0.0000054325	0.0000035142
3	7	7	0.0000006848	0.0000004295
4	6	5	0.0000007657	0.0000008795
5	5	6	0.0000012004	0.0000007131
6	1	1	0.0000158122	0.0000159841
7	4	4	0.0000019491	0.0000026241

3.2 *Example 2*

In the second example we consider computation of the sensitivities of a European basket equity derivative illustrating applicability of the methodology to sensitivity computation of the relatively simple instruments. The basket consists of four equities, each of which follows a GBM process

$$S_t^{(i)} = S_0^{(i)} \cdot e^{(r-\frac{1}{2}\sigma_i^2)t+\sigma_i W_t^{(i)}}, \quad (i = 1, 2, 3, 4), \quad t \geq 0,$$

with the initial values $\mathbf{S_0} = \left(S_0^{(1)}, S_0^{(2)}, S_0^{(3)}, S_0^{(4)}\right)$, where $S_0^{(1)} = 10, S_0^{(2)} = 11$, $S_0^{(3)} = 15$, and $S_0^{(4)} = 14.5$ and the volatilities σ_i are $\sigma_1 = 0.4, \sigma_2 = 0.2$, $\sigma_3 = 0.26, \sigma_4 = 0.18$.

The payoff of the option is $P_T = \max\left(\max_{1\leq l\leq 4} S_T^{(i)} - K, 0\right)$. The maturity of the option, $T = 2.5$ years. The strike, $K = 13.2729$. The interest rate is constant in this model, $r = 0.024$. The option value is $V = \mathbb{E}\left[e^{-rT} P_T\right]$.

The computation of the expectation in the pricing formula above is based on the QMC scenario generation. More precisely, we use the Sobol' points generator for production of the normally distributed random processes. As in Example 1, the rate of convergence of the QMC estimator is much higher then that of the MC estimators using a pseudo-random scenario generation. The surpassing efficiency of the QMC strategy, in general, is known [7]. In our particular case, it is illustrated in Fig. 1, where the option value is shown as a function of the logarithm of the number of scenarios, N. The number of scenarios $N \in \{2^k - 1, k = 5, 6, \ldots, 19.\}$. The benchmark value

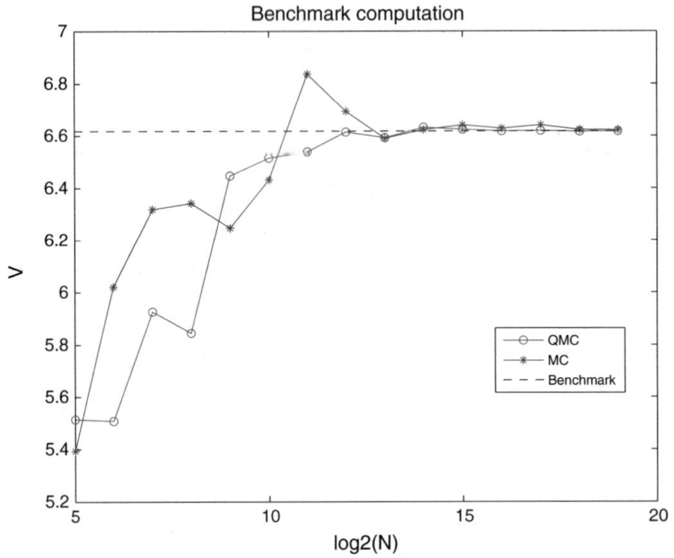

Fig. 1 Comparison of the QMC and MC estimators: option value as a function of $\log_2 N$, where N is the number of scenarios

of the option is obtained by MC with $N_B = 200 \cdot 2^{19}$ scenarios[4]: $V_B = 6.62$. The standard deviation of the MC results in this computation is $\sigma_B = 0.008$.

The vector of parameters includes 4 initial values $S_0^{(i)}$, $(i = 1, 2, 3, 4)$ and 4 volatilities σ_j, having indices $i = j + 4$. Thus the vector of sensitivities of the basket value has 8 components. Their ranks are shown in Table 8. The highest rank is rank_1, the lowest is rank_8. In particular, the value $\text{rank}_1 = 4$ in the second raw, corresponding to $k = 6$, means that the highest rank has sensitivity to the parameter 4. The stabilization of the ranks of the sensitivities is again obtained with relatively small number of scenarios.

4 Proof of the Result (14)

In this section, we provide the proof of the result (14), used in Theorem 1.

Lemma 1 *Let* $|\rho| < 1$ *and* $b > 0$. *The following asymptotics is valid as* $b \to \infty$.

$$\Phi_\rho^{(2)}(-b, -b) = \frac{(1+\rho)^2}{2\pi\sqrt{1-\rho^2}} \frac{e^{-b^2/(1+\rho)}}{b^2}(1 + o(1))$$

[4]We ran MC simulation 200 times with 2^{19} scenarios.

Table 8 Estimation of sensitivity ranking by QMC

k	Rank_1	Rank_2	Rank_3	Rank_4	Rank_5	Rank_6	Rank_7	Rank_8
5	1	5	4	2	3	6	8	7
6	4	1	3	2	6	5	8	7
7	2	1	5	3	6	4	8	7
8	3	1	4	2	6	5	8	7
9	5	1	4	2	6	3	8	7
10	5	1	4	2	6	3	8	7
11	5	1	4	2	6	3	8	7
12	4	1	5	2	6	3	8	7
13	4	1	5	2	6	3	8	7
14	4	1	5	2	6	3	8	7
15	4	1	5	2	6	3	8	7
16	4	1	5	2	6	3	8	7
17	4	1	5	2	6	3	8	7
18	4	1	5	2	6	3	8	7
19	4	1	5	2	6	3	8	7

Proof By symmetry, the asymptotics will be obtained for the equivalent probability,

$$\int_b^\infty \int_b^\infty \phi_\rho^{(2)}(x, y)\,dx\,dy, \qquad \phi_\rho^{(2)}(x, y) := \frac{e^{-\frac{1}{2(1-\rho^2)}(x^2-2\rho xy+y^2)}}{2\pi\sqrt{1-\rho^2}}.$$

Making the change of variables, $x \mapsto bx$, $y \mapsto by$, and setting $\theta = b^2$, transforms the integral to

$$\theta \int_1^\infty \int_1^\infty \frac{e^{-\frac{\theta}{2(1-\rho^2)}(x^2-2\rho xy+y^2)}}{2\pi\sqrt{1-\rho^2}}\,dx\,dy.$$

The dominant contribution to the integral comes from any neighbourhood of the point, in the region of integration, which minimizes the quadratic form in the exponent of the integrand. It is straightforward to check that the quadratic form has no critical points in the region and, along each of the two boundary lines, it is an increasing function. Therefore the minimum is attained at the corner point, (1, 1), and for the purposes of asymptotics, we change coordinates to make (1, 1) our new origin.

Express the quadratic form in terms of $x - 1$ and $y - 1$ (an exact Taylor expansion):

$$\begin{aligned} x^2 - 2\rho xy + y^2 &= 2(1-\rho) + 2(1-\rho)(x-1) + 2(1-\rho)(y-1) \\ &\quad + (x-1)^2 - 2\rho(x-1)(y-1) + (y-1)^2. \end{aligned}$$

Making the change of variables, $x \mapsto (x+1)/\theta$, $y \mapsto (y+1)/\theta$, and using the latter expansion, transforms the integral (and its preceding factor, θ) to

$$e^{-\theta/(1+\rho)}\theta^{-1}\int_0^\infty\int_0^\infty \frac{e^{-\frac{1}{1+\rho}(x+y)-\frac{1}{2\theta(1-\rho^2)}(x^2-2\rho xy+y^2)}}{2\pi\sqrt{1\ \ \rho^2}}\,dx\,dy.$$

As $\theta \to \infty$, the integral converges to the constant

$$\int_0^\infty\int_0^\infty \frac{e^{-\frac{1}{1+\rho}(x+y)}}{2\pi\sqrt{1-\rho^2}}\,dx\,dy = \frac{1}{2\pi\sqrt{1-\rho^2}}\left(\int_0^\infty e^{-x/(1+\rho)}\,dx\right)^2 = \frac{(1+\rho)^2}{2\pi\sqrt{1-\rho^2}}.$$

The lemma is now immediate, as $\theta = b^2$. □

Let us now provide the proof of the result $\gamma_* > 0$, in Theorem 2, and also present some provisional material concerning a lower bound, to compliment the upper bound given in that theorem.

Lemma 2 *Let X be a real-valued r.v. such that $M(\theta) := \mathbb{E}[\exp(\theta X)]$ is finite for $|\theta| < \theta_o$, for some $\theta_o > 0$. Denote $m := \mathbb{E}[X]$. Let I be the Legendre transform of $\log M$; so that $I(m) = 0$. Then $I(\beta) > 0$ for all $\beta > m$.*

Proof Without loss of generality, we may and do assume that $m = 0$. Now, $I(0) = 0$. Assume, for the sake of a contradiction, that I is identically 0 on some $[0, \beta^*]$, with $\beta^* > 0$. (I is nondecreasing on (m, ∞), so the previous statement is the negation of the conclusion in the theorem.)

Set $L(t) := \log M(\theta)$; then $L(0) = 0 = L'(0)$. Now, L is convex, so its derivative, L', is nondecreasing, hence nonnegative for $\theta > 0$. L' cannot be 0 along any sequence tending to 0 because $L' = M'/M$ being analytic near $\theta = 0$, would force L' to be identically 0 near $\theta = 0$; then L, and hence M, would then be constant there. M would have to be identically 1, forcing X to be identically 0. This is of course impossible because the rate function, I, would then be identically infinite above the mean, 0, and so never 0 anywhere above the mean. Thus $L'(\theta) > 0$ for $0 < \theta < \theta^*$, for some $\theta^* > 0$.

By continuity, we can assume θ^* is sufficiently small, that $0 < L'(\theta) < \beta^*$, for $0 < \theta < \theta^*$. By the fundamental connection between L and I:

$$\text{“}L(\theta) < \infty \text{ and } L'(\theta) = \beta\text{”} \implies I(\beta) = \theta\beta - L(\theta),$$

we conclude from $I(L'(\theta)) = 0$, $0 \le \theta < \theta^*$, that

$$0 = \theta L'(\theta) - L(\theta), \quad \text{for all } 0 \le \theta < \theta^*.$$

Solving this ODE, yields $L(t) = ct$, for some constant, c. Since $L(0) = 0$, c must be 0 and L is identically 0; i.e., M is identically 1, implying that X is identically 0—a contradiction, as in a previous argument in the proof. □

For a lower bound corresponding to the upper bound given in Theorem 2, we see from (18), that we require a lower bound for the innermost sum and an upper bound for the outermost sum. By a general result in Large Deviation theory (see e.g., [4]), for each k, we have

$$\mathbb{P}(S_{k,N} \geq \beta_k N) \geq \exp(-I_k(\beta_k)N + o(N)).$$

Therefore, by retaining only the slowest, decaying terms, we obtain

$$\sum_{k=1}^{K-1} \mathbb{P}(S_{k,N} \geq \beta_k N) \geq m_* \exp(-\gamma_* N + o(N)). \tag{24}$$

Next, we come to the bivariate probabilities in (18), $\mathbb{P}\big(S_{i,N} \geq \beta_i N,\ S_{j,N} \geq \beta_j N\big)$, which we can rewrite in vectorial form, $\mathbb{P}\big(S_{ij,N} \in N\,\mathscr{R}_{ij}\big)$, $\quad 1 \leq i < j < K$, where $S_{ij,N} := (S_{i,N}, S_{j,N})$ and $\mathscr{R}_{ij}$ is the closed rectangle,

$$\mathscr{R}_{ij} := \{(s_1, s_2) \in \mathbb{R}^2 : s_1 \geq \beta_i, s_1 \geq \beta_j\}.$$

Denote the moment generating function of the pair, $\varepsilon_{ij} := (\varepsilon_i - \varepsilon_{i+1}, \varepsilon_j - \varepsilon_{j+1})$, by[5]

$$M_{ij}(\theta) := \mathbb{E}[e^{\langle \theta, \varepsilon_{ij} \rangle}], \quad \theta = (\theta_1, \theta_2), \quad 1 \leq i < j < K,$$

which we assume is finite for all θ in a neighbourhood of the origin, $\mathbf{0} := (0, 0) \in \mathbb{R}^2$, and set

$$\begin{aligned} I_{ij}(\beta) &:= \sup_{\theta}[\langle \theta, \beta \rangle - \log M_{ij}(\theta)], \quad 1 \leq i < j < K, \\ J(\mathscr{R}_{ij}) &:= \inf_{\beta \in \mathscr{R}_{ij}} I_{ij}(\beta), \quad 1 \leq i < j < K. \end{aligned}$$

A general result from the theory of Large Deviations (Corollary 6.1.6 in [4]), states that, for all $\delta > 0$ there exists an N_0 such that for all $N \geq N_0$,

$$\mathbb{P}\big(S_{ij,N} \in N\,\mathscr{R}_{ij}\big) \leq \exp(-N(J(\mathscr{R}_{ij}) - \delta)).$$

However, this is too weak for our purpose, so we return to the simple argument using Chebyshev's inequality, to derive a better bound. Set $\beta_{ij} := (\beta_i, \beta_j)$. Then, for any $\theta = (\theta_1, \theta_2) \geq \mathbf{0}$ (componentwise):

$$\begin{aligned} \mathbb{P}\big(S_{i,N} \geq \beta_i N,\ S_{j,N} \geq \beta_j N\big) &\leq \mathbb{E}[e^{\theta_1 S_{i,N} - \theta_1 \beta_i N} e^{\theta_2 S_{j,N} - \theta_2 \beta_j N}] \\ &= \exp(-[\langle \theta, \beta_{ij} \rangle - \log M_{ij}(\theta)]N) \\ &\leq \exp(-I_{ij}(\beta_{ij})N). \end{aligned} \tag{25}$$

[5]The angled brackets $\langle \cdot, \cdot \rangle$ denote the Euclidean inner product on $\mathbb{R}^2$.

Remark 1 In the definition of I_{ij}, we only need to consider $\theta \geq \mathbf{0}$ (componentwise) for this derivation, when taking the supremum over θ, because $\beta_{ij} > \mathbf{0}$—the proof is identical to that in the one-dimensional case (see the proof of (2.2.6) in Lemma 2.2.5(b) of [4]).

The problem is now reduced to comparing the large-deviation rate function (I_{ij}) for a bivariate random vector with the rate functions (I_i and I_j) of the vector's components. There is a simple relation in the theory of Large Deviations, known as the *Contraction Principle* (Theorem 4.2.1 in [4]), which states:

Theorem 3 (Contraction Principle) *Let $f : X \to Y$ be a continuous mapping between two Hausdorff topological spaces, and let I be a good rate function*[6] *on X. Define*

$$I^f(y) := \inf\{I(x) : f(x) = y\}$$

with the convention that the infimum over the empty set is ∞. Then I^f is a good rate function on Y and if $X_1, X_2, \ldots$ satisfies a large deviation principle with rate function I, then $Y_1, Y_2, \ldots$ satisfies a large deviation principle with rate function I^f, where $Y_r = f(X_r)$, $r = 1, 2, \ldots$.

Each sequence, $(S_{ij,N} : N \geq 1)$, satisfies a Large Deviation principle with good rate function, I_{ij} (for the goodness, see Corollary 6.1.6 in [4]). Therefore, we can apply the Contraction Principle to each of the projections, f_1 and f_2, of $\mathbb{R}^2$ onto the first and second coordinate axes, respectively. We already know that $S_{i,N} \equiv f_1(S_{ij,N})$ and $S_{j,N} \equiv f_2(S_{ij,N})$ satisfy Large Deviation principles with rate functions I_i and I_j, respectively. This allows us to identify the latter with the rate functions, $I_{ij}^{f_1}$ and $I_{ij}^{f_2}$, respectively, coming from the Contraction Principle. Then we obtain the inequality

$$I_{ij}(\beta_{ij}) \geq \max(I_i(\beta_i), I_j(\beta_j)),$$

where $I_i(\beta_i) = \inf_{\beta \in \mathbb{R}} I_{ij}((\beta_i, \beta))$ and $I_j(\beta_j) = \inf_{\beta \in \mathbb{R}} I_{ij}((\beta, \beta_j))$. Substituting this lower bound into (25) and summing over index pairs $i < j$, yields the upper bound

$$\sum_{1 \leq i < j < K} \mathbb{P}\big(S_{i,N} \geq \beta_i N,\ S_{j,N} \geq \beta_j N\big) \leq \sum_{1 \leq i < j < K} e^{-\max(I_i(\beta_i), I_j(\beta_j)) \cdot N}. \tag{26}$$

Unfortunately, the right-hand side of this inequality might not be of smaller order than the right-hand side of the inequality, (24). Indeed, if $m_* > 1$, then

$$\sum_{1 \leq i < j < K} \exp\big(-\max(I_i(\beta_i), I_j(\beta_j))N\big) \sim \frac{m_*(m_* - 1)}{2} e^{-\gamma_* N}, \quad \text{as } N \to \infty.$$

[6] I.e., I is a non-negative, lower semicontinuous function such that its levels sets, $\{x \in X : I(x) \leq a\}$, are compact for every real a.

The only case in which this approach will yield a lower bound which is asymptotically equivalent to the upper bound in Theorem 2, is when $m_* = 1$.

Proposition 1 *Under the hypotheses of Theorem 2,*

$$1 - p_K(N) \geq m_* \exp(-\gamma_* N) - \sum_{1 \leq i < j < K} \mathbb{P}\big(S_{i,N} \geq \beta_i N,\ S_{j,N} \geq \beta_j N\big)$$

with

$$\sum_{1 \leq i < j < K} \mathbb{P}\big(S_{i,N} \geq \beta_i N,\ S_{j,N} \geq \beta_j N\big) \leq \sum_{1 \leq i < j < K} \exp\big(-\max(I_i(\beta_i), I_j(\beta_j))N\big);$$

so that if in addition, $m_* = 1$, *then* $1 - p_K(N) \sim \exp(-\gamma_* N)$, *as* $N \to \infty$.

Proof To see that the lower bound in the present proposition (cf. (18) and (26)), is asymptotically dominated by the first summation, it only remains to observe that, with $\gamma^* := \min(I_k(\beta_k) : I_k(\beta_k) \neq \gamma_*)$,

$$\sum_{1 \leq i < j < K} \exp\big(-\max(I_i(\beta_i), I_j(\beta_j))N\big) = \mathcal{O}(e^{-\gamma^* N}) = o(e^{-\gamma_* N}), \text{ as } N \to \infty.$$

Then the asymptotics for $p_K(N)$, as $N \to \infty$, follows from the asymptotics for the lower bound, combined with that of the upper bound obtained in Theorem 2. □

5 Conclusion

Estimation of the ranks of sensitivities has been found empirically to require a relatively small number of scenarios. In this paper, we developed a theoretical justification under a simple probabilistic model that helps to understand this phenomenon. Our numerical experiments demonstrate that fast convergence is observed both for sufficiently large portfolios as well as for relatively simple financial derivatives evaluated by QMC methods. Similar results can be obtained for the derivatives priced with MC scenarios but the rate of convergence, usually, is not that high.

The inversions of the rankings for small sample sizes correspond to the case of close sensitivity ranks. In this case the group of sensitivities remains invariant across different sizes. This observation leads us to an efficient numerical strategy for sensitivity calculations based on the idea of grouping the most significant sensitivities, instead of estimating them individually, in a stable manner, using only a very small number of scenarios; then re-estimate that smaller group of sensitivities more accurately.

Acknowledgements This paper was discussed with S. R. S. Varadhan. During the conference on MC and QMC in Rennes in July 2018, the results were discussed with Pierre L'Ecuyer, Barry Nelson, Sergei Kucherenko, Marvin Nakayama and Giray Öcten. We are very grateful to them and the referees for their comments and suggestions.

References

1. Brigo, D., Mercurio, F.: Interest Rate Models–Theory and Practice, With Smile, Inflation, and Credit, 2nd ed., 981 pp. Springer (2006)
2. Dai, L.: Convergence properties of ordinal comparison in the simulation of discrete event dynamic systems. J. Optim. Theory Appl. **91**(2), 363–388 (1996)
3. Glynn, P., Juneja, S.: A large deviations perspective on ordinal optimization. In: Proceedings of the 2004 Winter Simulation Conference
4. Dembo, A., Zeitouni, O.: Large Deviations Techniques and Applications, 2nd ed., 396 pp. Springer (1998)
5. Glasserman, P., Vakili, P.: Correlation of Markov chains simulated in parallel. In: Swain, J.J., Goldsman, D., Crainl, R.C., Wilson, J.R. (eds.) Proceedings of the 1992 Winter Simulation Conference, pp. 475–483
6. Glasserman, P., Vakili, P.: Comparing Markov chains simulated in parallel. Probab. Eng. Inf. Sci. **8**, 309–326 (1994)
7. Glasserman, P.: Monte Carlo Methods in Financial Engineering. Springer (2004)
8. Goldsman, D., Nelson, B.: Statistical selection of best system. In: Peters, B.A., Smith, J.S., Medeiros, D.J., Rohrer, M.W. (eds.) Proceedings of the 2001 Winter Simulation Conference, pp. 139–146. Institute of Electrical and Electronics Engineers, Piscataway, New Jersey

Lower Bounds on the L_p Discrepancy of Digital NUT Sequences

Ralph Kritzinger and Friedrich Pillichshammer

Abstract We study the L_p discrepancy of digital sequences generated by non-singular upper triangular (NUT) matrices which are an important sub-class of digital $(0, 1)$-sequences in the sense of Niederreiter. The main result is a lower bound for certain sub-classes of digital NUT sequences.

Keywords L_p discrepancy · van der Corput sequence · Digital $(0, 1)$ sequence

1 Introduction

For a set $\mathscr{P} = \{x_0, \ldots, x_{N-1}\}$ of N points in $[0, 1)$ the (non-normalized) L_p *discrepancy* for $p \in [1, \infty]$ is defined as

$$L_p(\mathscr{P}) = \|\Delta_{\mathscr{P}}\|_{L_p([0,1])} = \left(\int_0^1 |\Delta_{\mathscr{P}}(t)|^p \mathrm{d}t\right)^{\frac{1}{p}}$$

(with the usual modification if $p = \infty$), where

$$\Delta_{\mathscr{P}}(t) = \sum_{n=0}^{N-1} \mathbf{1}_{[0,t)}(x_n) - Nt \quad \text{for } t \in [0, 1]$$

is the (non-normalized) *discrepancy function* of $\mathscr{P}$.

We denote by $\mathbb{N}$ the set of positive integers and define $\mathbb{N}_0 = \mathbb{N} \cup \{0\}$. Let $X = (x_n)_{n \geq 0}$ be an infinite sequence in $[0, 1)$ and, for $N \in \mathbb{N}$, let $X_N = \{x_0, x_1, \ldots, x_{N-1}\}$

R. Kritzinger (✉) · F. Pillichshammer
Johannes Kepler University, Altenbergerstr. 69, Linz, Austria
e-mail: ralph.kritzinger@jku.at

F. Pillichshammer
e-mail: friedrich.pillichshammer@jku.at

B. Tuffin and P. L'Ecuyer (eds.), *Monte Carlo and Quasi-Monte Carlo Methods*, Springer Proceedings in Mathematics & Statistics 324,
https://doi.org/10.1007/978-3-030-43465-6_16

denote the set consisting of the first N elements of X. It is well known that for all $p \in [1, \infty)$ we have

$$L_p(X_N) \gtrsim \sqrt{\log N} \quad \text{for infinitely many } N \in \mathbb{N}$$

and

$$L_\infty(X_N) \gtrsim \log N \quad \text{for infinitely many } N \in \mathbb{N}. \tag{1}$$

(For functions $f, g : \mathbb{N} \to \mathbb{R}^+$, we write $g(N) \lesssim f(N)$ or $g(N) \gtrsim f(N)$, if there exists a positive constant C that is independent of N such that $g(N) \leq Cf(N)$ or $g(N) \geq Cf(N)$, respectively.) The lower estimate for finite p was first shown by Proĭnov [14] (see also [3]) based on famous results of Roth [15] and Schmidt [17] for finite point sets in dimension two. Using the method of Proĭnov in conjunction with a result of Halász [7] for finite point sets in dimension two the lower bound follows also for the L_1-discrepancy. The estimate for $p = \infty$ was first shown by Schmidt [16] in 1972 (see also [1, 10, 18]).

In this paper we investigate the L_p discrepancy of *digital* (0, 1)*-sequences*. Since we only deal with digital sequences over $\mathbb{Z}_2$ and in dimension 1 we restrict the necessary definitions to this case. For the general setting we refer to [2, 11, 12].

Let $\mathbb{Z}_2$ be the finite field of order 2, which we identify with the set $\{0, 1\}$ equipped with arithmetic operations modulo 2. For the generation of a digital (0, 1) sequence $(x_n)_{n \geq 0}$ over $\mathbb{Z}_2$ we require an infinite matrix $C = (c_{i,j})_{i,j \geq 1}$ over $\mathbb{Z}_2$ with the following property[1]: for every $n \in \mathbb{N}$ the left upper $n \times n$ submatrix $(c_{i,j})_{i,j=1}^n$ has full rank. In order to construct the nth element x_n for $n \in \mathbb{N}_0$ compute the base 2 expansion $n = n_0 + n_1 2 + n_2 2^2 + \cdots$ (which is actually finite), set $\mathbf{n} := (n_0, n_1, n_2, \ldots)^\top \in \mathbb{Z}_2^\infty$ and compute the matrix vector product

$$C\mathbf{n} =: (y_1^{(n)}, y_2^{(n)}, y_3^{(n)}, \ldots)^\top \in \mathbb{Z}_2^\infty$$

over $\mathbb{Z}_2$. Finally, set

$$x_n := \frac{y_1^{(n)}}{2} + \frac{y_2^{(n)}}{2^2} + \frac{y_3^{(n)}}{2^3} + \cdots .$$

We denote the digital (0, 1)-sequence[2] constructed in this way by X^C.

An important sub-class of digital (0, 1)-sequences which is studied in many papers (initiated by Faure [5]) are so-called *digital NUT sequences* whose generator matrices are of non-singular upper triangular (NUT) form

---

[1] A further technical condition which is sometimes required, see [12, p. 72, (S6)], is that for each $j \geq 1$ the sequence $(c_{i,j})_{i \geq 1}$ becomes eventually zero. Otherwise it could happen that one or more elements of the digital (0, 1)-sequence are 1 and therefore do not belong to [0, 1).

[2] In the general notation, the 1 refers to the dimension and the 0 refers to the full rank condition of the generator matrix C.

$$C = \begin{pmatrix} 1 & c_{1,2} & c_{1,3} & \cdots \\ 0 & 1 & c_{2,3} & \cdots \\ 0 & 0 & 1 & \cdots \\ \vdots & \vdots & \vdots & \ddots \end{pmatrix}. \tag{2}$$

For example, if $C = I$ is the identity matrix, then the corresponding digital NUT sequence is the *van der Corput sequence* in base 2. For information about digital NUT sequences and the van der Corput sequence see the survey [6] and the references therein.

For digital NUT sequences X^C it is known (see [4, Theorem 1]), that

$$L_2(X_N^C) \le L_2(X_N^I) \le \left(\left(\frac{\log N}{6 \log 2} \right)^2 + O(\log N) \right)^{1/2}$$

and for general $p \ge 1$ it is known (see [13, Theorem 2]) that

$$L_p(X_N^C) \le L_\infty(X_N^C) \le L_\infty(X_N^I) \le \frac{\log N}{3 \log 2} + 1. \tag{3}$$

Note that according to the lower bound (1) of Schmidt the upper bound for the L_∞ discrepancy in (3) is optimal in the order of magnitude in N. This is not the case for finite p, as for instance the symmetrized van der Corput sequence achieves an L_p discrepancy of order $\sqrt{\log N}$ for all $N \ge 2$ and all $p \in [1, \infty)$, see [8].

Concerning lower bounds on the L_p discrepancy of digital NUT sequences very few is known and only for very special cases. For the van der Corput sequence we have for all $p \in [1, \infty)$

$$\limsup_{N \to \infty} \frac{L_p(X_N^I)}{\log N} = \frac{1}{6 \log 2} \tag{4}$$

and hence $L_p(X_N^I) \gtrsim \log N$ for infinitely many $N \in \mathbb{N}$; see [13, Corollary 1].

For the so-called *upper-$\mathbf{1}$-sequence* X^U, which is generated by the matrix

$$U = \begin{pmatrix} 1 & 1 & 1 & \cdots \\ 0 & 1 & 1 & \cdots \\ 0 & 0 & 1 & \cdots \\ \vdots & \vdots & \vdots & \ddots \end{pmatrix} \tag{5}$$

it is known that for every $p \ge 1$ we have $L_p(X_N^U) \ge \frac{\log N}{20 \log 2} + O(1)$ for infinitely many $N \in \mathbb{N}$; see [4].

In [4] the authors study the L_p discrepancy of X^C for special types of NUT matrices C of the form

$$C = \begin{pmatrix} \boldsymbol{a}_1 & & & \\ 0 & \boldsymbol{a}_2 & & \\ 0 & 0 & \boldsymbol{a}_3 & \\ & & \dots & \end{pmatrix} \tag{6}$$

with

$$\boldsymbol{a}_i = (1, 0, 0, \dots) \quad \text{or} \quad \boldsymbol{a}_i = (1, 1, 1, \dots) \quad \text{for } i \in \mathbb{N}.$$

Note that these NUT sequences comprise the van der Corput sequence and the upper-**1**-sequence X^U as special cases. For $m \in \mathbb{N}$ let $h(m)$ denote the number of $(1, 0, 0, \dots)$ rows among the first m rows of C. For example, $h(m) = m$ in case of the van der Corput sequence and $h(m) = 0$ in case of the sequence X^U. Then it follows from [4, Lemma 4] that for every $m \in \mathbb{N}$ there exists an integer $N \in [2^m, 2^{m+1})$ such that $L_1(X_N^C) \gtrsim (m + h(m)^2)^{1/2}$. This implies that if $h(m) \gtrsim m$ we have for every $p \geq 1$

$$L_p(X_N^C) \gtrsim \log N \quad \text{for infinitely many } N \in \mathbb{N}$$

In general, however, it is a very difficult task to give precise lower bounds on the L_p discrepancy of digital NUT sequences. We strongly conjecture the following:

Conjecture 1 *For every digital NUT sequence X^C we have*

$$L_p(X_N^C) \gtrsim \log N \quad \textit{for infinitely many } N \in \mathbb{N}. \tag{7}$$

Note that for every digital NUT sequence and for every $p \geq 1$ we have

$$L_p(X_N^C) \leq L_\infty(X_N^C) \leq L_\infty(X_N^I) \leq s_2(N), \tag{8}$$

where $s_2 : \mathbb{N} \to \mathbb{N}$ denotes the binary sum-of-digits function which is defined as $s_2(N) = N_0 + N_1 + \dots + N_m$ whenever N has binary expansion $N = N_0 + N_1 2 + \dots + N_m 2^m$. The very last inequality in (8) follows from the proof of [9, Theorem 3.5 in Chap. 2].

Remark 1 The result in (8) can be generalized and improved in the following sense: For every $p \in [1, \infty]$ and for **every** digital $(0, 1)$-sequence X^C we have

$$L_p(X_N^C) \leq c_p s_2(N) \quad \text{for all } N \in \mathbb{N},$$

where

$$c_p = \begin{cases} 1/\sqrt{3} = 0.5773\dots & \text{if } p \in [1, 2], \\ 1 & \text{if } p \in (2, \infty]. \end{cases}$$

We omit the proof.

The sum-of-digits function is very fluctuating. For example we have $s_2(2^m) = 1$, but $s_2(2^m - 1) = m$. In any case we have $s_2(N) \leq \frac{\log N}{\log 2} + 1$.

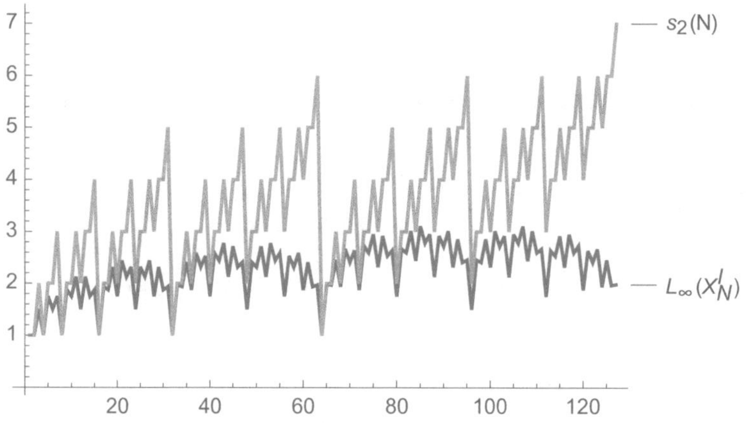

Fig. 1 The L_∞ discrepancy of the van der Corput sequence $L_\infty(X_N^I)$ and the binary sum-of-digits function $s_2(N)$ for $N = 1, 2, \dots, 127$

Remark 2 The inequalities in (8) show that having only very few non-zero binary digits is a sufficient condition on $N \in \mathbb{N}$ which guarantees that X_N^C has very low L_p discrepancy. For example we have

$$L_p(X_N^C) \leq 1 \quad \text{for all } N \text{ of the form } N = 2^m$$

or

$$L_p(X_N^C) \lesssim \sqrt{\log N} \quad \text{for all } N \text{ of the form } N = 2^m + 2^{\lfloor \sqrt{m} \rfloor - 1} - 1$$

or

$$L_p(X_N^C) \lesssim \log N \quad \text{for all } N \geq 2.$$

See Fig. 1 for a comparison for the van der Corput sequence.

However, the condition on N of having very few non-zero binary digits is not a necessary one for low discrepancy. For example, consider N of the form $N = 2^m - 1$. Then we have $s_2(N) = m = \lfloor \frac{\log N}{\log 2} + 1 \rfloor$ but: since the discrepancy of X_N^C and of X_{N+1}^C differ at most by 1 and since $L_p(X_{N+1}^C) = L_p(X_{2^m}^C) \leq 1$ we obtain $L_p(X_N^C) \lesssim 1$. Hence, while $s_2(N)$ is very large, the discrepancy $L_p(X_N^C)$ is low.

But in any case: the only possible candidates of N that satisfy (7) are required to have $s_2(N) \gtrsim \log N$.

In Sect. 2 we provide a lower bound for $L_p(X_N^C)$ for special types of NUT matrices.

2 Lower Bound on $L_p(X_N^C)$

We study two sub-classes of NUT matrices. The first class has a certain band structure. More detailed, the considered matrices are of the form $C(\alpha) = (c_{i,j})_{i,j\geq 1}$ where, for fixed $\alpha \in \mathbb{N}$,

$$c_{i,j} = \begin{cases} 1 \text{ if } i \leq j < i + \alpha, \\ 0 \text{ in all other cases.} \end{cases}$$

For example, if $\alpha = 1$, we obtain the identity matrix, i.e., $C(1) = I$.

Theorem 1 *For all $\alpha \in \mathbb{N}$ and $p \in [1, \infty]$ we have*

$$L_p(X_N^{C(\alpha)}) \geq \frac{2^{\alpha-1}}{2^{2\alpha}-1} \frac{\log N}{2\alpha \log 2} + O_\alpha(1) \quad \textit{for infinitely many } N \in \mathbb{N}.$$

The bound above is satisfied for N of the form

$$N = \sum_{\ell=1}^{r} 2^{2\alpha(r-\ell)} = \frac{2^{2\alpha r}-1}{2^{2\alpha}-1} \quad \textit{for arbitrary } r \in \mathbb{N}.$$

Remark 3 1. Following all the details in the proof the constant hidden in $O_\alpha(1)$ can be computed exactly.
2. For $\alpha = 1$ we have $C(1) = I$ and hence the resulting NUT sequence is the van der Corput sequence. Theorem 1 gives

$$L_p(X_N^I) \geq \frac{\log N}{6 \log 2} + O(1) \quad \text{for infinitely many } N \in \mathbb{N}.$$

This matches the corresponding value in (4).

We also study NUT matrices which have the same entries in each column above the diagonal; i.e. we deal with matrices of the form

$$C(\boldsymbol{a}) = \begin{pmatrix} 1 & a_1 & a_2 & a_3 & \cdots \\ 0 & 1 & a_2 & a_3 & \cdots \\ 0 & 0 & 1 & a_3 & \cdots \\ 0 & 0 & 0 & 1 & \cdots \\ \vdots & \vdots & \vdots & \vdots & \ddots \end{pmatrix}, \tag{9}$$

where $\boldsymbol{a} = (a_1, a_2, \ldots) \in \mathbb{Z}_2^{\mathbb{N}}$ is chosen arbitrarily. We set $l_0(m) := \#\{i \in \{1, \ldots, m\} : a_i = 0\}$ and $l_1(m) := \#\{i \in \{1, \ldots, m\} : a_i = 1\}$. For $m \geq 2$ let further $d_0(m)$ be the minimal distance of consecutive zeroes and $d_1(m)$ be the minimal distance of consecutive ones in the string $(a_1, \ldots, a_m)$, i.e. for $\ell \in \{0, 1\}$ we define

$$d_\ell(m) := \min_{1\le n\le m-1}\left\{\exists i \in \{1,\dots,m-n\} : a_i = a_{i+n} = \ell, a_{i+1} = \cdots = a_{i+n-1} \neq \ell\right\}.$$

Theorem 2 *For all $m \geq 2$, $p \in [1,\infty]$, and $N_{\boldsymbol{a}} = 1 + \sum_{i=1}^{m-1} 2^i(1-a_i) + 2^m$, we have*

$$L_p(X_{N_{\boldsymbol{a}}}^{C(\boldsymbol{a})}) \geq \frac{1}{3} l_0(m) + O(1) \tag{10}$$

if $d_0(m) \geq 2$, and

$$L_p(X_{N_{\boldsymbol{a}}}^{C(\boldsymbol{a})}) \geq \frac{1}{3} l_1(m) + O(1) \tag{11}$$

if $d_1(m) \geq 2$.

Corollary 1 *The first N elements of a NUT-sequence generated by a matrix of the form $C(\boldsymbol{a})$ satisfy*

$$L_p(X_N^{C(\boldsymbol{a})}) \geq c \log N \quad \text{for infinitely many } N \tag{12}$$

for some constant $c > 0$ if $l_1(m) \geq c_1 m$ for some $c_1 > 0$ and $d_1(m) \geq 2$ for all $m \geq 2$ or if $l_0(m) \geq c_2 m$ for some $c_2 > 0$ and $d_0(m) \geq 2$ for all $m \geq 2$.

One example for a generator matrix satisfying the hypotheses of Corollary 1 is

$$C(\boldsymbol{a}) = \begin{pmatrix} 1 & 0 & 1 & 0 & 1 & \cdots \\ 0 & 1 & 1 & 0 & 1 & \cdots \\ 0 & 0 & 1 & 0 & 1 & \cdots \\ 0 & 0 & 0 & 1 & 1 & \cdots \\ 0 & 0 & 0 & 0 & 1 & \cdots \\ \vdots & \vdots & \vdots & \vdots & \vdots & \ddots \end{pmatrix}.$$

3 The Proofs

The following auxiliary result will be the main tool of our proofs.

Lemma 1 *For every NUT digital sequence X^C and every $N \in \mathbb{N}$ of the form $N = 2^{n_1} + 2^{n_2} + \cdots + 2^{n_r}$ with $n_1 > n_2 > \cdots > n_r$ and $r \in \mathbb{N}$ we have*

$$\int_0^1 \Delta_{X_N^C}(t)\mathrm{d}t = \sum_{i=2}^{r} \sigma_{r,n_i+1} - \sum_{k=2}^{r} \sum_{j=n_k+1}^{n_{k-1}} \frac{\sigma_{r,j}}{2^j} \sum_{i=k}^{r} 2^{n_i} + O(1),$$

where the $\sigma_{r,j}$ are given by the following matrix-vector product over $\mathbb{Z}_2$:

$$\begin{pmatrix} \sigma_{r,n_r+1} \\ \vdots \\ \vdots \\ \vdots \\ \vdots \\ \sigma_{r,n_1+1} \end{pmatrix} = \begin{pmatrix} c_{n_r+1,n_r+1} & \cdots\cdots & c_{n_r+1,n_1+1} \\ & \cdots\cdots\cdots\cdots & \\ & \cdots\cdots\cdots\cdots & \\ & \cdots\cdots\cdots\cdots & \\ & \cdots\cdots\cdots\cdots & \\ 0 \ldots & \ldots 0 & c_{n_1+1,n_1+1} \end{pmatrix} \begin{pmatrix} 0 \\ \vdots \\ 0 \\ \hline 1 \\ 0 \\ \vdots \\ 0 \\ \hline \vdots \\ \hline 1 \\ 0 \\ \vdots \\ 0 \\ \hline 1 \end{pmatrix},$$

where the digits 1 in the latter vector are placed at positions $n_l - n_r + 1$ for $l \in \{1, \ldots, r-1\}$.

Proof Let $X^C = (x_n)_{n \geq 0}$ be the NUT digital sequence which is generated by the $\mathbb{N} \times \mathbb{N}$ matrix C. Let $N \in \mathbb{N}$ be of the form

$$N = 2^{n_1} + 2^{n_2} + \cdots + 2^{n_r},$$

where $n_1 > \cdots > n_r$. For $i = 1, \ldots, r$ consider

$$\mathscr{P}_i = \{x_{2^{n_1}+\cdots+2^{n_{i-1}}}, \ldots, x_{2^{n_1}+\cdots+2^{n_{i-1}}+2^{n_i}-1}\},$$

where for $i = 1$ we define $2^{n_1} + \cdots + 2^{n_{i-1}} = 0$. Every

$$n \in \{2^{n_1} + \cdots + 2^{n_{i-1}}, \ldots, 2^{n_1} + \cdots + 2^{n_{i-1}} + 2^{n_i} - 1\} \tag{13}$$

can be written as

$$n = 2^{n_1} + \cdots + 2^{n_{i-1}} + a = 2^{n_{i-1}} l_i + a,$$

where $a \in \{0, 1, \ldots, 2^{n_i} - 1\}$ and

$$l_i = \begin{cases} 0 & \text{if } i = 1, \\ 1 & \text{if } i = 2, \\ 1 + 2^{n_{i-2}-n_{i-1}} + \cdots + 2^{n_1-n_{i-1}} & \text{if } i > 2. \end{cases}$$

For fixed $i = 1, \ldots, r$ we decompose the matrix C in the form

$$\left(\begin{array}{c|c} C^{(n_i\times n_i)} & D^{(n_i\times\mathbb{N})} \\ \hline 0^{(\mathbb{N}\times n_i)} & F^{(\mathbb{N}\times\mathbb{N})} \end{array}\right) \in \mathbb{Z}_2^{\mathbb{N}\times\mathbb{N}},$$

where $C^{(n_i\times n_i)}$ is the left upper $n_i \times n_i$ sub-matrix of C. To n in (13) we associate

$$\mathbf{n} = (a_0, a_1, \ldots, a_{n_i-1}, \ell_0, \ell_1, \ell_2, \ldots)^\top =: \begin{pmatrix} \mathbf{a} \\ \mathbf{l}_i \end{pmatrix},$$

where $a_0, \ldots, a_{n_i-1}$ are the binary digits of a and $\ell_0, \ell_1, \ell_2, \ldots$ are the binary digits of l_i. With this notation for n in the range (13) we have

$$C\mathbf{n} = \begin{pmatrix} C^{(n_i\times n_i)}\mathbf{a} \\ 0 \\ 0 \\ \vdots \end{pmatrix} + \left(\begin{array}{c} D^{(n_i\times\mathbb{N})} \\ \hline F^{(\mathbb{N}\times\mathbb{N})} \end{array}\right) \mathbf{l}_i.$$

This shows that the point set $\mathscr{P}_i$ is a digitally shifted digital net with generating matrix $C^{(n_i\times n_i)}$ and with digital shift vector

$$\boldsymbol{\sigma}_i = (\sigma_{i,1}, \sigma_{i,2}, \ldots)^\top := \left(\begin{array}{c} D^{(n_i\times\mathbb{N})} \\ \hline F^{(\mathbb{N}\times\mathbb{N})} \end{array}\right) \mathbf{l}_i. \tag{14}$$

Since $F^{(\mathbb{N}\times\mathbb{N})}$ is also a NUT matrix we find that the shift is of the form

$$\boldsymbol{\sigma}_i = (\sigma_{i,1}, \sigma_{i,2}, \ldots, \sigma_{i,n_1+1}, 0, 0, \ldots)^\top \in \mathbb{Z}_2^\infty$$

Note that the matrix $C^{(n_i\times n_i)}$ has full rank, as X^C is a NUT digital sequence. Hence the shifted digital net $\mathscr{P}_i$ can be written as the set of points

$$\mathscr{P}_i = \left\{ \frac{b_1}{2} + \cdots + \frac{b_{n_i}}{2^{n_i}} + \sum_{j=1}^{\infty} \frac{\sigma_{i,n_i+j}}{2^{n_i+j}} : a_0, \ldots, a_{n_i-1} \in \{0,1\} \right\},$$

where $b_k = c_{k,1}a_0 \oplus \cdots \oplus c_{k,n_i}a_{n_i-1} \oplus \sigma_{i,k}$ for $1 \le k \le n_i$. Here and in the following $\oplus$ denotes addition in $\mathbb{Z}_2$.

We emphasize that $\sigma_{i,1}, \ldots, \sigma_{i,n_i}$ do not depend on the a_i's, whereas the components $\sigma_{i,j}$ for $j \ge n_i + 1$ may do so. Therefore we can also write

$$\mathscr{P}_i = \left\{ \frac{k_i}{2^{n_i}} + \delta_i : k_i \in \{0, 1, \ldots, 2^{n_i} - 1\} \right\},$$

where $\delta_i = \sum_{j=1}^{n_1 - n_i + 1} \frac{\sigma_{i,n_i+j}}{2^{n_i+j}}$ for $i > 1$ and $\delta_1 = 0$.

We have the following decomposition of X_N^C:

$$X_N^C = \bigcup_{i=1}^{r} \mathscr{P}_i.$$

Therefore and from the fact that

$$\int_0^1 \Delta_{\mathscr{P}_i}(t)\,\mathrm{d}t = \sum_{z \in \mathscr{P}_i} \left(\frac{1}{2} - z\right)$$

we obtain

$$\begin{aligned}
\int_0^1 \Delta_{X_N^C}(t)\,\mathrm{d}t &= \sum_{i=1}^{r} \int_0^1 \Delta_{\mathscr{P}_i}(t)\,\mathrm{d}t = \sum_{i=1}^{r} \sum_{\ell=0}^{2^{n_i}-1} \left(\frac{1}{2} - \left(\frac{\ell}{2^{n_i}} + \delta_i\right)\right) \\
&= \sum_{i=1}^{r} \left(\frac{1}{2} - 2^{n_i}\delta_i\right) = \frac{r}{2} - \sum_{i=1}^{r} 2^{n_i}\delta_i,
\end{aligned}$$

where

$$2^{n_i}\delta_i = \begin{cases} 0 & \text{if } i = 1, \\ \frac{\sigma_{i,n_i+1}}{2} + \frac{\sigma_{i,n_i+2}}{2^2} + \frac{\sigma_{i,n_i+3}}{2^3} + \cdots + \frac{\sigma_{i,n_1+1}}{2^{n_1-n_i+1}} & \text{if } i > 1 \end{cases}$$

and, for $k \geq 1$,

$$\begin{aligned}
\sigma_{i,n_i+k} &= \bigoplus_{j=0}^{n_1-n_i-k+1} c_{n_i+k,n_i+k+j}\, a_{n_i+k-1+j} \\
&= c_{n_i+k,n_i+k}\, a_{n_i+k-1} + c_{n_i+k,n_i+k+1}\, a_{n_i+k} + \cdots + c_{n_i+k,n_1+1}\, a_{n_1} \pmod 2,
\end{aligned}$$

where $a_{n_\ell} = 1$ for $\ell = 1, \ldots, i-1$ and all other a_r's are zero. Note that $\sigma_{i,n_i+k} \in \mathbb{Z}_2$.

We have

$$\sum_{i=1}^{r} 2^{n_i}\delta_i = \sum_{i=2}^{r} \sum_{j=1}^{n_1-n_i+1} \frac{\sigma_{i,n_i+j}}{2^j}.$$

Observe that

$$\begin{pmatrix} \sigma_{i,n_i+1} \\ \cdots \\ \sigma_{i,n_{i-1}} \\ \sigma_{i,n_{i-1}+1} \\ \cdots \\ \cdots \\ \cdots \\ \cdots \\ \sigma_{i,n_1+1} \end{pmatrix} = \begin{pmatrix} c_{n_i+1,n_i+1} \cdots c_{n_i+1,n_{i-1}+1} & \cdots\cdots & c_{n_i+1,n_1+1} \\ & \cdots\cdots\cdots\cdots\cdots\cdots\cdots\cdots & \\ & \cdots\cdots\cdots\cdots\cdots\cdots\cdots\cdots & \\ 0\ldots \quad \ldots c_{n_{i-1}+1,n_{i-1}+1} & \cdots\cdots & c_{n_{i-1}+1,n_1+1} \\ & \cdots\cdots\cdots\cdots\cdots\cdots\cdots\cdots & \\ & \cdots\cdots\cdots\cdots\cdots\cdots\cdots\cdots & \\ & \cdots\cdots\cdots\cdots\cdots\cdots\cdots\cdots & \\ & \cdots\cdots\cdots\cdots\cdots\cdots\cdots\cdots & \\ 0\ldots & \cdots\cdots\cdots\cdots & c_{n_1+1,n_1+1} \end{pmatrix} \begin{pmatrix} 0 \\ \vdots \\ 0 \\ 1 \\ 0 \\ \vdots \\ 0 \\ \vdots \\ 1 \end{pmatrix},$$

and

$$\begin{pmatrix} \sigma_{i,n_{i-1}+1} \\ \cdots \\ \cdots \\ \cdots \\ \cdots \\ \sigma_{i,n_1+1} \end{pmatrix} = \begin{pmatrix} c_{n_{i-1}+1,n_{i-1}+1} & \cdots\cdots & c_{n_{i-1}+1,n_1+1} \\ & \cdots\cdots\cdots\cdots\cdots\cdots & \\ & \cdots\cdots\cdots\cdots\cdots\cdots & \\ & \cdots\cdots\cdots\cdots\cdots\cdots & \\ & \cdots\cdots\cdots\cdots\cdots\cdots & \\ 0\ldots & \ldots 0 & c_{n_1+1,n_1+1} \end{pmatrix} \begin{pmatrix} 1 \\ 0 \\ \vdots \\ 0 \\ \vdots \\ 1 \end{pmatrix}.$$

Hence

$$\begin{pmatrix} \sigma_{i-1,n_{i-1}+1} \\ \sigma_{i-1,n_{i-1}+2} \\ \cdots \\ \cdots \\ \cdots \\ \sigma_{i-1,n_1+1} \end{pmatrix} = \begin{pmatrix} c_{n_{i-1}+1,n_{i-1}+1} & \cdots\cdots & c_{n_{i-1}+1,n_1+1} \\ & \cdots\cdots\cdots\cdots\cdots\cdots & \\ & \cdots\cdots\cdots\cdots\cdots\cdots & \\ & \cdots\cdots\cdots\cdots\cdots\cdots & \\ & \cdots\cdots\cdots\cdots\cdots\cdots & \\ 0\ldots & \ldots 0 & c_{n_1+1,n_1+1} \end{pmatrix} \begin{pmatrix} 0 \\ 0 \\ \vdots \\ 0 \\ \vdots \\ 1 \end{pmatrix} = \begin{pmatrix} \sigma_{i,n_{i-1}+1} \oplus 1 \\ \sigma_{i,n_{i-1}+2} \\ \cdots \\ \cdots \\ \cdots \\ \sigma_{i,n_1+1} \end{pmatrix}.$$

This shows that we have

$$\sigma_{i,k} = \begin{cases} \sigma_{i-1,k} & \text{for } k = n_{i-1}+2, n_{i-1}+3, \ldots, n_1+1, \\ \sigma_{i-1,k} \oplus 1 & \text{for } k = n_{i-1}+1. \end{cases}$$

From this we obtain for all $i \in \{2, 3, \ldots, r\}$ that

$$\sigma_{i,n_i+j} = \begin{cases} \sigma_{r,n_i+j} & \text{for } j = 2, 3, \ldots, n_1 - n_i + 1, \\ \sigma_{r,n_i+j} \oplus 1 = 1 - \sigma_{r,n_i+j} & \text{for } j = 1. \end{cases}$$

Hence

$$\sum_{i=1}^{r} 2^{n_i}\delta_i = \sum_{i=2}^{r} \frac{1-\sigma_{r,n_i+1}}{2} + \sum_{i=2}^{r}\sum_{j=2}^{n_1-n_i+1} \frac{\sigma_{r,n_i+j}}{2^j}$$
$$= \frac{r-1}{2} - \sum_{i=2}^{r}\sigma_{r,n_i+1} + \sum_{i=2}^{r}\sum_{j=1}^{n_1-n_i+1} \frac{\sigma_{r,n_i+j}}{2^j}.$$

For the very last double sum we have

$$\sum_{i=2}^{r}\sum_{j=1}^{n_1-n_i+1} \frac{\sigma_{r,n_i+j}}{2^j} = \sum_{i=2}^{r}\sum_{j=n_i+1}^{n_1+1} \frac{\sigma_{r,j}}{2^{j-n_i}} = \sum_{j=n_r+1}^{n_1+1} \frac{\sigma_{r,j}}{2^j} \sum_{\substack{i=2\\ n_i\le j-1}}^{r} 2^{n_i}$$
$$= \sum_{k=2}^{r}\sum_{j=n_k+1}^{n_{k-1}} \frac{\sigma_{r,j}}{2^j} \sum_{\substack{i=2\\ n_i\le j-1}}^{r} 2^{n_i} + \frac{\sigma_{r,n_1+1}}{2^{n_1+1}} \sum_{\substack{i=2\\ n_i\le n_1}}^{r} 2^{n_i}$$
$$= \sum_{k=2}^{r}\sum_{j=n_k+1}^{n_{k-1}} \frac{\sigma_{r,j}}{2^j} \sum_{i=k}^{r} 2^{n_i} + \frac{\sigma_{r,n_1+1}}{2^{n_1+1}}(N-2^{n_1}).$$

Hence

$$\sum_{i=1}^{r} 2^{n_i}\delta_i = \frac{r-1}{2} - \sum_{i=2}^{r}\sigma_{r,n_i+1} + \sum_{k=2}^{r}\sum_{j=n_k+1}^{n_{k-1}} \frac{\sigma_{r,j}}{2^j} \sum_{i=k}^{r} 2^{n_i} + \frac{\sigma_{r,n_1+1}}{2^{n_1+1}}(N-2^{n_1}).$$

This gives

$$\int_0^1 \Delta_{X_N^C}(t)\mathrm{d}t = \sum_{i=2}^{r}\sigma_{r,n_i+1} - \sum_{k=2}^{r}\sum_{j=n_k+1}^{n_{k-1}} \frac{\sigma_{r,j}}{2^j} \sum_{i=k}^{r} 2^{n_i} + O(1).$$

□

Now we give the proof of Theorem 1.

Proof In order to simplify the notation we will write C instead of $C(\alpha)$ in the following. For every N and p we have

$$L_p(X_N^C) \ge L_1(X_N^C) = \|\Delta_{X_N^C}\|_{L_1([0,1))} \ge \left|\int_0^1 \Delta_{X_N^C}(t)\mathrm{d}t\right|. \tag{15}$$

Now choose $n_i = 2\alpha(r-i)$ for $i \in \{1, 2, \ldots, r\}$, i.e.

$$N = \sum_{i=1}^{r} 2^{2\alpha(r-i)} = \frac{2^{2\alpha r}-1}{2^{2\alpha}-1} \quad \text{and hence} \quad r = \frac{\log((2^{2\alpha}-1)N+1)}{2\alpha\log 2}.$$

We have

$$\sum_{i=k}^{r} 2^{n_i} = \sum_{i=k}^{r} 2^{2\alpha(r-i)} = \frac{2^{2\alpha(r-k+1)}-1}{2^{2\alpha}-1}.$$

Therefore

$$\begin{aligned}\sum_{k=2}^{r}\sum_{j=n_k+1}^{n_{k-1}} \frac{\sigma_{r,j}}{2^j}\sum_{i=k}^{r}2^{n_i} &= \sum_{k=2}^{r}\sum_{j=2\alpha(r-k)+1}^{2\alpha(r-k)+2\alpha} \frac{\sigma_{r,j}}{2^j}\frac{2^{2\alpha(r-k+1)}-1}{2^{2\alpha}-1}\\ &= \frac{1}{2^{2\alpha}-1}\sum_{k=2}^{r}\sum_{j=1}^{2\alpha}\frac{\sigma_{r,2\alpha(r-k)+j}}{2^{2\alpha(r-k)+j}}(2^{2\alpha(r-k+1)}-1)\\ &= \frac{2^{2\alpha}}{2^{2\alpha}-1}\sum_{k=2}^{r}\sum_{j=1}^{2\alpha}\frac{\sigma_{r,2\alpha(r-k)+j}}{2^j}+O(1)\\ &= \frac{2^{2\alpha}}{2^{2\alpha}-1}\sum_{\ell=0}^{r-2}\sum_{j=1}^{2\alpha}\frac{\sigma_{r,2\alpha\ell+j}}{2^j}+O(1).\end{aligned}$$

Hence, using Lemma 1, we get

$$\int_0^1 \Delta_{X_N^{\mathcal{C}}}(t)\mathrm{d}t = \sum_{\ell=0}^{r-2}\sigma_{r,2\ell\alpha+1} - \frac{2^{2\alpha}}{2^{2\alpha}-1}\sum_{\ell=0}^{r-2}\sum_{j=1}^{2\alpha}\frac{\sigma_{r,2\alpha\ell+j}}{2^j}+O(1).$$

Now we have to determine the numbers $\sigma_{r,j}$. Observe that

$$\begin{pmatrix}\sigma_{r,1}\\ \vdots\\ \vdots\\ \vdots\\ \vdots\\ \sigma_{r,(2r-2)\alpha+1}\end{pmatrix} = \begin{pmatrix} c_{1,1} & \cdots\cdots & c_{1,(2r-2)\alpha+1}\\ & \cdots\cdots\cdots & \\ & \cdots\cdots\cdots & \\ & \cdots\cdots\cdots & \\ & \cdots\cdots\cdots & \\ 0\ldots\ldots 0 & & c_{(2r-2)\alpha+1,(2r-2)\alpha+1}\end{pmatrix}\begin{pmatrix}0\\ \vdots\\ 0\\ 1\\ 0\\ \vdots\\ 0\\ \hline \vdots\\ \hline 1\\ 0\\ \vdots\\ 0\\ \hline 1\end{pmatrix},$$

where the 1's in the latter vector are in positions $l\alpha+1$ for $l \in \{2, \ldots, 2r-2\}$. From the structure of the matrix we find that

$$\begin{aligned}\sigma_{r,1} = \cdots = \sigma_{r,\alpha+1} &= 0\\ \sigma_{r,\alpha+2} = \cdots = \sigma_{r,2\alpha+1} &= 1\\ \sigma_{r,2\alpha+2} = \cdots = \sigma_{r,3\alpha+1} &= 0\\ \sigma_{r,3\alpha+2} = \cdots = \sigma_{r,4\alpha+1} &= 1\\ \sigma_{r,4\alpha+2} = \cdots = \sigma_{r,5\alpha+1} &= 0\\ &\ldots\\ \sigma_{r,(2r-3)\alpha+2} = \cdots = \sigma_{r,(2r-2)\alpha+1} &= 1\end{aligned}$$

and therefore

$$\begin{aligned}\sum_{\ell=0}^{r-2}\sum_{j=1}^{2\alpha}\frac{\sigma_{r,2\alpha\ell+j}}{2^j} &= \frac{1}{2^{\alpha+2}}+\cdots+\frac{1}{2^{2\alpha}}+\sum_{\ell=1}^{r-2}\left(\frac{1}{2}+\frac{1}{2^{\alpha+2}}+\cdots+\frac{1}{2^{2\alpha}}\right)\\ &= \frac{r-2}{2}+(r-1)\frac{1}{2^{\alpha+2}}\frac{1-(1/2)^{\alpha-1}}{1/2}.\end{aligned}$$

Furthermore

$$\sum_{\ell=0}^{r-2}\sigma_{r,2\alpha\ell+1} = 0+1+1+\cdots+1 = r-2.$$

Putting all together we obtain

$$\begin{aligned}\int_0^1 \Delta_{X_N^C}(t)\mathrm{d}t &= r-2-\frac{2^{2\alpha}}{2^{2\alpha}-1}\left(\frac{r-2}{2}+(r-1)\frac{1}{2^{\alpha+2}}\frac{1-(1/2)^{\alpha-1}}{1/2}\right)+O(1)\\ &= r\frac{2^{\alpha-1}}{2^{2\alpha}-1}+O(1).\end{aligned}$$

Hence, using (15), we get

$$L_p(X_N^C) \geq \left|\int_0^1 \Delta_{X_N^C}(t)\mathrm{d}t\right| = \frac{2^{\alpha-1}}{2^{2\alpha}-1}\frac{\log((2^{2\alpha}-1)N+1)}{2\alpha\log 2}+O(1).$$

□

In the following, we give the proof of Theorem 2.

Proof Note that in the case $n_r = 0$ the numbers $\sigma_{r,j}$ appearing in Lemma 1 can also be understood in the following way: Let $N = 2^{n_1}+\sum_{i=0}^{n_1-1}N_i 2^i = \sum_{i=1}^r 2^{n_i}$ with $n_1 = m \in \mathbb{N}$, $N_i \in \mathbb{Z}_2$ for $i \in \{0,\ldots,n_1-1\}$ and $r = s_2(N)$. Let

$$\eta_j := c_{j,j+1}N_j \oplus \cdots \oplus c_{j,n_1}N_{n_1-1} \oplus c_{j,n_1+1}.$$

Then we have for $j \in \{1,\ldots,n_1+1\}$

$$\sigma_{r,j} = \begin{cases} 1 & \text{if } j = n_1 + 1, \\ \eta_j \oplus 1 & \text{if } j = n_k + 1 \text{ for some } k \in \{2, \ldots, r\}, \\ \eta_j & \text{otherwise.} \end{cases}$$

Now consider a matrix of the form $C(\boldsymbol{a})$ and set $N_{\boldsymbol{a}} = 2^{n_1} + \sum_{i=1}^{n_1-1}(1-a_i)2^i + 1 = \sum_{i=1}^{r} 2^{n_i}$, where $r = l_0(m) + 2$. Then we have

$$\eta_j = a_j N_j \oplus \cdots \oplus a_{n_1-1} N_{n_1-1} \oplus a_{n_1} = a_{n_1}.$$

We observe that for $N_{\boldsymbol{a}}$ and $j \in \{1, \ldots, n_1\}$ we have $\sigma_{r,j} = a_{n_1} \oplus 1$ if and only if $j = n_k + 1$ for some $k \in \{2, \ldots, r\}$, and $\sigma_{r,j} = a_{n_1}$ otherwise. Hence with Lemma 1 we find

$$\int_0^1 \Delta_{X_{N_{\boldsymbol{a}}}^{C(\boldsymbol{a})}}(t)\mathrm{d}t = (-1)^{a_{n_1}} \left(\frac{r}{2} - \frac{1}{2} \sum_{k=2}^{r} \frac{1}{2^{n_k}} \sum_{i=k+1}^{r} 2^{n_i} \right) + O(1).$$

The fact that $d_0(m) \geq 2$ implies $n_i - n_k \leq 2(k-i)$ and further

$$\left| \int_0^1 \Delta_{X_{N_{\boldsymbol{a}}}^{C(\boldsymbol{a})}}(t)\mathrm{d}t \right| \geq \frac{r}{2} - \frac{1}{2} \sum_{k=2}^{r} \sum_{i=k+1}^{r} 2^{2(k-i)} + O(1) = \frac{r}{3} + O(1).$$

This completes the proof of the first claim (10). To derive (11) from (10), we show that changing the tuple $\boldsymbol{a}$ which defines the matrix to $\tilde{\boldsymbol{a}} = (1-a_i)_{i\geq 1}$ does not change the integral of $\Delta_{X_{N_{\boldsymbol{a}}}^{C(\boldsymbol{a})}}$ much. We use the following argument: Let for $2^{n_1} \leq N \leq 2^{n_1+1} - 1$ with $N = 2^{n_1} + \sum_{i=0}^{n_1-1} N_i 2^i = \sum_{i=1}^{r} 2^{n_i}$

$$S(N) := \frac{r}{2} - \frac{1}{2} \sum_{k=2}^{r} \frac{1}{2^{n_k}} \sum_{i=k+1}^{r} 2^{n_i}.$$

It is not hard to show that $S(N) = \frac{1}{2} \sum_{\ell=0}^{n_1-1} \left\| \frac{N}{2^{\ell+1}} \right\| + O(1)$, where $\|x\|$ denotes the distance of a real number x to its nearest integer. For N as defined above we define the integer $N' := 2^{n_1} + \sum_{i=0}^{n_1-1}(1-N_i)2^i$ and prove $S(N') = S(N) + O(1)$. This is the case, since

$$\begin{aligned} S(N) - S(N') &= \sum_{\ell=0}^{m-1} \left\{ \left\| \frac{N_r}{2} + \cdots + \frac{N_0}{2^{r+1}} \right\| - \left\| \frac{1-N_r}{2} + \cdots + \frac{1-N_0}{2^{r+1}} \right\| \right\} + O(1) \\ &= \sum_{\substack{\ell=0 \\ N_r=0}}^{m-1} (-2^{-r-1}) + \sum_{\substack{\ell=0 \\ N_r=1}}^{m-1} 2^{-r-1} + O(1) = \sum_{\ell=0}^{m-1} (2N_r - 1)2^{-r-1} + O(1) \end{aligned}$$

and therefore

$$|S(N) - S(N')| \leq \sum_{r=0}^{m-1} 2^{-r-1} + O(1) = O(1).$$

This implies inequality (11). □

Acknowledgements The authors are supported by the Austrian Science Fund (FWF): Project F5509-N26, which is a part of the Special Research Program "Quasi-Monte Carlo Methods: Theory and Applications".

References

1. Béjian, R.: Minoration de la discrépance d'une suite quelconque sur T. Acta Arith. **41**, 185–202 (1982)
2. Dick, J., Pillichshammer, F.: Digital Nets and Sequences: Discrepancy Theory and Quasi-Monte Carlo Integration. Cambridge University Press, Cambridge (2010)
3. Dick, J., Pillichshammer F.: Explicit constructions of point sets and sequences with low discrepancy. In: Kritzer, P., Niederreiter, H., Pillichshammer, F., Winterhof, A. (eds.) Uniform Distribution and Quasi-Monte Carlo Methods. Radon Series on Computational and Applied Mathematics, vol. 15, pp. 63–86. DeGruyter, Berlin (2014)
4. Drmota, M., Larcher, G., Pillichshammer, F.: Precise distribution properties of the van der Corput sequence and related sequences. Manuscr. Math. **118**, 11–41 (2005)
5. Faure, H.: Discrepancy and diaphony of digital (0, 1)-sequences in prime bases. Acta Arith. **117**, 125–148 (2005)
6. Faure, H., Kritzer, P., Pillichshammer, F.: From van der Corput to modern constructions of sequences for quasi-Monte Carlo rules. Indag. Math. **26**, 760–822 (2015)
7. Halász, G.: On Roth's method in the theory of irregularities of point distributions. Recent progress in analytic number theory, vol. 2, pp. 79–94. Academic, London (1981)
8. Kritzinger, R., Pillichshammer, F.: L_p-discrepancy of the symmetrized van der Corput sequence. Arch. Math. **104**, 407–418 (2015)
9. Kuipers, L., Niederreiter, H.: Uniform distribution of sequences. Wiley, New York (1974); Reprint, Dover Publications, Mineola (2006)
10. Larcher, G.: On the star discrepancy of sequences in the unit interval. J. Complex. **31**, 474–485 (2015)
11. Niederreiter, H.: Point sets and sequences with small discrepancy. Monatshefte Math. **104**, 273–337 (1987)
12. Niederreiter, H.: Random Number Generation and Quasi-monte Carlo Methods. CBMS-NFS Series in Applied Mathematics, vol. 63. SIAM, Philadelphia (1992)
13. Pillichshammer, F.: On the discrepancy of (0, 1)-sequences. J. Number Theory **104**, 301–314 (2004)
14. Proĭnov, P.D.: On irregularities of distribution. Comptes Rendus Acad. Bulg. Sci. **39**, 31–34 (1986)
15. Roth, K.F.: On irregularities of distribution. Mathematika **1**, 73–79 (1954)
16. Schmidt, W.M.: Irregularities of distribution VII. Acta Arith. **21**, 45–50 (1972)
17. Schmidt, W.M.: Irregularities of distribution. X. Number Theory and Algebra, pp. 311–329. Academic, New York (1977)
18. Wagner, G.: On a problem of Erdős in diophantine approximation. Bull. Lond. Math. Soc. **12**, 81–88 (1980)

Randomized QMC Methods for Mixed-Integer Two-Stage Stochastic Programs with Application to Electricity Optimization

H. Leövey and W. Römisch

Abstract We consider randomized QMC methods for approximating the expected recourse in two-stage stochastic optimization problems containing mixed-integer decisions in the second stage. It is known that the second-stage optimal value function is piecewise linear-quadratic with possible kinks and discontinuities at the boundaries of certain convex polyhedral sets. This structure is exploited to provide conditions implying that first and second order ANOVA terms of the integrand have mixed first order partial derivatives in the sense of Sobolev. This shows that the integrand can be decomposed into a smooth part and a not well-behaved but small part if the effective dimension is low. This leads to good convergence properties of randomized QMC methods. In a case study we consider an optimization model for generating and trading electricity under normal load and price stochasticity. Our numerical experiments where we compare Monte Carlo and two randomized QMC methods indicate that the latter can be superior which confirms our analysis.

Keywords Two-stage stochastic programming · Mixed-integer · Randomized Quasi-Monte Carlo · Convergence rate · Electricity portfolio optimization

1 Introduction

Two-stage stochastic programming models represent a classical approach to deal with optimization problems containing random parameters in the constraints. Its idea is to introduce a two-stage decision process, where the first-stage decision x has to be decided before the randomness occurs, and the second-stage decision y satisfies the constraints that depend on x and the random parameter. Then the sum of the first-stage objective and the expected optimal value of the second-stage problem is optimized

H. Leövey
Structured Energy Management Team, Axpo, Baden, Switzerland
e-mail: hernaneugenio.leoevey@axpo.com

W. Römisch (✉)
Institute of Mathematics, Humboldt-University Berlin, Berlin, Germany
e-mail: romisch@math.hu-berlin.de

B. Tuffin and P. L'Ecuyer (eds.), *Monte Carlo and Quasi-Monte Carlo Methods*, Springer Proceedings in Mathematics & Statistics 324,
https://doi.org/10.1007/978-3-030-43465-6_17

with respect to x. If the second-stage problem contains also integer decisions, we arrive at *mixed-integer two-stage stochastic programs*. We refer to Sect. 2 for a formal mathematical description and for recalling some structural properties. For further information we refer to [22, 29] and to [30] for a recent monograph on stochastic programming. We also refer to Sect. 6 for a practical application from electricity management.

Mixed-integer two-stage stochastic programs belong to the most complicated optimization problems. For a long time it was believed that the only way to tackle the solution of such models is by Monte Carlo (MC) methods [13]. In this paper, we study the possibility of applying randomized Quasi-Monte Carlo (QMC) methods and thereby extending our earlier work [11, 20] on two-stage models without integer decisions. In the present paper we review in Sects. 2–4 theoretical results from [21], but discuss exclusively the error analysis in Sect. 5 and the numerical experiments on solving a practical optimization problem from electric power industry by using randomized QMC methods.

We consider two specific randomized QMC methods, namely, *randomly scrambled Sobol' point sets* [5, 27] and *randomly shifted lattice rules* [15, 31]. For further reading we refer to a survey of randomized QMC methods [19] and to the recent survey [4]. It is well known that such methods display their power and fast convergence in weighted tensor product Sobolev spaces of functions on $[0, 1]^d$ or $\mathbb{R}^d$ (see [4] and Sect. 5). However, there exist several attempts to study the convergence behavior also for functions with kinks [8] and discontinuities [9, 10]. The performance of randomized QMC methods may be significantly deteriorated for such functions. In [10] the authors derive convergence rates for functions of the form $g(x)\mathbb{1}_B(x)$, $x \in [0, 1]^d$, where the function g is smooth and B is a convex polyhedron. They show that the convergence rate can be improved can be improved if some of the discontinuity faces of B are parallel to some coordinate axes (best case being all faces parallel to some coordinate axes since then the function exhibits bounded HK variation).

Integrands of mixed-integer two-stage models are piecewise linear-quadratic with kinks and discontinuities at boundaries of convex polyhedral sets. However, the structure of the convex polyhedra is not known, but hidden in the problem data. Therefore, our approach is different and motivated by the work of [8]. We study the smoothness of lower order ANOVA terms of the integrands and show that they are indeed much smoother than the integrand itself under certain conditions (Sect. 4). Hence, the integrands my be decomposed into a smooth part consisting of lower order ANOVA terms and a nonsmooth part which is small if the effective dimension of the integrand is low (see Sect. 3). This fact indicates that randomized QMC methods can be applied to mixed-integer two-stage models if the integrand has low effective dimension relative to the underlying probability distribution. Details are discussed in the error analysis for randomly shifted lattice rules (see Sect. 5) where we derive an error estimate for the root mean square error of true and approximate optimal values. In our numerical experiments we consider a practical electricity optimization model under uncertainty with normal load and price processes (see Sect. 6). In that case the effective dimension of the integrand can be reduced by factorizing the covariance matrix using principal component analysis.

2 Mixed-Integer Two-Stage Stochastic Programs

We consider the mixed-integer two-stage stochastic optimization problem

$$\min\left\{\langle c, x\rangle + \int_{\mathbb{R}^d} \Phi(q(\xi), h(\xi) - Vx)P(d\xi) : x \in X\right\}, \tag{1}$$

where Φ denotes the parametric infimal function of the second-stage program

$$\Phi(u,t) := \inf\left\{\langle u_1, y_1\rangle + \langle u_2, y_2\rangle : W_1y_1 + W_2y_2 \le t,\, y_1 \in \mathbb{R}^{m_1},\, y_2 \in \mathbb{Z}^{m_2}\right\} \tag{2}$$

for all $(u,t) \in \mathbb{R}^{m_1+m_2} \times \mathbb{R}^r$, and $c \in \mathbb{R}^m$, a closed subset X of $\mathbb{R}^m$, (r, m_1) and (r, m_2)-matrices W_1 and W_2, (r, m)-matrix V, affine functions $q(\xi) \in \mathbb{R}^{m_1+m_2}$, $h(\xi) \in \mathbb{R}^r$, and a Borel probability measure P on $\mathbb{R}^d$. To characterize the domain of Φ we introduce

$$\begin{aligned}\mathscr{T} &= \left\{t \in \mathbb{R}^r : \exists (y_1, y_2) \in \mathbb{R}^{m_1} \times \mathbb{Z}^{m_2} \text{ such that } W_1y_1 + W_2y_2 \le t\right\}\\ \mathscr{U} &= \left\{u = (u_1, u_2) \in \mathbb{R}^{m_1+m_2} : \exists v \in \mathbb{R}^r_- \text{ such that } W_1^\top v = u_1,\ W_2^\top v = u_2\right\}\end{aligned}$$

the primal and dual feasible right-side sets of (2) and assume:

(A1) The matrices W_1 and W_2 have only rational elements.

(A2) The cardinality of the set

$$\bigcup_{t\in\mathscr{T}} \{y_2 \in \mathbb{Z}^{m_2} : \exists y_1 \in \mathbb{R}^{m_1} \text{ such that } W_1y_1 + W_2y_2 \le t\}$$

is finite, i.e., the number of integer decisions appearing in (2) is finite.

It is well known that the presence of integer decisions in (2) leads to discontinuities of Φ. By imposing conditions (A1) and (A2) the structure of the function Φ and of its discontinuity and nondifferentiability regions can be further characterized by utilizing results from parametric mixed-integer linear programming [1, Sect. 5.6].

Proposition 1 ([21]) *Assume (A1) and (A2). The function Φ is finite and lower semicontinuous on $\mathscr{U} \times \mathscr{T}$ and there exists a finite index set $\mathscr{N}$ and a decomposition of $\mathscr{U} \times \mathscr{T}$ consisting of Borel sets $U_\nu \times B_\nu$, $\nu \in \mathscr{N}$, such that their closure is convex polyhedral and Φ is bilinear in (u,t) on each $U_\nu \times B_\nu$. Φ may have kinks and discontinuities at the boundaries of $U_\nu \times B_\nu$.*

In order to have the integrand in (1) well defined we need the additional assumptions known as *relatively complete recourse* and *dual feasibility*:

(A3) For each pair $(x,\xi) \in X \times \mathbb{R}^d$ it holds that $h(\xi) - Vx \in \mathscr{T}$.

(A4) For each $\xi \in \mathbb{R}^d$ the recourse cost $q(\xi)$ belongs to the dual feasible set $\mathscr{U}$.

Proposition 2 ([21]) *Assume (A1)–(A4). Then the integrand*

$$f(x,\xi) = \langle c, x\rangle + \Phi(q(\xi), h(\xi) - Vx) \tag{3}$$

in (1) *is finite and lower semicontinuous on* $X \times \mathbb{R}^d$.
For fixed $x \in X$ *the function* $f(x,\cdot)$ *is linear-quadratic in* ξ *on the Borel sets*

$$\Xi_\nu(x) = \{\xi \in \mathbb{R}^d : q(\xi) \in U_\nu, h(\xi) \in Vx + B_\nu\},\ \nu \in \mathscr{N}, \tag{4}$$

that decompose $\mathbb{R}^d$ *and have convex polyhedral closures. Kinks and discontinuities of* $f(x,\cdot)$ *may appear at the boundaries of* $\Xi_\nu(x)$.

If the probability distribution P has at least finite second order moments, the objective function of (1) is finite and lower semicontinuous due to Fatou's lemma. Hence, the minimization problem (1) is well defined and solvable if the objective is inf-compact. Later we assume even a stronger moment condition in order to be able to use properties of the ANOVA decomposition which we recall next.

3 ANOVA Decomposition and Effective Dimension

We consider a nonlinear function $f : \mathbb{R}^d \to \mathbb{R}$ and intend to compute the expectation $\mathbb{E}[f(\xi)]$ with respect to a probability distribution P having a density ρ given in product form

$$\rho(\xi) = \prod_{k=1}^{d} \rho_k(\xi_k) \quad (\xi \in \mathbb{R}^d).$$

In this context, representations of f that are of interest are of the form

$$f(\xi) = f_0 + \sum_{i=1}^{d} f_i(\xi_i) + \sum_{\substack{i,j=1 \\ i<j}}^{d} f_{ij}(\xi_i, \xi_j) + \cdots + f_{12\cdots d}(\xi_1, \ldots, \xi_d).$$

Such representations can be written more compactly in the form

$$f(\xi) = \sum_{u \subseteq \mathfrak{D}} f_u(\xi^u), \tag{5}$$

where $\mathfrak{D} = \{1, \ldots, d\}$, f_u is defined on $\mathbb{R}^{|u|}$ and ξ^u belongs to $\mathbb{R}^{|u|}$ and contains only the components ξ_j with $j \in u$. Here and in what follows, $|u|$ denotes the cardinality of u and $-u$ the complement $\mathfrak{D} \setminus u$ of u.

Next we make use of the space $L_{2,\rho}(\mathbb{R}^d)$ of all real-valued square integrable functions with inner product

$$\langle f, \tilde{f} \rangle_{2,\rho} = \int_{\mathbb{R}^d} f(\xi)\tilde{f}(\xi)\rho(\xi)d\xi \,.$$

For each function $f \in L_{2,\rho}(\mathbb{R}^d)$ a representation of the form (5) is called ANOVA decomposition of f and the functions f_u are called ANOVA terms if

$$\int_{\mathbb{R}} f_u(\xi^u)\rho_k(\xi_k)d\xi_k = 0 \quad \text{holds for all } k \in u \text{ and } u \subseteq \mathfrak{D}.$$

The ANOVA terms f_u, $\emptyset \neq u \subseteq \mathfrak{D}$, are orthogonal in $L_{2,\rho}(\mathbb{R}^d)$, i.e.

$$\langle f_u, f_v \rangle_{2,\rho} = \int_{\mathbb{R}^d} f_u(\xi) f_v(\xi)\rho(\xi)d\xi = 0 \quad \text{if and only if} \quad u \neq v,$$

and allow a representation by means of (so-called) ANOVA projections. The latter are defined recursively as follows. The first and higher order projections $P_k = P_{-\{k\}}$, $k \in \mathfrak{D}$, and P_u, $u \subseteq \mathfrak{D}$, are given by

$$(P_k f)(\xi^k) = \int_{-\infty}^{\infty} f(\xi_1, \ldots, \xi_{k-1}, s, \xi_{k+1}, \ldots, \xi_d)\rho_k(s)ds$$
$$P_u f(\xi^u) = \Big(\prod_{k \in u} P_k f\Big)(\xi^u)$$

and it holds (see [17])

$$f_u = \Big(\prod_{j \in u}(I - P_j)\Big)P_{-u}(f) = P_{-u}(f) + \sum_{v \subsetneq u}(-1)^{|u|-|v|}P_{-v}(f). \tag{6}$$

To define the effective dimension we consider the variances of f and f_u

$$\sigma^2(f) = \|f - I_{d,\rho}(f)\|^2_{2,\rho} \quad \text{and} \quad \sigma_u^2(f) = \|f_u\|^2_{2,\rho}. \tag{7}$$

Due to the orthogonality of the ANOVA terms we obtain

$$\sigma^2(f) = \|f\|^2_{2,\rho} - (I_{d,\rho}(f))^2 = \sum_{\emptyset \neq u \subseteq \mathfrak{D}} \sigma_u^2(f)\,.$$

Since the quotients $\sigma_u^2(f)/\sigma^2(f)$ indicate for any $u \subseteq \mathfrak{D}$ the importance of the group ξ_j, $j \in u$, of variables of f relative to the underlying distribution P, we define for small $\varepsilon \in (0, 1)$ (e.g. $\varepsilon = 0.01$) the *effective (superposition) dimension* $d_S(\varepsilon)$ *of* f given P [26] as

$$d_S(\varepsilon) = \min\left\{ s \in \mathfrak{D} : \sum_{|u| \leq s} \sigma_u^2(f) \geq (1-\varepsilon)\sigma^2(f) \right\}. \tag{8}$$

An important property of the effective dimension consists in the estimate (see [32])

$$\left\| f - \sum_{|u| \le d_S(\varepsilon)} f_u \right\|_{2,\rho} \le \sqrt{\varepsilon}\sigma(f) \tag{9}$$

showing that the function f is approximated by a truncated ANOVA decomposition which contains all ANOVA terms f_u such that $|u| \le d_S(\varepsilon)$.

If the function f is nonsmooth, the ANOVA terms f_u, $|u| \le d_S(\varepsilon)$, are often smoother than f due to their relation to ANOVA projections and the smoothing effect of integration (see [8, 9]). Hence, the estimate (9) indicates that the main part of f can be smooth and the remaining nonsmooth part be small. Unfortunately, the effective superposition dimension is hardly computable in general, but an upper bound can be computed by finding the smallest $s \in \mathfrak{D}$ such that

$$\sum_{v \subseteq \{1,\ldots,s\}} \sigma_v^2(f) \ge (1 - \varepsilon)\sigma^2(f). \tag{10}$$

This relies on a particular integral representation of the left-hand side of (10), where the occuring integrals can be computed approximately by means of Monte Carlo or Quasi-Monte Carlo methods based on large samples. It should be mentioned, however, that the upper bound can be (extremely) conservative.

4 ANOVA Terms of Mixed-Integer Two-Stage Integrands

According to Proposition 2 mixed-integer two-stage integrands (3) are discontinuous and piecewise linear-quadratic and may be written in the form

$$f(x, \xi) = \langle A_\nu(x)\xi, \xi\rangle + \langle b_\nu(x), \xi\rangle + c_\nu(x) \tag{11}$$

for all $\xi \in \Xi_\nu(x)$, $\nu \in \mathcal{N}$, $x \in X$ if (A1)–(A4) are satisfied. Here, $A_\nu(\cdot)$ are (d, d)-matrices, $b_\nu(\cdot) \in \mathbb{R}^d$ and $c_\nu(\cdot) \in \mathbb{R}$, which are all affine functions of x. The sets $\Xi_\nu(x)$, $\nu \iota \mathcal{N}$, $x \in X$, are defined in (4). They decompose $\mathbb{R}^d$ and their closures are convex polyhedral.

In this section we need further assumptions to prove our main results:

(A5) The probability distribution P has finite fourth order absolute moments.

(A6) P has a density ρ with respect to the Lebesgue measure on $\mathbb{R}^d$ and ρ admits product form

$$\rho(\xi) = \prod_{i=1}^{d} \rho_i(\xi_i) \quad (\xi = (\xi_1, \ldots, \xi_d) \in \mathbb{R}^d,$$

where the densities ρ_i are positive and continuously differentiable, and ρ_i and its derivative are bounded on $\mathbb{R}$.

(A7) For each face F of dimension greater than zero of the convex polyhedral sets cl $\Xi_\nu(x)$, $\nu \in \mathcal{N}$, the affine hull aff(F) of F does not parallel any coordinate axis in $\mathbb{R}^d$ for each $x \in X$ (*geometric condition*).

Due to (A5) and (A6) we may use the concepts ANOVA decomposition and effective dimension for studying mixed-integer two-stage integrands. Using the representation (11) of f the structure of first and second order ANOVA projections can be computed explicitly. This allows conclusions also on the smoothness of higher order projections and, hence, of lower order ANOVA terms due to (6). Finally this leads to the main result of this section. It is proved in [21] and states that at least lower order ANOVA terms of $f = f(x, \cdot)$ for fixed $x \in X$ have all mixed first order partial derivatives in the sense of Sobolev.

Theorem 1 *Assume (A1)–(A7). For fixed $x \in X$ we consider $f = f(x, \cdot)$. Then the ANOVA terms f_u, $|u| \le 2$, $u \subset \mathfrak{D}$, of f are continuously differentiable and have partial mixed first Sobolev derivatives which belong to $L_{2,\rho}(\mathbb{R}^d)$.*

We recall that a real-valued function g on $\mathbb{R}^d$ is the partial *weak or Sobolev derivative* $D^\alpha f$ of a given function f if it is measurable on $\mathbb{R}^d$ and satisfies

$$\int_{\mathbb{R}^d} g(\xi)v(\xi)d\xi = (-1)^{|\alpha|} \int_{\mathbb{R}^d} f(\xi)(D^\alpha v)(\xi)d\xi \quad \text{for all } v \in C_0^\infty(\mathbb{R}^d), \tag{12}$$

where $C_0^\infty(\mathbb{R}^d)$ denotes the space of infinitely differentiable functions with compact support in $\mathbb{R}^d$ and

$$D^\alpha v = \frac{\partial^{|\alpha|} v}{\partial \xi_1^{\alpha_1} \cdots \partial \xi_d^{\alpha_d}} \tag{13}$$

is the classical derivative of v of order $|\alpha| = \sum_{i=1}^d \alpha_i$, where $\alpha = (\alpha_1, \ldots, \alpha_d)$ is a multi-index. The same symbol as in (13) is also used for partial Sobolev derivatives, since classical are also Sobolev derivatives. In the classical case Eq. (12) is just the classical multivariate integration by parts formula.

Remark 1 Theorem 1 shows that the second order ANOVA approximation

$$f^{(2)} = \sum_{\substack{|u| \le 2 \\ u \in \mathfrak{D}}} f_u \tag{14}$$

of the mixed-integer two-stage integrand f (see (3)) has all mixed first partial Sobolev derivatives. If the effective dimension $d_S(\varepsilon)$ of f (see (8)) is at most 2, the mean square distance between the integrand f and $f^{(2)}$ satisfies

$$\|f - f^{(2)}\|_{2,\rho}^2 \le \varepsilon\sigma^2(f)$$

due to (9). For a discussion of techniques for reducing the effective dimension we refer to [32, 33].

While the assumptions (A1)–(A6) are reasonable, assumption (A7) seems somewhat implicit and restrictive at first sight and needs further explanation. For a normal probability distribution P with nonsingular covariance matrix Σ, the orthogonal matrix Q of eigenvectors allows a transformation of Σ into a diagonal matrix D containing the eigenvalues in its main diagonal. This observation enables the following characterization of the geometric condition (A7) using the Haar measure over the topological group of orthogonal matrices. For its proof we refer to [21] and for further information on the Haar measure to [3, Chap. 9].

Theorem 2 *We consider* (1) *and assume (A1)–(A4). If P is multivariate normal on $\mathbb{R}^d$ with nonsingular covariance matrix Σ, the geometric condition (A7) is satisfied almost everywhere with respect to the Haar measure over the topological group of orthogonal (d, d) matrices needed to transform Σ into diagonal form.*

5 Error Analysis of Randomly Shifted Lattice Rules

In this section we provide an error analysis for randomly shifted lattice rules applied to solving mixed-integer two-stage stochastic programs (1). Since typical integrands in stochastic programming are defined on $\mathbb{R}^d$, we introduce first appropriate Sobolev spaces. Following [17, 25] we start with the weighted Sobolev spaces $W^1_{2,\gamma_i,\rho_i,\psi_i}(\mathbb{R})$ of functions $h \in L_{2,\rho_i}(\mathbb{R})$ that are absolutely continuous with derivatives $h' \in L_{2,\psi_i}(\mathbb{R})$ with positive continuous weight functions ψ_i, $i \in \mathfrak{D}$. They are endowed with the weighted inner product

$$\langle h, \tilde{h}\rangle_{\gamma_i,\psi_i} = \Big(\int_{\mathbb{R}} h(\xi)\rho_i(\xi)d\xi\Big)\Big(\int_{\mathbb{R}} \tilde{h}(\xi)\rho_i(\xi)d\xi\Big) + \frac{1}{\gamma_i}\int_{\mathbb{R}} h'(\xi)\tilde{h}'(\xi)\psi_i^2(\xi)d\xi\,,$$

where for each $i \in \mathfrak{D}$ the weight γ_i is positive and we assume that for any $x, \tilde{x} \in \mathbb{R}$

$$\int_x^{\tilde{x}} \psi_i^{-2}(t)dt < \infty\,.$$

The latter condition implies that the weighted Sobolev space is complete [14] and, thus, a Hilbert space. Then the weighted tensor product Sobolev space

$$\mathbb{F}_d = \mathscr{W}^{(1,\ldots,1)}_{2,\gamma,\rho,\psi,\mathrm{mix}}(\mathbb{R}^d) = \bigotimes_{i=1}^{d} W^1_{2,\gamma_i,\rho_i,\psi_i}(\mathbb{R})$$

is equipped with the inner product

$$\langle f, \tilde{f}\rangle_{\gamma,\psi} = \sum_{u\subseteq\mathfrak{D}} \gamma_u^{-1} \int_{\mathbb{R}^{|u|}} I_{u,\rho}(f)(\xi^u) I_{u,\rho}(\tilde{f})(\xi^u) \prod_{i\in u} \psi_i^2(\xi_i) d\xi^u,$$

where the integrands $I_{u,\rho}(f)(\xi^u)$ and the weights γ_u are defined by

$$I_{u,\rho}(f)(\xi^u) = \int_{\mathbb{R}^{|-u|}} \frac{\partial^{|u|} f}{\partial \xi^u}(\xi) \prod_{i\in -u} \rho_i(\xi_i) d\xi^{-u} \quad \text{and} \quad \gamma_u = \prod_{i\in u} \gamma_i\,, \quad \gamma_\emptyset = 1\,.$$

In the QMC literature, this is called the unanchored setting with product weights.

In order to apply QMC methods to the computation of integrals

$$I_\rho(f) = \int_{\mathbb{R}^d} f(\xi)\rho(\xi)d\xi = \int_{\mathbb{R}^d} f(\xi) \prod_{i=1}^{d} \rho_i(\xi_i) d\xi$$

with $f \in \mathbb{F}_d$, the Hilbert space $\mathbb{F}_d$ has to be transformed to a Hilbert space $\mathbb{G}_d$ of functions g on $[0, 1]^d$ by the isometry

$$f \in \mathbb{F}_d \Longleftrightarrow g(\cdot) = f(\Phi^{-1}(\cdot)) \in \mathbb{G}_d\,,$$

where $\Phi^{-1}(t) = (\phi_1^{-1}(t_1), \ldots, \phi_d^{-1}(t_d))$, $t \in [0, 1]^d$, and ϕ_i denotes the one-dimensional distribution function to the density ρ_i, $i \in \mathfrak{D}$. The inner product of $\mathbb{G}_d$ is

$$\langle g, \tilde{g}\rangle_\gamma = \langle f(\Phi^{-1}(\cdot)), \tilde{f}(\Phi^{-1}(\cdot))\rangle_\gamma = \langle f, \tilde{f}\rangle_{\gamma,\psi}.$$

The choice of the weight functions ψ_i depends on the marginal densities ρ_i, $i \in \mathfrak{D}$. We refer to [16, 25] for a discussion of this aspect and for a list of marginal densities and the corresponding weight functions.

Now we consider randomly shifted lattice rules for numerical integration in $\mathbb{G}_d$ (see [15, 31]). Let $Z_n = \{z \in \mathbb{N} : 1 \leq z \leq n,\ \gcd(z, n) = 1\}$ denote the set of natural numbers between 1 and n that are relatively prime to n. Given a generating vector $\mathbf{g} \in Z_n^d$ and a random shift vector Δ which is uniformly distributed in $[0, 1]^d$, the shifted lattice rule points are $t^j = \{\frac{j\mathbf{g}}{n} + \Delta\}$, $j = 1, \ldots, n$, where the braces indicate taking componentwise the fractional part. The corresponding randomized QMC method on $\mathbb{G}_d$ is of the form

$$Q_{n,d}(g) = \frac{1}{n}\sum_{j=1}^{n} g(t^j) \quad (g \in \mathbb{G}_d, n \in \mathbb{N}). \tag{15}$$

Let $\varphi(n)$ denote the cardinality of Z_n, thus, $\varphi(n) = n$ if n is prime, and let $\xi^j = \Phi^{-1}(t^j)$ for $j = 1, \ldots, n$. Then we obtain from [25, Theorem 8] that a generating vector $\mathbf{g} \in Z_n^d$ can be constructed by a component-by-component algorithm such that for each $\delta \in (0, \frac{1}{2}]$ there exists $C(\delta) > 0$ with

$$\left(\mathbb{E}\left|I_{d,\rho}(f) - Q_{n,d}(f(\Phi^{-1}(\cdot)))\right|^2\right)^{\frac{1}{2}} \leq C(\delta)\|f\|_{\gamma,\psi}\,\varphi(n)^{-1+\delta} \tag{16}$$

if the following condition

$$\sum_{i=1}^{\infty} \gamma_i^{\frac{1}{2(1-\delta)}} < \infty \tag{17}$$

on the weights is satisfied and f belongs to $\mathbb{F}_d$. To state our next result we denote by $v(P)$ the infimal value of (1) and by $v(Q_{n,d})$ the infimum if the integral in (1) is replaced by the randomly shifted lattice rule (15).

Theorem 3 *Let (A1)–(A7) be satisfied and X be compact. Assume that all integrands $f = f_x, x \in X$, of the form (3) have at most effective superposition dimension $d_S(\varepsilon) = 2$ for some $\varepsilon > 0$ and that the second order ANOVA approximation $f^{(2)}$ of f belongs to $\mathbb{F}_d$. Furthermore, we assume that $Q_{n,d}$ is a randomly shifted lattice rule* (15) *satisfying* (16). *For each $\delta \in (0, \frac{1}{2}]$ there exists $\hat{C}(\delta) > 0$ such that*

$$\left(\mathbb{E}\left|v(P) - v(Q_{n,d})\right|^2\right)^{\frac{1}{2}} \leq \hat{C}(\delta)\varphi(n)^{-1+\delta} + a_n\,, \tag{18}$$

where the sequence (a_n) converges to zero and allows the estimate

$$a_n \leq \sqrt{\varepsilon}\,\sigma(f) \tag{19}$$

with $\sigma(f)$ denoting the variance (7) *of f.*

Proof Let $x \in X$ be fixed and we consider $f = f_x$. The QMC error may be estimated using the ANOVA approximation $f^{(2)}$ of f of order 2 as follows:

$$\begin{aligned}\left|I_{d,\rho}(f) - Q_{n,d}(f(\Phi^{-1}(\cdot)))\right| \leq &\left|\int_{\mathbb{R}^d} f^{(2)}(\xi)\rho(\xi)d\xi - \frac{1}{n}\sum_{j=1}^{n} f^{(2)}(\xi^j)\right| \\ &+\left|\int_{\mathbb{R}^d} f^{-(2)}(\xi)\rho(\xi)d\xi - \frac{1}{n}\sum_{j=1}^{n} f^{-(2)}(\xi^j)\right|,\end{aligned}$$

where $f^{-(2)} = f - f^{(2)}$. For any $\delta \in (0, \frac{1}{2}]$ we continue

$$\begin{aligned}\left(\mathbb{E}\left|I_{d,\rho}(f) - Q_{n,d}(f(\Phi^{-1}(\cdot)))\right|^2\right)^{\frac{1}{2}} &\leq \left(\mathbb{E}\left|I_{d,\rho}(f^{(2)}) - Q_{n,d}(f^{(2)}(\Phi^{-1}(\cdot)))\right|^2\right)^{\frac{1}{2}} \quad (20)\\ &\quad+\left(\mathbb{E}\left|I_{d,\rho}(f^{-(2)}) - Q_{n,d}(f^{-(2)}(\Phi^{-1}(\cdot)))\right|^2\right)^{\frac{1}{2}} \\ &\leq C(\delta)\|f^{(2)}\|_{\gamma,\psi}\,\varphi(n)^{-1+\delta} + a_n\,, \quad (21)\end{aligned}$$

where we use (16) with $f = f^{(2)}$ to estimate the first term and denote the second term by a_n. Since the integrand $f^{-(2)}$ is Riemann-integrable, the sequence (a_n) converges to zero. Next we utilize [18, Proposition 4] on expressing the variance of randomly shifted lattice rules in terms of squared Fourier coefficients, Parseval's identity for $\|f - f^{(2)}\|_{2,\rho}^2$ and the estimate (9) to obtain

$$a_n \le \|f - f^{(2)}\|_{2,\rho} \le \sqrt{\varepsilon}\,\sigma(f).$$

Our next step is to study how the right-hand side in the estimate (20), (21) depends on $x \in X$. The only term depending on x is the $\mathbb{F}_d$-norm of $f^{(2)} = f_x^{(2)}$. Since $f^{(2)}$ contains only ANOVA terms of order 1 and 2, its norm is given by

$$\|f^{(2)}\|_{\gamma,\psi}^2 = \sum_{|u|\le 2} \gamma_u^{-1} \int_{\mathbb{R}^{|u|}} \left| \int_{\mathbb{R}^{|-u|}} \frac{\partial^{|u|} f^{(2)}}{\partial \xi^u}(\xi) \prod_{i\in -u} \rho_i(\xi_i) d\xi^{-u} \right|^2 \prod_{i\in u} \psi_i^2(\xi_i) d\xi^u .$$

Due to (14) and (6) the second order ANOVA approximation allows a representation in terms of ANOVA projections $P_u f$ with $d-2 \le |u| \le d$. The modulus of such ANOVA projections and of their first and second order derivatives can be bounded by some constant times $\max\{1, \|x\|\}\|\xi^{-u}\|^2$ (at least almost everywhere). Since X is compact, those bounds being continuous functions with respect to x are uniformly bounded on X. Using (A5) this implies that $\|f^{(2)}\|_{\gamma,\psi}$ can be bounded by some uniform constant $\bar{C}$. Now, it remains to appeal to a standard stability result for stochastic programs (see [28, Theorem 5]) to obtain

$$\begin{aligned}\left(\mathbb{E}\left|v(P) - v(Q_{n,d})\right|^2\right)^{\frac{1}{2}} &\le \sup_{x\in X} \left(\mathbb{E}\left|I_{d,\rho}(f_x) - Q_{n,d}(f_x(\Phi^{-1}(\cdot)))\right|^2\right)^{\frac{1}{2}} \\ &\le C(\delta)\bar{C}\varphi(n)^{-1+\delta} + a_n ,\end{aligned}$$

which completes the proof. □

We note that the differentiability properties of $f^{(2)}$ in Theorem 1 motivate the condition for $f^{(2)}$ imposed in Theorem 3.

6 Application to Electricity Optimization Under Uncertainty

We consider a model for the optimal operation of an electricity company in the presence of stochasticity of the electrical load ξ_λ and market price ξ_π. The company owns a number of thermal units and bilateral contracts with other power producers. In addition it trades at electricity markets. Load and price are components of the random vector

$$\xi = (\xi_{\lambda,1}, \dots, \xi_{\lambda,T}, \xi_{\pi,1}, \dots, \xi_{\pi,T})^\top .$$

The time horizon consists of T hourly intervals. At each time period $t \in \{1, \dots, T\}$ the load has to be covered. During peak load periods the production capacity based on their own m units does eventually not suffice to cover the load. Hence, it has to buy the necessary extra amounts from other m_1 markets and m_2 producers at prices

$$p_{1,j_1,t}(\xi) = \bar{p}_{1,j_1,t} + \xi_{\pi,t},\ p_{2,j_2,t} = \bar{p}_{2,j_2,t},\ t = 1,\dots,T,\ j_1 = 1,\dots,m_1,\ j_2 = 1,\dots,m_2,$$

where the vector $\xi_{\pi,t}$ represents the stochastic part of the prices $p_{1,j_1,t}$ at the markets, and $\bar{p}_{1,j_1,t}$, $\bar{p}_{2,j_2,t}$, $t = 1,\dots,T$, represent contractual fixed prices. The aim of the company consists in minimizing its expected costs in the presence of uncertain load and prices. The two-stage stochastic electricity optimization model is of the form

$$\min\left\{\sum_{t=1}^{T}\sum_{j=1}^{m} c_{j,t}\,x_{j,t} + \int_{\mathbb{R}^{2T}} \inf\{g(x,y,u,\xi) : (y,u) \in Y(x,\xi)\}\,P(d\xi) : x \in X\right\} \tag{22}$$

with the convex polyhedral feasible set

$$X := \left\{x \in \mathbb{R}^{mT} \;\middle|\; \begin{array}{l} a_{i,t} \le x_{i,t} \le b_{i,t}\ ,\, i = 1,\dots,m\ ,\ t = 1,\dots,T \\ |x_{i,t} - x_{i,t+1}| \le \delta_{i,t}\ ,\, i = 1,\dots,m\ ,\ t = 1,\dots,T-1 \end{array}\right\},$$

where the linear constraints model capacity limits and ramping constraints. The second-stage objective function g is given by

$$g(x,y,u,\xi) = \sum_{t=1}^{T}\left[\sum_{j_1=1}^{m_1} p_{1,j_1,t}(\xi)(y_{1,j_1,t} + \eta_{j_1}u_{j_1,t}) + \sum_{j_2=1}^{m_2} p_{2,j_2,t}\,y_{2,j_2,t}\right]$$

and the second-stage constraint set $Y(x,\xi)$ as subset of points $(y,u) \in \mathbb{R}^{(m_1+m_2)T} \times \{0,1\}^{m_1T}$ such that

$$\sum_{i=1}^{m} x_{i,t} + \sum_{j_1=1}^{m_1} y_{1,j_1,t} + \sum_{j_2=1}^{m_2} y_{2,j_2,t} \ge \xi_{\lambda,t}\ ,\quad t = 1,\dots,T,$$

$$\begin{array}{ll}
w_{2,j_2,t} \le y_{2,j_2,t}\ , & j_2 = 1,\dots,m_2\ ,\ t = 1,\dots,T, \\
|y_{2,j_2,t} - y_{2,j_2,t+1}| \le \rho_{j_2,t}\ , & j_2 = 1,\dots,m_2\ ,\ t = 1,\dots,T-1, \\
w_{1,j_1,t}u_{j_1,t} \le y_{j_1,t} \le z_{j_1,t}u_{j_1,t}\ , & j_1 = 1,\dots,m_1\ ,\ t = 1,\dots,T, \\
u_{j_1,\tau} - u_{j_1,\tau-1} \le u_{j_1,t}\ , & \tau = t-\overline{\tau},\dots,t-1,\ j_1 = 1,\dots,m_1\ ,\ t = 1,\dots,T, \\
u_{j_1,\tau-1} - u_{j_1,\tau} \le 1 - u_{j_1,t}\ , & \tau = t-\underline{\tau},\dots,t-1,\ j_1 = 1,\dots,m_1\ ,\ t = 1,\dots,T,
\end{array}$$

with fixed positive costs $c_{i,t}$, up/down price proportion η_{j_1}, bounds $a_{i,t}$, $b_{i,t}$, $\delta_{i,t}$, $w_{1,j_1,t}$,$w_{2,j_2,t}$, $z_{j_1,t}$, $\rho_{j_2,t}$ modeling capacity limits and ramp constraints. The variables $u_{j_1,t} \in \{0,1\}$, $j_1 = 1,\dots,m_1$, $t = 1,\dots,T$, model on/off decisions for external units and the bounds $\overline{\tau}$, $\underline{\tau}$ are their minimum up/down times.

We assume that the stochastic loads and prices $\xi_{\lambda,t}, \xi_{\pi,t}$ follow the condition

$$\begin{pmatrix}\xi_{\lambda,t}\\ \xi_{\pi,t}\end{pmatrix} = \begin{pmatrix}\bar{\xi}_{\lambda,t}\\ \bar{\xi}_{\pi,t}\end{pmatrix} + \begin{pmatrix}E_{1,t}\\ E_{2,t}\end{pmatrix}, \quad t = 1, \ldots, T,$$

$$\begin{pmatrix}\bar{\xi}_{\lambda,1}\\ \bar{\xi}_{\pi,1}\end{pmatrix} = B_1 \begin{pmatrix}\gamma_{1,1}\\ \gamma_{2,1}\end{pmatrix}, \ \begin{pmatrix}\bar{\xi}_{\lambda,t}\\ \bar{\xi}_{\pi,t}\end{pmatrix} = A \begin{pmatrix}\bar{\xi}_{\lambda,t-1}\\ \bar{\xi}_{\pi,t-1}\end{pmatrix} + B_1 \begin{pmatrix}\gamma_{1,t}\\ \gamma_{2,t}\end{pmatrix} + B_2 \begin{pmatrix}\gamma_{1,t-1}\\ \gamma_{2,t-1}\end{pmatrix}, \quad t = 2, \ldots, T,$$

where $(E_{1,1}, \ldots, E_{1,T})$ and $(E_{2,1}, \ldots, E_{2,T})$ are fixed mean vectors for loads and prices simulating the *trend* or *seasonality*, $A, B_1, B_2 \in \mathbb{R}^{2\times 2}$, and $\gamma_{1,t}, \gamma_{2,t} \sim \mathrm{N}(0,1)$ are independent standard normal random variables. The resulting stochastic process $\xi = \{(\xi_{\lambda,t}, \xi_{\pi,t})\}_{t=1}^{T}$ is thus a multivariate ARMA(1,1) process. Similar models have been considered for simulating prices and demands in the energy industry in the literature, see e.g. [6]. Note that since the model contains unbounded demands $\xi_{\lambda,1}, \ldots, \xi_{\lambda,T}$, no upper bounds on the variables $y_{2,j_2,t}$, $j_2 = 1, \ldots, m_2$, $t = 1, \ldots, T$ were imposed, allowing the latter to cover arbitrarily large demand values. We select in addition the prices $\bar{\pi}_{2,j_2,t}$ significantly higher than the prices $\bar{\pi}_{1,j_1,t}$, such that the variables $y_{2,j_2,t}$, $j_2 = 1, \ldots, m_2$, $t = 1, \ldots, T$ do not always represent the trivial choice for costs minimization. For our tests, we chose the time horizon $T = 100$, therefore the real dimension of the model is $d = 2T = 200$. Further model constants were set to

$$A = \begin{pmatrix}0.29 & 0.44\\ 0.44 & 0.70\end{pmatrix}, \quad B_1 = \begin{pmatrix}1 & 0\\ 0 & 1\end{pmatrix}, \quad B_2 = \begin{pmatrix}0.75 & 0.053\\ 0.053 & 0.43\end{pmatrix}.$$

We refer to [2, Sect. 7] for detailed information about modeling with multivariate ARMA processes. The resulting joint probability distribution P of the process is normal with dimension $d = 2T$ and covariance matrix Σ. The expectation integral is transformed by factorizing the covariance matrix $\Sigma = A\,A^\top$ as usually recommended in normal high-dimensional integration (see [7, Sect. 2.3.3]). We carry out our tests using the standard lower triangular Cholesky matrix for A (CH) and the principal component analysis factorization, in which $A = (\sqrt{\lambda_1}u_1, \ldots, \sqrt{\lambda_d}u_d)$ with the eigenvalues $\lambda_1 \geq \lambda_2 \geq \cdots, \lambda_d > 0$ of Σ in decreasing order and the corresponding orthonormal eigenvectors u_i, $i = 1, \ldots, d$. Another description of PCA is

$$\Sigma = Q\,\mathrm{diag}(\lambda_1, \ldots, \lambda_d)\,Q^\top,$$

where Q denotes the orthogonal matrix $Q = (u_1 \cdots u_d)$. While the Cholesky factorization seems to assign the same importance to every variable and, hence, is not suitable to reduce the effective dimension, several authors report an enormous reduction of the effective dimension in financial models if PCA is used (e.g., [32]).

A simulated demands and prices-path ξ can then be obtained by

$$\xi = A\,(\phi^{-1}(z_1), \ldots, \phi^{-1}(z_{2T}))^\top + (E_{1,1}, \ldots, E_{1,T}, E_{2,1}, \ldots, E_{2,T}),$$

where $Z = (z_1, \ldots, z_{2T}) \sim U([0, 1]^{2T})$ (i.e., the probability distribution of Z is the uniform distribution on $[0, 1]^{2T}$), and $\phi^{-1}(.)$ represents the inverse cumulative normal distribution function, which can be efficiently and accurately calculated by Moro's algorithm (see [7, Sect. 2.3.2]). The evaluation begins then with MC or randomized QMC points for the samples $Z \sim U([0, 1]^{2T})$. For MC points in $[0, 1]^{2T}$ we used the Mersenne Twister [24] as pseudo random number generator. For QMC, we use randomly scrambled Sobol' points with direction numbers given in [12] and randomly shifted lattice rules [15, 31]. As scrambling technique we used random linear scrambling described in [23]. For our tests, we considered cubic decaying weights $\gamma_j = \frac{1}{j^3}$ for constructing the lattice rules.

We chose the following parameters for the numerical experiments:

- $m = 8, m_1 = 3, m_2 = 4$.
- For all i, j_1, j_2, t, we select randomly $a_{i,t} \in [0.001, 0.003]$, $b_{i,t} \in [0.3, 0.6]$, $\delta_{i,t} \in [0.3, 0.35]$, $w_{1,j_1,t}, w_{2,j_2,t} \in [0.000001, 0.00002]$, $z_{j_1,t} \in [5, 7]$, $\gamma \in [0.1, 0.3]$, $\rho_{j_2,t} \in [1.0, 1.1]$, and $\overline{\tau} = \underline{\tau} = 2$.
- For all i, j_1, j_2, t, we select randomly $c_{i,t} \in [7, 9]$, $\bar{c}_{1,j_1,t} \in [8, 10]$, and $\bar{c}_{2,j_2,t} \in [11, 13]$. We fixed $(E_{1,1}, \ldots, E_{1,d}) = (6, 6, \ldots, 6)$, and $(E_{2,1}, \ldots, E_{2,d}) = (0, 0, \ldots, 0)$.

We performed the following computational experiments. We fixed N sampling points ξ^j and replaced the expected recourse costs by the corresponding equal-weight MC or randomized QMC quadrature rule. Then the resulting approximate stochastic program is of the form

$$\min_{x \in X} \left\{ \sum_{t=1}^{T} \sum_{i=1}^{m} c_{i,t} x_{i,t} + \frac{1}{N} \sum_{j=1}^{N} g(x, y^j, u^j, \xi^j) : (y^j, u^j) \in Y(x, \xi^j), j = 1, \ldots, N \right\}. \tag{23}$$

It represents a mixed-integer linear program comprising $(m + (m_1 + m_2)N)T$ continuous and $m_1 NT$ binary variables. Since N ranges between 2^7 and 2^9, the program (23) contains more than 30.000–150.000 binary variables. These large scale mixed-integer linear programs are solved by means of the standard solver ILOG CPLEX (2014). The aim of the experiments is to examine the convergence rate with respect to the sample size N of the estimated optimal value from (23) obtained by replacing the expectation with MC or randomized QMC quadrature rules. We performed 5 runs for all experiments by changing the set of randomly selected parameters. But the qualitative results remained very similar, therefore we only expose one of these results in the figures. Figure 1 summarizes the convergence behavior under PCA factorizations and Table 1 shows the mean and standard deviation of the estimated optimal values under PCA for each sampling method and each sample size over the 300 replications. We chose $N_1 = 128$, $N_2 = 256$, $N_3 = 512$ as sample sizes for the Mersenne Twister and for the scrambled Sobol' points. For randomly shifted lattices, we chose $N_1 = 127$, $N_2 = 257$, $N_3 = 509$. The random shifts were generated using the Mersenne Twister. We estimate the relative root mean square errors (RMSE) of the optimal values by taking 10 runs of every experiment, and repeat the process 30

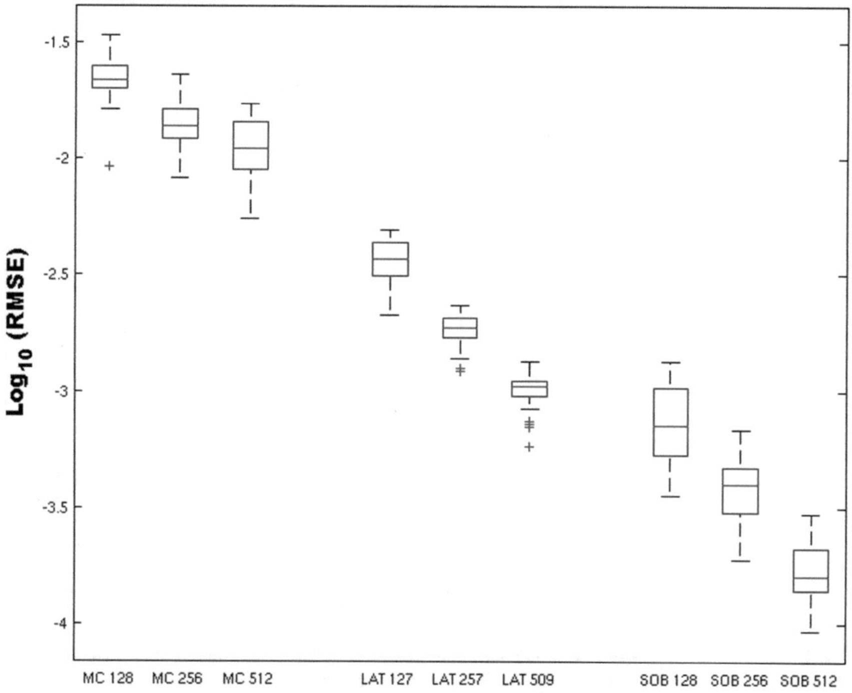

Fig. 1 Shown are the Log_{10} of the relative RMSE with PCA factorization of covariance matrix for computing the optimal value of (23) for parameters as stated above. Results for Mersenne Twister MC and randomly scrambled Sobol' QMC with $N_1 = 128$, $N_2 = 256$ and $N_3 = 512$ points (MC 128, ... or SOB 128, ...), and randomly shifted lattice rules QMC with $N_1 = 127$, $N_2 = 257$ and $N_3 = 509$ lattice points (LAT 127, ...)

Table 1 Mean and standard deviation of the estimated optimal values under PCA for different sampling methods and sample sizes

PCA	Mean			Standard deviation		
	N_1	N_2	N_3	N_1	N_2	N_3
MC	5022.61	5024.13	5026.24	121.86	77.53	62.28
LAT	5026.65	5026.79	5026.99	19.60	9.90	5.41
SOB	5027.14	5027.50	5027.53	4.34	2.16	0.96

times for the box plots in the figures. The box-plots show the median value (red line), first quartile (lower bound of the box) and third quartile (upper bound of the box). Outliers are marked in red and the rest of the results lie between the brackets.

The average of the estimated rates of convergence for the tests under PCA were approximately -0.91 for randomly shifted lattice rules, and -1.05 for the randomly scrambled Sobol' points, for different price- and bound-parameters as listed above. This is clearly superior to the MC convergence rate of -0.5. The upper bound for

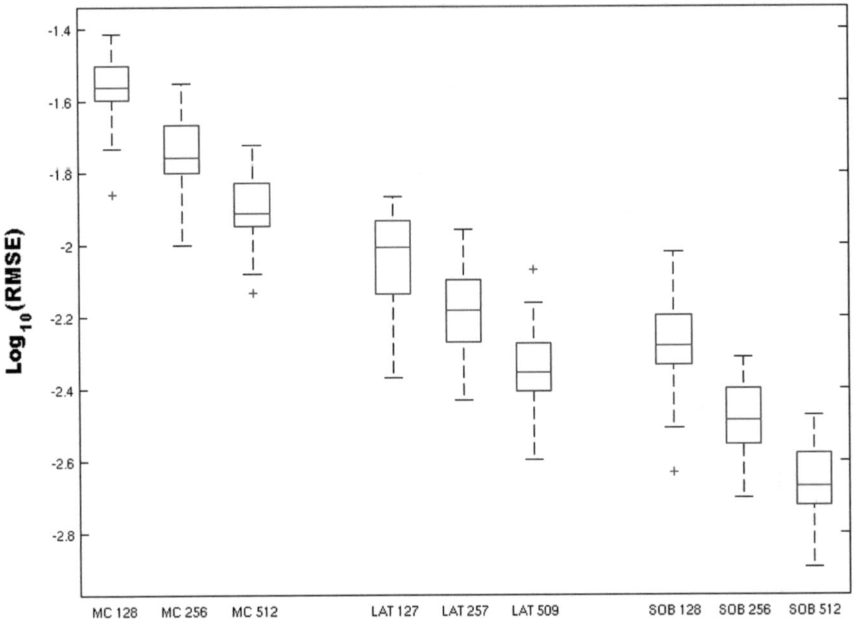

Fig. 2 Shown are the Log_{10} of relative RMSE with Cholesky factorization of covariance matrix for computing the optimal value of (23) for parameters as stated above. Results for Mersenne Twister MC and randomly scrambled Sobol' QMC with $N_1 = 128$, $N_2 = 256$ and $N_3 = 512$ points (MC 128, ... or SOB 128, ...), and randomly shifted lattice rules QMC with $N_1 = 127$, $N_2 = 257$ and $N_3 = 509$ lattice points (LAT 127, ...)

the effective dimension of the integrand $f(x, \cdot)$ in (22) was computed by means of (10) at 5 different feasible vertices x. We used the algorithm proposed in [32] with 2^{16} randomly scrambled Sobol' points ensuring that all results for the ANOVA total and partial variances were obtained with at least 3 digits accuracy. The upper bound of $d_S(\varepsilon)$ with $\varepsilon = 0.01$ is computed by using (10) and remained always equal to 2. We observed also that the first variable under PCA seems to accumulate always more than 90% of the total variance $\sigma^2(f(x, \cdot))$. Hence, PCA serves as excellent dimension reduction technique in this case. Additionally, we performed the same test runs by using the Cholesky decomposition CH instead of PCA for factorizing the covariance matrix. Using CH the observed results, see Fig. 2, were completely different than those under PCA. The average of the estimated rates of convergence of randomized QMC was approximately -0.5, which is the same as the expected MC rate, although the implied error constants seem to be smaller for randomly shifted lattice rules and randomly scrambled Sobol' points than for MC. The upper bound for the effective dimension of the integrand $f(x, \cdot)$ in (22) was estimated by using (10) to be 200 in all tests.

7 Conclusions

The theoretical and numerical results indicate that randomized QMC methods can be superior to MC for solving two-stage stochastic programming problems at least if the recourse cost function has low effective dimension and $\sqrt{\varepsilon}\sigma(f)$ is smaller than the target accuracy for solving the optimization problem. Then using randomized QMC methods instead of MC allows a reduction of sample sizes from N approximately to $\sqrt{N}$. This fact becomes especially important when solving practical mixed-integer stochastic programming models because it reduces the dimension of the large scale mixed-integer linear programs of type (23) and, hence, leads to a considerable reduction of running time. But, Fig. 2 shows that the error constants for randomized QMC methods tend to be smaller than for MC even if the effective dimension is not low. Hence, the use of randomized QMC methods for solving stochastic programs instead of MC seems to pay in any case.

Acknowledgements The authors wish to thank the Editors and the referee for their detailed and valuable comments on an earlier version of this paper.

References

1. Bank, B., Guddat, J., Klatte, D., Kummer, B., Tammer, K.: Non-Linear Parametric Optimization. Akadamie-Verlag, Berlin (1982)
2. Brockwell, P.J., Davis, R.A.: Introduction to Time Series and Forecasting, 2nd edn. Springer, New York (2002)
3. Cohn, D.L.: Measure Theory, 2nd edn. Springer, New York (2013)
4. Dick, J., Kuo, F.Y., Sloan, I.H.: High-dimensional integration - the Quasi-Monte Carlo way. Acta Numer. **22**, 133–288 (2013)
5. Dick, J., Pillichshammer, F.: Digital Nets and Sequences. Cambridge University Press, Cambridge (2010)
6. Eichhorn, A., Römisch, W., Wegner, I.: Mean-risk optimization of electricity portfolios using multiperiod polyhedral risk measures. IEEE St. Petersburg Power Tech (2005)
7. Glasserman, P.: Monte-Carlo Methods in Financial Engineering. Springer, New York (2003)
8. Griebel, M., Kuo, F.Y., Sloan, I.H.: The smoothing effect of integration in $\mathbb{R}^d$ and the ANOVA decomposition. Math. Comput. **82**, 383–400 (2013)
9. Griewank, A., Kuo, F.Y., Leövey, H., Sloan, I.H.: High dimensional integration of kinks and jumps - smoothing by preintegration. J. Comput. Appl. Math. **344**, 259–274 (2018)
10. He, Z., Wang, X.: On the convergence rate of randomized Quasi-Monte Carlo for discontinuous functions. SIAM J. Numer. Anal. **53**, 2488–2503 (2015)
11. Heitsch, H., Leövey, H., Römisch, W.: Are Quasi-Monte Carlo algorithms efficient for two-stage stochastic programs? Comput. Optim. Appl. **65**, 567–603 (2016)
12. Joe, S., Kuo, F.Y.: Remark on algorithm 659: implementing Sobol's quasirandom sequence generator. ACM Trans. Math. Softw. **29**, 49–57 (2003)
13. Kleywegt, A.J., Shapiro, A., Homem-de-Mello, T.: The sample average approximation method for stochastic discrete optimization. SIAM J. Optim. **12**, 479–502 (2001)
14. Kufner, A., Opic, B.: How to define reasonably weighted Sobolev spaces. Comment. Math. Univ. Carol. **25**, 537–554 (1984)
15. Kuo, F.Y.: Component-by-component constructions achieve the optimal rate of convergence in weighted Korobov and Sobolev spaces. J. Complex. **19**, 301–320 (2003)

16. Kuo, F.Y., Sloan, I.H., Wasilkowski, G.W., Waterhouse, B.J.: Randomly shifted lattice rules with the optimal rate of convergence for unbounded integrands. J. Complex. **26**, 135–160 (2010)
17. Kuo, F.Y., Sloan, I.H., Wasilkowski, G.W., Woźniakowski, H.: On decomposition of multivariate functions. Math. Comput. **79**, 953–966 (2010)
18. L'Ecuyer, P., Lemieux, Ch.: Variance reduction via lattice rules. Manag. Sci. **46**, 1214–1235 (2000)
19. L'Ecuyer, P., Lemieux, Ch.: Recent advances in randomized quasi-Monte Carlo methods. In: Dror, M., L'Ecuyer, P., Szidarovski, F. (eds.) Modeling Uncertainty, pp. 419–474. Kluwer, Boston (2002)
20. Leövey, H., Römisch, W.: Quasi-Monte Carlo methods for linear two-stage stochastic programming problems. Math. Program. **151**, 315–345 (2015)
21. Leövey, H., Römisch, W.: Quasi-Monte Carlo methods for two-stage stochastic programs: mixed-integer models. Optim. Online **12**, 7546 (2019)
22. Louveaux, F., Schultz, R.: Stochastic integer programming. In: Ruszczyński, A., Shapiro, A. (eds.) Stochastic Programming, Handbooks in Operations Research and Management Science, vol. 10, pp. 213–266. Elsevier, Amsterdam (2003)
23. Matoušek, J.: On the L_2-discrepancy for anchored boxes. J. Complex. **14**, 527–556 (1998)
24. Matsumoto, M., Nishimura, T.: Mersenne twister: a 623-dimensionally equidistributed uniform pseudo-random number generator. ACM Trans. Model. Comput. Simul. **8**, 3–30 (1998)
25. Nichols, J.A., Kuo, F.Y.: Fast CBC construction of randomly shifted lattice rules achieving $O(n^{-1+\delta})$ convergence for unbounded integrands over $\mathbb{R}^s$ in weighted spaces with POD weights. J. Complex. **30**, 444–468 (2014)
26. Owen, A.B.: The dimension distribution and quadrature test functions. Stat. Sin. **13**, 1–17 (2003)
27. Owen, A.B.: Scrambling Sobol' and Niederreiter-Xing points. J. Complex. **14**, 466–489 (1998)
28. Römisch, W.: Stability of stochastic programming problems. In: Ruszczyński, A., Shapiro, A. (eds.) Stochastic Programming, Handbooks in Operations Research and Management Science, vol. 10, pp. 483–554. Elsevier, Amsterdam (2003)
29. Schultz, R.: Stochastic programming with integer variables. Math. Program. **97**, 285–309 (2003)
30. Shapiro, A., Dentcheva, D., Ruszczyński, A.: Lectures on Stochastic Programming, 2nd edn. MPS-SIAM Series on Optimization, Philadelphia (2014)
31. Sloan, I.H., Kuo, F.Y., Joe, S.: Constructing randomly shifted lattice rules in weighted Sobolev spaces. SIAM J. Numer. Anal. **40**, 1650–1665 (2002)
32. Wang, X., Fang, K.-T.: The effective dimension and Quasi-Monte Carlo integration. J. Complex. **19**, 101–124 (2003)
33. Wang, X., Sloan, I.H.: Quasi-Monte Carlo methods in financial engineering: an equivalence principle and dimension reduction. Oper. Res. **59**, 80–95 (2011)

Approximating Gaussian Process Emulators with Linear Inequality Constraints and Noisy Observations via MC and MCMC

Andrés F. López-Lopera, François Bachoc, Nicolas Durrande, Jérémy Rohmer, Déborah Idier and Olivier Roustant

Abstract Adding inequality constraints (e.g. positivity, monotonicity, convexity) in Gaussian processes (GPs) leads to more realistic stochastic emulators. Due to the truncated Gaussianity of the posterior, its distribution has to be approximated. In this work, we consider Monte Carlo (MC) and Markov Chain MC (MCMC) methods. However, strictly interpolating the observations may entail expensive computations due to highly restrictive sample spaces. Furthermore, having emulators when data are actually noisy is also of interest for real-world applications. Hence, we introduce a noise term for the relaxation of the interpolation conditions, and we develop the corresponding approximation of GP emulators under linear inequality constraints. We demonstrate on various synthetic examples that the performance of MC and MCMC samplers improves when considering noisy observations. Finally, on 2D and 5D coastal flooding applications, we show that more flexible and realistic emulators are obtained by considering noise effects and by enforcing the inequality constraints.

A. F. López-Lopera (✉) · N. Durrande · O. Roustant
Mines Saint-Étienne, CNRS, UMR 6158 LIMOS, Institut Henri Fayol, Université Clermont Auvergne, 42023 Saint-Étienne, France
e-mail: andres-felipe.lopez@emse.fr; andres-felipe.lopez@math.univtoulouse.fr

N. Durrande
e-mail: nicolas.durrande@emse.fr; nicolas@prowler.io

O. Roustant
e-mail: roustant@emse.fr

F. Bachoc
Institut de Mathématiques de Toulouse, Université Paul Sabatier, 31062 Toulouse, France
e-mail: Francois.Bachoc@math.univ-toulouse.fr

N. Durrande
PROWLER.io, 72 Hills Road, Cambridge CB2 1LA, UK

J. Rohmer · D. Idier
BRGM, DRP/R3C, 3 avenue Claude Guillemin, 45060 Orléans cédex 2, France
e-mail: j.rohmer@brgm.fr

D. Idier
e-mail: d.idier@brgm.fr

B. Tuffin and P. L'Ecuyer (eds.), *Monte Carlo and Quasi-Monte Carlo Methods*, Springer Proceedings in Mathematics & Statistics 324,
https://doi.org/10.1007/978-3-030-43465-6_18

Keywords Linear Inequality Constraints · Gaussian Processes Emulators · Monte Carlo and Markov Chain · Monte Carlo Methods · Uncertainty Quantification with Noisy Observations

1 Introduction

Gaussian processes (GPs) have been applied in a great variety of real-world problems as stochastic emulators in fields such as physics, biology, finance and robotics [18, 20]. In the latter, they can be used for emulating the dynamics of robots when experiments become costly-to-evaluate (e.g. motion caption) [20].

Imposing inequality constraints (e.g. boundedness, monotonicity, convexity) into GP emulators leads to more realistic profiles guided by the physics of data [9, 15, 17]. Some applications where constrained GP emulators have been successfully used are computer networking (monotonicity) [9], econometrics (positivity or monotonicity) [5], and nuclear safety criticality assessment (positivity and monotonicity) [15].

In [15, 17], an approximation of GP emulators based on (first-order) regression splines was introduced in order to satisfy general sets of linear inequality constraints. Because of the piecewise linearity of the finite-dimensional approximation used there, the inequalities are satisfied everywhere in the input space. Furthermore, authors in [15, 17] showed that the resulting posterior distribution conditioned on both observations and inequality constraints is truncated Gaussian-distributed. Finally, it was shown in [3] that the resulting posterior mode converges uniformly to the spline interpolation when the number of knots of the spline goes to infinity.

Since the posterior of the model in [15, 17] is a truncated GP, its distribution cannot be computed in closed-form but can be approximated via Monte Carlo (MC) or Markov Chain MC (MCMC). Several MC/MCMC samplers have been tested in [15], leading to emulators that perform well up to 2D. Starting from the claim that allowing noisy observations could yield less constrained sample spaces for samplers, here we develop the approximation of constrained GP emulators when adding noise. Moreover, (constrained) GP emulators for observations that are truly noisy are also of interest for practical implementations. We test the efficiency of various MC and MCMC samplers under 1D synthetic examples where models without observation noise yield impractical sampling routines. We also show that, in monotonic examples, our framework can be applied up to 5D and/or for thousands of observations providing high-quality effective sample sizes within reasonable running times.

This paper is organised as follows. In Sect. 2, we introduce the finite-dimensional approximation of GP emulators with linear inequality constraints and noisy observations. In Sect. 3, we apply our framework to synthetic examples where the consideration of noise-free observations is unworkable. We also test it on 2D and 5D coastal flooding applications. Finally, in Sect. 4, we highlight the conclusions, as well as potential future works.

2 Gaussian Process Emulators with Linear Inequality Constraints and Noisy Observations

In this paper, we aim at imposing linear inequality constraints on GP emulators when observations are considered noisy. As an example, Fig. 1 shows three GP emulators Y satisfying different inequality conditions, with training points at $x_1 = 0.2$, $x_2 = 0.5$, $x_3 = 0.8$. We use a squared exponential (SE) covariance function,

$$k_{\boldsymbol{\theta}}(x, x') = \sigma^2 \exp\left\{ -\frac{(x - x')^2}{2\ell^2} \right\},$$

with covariance parameters $\boldsymbol{\theta} = (\sigma^2 = 0.5^2, \ell = 0.2)$. We set a noise variance to be equal to 0.5% of the variance parameter σ^2. One can note that different types of (constrained) Gaussian priors (top) yield different GP emulators (bottom) for the same training data. Observe also that the interpolation constraints are relaxed due to the noise effect, and that the inequality constraints are still satisfied everywhere.

Next, we formally introduce the corresponding model to obtain constrained GP emulators with linear inequality constraints and noisy observations as in Fig. 1.

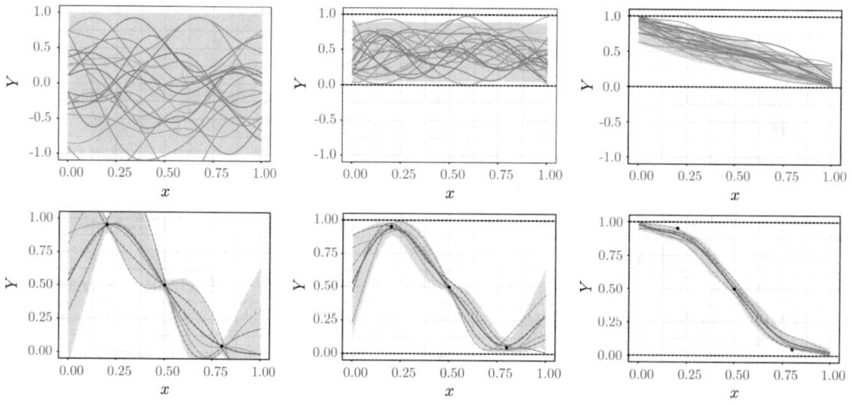

Fig. 1 GP emulators under no constraints (left), boundedness constraints $Y \in [0, 1]$ (centre), and $Y \in [0, 1]$ with non-increasing trajectories (right). Samples from the different types of (constrained) Gaussian priors and resulting GP emulators are shown in the first and second row, respectively. Each panel shows: conditional emulations (dashed lines), and the 95% prediction interval (grey region). For boundedness constraints, bounds at $l = 0$ and $u = 1$ correspond to horizontal dashed lines. For GP emulators, the conditional mean (blue solid line) and observations (black dots) are shown

2.1 Finite-Dimensional Approximation of Gaussian Process Emulators with Noisy Observations

Let Y be a GP on $\mathbb{R}$ with mean zero and covariance function k. Consider $x \in \mathcal{D}$, with compact input space $\mathcal{D} = [0, 1]$. Consider a spline decomposition with an equispaced set of knots $t_1, \ldots, t_m \in \mathcal{D}$ such that $t_j = (j-1)\Delta_m$, for $j = 1, \ldots, m$, with $\Delta_m = 1/(m-1)$. This assumption can be relaxed for non-equispaced designs as in [13], leading to similar developments as the ones in this paper but with slight differences when imposing some constraints (e.g. convexity condition). In contrast to [15], we consider noisy observations $y_i \in \mathbb{R}$, for $i = 1, \ldots, n$. Define Y_m as a stochastic emulator consisting of the piecewise-linear approximation of Y at knots $(t_1, \ldots, t_m)$:

$$Y_m(x) = \sum_{j=1}^{m} \phi_j(x) Y(t_j), \quad \text{s.t.} \quad Y_m(x_i) + \varepsilon_i = y_i, \tag{1}$$

where $x_i \in \mathcal{D}$, for $i = 1, \ldots, n$, $\varepsilon_i \sim \mathcal{N}\left(0, \tau^2\right)$ with noise variance τ^2, and $\phi_1, \ldots, \phi_m$ are hat basis functions given by

$$\phi_j(x) := \begin{cases} 1 - \left|\frac{x-t_j}{\Delta_m}\right| & \text{if } \left|\frac{x-t_j}{\Delta_m}\right| \leq 1, \\ 0 & \text{otherwise.} \end{cases} \tag{2}$$

As in many GP implementations, we assume that $\varepsilon_1, \ldots, \varepsilon_n$ are independent, and independent of Y. However, since the proposed framework does not have any restriction on the type of the covariance function, the extension to other noise distributions and/or noise with autocorrelation is achieved as in [18, 20].

The benefit of considering noisy observations in (1) is that, due to the "relaxation" of the interpolation conditions, the number of knots m does not have to be larger than the number of observations n (assumption required in [15, 17] for the interpolation of noise-free observations). Then, for $m \ll n$, the approximation in (1) would lead to less expensive procedures since the cost of the MC and MCMC samplers grow with the value of m rather than n (see Sects. 2.2 and 2.3).

2.2 Imposing Linear Inequality Constraints

We now assume that Y_m also satisfies inequality constraints everywhere in the input space (e.g. boundedness, monotonicity, convexity), i.e.

$$Y_m \in \mathcal{E}, \tag{3}$$

with $\mathcal{E}$ a convex set of functions defined by some constraints.

The benefit of using (1) is that, for many constraint sets $\mathscr{E}$, satisfying $Y_m \in \mathscr{E}$ is equivalent to satisfying only a finite number of inequalities at the knots $Y(t_1), \ldots, Y(t_m)$ [17]:

$$Y_m \in \mathscr{E} \quad \Leftrightarrow \quad \boldsymbol{\xi} \in \mathscr{C}, \tag{4}$$

where $\boldsymbol{\xi} = [\xi_1, \ldots, \xi_m]^\top$, with $\xi_j := Y(t_j)$ for $j = 1, \ldots, m$, and $\mathscr{C}$ is a convex set on $\mathbb{R}^m$. As an example, when we evaluate a GP with bounded trajectories $l \leq Y_m(x) \leq u$, $\mathscr{C}$ is defined by $\mathscr{C}_{[l,u]} := \{\mathbf{c} \in \mathbb{R}^m; \forall j = 1, \ldots, m : l \leq c_j \leq u\}$. In this paper, we consider the case where $\mathscr{C}$ is composed by a set of q linear inequalities of the form:

$$\mathscr{C} = \left\{\mathbf{c} \in \mathbb{R}^m; \forall p = 1, \ldots, q : l_p \leq \sum_{j=1}^{m} \lambda_{p,j} c_j \leq u_p\right\}, \tag{5}$$

where the $\lambda_{p,j}$'s encode the linear operations, the l_p's and u_p's represent the lower and upper bounds, respectively. Note that $\mathscr{C}_{[l,u]}$ is a particular case of $\mathscr{C}$ where $\lambda_{p,j} = 1$ if $p = j$ and zero otherwise, and with bounds $l_p = l$, $u_p = u$, for $p = 1, \ldots, m$.

We aim at computing the distribution of Y_m conditionally on the constraints in (1) and (3). Observe that the vector $\boldsymbol{\xi}$ is a Gaussian vector with mean zero and covariance matrix $\boldsymbol{\Gamma} = (k(t_i, t_j))_{1 \leq i,j \leq m}$. Denote $\boldsymbol{\Lambda} = (\lambda_{p,j})_{1 \leq p \leq q, 1 \leq j \leq m}$, $\boldsymbol{l} = (\ell_p)_{1 \leq p \leq q}$, $\boldsymbol{u} = (u_p)_{1 \leq p \leq q}$, $\boldsymbol{\Phi}$ the $n \times m$ matrix defined by $\boldsymbol{\Phi}_{i,j} = \phi_j(x_i)$, and $\boldsymbol{y} = [y_1, \ldots, y_n]^\top$ the vector of noisy observations at points $x_1, \ldots, x_n$. Then, the distribution of $\boldsymbol{\xi}$ conditioned on $\boldsymbol{\Phi}\boldsymbol{\xi} + \boldsymbol{\varepsilon} = \boldsymbol{y}$, with $\boldsymbol{\varepsilon} \sim \mathscr{N}\left(\mathbf{0}, \tau^2 \boldsymbol{I}\right)$, is given by

$$\boldsymbol{\xi} | \{\boldsymbol{\Phi}\boldsymbol{\xi} + \boldsymbol{\varepsilon} = \boldsymbol{y}\} \sim \mathscr{N}(\boldsymbol{\mu}, \boldsymbol{\Sigma}), \tag{6}$$

with conditional parameters,

$$\boldsymbol{\mu} = \boldsymbol{\Gamma}\boldsymbol{\Phi}^\top[\boldsymbol{\Phi}\boldsymbol{\Gamma}\boldsymbol{\Phi}^\top + \tau^2 \boldsymbol{I}]^{-1}\boldsymbol{y}, \quad \text{and} \quad \boldsymbol{\Sigma} = \boldsymbol{\Gamma} - \boldsymbol{\Gamma}\boldsymbol{\Phi}^\top[\boldsymbol{\Phi}\boldsymbol{\Gamma}\boldsymbol{\Phi}^\top + \tau^2 \boldsymbol{I}]^{-1}\boldsymbol{\Phi}\boldsymbol{\Gamma}. \tag{7}$$

Note that, in the limit as the noise variance $\tau^2 \to \infty$, then $\boldsymbol{\mu} \to \mathbf{0}$ and $\boldsymbol{\Sigma} \to \boldsymbol{\Gamma}$, and therefore the distribution in (6) ignores the observations $\boldsymbol{y}$. In that case, MC and MCMC samplers are performed in the sample space of the prior of $\boldsymbol{\xi}$, which is less restrictive than the one of $\boldsymbol{\xi} | \{\boldsymbol{\Phi}\boldsymbol{\xi} + \boldsymbol{\varepsilon} = \boldsymbol{y}\}$.

Since the constraints are on $\boldsymbol{\Lambda}\boldsymbol{\xi}$, one can first show that the posterior distribution of $\boldsymbol{\Lambda}\boldsymbol{\xi}$ conditioned on $\boldsymbol{\Phi}\boldsymbol{\xi} + \boldsymbol{\varepsilon} = \boldsymbol{y}$ and $\boldsymbol{l} \leq \boldsymbol{\Lambda}\boldsymbol{\xi} \leq \boldsymbol{u}$ is truncated Gaussian-distributed (see, e.g., [15] for a further discussion when noise-free observations are considered):

$$\boldsymbol{\Lambda}\boldsymbol{\xi} | \{\boldsymbol{\Phi}\boldsymbol{\xi} + \boldsymbol{\varepsilon} = \boldsymbol{y}, \boldsymbol{l} \leq \boldsymbol{\Lambda}\boldsymbol{\xi} \leq \boldsymbol{u}\} \sim \mathscr{T}\mathscr{N}\left(\boldsymbol{\Lambda}\boldsymbol{\mu},\ \boldsymbol{\Lambda}\boldsymbol{\Sigma}\boldsymbol{\Lambda}^\top,\ \boldsymbol{l},\ \boldsymbol{u}\right). \tag{8}$$

Notice that the inequality constraints are encoded in the posterior mean $\boldsymbol{\Lambda}\boldsymbol{\mu}$, the posterior covariance $\boldsymbol{\Lambda}\boldsymbol{\Sigma}\boldsymbol{\Lambda}^\top$, and bounds $(\boldsymbol{l}, \boldsymbol{u})$. Observe that, due to the "relaxation" of the interpolation conditions, constraints can also be imposed when the observations $y_1, \ldots, y_n$ do not fulfil the inequalities.

Algorithm 1 GP emulator with linear inequality constraints.

Require: $\boldsymbol{y} \in \mathbb{R}^n, \boldsymbol{\Gamma} \in \mathbb{R}^{m \times m}, \tau^2 \in \mathbb{R}^+, \boldsymbol{\Phi} \in \mathbb{R}^{n \times m}, \boldsymbol{\Lambda} \in \mathbb{R}^{q \times m}, \boldsymbol{l} \in \mathbb{R}^q, \boldsymbol{u} \in \mathbb{R}^q$
Ensure: Emulated samples from $\boldsymbol{\xi}|\{\boldsymbol{\Phi}\boldsymbol{\xi} + \boldsymbol{\varepsilon} = \boldsymbol{y}, \boldsymbol{l} \leq \boldsymbol{\Lambda}\boldsymbol{\xi} \leq \boldsymbol{u}\}$
1: Compute the conditional mean and covariance of $\boldsymbol{\xi}|\{\boldsymbol{\Phi}\boldsymbol{\xi} + \boldsymbol{\varepsilon} = \boldsymbol{y}\}$,
2: $\quad \boldsymbol{\mu} = \boldsymbol{\Gamma}\boldsymbol{\Phi}^\top(\boldsymbol{\Phi}\boldsymbol{\Gamma}\boldsymbol{\Phi}^\top + \tau^2\boldsymbol{I})^{-1}\boldsymbol{y}$,
3: $\quad \boldsymbol{\Sigma} = \boldsymbol{\Gamma} - \boldsymbol{\Gamma}\boldsymbol{\Phi}^\top(\boldsymbol{\Phi}\boldsymbol{\Gamma}\boldsymbol{\Phi}^\top + \tau^2\boldsymbol{I})^{-1}\boldsymbol{\Phi}\boldsymbol{\Gamma}$.
4: Sample z from the truncated Gaussian distribution via MC/MCMC,
5: $\quad z = \boldsymbol{\Lambda}\boldsymbol{\xi}|\{\boldsymbol{\Phi}\boldsymbol{\xi} + \boldsymbol{\varepsilon} = \boldsymbol{y}, \boldsymbol{l} \leq \boldsymbol{\Lambda}\boldsymbol{\xi} \leq \boldsymbol{u}\} \sim \mathcal{TN}\left(\boldsymbol{\Lambda}\boldsymbol{\mu}, \; \boldsymbol{\Lambda}\boldsymbol{\Sigma}\boldsymbol{\Lambda}^\top, \; \boldsymbol{l}, \; \boldsymbol{u}\right)$.
6: Compute $\boldsymbol{\xi}$ by solving the linear system $\boldsymbol{\Lambda}\boldsymbol{\xi} = z$.

Finally, the truncated Gaussian distribution in (8) does not have a closed-form expression but can be approximated via MC or MCMC. Hence, samples of $\boldsymbol{\xi}$ can be recovered from samples of $\boldsymbol{\Lambda}\boldsymbol{\xi}$, by solving a linear system. As discussed in [15], the number of inequalities q is usually larger than the number of knots m for many convex sets $\mathscr{C}$. If we further assume that $q \geq m$, and that $\mathrm{rank}(\boldsymbol{\Lambda}) = m$, then the solution of the linear system $\boldsymbol{\Lambda}\boldsymbol{\xi}$ exists and is unique (see [15] for a further discussion). Therefore, samples of Y_m are obtained from samples of $\boldsymbol{\xi}$, with the formula $Y_m(x) = \sum_{j=1}^{m} \phi_j(x)\xi_j$ for $x \in \mathscr{D}$. The implementation of the GP emulator Y_m is summarised in Algorithm 1.

2.3 Maximum a Posteriori Estimate via Quadratic Programming

In practice, the posterior mode (maximum a posteriori estimate, MAP) of (8) can be used as a point estimate of unobserved quantities [20], and as a starting state of MCMC samplers [18]. Let $\boldsymbol{\mu}^*$ be the posterior mode that maximises the probability density function (pdf) of $\boldsymbol{\xi}$ conditioned on $\boldsymbol{\Phi}\boldsymbol{\xi} + \boldsymbol{\varepsilon} = \boldsymbol{y}$ and $\boldsymbol{l} \leq \boldsymbol{\Lambda}\boldsymbol{\xi} \leq \boldsymbol{u}$. Then, maximising the pdf in (8) is equivalent to maximise the quadratic problem:

$$\boldsymbol{\mu}^* = \underset{\boldsymbol{\xi} \text{ s.t. } \boldsymbol{l} \leq \boldsymbol{\Lambda}\boldsymbol{\xi} \leq \boldsymbol{u}}{\arg\max} \; \{-[\boldsymbol{\xi} - \boldsymbol{\mu}]^\top \boldsymbol{\Sigma}^{-1} [\boldsymbol{\xi} - \boldsymbol{\mu}]\}, \tag{9}$$

with conditional parameters $\boldsymbol{\mu}$, $\boldsymbol{\Sigma}$ as in (7). By maximising (9), we are looking for the most likely vector $\boldsymbol{\xi}$ satisfying both the observation and inequality constraints. The optimisation problem in (9) is equivalent to

$$\boldsymbol{\mu}^* = \underset{\boldsymbol{\xi} \text{ s.t. } \boldsymbol{l} \leq \boldsymbol{\Lambda}\boldsymbol{\xi} \leq \boldsymbol{u}}{\arg\min} \; \{\boldsymbol{\xi}^\top \boldsymbol{\Sigma}^{-1} \boldsymbol{\xi} - 2\boldsymbol{\mu}^\top \boldsymbol{\Sigma}^{-1} \boldsymbol{\xi}\}, \tag{10}$$

which is solved via quadratic programming [10]. One must note that the mode of (8) converges uniformly to the spline solution when the number of knots $m \to \infty$ [3, 17].

2.4 Extension to Higher Dimensions

The GP emulator in Sect. 2.1 can be extended to d dimensions by tensorisation (see, e.g., [15, 17] for a further discussion on imposing constraints for $d \geq 2$). Consider $\boldsymbol{x} = (x_1, \ldots, x_d) \in \mathscr{D}$ with compact space $\mathscr{D} = [0, 1]^d$, and a set of knots per dimension $(t_1^1, \ldots, t_{m_1}^1), \ldots, (t_1^d, \ldots, t_{m_d}^d)$. Then, the GP emulator $Y_{m_1,\ldots,m_d}$ is given by

$$Y_{m_1,\ldots,m_d}(\boldsymbol{x}) = \sum_{j_1=1,\ldots,m_1} \cdots \sum_{j_d=1,\ldots,m_d} [\phi_{j_1}^1(x_1) \times \cdots \times \phi_{j_d}^d(x_d)]\xi_{j_1,\ldots,j_d}, \tag{11}$$

where $\xi_{j_1,\ldots,j_d} := Y(t_{j_1}, \ldots, t_{j_d})$, and $\phi_{j_1}^1, \ldots, \phi_{j_d}^d$ are hat functions as in (2). We aim at computing (11) subject to the observations $Y_{m_1,\ldots,m_d}(\boldsymbol{x}_i) + \varepsilon_i = y_i$, with $y_i \in \mathbb{R}$ and $\varepsilon_i \sim \mathscr{N}\left(0, \tau^2\right)$ for $i = 1, \ldots, n$; and inequality constraints $\boldsymbol{\xi} = [\xi_{1,\ldots,1}, \ldots, \xi_{m_1,\ldots,m_d}]^\top \in \mathscr{C}$ with $\mathscr{C}$ a convex set of $\mathbb{R}^{m_1 \times \cdots \times m_d}$. We assume that $\varepsilon_1, \ldots, \varepsilon_n$ are independent, independent of Y. Then, following a similar procedure as in Sect. 2.2, Algorithm 1 can be used with $\boldsymbol{\xi}$ a Gaussian vector with mean zero and covariance matrix $\boldsymbol{\Gamma}$.

Note that considering less knots than observations has a great impact since MC/MCMC samplers will then be performed in low dimensional spaces when $m = m_1 \times \cdots \times m_d \ll n$. In that case, the inversion of $(\boldsymbol{\Phi}\boldsymbol{\Gamma}\boldsymbol{\Phi}^\top + \tau^2\boldsymbol{I})$ is obtained more efficiently through the matrix inversion lemma (see, e.g. [20], Appendix A.3), reducing the computational complexity to the inversion of an $m \times m$ full-rank matrix. Thus, the computation of the conditional parameters in (7) and the estimation of the covariance parameter are achieved faster.

3 Numerical Experiments

The codes are implemented in the R programming language, based on the open source package `lineqGPR` [14]. This package is based on previous R developments produced by the Dice (Deep Inside Computer Experiments) and ReDice Consortiums (e.g. `DiceKriging` [22], `DiceDesign` [7], `kergp` [6]), but incorporating some structures of classic libraries for GP regression modelling from other platforms (e.g. the `GPmat` toolbox from MATLAB, and the `GPy` library from Python).

`lineqGPR` also contains implementations of various samplers for the approximation of truncated (multivariate) Gaussian distributions. Samplers are based on recent contributions on efficient MC and MCMC inference methods. Table 1 summarises some properties of the samplers used in this paper (see, e.g., [4, 16, 19, 23]).

Experiments are executed on a single core of an Intel® Core™ i7-6700HQ CPU.

Table 1 Comparison between the MC and MCMC samplers provided in `lineqGPR`: rejection sampling from the mode (RSM) [16], exponential tilting (ExpT) [4], Gibbs sampling [23], and Hamiltonian MC (HMC) [19]

Item	RSM	ExpT	Gibbs	HMC
Exact method	✓	✓	✗	✗
Non parametric	✓	✓	✓	✓
Acceptance rate	Low	High	100%	100%
Speed	Slow	Fast	Slow-fast	Fast
Uncorrelated samples	✓	✓	✗	✗
Previous R Implementations	`constrKriging`	`TruncatedNormal`	`tmvtnorm`	`tmg`

3.1 1D Toy Example Under Boundedness Constraints

Here, we use the GP framework introduced in Sect. 2 for emulating bounded trajectories $Y_m \in [-\alpha, \alpha]$ with constant $\alpha \in \mathbb{R}^+$. We aim at testing the resulting constrained GP emulator when noise-free or noisy observations are considered. The dataset is $(x_i, y_i)_{1 \leq i \leq 5}$: $(0, 0)$, $(0.2, -0.5)$, $(0.5, -0.3)$, $(0.75, 0.5)$, and $(1, 0.4)$. We use a Matérn 5/2 covariance function,

$$k_{\boldsymbol{\theta}}(x, x') = \sigma^2 \left(1 + \frac{\sqrt{5}|x - x'|}{\ell} + \frac{5}{3}\frac{(x - x')^2}{\ell^2}\right) \exp\left\{-\frac{\sqrt{5}|x - x'|}{\ell}\right\},$$

with $\boldsymbol{\theta} = (\sigma^2, \ell)$. We fix $\sigma^2 = 10$, leading to highly variable trajectories. The length-scale ℓ and the noise variance τ^2 are estimated via maximum likelihood (ML).

The effect of different bounds $[-\alpha, \alpha]$ on constrained GP emulators can be seen in Fig. 2. There, we set $m = 100$ for having emulations with high-quality of resolution, and we generated 10^4 constrained emulations via RSM [16]. One can observe that, since interpolation conditions are relaxed due to the influence of the noise variance τ^2, the prediction intervals are wider when bounds become closer to the observations. For the case $\alpha = 0.5$, the noise-free GP emulator yielded costly procedures due to a small acceptance rate equal to 0.1%. In contrast, when noisy observations are assumed, emulations are more likely to be accepted leading to an acceptance rate equal to 16.92%.

We now test the samplers in Table 1 for the approximation of (8). We consider the examples in Fig. 2. For the MCMC samplers, we use the mode in (10) as the starting state of the Markov chains. This initialises the chains in a high probability region. Thus, only few emulations are "burned" in order to have samples that appeared to be independent of the starting state. We burn the first 100 emulations. We evaluate the performance of MC/MCMC samplers in terms of the effective sample size (ESS):

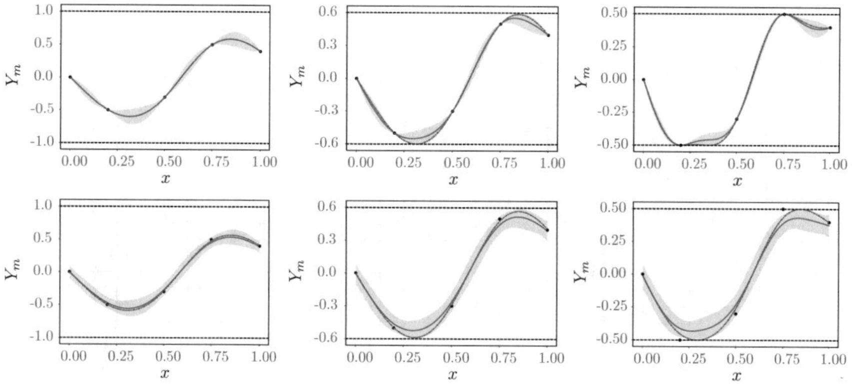

Fig. 2 GP emulators under boundedness constraints $Y_m \in [-\alpha, \alpha]$. Results are shown considering noise-free (top) and noisy observations (bottom): $\alpha = 1$ (left), $\alpha = 0.6$ (centre), and $\alpha = 0.5$ (right). Each panel shows: the observations (dots), the conditional mean (blue solid line), the conditional mode (green dot-dash line), the 95% prediction interval (grey region), and the bounds (dashed lines)

$$\text{ESS} = \frac{n_s}{1 + 2\sum_{k=1}^{n_s} \rho_k}, \tag{12}$$

with n_s the size of the sample path, and ρ_k the sample autocorrelation with lag k. The ESS indicator gives an intuition on how many emulations of the sample path are considered independent [11]. To obtain non-negative values of ρ_k, we use the estimator proposed in [8]. We compute the ESS of each coordinate of the vector $\boldsymbol{\xi}$, i.e. $\text{ESS}_j = \text{ESS}(\xi_j^1, \ldots, \xi_j^{n_s})$ for $j = 1, \ldots, m$, and we evaluate the quantiles $(q_{10\%}, q_{50\%}, q_{90\%})$ over the m resulting ESS values. The value of $n_s = 10^4$ is chosen to be larger than the minimum ESS required to obtain a proper estimation of $\boldsymbol{\xi}$ [11]. Finally, we test the efficiency of each sampler by computing the time normalised ESS (TN-ESS) [12] at $q_{10\%}$ (worst case): TN-ESS $= q_{10\%}(\text{ESS})/$ (CPU Time).

Table 2 displays the performance indicators obtained for each samplers from Table 1. Firstly, one can observe that RSM yielded the most expensive procedures due to its high rejection rate when sampling the constrained trajectories from the posterior mode. In particular, for $\alpha = 0.5$, and assuming noise-free observations, the prohibitively small acceptance rate of RSM led to costly procedures (about 7 h) making it impractical. Secondly, although the Gibbs sampler needed to discard intermediate samples (thinning effect), it provided accurate ESS values within a moderate running time (with effective sampling rates of $400\,\text{s}^{-1}$). Thirdly, due to the high acceptance rates obtained by ExpT, and good exploratory behaviour of the HMC, both samplers provided much more efficient TN-ESS values compared to their competitors, generating thousands of effective emulations each second. Finally, as we expected, the performance of some samplers were improved when adding a noise. For RSM, due to the relaxation of the interpolation conditions, we noted that emulations were more likely to be accepted leading quicker routines: more than 150 times faster with noise (see Table 2, $\alpha = 0.5$).

Table 2 Efficiency of MC and MCMC samplers from Table 1 for emulating bounded samples $Y_m \in [-\alpha, \alpha]$ of Fig. 2. Best results are shown in bold. For the Gibbs sampler, we set the thinning parameter to 200 emulations aiming to obtain competitive ESS values with respect to other samplers. [a]Results could not be obtained due to numerical instabilities

Bounds	Method	Without noise variance			With noise variance		
		CPU Time (s)	ESS [$\times 10^4$] ($q_{10\%}, q_{50\%}, q_{90\%}$)	TN-ESS [$\times 10^4\,s^{-1}$]	CPU Time (s)	ESS [$\times 10^4$] ($q_{10\%}, q_{50\%}, q_{90\%}$)	TN-ESS [$\times 10^4\,s^{-1}$]
[−1.0, 1.0]	RSM	61.30	(0.97, 1.00, 1.00)	0.02	57.64	(0.91, 1.00, 1.00)	0.02
	ExpT	2.30	(0.98, 1.00, 1.00)	0.43	2.83	(0.96, 1.00, 1.00)	0.34
	Gibbs	19.70	(0.84, 0.86, 0.91)	0.04	21.18	(0.75, 0.84, 0.91)	0.04
	HMC	**1.89**	(0.95, 0.99, 1.00)	**0.50**	**1.92**	(0.94, 0.99, 1.00)	**0.49**
[−0.75, 0.75]	RSM	63.59	(1.00, 1.00, 1.00)	0.02	48.66	(0.95, 0.99, 1.00)	0.02
	ExpT	3.22	(0.96, 0.99, 1.00)	0.30	3.24	(0.98, 1.00, 1.00)	0.30
	Gibbs	20.20	(0.83, 0.86, 0.91)	0.04	18.23	(0.74, 0.84, 0.93)	0.04
	HMC	**1.46**	(0.94, 1.00, 1.00)	**0.64**	**1.28**	(0.94, 0.97, 1.00)	**0.73**
[−0.6, 0.6]	RSM	242.34	(0.94, 0.97, 1.00)	0	101.20	(0.96, 1.00, 1.00)	0.01
	ExpT	2.94	(0.94, 1.00, 1.00)	0.32	2.80	(0.98, 1.00, 1.00)	0.35
	Gibbs	18.89	(0.80, 0.83, 0.94)	0.04	18.90	(0.77, 0.84, 0.92)	0.04
	HMC	**1.72**	(0.92, 0.99, 1.00)	**0.53**	**1.68**	(0.93, 0.96, 1.00)	**0.55**
[−0.5, 0.5]	RSM	25512.77	(0.98, 1.00, 1.00)	0	157.06	(0.96, 0.99, 1.00)	0.01
	ExpT	**2.50**	(0.99, 1.00, 1.00)	**0.40**	**2.69**	(0.97, 1.00, 1.00)	**0.36**
	Gibbs[a]	–	–	–	–	–	–
	HMC	6.20	(0.86, 0.90, 0.98)	0.14	2.14	(0.52, 0.85, 0.97)	0.24

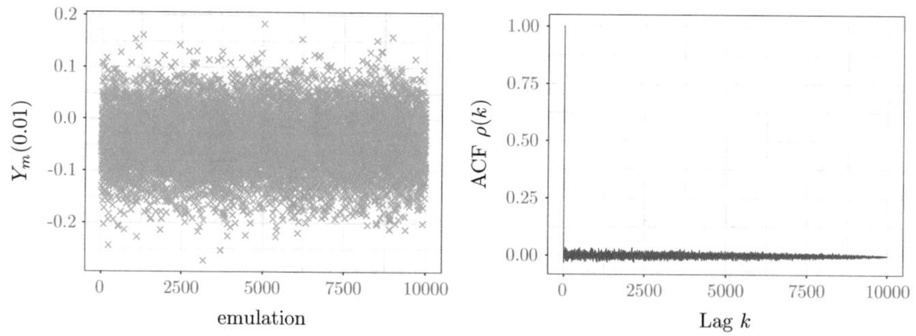

Fig. 3 Efficiency of the HMC sampler in terms of its mixing performance. Results are shown for the trace (left) and autocorrelation (right) plots at $Y_m(0.01)$

Finally, we assess the efficiency of the HMC sampler in terms of its mixing performance (see Fig. 3). We analyse the example of Fig. 2 using the noisy GP emulator with $\alpha = 0.5$. From both the trace and autocorrelation plots at $Y_m(0.01)$, one can conclude that the HMC sampler mixes well with small correlations.

3.2 1D Toy Example Under Multiple Constraints

In [15], numerical implementations were limited to noise-free observations that fulfilled the inequality constraints. In this example, we test the case when noisy observations do not necessarily satisfy the inequalities.

Consider the sigmoid function given by

$$x \mapsto \frac{1}{1 + \exp\left\{-10(x - \frac{1}{2})\right\}}, \quad \text{for } x \in [0, 1]. \tag{13}$$

We evaluate (13) at $n = 300$ random values of x, and we contaminate the function evaluations with an additive Gaussian white noise with a standard deviation equal to 10% of the sigmoid range. Since (13) exhibits both boundedness and non-decreasing conditions, we add those constraints in the GP emulator Y_m using the convex set:

$$\mathscr{C}^{\uparrow}_{[0,1]} = \Big\{\mathbf{c} \in \mathbb{R}^m;\, \forall j = 2, \ldots, m : c_j \geq c_{j-1},\ c_1 \geq 0,\ c_m \leq 1\Big\}.$$

Hence, MC/MCMC samplers will be performed on $\mathbb{R}^{m+1}$ (number of inequalities). As a covariance function, we use a SE kernel and estimate (σ^2, ℓ, τ^2) via ML.

Unlike [15], there is no need here to satisfy the condition $m \geq n$, due to the noise. Therefore, the finite approximation of Sect. 2 can be seen as a surrogate model of standard GP emulators for $m \ll n$. Figure 4 shows the performance of the constrained emulators via HMC for $m = 5, 25, 100$. Note that, for small values of m, the GP

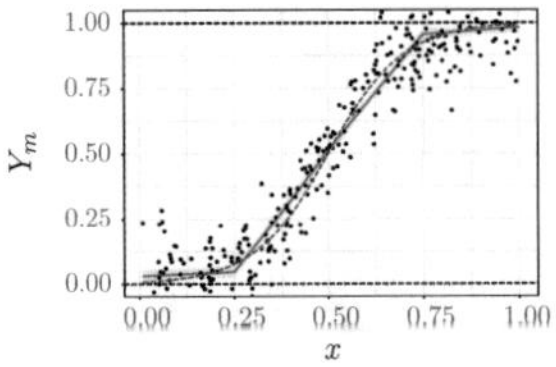

(a) $m = 5$, CPU Time = 0.03 s

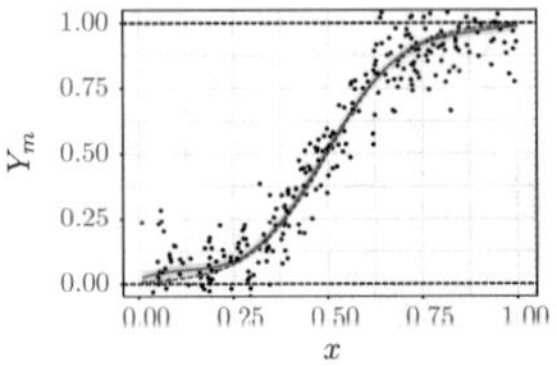

(b) $m = 25$, CPU Time = 0.09 s

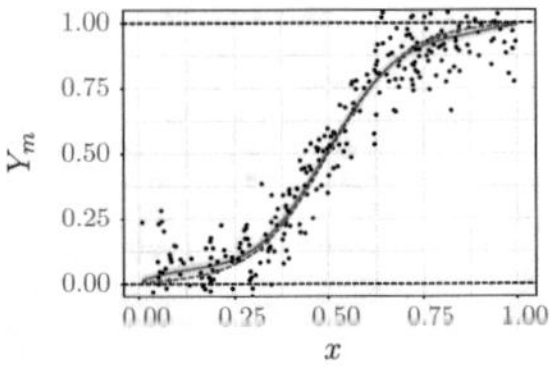

(c) $m = 100$, CPU Time = 1.20 s

Fig. 4 GP emulators under boundedness and monotonicity constraints. Results are shown for different number of knots m. Each panel shows: the target function (red dashed line), the noisy training points (black dots), the conditional mean (blue solid line), the 95% prediction interval (grey region), and the bounds (horizontal dashed lines)

emulator runs fast but with a low quality of resolution of the approximation. For example, for $m = 5$, because of the linearity assumption between knots, the predictive mean presents breakpoints at the knots. On the other hand, the GP emulator yields smoother (constrained) emulations as m increases ($m \geq 25$). In particular, one can observe that for $m = 25$, the emulator leads to a good trade-off between quality of resolution and running time (13 times faster than for $m = 100$).

We now test the performance of the constrained framework under various regularity assumptions, noise levels and constraints. For the example in Fig. 4, we fix $m = 200$ and use different choices of covariance functions: either a Matérn 3/2, a Matérn 5/2 or a SE kernel. Given a fixed noise level, we estimate $\boldsymbol{\theta} = (\sigma^2, \ell)$ via ML. The noise levels are chosen using different proportions of the sigmoid range. We assess GP emulators accounting for either boundedness constraints, monotonicity constraints or both. We compute the CPU time and the Q^2 criterion. The Q^2 criterion is given by $Q^2 = 1 - \text{SMSE}$, where SMSE is the standardised mean squared error [20], and is equal to one if the predictive mean is equal to the test data and lower than one otherwise. We use the 300 noise-free function evaluations from (13) as test data.

Results are shown in Table 3. Note that the introduction of noise let us also have emulations in the cases where the regularity of the GP prior is not in agreement with the regularity of data and constraints. In particular, expensive procedures were obtained for the Matérn 3/2 kernel when monotonicity was considered. In those cases, the high irregularity of the prior yielded more restrictive sample spaces ensuring monotonicity. Observe also that the computational cost of GP emulators can be attenuated by increasing the noise level but at the cost of the accuracy of predictions.

3.3 Coastal Flooding Applications

Coastal flooding models based on GPs have taken great attention regarding computational simplifications for estimating flooding indicators (like maximum water level at the coast, discharge, flood spatial extend, etc.) [2, 21]. However, since standard

Table 3 Performance of the GP emulators in Fig. 4 under different regularity assumptions, noise levels and inequality constraints. Noise levels were chosen using different proportions of the range of the sigmoid function in (13). CPU Time [s] and Q^2 [%] results are shown for various covariance function (i.e. Matérn 3/2, Matérn 5/2 and SE), and different inequality constraints

Noise level (%)	Boundedness constraints						Monotonicity constraints						Both constraints					
	Matérn $\frac{3}{2}$		Matérn $\frac{5}{2}$		SE		Matérn $\frac{3}{2}$		Matérn $\frac{5}{2}$		SE		Matérn $\frac{3}{2}$		Matérn $\frac{5}{2}$		SE	
	Time	Q^2	Time	Q^2	Time	Q^2	Time	Q^2	Time	Q^2	Time	Q^2	Time	Q^2	Time	Q^2	Time	Q^2
0	–	–	–	–	–	–	–	–	–	–	–	–	–	–	–	–	–	–
0.5	1.0	99.4	0.8	99.6	0.6	99.7	117.0	99.5	1.4	99.8	1.2	99.8	–	–	17.3	99.7	13.9	99.8
1.0	1.1	99.4	0.7	99.6	0.6	99.7	14.5	99.1	1.2	99.8	1.0	99.8	$>10^4$	99.4	15.2	99.6	10.4	99.6
5.0	1.0	98.9	0.8	99.3	0.6	99.5	7.4	95.6	1.0	99.3	0.8	99.3	251.8	96.7	13.3	98.6	8.6	98.3
10.0	0.9	98.2	0.8	98.9	0.6	99.2	6.3	91.9	1.0	98.7	0.6	98.9	246.1	94.6	13.3	97.5	8.6	97.0

GP emulators do not take into account the nature of many coastal flooding events satisfying positivity and/or monotonicity constraints, those approaches often require a large number of observations (commonly costly to obtain) in order to have reliable predictions. In those cases, GP emulators yield expensive procedures. Here we show that, by enforcing GP emulators to those inequality constraints, our framework leads to more reliable prediction also when a small amount of data is available.

We test the performance of the emulator in (11) on two coastal flooding datasets provided by the BRGM (the French Geological Survey, "Bureau de Recherches Géologiques et Minières" in French). The first dataset corresponds to a 2D coastal flooding application located on the Mediterranean coast, focusing on the water level at the coast [21]. The second one describes a 5D coastal flooding example induced by overflow on the Atlantic coast, focusing on the inland flooded surface [2]. We train GPs whether the constraints are considered or not. For the unconstrained emulators, we use the GP-based scheme provided by the R package `DiceKriging` [22].

3.3.1 2D Coastal Flooding Application on the Mediterranean Coast

The coastal study site is located on a lido, which has faced two flood events in the past [21]. The dataset used here contains 900 observations of the maximum water level at the coast ξ_m depending on two input parameters: the offshore water level (ξ_o) and the wave height (H_s), both in meters. The observations are taken within the domains $\xi_o \in [0.25, 1.50]$ and $H_s \in [0.5, 7]$ (with each dimension being discretised in 30 elements). Note that, on the domain considered for the input variables, ξ_m increases as ξ_o and H_s increase (see Fig. 5).

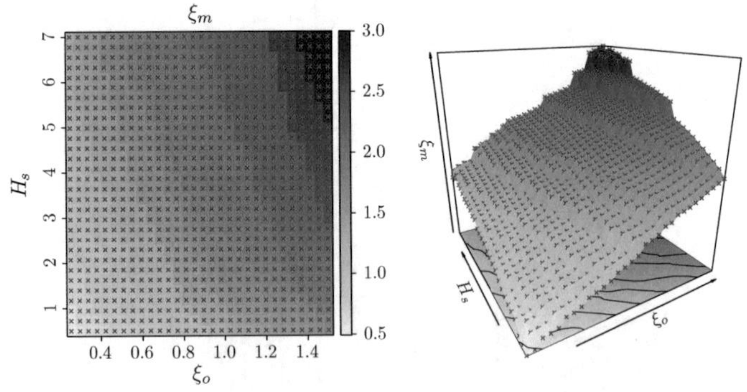

Fig. 5 2D coastal flooding application. (Left) 2D visualisation of the ξ_m values measured over a regular grid. (Right) 3D visualisation of the ξ_m data

Here, we normalise the input space to be in $[0, 1]^2$. As covariance function, we use the product of 1D SE kernels,

$$k_{\boldsymbol{\theta}}(\boldsymbol{x}, \boldsymbol{x}') = \sigma^2 \exp\left\{ - \frac{(x_1 - x_1')^2}{2\ell_1^2} \right\} \exp\left\{ - \frac{(x_2 - x_2')^2}{2\ell_2^2} \right\},$$

with parameters $\boldsymbol{\theta} = (\sigma^2, \ell_1, \ell_2)$. Both $\boldsymbol{\theta}$ and the noise variance τ^2 are estimated via ML. For the constrained model, we propose emulators accounting for both positivity and monotonicity constraints, and we manually fix the number of knots $m_1 = m_2 = 25$ aiming a trade-off between high quality of resolution and computational cost.

For illustrative purposes, we first train both unconstrained and constrained GP emulators using 5% of the data (equivalent to 45 training points chosen by a maximin Latin hypercube DoE [7]), and we aim at predicting the remaining 95%. Results are shown in Fig. 6a, b. In particular, one can observe that the constrained GP emulator slightly outperforms the prediction around the extreme values of ξ_m, leading to an absolute improvement of 4% of the Q^2 indicator.

We then repeat the experiment using twenty different sets of training data and different proportions of training points. According to Fig. 6c, one can observe that the constrained emulator often outperforms the unconstrained one, with significant Q^2 improvements for small training sets. As coastal flooding simulators are commonly costly-to-evaluate, the benefit of having accurate prediction with fewer observations becomes useful for practical implementations.

3.3.2 5D Coastal Flooding Application on the Atlantic Coast

As in [2], here we focus on the coastal flooding induced by overflow. We consider the "Boucholeurs" area located close to "La Rochelle", France. This area was flooded during the 2010 Xynthia storm, an event characterised by a high storm surge in phase with a high spring tide. We focus on those primary drivers, and on how they affect the resulting flooded surface. We refer to [2] for further details.

The dataset contains 200 observations of the flooded area Y in m^2 depending on five input parameters $\mathbf{x} = (T, S, \phi, t_+, t_-)$ detailing the offshore forcing conditions:

- The tide is simplified by a sinusoidal signal parametrised by its high tide level $T \in [0.95, 3.70]$ (m).
- The surge signal is described by a triangular model using four parameters: the peak amplitude $S \in [0.65, 2.50]$ (m), the phase difference $\phi \in [-6, 6]$ (hours), between the surge peak and the high tide, the time duration of the raising part $t_- \in [-12.0, -0.5]$ (hours), and the falling part $t_+ \in [0.5, 12.0]$ (hours).

One must note that the flooded area Y increases as T and S increase. The dataset is freely available in the R package `profExtrema` [1].

Before implementing the corresponding GP emulators, we first analyse the structure of the dataset. We test various standard linear regression models in order to understand the influence of each input variable $\mathbf{x} = (T, S, \phi, t_+, t_-)$. We assess the

Fig. 6 2D GP emulators for modelling the coastal flooding data in [21]. (Left) Prediction results using 5% of the dataset via maximin Latin hypercube DoE. Each panel shows: training and test points (black dots and red crosses), the conditional mean function (solid surface), and the Q^2 criterion (subcaptions). **c** Q^2 assessment using different proportions of training points (x-axis) and using twenty different random training sets. Results are shown for the unconstrained (red) and constrained (blue) GP emulators

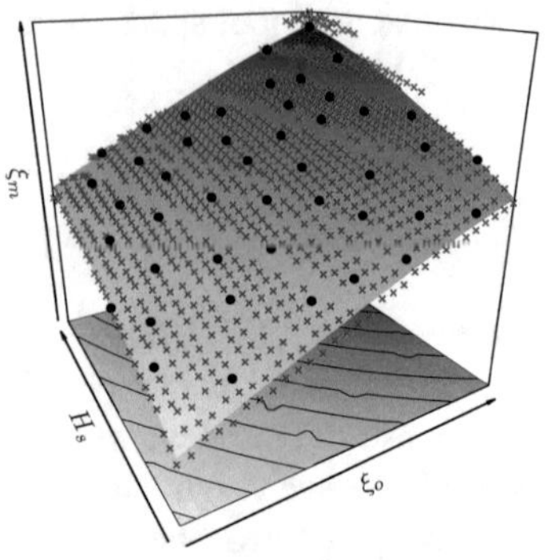

(a) Unconstrained GP: $Q^2 = 0.987$

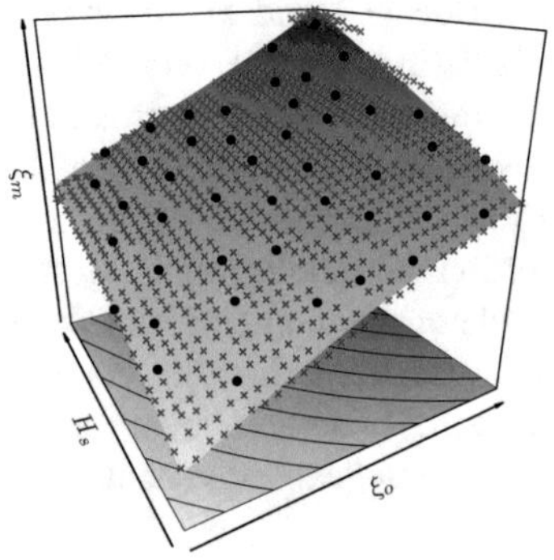

(b) Constrained GP: $Q^2 = 0.991$

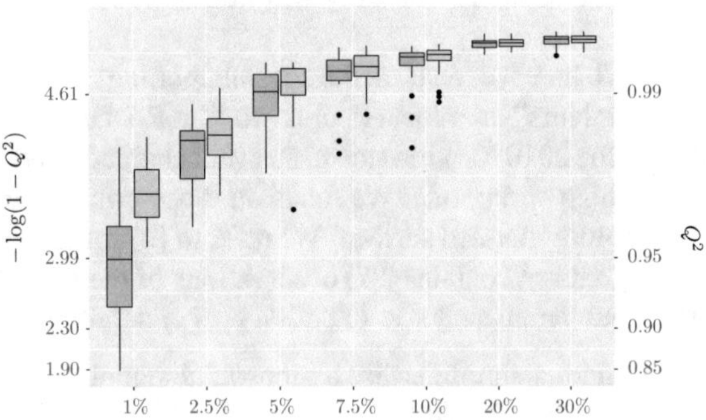

(c) Q^2 performance

quality of the linear models using the adjusted R^2 criterion. Similarly to the Q^2, the R^2 indicator evaluates the quality of predictions over all the observation points rather than only over the training data. Therefore, for noise-free observations, the R^2 indicator is equal to one if the predictors are exactly equal to the data. We also test various models considering different input variables (e.g. transformation of variables, or inclusion of interaction terms).

After testing different linear models, we observed that they were more sensitive to the inputs T and S rather than to other ones. We also noted that, by transforming the phase coordinate $\phi \mapsto \cos(2\pi\phi)$, an absolute improvement about 26% of the R^2 indicator was obtained, and the influence of both t_- and t_+ became more significant. Thus, we used these settings for the GP implementations.

We normalised the input space to be in $[0, 1]^5$, and we used a covariance function given by the product of 1D Matérn 5/2 kernels. The covariance parameters $\boldsymbol{\theta} = (\sigma^2, \ell_1, \ldots, \ell_5)$ and the noise variance τ^2 were estimated via ML. We also tested other types of covariance structures, including SE and Matérn 3/2 kernels, but less accurate predictions were obtained according to the Q^2 criterion. For the constrained model, we proposed GP emulators accounting for positivity constraints everywhere. We also imposed monotonicity constraints along the T and S input dimensions. Since the computational complexity of the constrained GP emulator increases with the number of knots m used in the piecewise-linear representation, we strategically fixed them in coordinates requiring high quality of resolution. Since we observed that the contribution of the inputs T, S, t_- and t_+ was almost linear (result in agreement with [2]), we placed fewer knots over those entries. In particular, we fixed as number of knots per dimension: $m_1 = m_2 = 4$, $m_3 = 5$ and $m_4 = m_5 = 3$.

As in Sect. 3.3.1, we trained GP emulators using twenty different sets of training data and different proportions of training points. According to Fig. 7, one can observe that the constrained GP emulator often outperforms the unconstrained one, with significant Q^2 improvements for small training sets. In particular, one can note that, by enforcing the GP emulators with both positivity and monotonicity constraints, accurate predictions were also provided by using only 10% of the observations as training points (equivalent to 20 observations).

4 Conclusions

We have introduced a constrained GP emulator with linear inequality conditions and noisy observations. By relaxing the interpolation of observations through a noise effect, MC and MCMC samplers are performed in less restrictive sample spaces. This leads to faster emulators while preserving high effective sampling rates. As seen in simulations, the HMC sampler in [19] usually outperformed its competitors, providing more efficient effective sample rates in high dimensional sample spaces.

Since there is no need of having more knots than observations ($m \geq n$), the computational complexity of MC and MCMC samplers is independent of n. Therefore, since the samplers are performed on $\mathbb{R}^m$, they can be used for large values of n by

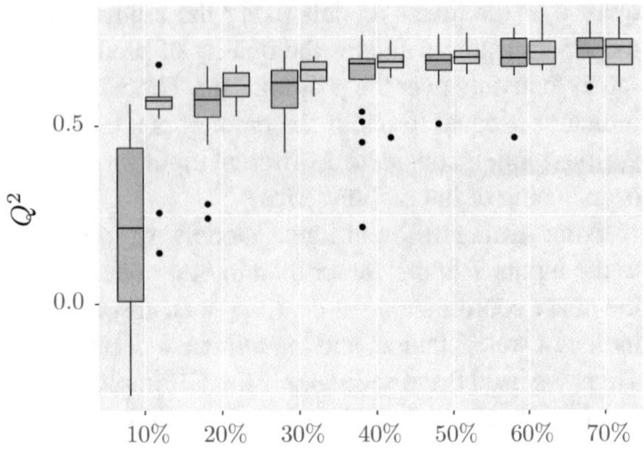

Fig. 7 5D GP emulators for modelling the coastal flooding data in [2]. The boxplots show the Q^2 results using different proportions of training points (x-axis) and using twenty different random training sets. Results are shown for the unconstrained (red) and constrained (blue) GP emulators

letting $m \ll n$. On 2D and 5D coastal flooding applications, we also showed that more flexible and realistic GP emulators were obtained by considering noise effects and by enforcing the (linear) inequality constraints.

Despite the improvements obtained here for scaling constrained GP emulators up to 5D, their tensor structure makes them impractical for tens of input variables. We believe that this limitation can be mitigated by using other types of designs of the knots (e.g. sparse designs). In addition, supplementary assumptions on the nature of the target function can also be made to reduce the dimensionality of the sample spaces where MC and MCMC samplers are performed (e.g. additivity).

Acknowledgements This research was conducted within the frame of the Chair in Applied Mathematics OQUAIDO, gathering partners in technological research (BRGM, CEA, IFPEN, IRSN, Safran, Storengy) and academia (CNRS, Ecole Centrale de Lyon, Mines Saint-Etienne, Univ. Grenoble, Univ. Nice, Univ. Toulouse) around advanced methods for computer experiments.

References

1. Azzimonti, D.: profExtrema: Compute and visualize profile extrema functions. R package version 0.2.0 (2018)
2. Azzimonti, D., Ginsbourger, D., Rohmer, J., Idier, D.: Profile extrema for visualizing and quantifying uncertainties on excursion regions. Application to coastal flooding. Technometrics **0**(ja), 1–26 (2019)
3. Bay, X., Grammont, L., Maatouk, H.: Generalization of the Kimeldorf-Wahba correspondence for constrained interpolation. Electron. J. Stat. **10**(1), 1580–1595 (2016)
4. Botev, Z.I.: The normal law under linear restrictions: simulation and estimation via minimax tilting. J. R. Stat. Soc.: Ser. B **79**(1), 125–148 (2017)
5. Cousin, A., Maatouk, H., Rullière, D.: Kriging of financial term-structures. Eur. J. Oper. Res. **255**(2), 631–648 (2016)
6. Deville, Y., Ginsbourger, D., Durrande, N., Roustant, O.: kergp: Gaussian process laboratory. R package version 0.2.0 (2015)
7. Dupuy, D., Helbert, C., Franco, J.: DiceDesign and DiceEval: two R packages for design and analysis of computer experiments. J. Stat. Softw. **65**(11), 1–38 (2015)

8. Geyer, C.J.: Practical Markov Chain Monte Carlo. Stat. Sci. **7**(4), 473–483 (1992)
9. Golchi, S., Bingham, D.R., Chipman, H., Campbell, D.A.: Monotone emulation of computer experiments. SIAM/ASA J. Uncertain. Quantif. **3**(1), 370–392 (2015)
10. Goldfarb, D., Idnani, A.: Dual and primal-dual methods for solving strictly convex quadratic programs. Numerical Analysis, pp. 226–239. Springer, New York (1982)
11. Gong, L., Flegal, J.M.: A practical sequential stopping rule for high-dimensional Markov Chain Monte Carlo. J. Comput. Graph. Stat. **25**(3), 684–700 (2016)
12. Lan, S., Shahbaba, B.: Sampling constrained probability distributions using spherical augmentation. Algorithmic Advances in Riemannian Geometry and Applications, pp. 25–71. Springer, New York (2016)
13. Larson, M.G., Bengzon, F.: The Finite Element Method: Theory, Implementation, and Applications. Springer, New York (2013)
14. López-Lopera, A.F.: lineqGPR: Gaussian process regression models with linear inequality constraints. R package version 0.0.3 (2018)
15. López-Lopera, A.F., Bachoc, F., Durrande, N., Roustant, O.: Finite-dimensional Gaussian approximation with linear inequality constraints. SIAM/ASA J. Uncertain. Quantif. **6**(3), 1224–1255 (2018)
16. Maatouk, H., Bay, X.: A new rejection sampling method for truncated multivariate Gaussian random variables restricted to convex sets. Monte Carlo and Quasi-Monte Carlo Methods, pp. 521–530. Springer, New York (2016)
17. Maatouk, H., Bay, X.: Gaussian process emulators for computer experiments with inequality constraints. Math. Geosci. **49**(5), 557–582 (2017)
18. Murphy, K.P.: Machine Learning: A Probabilistic Perspective. Adaptive Computation and Machine Learning Series. The MIT Press, Cambridge (2012)
19. Pakman, A., Paninski, L.: Exact Hamiltonian Monte Carlo for truncated multivariate Gaussians. J. Comput. Graph. Stat. **23**(2), 518–542 (2014)
20. Rasmussen, C.E., Williams, C.K.I.: Gaussian Processes for Machine Learning. Adaptive Computation and Machine Learning. The MIT Press, Cambridge (2005)
21. Rohmer, J., Idier, D.: A meta-modelling strategy to identify the critical offshore conditions for coastal flooding. Nat. Hazards Earth Syst. Sci. **12**(9), 2943–2955 (2012)
22. Roustant, O., Ginsbourger, D., Deville, Y.: DiceKriging, DiceOptim: two R packages for the analysis of computer experiments by Kriging-based metamodeling and optimization. J. Stat. Softw. **51**(1), 1–55 (2012)
23. Taylor, J., Benjamini, Y.: restrictedMVN: multivariate normal restricted by affine constraints. R package version 1.0 (2016)

A Multilevel Monte Carlo Asymptotic-Preserving Particle Method for Kinetic Equations in the Diffusion Limit

Emil Løvbak, Giovanni Samaey and Stefan Vandewalle

Abstract We propose a multilevel Monte Carlo method for a particle-based asymptotic-preserving scheme for kinetic equations. Kinetic equations model transport and collision of particles in a position-velocity phase-space. With a diffusive scaling, the kinetic equation converges to an advection-diffusion equation in the limit of zero mean free path. Classical particle-based techniques suffer from a strict time-step restriction to maintain stability in this limit. Asymptotic-preserving schemes provide a solution to this time step restriction, but introduce a first-order error in the time step size. We demonstrate how the multilevel Monte Carlo method can be used as a bias reduction technique to perform accurate simulations in the diffusive regime, while leveraging the reduced simulation cost given by the asymptotic-preserving scheme. We describe how to achieve the necessary correlation between simulation paths at different levels and demonstrate the potential of the approach via numerical experiments.

Keywords Multilevel Monte Carlo · Kinetic equations · Particle methods · Asymptotic-preserving schemes

1 Introduction

Kinetic equations, modeling particle behavior in a position-velocity phase space, occur in many domains. Examples are plasma physics [4], bacterial chemotaxis [33] and computational fluid dynamics [32]. Many of these applications exhibit a strong

---

E. Løvbak (✉) · G. Samaey · S. Vandewalle
Department of Computer Science, NUMA Section, Celestijnenlaan 200A box 2402, 3001 Leuven, Belgium
e-mail: emil.loevbak@cs.kuleuven.be

G. Samaey
e-mail: giovanni.samaey@cs.kuleuven.be

S. Vandewalle
e-mail: stefan.vandewalle@cs.kuleuven.be

B. Tuffin and P. L'Ecuyer (eds.), *Monte Carlo and Quasi-Monte Carlo Methods*, Springer Proceedings in Mathematics & Statistics 324,
https://doi.org/10.1007/978-3-030-43465-6_19

time-scale separation, leading to an unacceptably high simulation cost [7]. However, one typically is only interested in computing the evolution of some macroscopic quantities of interest. These are usually some moments of the particle distribution, which can be computed as averages over velocity space. The time-scale at which these quantities of interest change is often much slower than the time-scale governing the particle dynamics. The nature of the macroscopic dynamics depends on the scaling of the problem, which can be either *hyperbolic* or *diffusive* [15].

The model problem in this work is a one-dimensional kinetic equation of the form

$$\partial_t f(x, v, t) + v\partial_x f(x, v, t) = Q\left(f(x, v, t)\right), \tag{1}$$

where $f(x, v, t)$ represents the distribution of particles as a function of position $x \in \mathbb{R}$ and velocity $v \in \mathbb{R}$ as it evolves in time $t \in \mathbb{R}^+$. The left-hand side of (1) represents transport, while $Q(f(x, v, t))$ is a collision operator that results in discontinuous velocity changes. As the collision operator, we take the BGK model [3], which represents linear relaxation to an equilibrium distribution that only depends on the particle density

$$\rho(x, t) = \int f(x, v, t)dv. \tag{2}$$

We introduce a parameter ε that represents the mean free path. When decreasing ε, the average time between collisions decreases. In this paper, we consider the diffusive scaling. In that case, we simultaneously increase the time scale at which we observe the evolution of the particle distribution, arriving at

$$\varepsilon\partial_t f(x, v, t) + v\partial_x f(x, v, t) = \frac{1}{\varepsilon}\left(\mathscr{M}(v)\rho(x, t) - f(x, v, t)\right), \tag{3}$$

with $\mathscr{M}(v)$ the particles' steady state velocity distribution. It has been shown that when taking the limit $\varepsilon \to 0$, the behavior of equations of the form (3) is fully described by the diffusion equation [25]

$$\partial_t \rho(x, t) = \partial_{xx}\rho(x, t). \tag{4}$$

Kinetic equations can be simulated with deterministic methods, solving the partial differential equation (PDE) that describes the evolution of the particle distribution in the position-velocity phase space. Alternatively, one can use stochastic methods that simulate a large number of particle trajectories. Deterministic methods become prohibitively expensive for higher dimensional applications. Particle-based methods do not suffer from this curse of dimensionality, at the expense of introducing a statistical error in the computed solution. The issue of time-scale separation is present in both deterministic and stochastic methods.

One way to avoid the issue of time-scale separation is through the use of asymptotic-preserving methods, which aim at reproducing a scheme for the limiting macroscopic equation in the limit of infinite time-scale separation. For deterministic discretization methods, there is a long line of such methods. We refer to [2, 5, 6, 10, 14, 19–24, 26–28] as a representative sample of such methods in the diffusive scaling. The recent review paper [15] contains an overview of the state of the art on asymptotic-preserving methods for kinetic equations, and ample additional references. In the particle-based setting, only a few asymptotic-preserving methods have been developed, mostly in the hyperbolic scaling [11–13, 29–31]. In the diffusive scaling, there are only two works [9, 16] so far, to the best of our knowledge. Both methods avoid the time step restrictions caused by fast problem time-scales, at the expense of introducing a bias, which is of order one in the time step size.

The goal of the present paper is to combine the asymptotic-preserving scheme in [16] with the multilevel Monte Carlo method. Given a fixed computational budget, a trade-off typically has to be made between a small bias and a low variance. The former can be obtained by reducing the time step, the latter by simulating many trajectories with large time steps. The core idea behind the multilevel Monte Carlo method [17] is to reduce computational cost, by combining estimates computed with different time step sizes. The multilevel Monte Carlo method, originally developed in the context of stochastic processes, has been applied to problems across many fields, for example, finance [17] and biochemistry [1]. The method has successfully been applied to simulating large PDE's with random coefficients [8]. Recent work has also used multilevel Monte Carlo methods in an optimization context [34].

The remainder of this paper is organized as follows. In Sect. 2, we describe the model kinetic equation on which we will demonstrate our approach, as well as the asymptotic-preserving Monte Carlo scheme that was introduced in [16]. In Sect. 3, we cover the multilevel Monte Carlo method that is the core contribution of this paper. In Sect. 4, we present some preliminary experimental results, demonstrating the properties of the new scheme as well as its computational gain. Finally, in Sect. 5 we will summarize our main results and mention some possible future extensions.

2 Model Problem and Asymptotic-Preserving Scheme

2.1 Model Equation in the Diffusive Limit

The model problem considered in this work is a one-dimensional kinetic equation in the diffusive scaling of the form (3), which we rewrite as

$$\partial_t f(x, v, t) + \frac{v}{\varepsilon}\partial_x f(x, v, t) = \frac{1}{\varepsilon^2}\left(\mathcal{M}(v)\rho(x, t) - f(x, v, t)\right). \qquad (5)$$

For ease of exposition, we restrict ourselves to the case of two discrete velocities, $v = \pm 1$. Then, we can write $f_+(x, t)$ and $f_-(x, t)$ to represent the distribution of particles with, respectively, positive and negative velocities, and $\rho(x, t) = f_+(x, t) + f_-(x, t)$ represents the total density of particles. In this case, Eq. (5) simplifies to

$$\begin{cases} \partial_t f_+(x,t) + \frac{1}{\varepsilon}\partial_x f_+(x,t) = \frac{1}{\varepsilon^2}\left(\frac{\rho(x,t)}{2} - f_+(x,t)\right) \\ \partial_t f_-(x,t) - \frac{1}{\varepsilon}\partial_x f_-(x,t) = \frac{1}{\varepsilon^2}\left(\frac{\rho(x,t)}{2} - f_-(x,t)\right) \end{cases}. \tag{6}$$

Equation (6) is also known as the Goldstein-Taylor model, and can be solved using a particle scheme. For this, we introduce a time step Δt and an ensemble of P particles

$$\left\{\left(X^n_{p,\Delta t}, V^n_{p,\Delta t}\right)\right\}_{p=1}^P. \tag{7}$$

The particle state (position and velocity) is represented as (X, V), p is the particle index ($1 \leq p \leq P$), and n represents the time index, i.e., $X^n_{p,\Delta t} \approx X_p(n\Delta t)$. Equation (6) is then solved via operator splitting as

1. **Transport step**. The position of each particle is updated based on its velocity

$$X^{n+1}_{p,\Delta t} = X^n_{p,\Delta t} + V^n_{p,\Delta t}\Delta t. \tag{8}$$

2. **Collision step**. During collisions, each particle's velocity is updated as:

$$V^{n+1}_{p,\Delta t} = \begin{cases} \pm 1/\varepsilon, & \text{with probability } p_{c,\Delta t} = \Delta t/\varepsilon^2 \text{ and equal probability in the sign,} \\ V^n_{p,\Delta t}, & \text{otherwise.} \end{cases} \tag{9}$$

This approximation requires a time step restriction $\Delta t = \mathcal{O}(\varepsilon^2)$ as $\varepsilon \to 0$, both to ensure $p_{c,\Delta t} < 1$ in the collision phase, and to keep the increments in the transport phase finite. This leads to unacceptably high computational costs for small ε.

2.2 Asymptotic-Preserving Monte Carlo Scheme

Recently, an asymptotic-preserving Monte Carlo scheme was proposed [16], based on the simulation of a modified equation

$$\begin{cases} \partial_t f_+ + \frac{\varepsilon}{\varepsilon^2+\Delta t}\partial_x f_+ = \frac{\Delta t}{\varepsilon^2+\Delta t}\partial_{xx} f_+ + \frac{1}{\varepsilon^2+\Delta t}\left(\frac{\rho}{2} - f_+\right) \\ \partial_t f_- - \frac{\varepsilon}{\varepsilon^2+\Delta t}\partial_x f_- = \frac{\Delta t}{\varepsilon^2+\Delta t}\partial_{xx} f_- + \frac{1}{\varepsilon^2+\Delta t}\left(\frac{\rho}{2} - f_-\right) \end{cases}. \tag{10}$$

In (10) we have dropped the space and time dependency of $f_\pm$ and ρ, for conciseness. The model given by (10) reduces to (6) in the limit when Δt tends to zero and has an $\mathcal{O}(\Delta t)$ bias. In the limit when ε tends to zero, the equations reduce to (4).

Discretizing this equation, using operator splitting as above, again leads to a Monte Carlo scheme. For each particle X_p and for each time step n, one time step now consists of a transport-diffusion and a collision step:

1. **Transport-diffusion step**. The position of the particle is updated based on its velocity and a Brownian increment

$$
\begin{aligned}
X^{n+1}_{p,\Delta t} &= X^n_{p,\Delta t} \pm \frac{\varepsilon}{\varepsilon^2 + \Delta t}\Delta t + \sqrt{2\Delta t}\sqrt{\frac{\Delta t}{\varepsilon^2 + \Delta t}}\xi^n_p \\
&= X^n_{p,\Delta t} + V^n_{p,\Delta t}\Delta t + \sqrt{2\Delta t}\sqrt{D_{\Delta t}}\xi^n_p,
\end{aligned}
\tag{11}
$$

in which we have taken $\xi^n_p \sim \mathscr{N}(0, 1)$ and introduced a Δt-dependent velocity $V^n_{p,\Delta t}$ and diffusion coefficient $D_{\Delta t}$:

$$
V^n_{p,\Delta t} = \pm \frac{\varepsilon}{\varepsilon^2 + \Delta t}, \qquad D_{\Delta t} = \frac{\Delta t}{\varepsilon^2 + \Delta t}. \tag{12}
$$

2. **Collision step**. During collisions, each particle's velocity is updated as:

$$
V^{n+1}_{p,\Delta t} = \begin{cases} \pm \dfrac{\varepsilon}{\varepsilon^2 + \Delta t}, & \text{with probability } p_{c,\Delta t} = \dfrac{\Delta t}{\varepsilon^2 + \Delta t} \\ & \qquad \text{and equal probability in the sign,} \\ V^n_{p,\Delta t}, & \text{otherwise.} \end{cases} \tag{13}
$$

For more details, we refer the reader to [16].

3 Multilevel Monte Carlo Method

3.1 Method and Notation

We want to estimate some quantity of interest Y that is a function of the particle distribution $f(x, v, t)$ at some specific moment $t = t^*$ in time, i.e., we are interested in

$$
Y(t^*) = \mathbb{E}[F(X(t^*))] = \int\int F(x) f(x, v, t^*) dx dv. \tag{14}
$$

Note that, in Eq. (14), the function F only depends on the position x and not on velocity. This is a choice we make for notational convenience and is not essential for the method we present.

The classical Monte Carlo estimator $\hat{Y}(t^*)$ for (14) is given by

$$
\hat{Y}(t^*) = \frac{1}{P}\sum_{p=1}^{P} F\left(X^N_{p,\Delta t}\right), \quad t^* = N\Delta t. \tag{15}
$$

Here, P denotes the number of simulated trajectories, N the number of simulated time steps, Δt the time step size, and $X^N_{p,\Delta t}$ is generated by the time-discretised process (11)–(13). Given a constrained computational budget, a trade-off has to be made when selecting the time step size Δt. On the one hand, a small time step reduces the bias of the simulation of each sampled trajectory, and thus of the estimated quantity of interest. On the other hand, a large time step reduces the cost per trajectory, which increases the number of trajectories that can be simulated and thus reduces the resulting variance on the estimate. The key idea behind the Multilevel Monte Carlo method [17] is to generate a sequence of estimates with varying discretization accuracy and a varying number of realizations. The method achieves the bias of the finest discretization, with the variance of the coarsest discretization.

To apply the multilevel Monte Carlo method, we define a sequence of time step sizes, denoted by Δt_ℓ with $\ell = 0 \dots L$, with $\ell = L$ denoting the finest level of discretization (smallest time step), and $\ell = 0$ the coarsest level. We use a fixed ratio of time steps between subsequent levels, i.e., we set $\Delta t_{\ell-1} = M \Delta t_\ell$ for some integer M. At each level, we simulate a number P_ℓ of particle trajectories. An initial coarse estimator with a large number P_0 of sample trajectories is given by

$$\hat{Y}_0(t^*) = \frac{1}{P_0} \sum_{p=1}^{P_0} F\left(X_{p,\Delta t_0}^{N_0}\right), \quad t^* = N_0 \Delta t_0. \tag{16}$$

This initial estimate can be improved upon by a series of difference estimators $\hat{Y}_\ell(t^*)$, $\ell = 1 \dots L$, of the form

$$\hat{Y}_\ell(t^*) = \frac{1}{P_\ell} \sum_{p=1}^{P_\ell} \left(F\left(X_{p,\Delta t_\ell}^{N_\ell}\right) - F\left(X_{p,\Delta t_{\ell-1}}^{N_{\ell-1}}\right)\right), \tag{17}$$

with $N_\ell \Delta t_\ell = t^*$, for each value of ℓ, and P_ℓ the number of correlated sample trajectories at each level. The estimators (17) estimate the bias induced by sampling with a simulation time step size $\Delta t_{\ell-1}$ by comparing the sample results with a simulation using a time step size Δt_ℓ. The estimators (16)–(17) are then combined into a multilevel Monte Carlo estimator via a telescopic sum,

$$\hat{Y}(t^*) = \sum_{\ell=0}^{L} \hat{Y}_\ell(t^*). \tag{18}$$

It can easily be seen that the expected value of estimator (18) is the same as that of estimator (15) with the finest time step Δt_L. If the required number of particles P_ℓ at each level decreases sufficiently fast with increasing level ℓ, the multilevel estimator will result in a reduced computational cost for a given accuracy. For more details on the multilevel Monte Carlo method, we refer to [18].

3.2 Correlating Asymptotic-Preserving Monte Carlo Simulations

3.2.1 Coupled Trajectories and Notation

The differences in (17) will only have low variance if the simulated paths $X^{n,m}_{\Delta t_\ell,p}$ and $X^n_{\Delta t_{\ell-1},p}$ are correlated. To achieve this correlation, we will couple the different sources of randomness in the simulation at consecutive levels. In each time step using the asymptotic-preserving particle scheme (11)–(13), there are two sources of stochastic behavior. On the one hand, a new Brownian increment ξ^n_p is generated for each particle in each transport-diffusion step (11). On the other hand, in each collision step (13), a fraction of particles randomly get a new velocity V^n_p.

Particle trajectories can be coupled by separately correlating the random numbers used for the individual particles in the transport-diffusion and collision phase of each time step. To show how this is done, we introduce a pair of simulations spanning a time step with size $\Delta t_{\ell-1}$: (i) a simulation at level $\ell-1$, using a single time step of size $\Delta t_{\ell-1}$; and (ii) a simulation at level ℓ, using M time steps of size Δt_ℓ:

$$\begin{cases} X^{n+1}_{p,\Delta t_{\ell-1}} = X^n_{p,\Delta t_{\ell-1}} + \Delta t_{\ell-1} V^n_{p,\Delta t_{\ell-1}} + \sqrt{2\Delta t_{\ell-1}}\sqrt{D_{\Delta t_{\ell-1}}}\xi^n_{p,\ell-1}, & \xi^n_{p,\ell-1} \sim \mathcal{N}(0,1), \\ X^{n+1,0}_{p,\Delta t_\ell} = X^{n,0}_{p,\Delta t_\ell} + \sum_{m=1}^{M}\left(\Delta t_\ell V^{n,m}_{p,\Delta t_\ell} + \sqrt{2\Delta t_\ell}\sqrt{D_{\Delta t_\ell}}\xi^{n,m}_{p,\ell}\right), & \xi^{n,m}_{p,\ell} \sim \mathcal{N}(0,1), \end{cases} \tag{19}$$

with $m \in \{1,\ldots,M\}$ and $X^{n,m}_{p,\Delta t_\ell} \approx X_p(n\Delta t_{\ell-1} + m\Delta t_\ell) \equiv X_p((nM+m)\Delta t_\ell)$.

The key point of the algorithm is to compute the velocities $V^n_{p,\Delta t_{\ell-1}}$ and the Brownian increments $\xi^n_{p,\ell-1}$ at level $\ell-1$, based on the randomly generated values $\xi^{n,m}_{p,\ell}$ and $V^{n,m}_{p,\Delta t_\ell}$ at level ℓ, instead of generating these independently. The main difficulty lies in maximizing the correlation between the velocities and Brownian increments at levels ℓ and $\ell-1$ while avoiding the introduction of an extra bias at level $\ell-1$. Once the coupled simulation (19) at level $\ell-1$ is performed, we can insert the results in (17) to obtain a low-variance difference estimator. In the next two subsections, we explain how we correlate the Brownian increments during the transport phase (Sect. 3.2.2) and the velocities during the collision phase (Sect. 3.2.3). We present the complete algorithm in Sect. 3.2.4.

3.2.2 Coupling the Transport-Diffusion Phase

We first correlate the Brownian increments at levels ℓ and $\ell-1$. To this end, we first simulate the stochastic process at level ℓ, using i.d.d. increments $\xi^{n,m}_{p,\ell}$. Then, at level $\ell-1$, we compute the Brownian increments, $\xi^n_{p,\ell-1}$, from those at level ℓ, $\left\{\xi^{n,m}_{p,\ell}\right\}$, ensuring that $\xi^n_{p,\ell-1} \sim \mathcal{N}(0,1)$. This condition is clearly satisfied if we define $\xi^n_{p,\ell-1}$ as

$$\xi^n_{p,\ell-1} = \sum_{m=1}^{M} \frac{\xi^{n,m}_{p,\ell}}{\sqrt{M}}. \tag{20}$$

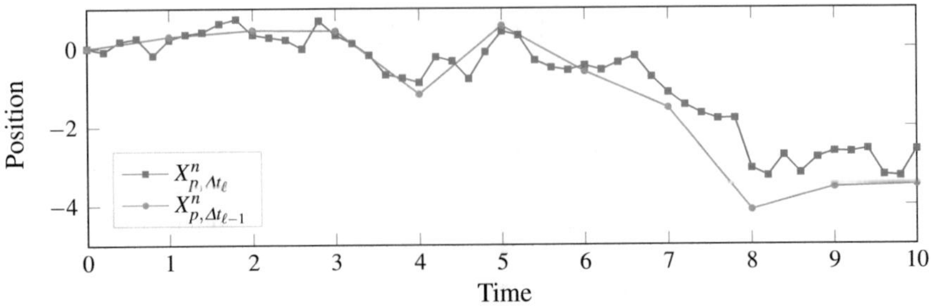

Fig. 1 Correlated diffusion steps with $\varepsilon = 0.5$, $\Delta t_\ell = 0.2$ and $\Delta t_{\ell-1} = 1$

Correlating the simulations in this way means that both levels use the same Brownian path, and differences in the diffusion part of the motion only result from differences in the diffusion coefficients D_ℓ and $D_{\ell-1}$ at different levels.

In Fig. 1, we show two particle trajectories, containing only diffusion behavior, i.e., (19) with $V^n_{p,\Delta t_{\ell-1}} = V^n_{p,\Delta t_\ell} = 0$, coupled as described in (20) with $\varepsilon = 0.5$, $\Delta t_\ell = 0.2$ and $M = 5$. We observe that the paths have similar behavior, i.e., if the fine simulation tends towards negative values, so does the coarse simulation and vice versa. Still, there is an observable difference between them. This is due to the bias caused by the paths having different diffusion coefficients.

3.2.3 Coupling the Collision Phase

While correlating the Brownian paths is relatively straightforward, the coupling of the velocities in the collision phase is more involved. Since we simulate level ℓ first, we have at our disposal the velocities $V^{n,m}_{p,\Delta t_\ell}$ at level ℓ, which are again i.i.d. Our goal is to compute the velocities $V^n_{p,\Delta t_{\ell-1}}$ at level $\ell - 1$ from those at level ℓ, to maximize correlation, while ensuring that the collision probability and post-collision velocity distribution at level $\ell - 1$ are satisfied. Note that, in the collision phase of the asymptotic-preserving particle scheme (13), both the value of the velocity and the probability of collision depend on the value of the time step Δt, and therefore depend on the level ℓ.

The computation of $V^n_{p,\Delta t_{\ell-1}}$ is done in two steps. First, we will couple the occurrence of a collision at level $\ell - 1$ to the occurrence of a collision in one of the M sub-steps of the correlated fine simulation. If we decide to perform a collision both at level ℓ and $\ell - 1$, we will correlate the new velocities generated in both simulations.

Let us first consider the simulation at level ℓ. When simulating the collision step, we decide whether a collision has occurred during a time step of length Δt_ℓ by drawing a random number $\alpha^{n,m}_{p,\ell} \sim \mathscr{U}([0, 1])$ and comparing it to the probability that no collision has occurred in the simulation, $p_{nc,\Delta t_\ell} = 1 - p_{c,\Delta t_\ell}$, with $p_{c,\Delta t_\ell}$, defined in Eq. (13). A collision takes place if and only if

$$\alpha_{p,\ell}^{n,m} \geq p_{nc,\Delta t_\ell} = \frac{\varepsilon^2}{\varepsilon^2 + \Delta t_\ell}. \tag{21}$$

Now consider M time steps of length Δt_ℓ. At least one collision has taken place if at least one of the generated $\alpha_{p,\ell}^{n,m}$, $m \in \{1, \ldots, M\}$, satisfies (21).

Deciding upon collision in the coarse simulation. At level $\ell - 1$, we want to use the values $\alpha_{p,\ell}^{n,m}$, $m \in \{1, \ldots, M\}$ to compute a uniformly distributed number $\alpha_{p,\ell-1}^{n}$, that is correlated with the largest of the generated $\alpha_{p,\ell}^{n,m}$

$$\alpha_{p,\ell}^{n,\max} = \max_m \alpha_{p,\ell}^{n,m}, \tag{22}$$

to compare with the collision probability $p_{nc,\Delta t_{\ell-1}}$. However, the maximum of a set of uniformly distributed random numbers is not uniformly distributed. The cumulative density function of $\alpha_{p,\ell}^{n,\max}$ is given by

$$\mathrm{CDF}\left(\alpha_{p,\ell}^{n,\max}\right) = \left(\alpha_{p,\ell}^{n,\max}\right)^M. \tag{23}$$

Hence, by the inverse transform method, $\left(\alpha_{p,\ell}^{n,\max}\right)^M \sim \mathscr{U}([0,1])$. Equation (23) implies that we can define this random number as

$$\alpha_{p,\ell-1}^{n} = \left(\alpha_{p,\ell}^{n,\max}\right)^M, \tag{24}$$

without affecting the simulation statistics at level $\ell - 1$.

It is possible to show that, given the relation in (24), a collision can occur in the fine simulation without a collision occurring in the coarse simulation. The inverse, i.e., a collision in the coarse simulation, without a fine simulation collision, is not possible.

Choosing a new velocity. If a collision takes place in both simulations in a given time step $\Delta t_{\ell-1}$, then we set the sign of the velocity of the coarse simulation, at the end of the time step to be equal in sign to the velocity of the last subdividing fine time step for which (21) holds,

$$\mathrm{sign}\left(V_{p,\Delta t_{\ell-1}}^{n+1}\right) = \mathrm{sign}\left(V_{p,\Delta t_\ell}^{n,i}\right), \quad i = \operatorname*{argmax}_{1\le m\le M}\left(m \middle| \alpha_{p,\ell}^{n,m} \geq p_{nc,\Delta t_\ell}\right). \tag{25}$$

Because the new velocities generated in the fine simulation are i.i.d., we are free to make this selection, without altering the statistics of the coarse simulation. This approach to selecting the sign of $V_{p,\Delta t_{\ell-1}}^{n+1}$ means that the velocities going into the next time step will have the same sign.

Two particle trajectories without diffusion behavior, i.e., (19) with $D_{\Delta t_{\ell-1}} = D_{\Delta t_\ell} = 0$ are shown in Fig. 2. In this figure, a number of interesting phenomena can be observed. First of all, the fact that the particle's characteristic velocity is

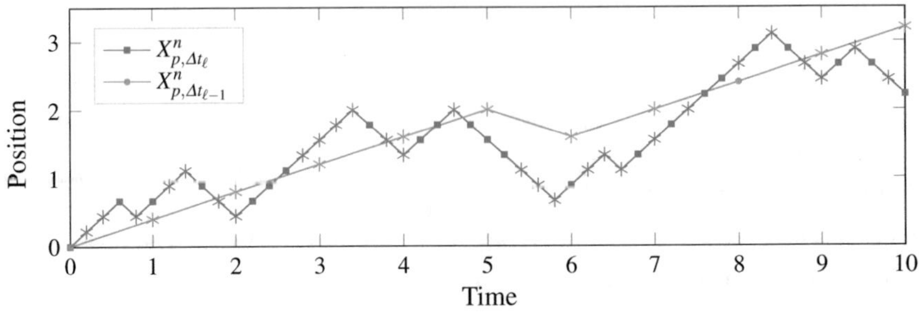

Fig. 2 Correlated transport steps with $\varepsilon = 0.5$, $\Delta t_\ell = 0.2$ and $\Delta t_{\ell-1} = 1$. Stars mark collisions

dependent on the time step sizes $\Delta t_{\ell-1}$ and Δt_ℓ results in different slopes in the curves. This is one source of the bias that we want to estimate using the multilevel Monte Carlo method. Second of all, the collision probability between the coupled trajectories does not match precisely, as this probability also depends on $\Delta t_{\ell-1}$ and Δt_ℓ. For instance, no collision occurs at $t = 8$ in the coarse simulation, while a collision takes place at time $t = 7.4$ and $t = 8$ in the fine simulation. By coincidence, the new velocity generated at $t = 8$ in the fine simulation has the same sign as the coarse simulation velocity. This mismatch is also part of the bias we wish to estimate.

3.2.4 The Complete Algorithm

Combining the correlation of the Brownian increments and velocities results in Algorithm 1. The correlation of the trajectories can be seen in Fig. 3 which shows the particle trajectory given by the sum of the behaviors in Figs. 1 and 2.

Algorithm 1 Performing correlated simulation steps.

1: **for** Each time step n **do**
2: **for** $m = 1 \ldots M$ **do**
3: Simulate (11)–(13) with Δt_ℓ, saving the $\xi^{n,m}_{p,\ell}$, $\alpha^{n,m}_{p,\ell}$ and $V^{n,m}_{p,\Delta t_\ell}$.
4: **end for**
5: Generate $\xi^n_{p,\ell-1}$ from the $\xi^{n,m}_{p,\ell}$ according to (20).
6: Generate $\alpha^n_{p,\ell-1}$ from the $\alpha^{n,m}_{p,\ell}$ according to (22) and (24).
7: Set $V^{n+1}_{p,\Delta t_{\ell-1}} = V^n_{p,\Delta t_{\ell-1}}$
8: **if** $\alpha^n_{p,\ell-1} \geq p_{nc,\Delta t_{\ell-1}}$ **then**
9: **for** $m = 0 \ldots M-1$ **do**
10: **if** $\alpha^{n,m}_{p,\ell} \geq p_{nc,\Delta t_\ell}$ **then**
11: Change the sign of $V^{n+1}_{p,\Delta t_{\ell-1}}$ to be equal to that of $V^{n,m}_{p,\Delta t_\ell}$.
12: **end if**
13: **end for**
14: **end if**
15: **end for**

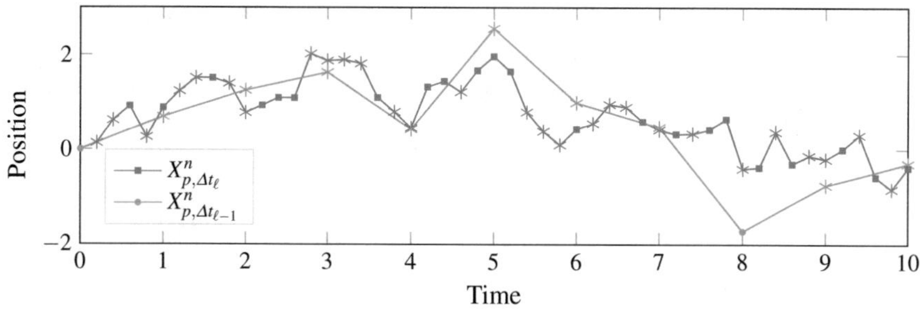

Fig. 3 Correlated paths steps with $\varepsilon = 0.5$, $\Delta t_\ell = 0.2$ and $\Delta t_{\ell-1} = 1$. Stars mark collisions

4 Experimental Results

We will now demonstrate the viability of the suggested approach through some numerical experiments. We will simulate the model given by (10), using the multilevel Monte Carlo method to estimate a selected quantity of interest, which is the expected value of the square of the particle position, at t^*. The ensemble of particles is initialized at the origin with equal probability of having a left and right velocity. When discussing results we will replace the full expression for a sample of the quantity of interest, based on an arbitrary particle p, $F\left(X^{N,0}_{\Delta t_\ell,p}\right)$, with the symbol F_ℓ to simplify notation.

4.1 Model Correlation Behavior

In a first test, we set $t^* = 5$ and investigate the variance of the difference estimators (17) as a function of the time step Δt_ℓ (or, equivalently) the level number. At level $\ell = 0$, we set $\Delta t_0 = 2.5$. All finer levels ($\ell \geq 1$) are defined by setting $\Delta t_\ell = \Delta t_{\ell-1}/M$ with $M = 2$. We fix the number of samples per difference estimator at 100 000. For a selection of values of ε, we calculate the expected value and variance as a function of Δt_ℓ, for $1 \leq \ell$. We compute both the variance of the function samples for a given Δt_ℓ, and the variance of the sampled differences (17), based on the coupled trajectories computed using $\Delta t_{\ell-1}$ and Δt_ℓ. We choose $\varepsilon = 10$ (Fig. 4), $\varepsilon = 1$ (Fig. 5), $\varepsilon = 0.1$ (Fig. 6) and $\varepsilon = 0.01$ (Fig. 7).

The regime $\Delta t \ll \varepsilon^2$. In Figs. 4, 5 and 6, we see that the slopes of both the mean and variance curves for the differences approach an asymptotic limit $\mathcal{O}(\Delta t)$ for $\Delta t \ll \varepsilon^2$. This matches the weak convergence order of the Euler-Maruyama scheme, used to simulate the model (11)–(13), as well as the expected behavior from the time step dependent bias in the asymptotic-preserving model. Given this asymptotic geometric convergence, it is possible to apply thecomplexity theorem in [17] to analyze the

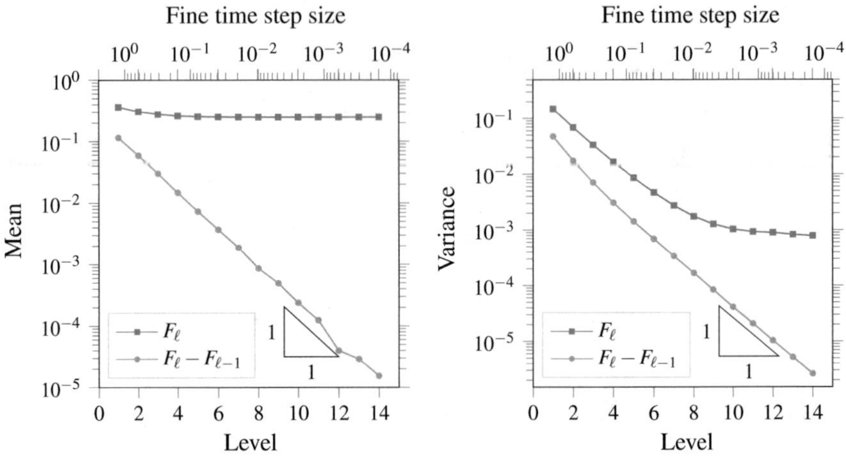

Fig. 4 Mean and variance of the squared particle position for $\varepsilon = 10$

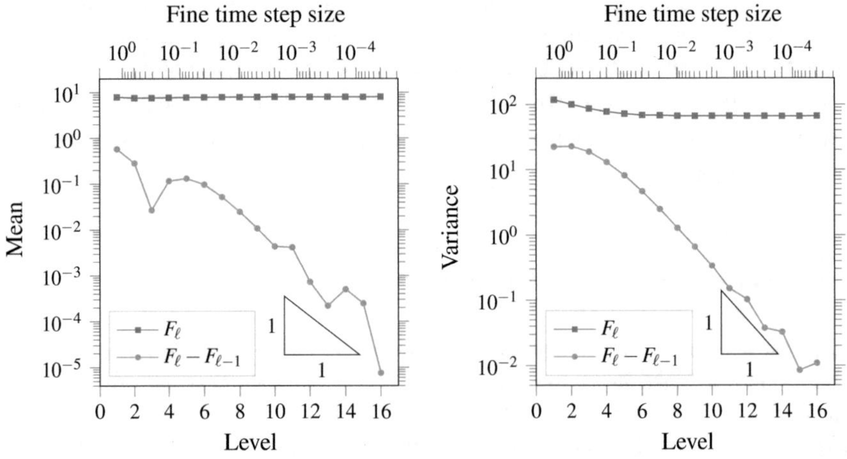

Fig. 5 Mean and variance of the squared particle position for $\varepsilon = 1$

method's computational cost and error bounds . This means that existing theory for multilevel Monte Carlo methods [18] concerning, e.g. samples per level, convergence criteria and conditions for adding levels, can be applied in this regime.

The regime $\Delta t \gg \varepsilon^2$. For time steps $\Delta t \gg \varepsilon^2$, however, we see in Figs. 6 and 7 that both the mean and the variance curves increase geometrically in terms of increasing level. To explain this perhaps counterintuitive result, we will look at the limit of the modified Goldstein-Taylor model when Δt tends to infinity. In this limit, the model (10) converges to the heat equation:

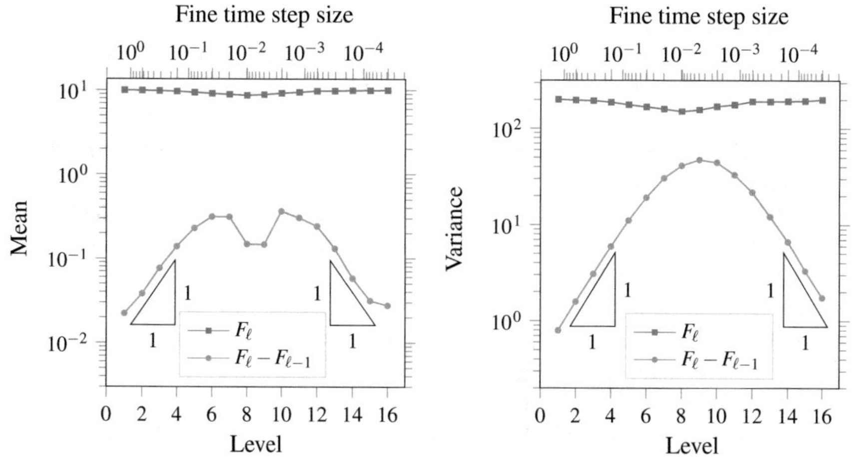

Fig. 6 Mean and variance of the squared particle position for $\varepsilon = 0.1$

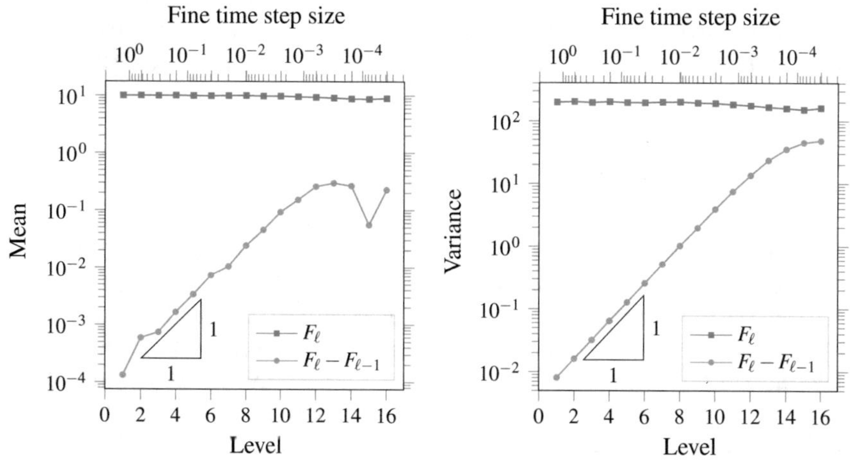

Fig. 7 Mean and variance of the squared particle position for $\varepsilon = 0.01$

$$\begin{cases} \partial_t f_+(x,t) = \partial_{xx} f_+(x,t) \\ \partial_t f_-(x,t) = \partial_{xx} f_-(x,t) \end{cases} \Rightarrow \partial_t \rho(x,t) = \partial_{xx} \rho(x,t). \tag{26}$$

This means that taking increasingly larger time steps in (10) is equivalent to taking the limit $\varepsilon \to 0$. This observation is precisely the asymptotic-preserving property of the particle scheme of Sect. 2.2.

The fact that the two limits approach different models can be seen most clearly in Figs. 4 and 6. In the right hand panel of Fig. 4 we see that the variance of the individual simulations at level ℓ (blue line with squares) changes drastically as a function of Δt_ℓ

in the region where it is of the same order of magnitude as ε^2. This is caused by the approximated models for large and small Δt having differences in behavior, which are significant enough to be observed when plotted. The scheme thus converges to different equations for the two limits in Δt. For small Δt, there is convergence to (6). For large Δt there is convergence to (4). In practice, the size of Δt is limited by the simulation time horizon, so it is not possible to get arbitrarily close to (4) by increasing the time step size, however. This phenomenon also has an effect on the curves in Fig. 6. The curves for the mean and variance of the differences $F_\ell - F_{\ell-1}$ (orange lines with dots) decrease for both small and large Δt, as the model converges to the two limits.

Combining the observations from the two limits in the time step size gives an intuitive interpretation to the multilevel Monte Carlo method in this setting: The method can be interpreted as correcting the result of a pure diffusion simulation by decreasing Δt to get a good approximation of the transport-diffusion equation that describes the behavior for a given value of ε. The peak of the variance of the differences lies near $\Delta t \approx \varepsilon^2$. This makes sense, as this is the region where the model parameters $D_{\Delta t}$ and $V^n_{p,\Delta t}$ vary the most in function of Δt. We also see a dip in the mean of the difference curves in the region of $\Delta t \approx \varepsilon^2$. A full analysis of the behavior that occurs in the transition between the asymptotic regimes is left for future work.

4.2 Comparison with Classical Monte Carlo

The analysis in Sect. 4.1 demonstrated a fast decay of the variance of the differences for increasingly fine levels in the region where $\Delta t \ll \varepsilon^2$. As such, one of the necessary requirements for convergence of the multilevel Monte Carlo method is present in this region. This is, however, not the case in the regime where $\Delta t \gg \varepsilon^2$. Here, the variance of the differences increases as the time step is refined. It is therefore highly non-trivial to perform an adequate selection of coarse levels in the regime $\Delta t \gg \varepsilon^2$. For the fine levels, a standard multilevel Monte Carlo approach can be applied. We therefore propose two simulation strategies:

1. A geometric sequence of levels $\Delta t_\ell = \varepsilon^2 M^{-\ell}$ for $\ell > 0$ starting with a coarse simulation time step $\Delta t_0 = \varepsilon^2$;
2. The same geometric sequence, preceded by a coarse simulation time step t^*, i.e., $\Delta t_0 = t^*$, $\Delta t_1 = \varepsilon^2$ and $\Delta t_\ell = \varepsilon^2 M^{1-\ell}$ for $\ell > 1$.

We compare these approaches in the following two sub-sections.

4.2.1 Standard MLMC Refinement

We will now compute the quantity of interest described at the beginning of this section to a range of prescribed error tolerances, to verify the reduced computational cost of the multilevel Monte Carlo method. We choose to set $M = 2$ and $\varepsilon = 0.1$, and

Table 1 Results of the simulation in Sect. 4.2 with a geometric level sequence for $E = 0.1$

Level	Δt_ℓ	P_ℓ	$\mathbb{V}[F_\ell]$	$\mathbb{E}[F_\ell - F_{\ell-1}]$	V_ℓ	$\mathbb{V}[\hat{Y}_\ell]$	C_ℓ	$P_\ell C_\ell$
0	1.00×10^{-2}	1 393	1.32	8.18×10^{-1}	1.32×10^{0}	9.45×10^{-4}	1	1 393
1	5.00×10^{-3}	395	1.52	7.91×10^{-3}	3.58×10^{-1}	9.07×10^{-4}	3	1 185
2	2.50×10^{-3}	296	1.59	2.18×10^{-2}	4.82×10^{-1}	1.59×10^{-3}	6	1 776
3	1.25×10^{-3}	229	2.22	-1.48×10^{-2}	3.22×10^{-1}	1.41×10^{-3}	12	2 748
4	6.25×10^{-4}	40	1.70	1.57×10^{-3}	4.56×10^{-2}	1.14×10^{-3}	24	960
Σ						6.00×10^{-3}		8 062

Table 2 Results of the simulation in Sect. 4.2 with a geometric level sequence for $E = 0.01$

Level	Δt_ℓ	P_ℓ	$\mathbb{V}[F_\ell]$	$\mathbb{E}[F_\ell - F_{\ell-1}]$	V_ℓ	$\mathbb{V}[\hat{Y}_\ell]$	C_ℓ	$P_\ell C_\ell$
0	1.00×10^{-2}	527920	1.47	8.65×10^{-1}	1.47×10^{0}	2.79×10^{-6}	1	527 920
1	5.00×10^{-3}	165386	1.49	1.06×10^{-2}	4.35×10^{-1}	2.63×10^{-6}	3	496158
2	2.50×10^{-3}	112208	1.59	2.98×10^{-2}	3.99×10^{-1}	3.55×10^{-6}	6	673248
3	1.25×10^{-3}	69135	1.64	2.84×10^{-2}	3.01×10^{-1}	4.36×10^{-6}	12	829620
4	6.25×10^{-4}	39146	1.73	2.00×10^{-2}	1.95×10^{-1}	4.98×10^{-6}	24	939504
5	3.13×10^{-4}	20670	1.76	7.53×10^{-3}	1.09×10^{-1}	5.28×10^{-6}	48	992160
6	1.56×10^{-4}	10842	1.75	9.55×10^{-3}	6.14×10^{-2}	5.67×10^{-6}	96	1040832
7	7.81×10^{-5}	4894	1.91	6.77×10^{-3}	2.42×10^{-2}	4.94×10^{-6}	192	939648
8	3.91×10^{-5}	3937	1.77	2.88×10^{-3}	1.21×10^{-2}	3.08×10^{-6}	384	1511808
9	1.95×10^{-5}	2721	1.81	2.35×10^{-3}	1.21×10^{-2}	4.46×10^{-6}	768	2089728
10	9.75×10^{-6}	40	1.47	2.00×10^{-3}	4.35×10^{-4}	1.09×10^{-5}	1536	61440
Σ						5.26×10^{-5}		10 102 066

reduce the time horizon to $t^* = 0.5$. This gives us an expensive, but computationally feasible problem. The number of samples per level is derived using the formula [18]

$$\left\lceil 2E^{-2}\sqrt{\frac{V_\ell}{C_\ell}}\left(\sum_{\ell=0}^{L}\sqrt{V_\ell C_\ell}\right)\right\rceil, \tag{27}$$

where E is the desired root mean square error, C_ℓ is the computational cost of the estimator at level ℓ, and V_ℓ is the estimated variance of the estimator at level ℓ, i.e., $V_\ell = \mathbb{V}\left[F_\ell - F_{\ell-1}\right]$, where we set $F_{-1} \equiv 0$. The criterion for adding levels and determining convergence are as described in [18]. The cost of a sample will be determined relative to the cost of a simulated trajectory with $\Delta t = \varepsilon^2$. The results of the simulations for E values 0.1, 0.01 and 0.001 can be found in Tables 1, 2 and 3.

In these tables, we list the time step size Δt_ℓ, number of samples P_ℓ, variance of the fine simulations $\mathbb{V}[F_\ell]$, expected value $\mathbb{E}\left[F_\ell - F_{\ell-1}\right]$ and variance V_ℓ of the differences of simulations, estimated variance of the estimator $\mathbb{V}[\hat{Y}_\ell]$, cost per sample C_ℓ and cost per level $P_\ell C_\ell$. The variance of the estimator at level ℓ is estimated as

Table 3 Results of the simulation in Sect. 4.2 with a geometric level sequence for $E = 0.001$

Level	Δt_ℓ	P_ℓ	$\mathbb{V}[F_\ell]$	$\mathbb{E}[F_\ell - F_{\ell-1}]$	V_ℓ	$\mathbb{V}[\hat{Y}_\ell]$	C_ℓ	$P_\ell C_\ell$
0	1.00×10^{-2}	71 593 376	1.47	8.65×10^{-1}	1.47×10^{0}	2.06×10^{-8}	1	71 593 376
1	5.00×10^{-3}	22 501 565	1.49	1.08×10^{-2}	4.36×10^{-1}	1.94×10^{-8}	3	67 504 695
2	2.50×10^{-3}	15 284 042	1.57	2.84×10^{-2}	4.03×10^{-1}	2.63×10^{-8}	6	91 704 252
3	1.25×10^{-3}	9 372 999	1.66	2.88×10^{-2}	3.03×10^{-1}	3.23×10^{-8}	12	112 475 988
4	6.25×10^{-4}	5 322 687	1.73	2.07×10^{-2}	1.95×10^{-1}	3.67×10^{-8}	24	127 744 488
5	3.13×10^{-4}	2 850 794	1.77	1.24×10^{-2}	1.12×10^{-1}	3.93×10^{-8}	48	136 838 112
6	1.56×10^{-4}	1 480 624	1.77	6.73×10^{-3}	6.03×10^{-2}	4.07×10^{-8}	96	142 139 904
7	7.81×10^{-5}	749 144	1.80	3.42×10^{-3}	3.09×10^{-2}	4.13×10^{-8}	192	143 835 648
8	3.91×10^{-5}	382 855	1.79	1.93×10^{-3}	1.61×10^{-2}	4.20×10^{-8}	384	147 016 320
9	1.95×10^{-5}	192 847	1.80	6.71×10^{-4}	8.14×10^{-3}	4.22×10^{-8}	768	148 106 496
10	9.75×10^{-6}	95 971	1.85	6.37×10^{-4}	4.01×10^{-3}	4.18×10^{-8}	1 536	147 411 456
11	4.88×10^{-6}	50 319	1.76	-1.48×10^{-4}	2.18×10^{-3}	4.33×10^{-8}	3 072	154 579 968
12	2.44×10^{-6}	16 002	1.78	3.65×10^{-4}	4.87×10^{-4}	3.04×10^{-8}	6 144	98 316 288
13	1.22×10^{-6}	8 373	1.85	-1.06×10^{-4}	1.75×10^{-3}	2.09×10^{-7}	12 288	102 887 424
14	6.10×10^{-7}	1974	1.97	-6.83×10^{-4}	1.78×10^{-3}	9.00×10^{-7}	24 576	48 513 024
15	3.05×10^{-7}	40	1.49	-1.05×10^{-3}	1.49×10^{-5}	3.72×10^{-7}	49 152	1 966 080
$\sum$						1.94×10^{-6}		1 742 633 519

Table 4 Cost comparison between classical and multilevel Monte Carlo

RMSE	Classical cost	Multilevel cost	Speedup
0.1	4 544	8 062	0.56
0.01	28 627 968	10 102 066	2.83
0.001	25 167 200 256	1 742 633 519	14.4

$$\mathbb{V}\left[\hat{Y}_\ell\right] = \frac{V_\ell}{P_\ell}. \tag{28}$$

We see that the experimental results match the expected behavior of the multilevel Monte Carlo method. The number of samples P_ℓ needed to keep $\sum_{\ell=0}^{L} \mathbb{V}\left[\hat{Y}_\ell\right] < E^2$ decreases drastically in function of ℓ. We also see that $\mathbb{E}\left[F_L - F_{L-1}\right] < E^2$. The cost per level $P_\ell C_\ell$ is also spread quite evenly over the levels, once the time step is a couple orders of magnitude smaller than ε^2. This is to be expected, as the geometric factor with which the cost increases with ℓ is asymptotically the same as that with which V_ℓ decreases. In short, we thus achieve the bias of the finest level, while a large amount of variance reduction is performed in the coarser levels.

The total cost of each multilevel simulation, relative to the cost of a single sample at the coarsest level is computed as the sum of the cost of each level. We can estimate the cost for an equivalent classical Monte Carlo simulation by considering that one needs to perform

$$P_C = \left\lceil \frac{\mathbb{V}\left[F_L\right]}{\sum_{\ell=0}^{L} \mathbb{V}\left[\hat{Y}_\ell\right]} \right\rceil \tag{29}$$

samples with the fine time step at level L, to achieve the same bias and variance as the multilevel estimator. The cost of each sample in the classic Monte Carlo estimator is $\frac{2}{3}C_L$, as we do not need to perform a correlated coarse simulation. Note that, for the numbers in Table 4, $\mathbb{V}\left[F_L\right]$ is estimated using very few samples, so these results should not be taken to literally. They do give the correct order of magnitude of the cost of the equivalent classical Monte Carlo method, however. We now compare the cost of the classical and multilevel Monte Carlo simulations in Table 4.

As can be concluded from Table 4 the multilevel Monte Carlo method gives a significant computational advantage when we want to compute low bias results in the setting of the modified Goldstein-Taylor model. This speedup increases as the requested accuracy of the simulation is increased.

4.2.2 Adding a Coarse Level

It makes little sense to add a full sequence of levels in the regime where $\Delta t \gg \varepsilon^2$. It could however make sense to add a single very coarse level to the simulation as the

Table 5 Results of the simulation in Sect. 4.2 with an extra coarse level for $E = 0.01$

Level	Δt_ℓ	P_ℓ	$\mathbb{V}[F_\ell]$	$\mathbb{E}[F_\ell - F_{\ell-1}]$	V_ℓ	$\mathbb{V}[\hat{Y}_\ell]$	C_ℓ	$P_\ell C_\ell$
0	5.00×10^{-1}	2978687	1.96	9.91×10^{-1}	1.96×10^{0}	6.60×10^{-7}	0.02	59574
1	1.00×10^{-2}	354282	1.47	-1.26×10^{-1}	1.42×10^{0}	4.01×10^{-6}	1.02	361368
2	5.00×10^{-3}	114863	1.47	1.06×10^{-2}	4.36×10^{-1}	3.79×10^{-6}	3	344589
3	2.50×10^{-3}	77905	1.57	3.14×10^{-2}	4.01×10^{-1}	5.15×10^{-6}	6	467430
4	1.25×10^{-3}	47439	1.68	2.92×10^{-2}	3.00×10^{-1}	6.32×10^{-6}	12	569268
5	6.25×10^{-4}	27466	1.78	2.57×10^{-2}	2.01×10^{-1}	7.32×10^{-6}	24	659184
6	3.13×10^{-4}	14599	1.75	9.28×10^{-3}	1.10×10^{-1}	7.56×10^{-6}	48	700752
7	1.56×10^{-4}	7666	1.71	3.99×10^{-3}	6.74×10^{-2}	8.79×10^{-6}	96	735936
8	7.81×10^{-5}	3195	2.04	4.04×10^{-3}	2.49×10^{-2}	7.80×10^{-6}	192	613440
9	3.91×10^{-5}	40	1.54	5.51×10^{-3}	3.65×10^{-3}	9.12×10^{-5}	384	15360
Σ						1.42×10^{-4}		4 526 900

variance of F_ℓ is consistently larger than that of $F_\ell - F_{\ell-1}$ in Figs. 4 through 7. To test this idea we repeat the experiment as before, for $E = 0.01$, with a coarse level at $\Delta t_0 = 0.5$. The results of this experiment can be seen in Table 5.

We see that the total cost of the simulation with the extra coarse level is lower than that of the simulation starting with $\Delta t = \varepsilon^2$ (4 526 900 as apposed to 10 102 066). Based on this initial experiment, it makes sense to include a very coarse level when using the multilevel Monte Carlo method in this context. For a detailed analysis and more extensive numerical results, we refer to future work.

5 Conclusion

In this work, we have derived a new multilevel scheme for asymptotic-preserving particle schemes of the form given in (10). We have demonstrated that this scheme has interesting convergence behavior as the time step is refined, which is apparent in the expected value and variance of sampled differences of the quantity of interest. On the one hand, we get the expected linear convergence to the exact model in terms of Δt for a fixed value of ε. On the other hand, we get convergence to pure diffusion in the limit for large values of Δt. This means that we can interpret the multilevel Monte Carlo method in this setting as refining upon an initial simulation of the heat equation, gradually including transport effects until the correct regime set by ε has been achieved. We have shown that a significant speedup over classical Monte Carlo simulation is achieved when applying a geometric sequence of levels starting from $\Delta t = \varepsilon^2$. We have also shown that adding an extra coarse level to the simulation further accelerates the computation in the considered test case.

The approach taken in developing the asymptotic-preserving scheme is general, and it is straightforward to apply the coupling described in Sect. 3.2 to other, more general, models. As such, we are confident that the ideas expressed in this paper will also be applicable to more general equations than the Goldstein-Taylor model studied here. In future work, this scheme can, for example, be extended to higher dimensional models, both in terms of position and velocity. More complicated models including, for example, absorption terms can also be studied. We intend to expand upon the results in Sect. 4.2.2, as well as considering varying ε together with Δt.

Acknowledgements We thank Pieterjan Robbe for many helpful discussions on the multilevel Monte Carlo method. We also thank the anonymous reviewers for their helpful suggestions for improving the quality of this work. The computational resources and services used in this work were provided by the VSC (Flemish Supercomputer Center), funded by the Research Foundation—Flanders (FWO) and the Flemish Government—department EWI.

References

1. Anderson, D.F., Higham, D.J.: Multilevel Monte Carlo for continuous time Markov chains, with applications in biochemical kinetics. Multiscale Model. Simul. **10**(1), 146–179 (2012)
2. Bennoune, M., Lemou, M., Mieussens, L.: Uniformly stable numerical schemes for the Boltzmann equation preserving the compressible Navier-Stokes asymptotics. J. Comput. Phys. **227**(8), 3781–3803 (2008)
3. Bhatnagar, P.L., Gross, E.P., Krook, M.: A model for collision processes in gases. I. Small amplitude processes in charged and neutral one-component systems. Phys. Rev. **94**(3), 511–525 (1954)
4. Birdsall, C.K., Langdon, A.B.: Plasma Physics via Computer Simulation. Series in Plasma Physics and Fluid Dynamics. Taylor & Francis (2004)
5. Boscarino, S., Pareschi, L., Russo, G.: Implicit-explicit Runge-Kutta schemes for hyperbolic systems and kinetic equations in the diffusion limit. SIAM J. Sci. Comput. **35**(1), A22–A51 (2013)
6. Buet, C., Cordier, S.: An asymptotic preserving scheme for hydrodynamics radiative transfer models. Numer. Math. **108**(2), 199–221 (2007)
7. Cercignani, C.: The Boltzmann Equation and Its Applications, Applied Mathematical Sciences, vol. 67. Springer, New York, NY (1988)
8. Cliffe, K.A., Giles, M.B., Scheichl, R., Teckentrup, A.L.: Multilevel Monte Carlo methods and applications to elliptic PDEs with random coefficients. Comput. Vis. Sci. **14**(1), 3–15 (2011)
9. Crestetto, A., Crouseilles, N., Lemou, M.: A particle micro-macro decomposition based numerical scheme for collisional kinetic equations in the diffusion scaling. Commun. Math. Sci. **16**(4), 887–911 (2018)
10. Crouseilles, N., Lemou, M.: An asymptotic preserving scheme based on a micro-macro decomposition for Collisional Vlasov equations: diffusion and high-field scaling limits. Kinet. Relat. Model. **4**(2), 441–477 (2011)
11. Degond, P., Dimarco, G., Pareschi, L.: The moment-guided Monte Carlo method. Int. J. Numer. Methods Fluids **67**(2), 189–213 (2011)
12. Dimarco, G., Pareschi, L.: Hybrid multiscale methods II. Kinetic equations. Multiscale Model. Simul. **6**(4), 1169–1197 (2008)
13. Dimarco, G., Pareschi, L.: Fluid solver independent hybrid methods for multiscale kinetic equations. SIAM J. Sci. Comput. **32**(2), 603–634 (2010)
14. Dimarco, G., Pareschi, L.: High order asymptotic-preserving schemes for the Boltzmann equation. Comptes Rendus Math. **350**(9–10), 481–486 (2012)
15. Dimarco, G., Pareschi, L.: Numerical methods for kinetic equations. Acta Numer. **23**, 369–520 (2014)

16. Dimarco, G., Pareschi, L., Samaey, G.: Asymptotic-Preserving Monte Carlo methods for transport equations in the diffusive limit. SIAM J. Sci. Comput. **40**, A504–A528 (2017)
17. Giles, M.B.: Multilevel Monte Carlo path simulation. Oper. Res. **56**(3), 607–617 (2008)
18. Giles, M.B.: Multilevel Monte Carlo methods. Acta Numer. **24**, 259–328 (2015)
19. Gosse, L., Toscani, G.: An asymptotic-preserving well-balanced scheme for the hyperbolic heat equations. Comptes Rendus Math. **334**(4), 337–342 (2002)
20. Jin, S.: Efficient asymptotic-preserving (AP) schemes for some multiscale kinetic equations. SIAM J. Sci. Comput. **21**(2), 441–454 (1999)
21. Jin, S., Pareschi, L., Toscani, G.: Diffusive relaxation schemes for multiscale discrete-velocity kinetic equations. SIAM J. Numer. Anal. **35**(6), 2405–2439 (1998)
22. Jin, S., Pareschi, L., Toscani, G.: Uniformly accurate diffusive relaxation schemes for multiscale transport equations. SIAM J. Numer. Anal. **38**(3), 913–936 (2000)
23. Klar, A.: An asymptotic-induced scheme for nonstationary transport equations in the diffusive limit. SIAM J. Numer. Anal. **35**(3), 1073–1094 (1998)
24. Klar, A.: A numerical method for kinetic semiconductor equations in the drift-diffusion limit. SIAM J. Sci. Comput. **20**(5), 1696–1712 (1999)
25. Lapeyre, B., Pardoux, É., Sentis, R., Craig, A.W., Craig, F.: Introduction to Monte Carlo Methods for Transport and Diffusion Equations, vol. 6. Oxford University Press (2003)
26. Larsen, E.W., Keller, J.B.: Asymptotic solution of neutron transport problems for small mean free paths. J. Math. Phys. **15**(1), 75–81 (1974)
27. Lemou, M., Mieussens, L.: A new asymptotic preserving scheme based on micro-macro formulation for linear kinetic equations in the diffusion limit. SIAM J. Sci. Comput. **31**(1), 334–368 (2008)
28. Naldi, G., Pareschi, L.: Numerical schemes for hyperbolic systems of conservation laws with stiff diffusive relaxation. SIAM J. Numer. Anal. **37**(4), 1246–1270 (2000)
29. Pareschi, L., Caflisch, R.E.: An implicit Monte Carlo method for rarefied gas dynamics. J. Comput. Phys. **154**(1), 90–116 (1999)
30. Pareschi, L., Russo, G.: An introduction to Monte Carlo method for the Boltzmann equation. ESAIM Proc. **10**, 35–75 (2001)
31. Pareschi, L., Trazzi, S.: Numerical solution of the Boltzmann equation by time relaxed Monte Carlo (TRMC) methods. Int. J. Numer. Methods Fluids **48**(9), 947–983 (2005)
32. Pope, S.B.: A Monte Carlo method for the PDF equations of turbulent reactive flow. Combust. Sci. Technol. **25**(5–6), 159–174 (1981)
33. Rousset, M., Samaey, G.: Simulating individual-based models of bacterial chemotaxis with asymptotic variance reduction. Math. Model. Methods Appl. Sci. **23**(12), 2155–2191 (2011)
34. Van Barel, A., Vandewalle, S.: Robust optimization of PDEs with random coefficients using a multilevel Monte Carlo method. SIAM J. Uncertain. Quantif. **7**(1), 174–202 (2019)

Randomized Global Sensitivity Analysis and Model Robustness

David Mandel and Giray Ökten

Abstract Global sensitivity analysis allows the modeler to assess the importance of a model parameter in terms of its impact on the variance of the model output. Parameters that are not important can be frozen, and the important ones can be treated with care. This information, however, could be sensitive to data used in estimation of the parameters. To address this, we develop a notion of robustness of a model using randomized Sobol' sensitivity indices. We use the robustness definition to compare some models from computational finance.

Keywords Global sensitivity analysis · Interest rate models · Model robustness · Sobol' sensitivity indices · Temperature derivatives

1 Introduction

Virtually all practical applications of mathematical models involve uncertainty. This uncertainty can arise from two sources: the values of the input parameters used in the model, and the error between the model output and the true value of the quantity of interest. It is the former source of uncertainty for which global sensitivity analysis (GSA) provides insights. GSA consists of a suite of techniques that attribute the uncertainty in the output of a model to each of its parameters; see Saltelli et al. [23] for a thorough introduction. The result of such an analysis is called a sensitivity pattern which makes it possible to rank model parameters in order of contribution to model uncertainty. Since parameters which do not contribute to model uncertainty can be regarded as variables to which the model is largely invariant, such parameters may be frozen at some nominal value, reducing model and calibration complexity. Moreover, additional resources may be allocated to obtain accurate estimates of the

D. Mandel · G. Ökten (✉)
Department of Mathematics, Florida State University, Tallahassee, FL, USA
e-mail: okten@math.fsu.edu

D. Mandel
e-mail: dmandel@math.fsu.edu

B. Tuffin and P. L'Ecuyer (eds.), *Monte Carlo and Quasi-Monte Carlo Methods*,
Springer Proceedings in Mathematics & Statistics 324,
https://doi.org/10.1007/978-3-030-43465-6_20

influential parameters, efficiently reducing the model uncertainty. Global sensitivity methods have enjoyed success in engineering and applied sciences; see, for example, [3, 6, 15–18, 24]. Using GSA to make inferences about the robustness of a model was first done by Göncü et al. [9] where a qualitative approach was used. In this paper, we suggest a different and a quantitative approach to measure robustness.

The first step in GSA is to assume distributions for the input parameters. The modeler may know the distribution of an input parameter through empirical measurements. If the input parameter is estimated from data using a statistical technique, there might be an associated sampling distribution that can be used. And in the absence of any prior information, it is not uncommon to see the assignment of an arbitrary distribution, usually normal or uniform, to the input parameters. In this paper, we are interested in problems where parameters are estimated from data, and there is a sampling distribution suggested by the statistical method, such as asymptotically normal sampling distributions for the maximum likelihood estimation technique.

Once distributions are specified for the input parameters, GSA can be used to obtain the sensitivity pattern of the model, and decisions regarding freezing of unimportant parameters or spending extra resources on better estimation of important parameters, can be made. We emphasize that these decisions are based on the sampling distributions obtained from specific data, i.e., based on one *application* of the model. Using examples from interest rate models, Mandel and Ökten [19] showed that the sensitivity patterns of models can vary significantly depending on the data set used in parameter estimation. Thus any decision to freeze parameters, or rank models based on uncertainty, could be highly sensitive to data. To accommodate this phenomenon of changing sensitivity patterns, a generalization of Sobol' sensitivity analysis, a variance-based GSA, was developed in [19]. In this paper, we present a quantitative description of model robustness based on generalized Sobol' sensitivity analysis, where robustness of a model means the robustness of its sensitivity pattern.

The paper is organized as follows. In Sect. 2 we present the classical Sobol' sensitivity analysis, along with a few new identities among sensitivity indices. In Sect. 3, the generalized Sobol' sensitivity indices developed in [19] are revisited with results from interest rates and temperature derivative modeling. In Sect. 4 we develop a quantitative measure of model robustness and apply it to interest rate and temperature derivative models. We conclude in Sect. 5.

2 Classical Sobol' Sensitivity Analysis

A global sensitivity analysis is accomplished in practice through a *sensitivity index*, of which there are a number of candidates. One of the most popular sensitivity indices was developed in the early 1990s by Sobol' [25, 26]. Known as *Sobol' sensitivity indices*, these are based on the functional ANOVA decomposition for a square integrable function $f : \mathbb{R}^d \to \mathbb{R}$, which is given by

$$f(\mathbf{x}) = \sum_{u \subseteq \mathscr{D}} f_u(\mathbf{x}_u) , \tag{1}$$

where f_u is a function of only $\mathbf{x}_u = (x_i)_{i \in u} \in \mathbb{R}^{|u|}$ and $u \subseteq \mathscr{D} = \{1, \ldots, d\}$. If we think of f as a mathematical model, or a specific numerical implementation of a model, then the variables $\{x_i\}_{i=1}^d$ are taken to be the parameters of the model. The parameters would then be estimated from data, a process sometimes called calibration, and the resulting estimators would be given as random variables $\mathbf{X} = (X_1, \ldots, X_d) \in \mathbb{R}^d$ with joint distribution $\Lambda : \mathscr{B}(\mathbb{R}^d) \to [0, 1]$, where $\mathscr{B}(\mathbb{R}^d)$ is the Borel sigma algebra on $\mathbb{R}^d$. If we impose that the collection $\{X_i\}_{i=1}^d$ is mutually independent, then the component functions $\{f_u\}_{u \subseteq \mathscr{D}}$ may be uniquely constructed using the recursive definitions

$$f_\emptyset = \int_{\mathbb{R}^d} f(\mathbf{x})\, \Lambda(\mathrm{d}\mathbf{x}) , \tag{2}$$

$$f_u(\mathbf{x}_u) = \int_{\mathbb{R}^{|\bar{u}|}} f(\mathbf{x})\, \Lambda(\mathrm{d}\mathbf{x}_{\bar{u}}) - \sum_{v \subsetneq u} f_v(x_v) , \tag{3}$$

where $\mathbf{x}_{\bar{u}} = (x_i)_{i \notin u} \in \mathbb{R}^{d-|u|}$ and $\Lambda(\mathrm{d}\mathbf{x}_u) = P(\mathbf{X}_u \in \mathrm{d}\mathbf{x}_u)$ for any $\emptyset \neq u \subseteq \mathscr{D}$. Furthermore, the component functions are orthogonal in $\mathscr{L}^2(\Lambda)$; that is,

$$\int_{\mathbb{R}^d} f_u(\mathbf{x}_u) f_v(\mathbf{x}_v)\, \Lambda(\mathrm{d}\mathbf{x}) = 0 \tag{4}$$

for any $u, v \subseteq \mathscr{D}$ with $u \neq v$.

For square integrable f and mutually independent parameters $\{X_i\}_{i=1}^d$, the orthogonality result (4) permits the variance decomposition

$$\sigma^2 = \sum_{u \subseteq \mathscr{D}} \sigma_u^2 , \tag{5}$$

where $\sigma^2 = \mathrm{Var}(f(\mathbf{X}))$ and $\sigma_u^2 = \mathrm{Var}(f_u(\mathbf{X}_u))$ for $u \subseteq \mathscr{D}$. It is assumed throughout that $\sigma^2 > 0$. Equation 5 reveals the namesake of the ANOVA decomposition—the variance of f is decomposed into an additive combination of component variances, each of which depends on a unique subset of parameters. If we regard variance as uncertainty, then (5) provides a means to recover total model uncertainty from the uncertainty in its parameters.

For any subset u of the index set $\mathscr{D}$, Sobol' sensitivity indices are defined as

$$\underline{S}_u = \frac{1}{\sigma^2} \sum_{v \subseteq u} \sigma_v^2 , \tag{6}$$

$$\overline{S}_u = \frac{1}{\sigma^2} \sum_{v \cap u \neq \emptyset} \sigma_v^2 . \tag{7}$$

The quantity $\underline{S}_u$ in (6) is called the *lower sensitivity index* or *closed Sobol' index*, whereas $\overline{S}_u$ in (7) is called the *upper sensitivity index* or *total effect*. It is clear that $0 \leq \underline{S}_u \leq \overline{S}_u \leq 1$. The indices also satisfy the identity $\overline{S}_u + \underline{S}_{(-u)} = 1$.

Let us consider the variance, σ^2, as a measure of model uncertainty. Intuitively, $\underline{S}_u$ measures the proportion of model uncertainty that is caused by the parameter vector $\mathbf{X}_u$. If $\underline{S}_u \approx 1$ then the parameters comprising the vector $\mathbf{X}_u$ are considered important, since they contribute significantly to the model uncertainty. When applied to singleton sets $i, j \in \mathscr{D}$ for $i \neq j$ we have an intuitive interpretation: if $\underline{S}_{\{i\}} < \underline{S}_{\{j\}}$ we consider X_j to be more important than X_i, because X_j contributes more to model uncertainty than does X_i; hence, X_j should receive more attention when attempting to reduce model uncertainty. The lower sensitivity index therefore provides a means of parameter ranking. The upper sensitivity index, $\overline{S}_u$, on the other hand, provides a criterion for neglecting unimportant parameters. Indeed, if $\overline{S}_u \approx 0$ then there is no component of the model that depends on the parameters in $\mathbf{X}_u$ that affects the model uncertainty. Thus it is of little consequence to forgo estimation of the parameters $\mathbf{X}_u$ and instead fix them at some nominal value in their domain; this process is known as *freezing* parameters. Sobol' et al. [27] provide an approximation to the model error associated with such a freezing of parameters. Formulas for a Monte Carlo approximation of $\underline{S}_u$ and $\overline{S}_u$ are given in Sobol' [25].

A third sensitivity index, studied by Liu and Owen [17], is given by

$$\Upsilon_u^2 = \sum_{w \supseteq u} \sigma_w^2 \,. \tag{8}$$

Υ_u^2 measures the loss in model accuracy caused by ignoring the component functions f_w for $u \subseteq w$. The following inverse relationship was proved in [17]: $\sigma_u^2 = \sum_{w \supseteq u} (-1)^{|w-u|} \Upsilon_w^2$. Here $w - u$ denotes the set difference. The notation $-u$ denotes the complement $\mathscr{D} - u$, and $|u|$ is the cardinality of the set u. For a given $u \subseteq \mathscr{D}$, some additional relationships between Υ_u^2, $\underline{S}_u$ and $\overline{S}_u$ are provided below. The following two standard results are needed in the proof.

Lemma 1 *Let* $u \subseteq \mathscr{D} = \{1, \ldots, d\}$. *Then* $\sum_{v \subseteq u} (-1)^{|v|} = 1$ *if* $u = \emptyset$, *and 0 otherwise.*

Proof If $u = \emptyset$ the result is immediate. Let $u = \{i_1, \ldots, i_s\}$ for $s \in \{1, \ldots, d\}$. Then

$$\sum_{v \subseteq u} (-1)^{|v|} = (-1)^{|\emptyset|} + \sum_{k=1}^{s} (-1)^{|\{i_k\}|} + \sum_{k=1}^{s} \sum_{l=k+1}^{s} (-1)^{|\{i_k, i_l\}|} + \cdots + (-1)^{|u|}$$

$$= \binom{s}{0}(-1)^0 + \binom{s}{1}(-1)^1 + \binom{s}{2}(-1)^2 - \cdots + \binom{s}{s}(-1)^s = (1 + (-1))^s = 0.$$

□

Lemma 2 *Let $\mathscr{D}$ be a finite set, and f, g be functions on $2^{\mathscr{D}}$ such that $f(u) = \sum_{v\subseteq u} g(v)$ for all $u \subseteq \mathscr{D}$. Then, for all $u \subseteq \mathscr{D}$, $g(u) = \sum_{v\subseteq u}(-1)^{|u-v|} f(v)$.*

Proof This is a consequence of the Möbius inversion theorem (see Hall [10]). □

Theorem 1 *Let $u \subseteq \mathscr{D} = \{1, \ldots, d\}$. Then*

1. $\underline{S}_u = \frac{1}{\sigma^2}\sum_{v\subseteq -u}(-1)^{|v|}\Upsilon_v^2$,
2. $\Upsilon_u^2 = \sigma^2\sum_{v\subseteq u}(-1)^{|v|}\underline{S}_{(-v)}$,
3. $\overline{S}_u = \frac{1}{\sigma^2}\sum_{\emptyset\neq v\subseteq u}(-1)^{|v|+1}\Upsilon_v^2$, *if* $u \neq \emptyset$,
4. $\Upsilon_u^2 = \sigma^2\sum_{v\subseteq u}(-1)^{|v|+1}\overline{S}_v$, *if* $u \neq \emptyset$.

Proof 1. For $u \subseteq \mathscr{D}$ we have

$$\frac{1}{\sigma^2}\sum_{v\subseteq -u}(-1)^{|v|}\Upsilon_v^2 = \frac{1}{\sigma^2}\sum_{v\subseteq -u}(-1)^{|v|}\sum_{w\supseteq v}\sigma_w^2 = \frac{1}{\sigma^2}\sum_{w\subseteq\mathscr{D}}\sigma_w^2\sum_{v\subseteq w\cap(-u)}(-1)^{|v|}$$
$$= \frac{1}{\sigma^2}\sum_{w\cap(-u)=\emptyset}\sigma_w^2 = \frac{1}{\sigma^2}\sum_{w\subseteq u}\sigma_w^2 = \underline{S}_u,$$

where the third equality is by Lemma 1.

2. Let $f(u) = \underline{S}_{(-u)}$ and $g(u) = (-1)^{|u|}\Upsilon_u^2$ for all $u \subseteq \mathscr{D}$. From part 1, $f(-u) = \frac{1}{\sigma^2}\sum_{v\subseteq -u} g(v)$. Then, by Lemma 2,

$$\frac{1}{\sigma^2}g(-u) = \sum_{v\subseteq -u}(-1)^{|(-u)-v|}f(v) = \sum_{v\subseteq -u}(-1)^{|-u|-|v|}f(v),$$

(since $v \subseteq -u$) or equivalently,

$$\frac{1}{\sigma^2}(-1)^{|-u|}\Upsilon_{-u}^2 = \sum_{v\subseteq -u}(-1)^{|-u|-|v|}\underline{S}_{(-v)}\,.$$

This gives $\Upsilon_{-u}^2 = \sigma^2\sum_{v\subseteq -u}(-1)^{|v|}\underline{S}_{(-v)}$, and setting $-u = u$ concludes the proof.

3. For $u \subseteq \mathscr{D}$ we have

$$\frac{1}{\sigma^2}\sum_{\emptyset\neq v\subseteq u}(-1)^{|v|+1}\Upsilon_v^2 = \frac{1}{\sigma^2}\left(\Upsilon_\emptyset^2 + \sum_{v\subseteq u}(-1)^{|v|+1}\Upsilon_v^2\right) = 1 - \frac{1}{\sigma^2}\sum_{v\subseteq u}(-1)^{|v|}\Upsilon_v^2.$$

By part 1, the expression on the right-hand side equals $1 - \underline{S}_{(-u)} = \overline{S}_u$.

4. Follows from part 2, Lemma 1, and the identity $\overline{S}_u + \underline{S}_{(-u)} = 1$. □

3 Generalized Sobol' Sensitivity Indices

Sobol' sensitivity indices provide the modeler with practical information about the influence of parameter uncertainty on a model. Additional resources, such as computer time or experimental work, may be allocated to mitigating uncertainty in estimation of important parameters. One must, however, be mindful of the fact that these decisions are based on the sampling distribution, Λ, of parameter estimators obtained from a specific data. The central question we investigate in this paper is the following: how would the sensitivity pattern of the model parameters change as the sampling distribution, Λ, changes when one considers different data sets (or, calibration methods)? In other words, how robust is the quantitative information about the sensitivity of parameters to changing data sets? In this section we provide two examples—one from interest rates modeling and another from temperature derivative modeling—where we observe changing sensitivity patterns with data. In both examples, the higher order ANOVA terms are negligible and $\underline{S}_{\{i\}} \approx \overline{S}_{\{i\}}$. These examples motivate the introduction of randomized Sobol' sensitivity indices discussed in this section, and a definition of model robustness that incorporates sensitivity indices in Sect. 4.

3.1 Motivation: Short Rate Models

As a first example, consider pricing a one-year US Treasury bill (T-bill) assuming a stochastic short rate. There are a number of models in the literature for the task, but two of the most-studied and simplest are the Vasicek [29] and CIR [7] models (see [4] for an excellent introduction to either model). Vasicek and CIR models describe the evolution of the short rate by the following stochastic differential equations:

$$\text{Vasicek:}\quad \mathrm{d}r_t = a(b - r_t)\mathrm{d}t + \sigma \mathrm{d}W_t\ , \tag{9}$$

$$\text{CIR:}\quad \mathrm{d}r_t = a(b - r_t)\mathrm{d}t + \sigma\sqrt{r_t}\mathrm{d}W_t\ . \tag{10}$$

Each model has three parameters: $a > 0$, the mean reversion speed; $b > 0$, the long-term mean; and $\sigma > 0$, the volatility. In practice, an optimization method is performed using historical data that returns the parameter estimates for which the models "best fit" the data; this process is called calibration. The closed-form bond price for either model may then be computed using the parameter estimates.

For each of the years 1962–2015, such a calibration (using maximum likelihood estimation) was applied to one year's worth of daily observation of yields on one-year US T-bills. Specifically, we use the method of Duan [8] by observing yields and estimating the parameters for the short rate distribution, which we know to be asymptotically normal from MLE theory. For each year, the output of both models is the price of a one-year zero coupon bond maturing on December 31 of the year following the year over which the data span. For example, in the year 2006 both

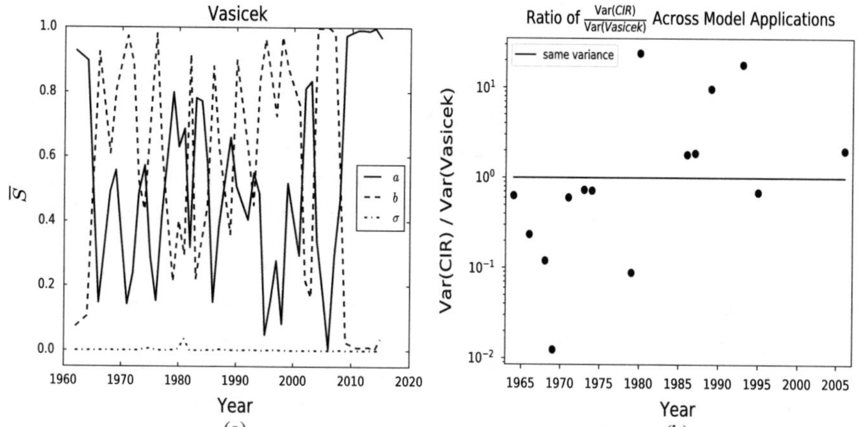

(a) (b)

Fig. 1 Upper Sobol' indices and ratio of variances in bond prices across model applications

models are calibrated to data spanning Jan 1, 2006–Dec 31, 2006, and the model output is the price of a zero coupon bond maturing on Dec 31, 2007.

We assume that the model parameters are independent throughout the paper. Although this assumption may be violated in practice, we concluded the assumption is not unreasonable after inspecting the Fisher information matrix obtained from the joint estimation of the parameters. In our numerical results the covariances of the parameters were several orders of magnitude smaller than the variances.

The disparate sensitivity indices are illustrated in Fig. 1a, where the upper Sobol' indices for bond prices under the Vasicek model are plotted as a function of data set to which the model was calibrated. First, we observe that the sensitivity to the volatility parameter, σ, is near zero for all calibrations (i.e., model applications). This suggests that volatility is a negligible parameter (or rather, a well estimated parameter) and is not causing noticeable uncertainty on the bond price. Perhaps a more important observation is about the behavior of the sensitivity indices for the parameters a and b in Fig. 1a. The upper Sobol' index switches order across model applications, sometimes in an extreme way, so that decisions based on a single year's sensitivity pattern may be wrong for other years. For example, if a modeler happens to consider the sensitivity results only from 2006, he may conclude the mean reversion speed, a, is negligible and may be frozen. However, from 2009 onward, the parameter a is the most important one, and therefore freezing it would have been a substantial error. A similar sensitivity pattern was observed for the CIR model; see Mandel and Ökten [19].

The order of model variances for Vasicek and CIR also change from year to year. Figure 1b plots the ratio of the variance of bond prices under the CIR model to the variance of bond prices under the Vasicek model, Var(CIR)/Var(Vasicek), for different years. Bonds priced under the CIR model have a higher variance for instances above the horizontal line at $y = 1$, whereas bonds priced under the Vasicek model have a higher variance in instances below the line.

Together, Fig. 1a, b underline a crucial limitation of classical sensitivity analysis: conclusions regarding ranking or freezing parameters, or model selection based on variance, can greatly vary depending on the particular application of the model. Thus, one cannot make sweeping assumptions about the sensitivity of a model in general; rather, one is limited to a particular application for which the parameter distributions have been fixed.

These observations lead us to a framework where Sobol' indices are not thought of as constant quantities, but as random variables that take on values for each particular model application. The sensitivity pattern of a model will then be about the ordering of random variables. This proposed framework is called *randomized Sobol' sensitivity analysis*, and will be used as the foundation of our notion of *model robustness*. The consideration of Sobol' indices as random variables has appeared in the literature before, for example, in Hart et al. [12] and Marrel et al. [22].[1] In these papers, the model is replaced by a surrogate model, which is easier to analyze, and then the statistical properties of the Sobol' indices are computed by simulation. In this paper and [19], we directly access the distribution of the Sobol' indices and do not employ surrogate models.

To randomize the Sobol' sensitivity indices, we model the *hyperparameters* as random variables; these are the variables that describe the sampling distribution of the model parameter estimators. It is the hyperparameters that drive the disparate sensitivity indices apparent in Fig. 1a, so they are a natural starting point to randomize the sensitivity indices. Let $m \in \mathbb{N}$ denote the number of such hyperparameters for a model. For example, in the short rate models (9), (10), there are $d = 3$ model parameters, and if each parameter estimator is normally distributed, there are $m = 6$ hyperparameters—the mean and standard deviation of each estimator. Let $\mathbf{Y} : \Omega \to \mathbb{R}^m$ denote the random vector of hyperparameters. Instead of specifying the sampling distribution of the parameter estimators, Λ, through a calibration technique on a single data set, we specify it only in term of each realization of the hyperparameters, $\Lambda(\cdot \mid \mathbf{y})$ for $\mathbf{y} \in \mathbb{R}^m$. This provides a theoretical foundation in which the sampling distributions vary when parameters are estimated on different sets of data. The following result from [19] shows that the ANOVA decomposition holds in this framework as well.

Theorem 2 *Let* $\mathbf{Y} : \Omega \to \mathbb{R}^m$ *be a random vector on a probability space* $(\Omega, \mathscr{F}, P)$*. For fixed* $\mathbf{y} \in \mathbb{R}^m$*, let* $f^{\mathbf{y}} : \mathbb{R}^d \to \mathbb{R}$ *be square-integrable. Finally, for fixed* $\mathbf{y} \in \mathbb{R}^m$*, let* $X_1, \ldots, X_d$ *be mutually independent, real-valued random variables with finite variance, where independence is with respect to the joint distribution* $\Lambda(\cdot \mid \mathbf{y})$*. Then for each* $\mathbf{y} \in \mathbb{R}^m$*, there exists a unique representation of* f *as*

$$f^{\mathbf{y}}(\mathbf{X}) = \sum_{u \subseteq \mathscr{D}} f_u^{\mathbf{y}}(\mathbf{X}_u) \tag{11}$$

[1] We thank the referee for bringing these papers to our attention.

where each component function $f_u^{\mathbf{y}}$ satisfies, provided $j \in u$,

$$\int_{\mathbb{R}} f_u^{\mathbf{y}}(x_j, \mathbf{X}_{-\{j\}})\, \Lambda(\mathrm{d}x_j \mid \mathbf{y}) = 0. \tag{12}$$

Proof Once $\mathbf{Y} = \mathbf{y}$, the construction of the components functions is identical to the classical construction [25] if the uniform measure is replaced with the conditional distributions $\Lambda(\cdot \mid \mathbf{y})$. □

It can be shown that (12) is sufficient for the analogue of the orthogonality result (4) in which the measure is the conditional distribution $\Lambda(\mathrm{d}\mathbf{x} \mid \mathbf{y})$. For each $u \subseteq \mathscr{D}$, let $(\sigma_u^{\mathbf{Y}})^2 = \mathrm{Var}(f_u^{\mathbf{Y}} \mid \mathbf{Y})$; this is the analogue of the component variances $(\sigma_u)^2$ in which the hyperparameters are random variables. In this setting, each $\sigma_u^{\mathbf{Y}}$ is a random variable, as well. It is assumed throughout that $\sigma_u^{\mathbf{Y}} > 0$ almost surely. The randomized Sobol' sensitivity indices are defined as follows.

Definition 1 (*Randomized Sobol' Sensitivity Indices*) Let $f : \mathbb{R}^d \to \mathbb{R}$ be a square-integrable function with the random vector of parameters $\mathbf{X} \in \mathbb{R}^d$. Assume $\{X_i\}_{i=1}^d$ is mutually independent with respect to the conditional joint distribution $\Lambda(\cdot \mid \mathbf{y})$ for each $\mathbf{y} \in \mathbb{R}^m$. Then for any $u \subseteq \mathscr{D}$, the *randomized lower and upper Sobol' indices* are

$$\underline{S}_u^{\mathbf{Y}} = \frac{1}{(\sigma^{\mathbf{Y}})^2} \sum_{v \subseteq u} (\sigma_v^{\mathbf{Y}})^2 , \tag{13}$$

$$\overline{S}_u^{\mathbf{Y}} = \frac{1}{(\sigma^{\mathbf{Y}})^2} \sum_{v \cap u \neq \emptyset} (\sigma_v^{\mathbf{Y}})^2 . \tag{14}$$

The randomization of Sobol' sensitivity indices accommodates the empirical behavior in Fig. 1a. In the new framework, the indices are permitted to vary depending on the application of the model, allowing averages and other statistics to be analyzed as aggregate sensitivity measures.

3.2 Temperature Derivatives

In this section, we apply randomized Sobol' sensitivity analysis to temperature derivative pricing. For a given day of the year, define a *heating degree day* as the degrees Fahrenheit in which the average temperature for that day is below sixty-five degrees Fahrenheit; if the average temperature is above sixty-five, the heating degrees for that day is zero. The sixty-five degree reference temperature is chosen as an international standard proxy to measure the extent with which consumers run their heaters. For a given geographic location, a call option may be written on the total number of heating degree days in a period of $n \in \mathbb{N}$ days,

$$H_n = \sum_{i=1}^{n} (65 - T_i)^+ , \tag{15}$$

where T_i is the average temperature for day $i = 1, \ldots, n$. The time t_0 price of such a call option with strike K is given by

$$C = e^{-rt_n} E\left((H_n - K)^+\right) , \tag{16}$$

where the expectation is with respect to the physical probability measure of the temperature process (see Hull [14], Chap. 34). The call price (16) requires a model for the daily average temperature, T. We will consider three prominent models in the literature, each of which describes the evolution of T as a stochastic differential equation (SDE). The first two models are given by Alaton et al. [1], and Benth and Benth [2]. The models describe the daily average temperature by the SDE

$$dT_t = ds(t) + a(s(t) - T_t)dt + \sigma(t)dW_t , \tag{17}$$

where

$$s(t) = A + Bt + C \sin(\omega t) + D \cos(\omega t) . \tag{18}$$

In (17), a is the mean reversion parameter, $\sigma(t)$ is the volatility, and W_t is a Brownian motion. In (18), A, B, C, and D are constants to be estimated from data, and $\omega = 2\pi/365$. Here $s(t)$ models the long-term dynamics of the daily average temperatures deterministically, where the sin and cos functions capture the seasonality, and the linear term captures the trend in temperatures. Referring to each model by its first author, the model by Alaton assumes the volatility is given by a piecewise function that is constant in each month:

$$\sigma(t) \in \{\sigma_{Jan}, \sigma_{Feb}, \ldots, \sigma_{Dec}\} . \tag{19}$$

The model by Benth assumes a truncated Fourier series for the variance:

$$\sigma^2(t) = c_0 + \sum_{i=1}^{4} c_i \sin(\omega i t) + \sum_{j=1}^{4} d_j \cos(\omega j t) . \tag{20}$$

The third model is given by Brody et al. [5], and assumes the daily average temperature follows

$$dT_t = ds(t) + a(s(t) - T_t)dt + \sigma(t)dW_t^H , \tag{21}$$

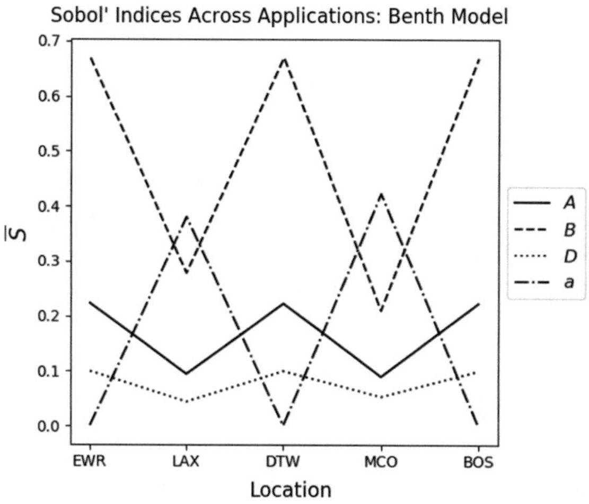

Fig. 2 Upper Sobol' indices for the Benth temperature model as a function of model application

where $H \in [0, 1]$ is the Hurst parameter for a fractional Brownian motion W_t^H, $s(t)$ is the same as in (18), and $\sigma(t)$ is the same piecewise constant function as in (19).[2] The benefits of assuming a fractional Brownian motion process are persuasively argued in Mandelbrot [21].

There are a total of seventeen parameters in the Alaton model, fourteen in the Benth model and eighteen parameters in the Brody model. For twenty-five locations across the US, the parameters of each model were estimated using daily average temperatures from 1985–2016. Call prices at each location were computed using formula (16) and $K = 300$, $t_n = 31$ days, and $r = 0.05$. The initial temperature was taken as the average temperature on the day on which the call price was computed. Specifically, we assumed it was currently Dec 31, 2016, and computed call prices that expire on Jan 31, 2017.

The temperature derivative models have input parameters that differ, but they all have the same output: the call price at each location. We want to investigate the sensitivity of the model output to its input parameters. In Fig. 2, the upper Sobol' sensitivity indices for four parameters in the Benth model are shown for a subset of the locations. The remaining ten parameters' sensitivity indices are not shown as they were near zero across locations. The behavior is similar to the one observed with the short rate models: sensitivities can vary, sometimes significantly, depending on the particular data set used to estimate the model parameters. For example, the mean reversion parameter a is the least important parameter at Detroit (DTW), but the most important at Orlando (MCO). Similar behavior was observed in the Alaton and Brody models.

Using the randomized Sobol' sensitivity analysis framework discussed earlier, we could, for example, compute the mean and variance of the randomized Sobol' sensi-

[2]In Brody et al. [5], the volatility function is left unspecified; we choose the same function as in Alaton [1] for convenience.

tivity indices, and make decisions regarding the sensitivity of the models accordingly (such as ranking parameters in importance, freezing, etc.). We will, however, explore another related question next. Imagine having two competing models—they have the same output, and possibly different inputs—and assume that one model's sensitivity pattern (ranking of parameters in importance) is relatively more robust (i.e., stable across applications) compared to the other model. So, important parameters tend to stay important for one model, more so than the other. This type of robustness would be a positive feature of a model. In the next section, we will introduce a definition for the robustness of a model based on this thought experiment.

4 Model Robustness

There are numerous notions of robustness in statistics literature, each crafted to address specific concerns of an application for which robustness is desired. In one instance, Huber [13] states that "robustness signifies insensitivity to small deviations from the assumptions." Here, concern is with respect to *distributional robustness* in the sense that the true underlying distribution of some random quantity deviates only slightly from the assumed distribution. The interest was in designing statistics that performed well (as measured by asymptotic efficiency) in such deviations from the distributional assumptions. This work in robust statistics, together with Hampel [11], resulted in estimator designs that were insensitive to outliers and that retained efficiency even under departures from the underlying distributional assumption. A robust design of experiment technique was developed by Taguchi [28] which challenged the use of relying only on mean-unbiased estimators to measure the robustness of manufactured products. Here, the context was manufacturing, where it was argued it is more important to incorporate a manufacturing technique in which products consistently hit a target, rather than one that is theoretically capable of yielding products closer to a target but does so inconsistently. Thus the emphasis was on *variance* of a product, which is a similar notion to model variance. More recently, Göncü et al. [9] developed a qualitative description of model robustness through analyzing the rate of increase in Sobol' sensitivity indices as the parameter uncertainty increased.

A quantitative description of *model robustness* is developed in this section. We have already discussed the importance of understanding the effects of parameter uncertainty on model output for a given model application. What we want to distinguish here is how this sensitivity may change across different applications of the model. That is, if one were to assume a different distribution for the parameters, how does the sensitivity pattern of the model change? Do important parameters stay important? Is the overall model uncertainty (variance) affected? If the answer to any of these questions is "no", how should we treat such a fickle model? Based on the previous notions of robustness, it should be clear that such a model should not be considered robust. In light of the results from Sect. 3, robustness will take into account two general properties of a model. First, all else being equal, a model with lower average variance across applications will be considered more robust than a model with higher variance. The second property is more nuanced and it penalizes models with fickle sensitivity patterns.

We need some notation to introduce our robustness definition. Let $f : \mathbb{R}^d \to \mathbb{R}$ be a mathematical model. The marginal distributions of its parameters $\mathbf{X} \in \mathbb{R}^d$ are themselves parametrized by the hyperparameters $\mathbf{Y} \in \mathbb{R}^m$, where m is the number of hyperparameters. Let $\mathscr{D}!$ denote the set of all permutations of the index set $\mathscr{D}$. For each $I \in \mathscr{D}!$, let $\overline{S}_I^{\mathbf{Y}}$ denote the event $(\overline{S}_{I_1}^{\mathbf{Y}} < \overline{S}_{I_2}^{\mathbf{Y}} < \ldots, \overline{S}_{I_d}^{\mathbf{Y}})$, and $P(\overline{S}_I^{\mathbf{Y}})$ denote its probability. For example, if $\mathscr{D} = \{1, 2, 3\}$ and $I = \{2, 1, 3\}$, then $P(\overline{S}_I^{\mathbf{Y}}) = P(\overline{S}_2^{\mathbf{Y}} < \overline{S}_1^{\mathbf{Y}} < \overline{S}_3^{\mathbf{Y}})$. Finally, let

$$CV = \frac{\sqrt{E(\mathrm{Var}(f(\mathbf{X}) \mid \mathbf{Y}))}}{|E(f(\mathbf{X}))|} \tag{22}$$

denote the coefficient of variation, where the denominator is assumed nonzero. CV describes the standard deviation of a model relative to its average output across model applications.[3] The absolute value around the denominator is to ensure $CV > 0$. Robustness is defined next.

Definition 2 The *robustness*, R, of the mathematical model f is

$$R = \frac{\max\limits_{I \in \mathscr{D}!} P(\overline{S}_I^{\mathbf{Y}})}{CV}. \tag{23}$$

For a fixed value of CV, the quantity R is largest when the randomized upper Sobol' sensitivity indices respect a fixed ordering across all model applications; this is given by $\max_{I \in \mathscr{D}!} P(\overline{S}_I^{\mathbf{Y}})$ in the numerator. This means a model is more robust if its parameters have a specific ordering which tends to stay fixed with a higher probability across applications. R is inversely proportional to the coefficient of variation across model applications, so that for a fixed maximal order probability, a model which is less volatile in its output, measured across applications, is more robust than one which is more volatile. Put together, R captures the two behaviors with which we would like to compare competing models.

It is worth mentioning the functional form of the robustness definition we have proposed—namely the ratio of maximal order probability to coefficient of variation—is not unique in its ability to capture the desired behavior described above. Indeed, an additive version, for example, $R = \alpha_1 \max\limits_{I \in \mathscr{D}!} P(\overline{S}_I^{\mathbf{Y}}) + \frac{\alpha_2}{CV}$, would also provide the desired behavior described above. Here the coefficients α_1 and α_2 provide freedom in expressing the modeler's view on the importance to robustness of either term.

There is still the issue of determining a joint probability distribution for the randomized upper Sobol' indices to use in the definition of robustness. Given that we have a collection of joint realizations of the upper indices (see Figs. 1a and 2), we

[3]The coefficient of variation is typically used for data measured on ratio scale. Measurement data in the physical sciences are often, but not always, on a ratio scale.

used the kernel estimation method to approximate their joint distribution function. We used the popular and well-studied Gaussian kernels in this estimation. For the examples in the next section, it is from this fitted kernel density that we sample from to estimate the probabilities in (23).

Let us summarize the computational framework used in computing robustness in this paper. Let s be the total number of model applications in a problem. In the interest rate models s is the number of years, and in the temperature models it is the number of locations. Corresponding to each model application, there is a vector of hyperparameters y estimated from data. Let $\mathscr{H} = \{y_1, \ldots, y_s\}$ be the set of all hyperparameters. Then one can proceed as follows:

1. Select k distinct hyperparameters $y_1, \ldots, y_k$, where $k < s$, corresponding to k model applications, uniformly at random from $\mathscr{H}$.
2. Assume the model parameters are $a_1, \ldots, a_p$. Compute Sobol' sensitivity indices $(\bar{S}_{a_1}, \ldots, \bar{S}_{a_p})$, for each $y_i, i = 1, \ldots, k$, using N Monte Carlo samples. Estimate the coefficient of variation CV.
3. Compute the Gaussian kernel density for the data $(\bar{S}_{a_1}, \ldots, \bar{S}_{a_p})$ obtained in step 2.
4. Generate M samples from the kernel density to estimate $\max_{I \in \mathscr{D}!} P(\bar{S}_I^Y)$. Compute robustness $R = \max_{I \in \mathscr{D}!} P(\bar{S}_I^Y)/CV$.

Remark 1 There are two main Monte Carlo simulations in our proposed framework. The first one is the computation of Sobol' sensitivity indices, and uses pkN samples. In our numerical results $p = 3$ for the interest rate problem, and $p = 5$ or 6, in the temperature derivatives problem. The number of model applications k is 46 for the Vasicek model, 16 for the CIR model, and 25 for the temperature models. Finally, $N = 20,000$ in all problems. The second Monte Carlo simulation samples from the kernel density to estimate the probabilities of different orderings of Sobol' indices, and the maximum probability. Here we used $M = 10,000$ Monte Carlo samples for the interest rate problem, and $M = 50,000$ for the temperature derivatives problem.

Remark 2 The accuracy of the robustness estimates can be assessed using bootstrapping. One way to do this is as follows: repeat steps 1–4, each time sampling distinct $\{y_1, \ldots, y_k\}$ from the set $\mathscr{H}$, with replacement. In other words, we sample with replacement from $\binom{s}{k}$ subsets of $\mathscr{H}$ of size k. If we sample j times, we obtain bootstrap estimates $R_1, \ldots, R_j$ for the robustness measure. We can then compute the standard deviation of these, or construct bootstrap confidence intervals. This will give us the necessary assessment of the robustness measure. Step 1 is essential in obtaining the randomization for bootstrapping. In our numerical results, we did not carry out this bootstrapping approach, and thus in step 1, we did not randomly select hyperparameters. Instead, we considered the entire population of hyperparameters $\mathscr{H} = \{y_1, \ldots, y_s\}$, and carried out the analysis of steps 2–4 for each $y_i \in \mathscr{H}$.

Table 1 Likelihood of orderings of randomized upper Sobol' indices

Event	Probability	
	Vasicek	CIR
$\overline{S}_a < \overline{S}_b < \overline{S}_\sigma$	0.0	0.0
$\overline{S}_a < \overline{S}_\sigma < \overline{S}_b$	0.001	0.0
$\overline{S}_b < \overline{S}_a < \overline{S}_\sigma$	0.0	0.0
$\overline{S}_b < \overline{S}_\sigma < \overline{S}_a$	0.003	0.0
$\overline{S}_\sigma < \overline{S}_a < \overline{S}_b$	0.53	0.69
$\overline{S}_\sigma < \overline{S}_b < \overline{S}_a$	0.47	0.31

4.1 Robustness of Interest Rate and Temperature Models

There are three parameters for the interest rate models—a, b, and σ (see (9) and (10))—which were calibrated to each year of data from 1962–2015. To make notation explicit we let $\overline{S}_a^{\mathbf{Y}}$ denote the randomized upper sensitivity index for the parameter a, for example, instead of specifying an index 1, 2, 3 for each parameter.[4] The estimated joint probability density function (pdf) of $(\overline{S}_a^{\mathbf{Y}}, \overline{S}_b^{\mathbf{Y}}, \overline{S}_\sigma^{\mathbf{Y}})$ was used to approximate the probability $P(\overline{S}_I)$ for all $I \in \mathscr{D}!$. By generating 10, 000 samples, we estimate the probability of the orderings in Table 1 (note we drop the $\mathbf{Y}$ superscript for readability).

The maximal probabilities for both models occur for the event $\overline{S}_\sigma < \overline{S}_a < \overline{S}_b$, and the probabilities are 0.53 and 0.69, for the Vasicek and CIR models, respectively. The average bond price under both models was \$0.97. The average variance of the Vasicek model was 0.00266, whereas the average variance of the CIR model was 0.00183. Putting this all together, we compute the robustness of each model as

$$R_{Vasicek} = \frac{0.53}{\sqrt{0.00226/0.97}} = 10.86\ , \tag{24}$$

$$R_{CIR} = \frac{0.69}{\sqrt{0.00183/0.97}} = 15.7\ . \tag{25}$$

Thus, for one-year US T-bills, the CIR model is approximately $15.7/10.86 = 1.45$ times more robust than the Vasicek model, and so the CIR model is preferred. However, a more statistically sound conclusion on which model has lower or higher robustness is possible when the accuracy of the robustness values is calculated (see Remark 2). In this paper we present a proof-of-concept for comparing models based on their robustness, but not the final verdict.

[4]To avoid obtaining negative values, maximum likelihood estimation was used to estimate $\log a$, $\log b$, and $\log \sigma$ from one year of data, for each interest rate model. Therefore a hyperparameter $y \in \mathbf{Y}$ is the mean and the variance parameter of the corresponding normal distribution. For details see Mandel [20].

Table 2 Probability of each randomized upper Sobol' index exceeding 0.1

Parameter	$P\left(\overline{S}_i^{\mathbf{Y}} > 0.1\right)$		
	Alaton	Benth	Brody
A	0.88	0.92	0.80
B	0.98	0.99	0.94
C	<0.05	<0.05	<0.05
D	0.061	0.12	0.15
a	0.71	0.73	0.77
H	–	–	0.16
σ_1	0.48	–	0.56
$\sigma_2, \ldots, \sigma_{12}$	<0.05	–	<0.05
$c_0, \ldots, c_4$	–	<0.05	–
d_1	–	0.067	–
d_2, d_3, d_4	–	<0.05	–

We now turn our attention to the robustness of the temperature models introduced in Sect. 3[5] Because of the multitude of parameters inherent in these models, we consider only a subset of the parameters when computing order probabilities of the randomized upper Sobol' indices. Specifically, we include only those parameters whose randomized upper Sobol' index exceeds 0.1, with a probability of at least 0.05. This probability is estimated by generating 10,000 samples from the Gaussian kernel density estimator of the joint pdf of the randomized Sobol' indices.

The reason we ignore the other parameters is for computational efficiency. For example, in the case of the Benth model which has 14 parameters, we have to compute the probabilities of 14! orderings, which is not feasible. However, Table 2 shows that the randomized upper Sobol' index of only 5 parameters for the Benth model exceed 0.1, with a probability larger than 0.05. These parameters are A, B, D, a, d_1, and the corresponding probabilities are 0.92, 0.99, 0.12, 0.73, and 0.067. Note that when deciding which parameters to ignore, we are using the Gaussian kernel density obtained from the entire data set available to us. In other words, the parameters we ignore have a small upper Sobol' index not just in one application (calibration) but across all applications.

The most likely orderings of upper Sobol' indices are depicted graphically in Fig. 3 for the Benth model. The x-axis labels denote the five most-likely orderings; for example, the event $(\overline{S}_{d_1} < \overline{S}_D < \overline{S}_a < \overline{S}_A < \overline{S}_B)$ occurred with probability 0.26.

The average call prices were \$64.57 under Alaton, \$64.72 under Benth and \$66.44 under Brody. The average variances were 73.67 under Alaton, 73.86 under Benth and

[5]See Mandel [20] for details on how the parameters are estimated from data obtained from each weather station. The sampling distribution of each parameter is normal. A hyperparameter is the vector whose components are the mean and variance of all the parameters, corresponding to one location.

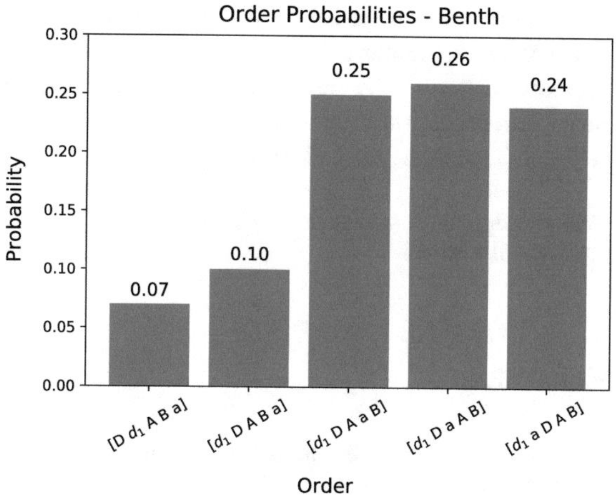

Fig. 3 Order probabilities of upper Sobol' indices for the Benth model. The parameters are written from the smallest to largest upper Sobol' index

74.69 under Brody. Putting these together gives the following robustness calculations.

$$R_{Alaton} = \frac{0.18}{\sqrt{73.67/64.57}} = 1.32 \,, \tag{26}$$

$$R_{Benth} = \frac{0.26}{\sqrt{73.86/64.72}} = 1.92 \,, \tag{27}$$

$$R_{Brody} = \frac{0.22}{\sqrt{74.69/66.44}} = 1.67 \,. \tag{28}$$

Thus the Benth model is 15% more robust than the Brody model, which in turn is 27% more robust than the Alaton model. The Benth model is 45% more robust than the Alaton model. In this case the order probability of upper Sobol' indices was the major factor in robustness rankings. Indeed, the CV for each model (the denominators in (26), (27), and (28)) are nearly equal; it is the maximal order probability in the numerator that differentiates the three models.

5 Conclusions

There are various notions for robustness proposed in the statistics literature. In this paper, we suggested one notion for robustness based on the sensitivity pattern of a model. We argued this notion becomes especially relevant if a modeler wants to distinguish between important and unimportant parameters in order to freeze parameters, and thus reduce the complexity of the model. Another scenario is where a modeler invests more resources in estimating the more sensitive parameters—this could be in the form of expensive experiments— and a model where important parameters tend to stay important across different applications will be desirable.

We discussed the robustness of some models for two problems from computational finance. In our numerical results we did not assess the accuracy of the robustness estimates, however we outlined a bootstrapping based approach in Remark 2. The further investigation of other estimators for the robustness measure and methods to estimate its accuracy is left for future research.

Acknowledgements We thank Fred Huffer, Sergei Kucherenko, Art Owen, and the referees for their helpful comments.

References

1. Alaton, P., Djehiche, B., Stillberger, D.: On modelling and pricing weather derivatives. Appl. Math. Financ. **9**(1), 1–20 (2002)
2. Benth, F.E., Benth, J.Š.: The volatility of temperature and pricing of weather derivatives. Quant. Financ. **7**(5), 553–561 (2007)
3. Bianchetti, M., Kucherenko, S., Scoleri, S.: Pricing and risk management with high-dimensional quasi-Monte Carlo and global sensitivity analysis. Wilmott **July**(78), 46–70 (2015)
4. Brigo, D., Mercurio, F.: Interest Rate Models: Theory and Practice. Springer, New York (2001)
5. Brody, D.C., Syroka, J., Zervos, M.: Dynamical pricing of weather derivatives. Quant. Financ **2**(3), 189–198 (2002)
6. Campolongo, F., Jönsson, H., Schoutens, W.: Quantitative Assessment of Securitisation Deals. Springer Science & Business Media (2012)
7. Cox, J.C., Ingersoll Jr, J.E., Ross, S.A.: A theory of the term structure of interest rates. Econometrica: J. Econom. Soc. 385–407 (1985)
8. Duan, J.C.: Maximum likelihood estimation using price data of the derivative contract. Math. Financ. **4**(2), 155–167 (1994)
9. Göncü, A., Liu, Y., Ökten, G., Hussaini, M.Y.: Uncertainty and robustness in weather derivative models. In: Cools, R., Nuyens, D. (eds.) Monte Carlo and Quasi-Monte Carlo Methods, vol. 163, pp. 351–365. Springer International Publishing, Cham (2016)
10. Hall, M.: Combinatorial Theory, vol. 71. Wiley (1998)
11. Hampel, F.: Robust Statistics: The Approach Based on Influence Functions. Wiley, New York (1986)
12. Hart, J.L., Alexanderian, A., Gremaud, P.A.: Efficient computation of Sobol' indices for stochastic models. SIAM J. Sci. Comput. **39**(4), A1514–A1530 (2017)
13. Huber, P.: Robust Statistics. Wiley, Hoboken, NJ (2009)
14. Hull, J.: Options, Futures, and Other Derivatives, 9th edn. Pearson, Boston (2015)
15. Kent, E., Neumann, S., Kummer, U., Mendes, P.: What can we learn from global sensitivity analysis of biochemical systems? PLoS One **8**(11), e79244 (2013)
16. Kucherenko, S., Shah, N.: The importance of being global. Application of global sensitivity analysis in Monte Carlo option pricing. Wilmott Mag. **4**, 2–10 (2007)
17. Liu, R., Owen, A.B.: Estimating mean dimensionality of analysis of variance decompositions. J. Am. Stat. Assoc. **101**(474), 712–721 (2006)
18. Liu, Y., Hussaini, M.Y., Ökten, G.: Global sensitivity analysis for the Rothermel model based on high-dimensional model representation. Can. J. For. Res. **45**(11), 1474–1479 (2015)
19. Mandel, D., Ökten, G.: Randomized Sobol' sensitivity indices. In: Owen, A.B., Glynn, P.W. (eds.) Monte Carlo and Quasi-Monte Carlo Methods, vol. 241, pp. 395–408. Springer International Publishing, Cham (2018)
20. Mandel, D.: Random Sobol' sensitivity analysis and model robustness. Ph.D. Dissertation, Florida State University (2017)

21. Mandelbrot, B.: The (Mis)behavior of Markets: A Fractal View of Financial Turbulence. Published by Basic Books, New York (2004)
22. Marrel, A., Iooss, B., Laurent, B., Roustant, O.: Calculations of Sobol' indices for the Gaussian process metamodel. Reliab. Eng. Syst. Saf. **94**(3), 742–751 (2009)
23. Saltelli, A., Ratto, M., Andres, T., Campolongo, F., Cariboni, J., Gatelli, D., Saisana, M., Tarantola, S.: Global Sensitivity Analysis: The Primer. Wiley (2008)
24. Saltelli, A., Tarantola, S., Campolongo, F.: Sensitivity analysis as an ingredient of modeling. Stat. Sci. **15**(4), 377–395 (2000)
25. Sobol', I.M.: Sensitivity estimates for nonlinear mathematical models. Math. Model. Comput. Exp. **1**(4), 407–414 (1993)
26. Sobol', I.M.: Global sensitivity indices for nonlinear mathematical models and their Monte Carlo estimates. Math. Comput. Simul. **55**(1), 271–280 (2001)
27. Sobol', I.M., Tarantola, S., Gatelli, D., Kucherenko, S., Mauntz, W.: Estimating the approximation error when fixing unessential factors in global sensitivity analysis. Reliab. Eng. Syst. Saf. **92**(7), 957–960 (2007)
28. Taguchi, G., Chowdhury, S., Wu, Y.: Taguchi's Quality Engineering Handbook, vol. 1736. Wiley, Hoboken, NJ (2005)
29. Vasicek, O.: An equilibrium characterization of the term structure. J. Financ. Econom. **5**(2), 177–188 (1977)

Rapid Covariance-Based Sampling of Linear SPDE Approximations in the Multilevel Monte Carlo Method

Andreas Petersson

Abstract The efficient simulation of the mean value of a non-linear functional of the solution to a linear stochastic partial differential equation (SPDE) with additive Gaussian noise is considered. A Galerkin finite element method is employed along with an implicit Euler scheme to arrive at a fully discrete approximation of the mild solution to the equation. A scheme is presented to compute the covariance of this approximation, which allows for rapid sampling in a Monte Carlo method. This is then extended to a multilevel Monte Carlo method, for which a scheme to compute the cross-covariance between the approximations at different levels is presented. In contrast to traditional path-based methods it is not assumed that the Galerkin subspaces at these levels are nested. The computational complexities of the presented schemes are compared to traditional methods and simulations confirm that, under suitable assumptions, the costs of the new schemes are significantly lower.

Keywords Stochastic partial differential equations · Finite element method · Monte Carlo · Multilevel Monte Carlo · Covariance operators

1 Introduction

Stochastic partial differential equations (SPDE) have many applications in engineering, finance, biology and meteorology. These include filtering problems, pricing of energy derivative contracts and modeling of sea surface temperature. For an overview of applications, we refer to [10, 22]. A natural quantity of interest for an SPDE is the expected value of a non-linear functional of the solution of the equation at a fixed time. This includes moments of the solution but also more concrete quantities, such as, in the case that the SPDE models sea surface temperature, the average amount of area in which the temperature exceeds a given temperature distribution.

A. Petersson (✉)
Department of Mathematical Sciences, Chalmers University of Technology & University of Gothenburg, 412 96 Göteborg, Sweden
e-mail: andreas.petersson@chalmers.se

B. Tuffin and P. L'Ecuyer (eds.), *Monte Carlo and Quasi-Monte Carlo Methods*, Springer Proceedings in Mathematics & Statistics 324,
https://doi.org/10.1007/978-3-030-43465-6_21

In order to determine such quantities, numerical approximations of the SPDE have to be considered, since analytical solutions are in general unavailable.

The field of numerical analysis of SPDE is very active and a multitude of approximations have been considered in the literature, see e.g., [15] and [21, Sect. 10.9] for an overview. In this paper we take the approach of [17], where the author considers an SPDE of evolutionary type and employs a Galerkin method for the spatial discretization of the equation (which includes both spectral and finite element methods) along with a drift-implicit Euler–Maruyama scheme for the temporal discretization. The finite element method in particular is useful and flexible as no explicit knowledge of eigenfunctions or eigenvalues is needed. The author of [17] does, however, omit the problem of how to, given this approximation, efficiently estimate expected values. In this paper we formulate methods for this problem that, under suitable assumptions, outperform standard methods based on the discretization considered in [17].

Typically, the approximation of expected values is accomplished by a *Monte Carlo* method (MC), i.e., by computing a large number of sample paths of the approximate solution and taking the average of the functional of interest applied to each path. This is however quite expensive. Starting with the publication of [11], the *multilevel Monte Carlo method* (MLMC) has become popular, since it can reduce computational cost while retaining accuracy. The method was first considered in [14] for the evaluation of functionals arising from the solution of integral equations. We refer to [12] for an introduction to this active field and to [5, 6] for the first applications of MLMC to finite element approximations of SPDE. MLMC was first considered for SPDE in the thesis [13], where a spectral Galerkin discretization was used.

Even though the MLMC method decreases the computational cost of the approximation of expected values, it is still fairly expensive. In this paper we formulate covariance-based variants of the MC and MLMC methods. The idea is to exploit the fact that as long as the considered SPDE has additive noise and is linear, then the approximation from [17] of the end-time solution is Gaussian. Since a Gaussian random variable is completely determined by its mean and covariance, calculating these parameters provides an efficient way of sampling the approximation (Algorithm 2 below). To incorporate this idea in an MLMC method (Algorithm 4) we calculate the cross-covariance between two SPDE approximations in different Galerkin subspaces. In contrast to [5, 6], the subspace sequence is not assumed to be nested. We demonstrate, using theoretical computations and numerical simulations, that the computational costs of these new algorithms are, under mild assumptions, substantially lower than their traditional path-based alternatives (Algorithms 1 and 3 respectively).

The paper is organized as follows: In Sect. 2 we recapitulate the theoretical setting and approximation results of [17]. We also introduce the assumptions we make along with a stochastic advection-diffusion equation as a concrete example that fulfills these. In Sect. 3 we introduce a covariance-based method for computing samples of SPDE approximations in an MC setting and compare the complexity of it to the traditional path-based method. We extend this in Sect. 4 to the setting of the MLMC method. Section 5 contains a description of the numerical implementation of our methods and a discussion of our assumptions. Finally in Sect. 6 we demonstrate the efficiency of our approach by simulation of the stochastic heat equation.

2 Stochastic Partial Differential Equations and Their Approximations

Let $(H, \langle\cdot,\cdot\rangle, \|\cdot\|)$ be a real separable Hilbert space and let $-A\colon \mathrm{dom}(-A) \subset H \to H$ be a positive definite, self-adjoint operator with a compact inverse on H. For a fixed time $T < \infty$, let $(\Omega, \mathscr{A}, (\mathscr{F}_t)_{t\in[0,T]}, P)$ be a complete filtered probability space satisfying the usual conditions. In this context we consider the linear SPDE

$$\begin{aligned} \mathrm{d}X(t) &= \big(AX(t) + F(t, X(t))\big)\,\mathrm{d}t + G(t)\,\mathrm{d}W(t), \\ X(0) &= x_0, \end{aligned} \tag{1}$$

for $t \in [0, T]$. Here W is an H-valued cylindrical Q-Wiener process, x_0 is a random member of a subspace of H, while F and G are mappings that fulfill Assumption 1 below. The solution $X = (X(t))_{t\in[0,T]}$ is then an H-valued stochastic process. Equation (1) is treated with the semigroup approach of [10, Chap. 7], resulting in a so-called mild solution of the equation. In order to introduce this notion, we start by describing the spectral structure induced by A on H.

By the spectral theorem applied to $(-A)^{-1}$, there is an orthonormal eigenbasis $(e_i)_{i\in\mathbb{N}}$ of H and a positive sequence $(\lambda_i)_{i\in\mathbb{N}}$ of eigenvalues of $-A$ that is increasing and for which $\lim_{i\to\infty}\lambda_i = \infty$. For $r \in \mathbb{R}$, fractional powers of $-A$ are defined by

$$(-A)^{r/2} f = \sum_{i=1}^{\infty} \lambda_i^{r/2} \langle f, e_i\rangle\, e_i$$

for $f \in \dot{H}^r = \mathrm{dom}((-A)^{r/2})$, which is characterized by

$$\dot{H}^r = \left\{ f = \sum_{i=1}^{\infty} f_i e_i : f_i \in \mathbb{R} \text{ for all } i \in \mathbb{N} \text{ and } \|f\|_r^2 = \sum_{i=1}^{\infty} \lambda_i^r f_i^2 < \infty \right\}.$$

This is a separable Hilbert space when equipped with the inner product

$$\langle\cdot,\cdot\rangle_r = \big\langle (-A)^{r/2}\,\cdot, (-A)^{r/2}\,\cdot \big\rangle.$$

For $r > 0$, we have the Gelfand triple $\dot{H}^r \subseteq H \subseteq \dot{H}^{-r}$ since $\dot{H}^{-r} \cong (\dot{H}^r)^*$, the dual of $\dot{H}^r$. The operator A is the generator of an analytic semigroup $E = (E(t))_{t\geq 0} \subset \mathscr{L}(H)$.

Next, we briefly recapitulate some notions from functional analysis and probability theory. For two real separable Hilbert spaces H_1 and H_2, we denote by $H_1 \oplus H_2 = \{[f, u]' : f \in H_1, u \in H_2\}$ the Hilbert (external) direct sum of H_1 and H_2, with an inner product defined by $\big\langle [f_1, u_1]', [f_2, u_2]' \big\rangle_{H_1\oplus H_2} = \langle f_1, f_2\rangle_{H_1} + \langle u_1, u_2\rangle_{H_2}$, $f_1, f_2 \in H_1, u_1, u_2 \in H_2$. Similarly, by $H_1 \otimes H_2$ we denote the Hilbert tensor product, i.e., the completion of the algebraic tensor product of H_1 and H_2 under the norm induced by the inner product $\langle f_1 \otimes u_1, f_2 \otimes u_2\rangle_{H_1\otimes H_2} = \langle f_1, f_2\rangle_{H_1} \langle u_1, u_2\rangle_{H_2}$, $f_1, f_2 \in H_1$, $u_1, u_2 \in H_2$. For $H_1 = H_2$ we write $H_1^{\otimes 2} = H_1 \otimes H_1$ and $f^{\otimes 2} =$

$f \otimes f$ for $f \in H_1$. We denote by $\mathscr{L}(H_1, H_2)$ and $\mathscr{L}_2(H_1, H_2)$, or $\mathscr{L}(H)$ respectively $\mathscr{L}_2(H)$ when $H_1 = H_2 = H$, the spaces of linear respectively Hilbert–Schmidt operators from H_1 to H_2. Similarly, we denote by $\mathscr{L}_1^{\mathrm{s}}(H)$ the family of operators $K \in \mathscr{L}(H)$ that are positive semidefinite, self-adjoint, and of *trace class*, i.e., such operators for which the trace $\operatorname{Tr} K = \sum_{i=1}^{\infty} \langle K e_i, e_i \rangle$ is finite.

For an H-valued random variable $X \in L^1(\Omega; H)$, i.e., $\mathbb{E}[\|X\|] < \infty$, the *expected value*, or *mean*, of X is defined by the Bochner integral $\mathbb{E}[X] = \int_\Omega X(\omega)\,\mathrm{d}P(\omega)$. If $X \in L^2(\Omega; H)$, we define the *covariance* or *covariance operator* of X by

$$\operatorname{Cov}(X) = \mathbb{E}[(X - \mathbb{E}[X])^{\otimes 2}] = \mathbb{E}[X^{\otimes 2}] - \mathbb{E}[X]^{\otimes 2} \in H^{\otimes 2}.$$

More generally, for Hilbert spaces H_1 and H_2, we define the *cross-covariance* or *cross-covariance operator* of $Y \in L^2(\Omega; H_2)$ and $Z \in L^2(\Omega; H_1)$ by

$$\operatorname{Cov}(Y, Z) = \mathbb{E}[(Y - \mathbb{E}[Y]) \otimes (Z - \mathbb{E}[Z])] = \mathbb{E}[Y \otimes Z] - \mathbb{E}[Y] \otimes \mathbb{E}[Z] \in H_2 \otimes H_1$$

so that $\operatorname{Cov}(X) = \operatorname{Cov}(X, X)$. Calling these quantities operators is justified by the fact that $H_2 \otimes H_1 \simeq \mathscr{L}_2(H_1, H_2) \subseteq \mathscr{L}(H_1, H_2)$. The action of the cross-covariance is given by $\operatorname{Cov}(Y, Z) = \mathbb{E}[\langle Z - \mathbb{E}[Z], \cdot \rangle_{H_1} (Y - \mathbb{E}[Y])]$ which means that $\operatorname{Cov}(X) \in \mathscr{L}_1(H)$ and also that $\operatorname{Cov}(X) \in \mathscr{L}_1^{\mathrm{s}}(H)$, implying in particular that it has a unique square root.

We next recall that an H-valued random variable X is said to be Gaussian if $X \in H$ P-a.s. and $\langle X, f \rangle$ is a real-valued Gaussian random variable for all $f \in H$. In this case $X \in L^p(\Omega; H)$ for all $p \geq 1$ so $\operatorname{Cov}(X)$ is well-defined. Now, a stochastic process $W : [0, T] \times \Omega \to H$ is said to be an *H-valued Q-Wiener process* adapted to $(\mathscr{F}_t)_{t \in [0,T]}$ if $W(0) = 0$, W has P-a.s. continuous trajectories, and if there exists a self-adjoint trace class operator $Q \in \mathscr{L}(H)$ such that for each $0 \leq s < t \leq T$, $W(t) - W(s)$ is Gaussian with zero mean and covariance $(t - s)Q$ and $W(t) - W(s)$ is independent of $\mathscr{F}_s$. Below we consider *cylindrical Q-Wiener processes* (see, e.g., [25, Sect. 2.5], [10, Chaps. 2–4]). These can be formally defined as Q-Wiener processes in $\dot{H}^{-r}$ for large enough $r > 0$, allowing for $\operatorname{Tr}(Q) = \infty$. In this case it is no longer true that $W(t) - W(s) \in H$ P-a.s., but $\langle W(t) - W(s), f \rangle$ is still a real-valued Gaussian random variable for all $f \in H$ and $\mathbb{E}[\langle W(t) - W(s), f \rangle \langle W(t) - W(s), g \rangle] = \langle (t - s)Qf, g \rangle$ for all $f, g \in H$.

With this in mind, a predictable process $X = (X(t))_{t \in [0,T]}$, with $T < \infty$ fixed, is called a *mild solution* to (1) if $\sup_{t \in [0,T]} \|X(t)\|_{L^2(\Omega; H)} < \infty$ and for all $t \in [0, T]$,

$$X(t) = E(t)x_0 + \int_0^t E(t - s)F(s, X(s))\,\mathrm{d}s + \int_0^t E(t - s)G(s)\,\mathrm{d}W(s), \quad P\text{-a.s.}$$

The first integral is of Bochner type while the second is an H-valued Itô-integral. For this to be well defined, we need G to map into $\mathscr{L}_2^0 = \mathscr{L}_2(Q^{1/2}(H), H)$. Here $Q^{1/2}(H)$ is a Hilbert space equipped with the inner product $\langle Q^{-1/2}\cdot, Q^{-1/2}\cdot \rangle$, where $Q^{-1/2}$ denotes the pseudoinverse of $Q^{1/2}$. We make this explicit in the assumption below, which by [17, Theorem 2.25] guarantees the existence of a mild solution to (1). The assumption also implies that the approximation we consider below is Gaussian.

Assumption 1 The parameters of (1) fulfill the following requirements.

(i) $W = (W(t))_{t\in[0,T]}$ is an $(\mathscr{F}_t)_{t\in[0,T]}$-adapted cylindrical Q-Wiener process, where $Q \in \mathscr{L}(H)$ is self-adjoint and positive semidefinite, not necessarily of trace class.

(ii) There is a constant $C > 0$ such that $G : [0, T] \to \mathscr{L}_2^0$ satisfies

$$\|G(t_1) - G(t_2)\|_{\mathscr{L}_2^0} \leq C|t_1 - t_2|^{1/2}, \text{ for all } t_1, t_2 \in [0, T].$$

(iii) The function $F : [0, T] \times H \to \dot{H}^{-1}$ is affine in H, i.e., for each $t \in [0, T]$ there exists an operator $F_t^1 \in \mathscr{L}(H, \dot{H}^{-1})$ and an element $F_t^2 \in \dot{H}^{-1}$ such that $F(t, f) = F_t^1 f + F_t^2$ for all $f \in H$. Furthermore, there exists a constant $C > 0$ such that $F : [0, T] \times H \to \dot{H}^{-1}$ satisfies

$$\|F(t_1, f) - F(t_2, f)\|_{-1} \leq C(1 + \|f\|)|t_1 - t_2|^{1/2}$$

for all $f \in H$, $t_1, t_2 \in [0, T]$, and $\|F_t^1\|_{\mathscr{L}(H,\dot{H}^{-1})} \leq C$ for all $t \in [0, T]$.

(iv) The initial value x_0 is a (possibly degenerate) $\mathscr{F}_0$-measurable $\dot{H}^1$-valued Gaussian random variable.

By degenerate, we mean that $\mathrm{Cov}(x_0)$ may only be positive semidefinite, allowing for a deterministic initial value. Since $x_0 \in L^p(\Omega; H)$ for all $p \geq 1$ we have $\sup_{t\in[0,T]} \|X(t)\|_{L^p(\Omega;\dot{H}^s)} < \infty$ for all $p \geq 1$ and $s \in [0, 1)$.

As a model problem in this context, we consider a stochastic advection-diffusion equation.

Example 1 For a convex polygonal domain $D \subset \mathbb{R}^d$, $d = 1, 2, 3$, let $H = L^2(D)$ and for a function f on D, let the operator $-A$: $\mathrm{dom}(-A) \to H$ be given by $Af - \nabla \cdot (a\nabla f)$ with Dirichlet zero boundary conditions, where $a : D \to \mathbb{R}$ is a sufficiently smooth strictly positive function. In this setting it holds (cf. [26, Chap. 3]) that $\dot{H}^1 = H_0^1(D)$ and $\dot{H}^2 = H^2(D) \cap H_0^1(D)$, where $H^k(D)$ is the Sobolev space of order k on D and $H_0^1(D)$ consists of all $f \in H^1(D)$ such that $f(x) = 0$ for $x \in \partial D$, the boundary of D. Let F be given by $F(t, f) = b(t, \cdot) \cdot \nabla f(\cdot) + c(t, \cdot) f(\cdot) + d(t, \cdot)$ for a function f on D. When $b : D \times [0, T] \to \mathbb{R}^d$ and $c, d : D \times [0, T] \to \mathbb{R}$ are smooth, $F(t, \cdot)$ is indeed a member of $\mathscr{L}(H, \dot{H}^{-1})$ (cf. [17, Example 2.22]). Choosing $G = g(t)\cdot$, where $g : [0, T] \to \mathbb{R}$ is smooth, Q such that $\mathrm{Tr}(Q) < \infty$ and x_0 smooth, Equation (1) is interpreted as the problem to find a function-valued stochastic process X such that

$$\begin{aligned} \mathrm{d}X(t, x) = {} & (\nabla \cdot (a(x)\nabla X(t, x)) + b(t, x) \cdot \nabla X(t, x) + c(t, x)X(t, x) + d(t, x))\, \mathrm{d}t \\ & + g(t)\, \mathrm{d}W(t, x) \end{aligned}$$

for all $t \in (0, T]$, $x \in D$, with $X(t, x) = 0$ for all $t \in (0, T]$, $x \in \partial D$ and $X(0, x) = x_0(x)$ for all $x \in D$. Moreover, the covariance of the noise has a more concrete meaning in this setting. A consequence of $\mathrm{Tr}(Q) < \infty$ is, by [23, Proposition A.7], the existence of a symmetric square integrable function $q : D \times D \to \mathbb{R}$ such that

$$Qf = \int_D q(\cdot, y) f(y)\, \mathrm{d}y \tag{2}$$

for $f \in H$. Similarly, if q is a symmetric, positive semidefinite continuous function on $D \times D$, then (2) defines a covariance operator by [23, Theorem A.8]. If we take $W(t, \cdot)$, $t \in [0, T]$, to be a random field (which in this setting is to say that it is defined in all of D and jointly $\mathscr{B}(D) \otimes \mathscr{F}_t$-measurable, where $\mathscr{B}(D)$ denotes the Borel σ-algebra on D), then tq is the covariance function of the field.

For the spatial discretization of (1), we assume the setting of [17, Chap. 3.2]. Let $(V_h)_{h \in (0,1]}$ be a family of subspaces of $\dot{H}^1$ equipped with $\langle \cdot, \cdot \rangle$ such that $N_h = \dim(V_h) < \infty$. By $P_h : H^{-1} \to V_h$ we denote the generalized orthogonal projector onto V_h, defined by $\langle P_h f, \Phi_h \rangle = {}_{\dot{H}^{-1}}\langle f, \Phi_h \rangle_{\dot{H}^1}$ for all $f \in \dot{H}^{-1}$ and $\Phi_h \in V_h$, where ${}_{\dot{H}^{-1}}\langle \cdot, \cdot \rangle_{\dot{H}^1}$ denotes the dual pairing. By R_h we denote the Ritz projector, i.e., the orthogonal projector $R_h : \dot{H}^1 \to V_h$ with respect to $\langle \cdot, \cdot \rangle_1$. We assume that there is a constant $C > 0$ such that $\|P_h f\|_1 \leq C \|f\|_1$ and $\|R_h f - f\| \leq C \|f\|_s h^s$ for all $h \in (0, 1]$ and all f in $\dot{H}^1$ and $\dot{H}^s$, $s \in \{1, 2\}$, respectively. This setting includes both finite element and spectral methods (see [17, Examples 3.6–3.7] for details on when these relatively mild assumptions hold). The operator $-A_h : V_h \to V_h$ is now defined by $\langle -A_h f_h, g_h \rangle = \langle f_h, g_h \rangle_1 = \left\langle (-A)^{1/2} f_h, (-A)^{1/2} g_h \right\rangle$, for all $f_h, g_h \in V_h$. For the time discretization, we use the drift-implicit Euler method. Let a uniform time grid be given by $t_j = j\Delta t$ for $j = 0, 1, \ldots, N_{\Delta t} = T/\Delta t \in \mathbb{N}$. A fully discrete approximation $(X_{h,\Delta t}^{t_j})_{j=0}^{N_{\Delta t}}$ is then given by

$$X_{h,\Delta t}^{t_{j+1}} - X_{h,\Delta t}^{t_j} = \left(A_h X_{h,\Delta t}^{t_{j+1}} + P_h F(t_j, X_{h,\Delta t}^{t_j}) \right) \Delta t + P_h G(t_j) \Delta W^j, \tag{3}$$

where $\Delta W^j = W(t_{j+1}) - W(t_j)$ and $j = 0, \ldots, N_{\Delta t} - 1$. It converges strongly to the solution of (1) in the sense of the following theorem.

Theorem 1 ([17, Theorem 3.14]) *Let the terms of* (1) *satisfy Assumption 1 and let* $(X_{h,\Delta t}^T)_{h,\Delta t}$ *be a family of approximations of* $X(T)$ *given by* (3). *Then, for all* $p \geq 1$, $\sup_{h,\Delta t}(\|X_{h,\Delta t}^T\|_{L^p(\Omega;H)}) < \infty$ *and there is a constant* $C > 0$ *such that*

$$\|X(T) - X_{h,\Delta t}^T\|_{L^p(\Omega;H)} \leq C \left(h + \Delta t^{1/2} \right), \text{ for all } h, \Delta t \in (0, 1].$$

Since the goal of this paper is the approximation of the quantity $\mathbb{E}[\phi(X(T))]$, where ϕ is a smooth functional, the concept of weak convergence, i.e., convergence with respect to the expectation of functionals of the solution, is vital as it allows for the efficient tuning of the MC estimators in Sects. 3 and 4. In order to use a result from [17], we need a stronger assumption. In particular, F is a function of time only and ϕ is smooth. This is formalized below, see also Remark 4.

Assumption 2 The parameters of (1) fulfill the following requirements.

(i) For some $\delta \in [1/2, 1]$, there is a constant $C > 0$ such that $G : [0, T] \to \mathscr{L}_2^0$ and $F : [0, T] \to H$ satisfy

$$\|G(t_1) - G(t_2)\|_{\mathscr{L}_2^0} \leq C|t_1 - t_2|^\delta, \text{ for all } t_1, t_2 \in [0, T]$$

and

$$\|F(t_1) - F(t_2)\| \leq C|t_1 - t_2|^\delta, \text{ for all } t_1, t_2 \in [0, T].$$

(ii) The functional ϕ is a member of $C_{\mathrm{p}}^2(H; \mathbb{R})$, the space of all continuous mappings from H to $\mathbb{R}$ which are twice continuously Fréchet-differentiable with at most polynomially growing derivatives.
(iii) The initial value $x_0 \in \dot{H}^1$ is deterministic.

With this assumption in place, we cite the following weak convergence result.

Theorem 2 ([17, Theorem 5.12]) *Under the assumptions of Theorem 1 and Assumption 2, there is a constant* $C > 0$ *such that*

$$\left|\mathbb{E}\left[\phi(X(T)) - \phi(X_{h,\Delta t}^T)\right]\right| \leq C\left(1 + |\log(h)|\right)\left(h^2 + \Delta t^\delta\right), \text{ for all } h, \Delta t \in (0, 1].$$

3 Covariance-Based Sampling in a Monte Carlo Setting

For $Y \in L^1(\Omega; \mathbb{R})$, the MC approximation of $\mathbb{E}[Y]$ is given by

$$E_N[Y] = \frac{1}{N}\sum_{i=1}^{N} Y^{(i)},$$

where $N \in \mathbb{N}$ is the number of independent realizations, $Y^{(i)}$, of Y. Any $Y \in L^2(\Omega; \mathbb{R})$ satisfies for $N \in \mathbb{N}$ the inequality

$$\|\mathbb{E}[Y] - E_N[Y]\|_{L^2(\Omega;\mathbb{R})} = \frac{1}{\sqrt{N}} \operatorname{Var}(Y)^{1/2} \leq \frac{1}{\sqrt{N}}\|Y\|_{L^2(\Omega;\mathbb{R})}. \quad (4)$$

In order to accurately approximate $\mathbb{E}[\phi(X(T))]$ by MC using our fully discrete approximation $X_{h,\Delta t}^T$ of $X(T)$, we must therefore generate many samples of $X_{h,\Delta t}^T$. In practice one samples the vector $\bar{\boldsymbol{x}}_h^T = [x_1, x_2, \ldots, x_{N_h}]'$ of coefficients of the expansion $X_{h,\Delta t}^T = \sum_{k=1}^{N_h} x_k \Phi_k^h$, where $\boldsymbol{\Phi}^h = (\Phi_k^h)_{k=1}^{N_h}$ is a basis of V_h.

The classical approach to this is that of *path-based sampling*, i.e., solving the $N_{\Delta t}$ matrix equations corresponding to (3) once for each sample $i = 1, 2, \ldots, N$ (Algorithm 1). These systems are obtained by expanding (3) on $\boldsymbol{\Phi}^h$ and applying $\langle \Phi_i^h, \cdot \rangle$ to each side of this equality for $i = 1, 2, \ldots, N_h$.

Algorithm 1 Path-based MC method of computing an estimate $E_N[\phi(X^T_{h,\Delta t})]$ of $\mathbb{E}[\phi(X(T))]$

1: $result = 0$
2: **for** $i = 1$ to N **do**
3: Sample increments of a realization $W^{(i)}$ of the Q-Wiener process W
4: Compute $\bar{\boldsymbol{x}}^T_h = [x_1, x_2, \ldots, x_{N_h}]'$ directly by solving the matrix equations corresponding to the drift-implicit Euler–Maruyama system (3) driven by $W^{(i)}$
5: Compute $\phi(X^T_{h,\Delta t}) = \phi\left(\sum_{k=1}^{N_h} x_k \Phi^h_k\right)$
6: $result = result + \phi(X^T_{h,\Delta t})^{(i)}/N$
7: **end for**
8: $E_N\left[\phi(X^T_{h,\Delta t})\right] = result$

Our alternative approach is that of *covariance-based sampling* where $\mathrm{Cov}(X_{h,\Delta t})$ is computed, yielding the covariance matrix of $\bar{\boldsymbol{x}}^T_h$ which is used to generate samples of $\bar{\boldsymbol{x}}^T_h$ directly (Algorithm 2). This is possible since Assumption 1 ensures that $\bar{\boldsymbol{x}}^T_h$ is Gaussian. To see this, we first introduce the abbreviations $R_{h,\Delta t} = (I_H - \Delta t A_h)$, $F^{1,j}_{h,\Delta t} = \left(I_H + \Delta t P_h F^1_{t_j}\right)$, and $F^{2,j}_{h,\Delta t} = \Delta t P_h F^2_{t_j}$, so that (3) can be written as

$$R_{h,\Delta t} X^{t_{j+1}}_{h,\Delta t} = F^{1,j}_{h,\Delta t} X^{t_j}_{h,\Delta t} + F^{2,j}_{h,\Delta t} + P_h G(t_j) \Delta W^j, \tag{5}$$

for $j = 0, 1, \ldots, N_{\Delta t} - 1$. The V_h-valued random variable $X^{t_{j+1}}_{h,\Delta t}$ is Gaussian by induction, due to the facts that affine transformations of Gaussian random variables remain Gaussian, that $X^{t_j}_{h,\Delta t}$ and $P_h G(t_j)\Delta W^j$ are independent, that the recursion is started at a (possibly degenerate) Gaussian random variable, and that $P_h G(t_j)\Delta W^j$ itself is Gaussian, which can be seen by, for example, [10, Theorems 4.6, 4.27]. That $\bar{\boldsymbol{x}}^T_h$ is an $\mathbb{R}^{N_h}$-valued Gaussian random variable is then a consequence of the equality $\langle \bar{\boldsymbol{x}}^T_h, \boldsymbol{a}\rangle_{\mathbb{R}^{N_h}} = \langle X^T_{h,\Delta t}, a_h\rangle_H$, where $\boldsymbol{a} = [a_1, a_2, \ldots, a_{N_h}]' \in \mathbb{R}^{N_h}$ is arbitrary and $a_h = \sum_{i=1}^{N_h} (\boldsymbol{M}_h^{-1}\boldsymbol{a})_i \phi_i$, where $\boldsymbol{M}_h$ is the symmetric positive definite matrix with entries $m_{i,j} = \langle \phi_i, \phi_j \rangle$.

In the next theorem we introduce a scheme for the calculation of $\mathrm{Cov}(X_{h,\Delta t})$, inspired by the derivation of stability properties of SPDE approximation schemes in [20], see also [9].

Theorem 3 *Let the terms of* (1) *satisfy Assumption 1 and let* $(X^{t_j}_{h,\Delta t})_{j=0}^{N_{\Delta t}}$ *be given by* (3). *Then* $\mu^T = \mathbb{E}[X^T_{h,\Delta t}] \in V_h$ *and* $\Sigma^T = \mathrm{Cov}(X^T_{h,\Delta t}) \in V_h^{\otimes 2}$ *are given by the recursions*

$$R_{h,\Delta t} \mu^{t_{j+1}} = F^{1,j}_{h,\Delta t} \mu^{t_j} + F^{2,j}_{h,\Delta t}, \tag{6}$$

$$\left(R_{h,\Delta t}\right)^{\otimes 2} \Sigma^{t_{j+1}} = \left(F^{1,j}_{h,\Delta t}\right)^{\otimes 2} \Sigma^{t_j} + \mathbb{E}\left[\left(P_h G(t_j) \Delta W^j\right)^{\otimes 2}\right] \tag{7}$$

for $j = 0, 1, \ldots, N_{\Delta t} - 1$.

Proof We first prove the result assuming $F_t^2 = 0$ for all $t \in [0, T]$. The recursion scheme (6) for the mean follows by applying $\mathbb{E}[\cdot]$ to both sides of (5), noting that $\mathbb{E}[\cdot]$ commutes with linear operators and that ΔW^j has zero mean.

For the covariance recursion scheme (7), we first tensorize (5) to get

$$\begin{aligned}\left(R_{h,\Delta t}\right)^{\otimes 2}\left(X_{h,\Delta t}^{t_{j+1}}\right)^{\otimes 2} = \left(F_{h,\Delta t}^{1,j}\right)^{\otimes 2}\left(X_{h,\Delta t}^{t_j}\right)^{\otimes 2} + F_{h,\Delta t}^{1,j} X_{h,\Delta t}^{t_j} \otimes P_h G(t_j)\Delta W^j \\ + P_h G(t_j)\Delta W^j \otimes F_{h,\Delta t}^{1,j} X_{h,\Delta t}^{t_j} + \left(P_h G(t_j)\Delta W^j\right)^{\otimes 2}.\end{aligned} \tag{8}$$

Since ΔW^j is independent of $X_{h,\Delta t}^{t_j}$ and has zero mean,

$$\mathbb{E}\left[F_{h,\Delta t}^{1,j} X_{h,\Delta t}^{t_j} \otimes P_h G(t_j)\Delta W^j\right] = \mathbb{E}\left[F_{h,\Delta t}^{1,j} X_{h,\Delta t}^{t_j}\right] \otimes \mathbb{E}\left[P_h G(t_j)\Delta W^j\right] = 0$$

and similarly, the third term has zero mean. Thus, the mean of (8) is given by

$$\left(R_{h,\Delta t}\right)^{\otimes 2}\mathbb{E}\left[\left(X_{h,\Delta t}^{t_{j+1}}\right)^{\otimes 2}\right] = \left(F_{h,\Delta t}^{1,j}\right)^{\otimes 2}\mathbb{E}\left[\left(X_{h,\Delta t}^{t_j}\right)^{\otimes 2}\right] + \mathbb{E}\left[\left(P_h G(t_j)\Delta W^j\right)^{\otimes 2}\right].$$

Tensorizing (6), the recursion scheme for the mean, we obtain

$$\left(R_{h,\Delta t}\right)^{\otimes 2}\left(\mu^{t_{j+1}}\right)^{\otimes 2} = \left(F_{h,\Delta t}^{1,j}\right)^{\otimes 2}\left(\mu^{t_j}\right)^{\otimes 2}$$

and by subtracting this from the previous equation we end up with (7). The general case of a non-zero $F_{h,\Delta t}^{2,j}$ term is proven in the same way, noting that all terms involving this disappears from (7) when subtracting the tensorized mean in the last step. □

Remark 1 Note that this computation can easily be extended to the case of linear multiplicative noise. However, $X_{h,\Delta t}^T$ is then non-Gaussian, so knowledge of its covariance is not sufficient to compute samples of it.

Algorithm 2 Covariance-based MC method of computing an estimate $E_N[\phi(X_{h,\Delta t}^T)]$ of $\mathbb{E}[\phi(X(T))]$

1: Form the mean vector $\boldsymbol{\mu}$ and covariance matrix $\boldsymbol{\Sigma}$ of $\bar{\boldsymbol{x}}_h^T$ by solving the matrix equations corresponding to (6) and (7)
2: $result = 0$
3: **for** $i = 1$ to N **do**
4: Sample $\bar{\boldsymbol{x}}_h^T = [x_1, x_2, \ldots, x_{N_h}]' \sim N(\boldsymbol{\mu}, \boldsymbol{\Sigma})$
5: Compute $\phi(X_{h,\Delta t}^T) = \phi\left(\sum_{k=1}^{N_h} x_k \Phi_k^h\right)$
6: $result = result + \phi(X_{h,\Delta t}^T)^{(i)}/N$
7: **end for**
8: $E_N\left[\phi(X_{h,\Delta t}^T)\right] = result$

Next, we compare the computational complexities of Algorithms 1 and 2. Combining (4) with Theorem 2 yields, using the triangle inequality, for a constant $C > 0$,

$$
\begin{aligned}
&\|\mathbb{E}[\phi(X(T))] - E_N[\phi(X^T_{h,\Delta t})]\|_{L^2(\Omega;\mathbb{R})} \\
&\quad \le \|\mathbb{E}[\phi(X(T))] - \mathbb{E}[\phi(X^T_{h,\Delta t})]\|_{L^2(\Omega;\mathbb{R})} + \|\mathbb{E}[\phi(X^T_{h,\Delta t})] - E_N[\phi(X^T_{h,\Delta t})]\|_{L^2(\Omega;\mathbb{R})} \\
&\quad \le C\left((1 + |\log(h)|)\left(h^2 + \Delta t^\delta\right) + N^{-1/2}\right).
\end{aligned}
$$

To balance this we couple Δt and N by $\Delta t^\delta \simeq N^{-1/2} \simeq h^2$. We make the following assumption for the computational complexities of the algorithms.

Assumption 3 There is a $d \in \mathbb{N}$ such that the cost of computing one step of (3) is $\mathrm{O}(h^{-\alpha d})$, where $\alpha \in [1, 2]$, while the cost of one step of the tensorized system (7) is $\mathrm{O}(h^{-2d})$. Moreover, the cost of sampling a Gaussian V_h-valued random variable with covariance given by (7) is $\mathrm{O}(h^{-2d})$. Finally, for Algorithm i, $i = 1, 2, 3, 4$, any additional offline cost is $\mathrm{O}(h^{-\omega_i d})$ for some $\omega_i \in \mathbb{N}$.

Remark 2 In our model problem Example 1, d is the dimension of the space $\mathbb{R}^d$ in which the domain D is contained and $\dim(V_h) = \mathrm{O}(h^{-d})$. See Sect. 5 for a discussion of this assumption.

In order to compute an approximation of $\mathbb{E}[\phi(X(T))]$, if we use Algorithm 1, we need to solve the drift-implicit Euler–Maruyama system (3) $N \cdot N_{\Delta t} = T N \Delta t^{-1}$ times, making the total (online) cost $\mathrm{O}(N\Delta t^{-1} h^{-\alpha d}) = \mathrm{O}(h^{-4-\alpha d-2/\delta})$. If we use Algorithm 2 instead, we need to solve (7) $N_{\Delta t} = T\Delta t^{-1}$ times and then sample from the resulting covariance N times, making the total cost $\mathrm{O}(h^{-2d-2/\delta}) + \mathrm{O}(h^{-2d-4}) = \mathrm{O}(h^{-2d-4})$. We collect these observations in the following proposition, which ends this section.

Proposition 1 *Let ϕ and the terms of* (1) *satisfy Assumptions 1 and 2. Assume that $X^T_{h,\Delta t}$ is given by* (3) *and that $X(T)$ is the solution to* (1) *at time $T < \infty$. If $N \simeq h^{-4}$ and $\Delta t \simeq h^{2/\delta}$ then there is a constant $C > 0$ such that*

$$
\|\mathbb{E}[\phi(X(T))] - E_N[\phi(X^T_{h,\Delta t})]\|_{L^2(\Omega;\mathbb{R})} \le C\,(1 + |\log(h)|)\,h^2, \text{ for all } h > 0.
$$

Under Assumption 3, the cost of computing $E_N[X^T_{h,\Delta t}]$ with Algorithm 1 is bounded by $\mathrm{O}(\max(h^{-4-\alpha d-2/\delta}, h^{-\omega_1 d}))$ and with Algorithm 2 by $\mathrm{O}(\max(h^{-2d-4}, h^{-\omega_2 d}))$.

4 Covariance-Based Sampling in a Multilevel Monte Carlo Setting

For our goal of estimating $\mathbb{E}[\phi(X(T))]$, the MLMC algorithm can be a more efficient alternative to the standard MC algorithm. For a sequence $(Y_\ell)_{\ell\in\mathbb{N}_0}$ of random variables in $L^2(\Omega;\mathbb{R})$ approximating $Y \in L^2(\Omega;\mathbb{R})$, where the index $\ell \in \mathbb{N}_0$ is referred to as a level, the *MLMC estimator* $E^L[Y_L]$ of $\mathbb{E}[Y_L]$ is, for $L \in \mathbb{N}$, defined by

$$E^L[Y_L] = E_{N_0}[Y_0] + \sum_{\ell=1}^{L} E_{N_\ell}[Y_\ell - Y_{\ell-1}],$$

where $(N_\ell)_{\ell=0}^L$ are level specific numbers of samples in the respective MC estimators.

To apply this algorithm in our setting, we take a sequence $(X_\ell^T)_{\ell\in\mathbb{N}_0}$ of approximations of $X(T)$, given by $X_\ell^T = X_{h_\ell,\Delta t_\ell}^T$, where $(h_\ell)_{\ell\in\mathbb{N}_0}$ is a decreasing sequence of mesh sizes and $\Delta t_\ell^\delta \simeq h_\ell^2$, so that $(\phi(X_\ell^T))_{\ell\in\mathbb{N}_0}$ becomes a sequence approximating $\phi(X(T))$. For notational convenience, we set $\phi(X_{-1}^T) = 0$. Computing $E^L\left[\phi\left(X_L^T\right)\right]$ involves, for each $\ell = 1, 2, \ldots, L$, sampling $\phi\left(X_\ell^T\right) - \phi\left(X_{\ell-1}^T\right)$ N_ℓ times (we specify how to choose the sample sizes below). For this it is key that X_ℓ^T on the *fine* level ℓ and $X_{\ell-1}^T$ on the *coarse* level $\ell - 1$ are positively correlated. In the classical path-based method (Algorithm 3, see also [2, 5, 6, 19]), this is achieved by computing them on the same discrete realization of W, assuming that the family $(V_h)_{h\in(0,1]}$ is nested.

Algorithm 3 Path-based MLMC method of computing an estimate $E^L\left[\phi\left(X_L^T\right)\right]$ of $\mathbb{E}[\phi(X(T))]$

$result = 0$
for $\ell = 0$ to L **do**
  for $i = 1$ to N_ℓ **do**
    Sample increments of a realization $W^{(i)}$ of the Q-Wiener process W
    Compute $\bar{\boldsymbol{x}}_{h_{\ell-1}}^T = [x_1^{\ell-1}, x_2^{\ell-1}, \ldots, x_{N_{h_{\ell-1}}}^{\ell-1}]'$ by solving the matrix equations corresponding to (3) driven by $W^{(i)}$
    Compute $\bar{\boldsymbol{x}}_{h_\ell}^T = [x_1^\ell, x_2^\ell, \ldots, x_{N_{h_\ell}}^\ell]'$ by solving the matrix equations corresponding to (3) driven by $W^{(i)}$
    Compute $\phi(X_\ell^T) - \phi(X_{\ell-1}^T) = \phi\left(\sum_{k=1}^{N_{h_\ell}} x_k^\ell \Phi_k^{h_\ell}\right) - \phi\left(\sum_{k=1}^{N_{h_{\ell-1}}} x_k^{\ell-1} \Phi_k^{h_{\ell-1}}\right)$
    $result = result + \left(\phi(X_\ell^T) - \phi(X_{\ell-1}^T)\right)/N_\ell$
  end for
end for
$E^L\left[\phi(X_L^T)\right] = result$

To introduce our alternative covariance-based method, the path-based sampling is rewritten as a system on $V_{h'} \oplus V_h$. Consider to this end, for $h, h', \Delta t, \Delta t' \in (0, 1]$, a pair $\left((X_{h',\Delta t'}^{t'_j})_{j=0}^{N_{\Delta t'}}, (X_{h,\Delta t}^{t_j})_{j=0}^{N_{\Delta t}}\right)$ of approximations of X, given by the drift-implicit Euler–Maruyama scheme (3). Assume further that they are nested in time, i.e., that $\Delta t' = K\Delta t$ for some $K \in \mathbb{N}$ with $K > 1$. We create an extension $(\hat{X}_{h',\Delta t}^{t_j})_{j=0}^{N_{\Delta t}}$ of the coarse approximation $(X_{h',\Delta t'}^{t'_j})_{j=0}^{N_{\Delta t'}}$ to the finer time grid by $\hat{X}_{h',\Delta t}^{t_0} = X_{h',\Delta t'}^{t'_0}$ and

$$\hat{R}_{h',\Delta t}^j \hat{X}_{h',\Delta t}^{t_{j+1}} = \hat{F}_{h',\Delta t}^{1,j} \hat{X}_{h',\Delta t}^{t_j} + \hat{F}_{h',\Delta t}^{2,j} + P_{h'}\hat{G}(t_j)\Delta W^j,$$

for $j = 0, 1, \ldots, N_{\Delta t} - 1$, where $\Delta W^j = W(t_{j+1}) - W(t_j)$. The operators are given by

$$\hat{R}^j_{h',\Delta t} = \begin{cases} R_{h',\Delta t'} & \text{if } j+1 = 0 \mod K, \\ I_H & \text{otherwise}, \end{cases} \quad \hat{F}^{1,j}_{h',\Delta t} = \begin{cases} F^{1,j/K}_{h',\Delta t'} & \text{if } j+1 = 1 \mod K, \\ I_H & \text{otherwise}, \end{cases}$$

$$\hat{F}^{2,j}_{h',\Delta t} = \begin{cases} F^{2,j/K}_{h',\Delta t'} & \text{if } j+1 = 1 \mod K, \\ 0 & \text{otherwise}, \end{cases} \quad \text{and } \hat{G}(t_j) = G(t_{j-(j \bmod K)}).$$

Note that $\hat{X}^{t_{j+1}}_{h',\Delta t} = X^{t'_{(j+1)/K}}_{h',\Delta t'}$ when $j+1 = 0 \mod K$ since then

$$\begin{aligned} \hat{R}^j_{h',\Delta t}\hat{X}^{t_{j+1}}_{h',\Delta t} &= \hat{F}^{1,j}_{h',\Delta t}\hat{X}^{t_j}_{h',\Delta t} + \hat{F}^{2,j}_{h',\Delta t} + P_{h'}\hat{G}(t_j)\Delta W^j = \hat{X}^{t_j}_{h',\Delta t} + P_{h'}G(t_{j-(K-1)})\Delta W^j \\ &= \hat{X}^{t_{j-1}}_{h',\Delta t} + P_{h'}G(t_{j-1-(K-2)})\Delta W^{j-1} + P_{h'}G(t_{j-(K-1)})\Delta W^j = \cdots \\ &= \hat{F}^{1,j+1-K}_{h',\Delta t}\hat{X}^{t_{j+1-K}}_{h',\Delta t} + \hat{F}^{2,j+1-K}_{h',\Delta t} + P_h G(t_{j+1-K})\sum_{i=1}^{K}\Delta W^{j-(i-1)} \\ &= F^{1,(j+1)/K-1}_{h',\Delta t'}\hat{X}^{t_{j+1-K}}_{h',\Delta t} + F^{2,(j+1)/K-1}_{h',\Delta t'} + P_h G(t'_{(j+1)/K-1})\Delta W'^{(j+1)/K-1}, \end{aligned}$$

where $\Delta W'^{(j+1)/K-1} = W(t'_{(j+1)/K}) - W(t'_{(j+1)/K-1})$. Hence, sampling the pair of discretizations $\left((X^{t_j}_{h,\Delta t})_{j=0}^{N_{\Delta t}}, (X^{t'_j}_{h',\Delta t'})_{j=0}^{N_{\Delta t'}}\right)$ on the same realization of the driving Q-Wiener process is equivalent to solving the system

$$\begin{bmatrix} \hat{R}^j_{h',\Delta t} & 0 \\ 0 & R^j_{h,\Delta t} \end{bmatrix} \begin{bmatrix} \hat{X}^{t_{j+1}}_{h',\Delta t} \\ X^{t_{j+1}}_{h,\Delta t} \end{bmatrix} = \begin{bmatrix} \hat{F}^{1,j}_{h',\Delta t} & 0 \\ 0 & F^{1,j}_{h,\Delta t} \end{bmatrix} \begin{bmatrix} \hat{X}^{t_j}_{h',\Delta t} \\ X^{t_j}_{h,\Delta t} \end{bmatrix} + \begin{bmatrix} \hat{F}^{2,j}_{h',\Delta t} \\ F^{2,j}_{h,\Delta t} \end{bmatrix} + \begin{bmatrix} P_{h'}\hat{G}(t_j) \\ P_h G(t_j) \end{bmatrix} \Delta W^j$$

in $V_{h'} \oplus V_h$ for $j = 0, 1, \ldots, N_{\Delta t} - 1$. We note that $[\hat{X}^{t_j}_{h',\Delta t}, X^{t_j}_{h,\Delta t}]'$ is a Gaussian $V_{h'} \oplus V_h$-valued random variable for all $j = 0, 1, \ldots, N_{\Delta t} - 1$. Therefore, a covariance-based approach for sampling $\left(X^T_{h,\Delta t}, X^T_{h',\Delta t'}\right)$ could be obtained by directly computing $\operatorname{Cov}\left([X^T_{h,\Delta t}, X^T_{h',\Delta t'}]'\right) = \operatorname{Cov}\left([X^T_{h,\Delta t}, \hat{X}^T_{h',\Delta t}]'\right)$. However, to save computational work, we base Algorithm 4 on computing $\operatorname{Cov}\left(X^T_{h',\Delta t'}, X^T_{h,\Delta t}\right)$ instead. The following theorem gives the scheme for this, which is derived analogously to Theorem 3.

Theorem 4 *Let the terms of* (1) *satisfy Assumption 1 and let, for* $h, h', \Delta t, \Delta t' > 0$, $(X^{t_j}_{h,\Delta t})_{j=0}^{N_{\Delta t}}$ *and* $(X^{t'_j}_{h',\Delta t'})_{j=0}^{N_{\Delta t'}}$ *be given by the drift-implicit Euler–Maruyama scheme* (3)*. Assume further that* $\Delta t' = K\Delta t$ *for some* $K \in \mathbb{N}$ *with* $K > 1$*. Then the cross-covariance* $\operatorname{Cov}\left(X^T_{h',\Delta t'}, X^T_{h,\Delta t}\right)$ *is given by* $\Sigma_T \in V_{h'} \otimes V_h$*, where the sequence* $(\Sigma_{t_j})_{j=0}^{N_{\Delta t}}$ *fulfills*

$$\left(\hat{R}^j_{h',\Delta t'} \otimes R_{h,\Delta t}\right)\Sigma_{t_{j+1}} = \left(\hat{F}^{1,j}_{h,\Delta t} \otimes F^{1,j}_{h,\Delta t}\right)\Sigma_{t_j} + \mathbb{E}\left[P_{h'}\hat{G}(t_j)\Delta W^j \otimes P_h G(t_j)\Delta W^j\right]. \quad (9)$$

The following proposition, which is an adaptation of [18, Theorem 1] to our setting, shows how one should choose the sample sizes in an MLMC algorithm and provides bounds on the overall computational work for Algorithms 3 and 4.

Proposition 2 *Let ϕ and the terms of* (1) *satisfy Assumptions 1 and 2. Let $(h_\ell)_{\ell\in\mathbb{N}_0}$ be a sequence of maximal mesh sizes that satisfy $h_\ell \simeq a^{-\ell}$ for some $a > 1$ and all $\ell \in \mathbb{N}_0$. Let $(X_\ell^T)_{\ell\in\mathbb{N}_0}$ be a sequence of approximations of $X(T)$, where $X_\ell^T = X_{h_\ell,\Delta t_\ell}^T$ is given by the recursion* (3) *with $\Delta t_\ell^\delta \simeq h_\ell^2$.*

Algorithm 4 Covariance-based MLMC method of computing an estimate $E^L\left[\phi\left(X_L^T\right)\right]$ of $\mathbb{E}[\phi(X(T))]$

---

result $= 0$
for $\ell = 0$ to L **do**
  Compute the covariance matrix $\boldsymbol{\Sigma}$ and mean vector $\boldsymbol{\mu}$ of

$$\bar{\boldsymbol{x}}_\ell = \left[[\bar{\boldsymbol{x}}_{h_{\ell-1}}^T]', [\bar{\boldsymbol{x}}_{h_\ell}^T]'\right]' = \left[x_1^{\ell-1}, x_2^{\ell-1}, \ldots, x_{N_{h_{\ell-1}}}^{\ell-1}, x_1^\ell, x_2^\ell, \ldots, x_{N_{h_\ell}}^\ell\right]'$$

  by computing the means, covariances and cross-covariances of the pair $(X_{\ell-1}^T, X_\ell^T)$ via the solution of the matrix equations corresponding to (6), (7) and (9)
  for $i = 1$ to N_ℓ **do**
    Sample $\bar{\boldsymbol{x}}_\ell \sim N(\boldsymbol{\mu}, \boldsymbol{\Sigma})$
    Compute $\phi(X_\ell^T) - \phi(X_{\ell-1}^T) = \phi\left(\sum_{k=1}^{N_{h_\ell}} x_k^\ell \Phi_k^{h_\ell}\right) - \phi\left(\sum_{k=1}^{N_{h_{\ell-1}}} x_k^{\ell-1} \Phi_k^{h_{\ell-1}}\right)$
    result $=$ *result* $+ \left(\phi(X_\ell^T) - \phi(X_{\ell-1}^T)\right)/N_\ell$
  end for
end for
$E^L\left[\phi(X_L^T)\right] =$ *result*

For $L \in \mathbb{N}$, $\ell = 1, \ldots, L$, $\varepsilon > 0$, set $N_\ell = \lceil h_L^{-4} h_\ell^2 \ell^{1+\varepsilon} \rceil$, where $\lceil\cdot\rceil$ is the ceiling function, and $N_0 = \lceil h_L^{-4} \rceil$. Then there exists a constant $C > 0$ such that, for all $L \in \mathbb{N}$,

$$\|\mathbb{E}\left[\phi\left(X(T)\right)\right] - E^L\left[\phi\left(X_L^T\right)\right]\|_{L^2(\Omega;\mathbb{R})} \leq C(1 + |\log(h_L)|)h_L^2.$$

Under Assumption 3, the cost of finding $E^L\left[\phi\left(X_L^T\right)\right]$ with Algorithm 3 is bounded by $\mathrm{O}(\max(h_L^{-2-\alpha d-2/\delta} L^{2+\varepsilon}, h_L^{-\omega_3 d}))$. With Algorithm 4 the cost is bounded by $\mathrm{O}(\max(h_L^{-2d-2/\delta} L, h_L^{-2-2d} L^{2+\varepsilon}, h_L^{-\omega_4 d}))$.

Proof By [18, Lemma 2],

$$\begin{aligned}
&\|\mathbb{E}\left[\phi\left(X(T)\right)\right] - E^L\left[\phi\left(X_L^T\right)\right]\|_{L^2(\Omega;\mathbb{R})} \\
&\leq \left|\mathbb{E}\left[\phi\left(X(T)\right) - \phi\left(X_L^T\right)\right]\right| \\
&\quad + \left(N_0^{-1}\|\phi\left(X_0^T\right)\|_{L^2(\Omega;\mathbb{R})}^2 + \sum_{\ell=1}^{L} N_\ell^{-1}\|\phi\left(X_\ell^T\right) - \phi\left(X_{\ell-1}^T\right)\|_{L^2(\Omega;\mathbb{R})}^2\right)^{1/2}.
\end{aligned} \tag{10}$$

By the fact that the Fréchet derivative ϕ' is at most polynomially growing and by the uniform bound on X_ℓ^T from Theorem 1, there is a constant $C > 0$ such that $\|\phi\left(X_0^T\right)\|^2_{L^2(\Omega;\mathbb{R})} < C$. Moreover, the mean-value theorem for Fréchet differentiable mappings (cf. [24, Example 4.2]) shows that there exist $p \geq 2$ and $C > 0$ such that

$$\|\phi\left(X_\ell^T\right) - \phi\left(X(T)\right)\|^2_{L^2(\Omega;\mathbb{R})} \leq C\|X_\ell^T - X(T)\|^2_{L^p(\Omega;H)} \text{ for all } \ell \in \mathbb{N}_0,$$

so that, using Theorem 1, we get a constant $C > 0$ such that

$$\begin{aligned}\|\phi\left(X_\ell^T\right) - \phi\left(X_{\ell-1}^T\right)\|^2_{L^2(\Omega;\mathbb{R})} &\leq C\left(\|X_\ell^T - X(T)\|^2_{L^p(\Omega;H)} + \|X_{\ell-1}^T - X(T)\|^2_{L^p(\Omega;H)}\right)\\ &\leq C(h_\ell^2 + h_{\ell-1}^2) \leq C(1+a^2)h_\ell^2.\end{aligned}$$

Hence, using Theorem 2 in (10) yields the existence of a constant $C > 0$ such that

$$\begin{aligned}&\|\mathbb{E}\left[\phi\left(X(T)\right)\right] - E^L\left[\phi\left(X_L^T\right)\right]\|_{L^2(\Omega;\mathbb{R})}\\ &\leq C\left((1+|\log(h_L)|)h_L^2 + \left(N_0^{-1} + \sum_{\ell=1}^{L} N_\ell^{-1}h_\ell^2\right)^{1/2}\right)\\ &\leq Ch_L^2\left(1 + |\log(h_L)| + (1+\zeta(1+\varepsilon))^{1/2}\right),\end{aligned}$$

where ζ denotes the Riemann zeta function and the last inequality follows from the choice of sample sizes. This shows the first part of the theorem.

If we use Algorithm 3 to compute $E^L[\phi\left(X_L^T\right)]$, the cost of sampling X_0^T N_0 times is by Proposition 1 $\mathrm{O}(N_0 h_0^{-\alpha d - 2/\delta}) = \mathrm{O}(h_L^{-4})$. Similarly, since the computation of $\phi\left(X_\ell^T\right) - \phi\left(X_{\ell-1}^T\right)$, $1 \leq \ell \leq L$, is dominated by the sampling of X_ℓ^T, the cost for the rest of the terms is bounded by a constant times

$$\sum_{\ell=1}^{L} N_\ell h_\ell^{-\alpha d - 2/\delta} = \sum_{\ell=1}^{L} h_L^{-4} h_\ell^{2-\alpha d - 2/\delta} \ell^{1+\varepsilon} \leq h_L^{-2-\alpha d - 2/\delta} L^{2+\varepsilon}.$$

For Algorithm 4, the cost of sampling X_0^T N_0 times is still $\mathrm{O}(h_L^{-4})$. For $1 \leq \ell \leq L$, the cost of computing the covariance of X_ℓ^T, $X_{\ell-1}^T$ and their cross-covariance is dominated by the cost of computing the covariance of X_ℓ^T, which is, by the same reasoning as that preceding Proposition 1, $\mathrm{O}(h_\ell^{-2d-2/\delta})$. The cost of sampling a positively correlated pair of X_ℓ^T and $X_{\ell-1}^T$, given that all covariances have been computed, is $\mathrm{O}(h_\ell^{-2d}) + \mathrm{O}(h_{\ell-1}^{-2d}) = \mathrm{O}(h_\ell^{-2d})$ so the total cost for sampling $\phi\left(X_\ell^T\right) - \phi\left(X_{\ell-1}^T\right)$ for all $\ell = 1, 2, \ldots, L$ is bounded by a constant times

$$\begin{aligned}\sum_{\ell=1}^{L}\left(h_\ell^{-2d-2/\delta} + N_\ell h_\ell^{-2d}\right) &= \sum_{\ell=1}^{L}\left(h_\ell^{-2d-2/\delta} + h_L^{-4}h_\ell^{2-2d}\ell^{1+\varepsilon}\right)\\ &\leq 2\max(h_L^{-2d-2/\delta}L, h_L^{-2-2d}L^{2+\varepsilon}).\end{aligned}$$

This finishes the proof. □

Remark 3 We note that this is a suboptimal choice of sample sizes compared to the standard $N_\ell \simeq \sqrt{V_\ell / C_\ell}$, see [12], where V_ℓ and C_ℓ are the variance and computational cost, respectively, of $\phi(X_\ell^T) - \phi(X_{\ell-1}^T)$. The reason for our choice is to avoid the estimation of the additional error resulting from the estimation of V_ℓ, cf. [5, 6].

Remark 4 Note that Assumption 2 is only used to tune the MC and MLMC estimators, using the weak convergence result Theorem 2. Assuming only Assumption 1, Theorem 1 can be used for the tuning if ϕ is Lipschitz. However, the result can be suboptimal, cf. [18]. Moreover, there exist results on weak convergence with different assumptions on the parameters of (1) (including rougher noise), for example [3]. If these are used for tuning, the conclusions regarding which methods are best may be different. The same is true when the parameters of (1) are such that the drift-implicit Euler scheme coincides with a Milstein scheme, see [4].

5 Implementation

In this section we describe the implementation of the algorithms of Sects. 3 and 4 in the setting of Example 1, and motivate Assumption 3 along the way. For simplicity, we conform to Assumption 2 by setting $b = c = 0$. For the spatial discretization, let V_h be the space of piecewise linear polynomials on a mesh of $D \subset \mathbb{R}^d$, $d = 1, 2, 3$ with maximal mesh size $h > 0$, with a basis $\boldsymbol{\Phi}^h = (\Phi_i^h)_{i=1}^{N_h}$.

Recall that the system of equations in $\mathbb{R}^{N_h}$ corresponding to (3) that we must solve in Algorithms 1 and 3 is given by

$$(\boldsymbol{M}_h + \Delta t \boldsymbol{A}_h)\bar{\boldsymbol{x}}_h^{t_j+1} = \boldsymbol{M}_h \bar{\boldsymbol{x}}_h^{t_j} + \boldsymbol{f}_h^{t_j} + g(t_j)\boldsymbol{\Delta W}_h^j, \tag{11}$$

for $j = 0, 1, \ldots, N_{\Delta t} - 1$. This gives the vector $\bar{\boldsymbol{x}}_h^T$. Here $\boldsymbol{M}_h$ is the mass matrix, $\boldsymbol{A}_h$ the stiffness matrix, $\boldsymbol{f}_h^{t_j}$ a vector corresponding to the function $d(t_j, \cdot)$ and $(\boldsymbol{\Delta W}_h^j)_{j=0}^{N_{\Delta t}-1}$ a family of iid Gaussian $\mathbb{R}^{N_h}$-valued random vectors with covariance matrix $\boldsymbol{\Sigma}_{h,\Delta W}$, having entries

$$s_{i,j} = \Delta t \int_{D^2} q(x, y)\Phi_i^h(x)\Phi_j^h(y)\,\mathrm{d}x\;\mathrm{d}y, \quad i, j \in \{1, 2, \ldots, N_h\}.$$

Solving the discretized elliptic problem (11) can be accomplished by direct or iterative solvers. Like the authors of [1, 5–7] we assume that we have access to a solver such that the inversion of $\boldsymbol{M}_h + \Delta t \boldsymbol{A}_h$ costs $\mathrm{O}(h^{-d})$ and any additional offline costs are $\mathrm{O}(h^{-\omega_1 d})$. Whether this is true will depend on the parameters, the mesh and the dimension of the problem. For $d = 1$ it is immediately true, for $d > 1$ we mention multigrid methods, see, e.g., [8]. Given this, the cost of solving (11) is dominated by the generation of the stochastic term $\boldsymbol{\Delta W}_h^j$. In special cases, for example, if q is piecewise analytic (see [16]) or if Q is specified via a truncated Karhunen–Loève expansion where the truncation does not depend on h (see also Sect. 6), the complexity is linear, i.e., $\alpha = 1$ in Assumption 3. If no special assumptions are made on q, the standard option is to use a Cholesky or eigenvalue decomposition of the covariance matrix for $P_h \Delta W$ (cf. [21, Chap. 7]). Since the cost of the decomposition is cubic and the matrix multiplication cost of generating $P_h \Delta W$ is quadratic, we then have $\omega_1, \omega_3 \geq 3$ and $\alpha = 2$ in Assumption 3.

For Algorithms 2 and 4 we must solve tensorized systems. By expanding (7) and applying $\left\langle \Phi_i^{2,h}, \cdot \right\rangle$ to each side for $i = 1, 2, \ldots, N_h^2$, where $\boldsymbol{\Phi}^{2,h} = (\Phi_i^{2,h})_{i=1}^{N_h^2}$ is a basis of $V_h^{\otimes 2}$, a system of equations in $\mathbb{R}^{2N_h}$ for the covariance recursion scheme of Theorem 3 is obtained. Choosing $\Phi_i^{2,h} = \Phi^h_{\lfloor (i-1)/N_h \rfloor + 1} \otimes \Phi^h_{i - \lfloor (i-1)/N_h \rfloor N_h}$ for $i = 1, 2, \ldots, N_h^2$, the matrices corresponding to $(R_{h,\Delta t})^{\otimes 2}$ and $(F^{1,j}_{h,\Delta t})^{\otimes 2}$ will be Kronecker products of the matrices corresponding to $R_{h,\Delta t}$ and $F^{1,j}_{h,\Delta t}$, $j = 0, 1, \ldots, N_{\Delta t} - 1$. In this setting, the resulting system at time t_{j+1} is

$$(\boldsymbol{M}_h + \Delta t \boldsymbol{A}_h)^{\otimes_K 2} \, \bar{\boldsymbol{y}}_h^{t_j+1} = \boldsymbol{M}_h^{\otimes_K 2} \bar{\boldsymbol{y}}_h^{t_j} + g(t_j)^2 \, \mathsf{Vec}\left(\boldsymbol{\Sigma}_{h,\Delta W}\right),$$

where $\otimes_K$ denotes the Kronecker product and Vec the vectorization operator. Here $\bar{\boldsymbol{y}}_h^{t_j} = \mathsf{Vec}(\boldsymbol{\Sigma}_{\bar{\boldsymbol{x}}_h^{t_j}})$, the covariance matrix of $\bar{\boldsymbol{x}}_h^{t_j}$. By the identity $\mathsf{Vec}\,(ABC) = (C' \otimes_K A)\, \mathsf{Vec}(B)$, where A, B and C are matrices such that ABC is well-defined, this is equivalent to the matrix system

$$(\boldsymbol{M}_h + \Delta t \boldsymbol{A}_h)\, \boldsymbol{\Sigma}_{\bar{\boldsymbol{x}}_h^{t_{j+1}}} \, (\boldsymbol{M}_h + \Delta t \boldsymbol{A}_h) = \boldsymbol{M}_h \boldsymbol{\Sigma}_{\bar{\boldsymbol{x}}_h^{t_j}} \boldsymbol{M}_h + g(t_j)^2 \boldsymbol{\Sigma}_{h,\Delta W}, \tag{12}$$

where we have used symmetry of the matrices involved. Assuming that the solver of (11) is used, the cost of this system is $\mathrm{O}(h^{-2d})$. To see this, note that since $\boldsymbol{M}_h$ has $\mathrm{O}(h^{-d})$ nonzero entries, the right hand side can be formed in $\mathrm{O}(h^{-2d})$. One then solves for the symmetric matrix $\boldsymbol{U} = \boldsymbol{\Sigma}_{\bar{\boldsymbol{x}}_h^{t_{j+1}}} (\boldsymbol{M}_h + \Delta t \boldsymbol{A}_h)$ by employing the solver for each of its $\mathrm{O}(h^{-d})$ columns, and then one similarly solves for the symmetric matrix $\boldsymbol{\Sigma}_{\bar{\boldsymbol{x}}_h^{t_{j+1}}}$ in the system $(\boldsymbol{M}_h + \Delta t \boldsymbol{A}_h)\, \boldsymbol{\Sigma}_{\bar{\boldsymbol{x}}_h^{t_{j+1}}} = \boldsymbol{U}$. Having computed $\boldsymbol{\Sigma}_{\bar{\boldsymbol{x}}_h^T}$, sampling using this costs $\mathrm{O}(h^{-2d})$, corresponding to a matrix-vector multiplication, assuming that a Cholesky or eigenvalue decomposition has been used, yielding $\omega_2, \omega_4 \geq 3$.

In Algorithm 4, we also need to solve for the cross-covariance. Given the approximations $(X^{t_j}_{h,\Delta t})_{j=0}^{N_{\Delta t}}$ and $(X^{t'_j}_{h',\Delta t'})_{j=0}^{N_{\Delta t'}}$ of Theorem 4, choosing a basis $\boldsymbol{\Phi}^{2,h',h} = (\Phi_i^{2,h',h})_{i=1}^{N_{h'} N_h}$ of $V_{h'} \otimes V_h$ by $\Phi_i^{2,h',h} = \Phi^{h'}_{\lfloor (i-1)/N_h \rfloor + 1} \otimes \Phi^h_{i - \lfloor (i-1)/N_h \rfloor N_h}$ yields, in the same way as above, a matrix system

$$\begin{aligned}(\boldsymbol{M}_h + \Delta t \boldsymbol{A}_h)\, &\boldsymbol{\Sigma}_{\bar{\boldsymbol{x}}_h^{t_{j+1}}, \bar{\boldsymbol{x}}_{h'}^{t_{j+1}}} \left(\boldsymbol{M}_{h'} + \Delta t' \boldsymbol{A}_{h'}\right) \\ &= \boldsymbol{M}_h \boldsymbol{\Sigma}_{\bar{\boldsymbol{x}}_h^{t_j}, \bar{\boldsymbol{x}}_{h'}^{t_j}} \boldsymbol{M}_{h'} + g(t_j) g(t_{j-(j \bmod K)}) \boldsymbol{\Sigma}_{h,h',\Delta W},\end{aligned}$$

corresponding to (9) at time t_{j+1} with $j + 1 = 0 \mod K$. Here $\boldsymbol{\Sigma}_{\bar{\boldsymbol{x}}_h^{t_j}, \bar{\boldsymbol{x}}_{h'}^{t_j}}$ is the cross-covariance matrix of $\bar{\boldsymbol{x}}_h^{t_j}$ and $\bar{\boldsymbol{x}}_{h'}^{t_j}$ and $\boldsymbol{\Sigma}_{h,h',\Delta W}$ a matrix with entries

$$s_{i,j} = \Delta t \int_{D^2} q(x, y)\Phi_i^h(x)\Phi_j^{h'}(y)\,\mathrm{d}x\,\mathrm{d}y, \quad i \in \{1, \dots, N_h\},\ j \in \{1, \dots, N_{h'}\}.$$

By the same reasoning as above, the cost of solving this system is $\mathrm{O}(\min(h', h)^{-2d})$. Having solved for the cross-covariance matrix, the vector $\bar{\boldsymbol{x}}_{h,h'}^T = \left[[\bar{\boldsymbol{x}}_{h'}^T]', [\bar{\boldsymbol{x}}_h^T]'\right]'$ is sampled at a cost of $\mathrm{O}(\min(h', h)^{-2d})$ using the matrix

$$\boldsymbol{\Sigma}_{\bar{\boldsymbol{x}}_{h,h'}^T} = \begin{bmatrix} \boldsymbol{\Sigma}_{\bar{\boldsymbol{x}}_{h'}^T} & \boldsymbol{\Sigma}_{\bar{\boldsymbol{x}}_{h'}^T, \bar{\boldsymbol{x}}_h^T} \\ \boldsymbol{\Sigma}'_{\bar{\boldsymbol{x}}_{h'}^T, \bar{\boldsymbol{x}}_h^T} & \boldsymbol{\Sigma}_{\bar{\boldsymbol{x}}_h^T} \end{bmatrix},$$

where the diagonal entries are obtained via (12) as before.

Having motivated Assumption 3, we compare the costs of the algorithms. Looking at Propositions 1 and 2, we see that if $\alpha = 2$, Algorithm 2 should be used for all $d \in \{1, 2, 3\}$ in the case that $\delta = 1/2$, while if $\delta = 1$, Algorithm 4 should be preferred for all $d \in \{1, 2, 3\}$. If $\alpha = 1$, for $d = 1, 2$ we are in the same case as before, that is to say, Algorithm 2 is preferred for $\delta = 1/2$ and Algorithm 4 for $\delta = 1$. For $d = 3$, however, Algorithm 3 outperforms both covariance-based algorithms. Note that we only consider the work needed to solve the problem and not the memory. If the stochastic terms can be generated in linear complexity (i.e. $\alpha = 1$), we may not need to store a covariance matrix of size $\mathrm{O}(h^{-2d})$, which implies that Algorithms 1 and 3 are preferred when lack of memory is a concern.

6 Simulation

In this section we illustrate our result numerically, employing the discretization of the previous section to the stochastic heat equation driven by additive noise,

$$\mathrm{d}X(t) = \Delta X(t)\,\mathrm{d}t + \mathrm{d}W(t),$$

on $H = L^2(D)$ with $D = (0, 1)$, for $t \in (0, T] = (0, 1]$ with initial value $X(0) = x_0 = x\chi_{(0,1/2)}(x) + (1 - x)\chi_{(1/2,1)}(x)$, where χ denotes the indicator function, and Dirichlet zero boundary conditions. We choose a simple Q-Wiener process

$$W(t, x) = \frac{5\beta_1(t)}{x^{0.45}} + \frac{5\beta_2(t)}{(1 - x)^{0.45}},$$

$t \in [0, T]$ and $x \in D$, where β_1, β_2 are two independent standard real-valued Wiener processes on $[0, T]$. This can be generated in linear complexity (i.e., $\alpha = 1$ in Assumption 3). This means that the covariance function corresponding to Q is, for $x, y \in D$, given by

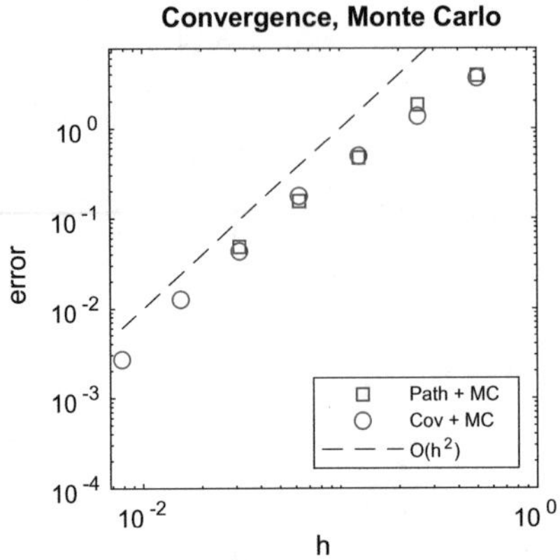

(a) Mean square errors for Algorithm 1 (Path+MC) and Algorithm 2 (Cov+MC).

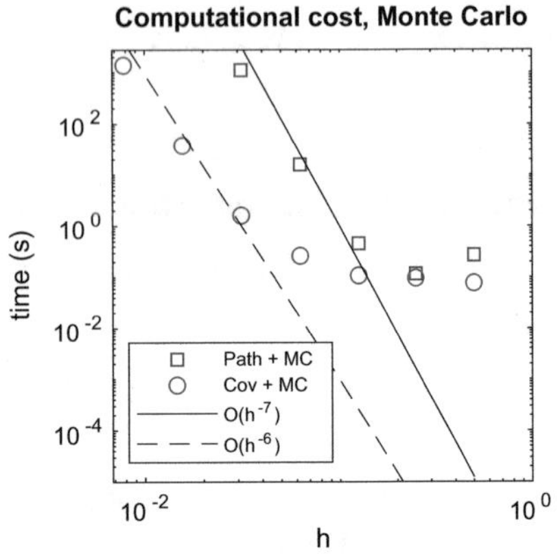

(b) Simulation times (in seconds) and bounds from Proposition 1.

Fig. 1 Convergence and computational costs of Algorithms 1 and 2

$$q(x,y) = \frac{25}{(xy)^{0.45}} + \frac{25}{((1-x)(1-y))^{0.45}}.$$

A uniform spatial mesh is used in our finite element discretization.

We now compute approximations of $\mathbb{E}[\phi(X(T))]$, with $\phi(\cdot) = \|\cdot\|^2 \in C^2_{\mathrm{p}}(H;\mathbb{R})$. Figure 1a shows estimates

$$\left(\sum_{i=1}^{5} \left(\mathbb{E}[\phi(X(T))] - E_N[\phi(X^T_{h,\Delta t})]^{(i)}\right)^2\right)^{1/2}$$

for 5 different realizations of $E_N[\phi(X^T_{h,\Delta t})]$, of the mean squared errors $\|\mathbb{E}[\phi(X(T))] - E_N[\phi(X^T_{h,\Delta t})]\|_{L^2(\Omega;\mathbb{R})}$ for $h = 2^{-1}, 2^{-2}, \ldots, 2^{-5}$, computed with Algorithms 1 and 2, choosing Δt and N according to Proposition 1. In the case of the covariance-based method, we also include $h = 2^{-6}$ and $h = 2^{-7}$. The quantity $\mathbb{E}[\phi(X(T))]$ is replaced by a reference solution $\mathbb{E}[\phi(X^T_{h,\Delta t})]$, with $h = 2^{-8}$, computed with a deterministic method, cf. [24, Sect. 6]. An eigenvalue decomposition was used for the covariance-based method. As expected, the order of convergence is $\mathrm{O}(h^2)$. In Fig. 1b we show the computational costs in seconds of the realizations of $E_N[\phi(X^T_{h,\Delta t})]$ along with the upper bounds on the costs from Proposition 1. The costs appear to asymptotically follow these bounds.

For the MLMC estimator $E^L[\phi(X^T_L)]$ we set, for $\ell = 0, \ldots, L$, $h_\ell = 2^{-\ell-1}$ and choose the temporal step sizes and sample 5 sizes according to Proposition 2. In Fig. 2a we show estimates

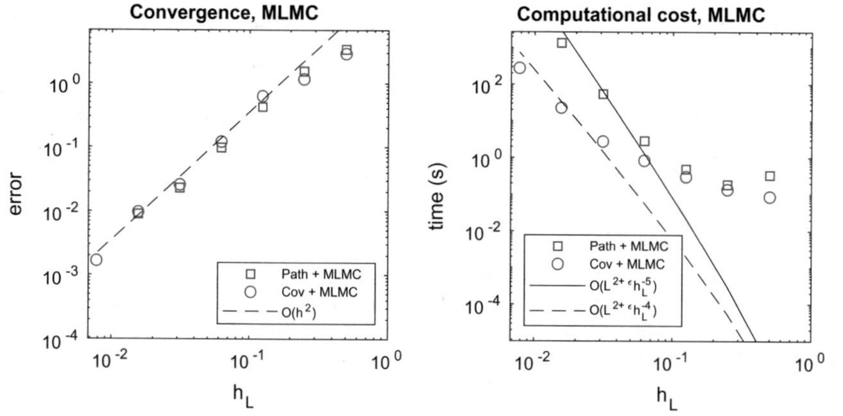

(a) Mean square errors for Algorithm 3 (Path+MLMC) and Algorithm 4 (Cov+MLMC).

(b) Simulation times (in seconds) and bounds from Proposition 2.

Fig. 2 Convergence and computational costs of the MLMC estimator (Algorithms 3 and 4)

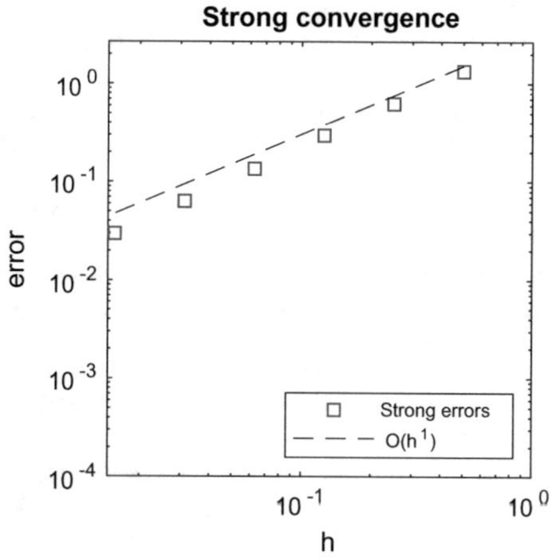

Fig. 3 A Monte Carlo estimate of the strong error $\|X(T) - X^T_{h,\Delta t}\|_{L^2(\Omega;H)}$

$$\left(\sum_{i=1}^{5} \left(\mathbb{E}[\phi(X(T))] - E^L[\phi(X_L^T)]^{(i)}\right)^2\right)^{1/2}$$

for 5 different realizations of $E^L[\phi(X_L^T)]$, of the mean squared errors $\|\mathbb{E}[\phi(X(T))] - E^L[\phi(X_L^T)]\|_{L^2(\Omega;\mathbb{R})}$ for $L = 0, 1, 2, \ldots, 5$ and for Algorithm 4 also for $L = 6$, using the reference solution from before. The order of convergence is again as expected. In Fig. 2b we show the computational costs with the upper bounds on the costs from Proposition 2. Both methods appear to follow the derived complexity bounds.

Finally, for completeness, we show in Fig. 3 a Monte Carlo approximation $E_N[\|X(T) - X^T_{h,\Delta t}\|^2]^{1/2}$ of the strong error $\|X(T) - X^T_{h,\Delta t}\|_{L^2(\Omega;H)}$ for $h = 2^{-1}$, $2^{-2}, \ldots, 2^{-6}$, $\Delta t = h^2$, with a reference solution at $h = 2^{-8}$ used in place of $X(T)$. $N = 100$ samples were used. The order of convergence is as expected from Theorem 1.

The computations in this section were performed in MATLAB® R2017b on a laptop with a dual-core Intel® Core™ i7-5600U 2.60GHz CPU.

Acknowledgements The author wishes to express many thanks to Annika Lang, Stig Larsson for support and fruitful discussions, to Michael B. Giles for helpful comments and to three anonymous referees who helped to improve the results and the presentation. The work was supported in part by the Swedish Research Council under Reg. No. 621-2014-3995 and the Knut and Alice Wallenberg foundation.

References

1. Abdulle, A., Barth, A., Schwab, C.: Multilevel Monte Carlo methods for stochastic elliptic multiscale PDEs. Multiscale Model. Simul. **11**(4), 1033–1070 (2013). https://doi.org/10.1137/120894725
2. Abdulle, A., Blumenthal, A.: Stabilized multilevel Monte Carlo method for stiff stochastic differential equations. J. Comput. Phys. **251**, 445–460 (2013). https://doi.org/10.1016/j.jcp.2013.05.039
3. Andersson, A., Kruse, R., Larsson, S.: Duality in refined Sobolev-Malliavin spaces and weak approximation of SPDE. Stoch. Partial Differ. Equ. Anal. Comput. **4**(1), 113–149 (2016)
4. Barth, A., Lang, A.: Milstein approximation for advection-diffusion equations driven by multiplicative noncontinuous martingale noises. Appl. Math. Optim. **66**(3), 387–413 (2012). https://doi.org/10.1007/s00245-012-9176-y
5. Barth, A., Lang, A.: Multilevel Monte Carlo method with applications to stochastic partial differential equations. Int. J. Comput. Math. **89**(18), 2479–2498 (2012). https://doi.org/10.1080/00207160.2012.701735
6. Barth, A., Lang, A., Schwab, C.: Multilevel Monte Carlo method for parabolic stochastic partial differential equations. BIT **53**(1), 3–27 (2013). https://doi.org/10.1007/s10543-012-0401-5
7. Barth, A., Schwab, C., Zollinger, N.: Multi-level Monte Carlo finite element method for elliptic PDEs with stochastic coefficients. Numer. Math. **119**(1), 123–161 (2011). https://doi.org/10.1007/s00211-011-0377-0
8. Brenner, S.C., Scott, L.R.: The Mathematical Theory of Finite Element Methods. Texts in Applied Mathematics, vol. 15, 3rd edn. Springer, New York (2008). https://doi.org/10.1007/978-0-387-75934-0
9. Buckwar, E., Sickenberger, T.: A structural analysis of asymptotic mean-square stability for multi-dimensional linear stochastic differential systems. Appl. Numer. Math. **62**(7), 842–859 (2012). https://doi.org/10.1016/j.apnum.2012.03.002
10. Da Prato, G., Zabczyk, J.: Stochastic Equations in Infinite Dimensions, vol. 152, 2nd edn. Cambridge University Press, Cambridge (2014)
11. Giles, M.B.: Multilevel Monte Carlo path simulation. Oper. Res. **56**(3), 607–617 (2008)
12. Giles, M.B.: Multilevel Monte Carlo methods. Acta Numer. **24**, 259–328 (2015)
13. Graubner, S.: Multi-level Monte Carlo Methoden für stochastische partielle Differentialgleichungen. Master's thesis, TU Darmstadt (2008)
14. Heinrich, S.: Monte Carlo complexity of global solution of integral equations. J. Complex. **14**, 151–175 (1998). https://doi.org/10.1006/jcom.1998.0471

15. Jentzen, A., Kloeden, P.E.: The numerical approximation of stochastic partial differential equations. Milan J. Math. **77**(1), 205–244 (2009)
16. Kovács, M., Larsson, S., Lindgren, F.: Strong convergence of the finite element method with truncated noise for semilinear parabolic stochastic equations with additive noise. Numer. Algorithms **53**(2), 309–320 (2010). https://doi.org/10.1007/s11075-009-9281-4
17. Kruse, R.: Strong and Weak Approximation of Semilinear Stochastic Evolution Equations. Lecture Notes in Mathematics, vol. 2093. Springer, Berlin (2014). https://doi.org/10.1007/978-3-319-02231-4
18. Lang, A.: A note on the importance of weak convergence rates for SPDE approximations in multilevel Monte Carlo schemes. In: R. Cools, D. Nuyens (eds.) Monte Carlo and Quasi-Monte Carlo Methods, MCQMC, Leuven, Belgium, April 2014, Springer Proceedings in Mathematics & Statistics, vol. 163, pp. 489–505 (2016). https://doi.org/10.1007/978-3-319-33507-0_25
19. Lang, A., Petersson, A.: Monte Carlo versus multilevel Monte Carlo in weak error simulations of SPDE approximations. Math. Comput. Simul. (2017). https://doi.org/10.1016/j.matcom.2017.05.002
20. Lang, A., Petersson, A., Thalhammer, A.: Mean-square stability analysis of approximations of stochastic differential equations in infinite dimensions. BIT Numer. Math. **57**(4), 963–990 (2017). https://doi.org/10.1007/s10543-017-0684-7
21. Lord, G.J., Powell, C.E., Shardlow, T.: An Introduction to Computational Stochastic PDEs. Cambridge Texts in Applied Mathematics. Cambridge University Press, Cambridge (2014). https://doi.org/10.1017/CBO9781139017329
22. Lototsky, S.V., Rozovsky, B.L.: Stochastic Partial Differential Equations. Springer International Publishing, Cham (2017)
23. Peszat, S., Zabczyk, J.: Stochastic Partial Differential Equations with Lévy Noise. An Evolution Equation Approach. Encyclopedia of Mathematics and Its Applications, vol. 113. Cambridge University Press, Cambridge (2007)
24. Petersson, A.: Computational aspects of Lévy-driven SPDE approximations. Licentiate Thesis, Chalmers University of Technology, Gothenburg, Sweden (2017). https://research.chalmers.se/en/publication/253417
25. Prévôt, C., Röckner, M.: A Concise Course on Stochastic Partial Differential Equations. Lecture Notes in Mathematics, vol. 1905. Springer, Berlin (2007)
26. Thomée, V.: Galerkin Finite Element Methods for Parabolic Problems. Springer Series in Computational Mathematics, vol. 25, 2nd edn. Springer, Berlin (2006). https://doi.org/10.1007/3-540-33122-0

A Multilevel Monte Carlo Algorithm for Parabolic Advection-Diffusion Problems with Discontinuous Coefficients

Andreas Stein and Andrea Barth

Abstract The Richards' equation is a model for flow of water in unsaturated soils. The coefficients of this (nonlinear) partial differential equation describe the permeability of the medium. Insufficient or uncertain measurements are commonly modeled by random coefficients. For flows in heterogeneous\fractured\porous media, the coefficients are modeled as discontinuous random fields, where the interfaces along the stochastic discontinuities represent transitions in the media. More precisely, the random coefficient is given by the sum of a (continuous) Gaussian random field and a (discontinuous) jump part. In this work moments of the solution to the random partial differential equation are calculated using a path-wise numerical approximation combined with multilevel Monte Carlo sampling. The discontinuities dictate the spatial discretization, which leads to a stochastic grid. Hence, the refinement parameter and problem-dependent constants in the error analysis are random variables and we derive (optimal) a-priori convergence rates in a mean-square sense.

Keywords Multilevel Monte Carlo method · Flow in heterogeneous media · Fractured media · Porous media · Jump-diffusion coefficient · Non-continuous random fields · Parabolic equation · Advection-diffusion equation

1 Introduction

We consider a linear (diffusion-dominated) advection-diffusion equation with random Lévy fields as coefficients. Adopting the term from stochastic analysis, by a Lévy field we mean a random field which is built from a (continuous) Gaussian random field and a (discontinuous) jump part (following a certain jump measure). In the last decade various ways to approximate the distribution or moments of the solution

A. Stein (✉) · A. Barth
SimTech, University of Stuttgart, Allmandring 5b, 70569 Stuttgart, Germany
e-mail: andreas.stein@mathematik.uni-stuttgart.de

A. Barth
e-mail: andrea.barth@mathematik.uni-stuttgart.de

B. Tuffin and P. L'Ecuyer (eds.), *Monte Carlo and Quasi-Monte Carlo Methods*, Springer Proceedings in Mathematics & Statistics 324,
https://doi.org/10.1007/978-3-030-43465-6_22

to a random equation were introduced. Next to classical Monte Carlo methods, their multilevel variants and further variance reduction techniques have been applied. Due to their low regularity constraints, multilevel Monte Carlo techniques have been successfully applied to various problems, for instance in the context of elliptic random PDEs in [1, 3, 5, 8, 16, 22] to just name a few. These sampling approaches differ fundamentally from Polynomial-Chaos-based methods. The latter suffer from high regularity assumptions. While in the case of continuous fields these algorithms can outperform sampling strategies, approaches—like stochastic Galerkin methods—are less promising in our discontinuous setting. In fact, it is even an open problem to define them for Lévy fields. While Richards' equation formulated as a deterministic interface problem was considered in numerous publications (see [10, 13] and the references therein), there is up-to-date no stochastic formulation.

After introducing the necessary basic notation, in this paper we show in Sect. 2 existence and uniqueness of a path-wise weak solution to the random advection-diffusion equation and prove an energy estimate which allows for a moment estimate. Next to space- and time-discretizations, the Lévy field has to be approximated, resulting in an approximated path-wise weak solution. In Sect. 3 we show convergence of this approximated path-wise weak solution, before we introduce a sample-adapted (path-wise) Galerkin approximation. Only if the discretization is adapted to the random discontinuities can we expect full convergence rates. As the main result of this article, we prove the error estimate of the spatial discretization in the L^2-norm. To this end, we utilize the corresponding results with respect to the H^1-norm from [6] and consider the parabolic dual problem. Finally, we combine the sample-adapted spatial discretization with a suitable time stepping method to obtain a fully discrete path-wise scheme. The path-wise approximations are used in Sect. 4 to estimate quantities of interest using a (coupled) multilevel Monte Carlo method. Naturally, the optimal sample numbers on each level depend on the sample-dependent convergence rate. The term *coupled* refers to a simplified version of *Multifidelity Monte Carlo* sampling (see [20]) that reuses samples across levels and is preferred when sampling from a certain distribution is computationally expensive. In Sect. 5, a numerical example confirms our theoretical results from Sect. 3 and shows that the sample-adapted strategy vastly outperforms a multilevel Monte Carlo estimator with a standard Finite Element discretization in space.

2 Parabolic Problems with Random Discontinuous Coefficients

Let $(\Omega, \mathscr{A}, \mathbb{P})$ be a complete probability space, $\mathbb{T} = [0, T]$ be a time interval for some $T > 0$ and $\mathbb{D} \subset \mathbb{R}^d$, $d \in \{1, 2\}$, be a polygonal and convex domain. We consider the linear, random initial-boundary value problem

$$\begin{aligned} \partial_t u(\omega, x, t) + [Lu](\omega, x, t) &= f(\omega, x, t) \quad \text{in } \Omega \times \mathbb{D} \times (0, T], \\ u(\omega, x, 0) &= u_0(\omega, x) \quad \text{in } \Omega \times \mathbb{D} \times \{0\}, \\ u(\omega, x, t) &= 0 \quad \text{on } \Omega \times \partial\mathbb{D} \times \mathbb{T}, \end{aligned} \tag{1}$$

where $f : \Omega \times \mathbb{D} \times \mathbb{T} \to \mathbb{R}$ is a random source function and $u_0 : \Omega \times \mathbb{D}$ denotes the initial condition of the above PDE. Furthermore, L is the second order partial differential operator given by

$$[Lu](\omega, x, t) = -\nabla \cdot (a(\omega, x)\nabla u(\omega, x, t)) + b(\omega, x)\mathbf{1}^T \nabla u(\omega, x, t) \tag{2}$$

for $(\omega, x, t) \in \Omega \times \mathbb{D} \times \mathbb{T}$ with ∇ operating on the second argument of u. In Eq. (2), we set $\mathbf{1} := (1, \dots, 1)^T \in \mathbb{R}^n$, such that $\mathbf{1}^T \nabla u = \sum_{i=1}^n \partial_{x_i} u$, and consider

- a stochastic jump-diffusion coefficient $a : \Omega \times \mathbb{D} \to \mathbb{R}$ and
- a random discontinuous convection term $b : \Omega \times \mathbb{D} \to \mathbb{R}$ coupled to a.

Throughout this article, we denote by C a generic positive constant which may change from one line to the next. Whenever helpful, the dependence of C on certain parameters is made explicit. To obtain a path-wise variational formulation, we use the standard Sobolev space $H^s(\mathbb{D})$ with norm $\|\cdot\|_{H^s(\mathbb{D})}$ for any $s > 0$, see for instance [2, 12]. Since $\mathbb{D}$ has a Lipschitz boundary, for $s \in (1/2, 3/2)$, the existence of a bounded, linear trace operator $\gamma : H^s(\mathbb{D}) \to H^{s-1/2}(\partial\mathbb{D})$ is ensured by the trace theorem, see [11]. We only consider homogeneous Dirichlet boundary conditions on $\partial\mathbb{D}$, hence we may treat γ independently of $\omega \in \Omega$ and define the suitable solution space V as

$$V := H_0^1(\mathbb{D}) = \{v \in H^1(\mathbb{D}) |\ \gamma v \equiv 0\},$$

equipped with the $H^1(\mathbb{D})$-norm $\|v\|_V := \|v\|_{H^1(\mathbb{D})}$. With $H := L^2(\mathbb{D})$, we work on the Gelfand triplet $V \subset H \subset V' = H^{-1}(\mathbb{D})$, where V' denotes the topological dual of V, i.e. the space of all bounded, linear functionals on V. In the variational version of Problem (1), $\partial_t u$ denotes the weak time derivative of u. Throughout this article, we may as well consider $\partial_t u$ as derivative in a strong sense (also with regard to its approximation at the end of Sect. 3) as we will always assume sufficient temporal regularity. As the coefficients a and b are random functions, any solution u to Problem (1) is a time-dependent V-valued random variable. To investigate the regularity of the solution u with respect to $\mathbb{T}$ and the underlying probability measure $\mathbb{P}$ on Ω, we need to introduce the corresponding Lebesgue–Bochner spaces. To this end, let $p \in [1, \infty)$ and$(\mathbb{X}, \|\cdot\|_{\mathbb{X}})$ be an arbitrary Banach space. For $Y \in \{\mathbb{T}, \Omega\}$, the Lebesgue–Bochner space $L^p(Y; \mathbb{X})$ is defined as

$$L^p(Y; \mathbb{X}) := \{\varphi : Y \to \mathbb{X} \text{ is strongly measurable and } \|\varphi\|_{L^p(Y;\mathbb{X})} < +\infty\},$$

with the norm

$$\|\varphi\|_{L^p(Y;\mathbb{X})} := \begin{cases} \left(\int_{\mathbb{T}} \|\varphi(t)\|_{\mathbb{X}}^p dt\right)^{1/p} & \text{for } Y = \mathbb{T}, \\ \mathbb{E}(\|\varphi\|^p)^{1/p} = \left(\int_{\Omega} \|\varphi(\omega)\|_{\mathbb{X}}^p d\mathbb{P}(d\omega)\right)^{1/p} & \text{for } Y = \Omega. \end{cases}$$

The bilinear form associated to L is introduced to derive a weak formulation of the initial-boundary value problem (1). For fixed $\omega \in \Omega$ and $t \in \mathbb{T}$, multiplying Eq. (1) with a test function $v \in V$ and integrating by parts yields

$$_{V'}\langle \partial_t u(\omega, \cdot, t), v\rangle_V + B_\omega(u(\omega, \cdot, t), v) = {}_{V'}\langle f(\omega, \cdot, t), v\rangle_V. \tag{3}$$

The bilinear form $B_\omega : V \times V \to \mathbb{R}$ is given by

$$B_\omega(u, v) = \int_{\mathbb{D}} a(\omega, x)\nabla u(x) \cdot \nabla v(x) + b(\omega, x)\mathbf{1}^T \nabla u(x)v(x)dx,$$

and $_{V'}\langle \cdot, \cdot\rangle_V$ denotes the (V', V)-duality pairing.

Definition 1 For fixed $\omega \in \Omega$, the *path-wise weak solution* to Problem (1) is a function $u(\omega, \cdot, \cdot) \in L^2(\mathbb{T}; V)$ with $\partial_t u(\omega, \cdot, \cdot) \in L^2(\mathbb{T}; V')$ such that, for $t \in \mathbb{T}$,

$$_{V'}\langle \partial_t u(\omega, \cdot, t), v\rangle_V + B_\omega(u(\omega, \cdot, t), v) = {}_{V'}\langle f(\omega, \cdot, t), v\rangle_V, \quad \text{for all } v \in V$$

and $u(\omega, \cdot, 0) = u_0(\omega, \cdot)$. Furthermore, we define the path-wise parabolic norm by

$$\begin{aligned} \|u(\omega, \cdot, \cdot)\|_{*,t} &:= \left(\|u(\omega, \cdot, t)\|_H^2 + \int_0^t \int_{\mathbb{D}} \nabla u(\omega, x, z) \cdot \nabla u(\omega, x, z)dxdz\right)^{1/2} \\ &= \left(\|u(\omega, \cdot, t)\|_H^2 + \|\|\nabla u(\omega, x, z)\|_2\|_{L^2([0,t];H)}^2\right)^{1/2}, \end{aligned} \tag{4}$$

where $\|\cdot\|_2$ is the Euclidean norm on $\mathbb{R}^d$.

To represent the (uncertain) permeability in a subsurface flow model, we use the random jump coefficients a, b from the elliptic/parabolic problems in [5, 6]. The diffusion coefficient is then given by a (spatial) Gaussian random field with additive discontinuities on random areas of $\mathbb{D}$. Its specific structure may be utilized to model the hydraulic conductivity within heterogeneous and/or fractured media and thus a is considered time-independent. The advection term in this model is driven by the same random field and inherits the same discontinuous structure as the diffusion, hence we consider the coefficient b as a linear mapping of a.

Definition 2 The *jump-diffusion coefficient* a is defined as

$$a : \Omega \times \mathbb{D} \to \mathbb{R}_{>0}, \quad (\omega, x) \mapsto \overline{a}(x) + \Phi(W(\omega, x)) + P(\omega, x),$$

where

- $\overline{a} \in C^1(\overline{\mathbb{D}}; \mathbb{R}_{\geq 0})$ is non-negative, continuous, and bounded.
- $\Phi \in C^1(\mathbb{R}; \mathbb{R}_{>0})$ is a continuously differentiable, positive mapping.
- $W \in L^2(\Omega; H)$ is a (zero-mean) Gaussian random field associated to a non-negative, symmetric trace class operator $Q : H \to H$.
- $\mathcal{T} : \Omega \to \mathcal{B}(\mathbb{D})$, $\omega \mapsto \{\mathcal{T}_1, \ldots, \mathcal{T}_\tau\}$ is a random partition of $\mathbb{D}$, i.e. the $\mathcal{T}_i$ are disjoint open subsets of $\mathbb{D}$ such that $|\mathcal{T}_i| > 0$ with $\overline{\mathbb{D}} = \bigcup_{i=1}^{\tau} \overline{\mathcal{T}}_i$, and $\mathcal{B}(\mathbb{D})$ denotes the Borel-σ-algebra on $\mathbb{D}$. The number of elements in $\mathcal{T}$, τ, is a random variable on $(\Omega, \mathcal{A}, \mathbb{P})$, i.e. $\tau : \Omega \to \mathbb{N}$.
- $(P_i, i \in \mathbb{N})$ is a sequence of non-negative random variables on $(\Omega, \mathcal{A}, \mathbb{P})$ and

$$P : \Omega \times \mathbb{D} \to \mathbb{R}_{\geq 0}, \quad (\omega, x) \mapsto \sum_{i=1}^{\tau(\omega)} \mathbf{1}_{\{\mathcal{T}_i\}}(x) P_i(\omega).$$

The sequence $(P_i, i \in \mathbb{N})$ is independent of τ (but not necessarily i.i.d.).

Based on a, the *jump-advection coefficient* b is given for $b_1, b_2 \in L^\infty(\mathbb{D})$ by

$$b : \Omega \times \mathbb{D} \to \mathbb{R}, \quad (\omega, x) \mapsto min(b1(x)a(\omega, x), b_2(x)).$$

The definition of the random partition $\mathcal{T}$ above is rather general and does not yet assume any structure on the discontinuities. A more specific class of random partitions is considered in our numerical experiment in Sect. 5. We assumed in Definition 2 that τ and P_i are independent due to technical reasons, i.e. to control for a possible sampling bias in P_i, see [5, Theorem 3.11]. On a further note, we do not require stochastic independence of W and P. In general, our aim is to estimate moments of a *quantity of interest* (QoI) $\Psi(\omega) := \psi(u(\omega, \cdot, \cdot))$ of the weak solution, where $\psi : L^2(\mathbb{T}; V) \to \mathbb{R}$ is a deterministic functional. To ensure existence and a certain regularity of u, and therefore of Ψ, we fix the following set of assumptions.

Assumption 1 1. Let $\eta_1 \geq \eta_2 \geq \cdots \geq 0$ denote the eigenvalues of Q in descending order and $(e_i, i \in \mathbb{N}) \subset H$ be the corresponding eigenfunctions. The e_i are continuously differentiable on $\mathbb{D}$ and there exist constants $\alpha, \beta, C_e, C_\eta > 0$ such that $2\alpha \leq \beta$ and for any $i \in \mathbb{N}$

$$\|e_i\|_{L^\infty(\mathbb{D})} \leq C_e, \quad \max_{j=1,\ldots,d} \|\partial_{x_j} e_i\|_{L^\infty(\mathbb{D})} \leq C_e i^\alpha \quad \text{and} \quad \sum_{i=1}^{\infty} \eta_i i^\beta \leq C_\eta < +\infty.$$

2. Furthermore, the mapping Φ as in Definition 2 and its derivative are bounded by

$$\phi_1 \exp(\phi_2|w|) \geq \Phi(w) \geq \phi_1 \exp(-\phi_2|w|), \quad |\frac{d}{dx}\Phi(w)| \leq \phi_3 \exp(\phi_4|w|), \quad w \in \mathbb{R},$$

where $\phi_1, \ldots, \phi_4 > 0$ are arbitrary constants.

3. For some $p > 2$, $f, \partial_t f \in L^p(\Omega; L^2(\mathbb{T}; H))$, $u_0 \in L^p(\Omega; H^2(\mathbb{D}) \cap V)$ and u_0 and f are stochastically independent of $\mathcal{T}$.
4. The partition elements $\mathcal{T}_i$ are almost surely polygons with piecewise linear boundary and "$\mathbb{E}(\tau^n) < +\infty$" for all $n \in \mathbb{N}$.
5. The sequence $(P_i, i \in \mathbb{N})$ consists of nonnegative and bounded random variables $P_i \in [0, \overline{P}]$ for some $\overline{P} > 0$.
6. The functional ψ is Lipschitz continuous on $L^2(\mathbb{T}; H)$, i.e. there exists $C_\psi > 0$ such that

$$|\psi(v) - \psi(w)| \leq C_\psi \|v - w\|_{L^2(\mathbb{T};H)} \quad \forall v, u \in L^2(\mathbb{T}; H).$$

Remark 1 The above assumptions are natural and cannot be relaxed significantly to derive the results in Sect. 3. The condition $2\alpha \leq \beta$ implies that W has almost surely Lipschitz continuous paths on $\mathbb{D}$, thus a is piecewise Lipschitz continuous. This is in turn necessary to derive the error estimates of orders $\mathcal{O}(\overline{h}_\ell^\kappa)$ and $\mathcal{O}(\overline{h}_\ell^{2\kappa})$ in Theorems 3 and 4, respectively, for some $\kappa \in (1/2, 1]$ that is independent of W. The parameter $\overline{h}_\ell$ denotes the Finite Element (FE) refinement and κ should only be influenced by the law of the random jump field P. If any of this assumptions were violated, however, κ may depend on other parameters of the random PDE, e.g. W or Ψ. For instance, if $\beta/2\alpha < \kappa \leq 1$, we would only obtain an error of approximate order $\mathcal{O}(\overline{h}_\ell^{\beta/2\alpha})$ in Theorem 3, see [6] for a detailed discussion. The remaining points in Assumption 1 are necessary to ensure that all estimates hold in the mean-square sense, i.e. the second moments of all estimates exist and can be bounded with respect to $\overline{h}_\ell$.

We have the following estimate on a and its piecewise Lipschitz norm.

Lemma 1 ([6, Lemma 3.6 and 4.8]) *Let Assumption 1 hold and define $a_-(\omega) := ess\inf_{x\in\mathbb{D}} a(\omega, x)$ and $a_+(\omega) := ess\sup_{x\in\mathbb{D}} a(\omega, x)$. Then, for any $q \in [1, \infty)$*

$$1/a_-,\ a_+,\ \max_{i=1,\dots,\tau} \sum_{j=1}^{d} \|\partial_{x_j} a\|_{L^\infty(\mathcal{T}_i)} \in L^q(\Omega; \mathbb{R}).$$

Theorem 1 *Under Assumption 1 there exists almost surely a unique path-wise weak solution $u(\omega, \cdot, \cdot) \in L^2(\mathbb{T}; V)$ to Problem* (1) *satisfying the estimate*

$$\sup_{t\in\mathbb{T}} \|u(\omega, \cdot, \cdot)\|_{*,t}^2 \leq C/a_-(\omega)\Big(\|u_0(\omega, \cdot)\|_H^2 + \|f(\omega, \cdot, \cdot)\|_{L^2(T;H)}^2\Big) < +\infty. \quad (5)$$

In addition, for any $r \in [1, p)$ (with p as in Ass. 1), u is bounded in expectation by

$$\mathbb{E}\Big(\sup_{t\in\mathbb{T}} \|u\|_{*,t}^r\Big)^{1/r} \leq C\|1/a_-\|_{L^{\widetilde{q}}(\Omega;\mathbb{R})} \left(\|u_0\|_{L^p(\Omega;H)} + \|f\|_{L^p(\Omega;L^2(\mathbb{T};V'))}\right) < +\infty. \quad (6)$$

with $C = C(r)$ and $\widetilde{q} := (1/r - 1/p)^{-1}$. Furthermore, it holds $\Psi \in L^r(\Omega; \mathbb{R})$.

Proof The estimates in Eqs. (5) and (6) follow from [6, Theorem 3.7]. To show that $\Psi \in L^r(\Omega;\mathbb{R})$, we use Assumption 1 to see that ψ fulfills the linear growth condition $|\psi(v)| \le C(1+\|v\|_{L^2(\mathbb{T};H)})$ for some deterministic constant $C = C(\psi) > 0$ and all $v \in L^2(\mathbb{T};H)$. Hence, we have

$$\mathbb{E}(\Psi^r) \le \mathbb{E}\Big(C^r(1+\|u\|_{L^2(\mathbb{T};V)})^r\Big) \le C^r 2^{r-1}\Big(1+\mathbb{E}\Big(\sup_{t\in\mathbb{T}}\|u\|_{*,t}^r\Big)\Big) < +\infty.$$

□

3 Numerical Approximation of the Solution

In general, the (exact) weak solution u to Problem (1) is out of reach and we have to find tractable approximations of u to apply Monte Carlo algorithms for the estimation of $\mathbb{E}(\Psi)$. A common approach is to use a FE discretization of V combined with a time marching scheme to sample path-wise approximations of u. For this, however, it is necessary to evaluate a and b at certain points in $\mathbb{D}$. This is in general infeasible, since the Gaussian field W usually involves an infinite series and/or the jump heights P_i might not be sampled without bias. The latter issue may arise if P_i has non-standard law, e.g. the *generalized inverse Gaussian* distribution, for more details we refer to [5, 6]. We may circumvent this issue by constructing suitable approximations of a and b, for instance by truncated Karhunen–Loève expansions ([7, 9]), circulant embedding methods ([18, 23]) or Fourier inversion techniques for the sampling of P_i ([4, 5]). Hence, we obtain a modified problem with approximated coefficients which may then be discretized in the spatial and temporal domain. To increase the order of convergence in the spatial discretization, we introduce a FE scheme in the second part of this section where we choose the FE grids adapted with respect to the discontinuities in each sample of a and b. Under mild assumptions on the coefficients we then derive errors on the semi- and fully discrete approximations of u.

3.1 Approximated Diffusion Coefficients

As discussed above, there are several methods available to obtain tractable approximations of the diffusion coefficient a, thus we consider a rather general setting here. For some $\varepsilon > 0$, let $a_\varepsilon : \Omega \times \mathbb{D} \to \mathbb{R}_{>0}$ be an arbitrary approximation of the diffusion coefficient and let (according to Definition 2)

$$b_\varepsilon : \Omega \times \mathbb{D} \to \mathbb{R}, \quad (\omega, x) \mapsto \min(b1(x)a_\varepsilon(\omega,x), b_2(x)),$$

be the canonical approximation of b. Substituting a_ε and b_ε into Problem (1) yields

$$
\begin{aligned}
\partial_t u_\varepsilon(\omega,x,t)+[L_\varepsilon u_\varepsilon](\omega,x,t) &= f(\omega,x,t) \quad \text{in } \Omega\times\mathbb{D}\times(0,T],\\
u_\varepsilon(\omega,x,0) &= u_0(\omega,x) \quad \text{in } \Omega\times\mathbb{D}\times\{0\} \\
u_\varepsilon(\omega,x,t) &= 0 \quad \text{on } \Omega\times\partial\mathbb{D}\times\mathbb{T},
\end{aligned} \tag{7}
$$

where the approximated second order differential operator L_ε is given by

$$[L_\varepsilon u](\omega,x,t) = -\nabla\cdot(a_\varepsilon(\omega,x)\nabla u(\omega,x,t)) + b_\varepsilon(\omega,x)\mathbf{1}^T\nabla u(\omega,x,t).$$

The path-wise variational formulation of Eq. (7) is then (analogous to Eq. (3)) given by: For almost all $\omega\in\Omega$ with given $f(\omega,\cdot,\cdot)$, find $u_\varepsilon(\omega,\cdot,\cdot)\in L^2(\mathbb{T};V)$ with $\partial_t u(\omega,\cdot,\cdot)\in L^2(\mathbb{T};V')$ such that, for $t\in\mathbb{T}$,

$$_{V'}\langle\partial_t u_\varepsilon(\omega,\cdot,t),v\rangle_V + B_{\varepsilon,\omega}(u_\varepsilon(\omega,\cdot,t),v) = F_{\omega,t}(v), \tag{8}$$

holds for all $v\in V$ with respect to the approximated bilinear form

$$B_{\varepsilon,\omega}(v,w) := \int_{\mathbb{D}} a_\varepsilon(\omega,x)\nabla v(x)\cdot\nabla w(x) + b_\varepsilon(\omega,x)\mathbf{1}^T\nabla v(x)w(x)dx, \quad v,w\in V.$$

The following assumption guarantees existence and uniqueness of u_ε and allows us to bound $u-u_\varepsilon$ in a mean-square sense.

Assumption 2 Let Assumption 1 hold and let $a_\varepsilon:\Omega\times\mathbb{D}\to\mathbb{R}_{>0}$ be an approximation of a for some fixed $\varepsilon>0$. Define $a_{\varepsilon,-}(\omega) := ess\inf_{x\in\mathbb{D}} a_\varepsilon(\omega,x)$ and $a_{\varepsilon,+}(\omega) := ess\sup_{x\in\mathbb{D}} a_\varepsilon(\omega,x)$. Assume that for some $s>(1/2-1/p)^{-1}$ and any $q\in[1,\infty)$, there are constants $C_i>0$, for $i=1,\dots,4$, independent of ε, such that

- $\|a-a_\varepsilon\|_{L^s(\Omega;L^\infty(\mathbb{D}))}\le C_1\varepsilon$,
- $\|1/a_{\varepsilon,-}\|_{L^q(\Omega;\mathbb{R})}\le C_2\|1/a_-\|_{L^q(\Omega;\mathbb{R})}<+\infty$,
- $\|a_{\varepsilon,+}\|_{L^q(\Omega;\mathbb{R})}\le C_3\|a_+\|_{L^q(\Omega;\mathbb{R})}<+\infty$ and
- $\|\max\limits_{i=1,\dots,\tau}\sum_{j=1}^d\|\partial_{x_j}a_\varepsilon\|_{L^\infty(\mathcal{T}_i)}\|_{L^q(\Omega;\mathbb{R})}\le C_4\|\max\limits_{i=1,\dots,\tau}\sum_{j=1}^d\|\partial_{x_j}a\|_{L^\infty(\mathcal{T}_i)}\|_{L^q(\Omega;\mathbb{R})}$ $<+\infty$.

At this point we remark that Assumption 2 is natural and essentially states that a_ε has the same regularity as a. Furthermore, the moments of $a-a_\varepsilon$ are controlled by the parameter ε and we may achieve an arbitrary good approximation by choosing ε sufficiently small. This holds for instance (with $C_2=C_3=C_4=1$) if W is approximated by a truncated Karhunen–Loève expansion (see [5, 6]) or if a_ε stems from linear interpolation of discrete sample points of W as we explain in Sect. 5.

Theorem 2 *Let Assumption 2 hold and let u_ε be the weak solution to Problem* (7)*. Then, the root-mean-squared approximation error is bounded by*

$$\mathbb{E}\Big(\sup_{t\in\mathbb{T}}\|u(\cdot,\cdot,t)-u_\varepsilon(\cdot,\cdot,t)\|_{*,t}^2\Big)^{1/2}\le C\varepsilon.$$

Proof By Theorem 1, we have existence of unique solutions u and u_ε to Eqs. (3) resp. (8) almost surely. Thus, we obtain the variational problem: Find $u - u_\varepsilon$ such that

$$ {}_{V'}\langle \partial_t (u(\omega,\cdot,t) - u_\varepsilon(\omega,\cdot,t)), v\rangle_V + B_\omega(u(\omega,\cdot,t) - u_\varepsilon(\omega,\cdot,t), v) = {}_{V'}\langle \widetilde{f}(\omega,\cdot,t), v\rangle_V $$

for all $t \in \mathbb{T}$ and $v \in V$ with initial condition $(u - u_\varepsilon)(\cdot,\cdot,0) \equiv 0$ and right hand side

$$ \widetilde{f}(\omega,\cdot,t) := \nabla \cdot ((a_\varepsilon - a)(\omega,\cdot)\nabla u_\varepsilon(\omega,\cdot,t)) + (b_\varepsilon - b)(\omega,\cdot)\mathbf{1}^T \nabla u_\varepsilon(\omega,\cdot,t) \in V'. $$

By Hölder's inequality it holds

$$ \begin{aligned} \|\widetilde{f}(\omega,\cdot,\cdot)\|_{L^2(\mathbb{T};V')} &\le \|(a - a_\varepsilon)(\omega,\cdot)\|_{L^\infty(\mathbb{D})} \| \|\nabla u(\omega,\cdot,\cdot)\|_2 \|_{L^2(\mathbb{T};H)} \\ &\quad + \|(b - b_\varepsilon)(\omega,\cdot)\|_{L^\infty(\mathbb{D})} \|\mathbf{1}^T \nabla u(\omega,\cdot,\cdot)\|_{L^2(\mathbb{T};H)} \\ &\le C(1 + |b_1|) \|(a - a_\varepsilon)(\omega,\cdot)\|_{L^\infty(\mathbb{D})} \| \|\nabla u(\omega,\cdot,\cdot)\|_2 \|_{L^2(\mathbb{T};H)}, \end{aligned} $$

which yields with Assumption 2 and Theorem 1

$$ \begin{aligned} \|\widetilde{f}(\omega,\cdot,\cdot)\|_{L^{p_1}(\Omega;L^2(\mathbb{T};V'))} &\le C(1 + |b_1|) \|(a - a_\varepsilon)\|_{L^s(\Omega;L^\infty(\mathbb{D}))} \mathbb{E}\Big(\sup_{t\in\mathbb{T}} \|u\|_{*,t}^r \Big)^{1/r} \\ &\le C\varepsilon \end{aligned} $$

for $r \in ((1/2 - 1/s)^{-1}, p)$ and $p_1 := (1/s + 1/r)^{-1} > 2$. We may now use Theorem 1 with $q = (1/2 - 1/p_1)^{-1}$ to estimate $u - u_\varepsilon$ via

$$ \mathbb{E}\Big(\sup_{t\in\mathbb{T}} \|u - u_\varepsilon\|_{*,t}^2 \Big)^{1/2} \le C \|1/a_-\|_{L^q(\Omega;\mathbb{R})} \|\widetilde{f}\|_{L^{p_1}(\Omega;L^2(\mathbb{T};V'))} \le C\varepsilon. $$

□

3.2 *Semi-discretization by Adaptive Finite Elements*

Given a suitable approximation a_ε of the diffusion coefficient, we discretize the (approximate) solution u_ε in the spatial domain. As a first step, we replace the (infinite-dimensional) solution space V by a sequence $\mathbb{V} = (V_\ell, \ell \in \mathbb{N}_0)$ of finite dimensional subspaces $V_\ell \subset V$. In general, V_ℓ are standard FE spaces of piecewise linear functions with respect to some given triangulation K_ℓ of $\mathbb{D}$ and h_ℓ represents the maximum diameter of K_ℓ. As indicated in [5, 6] using standard FE spaces will not yield the full order of convergence with respect to h_ℓ due to the discontinuities in a_ε and b_ε. Thus, we follow the same approach as in [5] for Problem (8) and utilize path-dependent meshes to match the interfaces generated by the jump-diffusion and advection coefficients. As this entails changing varying approximation spaces

V_ℓ with each sample of a_ε resp. b_ε, we have to formulate a semi-discrete version of problem (8) with respect to $\omega \in \Omega$:

Given a fixed $\omega \in \Omega$ and $\ell \in \mathbb{N}_0$, we consider a (stochastic) finite dimensional subspace $V_\ell(\omega) \subset V$ with sample-dependent basis $\{v_1(\omega), \dots, v_{d_\ell}(\omega)\} \subset V$ and stochastic dimension $d_\ell = d_\ell(\omega) \in \mathbb{N}$. For a given random partition $\mathscr{T}(\omega) = (\mathscr{T}_i, i = 1 \dots, \tau(\omega))$ of polygons on $\mathbb{D}$, we choose a conforming triangulation $\mathrm{K}_\ell(\omega)$ such that

$$\mathscr{T}(\omega) \subset \mathrm{K}_\ell(\omega) \;\text{ and }\; h_\ell(\omega) := \max_{K \in \mathrm{K}_\ell(\omega)} \operatorname{diam}(K) \leq \overline{h}_\ell \;\text{ for } \ell \in \mathbb{N}_0,$$

holds almost surely. The inclusion $\mathscr{T}(\omega) \subset \mathrm{K}_\ell(\omega)$ states that the triangles in $\mathrm{K}_\ell(\omega)$ are chosen to match and fully cover the polygonal partition elements in $\mathscr{T}(\omega)$. Furthermore, $(\overline{h}_\ell, \ell \in \mathbb{N}_0)$ is a sequence of positive, deterministic refinement thresholds, decreasing monotonically to zero. This guarantees that $h_\ell(\omega) \to 0$ for $\ell \to \infty$ almost surely, although the absolute speed of convergence varies for each ω. We assume shape-regularity of the triangulation uniform in Ω, i.e. there exist a $\vartheta \in (0,1)$ such that

$$0 < \vartheta \leq \sup_{\ell \in \mathbb{N}_0} \sup_{K \in K_\ell(\omega)} \frac{\operatorname{diam}(K)}{\iota_K} \leq \vartheta^{-1} < +\infty \quad \text{almost surely.}$$

In Eq. (3.2), ι_T denotes the diameter of the inscribed circle of the triangle K. For given $\{v_1(\omega), \dots, v_{d_\ell}(\omega)\}$, the semi-discrete version of the variational formulation (8) is then to find $u_{\varepsilon,\ell}(\omega, \cdot, t) \in V_\ell(\omega)$ such that for $t \in \mathbb{T}$ and $v_\ell(\omega) \in V_\ell(\omega)$

$$\begin{aligned} {}_{V'}\langle \partial_t u_{\varepsilon,\ell}(\omega,\cdot,t), v_\ell(\omega)\rangle_V + B_{\varepsilon,\omega}(u_{\varepsilon,\ell}(\omega,\cdot,t), v_\ell(\omega)) &= {}_{V'}\langle f(\omega,\cdot,t), v_\ell(\omega)\rangle_V, \\ u_{\varepsilon,\ell}(\omega,\cdot,0) &= u_{0,\ell}(\omega,\cdot), \end{aligned} \tag{9}$$

where $u_{0,\ell}(\omega,\cdot) \in V_\ell(\omega)$ is a suitable approximation of $u_0(\omega,\cdot)$, for instance the nodal interpolation of u_0 in $V_\ell(\omega)$. The function $u_{\varepsilon,\ell}(\omega,\cdot,t)$ may be expanded as

$$u_{\varepsilon,\ell}(\omega,\cdot,t) = \sum_{j=1}^{d_\ell(\omega)} c_j(\omega,t) v_j(\omega),$$

where the coefficients $c_1(\omega,t), \dots, c_{d_\ell}(\omega,t) \in \mathbb{R}$ depend on $(\omega,t) \in \Omega \times \mathbb{T}$ and the respective coefficient (column-)vector is $\mathbf{c}(\omega,\mathbf{t}) := (c_1(\omega,t), \dots, c_{d_\ell}(\omega,t))^T$. With this, the semi-discrete variational problem in the (stochastic) finite dimensional space $V_\ell(\omega)$ is equivalent to solving the system of ordinary differential equations

$$\frac{d}{dt}\mathbf{c}(\omega,t) + \mathbf{A}(\omega)\mathbf{c}(\omega,\mathbf{t}) = \mathbf{F}(\omega,t), \quad \mathbf{t} \in \mathbb{T} \tag{10}$$

for $\mathbf{c}$ with stochastic stiffness matrix $(\mathbf{A}(\omega))_{jk} = B_{\varepsilon,\omega}(v_j(\omega), v_k(\omega))$ and time-dependent load vector $(\mathbf{F}(\omega,t))_j = {}_{V'}\langle f(\omega,\cdot,t), v_j(\omega)\rangle_V$ for $j,k \in \{1,\dots,d_\ell(\omega)\}$. The following result gives an error estimate in the energy norm for $u_\varepsilon - u_{\varepsilon,\ell}$.

Theorem 3 ([6, Theorem 4.7]) *Let Assumption 2 hold such that for some* $\kappa \in (1/2, 1]$ *it holds that* $\mathbb{E}(\max_{i=1,\dots,\tau} \|u\|^2_{H^{1+\kappa}(\mathcal{T}_i)}) < +\infty$. *Let* $u_{\varepsilon,\ell}$ *be the semi-discrete sample-adapted approximation of* u_ε *as in Eq.* (9) *and let* $\|(u_0 - u_{\ell,0})(\omega, \cdot)\|_H \leq C\|u_0(\omega, \cdot)\|_V \overline{h}_\ell$ *almost surely for all* $\ell \in \mathbb{N}_0$. *Then, there holds almost surely the path-wise estimate*

$$\sup_{t\in\mathbb{T}} \|(u_\varepsilon - u_{\varepsilon,\ell})(\omega, \cdot, \cdot)\|_{*,t} \leq C/(a_{\varepsilon,-}(\omega))^{1/2}\Big(\|f(\omega, \cdot, \cdot)\|_{L^2(\mathbb{T};H)} + \|u_0(\omega, \cdot)\|_V\Big)\overline{h}_\ell^{\kappa}$$

and, for any $r \in [1, p)$ *(with* p *as in Assumption 1), the expected parabolic estimate*

$$\mathbb{E}(\sup_{t\in\mathbb{T}} \|u_\varepsilon - u_{\varepsilon,\ell}\|^r_{*,t})^{1/r} \leq C(\|f\|_{L^p(\Omega;L^2(\mathbb{T};H))} + \|u_0\|_{L^p(\Omega;V)})\overline{h}_\ell^{\kappa}.$$

The above statement gives a bound on the error in the $L^2(\mathbb{T}; V)$-norm. The functional Ψ however is defined on $L^2(\mathbb{T}; H)$, thus it is favorable to derive an error bound with respect to the weaker $L^2(\mathbb{T}; H)$-norm.

Theorem 4 *Let Assumption 2 hold such that for some* $\kappa \in (1/2, 1]$ *there holds* $\mathbb{E}(\max_{i=1,\dots,\tau} \|u\|^2_{H^{1+\kappa}(\mathcal{T}_i)}) < +\infty$ *and let* $\|(u_0 - u_{\ell,0})(\omega, \cdot)\|_H \leq C\|u_0(\omega, \cdot)\|_{H^2(\mathbb{D})}\overline{h}_\ell^2$ *almost surely. Then,*

$$\mathbb{E}(\|u_\varepsilon - u_{\ell,\varepsilon}\|^2_{L^2(\mathbb{T};H)})^{1/2} \leq C\overline{h}_\ell^{2\kappa}.$$

Proof For fixed ω, we consider the path-wise parabolic dual problem to find $w(\omega, \cdot, \cdot) \in L^2(\mathbb{T}; V)$ with $\partial_t w(\omega, \cdot, \cdot) \in L^2(\mathbb{T}; V')$ such that, for $t \in \mathbb{T}$,

$$_{V'}\langle \partial_t w(\omega, \cdot, t), v\rangle_V + B_{\varepsilon,\omega}(w(\omega, \cdot, t), v) = {}_{V'}\langle g(\omega, \cdot, t), v\rangle_V, \quad \text{for all } v \in V, \tag{11}$$

where $w(\omega, \cdot, 0) = w_0(\omega, \cdot) := 0$ and $g(\omega, \cdot, t) := (u_\varepsilon - u_{\varepsilon,\ell})(\omega, \cdot, T - t) \in V$ almost surely for any $t \in \mathbb{T}$ by Theorem 1. Hence, we may test against $v = g(\omega, \cdot, t)$ in Eq. (11) to obtain

$$\|g(\omega, \cdot, t)\|^2_H = {}_{V'}\langle \partial_t w(\omega, \cdot, t), g(\omega, \cdot, t)\rangle_V + B_{\varepsilon,\omega}(w(\omega, \cdot, t), g(\omega, \cdot, t)). \tag{12}$$

Furthermore, for any $v_\ell(\omega) \in V_\ell(\omega)$ it holds by Eqs. (8), (9)

$$_{V'}\langle \partial_t (u_\varepsilon - u_{\varepsilon,\ell})(\omega, \cdot, t), v_\ell(\omega)\rangle_V = -B_{\varepsilon,\omega}((u_\varepsilon - u_{\varepsilon,\ell})(\omega, \cdot, t), v_\ell(\omega)) \tag{13}$$

and thus

$$B_{\varepsilon,\omega}(g(\omega,\cdot,t), w(\omega,\cdot,t)) = {}_{V'}\langle \partial_t g(\omega,\cdot,t), v_\ell(\omega) - w(\omega,\cdot,t) + w(\omega,\cdot,t)\rangle_V \\ + B_{\varepsilon,\omega}(g(\omega,\cdot,t), w(\omega,\cdot,t) - v_\ell(\omega)), \tag{14}$$

where we have used the that $\partial_t g(\omega,\cdot,t) = -(\partial_t u_\varepsilon - \partial_t u_{\varepsilon,\ell})(\omega,\cdot,T-t)$ by the chain rule. Substituting Eq. (14) in Eq. (12) and integrating over $\mathbb{T}$ yields

$$\begin{aligned}\|g(\omega,\cdot,\cdot)\|^2_{L^2(\mathbb{T};H)} &= \int_0^T {}_{V'}\langle \partial_t w(\omega,\cdot,t), g(\omega,\cdot,t)\rangle_V + {}_{V'}\langle \partial_t g(\omega,\cdot,t), w(\omega,\cdot,t)\rangle_V dt \\ &+ \int_0^T {}_{V'}\langle \partial_t g(\omega,\cdot,t), v_\ell(\omega) - w(\omega,\cdot,t)\rangle_V dt \\ &+ \int_0^T B_{\varepsilon,\omega}(g(\omega,\cdot,t), w(\omega,\cdot,t) - v_\ell(\omega))dt \\ &=: I + II + III.\end{aligned}$$

Integration by parts and the path-wise estimate in Theorem 1 yield for I

$$\begin{aligned}I &= (w(\omega,\cdot,T), g(\omega,\cdot,T))_H - (w_0(\omega,\cdot), g(\omega,\cdot,0))_H \\ &\le \|w(\omega,\cdot,T)\|_H \|u_0(\omega,\cdot) - u_{0,\ell}(\omega,\cdot)\|_H \\ &\le C\frac{1}{a_{\varepsilon,-}(\omega)}\|g(\omega,\cdot,\cdot)\|_{L^2(\mathbb{T};H)}\|u_0(\omega,\cdot)\|_{H^2(\mathbb{D})}\overline{h}_\ell^2,\end{aligned}$$

where we have used $\|(u_0 - u_{\ell,0})(\omega,\cdot)\|_H \le C\|u_0(\omega,\cdot)\|_{H^2(\mathbb{D})}\overline{h}_\ell^2$ in the last step. To bound the second term, we choose $v_\ell = v_\ell(\omega,\cdot,t)$ to be the semi-discrete FE approximation of $w(\omega,\cdot,t)$ in $V_\ell(\omega)$. Since $w_0 \equiv 0$, there is no approximation error in the initial condition and with the path-wise estimate from Theorem 3 it follows that

$$\begin{aligned}II &\le \|\partial_t g(\omega,\cdot,\cdot)\|_{L^2(\mathbb{T};V')}\|v_\ell(\omega,\cdot,\cdot) - w(\omega,\cdot,\cdot)\|_{L^2(\mathbb{T};V)} \\ &\le C\frac{1}{(a_{\varepsilon,-}(\omega))^{1/2}}\|\partial_t g(\omega,\cdot,\cdot)\|_{L^2(\mathbb{T};V')}\|g(\omega,\cdot,\cdot)\|_{L^2(\mathbb{T};H)}\overline{h}_\ell^\kappa.\end{aligned}$$

From Eq. (13) and Theorem 3 we also see that

$$\|\partial_t g(\omega,\cdot,\cdot)\|_{L^2(\mathbb{T};V')} \le C\frac{a_{\varepsilon,+}(\omega)}{(a_{\varepsilon,-}(\omega))^{1/2}}\Big(\|f(\omega,\cdot,\cdot)\|_{L^2(\mathbb{T};H)} + \|u_0(\omega,\cdot)\|_V\Big)\overline{h}_\ell^\kappa$$

and thus

$$II \le C\frac{a_{\varepsilon,+}(\omega)}{a_{\varepsilon,-}(\omega)}\Big(\|f(\omega,\cdot,\cdot)\|_{L^2(\mathbb{T};H)} + \|u_0(\omega,\cdot)\|_V\Big)\|g(\omega,\cdot,\cdot)\|_{L^2(\mathbb{T};H)}\overline{h}_\ell^{2\kappa}.$$

Similarly, we bound the last term again with Theorem 3 via

$$
\begin{aligned}
III &\leq C a_{\varepsilon,+}(\omega)\|g(\omega,\cdot,\cdot)\|_{L^2(\mathbb{T};V)}\|v_\ell(\omega,\cdot,\cdot)-w(\omega,\cdot,\cdot)\|_{L^2(\mathbb{T};V)}\\
&\leq C\frac{a_{\varepsilon,+}(\omega)}{a_{\varepsilon,-}(\omega)}\Big(\|f(\omega,\cdot,\cdot)\|_{L^2(\mathbb{T};H)}+\|(u_0(\omega,\cdot)\|_V\Big)\|g(\omega,\cdot,\cdot)\|_{L^2(\mathbb{T};H)}\overline{h}_\ell^{2\kappa}.
\end{aligned}
$$

The estimates on $I-III$ now show that

$$
\|g(\omega,\cdot,\cdot)\|_{L^2(\mathbb{T};H)} \leq C\frac{a_{\varepsilon,+}(\omega)}{a_{\varepsilon,-}(\omega)}\Big(\|f(\omega,\cdot,\cdot)\|_{L^2(\mathbb{T};H)}+\|(u_0(\omega,\cdot)\|_{H^2(\mathbb{D})}\Big)\overline{h}_\ell^{2\kappa}.
$$

and the claim follows by Assumption 2 and Hölder's inequality. □

Remark 2 We remark that the additional condition on the initial data approximation in Theorem 4 is fulfilled if u_0 has almost surely continuous paths and $u_{\ell,0}$ is chosen as the path-wise nodal interpolation with respect to the sample-adapted FE basis.

3.3 *Fully Discrete Pathwise Approximation*

For a fully discrete formulation of Problem (9), we consider a time grid $0=t_0<t_1<\cdots<t_n=T$ in $\mathbb{T}$ for some $n\in\mathbb{N}$ and assume the grid is equidistant with fixed time step $\Delta t := t_i - t_{i-1} > 0$. The temporal derivative at t_i is approximated by the backward difference

$$
\partial_t u_{\varepsilon,\ell}(\omega,\cdot,t_i) = (u_{\varepsilon,\ell}(\omega,\cdot,t_i)-u_{\varepsilon,\ell}(\omega,\cdot,t_{i-1}))/\Delta t, \quad i=1,\ldots,n.
$$

We emphasize again that in our model problem the weak and strong temporal derivative of $u_{\varepsilon,\ell}$ coincide due to the temporal regularity of the solution. Hence, the backward difference as an approximation scheme in a strong sense is justified. This yields the fully discrete problem to find $(u_{\varepsilon,\ell}^{(i)}(\omega,\cdot), i=0,\ldots,n)\subset V_\ell(\omega)$ such that for all $v_\ell(\omega)\in V_\ell(\omega)$ and $i=1,\ldots,n$

$$
\begin{aligned}
\frac{((u_{\varepsilon,\ell}^{(i)}-u_{\varepsilon,\ell}^{(i-1)})(\omega,\cdot),v_\ell(\omega))_H}{\Delta t}+B_{\varepsilon,\omega}(u_{\varepsilon,\ell}^{(i)}(\omega,\cdot),v_\ell(\omega)) &= {}_{V'}\langle f(\omega,\cdot,t_i),v_\ell(\omega)\rangle_V,\\
u_{\varepsilon,\ell}^{(0)}(\omega,\cdot) &= u_{0,\ell}(\omega,\cdot).
\end{aligned}
$$

The fully discrete solution is given by

$$
u_{\varepsilon,\ell}^{(i)}(\omega,\cdot) = \sum_{j=1}^{d_\ell(\omega)} c_{i,j}(\omega)v_j(\omega), \quad i=1,\ldots,n,
$$

where the coefficient vector $\mathbf{c_i}(\omega) = (c_{i,1}(\omega), \ldots, c_{i,d_\ell}(\omega))$ solves the linear system of equations

$$(\mathbf{M} + \Delta t\mathbf{A}(\omega))\mathbf{c_i}(\omega) = \Delta t\mathbf{F}(\omega, t_i) + \mathbf{M}\mathbf{c_{i-1}}(\omega)$$

in every discrete point in time t_i, and $\mathbf{A}$ and $\mathbf{F}$ are as in Eq. (10). The mass matrix is given by $(\mathbf{M})_{jk} := (v_j(\omega), v_k(\omega))_H$ and $\mathbf{c_0}$ consists of the basis coefficients of $u_{0,\ell} \in V_\ell(\omega)$ with respect to $\{v_1(\omega), \ldots, v_{d_\ell}(\omega)\}$. We extend the discrete solution to the whole temporal domain by the linear interpolation

$$\overline{u}_{\varepsilon,\ell}(\cdot,\cdot,t) := (u_{\varepsilon,\ell}^{(i)} - u_{\varepsilon,\ell}^{(i-1)})\frac{(t-t_{i-1})}{\Delta t} + u_{\varepsilon,\ell}^{(i-1)}, \quad t \in [t_{i-1}, t_i], \quad i = 1, \ldots, n.$$

Theorem 5 ([6, Theorem 4.12]) *Let Assumption 2 hold, let* $(u_{\varepsilon,\ell}^{(i)}, i = 0, \ldots, n)$ *be the fully discrete sample-adapted approximation of* $u_{N,\varepsilon}$, *and let* $\overline{u}_{\varepsilon,\ell}$ *be the linear interpolation of* $(u_{\varepsilon,\ell}^{(i)}, i = 0, \ldots, n)$ *in* $\mathbb{T}$. *Then, for* $C > 0$ *independent of* ε, h_ℓ *and* Δt, *it holds*

$$\mathbb{E}(\sup_{t\in\mathbb{T}} \|u_{\varepsilon,\ell} - \overline{u}_{\varepsilon,\ell}\|_{*,t}^2)^{1/2} \le C\Delta t.$$

The final corollary on the overall approximation error is now an immediate consequence of Theorems 2, 4 and 5 and the Lipschitz condition on ψ.

Corollary 1 *Let Assumption 2 hold such that for some* $\kappa \in (1/2, 1]$ *there holds* $\mathbb{E}(\max_{i=1,\ldots,\tau} \|u\|_{H^{1+\kappa}(\mathcal{T}_i)}^2) < \infty$ *and let* $\|(u_0 - u_{\ell,0})(\omega,\cdot)\|_H \le C\|u_0(\omega,\cdot)\|_{H^2(\mathbb{D})}\overline{h}_\ell^2$ *almost surely. The (fully) approximated QoI is defined by* $\Psi_{\varepsilon,\ell,\Delta t} := \psi(\overline{u}_{\varepsilon,\ell})$. *Then, there holds the error bound*

$$\mathbb{E}(|\Psi - \Psi_{\varepsilon,\ell,\Delta t}|^2)^{1/2} \le C(\varepsilon + \overline{h}_\ell^{2\kappa} + \Delta t).$$

Given a sequence of discretization thresholds $\overline{h}_\ell > 0$ for $\ell \in \mathbb{N}_0$, one should adjust ε and Δt such that $\overline{h}_\ell^{2\kappa} \simeq \varepsilon \simeq \Delta t$ to achieve an error equilibrium. Hence, we denote the adjusted parameters on level ℓ by ε_ℓ and Δt_ℓ and assume that all errors are equilibrated in the sense that $c\overline{h}_\ell^{2\kappa} \le \varepsilon_\ell, \Delta t_\ell \le C\overline{h}_\ell^{2\kappa}$ holds for constants $c, C > 0$ independent of ℓ. We further define $\Psi_\ell := \Psi_{\varepsilon_\ell,\ell,\Delta t_\ell} = \psi(\overline{u}_{\varepsilon_\ell,\ell})$ and obtain with Corollary 1

$$\mathbb{E}((\Psi - \Psi_\ell)^2)^{1/2} \le C\overline{h}_\ell^{2\kappa}. \tag{15}$$

4 Estimation of Moments by Multilevel Monte Carlo Methods

As we are able to generate samples from $\Psi_\ell = \psi(\overline{u}_{\varepsilon_\ell,\ell})$ and control for the discretization error in each sample, we may estimate the expectation $\mathbb{E}(\Psi)$ by Monte Carlo methods. For convenience, we restrict ourselves to the estimation of $\mathbb{E}(\Psi)$, but we note that all results from this section are valid when estimating higher moments of Ψ, given that $u \in L^r(\Omega; L^2(\mathbb{T}; V))$ for sufficiently high r (cf. Theorem 1). Our focus is on multilevel Monte Carlo (MLMC) estimators, since they are easily implemented, do not require much regularity of Ψ and are significantly more efficient than standard Monte Carlo estimators. The main idea of the MLMC estimation has been developed in [21] and later been rediscovered and popularized in [14]. In this section, we briefly recall the MLMC method and then show how we achieve a desired error rate by adjusting the number of samples on each level to the discretization bias. We also suggest a modification of the MLMC algorithm to increase computational efficiency before we verify our results in Sect. 5.

Let $L \in \mathbb{N}$ be a fixed (maximum) discretization level and assume that the approximation parameters on each level $\ell = 0, \ldots, L$ satisfy $\overline{h}_\ell^{2\kappa} \simeq \varepsilon_\ell \simeq \Delta t_\ell$ (see Sect. 3). This yields a sequence $\Psi_0, \ldots, \Psi_L$ of approximated QoIs, hence the *MLMC estimator* of $\mathbb{E}(\Psi_L)$ is given by

$$E^L(\Psi_L) = \sum_{\ell=0}^{L} \frac{1}{M_\ell} \sum_{i=1}^{M_\ell} \Psi_\ell^{(i,\ell)} - \Psi_{\ell-1}^{(i,\ell)}, \tag{16}$$

where we have set $\Psi_{-1} := 0$. Above, $(\Psi_\ell^{(i,\ell)} - \Psi_{\ell-1}^{(i,\ell)}, i \in \mathbb{N})$ is a sequence of independent copies of $\Psi_\ell - \Psi_{\ell-1}$ and $M_\ell \in \mathbb{N}$ denotes the number of samples on each level. To achieve a desired target root mean-squared error (RMSE), this estimator requires less computational effort than the standard Monte Carlo approach under certain assumptions. This, by now, classical result was proven in [14, Theorem 3.1] for functionals of stochastic differential equations. The proof is rather general and may readily be transferred to other applications, for instance the estimation of functionals or moments of random PDEs, see [3, 15].

Theorem 6 *Let Assumption 2 hold such that for some* $\kappa \in (1/2, 1]$ *there holds* $\mathbb{E}(\max_{i=1,\ldots,\mathcal{T}} \|u\|^2_{H^{1+\kappa}(\mathcal{T}_i)}) < +\infty$ *and let* $\overline{h}_{\ell-1} \le C_1 \overline{h}_\ell$ *for some* $C_1 > 0$ *for all* $\ell \in \mathbb{N}_0$. *For* $L \in \mathbb{N}$ *and given refinement parameters* $h_0 > \cdots > \overline{h}_L > 0$ *choose* $\Delta t_\ell, \varepsilon_\ell > 0$ *such that* $\varepsilon_\ell, \Delta t_\ell \le C_2 \overline{h}_\ell^{2\kappa}$ *holds for fixed* $C_2 > 0$ *and* $\ell = 0, \ldots, L$. *Furthermore, let* $(\rho_\ell, \ell = 1, \ldots, L) \in (0, 1)^L$ *be a set of positive weights such that* $\sum_{\ell=1}^L \rho_\ell = C_\rho$, *with a constant* $C_\rho > 0$ *independent of L, and set*

$$M_0^{-1} := \left\lceil \overline{h}_L^{4\kappa} \right\rceil \quad \textit{and} \quad M_\ell^{-1} := \left\lceil \frac{\overline{h}_L^{4\kappa}}{\overline{h}_\ell^{4\kappa}} \rho_\ell^{-2} \right\rceil \quad \textit{for } \ell = 1, \ldots, L.$$

Then, there is a $C > 0$, *independent of L and* κ, *such that*

$$\|\mathbb{E}(\Psi) - E^L(\Psi_L)\|_{L^2(\Omega;\mathbb{R})} \le C \overline{h}_L^{2\kappa}.$$

Proof As all error contributions ε_ℓ, Δt_ℓ are adjusted to $\overline{h}_\ell$, we obtain by the triangle inequality and Eq. (15)

$$\begin{aligned}\|\mathbb{E}(\Psi) - E^L(\Psi_L)\|_{L^2(\Omega;\mathbb{R})} &\leq \|\mathbb{E}(\Psi) - \mathbb{E}(\Psi_L)\|_{L^2(\Omega;\mathbb{R})} + \|\mathbb{E}(\Psi_L) - E^L(\Psi_L)\|_{L^2(\Omega;\mathbb{R})} \\ &\leq \|\Psi - \Psi_L\|_{L^2(\Omega;\mathbb{R})} \\ &\quad + \|\sum_{\ell=0}^{L} \mathbb{E}(\Psi_\ell - \Psi_{\ell-1}) - \frac{1}{M_\ell}\sum_{i=1}^{M_\ell}(\Psi_\ell^{(i,\ell)} - \Psi_{\ell-1}^{(i,\ell)})\|_{L^2(\Omega;\mathbb{R})} \\ &\leq C\overline{h}_L^{2\kappa} + \sum_{\ell=0}^{L} \frac{1}{\sqrt{M_\ell}} \|\Psi_\ell - \Psi_{\ell-1}\|_{L^2(\Omega;\mathbb{R})}.\end{aligned}$$

At this point we emphasize that we did not use the independence of $\Psi_\ell^{(i,\ell)} - \Psi_{\ell-1}^{(i,\ell)}$ across the levels $\ell = 1, \ldots, L$ in the last inequality. We note that

$$\|\Psi_\ell - \Psi_{\ell-1}\|_{L^2(\Omega;\mathbb{R})} \leq \|\Psi - \Psi_\ell\|_{L^2(\Omega;\mathbb{R})} + \|\Psi - \Psi_{\ell-1}\|_{L^2(\Omega;\mathbb{R})} \leq C(1 + C_1)\overline{h}_\ell^{2\kappa}$$

for $\ell \geq 1$ and hence

$$\|\mathbb{E}(\Psi) - E^L(\Psi_L)\|_{L^2(\Omega;\mathbb{R})} \leq C\overline{h}_L^{2\kappa} + \|\Psi_0\|_{L^2(\Omega;\mathbb{R})}\overline{h}_L^{2\kappa} + C(1 + C_1)\overline{h}_L^{2\kappa} \sum_{\ell=1}^{L} \rho_\ell \leq C\overline{h}_L^{2\kappa}.$$

□

We remark that $C_\rho > 0$ may act as a normalizing constant if MLMC estimators based on different discretization techniques are compared, an example is provided in Sect. 5. To conclude this section, we briefly present a modified MLMC method to accelerate the estimation of $\mathbb{E}(\Psi_L)$. In the definition of the MLMC estimator from Eq. (16), the terms in the second sum are independent copies of the corrections $\Psi_\ell - \Psi_{\ell-1}$. Hence, one has to generate a total of $M_\ell + M_{\ell+1}$ samples of Ψ_ℓ for each $\ell = 0, \ldots, L$ (where we have set $M_{L+1} := 0$). This effort may be reduced if we "recycle" the already available samples and generate the differences $\Psi_\ell^{(i,\ell)} - \Psi_{\ell-1}^{(i,\ell)}$ and $\Psi_{\ell+1}^{(i,\ell)} - \Psi_\ell^{(i,\ell)}$ based on the same realization $\Psi_\ell^{(i,\ell)}$. That is, we drop the second superscript ℓ above and arrive at the *coupled MLMC estimator*

$$E_C^L(\Psi_L) := \sum_{\ell=0}^{L} \frac{1}{M_\ell} \sum_{i=1}^{M_\ell} \Psi_\ell^{(i)} - \Psi_{\ell-1}^{(i)}. \tag{17}$$

Instead of $M_\ell + M_{\ell+1}$ realizations of Ψ_ℓ, the coupled MLMC estimator requires only M_ℓ samples of Ψ_ℓ. The copies $\Psi_\ell^{(i)}$ are still independent in i, but not anymore across all levels ℓ for a fixed index i. Clearly, $\mathbb{E}(E_C^L(\Psi_L)) = \mathbb{E}(\Psi_L)$, and it holds

$$\lim_{L\to+\infty} \mathbb{E}(E_C^L(\Psi_L)) = \lim_{L\to+\infty} \mathbb{E}(E^L(\Psi_L)) = \lim_{L\to+\infty} \mathbb{E}(\Psi_L) = \mathbb{E}(u).$$

The introduced modification is a simplified version of the *Multifidelity Monte Carlo estimator* (see [20]), where the weighting coefficients for all level corrections $\Psi_\ell - \Psi_{\ell-1}$ are set equal to one. An estimator similar to (17) with coupled correction terms has also been introduced in the context of SDEs in [24]. As we mentioned in the proof of Theorem 6, independence of the sampled differences $\Psi_\ell - \Psi_{\ell-1}$ across ℓ is not required for the error estimate, thus, the asymptotic order of convergence also holds for the coupled estimator. To compare RMSEs of the estimators from Eq. (16) and (17), we calculate

$$\begin{aligned}
\mathrm{Var}(E_C^L(\Psi_L)) &= \mathrm{Var}\left(\sum_{\ell=0}^{L}\sum_{i=M_{\ell+1}+1}^{M_\ell}\sum_{k=0}^{\ell}\frac{\Psi_k^{(i)}-\Psi_{k-1}^{(i)}}{M_k}\right)\\
&=\sum_{\ell=0}^{L}(M_\ell - M_{\ell+1})\,\mathrm{Var}\left(\sum_{k=0}^{\ell}\frac{\Psi_k-\Psi_{k-1}}{M_k}\right)\\
&=\sum_{\ell=0}^{L}(M_\ell - M_{\ell+1})\left(\sum_{k=0}^{\ell}\frac{\mathbb{V}_k}{M_k^2}+2\sum_{k=0}^{\ell}\sum_{j=0}^{k-1}\frac{\mathbb{C}_{j,k}}{M_jM_k}\right)\\
&=\sum_{k=0}^{L}\left(\frac{\mathbb{V}_k}{M_k^2}+2\sum_{j=0}^{k-1}\frac{\mathbb{C}_{j,k}}{M_jM_k}\right)\sum_{\ell=k}^{L}(M_\ell - M_{\ell+1})\\
&=\mathrm{Var}(E^L(\Psi_L))+2\sum_{k=0}^{L}\sum_{j=0}^{k-1}\frac{\mathbb{C}_{j,k}}{M_j},
\end{aligned}$$

where $\mathbb{V}_k := \mathrm{Var}(\Psi_k - \Psi_{k-1})$ and $\mathbb{C}_{j,k} := \mathrm{Cov}(\Psi_j - \Psi_{j-1}, \Psi_k - \Psi_{k-1})$. Hence, the coupled estimator introduces a higher RMSE if the corrections $\Psi_\ell - \Psi_{\ell-1}$ are positively correlated across the levels. In this case, we trade in variance for simulation time and the ratio of this trade-off is problem-dependent and hard to assess in advance.

5 Numerical Results

For our numerical experiment we consider $\mathbb{D} = (0,1)^2$ with $T = 1$, initial data $u_0(x_1,x_2) = \frac{1}{10}\sin(\pi x_1)\sin(\pi x_2)$, source term $f \equiv 1$ and set $\bar{a} \equiv 0$. The covariance operator Q of W is given by the *Matérn covariance function*

$$[Q\varphi](y) := \int_{\mathbb{D}} \sigma^2\frac{2^{1-\nu}}{\Gamma(\nu)}\Big(\sqrt{2\nu}\frac{\|x-y\|_2}{\chi}\Big)^{\nu} K_\nu\Big(\sqrt{2\nu}\frac{\|x-y\|_2}{\chi}\Big)\varphi(x)dx, \quad \varphi \in H,$$

with smoothness parameter $\nu > 0$, variance $\sigma^2 > 0$ and correlation length $\chi > 0$. Above, Γ denotes the Gamma function, $\|\cdot\|_2$ is the Euclidean norm in $\mathbb{R}^2$ and K_ν is the modified Bessel function of the second kind with ν degrees of freedom. We set the covariance parameters as $\nu = 1.5$, $\sigma = 0.5$ and $\chi = 0.1$, hence Assumption 1 is fulfilled, see [17]. To approximate the Gaussian field, we use the circulant embedding method from [18] to draw samples of W at a grid of discrete points in $\mathbb{D}$ and then use linear interpolation to obtain an extension to $\overline{\mathbb{D}}$. We choose a maximum distance of $\varepsilon > 0$ for the grid points and denote the corresponding approximation by W_ε. Furthermore, we set $\Phi(\cdot) = \exp(\cdot)$ and observe that for any $s \in [1, \infty)$

$$\|\Phi(W) - \Phi(W_\varepsilon)\|_{L^s(\Omega;L^\infty(\mathbb{D}))} \le C\mathbb{E}\left(\left(\sum_{j=1}^{d} \|\partial_{x_j}\Phi(W)\|_{L^\infty(\mathbb{D})}\varepsilon\right)^s\right)^{1/s} \le C\varepsilon$$

holds by the path-wise Lipschitz regularity of W and Lemma 1 (cf. Assumption 2).

For the discontinuous random field P, we denote by $\mathscr{U}((c_1, c_2))$ the uniform distribution on the interval $(c_1, c_2) \subset \mathbb{R}$, sample four i.i.d. $\mathscr{U}((0.2, 0.8))$-distributed random variables $U_1, \ldots, U_4$ and assign one U_i to each side of the square $\partial\mathbb{D}$. We then connect the points on two opposing edges by a straight line to obtain a random partition $\mathscr{T}$ consisting of $\tau = 4$ convex quadrangles. Finally, we assign independent jump heights $P_1, P_2 \sim \mathscr{U}((0, 1))$, $P_3 \sim \mathscr{U}((5, 6))$ and $P_4 \sim \mathscr{U}((10, 11))$ to the partition elements, such that two adjacent elements do not have the same jump distribution. This guarantees rather steep discontinuities across the interfaces in $\mathscr{T}$, see Fig. 1. We do not need any approximation procedure for P and obtain $a_\varepsilon := \exp(W_\varepsilon) + P$. Clearly, a_ε satisfies Assumption 2 and we define $b_\varepsilon := \max(-2a_\varepsilon, -5)$. The QoI is given by

$$\Psi(u) := \int_{\mathbb{D}} u(x) \exp(-0.25\|(0.25, 0.75) - x\|_2^2)dx.$$

For the sample-adapted FE approach, we set the refinement parameters to $\overline{h}_\ell^{(a)} = \frac{1}{4}2^{-\ell/2}$ for $\ell \in \mathbb{N}_0$ and choose $\varepsilon_\ell^{(a)} = \Delta t_\ell^{(a)} = (\overline{h}_\ell^{(a)})^2$. While this choice gives an error equilibrium for $\kappa = 1$, it ensures that for any $\kappa < 1$ the RMSE is dominated solely by the spatial discretization error. Thus, we may infer the true value of κ from the numerical experiment. We also consider a non-adapted FE method with fixed and deterministic triangulations on $\mathbb{D}$. For given approximation parameters ε, $\overline{h}_\ell^{(na)}$ and Δt in the non-adapted setting, we may not expect a better error bound than

$$\mathbb{E}(|\Psi - \Psi_{\varepsilon,\ell,\Delta t}|^2)^{1/2} \le C(\varepsilon + \overline{h}_\ell^{(na)} + \Delta t)$$

in Corollary 1. This is due to the fact that the standard FE method for elliptic problems with discontinuous coefficients does not converge at a better rate than $\mathscr{O}((\overline{h}^{(na)})^{1/2})$ in the V-norm, see [5, Remark 4.2]. Thus, if we consider again the dual problem as in Theorem 4, we may not expect a better rate than $\mathscr{O}(\overline{h}^{(na)})$ with respect to the H-norm. We choose the non-adapted FE grid with diameter $\overline{h}_\ell^{(na)} := \frac{1}{4}2^{-\ell}$ and set accordingly $\varepsilon_\ell^{(na)} = \Delta t_\ell^{(na)} = \overline{h}_\ell^{(na)}$. In both FE methods, we use the midpoint rule on each triangle to approximate the entries of the stiffness matrix. The resulting quadrature error is of order $\mathscr{O}(\overline{h}_\ell^2)$ with respect to the H-norm in the sample-adapted case and hence does not dominate the overall approximation error, see [19, Sect. 2]. For non-adapted FE, no a-priori estimate on the quadrature error is possible due to the discontinuities in a and b, but our results suggest that this bias also in line with the overall approximation error. As $\varepsilon_{\ell-1} = 2\varepsilon_\ell$, the circulant embedding grids (to sample W_ε) are nested and we may achieve the MLMC coupling by first generating the discrete set of points on level ℓ and then taking the appropriate subset of points for level $\ell - 1$.

In the sample-adapted MLMC algorithm, we choose the number of samples via

$$(M_0^{(a)})^{-1} = \left\lceil (\overline{h}_L^{(a)})^4 \right\rceil \quad \text{and} \quad (M_\ell^{(a)})^{-1} = \left\lceil \frac{1}{4} \frac{(\overline{h}_L^{(a)})^4}{(\overline{h}_\ell^{(a)})^4} \left(\frac{(\ell+1)^{-1.001}}{\sum_{k=1}^L (k+1)^{-1.001}} \right)^{-2} \right\rceil$$

for $\ell = 1, \dots, L$, whereas, we choose

$$(M_0^{(na)})^{-1} = \left\lceil (\overline{h}_L^{(na)})^2 \right\rceil \quad \text{and} \quad (M_\ell^{(na)})^{-1} = \left\lceil \frac{(\overline{h}_L^{(na)})^2}{(\overline{h}_\ell^{(na)})^2} \left(\frac{(\ell+1)^{-1.001}}{\sum_{k=1}^L (k+1)^{-1.001}} \right)^{-2} \right\rceil$$

in the non-adapted MLMC approach. Basically, we choose $1/M_\ell$ proportional to $\mathbb{V}_\ell = \mathrm{Var}(\Psi_\ell - \Psi_{\ell-1})$ on each level and thus distribute the errors equally across all levels. Another possibility would be to distribute the computational effort equally (see [15]), which requires estimates on the cost of a single sample on each level. The sequence $(\ell^{-c}, \ell \in \mathbb{N})$ decreases rapidly for $c > 1$ and sums up to $\zeta(c) < +\infty$, where $\zeta(\cdot)$ is the Riemann ζ-function. Hence, the above choice of ρ_i ensures that only a few expensive samples on high levels are necessary and, due to the uniform bound $\sum_{\ell=1}^L \rho_\ell < \zeta(c)$, it is well suited to compare estimators for a varying choice of L. In terms of Theorem 6, we have chosen $C_\rho = 2$ for the number of samples in the sample-adapted method, whereas $C_\rho = 1$ for standard FE. Similar calculations as in Theorem 6 show that this choice leads to $\|\Psi - E^L(\Psi_L)\|_{L^2(\Omega,\mathbb{R})} \le C(2^{-2-L})$ in either case, where the constant C is the same for adapted and non-adapted FE. Hence, C_ρ is merely a normalizing constant and the above choice of M_ℓ ensures that both approaches produce a comparable error for fixed L. Finally, we calculate a reference QoI $\Psi_{ref} := E^L(\Psi_L)$ with $L = 7$ and the sample-adapted method and estimate the relative RMSE $\|\Psi_{ref} - E^L(\Psi_L)\|_{L^2(\Omega,\mathbb{R})}/\Psi_{ref}$ for $L = 0, \dots, 5$ based on 50 independent samples of $E^L(\Psi_L)$ for the sample-adapted and non-adapted MLMC algorithm. For each approach, we use adapted/non-adapted FE combined with a standard/coupled MLMC estimator, thus we compare a total of four algorithms regarding their error decay and efficiency.

Figure 1 confirms our theoretical results from Sect. 3, i.e. the sample-adapted spatial discretization yields rate $\mathscr{O}(\overline{h}_\ell^2)$ compared to $\mathscr{O}(\overline{h}_\ell)$ in the non-adapted setting. Hence, we are able to choose coarser spatial grids in the first approach which entails a better time-to-error ratio for both sample-adapted methods. The results also indicate

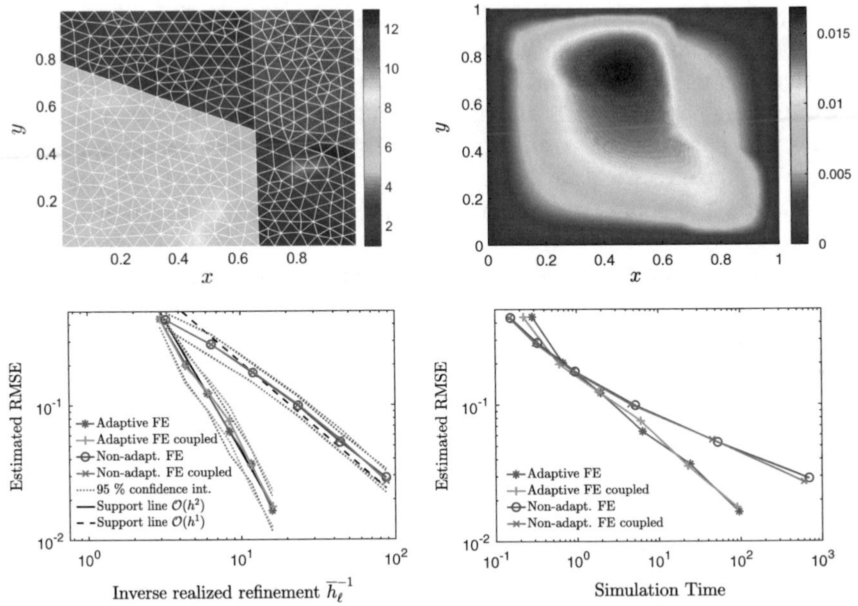

Fig. 1 Top: Sample of the diffusion coefficient with sample-adapted FE grid (left) and FE solution at $T = 1$ (right). Bottom: RMSE versus refinement (left) and RMSE versus simulation time (right)

that $\kappa \approx 1$ holds for this particular example, otherwise we would see a lower rate of convergence than $\mathcal{O}(\overline{h}_\ell^{-2})$ for the sample-adapted methods. While the sample-adapted FE grids have to be generated new for each sample, the $L + 1$ deterministic grids for the non-adapted FE method are generated and stored before the Monte Carlo loop. However, as we see from the time-to-error plot, the extra work of renewing the FE meshes for each sample in the sample-adapted method is more than compensated by the increased order of convergence. The computational cost of the sample-adapted MLMC estimators are (roughly) inversely proportional to the squared errors, which is the best possible results one may achieve with MLMC, see [15] and the references therein. To conclude, we remark that the coupled MLMC estimator yields a slight gain in efficiency if combined with non-adapted FE, whereas it produces similar results when using the sample-adapted discretization. We emphasize that there are scenarios where the coupled estimator outperforms standard MLMC and, on the other hand, there are examples were coupling performs worse due to high correlation terms $\mathbb{C}_{j,k}$ (for both, we refer to numerical examples in [5]). Hence, even though performance is similar to standard MLMC, it makes sense to consider the coupled estimator in our scenario. As we have mentioned at the end of Sect. 4, this behavior may not be expected a-priori.

Acknowledgements The research leading to these results has received funding from the German Research Foundation (DFG) as part of the Cluster of Excellence in Simulation Technology (EXC 310/2) at the University of Stuttgart and it is gratefully acknowledged.

References

1. Abdulle, A., Barth, A., Schwab, C.: Multilevel Monte Carlo methods for stochastic elliptic multiscale PDEs. Multiscale Model. Simul. **11**(4), 1033–1070 (2013). SIAM
2. Adams, R., Fournier, J.: Sobolev Spaces, 2nd edn. Elsevier, Amsterdam (2003)
3. Barth, A., Schwab, C., Zollinger, N.: Multi-level Monte Carlo finite element method for elliptic PDEs with stochastic coefficients. Numer. Math. **119**(1), 123–161 (2011). Springer
4. Barth, A., Stein, A.: Approximation and simulation of infinite-dimensional Lévy processes. Stoch. Partial Differ. Equ. **6**(2), 286–334 (2018). Springer
5. Barth, A., Stein, A.: A study of elliptic partial differential equations with jump-diffusion coefficients. SIAM/ASA J. Uncertain. Quantif. **6**(4), 1707-1743 (2018). SIAM
6. Barth, A., Stein, A.: Numerical analysis for time-dependent advection-diffusion problems with random discontinuous coefficients. Preprint, Arxiv (2019)
7. Charrier, J.: Strong and weak error estimates for elliptic partial differential equations with random coefficients. SIAM J. Numer. Anal. **50**(1), 216–246 (2012) SIAM
8. Cliffe, K., Giles, M., Scheichl, R., Teckentrup, A.: Multilevel Monte Carlo methods and applications to elliptic PDEs with random coefficients. Computing and Visualization in Science **14**(1), 3–15 (2011)
9. Charrier, J., Scheichl, R., Teckentrup, A.: Finite element error analysis of elliptic PDEs with random coefficients and its application to multilevel Monte Carlo methods. SIAM J. Numer. Anal. **51**(1), 322–352 (2013). SIAM
10. Dagan, G., Hornung, U., Knabner, P.: Mathematical Modeling for Flow and Transport Through Porous Media. Springer, Berlin (1991)
11. Ding, Z.: A proof of the trace theorem of Sobolev spaces on Lipschitz domains. Proc. Am. Math. Soc. **124**(2), 591–600 (1996)
12. Di Nezza, E., Palatucci, G., Valdinoci, E.: Hitchhiker's guide to the fractional Sobolev spaces. Bull. des Sci. Math. **136**(5), 521–573 (2012). Elsevier
13. Farthing, M., Ogden, F.: Numerical solution of Richard's equation: a review of advances and challenges. Soil Sci. Soc. Am. J. **81**(6), 1257 1269 (2017)
14. Giles, M.: Multilevel Monte Carlo path simulation. Oper. Res. **56**(3), 607-617 (2008). INFORMS
15. Giles, M.: Multilevel Monte Carlo methods. Acta Numer. **24**, 259–328 (2015). Cambridge University Press
16. Giles, M., Scheichl, R., Teckentrup, A., Ullmann, E.: Further analysis of multilevel Monte Carlo methods for elliptic PDEs with random coefficients. Numer. Math. **125**(3), 569–600 (2013). Springer
17. Graham, I.G., Kuo, F., Nichols, F.Y., Scheichl, R., Schwab, C., Sloan, I.H.: Quasi-Monte Carlo finite element methods for elliptic PDEs with lognormal random coefficients. Numer. Math. **131**(2), 329–368 (2015). Springer
18. Graham, I.G., Kuo, F., Nuyens, D., Scheichl, R., Sloan, I.H.: Analysis of circulant embedding methods for sampling stationary random fields. SIAM J. Numer. Anal. **56**(3), 1871–1895 (2018). SIAM
19. Graham, I.G., Kuo, F., Nuyens, D., Scheichl, R., Sloan, I.H.: Circulant embedding with QMC: analysis for elliptic PDE with lognormal coefficients. Numer. Math. **140**, 479–511 (2018). Springer
20. Gunzburger, M., Peherstorfer, B., Willcox, K.: Optimal model management for multifidelity Monte Carlo estimation. SIAM J. Sci. Comput. **38**(5), 3163–3194 (2016). SIAM

21. Heinrich, S.: Multilevel Monte Carlo methods. *Large-scale scientific computing. 3rd international conference, LSSC 2001, Sozopol, Bulgaria, June 6-10, 2001. Revised papers.* , Comput. Sci. 2179, 58–67, Springer, 2001
22. Li, J., Wang, X., Zhang, K.: Multi-level Monte Carlo weak Galerkin method for elliptic equations with stochastic jump coefficients. Appl. Math. Comput. **275**, 181–194 (2016). Elsevier
23. Lang, A., Potthoff, J.: Fast simulation of Gaussian random fields. Monte Carlo Methods Appl. **17**(3), 195–214 (2011). de Gruyter
24. Rhee, C.-H., Glynn, P.W.: Unbiased estimation with square root convergence for SDE models. Oper. Res. **63**(5), 1026–1043 (2015)

Estimates for Logarithmic and Riesz Energies of Spherical t-Designs

Tetiana A. Stepanyuk

Abstract In this paper we find asymptotic equalities for the discrete logarithmic energy of sequences of well separated spherical t-designs on the unit sphere $\mathbb{S}^d \subset \mathbb{R}^{d+1}$, $d \geq 2$. Also we establish exact order estimates for discrete Riesz s-energy, $s \geq d$, of sequences of well separated spherical t-designs.

Keywords Sphere · Well separated spherical t-design · Logarithmic energy · Riesz energy

1 Introduction

Let $\mathbb{S}^d := \{\mathbf{x} \in \mathbb{R}^{d+1} : |\mathbf{x}| = 1\}$, where $d \geq 2$, be the unit sphere in the Euclidean space $\mathbb{R}^{d+1}$, equipped with the Lebesgue measure σ_d normalized by $\sigma_d(\mathbb{S}^d) = 1$.

Definition 1 A spherical t-design is a finite subset $X_N \subset \mathbb{S}^d$ with a characterising property that an equal weight integration rule with node set X_N integrates all spherical polynomials p of total degree at most t exactly; that is,

$$\frac{1}{N} \sum_{\mathbf{x} \in X_N} p(\mathbf{x}) = \int_{\mathbb{S}^d} p(\mathbf{x}) d\sigma_d(\mathbf{x}), \quad \deg(p) \leq t.$$

Here N is the cardinality of X_N or the number of points of spherical design.

The concept of spherical t-designs was introduced by Delsarte, Goethals and Seidel in the groundbreaking paper [10], since then they attracted a lot of interest from scientific community (see, e.g., [7]).

T. A. Stepanyuk (✉)
Graz University of Technology, Kopernikusgasse 24, Graz, Austria
e-mail: tetiana.stepaniuk@ricam.oeaw.ac.at

Institute of Mathematics of Ukrainian National Academy of Sciences,
3, Tereshchenkivska st., 01601 Kyiv-4, Ukraine

B. Tuffin and P. L'Ecuyer (eds.), *Monte Carlo and Quasi-Monte Carlo Methods*,
Springer Proceedings in Mathematics & Statistics 324,
https://doi.org/10.1007/978-3-030-43465-6_23

The logarithmic energy of a set of N distinct points (or an N-point set) $X_N := \{\mathbf{x}_1, \ldots, \mathbf{x}_N\}$ on $\mathbb{S}^d$ is defined as

$$E_{\log}^{(d)}(X_N) := \sum_{\substack{i,j=1,\\ i\neq j}}^{N} \log \frac{1}{|\mathbf{x}_i - \mathbf{x}_j|}. \tag{1}$$

This paper investigates the logarithmic energy of spherical t-designs. Spherical t-designs can have points arbitrary close together (see, e.g. [13]), hence the logarithmic energy of N-point spherical t-designs can have no asymptotic bounds in terms of t and N. Hence we shall impose additional conditions and consider sequences of well-separated spherical t-designs.

Definition 2 A sequence of N-point sets (X_N), $X_N = \{\mathbf{x}_1, \ldots, \mathbf{x}_N\}$, is called well-separated if there exists a positive constant c_1 such that

$$\min_{i\neq j} |\mathbf{x}_i - \mathbf{x}_j| \geq \frac{c_1}{N^{\frac{1}{d}}} \qquad \text{for all } N. \tag{2}$$

The existence of N-point spherical t-designs with $N = N(t) \asymp t^d$[1] was proven by Bondarenko, Radchenko and Viazovska [3]. They showed that for $d \geq 2$ there exists a constant c_d, which depends only on d, such that for every $N \geq c_d t^d$ there exists a spherical t-design on $\mathbb{S}^d$ with N points. Later the existence of N-point well-separated spherical t-designs with $N(t) \asymp t^d$ points was also proven by these authors in [4]. Namely, they showed that for each $d \geq 2$, $t \in \mathbb{N}$ there exist positive constants c_d and λ_d, depending only on d, such that for every $N \geq c_d t^d$ there exists a spherical t-design on $\mathbb{S}^d$, consisting of N points $\{\mathbf{x}_i\}_{i=1}^N$ with $|\mathbf{x}_i - \mathbf{x}_j| \geq \lambda_d N^{-\frac{1}{d}}$ for $i \neq j$.

On the basis of these results we always assume that $N = N(t) \asymp t^d$.

Denote by $\mathscr{E}_{\log}^{(d)}(N)$ the minimal discrete logarithmic energy of N-points on the sphere

$$\mathscr{E}_{\log}^{(d)}(N) := \inf_{X_N} E_{\log}^{(d)}(X_N), \tag{3}$$

where the infimum is taken over all N-points subsets of $\mathbb{S}^d$.

From the papers of Wagner [18], Kuijlaars and Saff [14] and Brauchart [6] it follows that for $d \geq 2$ and as $N \to \infty$ the following asymptotic equality holds

$$\mathscr{E}_{\log}^{(d)}(N) = N^2 \int\limits_{\mathbb{S}^d}\int\limits_{\mathbb{S}^d} \log \frac{1}{|\mathbf{x}-\mathbf{y}|} d\sigma_d(\mathbf{x}) d\sigma_d(\mathbf{y}) - \frac{1}{d} N \log N + \mathscr{O}(N). \tag{4}$$

[1]We write $a_n \asymp b_n$ to mean that there exist positive constants C_1 and C_2 independent of n such that $C_1 a_n \leq b_n \leq C_2 a_n$ for all n.

Moreover, for the minimal logarithmic energy on the sphere $\mathbb{S}^2$ it was known (lower bound by Wagner [18], and upper bound by [14]) that

$$\mathscr{E}_{\log}^{(2)}(N) = \left(\frac{1}{2} - \log 2\right) N^2 - \frac{1}{2} N \log N + CN + o(N). \tag{5}$$

Brauchart, Hardin and Saff [8] made a conjecture that the constant C in (5) is equal to C_{BHS}, where

$$C_{\mathrm{BHS}} := 2\log 2 + \frac{1}{2}\log\frac{2}{3} + 3\log\frac{\sqrt{\pi}}{\Gamma(1/3)}.$$

Betermin and Sandier [2] proved that

$$\lim_{N\to+\infty} \frac{1}{N}\left(\mathscr{E}_{\log}^{(2)}(N) - \left(\frac{1}{2} - \log 2\right) N^2 + \frac{1}{2} N \log N\right) \leq C_{\mathrm{BHS}}.$$

Also in [5] some general upper and lower bounds for the potential energy of spherical designs were found.

We show that for every well-separated sequence of N-point spherical t-designs on $\mathbb{S}^d$, $d \geq 2$, with $N \asymp t^d$ the following asymptotic equality holds

$$E_{\log}^{(d)}(X_N) = N^2 \int\limits_{\mathbb{S}^d}\int\limits_{\mathbb{S}^d} \log\frac{1}{|\mathbf{x}-\mathbf{y}|} d\sigma_d(\mathbf{x}) d\sigma_d(\mathbf{y}) - \frac{1}{d} N \log N + \mathscr{O}(N).$$

Comparison with (4) gives that the leading and second terms in asymptotic expansion of minimal discrete logarithmic energy are exactly the same, and third terms are of the same order. So, we can summarize that for logarithmic energy well-separated spherical t-designs are asymptotically as good as point sets, which minimize the logarithmic energy.

For given $s > 0$ the discrete Riesz s-energy of a set of N distinct points (or an N-point set) X_N on $\mathbb{S}^d$ is defined as

$$E_s^{(d)}(X_N) := \sum_{\substack{i,j=1,\\ i\neq j}}^{N} |\mathbf{x}_i - \mathbf{x}_j|^{-s}, \tag{6}$$

where $|\mathbf{x}|$ denotes the Euclidean norm in $\mathbb{R}^{d+1}$ of the vector $\mathbf{x}$. In the case $s = d-1$ the energy (6) is Coulomb energy.

Hesse [12] showed that if spherical t-designs with $N = \mathscr{O}(t^2)$ exist, then for $s > 2$ there exists a positive constant c_s, such that for every well-separated sequence N-point spherical t-designs the following estimate holds

$$E_s^{(2)}(X_N) \leq c_s N^{1+\frac{s}{2}}, \tag{7}$$

and for $s = 2$, there exists a positive constant c_2, such that

$$E_s^{(2)}(X_N) \leq \frac{\sum\limits_{k=0}^{t} \frac{1}{k+1}}{2} N^2 + c_2 N^2, \tag{8}$$

and

$$\lim_{N\to\infty} \frac{E_s^{(2)}(X_N)}{N^2 \log N} = \frac{1}{4}. \tag{9}$$

Denote by $\mathscr{E}_s^{(d)}(X_N)$ the minimal discrete s-energy for N-points on the sphere $\mathbb{S}^d$

$$\mathscr{E}_s^{(d)}(N) := \inf_{X_N} E_s^{(d)}(X_N), \tag{10}$$

where the infimum is taken over all N-points subsets of $\mathbb{S}^d$.

Kuijlaars and Saff [14] proved that for $d \geq 2$ and $s > d$, there exist constants $C_{d,s}^{(1)}, C_{d,s}^{(2)} > 0$, such that

$$C_{d,s}^{(1)} N^{1+\frac{s}{d}} \leq \mathscr{E}_s^{(d)}(N) \leq C_{d,s}^{(2)} N^{1+\frac{s}{d}}. \tag{11}$$

Also in [14] it was shown that for $s = d$ the following formula holds

$$\lim_{N\to\infty} \frac{\mathscr{E}_s^{(d)}(N)}{N^2 \log N} = \frac{1}{d} \frac{\Gamma\left(\frac{d+1}{2}\right)}{\Gamma\left(\frac{d}{2}\right)\Gamma\left(\frac{1}{2}\right)}. \tag{12}$$

We show that for every well-separated sequence of N-point spherical t-designs on $\mathbb{S}^d$, $d \geq 2$, with $N \asymp t^d$ the following relations are true:

$$E_s^{(d)}(X_N) \ll N^{1+\frac{s}{d}}, \qquad s > d$$

and

$$\lim_{N\to\infty} \frac{E_s^{(d)}(X_N)}{N^2 \log N} = \frac{1}{d} \frac{\Gamma\left(\frac{d+1}{2}\right)}{\Gamma\left(\frac{d}{2}\right)\Gamma\left(\frac{1}{2}\right)}, \qquad s = d.$$

Here and throughout the paper we use the Vinogradov notation $a_n \ll b_n$ to mean that there exists a positive constant C independent of n such that $a_n \leq C b_n$ for all n.

First, we observe that since $\mathscr{E}_s^{(d)}(N) \leq E_s^{(d)}(X_N)$ for any N-point set, the lower bound in (11) provides a lower bound for the s-energy of any N-point set. So, the Riesz s-energy, $s \geq d$ of well-separated spherical t-designs has the same asymptotic order as the minimal Riesz s-energy.

This paper is organised as follows: Sect. 2 provides basic notations and necessary background for Jacobi polynomials, Sect. 3 collects our main results and their proofs.

2 Preliminaries

We use the Pochhammer symbol $(a)_n$, where $n \in \mathbb{N}_0$ and $a \in \mathbb{R}$, $a \neq 0, -1, -2, \ldots$, defined by

$$(a)_0 := 1, \qquad (a)_n := a(a+1)\cdots(a+n-1), \;\; n \in \mathbb{N},$$

which can be written in the terms of the gamma function $\Gamma(z)$ by means of

$$(a)_\ell = \frac{\Gamma(\ell + a)}{\Gamma(a)}, \qquad a \neq 0, -1, -2, \ldots. \tag{13}$$

For fixed a, b the following asymptotic equality is true

$$\frac{\Gamma(n+a)}{\Gamma(n+b)} = n^{a-b}\Big(1 + \mathcal{O}\Big(\frac{1}{n}\Big)\Big) \qquad \text{as } n \to \infty. \tag{14}$$

For any integrable function $f : [-1, 1] \to \mathbb{R}$ (see, e.g., [16]) we have

$$\int_{\mathbb{S}^d} f(\langle \mathbf{x}, \mathbf{y} \rangle) d\sigma_d(\mathbf{x}) = \gamma_d \int_{-1}^{1} f(t)(1-t^2)^{\frac{d}{2}-1} dt, \qquad \mathbf{y} \in \mathbb{S}^d, \tag{15}$$

where $\gamma_d := \frac{\Gamma(\frac{d+1}{2})}{\sqrt{\pi}\,\Gamma(\frac{d}{2})}$.

The Jacobi polynomials $P_\ell^{(\alpha,\beta)}$, $\alpha, \beta > -1$, are orthogonal on the interval $[-1, 1]$ with respect to the weight function $(1-x)^\alpha(1+x)^\beta$ and normalised by (see, e.g., [15, (5.2.1)])

$$P_\ell^{(\alpha,\beta)}(1) = \binom{\ell+\alpha}{\ell} = \frac{(1+\alpha)_\ell}{\ell!} = \frac{1}{\Gamma(1+\alpha)}\ell^\alpha \left(1 + \mathcal{O}\Big(\frac{1}{\ell}\Big)\right). \tag{16}$$

We will also use formula

$$P_\ell^{(\alpha,\beta)}(-x) = (-1)^\ell P_\ell^{(\alpha,\beta)}(x) \tag{17}$$

and the connection coefficient formula (see, e.g., [1, Theorem 7.1.4])

$$P_m^{(\gamma,\gamma)}(x) = \frac{(\gamma+1)_m}{(2\gamma+1)_m} \sum_{k=0}^{[\frac{m}{2}]} \frac{(2\alpha+1)_{m-2k}}{(\alpha+1)_{m-2k}} \frac{\left(\gamma+\frac{1}{2}\right)_{m-k}\left(\alpha+\frac{3}{2}\right)_{m-2k}(\gamma-\alpha)_k}{\left(\alpha+\frac{3}{2}\right)_{m-k}\left(\alpha+\frac{1}{2}\right)_{m-2k} k!} P_{m-2k}^{\alpha,\alpha}(x). \tag{18}$$

For fixed $\alpha, \beta > -1$ and $0 < \theta < \pi$, the following relation gives an asymptotic approximation for $\ell \to \infty$ (see, e.g., [17, Theorem 8.21.13])

$$P_\ell^{(\alpha,\beta)}(\cos\theta) = \frac{1}{\sqrt{\pi}}\ell^{-1/2}\Big(\sin\frac{\theta}{2}\Big)^{-\alpha-1/2}\Big(\cos\frac{\theta}{2}\Big)^{-\beta-1/2} \times\Big\{\cos\Big(\Big(\ell+\frac{\alpha+\beta+1}{2}\Big)\theta-\frac{2\alpha+1}{4}\pi\Big)+\mathcal{O}(\ell\sin\theta)^{-1}\Big\}.$$

Thus, for $c_{\alpha,\beta}\ell^{-1}\le\theta\le\pi-c_{\alpha,\beta}\ell^{-1}$ the last asymptotic equality yields

$$|P_\ell^{(\alpha,\beta)}(\cos\theta)|\le\tilde{c}_{\alpha,\beta}\ell^{-1/2}(\sin\theta)^{-\alpha-1/2}+\tilde{c}_{\alpha,\beta}\ell^{-3/2}(\sin\theta)^{-\alpha-3/2},\quad \alpha\ge\beta. \tag{19}$$

The following differentiation formula holds

$$\frac{d}{dx}P_n^{(\alpha,\beta)}(x)=\frac{n+\alpha+\beta+1}{2}P_{n-1}^{(\alpha+1,\beta+1)}(x). \tag{20}$$

If $\lambda > s-1$, $s\ge d$, then taking into account formula [15, (5.3.4)] and the fact that the Gegenbauer polynomials are a special case of the Jacobi polynomials $P_n^{(\alpha,\beta)}$ (see, e.g., [15, (5.3.1)]), we have that for $-1<x<1$ the following expansion holds

$$(1-x)^{-\frac{s}{2}}=2^{2\lambda-\frac{s}{2}}\pi^{-\frac{1}{2}}\Gamma(\lambda)\Gamma\Big(\lambda-\frac{s}{2}+\frac{1}{2}\Big) \times\sum_{n=0}^{\infty}\frac{(n+\lambda)\left(\frac{s}{2}\right)_n}{\Gamma\left(n+2\lambda-\frac{s}{2}+1\right)}\frac{(2\lambda)_n}{\left(\lambda+\frac{1}{2}\right)_n}P_n^{(\lambda-\frac{1}{2},\lambda-\frac{1}{2})}(x). \tag{21}$$

3 Main Results

By a spherical cap $S(\mathbf{x};\varphi)$ of centre $\mathbf{x}$ and angular radius φ we mean

$$S(\mathbf{x};\varphi):=\big\{\mathbf{y}\in\mathbb{S}^d:\ \langle\mathbf{x},\mathbf{y}\rangle\ge\cos\varphi\big\}.$$

The normalised surface area of a spherical cap is given by

$$|S(\mathbf{x};\varphi)|=\gamma_d\int_{\cos\varphi}^{1}(1-t^2)^{\frac{d}{2}-1}dt\asymp(1-\cos\varphi)^{\frac{d}{2}}\quad\text{as }\varphi\to 0. \tag{22}$$

If condition (2) holds for the sequence (X_N), then any spherical cap $S(\mathbf{x};\alpha_N)$, $\mathbf{x}\in\mathbb{S}^d$, where

$$\alpha_N:=\arccos\left(1-\frac{c_1^2}{8N^{\frac{2}{d}}}\right), \tag{23}$$

contains at most one point of the set X_N.

From the elementary estimates

$$\sin\theta \le \theta \le \frac{\pi}{2}\sin\theta, \quad 0 \le \theta \le \frac{\pi}{2}, \tag{24}$$

we obtain

$$\left(1 - \frac{c_1^2}{16N^{\frac{2}{d}}}\right)^{\frac{1}{2}} \frac{c_1}{2N^{\frac{1}{d}}} \le \alpha_N \le \frac{\pi}{4}\left(1 - \frac{c_1^2}{16N^{\frac{2}{d}}}\right)^{\frac{1}{2}} \frac{c_1}{N^{\frac{1}{d}}}. \tag{25}$$

The following two theorems are the main result of this paper.

Theorem 1 *Let $d \ge 2$ be fixed, $(X_{N(t)})_t$ be a sequence of well-separated spherical t-designs on $\mathbb{S}^d$ with $N(t) \asymp t^d$. Then for the logarithmic energy $E_{\log}^{(d)}(X_N)$ the following estimate holds*

$$E_{\log}^{(d)}(X_N) = N^2 \int_{\mathbb{S}^d}\int_{\mathbb{S}^d} \log\frac{1}{|\mathbf{x}-\mathbf{y}|} d\sigma_d(\mathbf{x}) d\sigma_d(\mathbf{y}) - \frac{1}{d} N \log N + \mathcal{O}(N). \tag{26}$$

Theorem 2 *Let $d \ge 2$ be fixed, and $(X_{N(t)})_t$ be a sequence of well-separated spherical t-designs on $\mathbb{S}^d$ with $N(t) \asymp t^d$. Then for $s > d$ for the s-energy $E_s^{(d)}(X_N)$ the following estimate holds*

$$E_s^{(d)}(X_N) \ll N^{1+\frac{s}{d}}, \tag{27}$$

and for $s = d$, the s-energy $E_s^{(d)}(X_N)$ satisfies following estimates

$$E_s^{(d)}(X_N) = \gamma_d \sum_{n=1}^{[\frac{t}{2}]} \frac{1}{n} N^2 + \mathcal{O}(N^2) \tag{28}$$

and

$$\lim_{N\to\infty} \frac{E_s^{(d)}(X_N)}{N^2 \log N} = \frac{\gamma_d}{d}. \tag{29}$$

3.1 Proof of Theorem 1

For each $i \in \{1, \ldots, N\}$ we divide the sphere $\mathbb{S}^d$ into an upper hemisphere H_i^+ with 'North Pole' $\mathbf{x}_i$ and a lower hemisphere H_i^-:

$$H_i^+ := \left\{\mathbf{x} \in \mathbb{S}^d \,\middle|\, \langle \mathbf{x}_i, \mathbf{x}\rangle \ge 0\right\},$$

$$H_i^- := \mathbb{S}^d \setminus H_i^+.$$

Noting that

$$|\mathbf{x}_i - \mathbf{x}_j|^{-1} = \frac{1}{\sqrt{2}}(1 - \langle \mathbf{x}_i, \mathbf{x}_j \rangle)^{-\frac{1}{2}}, \tag{30}$$

the logarithmic energy can be written in the form

$$E_{\log}^{(d)}(X_N) = \sum_{\substack{i,j=1,\\ i\neq j}}^{N} \log \frac{1}{|\mathbf{x}_i - \mathbf{x}_j|} = \frac{1}{2} \sum_{\substack{i,j=1,\\ i\neq j}}^{N} \left(\log \frac{1}{1 - \langle \mathbf{x}_i, \mathbf{x}_j \rangle} - \log 2 \right). \tag{31}$$

Let $\lambda > d + 1$. Then, putting $s = 2$ in (21) and using the relation $\Gamma(n + \frac{1}{2}) = \frac{(2n)!}{2^{2n} n!}\sqrt{\pi}$, we get

$$\frac{1}{1-x} = 2^{2\lambda-1}\pi^{-\frac{1}{2}}\Gamma(\lambda)\Gamma\left(\lambda - \frac{1}{2}\right) \sum_{n=0}^{\infty} \frac{(n+\lambda)\Gamma(n+1)}{\Gamma(n+2\lambda)} \frac{(2\lambda)_n}{\left(\lambda + \frac{1}{2}\right)_n} P_n^{\left(\lambda-\frac{1}{2},\lambda-\frac{1}{2}\right)}(x)$$

$$= \Gamma\left(\lambda - \frac{1}{2}\right) \sum_{n=0}^{\infty} \frac{(n+\lambda)\Gamma(n+1)}{\Gamma\left(n+\lambda+\frac{1}{2}\right)} P_n^{\left(\lambda-\frac{1}{2},\lambda-\frac{1}{2}\right)}(x), \quad -1 < x < 1. \tag{32}$$

Formula (20) implies that

$$\int P_n^{\left(\lambda-\frac{1}{2},\lambda-\frac{1}{2}\right)}(x)dx = \frac{2}{n+2\lambda-1} P_{n+1}^{\left(\lambda-\frac{3}{2},\lambda-\frac{3}{2}\right)}(x). \tag{33}$$

Integrating from 0 to x, we have

$$\frac{1}{2}\log\frac{1}{1-x}$$

$$= \Gamma\left(\lambda - \frac{1}{2}\right) \sum_{n=1}^{\infty} \frac{(n+\lambda-1)\Gamma(n)}{(n+2\lambda-2)\Gamma\left(n+\lambda-\frac{1}{2}\right)} \left(P_n^{\left(\lambda-\frac{3}{2},\lambda-\frac{3}{2}\right)}(x) - P_n^{\left(\lambda-\frac{3}{2},\lambda-\frac{3}{2}\right)}(0)\right). \tag{34}$$

We split the log-energy into two parts

$$E_{\log}^{(d)}(X_N) = \sum_{j=1}^{N} \sum_{\substack{i=1,\\ \mathbf{x}_i \in H_j^{\pm} \setminus S(\pm \mathbf{x}_j;\alpha_N)}}^{N} \log \frac{1}{|\mathbf{x}_i - \mathbf{x}_j|} + \sum_{j=1}^{N} \sum_{\substack{i=1,\\ \mathbf{x}_i \in S(-\mathbf{x}_j;\alpha_N)}}^{N} \log \frac{1}{|\mathbf{x}_i - \mathbf{x}_j|}. \tag{35}$$

From (2) and the fact the spherical cap $S(-\mathbf{x}_j; \alpha_N)$ contains at most one point of X_N, the second term in (35), where the scalar product is close to -1, can be bounded from above by

$$\sum_{j=1}^{N} \sum_{\substack{i=1, \\ \mathbf{x}_i \in S(-\mathbf{x}_j;\alpha_N)}}^{N} \log \frac{1}{|\mathbf{x}_i - \mathbf{x}_j|} = \mathcal{O}(N). \tag{36}$$

Taking into account (31), (34)–(36), we deduce

$$E_{\log}^{(d)}(X_N) = \frac{1}{2} E_{H_{\log,t}}(X_N) + \frac{1}{2} E_{R_{\log,t}}(X_N) - \frac{1}{2} N^2 \log 2 + \mathcal{O}(N), \tag{37}$$

where

$$E_U(X_N) := \sum_{j=1}^{N} \sum_{\substack{i=1, \\ \mathbf{x}_i \in H_j^{\pm} \setminus S(\pm x_j;\alpha_N)}}^{N} U(\langle \mathbf{x}_i, \mathbf{x}_j \rangle) \tag{38}$$

and

$$\begin{aligned} H_{\log,t}(x) &= H_{\log,t}(d,\lambda,x) \\ &:= -2\Gamma\left(\lambda - \frac{1}{2}\right) \sum_{n=1}^{\infty} \frac{(n+\lambda-1)\Gamma(n)}{(n+2\lambda-2)\Gamma\left(n+\lambda-\frac{1}{2}\right)} P_n^{\left(\lambda-\frac{3}{2},\lambda-\frac{3}{2}\right)}(0) \\ &+ 2\Gamma\left(\lambda - \frac{1}{2}\right) \sum_{n=1}^{t} \frac{(n+\lambda-1)\Gamma(n)}{(n+2\lambda-2)\Gamma\left(n+\lambda-\frac{1}{2}\right)} P_n^{\left(\lambda-\frac{3}{2},\lambda-\frac{3}{2}\right)}(x), \end{aligned} \tag{39}$$

$$\begin{aligned} R_{\log,t}(x) &= R_{\log,t}(d,\lambda,x) \\ &:= 2\Gamma\left(\lambda - \frac{1}{2}\right) \sum_{n=t+1}^{\infty} \frac{(n+\lambda-1)\Gamma(n)}{(n+2\lambda-2)\Gamma\left(n+\lambda-\frac{1}{2}\right)} P_n^{\left(\lambda-\frac{3}{2},\lambda-\frac{3}{2}\right)}(x). \end{aligned} \tag{40}$$

Let us show that

$$E_{R_{\log,t}}(X_N) = \mathcal{O}(N). \tag{41}$$

Applying (13), (14) and (19) to (40), we have

$$\begin{aligned} |R_{\log,t}(\cos\theta)| &\ll \sum_{n=t+1}^{\infty} n^{-\lambda+\frac{1}{2}} |P_n^{\left(\lambda-\frac{3}{2},\lambda-\frac{3}{2}\right)}(\cos\theta)| \\ &\ll \sum_{n=t+1}^{\infty} n^{-\lambda+\frac{1}{2}} \left(n^{-\frac{1}{2}}(\sin\theta)^{-\lambda+1} + n^{-\frac{3}{2}}(\sin\theta)^{-\lambda}\right) \\ &\ll (\sin\theta)^{-\lambda+1} t^{-\lambda+1} + (\sin\theta)^{-\lambda} t^{-\lambda}. \end{aligned} \tag{42}$$

We define $\theta_{ij}^{\pm} \in [0, \pi]$ by $\cos\theta_{ij}^{\pm} := \langle \mathbf{x}_i, \pm\mathbf{x}_j \rangle$. Then $\sin\theta_{ij}^{+} = \sin\theta_{ij}^{-}$. From [9, (3.30) and (3.33)], it follows that

$$\sum_{j=1}^{N} \sum_{\substack{i=1, \\ \mathbf{x}_i \in H_j^{\pm} \setminus S(\pm\mathbf{x}_j; \frac{c}{n})}}^{N} (\sin\theta_{ij}^{\pm})^{-\frac{d}{2}+\frac{1}{2}-k-L}$$

$$\ll N^2(1 + n^{L+k-(d+1)/2}), \quad k = 0, 1, \ldots \quad \text{for } L > \frac{d+1}{2}. \tag{43}$$

Estimates (25) and (43) imply

$$E_{R_{\log,t}}(X_N) \ll t^{-\lambda+1} \sum_{j=1}^{N} \sum_{\substack{i=1, \\ \mathbf{x}_i \in H_j^{\pm} \setminus S(\pm\mathbf{x}_j; \alpha_N)}}^{N} (\sin\theta^{\pm})^{-\lambda+1} + t^{-\lambda} \sum_{j=1}^{N} \sum_{\substack{i=1, \\ \mathbf{x}_i \in H_j^{\pm} \setminus S(\pm\mathbf{x}_j; \alpha_N)}}^{N} (\sin\theta^{\pm})^{-\lambda}$$

$$\ll N^2 t^{-d} \ll N, \quad \lambda > d+1. \tag{44}$$

This proves (41).

Now let us find the estimate for $E_{H_{\log,t}}(X_N)$. The polynomial $H_{\log,t}$ is a spherical polynomial of degree t and X_N is a spherical t-design. Thus, an equal weight integration rule with node set X_N integrates $H_{\log,t}$ exactly and

$$E_{H_{\log,t}}(X_N) = \sum_{j=1}^{N} \sum_{\substack{i=1, \\ \mathbf{x}_i \in H_j^{\pm} \setminus S(\pm\mathbf{x}_j; \alpha_N)}}^{N} H_{\log,t}(\langle \mathbf{x}_i, \mathbf{x}_j \rangle)$$

$$= N^2 \int_{\mathbb{S}^d} H_{\log,t}(\langle \mathbf{x}, \mathbf{y} \rangle) d\sigma_d(\mathbf{x}) - N H_{\log,t}(1) - \sum_{j=1}^{N} \sum_{\substack{i=1, \\ \mathbf{x}_i \in S(-\mathbf{x}_j; \alpha_N)}}^{N} H_{\log,t}(\langle \mathbf{x}_i, \mathbf{x}_j \rangle), \ \mathbf{y} \in \mathbb{S}^d. \tag{45}$$

Let $b_0 = b_0(d, N) \in \mathbb{R}_+$ be such that $b_0 = \mathcal{O}(1)$ as $N \to \infty$ and for $\beta_N := \arccos(1 - b_0 N^{-\frac{2}{d}})$ the following relation holds

$$\int_{S(\mathbf{y}; \beta_N)} d\sigma_d(\mathbf{x}) = \gamma_d \int_{1-b_0 N^{-\frac{2}{d}}}^{1} (1 - x^2)^{\frac{d}{2}-1} dx = \frac{1}{N}, \quad \mathbf{y} \in \mathbb{S}^d. \tag{46}$$

It is clear that

$$\beta_N \asymp N^{-\frac{1}{d}}. \tag{47}$$

Then

$$E_{H_{\log,t}}(X_N) = N^2 \int_{\mathbb{S}^d} \log \frac{1}{1 - \langle \mathbf{x}, \mathbf{y} \rangle} d\sigma_d(\mathbf{x}) + Q_t(X_N), \tag{48}$$

where

$$Q_t(X_N) = Q_t(d, X_N) := -N^2 \int\limits_{S(\pm\mathbf{y};\beta_N)} \log \frac{1}{1-\langle \mathbf{x}, \mathbf{y}\rangle} d\sigma_d(\mathbf{x})$$

$$- N^2 \int\limits_{\mathbb{S}^d \setminus S(\pm\mathbf{y};\beta_N)} R_{\log,t}(\langle \mathbf{x}, \mathbf{y}\rangle) d\sigma_d(\mathbf{x}) + N^2 \int\limits_{S(\pm\mathbf{y};\beta_N)} H_{\log,t}(\langle \mathbf{x}, \mathbf{y}\rangle) d\sigma_d(\mathbf{x})$$

$$- N H_{\log,t}(1) - \sum_{j=1}^{N} \sum_{\substack{i=1,\\ \mathbf{x}_i \in S(-\mathbf{x}_j;\alpha_N)}}^{N} H_{\log,t}(\langle \mathbf{x}_i, \mathbf{x}_j\rangle), \quad \mathbf{y} \in \mathbb{S}^d. \tag{49}$$

Now we shall prove that

$$Q_t(X_N) = -N^2 \int\limits_{S(\mathbf{y};\beta_N)} \log \frac{1}{1-\langle \mathbf{x}, \mathbf{y}\rangle} d\sigma_d(\mathbf{x}) + \mathscr{O}(N), \quad \mathbf{y} \in \mathbb{S}^d. \tag{50}$$

Using (15), (42) and (47), we get

$$N^2 \left| \int\limits_{\mathbb{S}^d \setminus S(\pm\mathbf{y};\beta_N)} R_{\log,t}(\langle \mathbf{x}, \mathbf{y}\rangle) d\sigma_d(\mathbf{x}) \right| \ll N^2 \int\limits_{-1+b_0 N^{-\frac{2}{d}}}^{1-b_0 N^{-\frac{2}{d}}} |R_{\log,t}(x)| (1-x^2)^{\frac{d}{2}-1} dx$$

$$\ll N^2 \int\limits_{-1+b_0 N^{-\frac{2}{d}}}^{1-b_0 N^{-\frac{2}{d}}} \left(t^{-\lambda+1} (\sqrt{1-x^2})^{-\lambda+1} + t^{-\lambda} (\sqrt{1-x^2})^{-\lambda} \right) (1-x^2)^{\frac{d}{2}-1} dx$$

$$= 2N^2 \int\limits_{\beta_N}^{\frac{\pi}{2}} \left(t^{-\lambda+1} (\sin y)^{-\lambda+1} + t^{-\lambda} (\sin y)^{-\lambda} \right) (\sin y)^{d-1} dy$$

$$\ll N^2 \int\limits_{\beta_N}^{\frac{\pi}{2}} \left(t^{-\lambda+1} y^{-\lambda+d} + t^{-\lambda} y^{-\lambda+d-1} \right) dy \ll N. \tag{51}$$

From the definition of β_n it is easy to see that

$$\left| N^2 \int\limits_{S(-\mathbf{y};\beta_N)} \log \frac{1}{1-\langle \mathbf{x}, \mathbf{y}\rangle} d\sigma_d(\mathbf{x}) \right| \ll N^2 |S(-\mathbf{y};\beta_N)| \ll N, \ \mathbf{y} \in \mathbb{S}^d. \tag{52}$$

According to the definition of β_N (46) we deduce

$$\left| N^2 \int\limits_{S(\mathbf{y};\beta_N)} H_{\log,t}(\langle \mathbf{x},\mathbf{y}\rangle) d\sigma_d(\mathbf{x}) - N H_{\log,t}(1) \right|$$
$$= \left| N^2 \gamma_d \int\limits_{1-b_0 N^{-\frac{2}{d}}}^{1} (H_{\log,t}(x) - H_{\log,t}(1))(1-x^2)^{\frac{d}{2}-1} dx \right|$$
$$\ll N \max_{x \in [1-b_0 N^{-\frac{2}{d}},1]} \Big(H_{\log,t}(1) - H_{\log,t}(x) \Big) \ll N^{1-\frac{2}{d}} |H'_{\log,t}(1)|. \tag{53}$$

Formulas (16), (20) and (39) imply

$$H'_{\log,t}(1) = \Gamma\left(\lambda - \frac{1}{2}\right) \sum_{n=1}^{t} \frac{(n+\lambda-1)\Gamma(n)}{\Gamma\left(n+\lambda-\frac{1}{2}\right)} P_{n-1}^{\left(\lambda-\frac{1}{2},\lambda-\frac{1}{2}\right)}(1)$$
$$= \Gamma\left(\lambda - \frac{1}{2}\right) \sum_{n=1}^{t} \frac{(n+\lambda-1)\Gamma(n)}{\Gamma\left(n+\lambda-\frac{1}{2}\right)} \frac{\left(\lambda+\frac{1}{2}\right)_{n-1}}{(n-1)!}$$
$$= \frac{1}{\lambda-\frac{1}{2}} \sum_{n=1}^{t} (n+\lambda-1) = t + \frac{t^2}{2\lambda-1} \asymp t^2 \asymp N^{\frac{2}{d}}. \tag{54}$$

From (17), (16), (19) and (39) it follows that

$$|H_{\log,t}(-1)| \ll \sum_{n=1}^{\infty} \frac{(n+\lambda-1)\Gamma(n)}{(n+2\lambda-2)\Gamma\left(n+\lambda-\frac{1}{2}\right)} \frac{1}{\sqrt{n}}$$
$$+ \left| \sum_{n=1}^{t} (-1)^n \frac{(n+\lambda-1)\Gamma(n)}{(n+2\lambda-2)\Gamma\left(n+\lambda-\frac{1}{2}\right)} P_n^{\left(\lambda-\frac{3}{2},\lambda-\frac{3}{2}\right)}(1) \right|$$
$$\ll 1 + \left| \sum_{n=1}^{t} (-1)^n \frac{n+\lambda-1}{n+2\lambda-2} \frac{1}{n} \right|. \tag{55}$$

Thus, (55) enables us to obtain

$$|H_{\log,t}(-1)| = \mathcal{O}(1). \tag{56}$$

Using (15), (54) and (56), we deduce

$$\left| N^2 \int\limits_{S(-\mathbf{y};\beta_N)} H_{\log,t}(\langle \mathbf{x}, \mathbf{y}\rangle) d\sigma_d(\mathbf{x}) \right|$$
$$= \left| N^2 \gamma_d \int\limits_{1-b_0 N^{-\frac{2}{d}}}^{1} (H_{\log,t}(-x) - H_{\log,t}(-1))(1-x^2)^{\frac{d}{2}-1} dx + N H_{\log,t}(-1) \right|$$
$$\ll N^{1-\frac{2}{d}} |H'_{\log,t}(1)| + N \ll N. \tag{57}$$

Applying (56), we have

$$\left| \sum_{j=1}^{N} \sum_{\substack{i=1, \\ \mathbf{x}_i \in S(-\mathbf{x}_j;\alpha_N)}}^{N} H_{\log,t}(\langle \mathbf{x}_i, \mathbf{x}_j \rangle) \right| \ll N |H_{\log,t}(\xi)|$$
$$= N |H_{\log,t}(\xi) - H_{\log,t}(-1) + H_{\log,t}(-1)| = \mathscr{O}(N), \tag{58}$$

where $\xi \in [-1, -1 + b_0 N^{-\frac{2}{d}}]$.

Relations (51)–(54), (57) and (58) prove (50).

Integrating by parts, we obtain

$$N^2 \int\limits_{S(\mathbf{y};\beta_N)} \log \frac{1}{1 - \langle \mathbf{x}, \mathbf{y} \rangle} d\sigma_d(\mathbf{x}) = \gamma_d N^2 \int\limits_{1-b_0 N^{-\frac{2}{d}}}^{1} \log \frac{1}{1-x} (1-x^2)^{\frac{d}{2}-1} dx$$
$$= \gamma_d N^2 \int\limits_{1-b_0 N^{-\frac{2}{d}}}^{1} \log \frac{1}{1-x} \frac{d}{dx} \left(- \int\limits_{x}^{1} (1-t^2)^{\frac{d}{2}-1} dt \right) dx$$
$$= N \log \left(\frac{N^{\frac{2}{d}}}{b_0} \right) + N^2 \gamma_d \int\limits_{1-b_0 N^{-\frac{2}{d}}}^{1} \frac{1}{1-x} \int\limits_{x}^{1} (1-t^2)^{\frac{d}{2}-1} dt dx$$
$$= \frac{2}{d} N \log N + \mathscr{O}(N). \tag{59}$$

So, combining (37), (41), (48), (50) and (59), we get

$$E_{\log}^{(d)}(X_N) = \frac{1}{2} N^2 \int\limits_{\mathbb{S}^d} \int\limits_{\mathbb{S}^d} \log \frac{1}{1 - \langle \mathbf{x}, \mathbf{y} \rangle} d\sigma_d(\mathbf{x}) d\sigma_d(\mathbf{y}) - \frac{1}{d} N \log N - \frac{1}{2} N^2 \log 2 + \mathscr{O}(N)$$
$$= N^2 \int\limits_{\mathbb{S}^d} \int\limits_{\mathbb{S}^d} \log \frac{1}{|\mathbf{x} - \mathbf{y}|} d\sigma_d(\mathbf{x}) d\sigma_d(\mathbf{y}) - \frac{1}{d} N \log N + \mathscr{O}(N). \tag{60}$$

This implies (26). Theorem 1 is proved. □

3.2 Proof of Theorem 2

In the same way as in the case for logarithmic energy, we split the s-energy into two parts

$$E_s^{(d)}(X_N) = \sum_{j=1}^{N} \sum_{\substack{i=1,\\ \mathbf{x}_i \in H_j^{\pm} \setminus S(\pm \mathbf{x}_j;\alpha_N)}}^{N} |\mathbf{x}_i - \mathbf{x}_j|^{-s} + \sum_{j=1}^{N} \sum_{\substack{i=1,\\ \mathbf{x}_i \in S(-\mathbf{x}_j;\alpha_N)}}^{N} |\mathbf{x}_i - \mathbf{x}_j|^{-s}$$

$$= \sum_{j=1}^{N} \sum_{\substack{i=1,\\ \mathbf{x}_i \in H_j^{\pm} \setminus S(\pm \mathbf{x}_j;\alpha_N)}}^{N} |\mathbf{x}_i - \mathbf{x}_j|^{-s} + \mathcal{O}(N). \tag{61}$$

Taking into account that the Jacobi series (21) converges uniformly in $\left[-1+\frac{c_1^2}{8N^{\frac{2}{d}}}, 1-\frac{c_1^2}{8N^{\frac{2}{d}}}\right]$, for $\lambda > s-1$ we get that

$$\sum_{j=1}^{N} \sum_{\substack{i=1,\\ \mathbf{x}_i \in H_j^{\pm} \setminus S(\pm \mathbf{x}_j;\alpha_N)}}^{N} |\mathbf{x}_i - \mathbf{x}_j|^{-s} = \frac{1}{2^{\frac{s}{2}}} \sum_{j=1}^{N} \sum_{\substack{i=1,\\ \mathbf{x}_i \in H_j^{\pm} \setminus S(\pm \mathbf{x}_j;\alpha_N)}}^{N} (1-\langle \mathbf{x}_i, \mathbf{x}_j\rangle)^{-\frac{s}{2}}$$

$$= 2^{2\lambda - s}\pi^{-\frac{1}{2}}\Gamma(\lambda)\Gamma\left(\lambda - \frac{s}{2} + \frac{1}{2}\right) \sum_{n=0}^{\infty} \frac{(n+\lambda)(\frac{s}{2})_n}{\Gamma\left(n+2\lambda-\frac{s}{2}+1\right)} \frac{(2\lambda)_n}{\left(\lambda+\frac{1}{2}\right)_n} P_n^{\left(\lambda-\frac{1}{2},\, \lambda-\frac{1}{2}\right)}(x)$$

$$= 2^{1-s}\Gamma\left(\lambda - \frac{s}{2} + \frac{1}{2}\right) \sum_{n=0}^{\infty} \frac{(n+\lambda)\Gamma\left(n+\frac{s}{2}\right)\Gamma(n+2\lambda)}{\Gamma(n+2\lambda-\frac{s}{2}+1)\Gamma\left(\frac{s}{2}\right)\Gamma\left(n+\lambda+\frac{1}{2}\right)} P_n^{\left(\lambda-\frac{1}{2},\, \lambda-\frac{1}{2}\right)}(x)$$

$$= E_{H_{s,t}}(X_N) + E_{R_{s,t}}(X_N), \tag{62}$$

where

$$H_{s,t}(x) = H_{s,t}(d, \lambda, x)$$
$$:= 2^{1-s}\Gamma\left(\lambda - \frac{s}{2} + \frac{1}{2}\right) \sum_{n=0}^{t-1} \frac{(n+\lambda)\Gamma\left(n+\frac{s}{2}\right)\Gamma(n+2\lambda)}{\Gamma\left(n+2\lambda-\frac{s}{2}+1\right)\Gamma\left(\frac{s}{2}\right)\Gamma\left(n+\lambda+\frac{1}{2}\right)} P_n^{\left(\lambda-\frac{1}{2},\, \lambda-\frac{1}{2}\right)}(x), \tag{63}$$

$$R_{s,t}(x) = R_{s,t}(d, \lambda, x)$$
$$:= 2^{1-s}\Gamma\left(\lambda - \frac{s}{2} + \frac{1}{2}\right) \sum_{n=t}^{\infty} \frac{(n+\lambda)\Gamma\left(n+\frac{s}{2}\right)\Gamma(n+2\lambda)}{\Gamma\left(n+2\lambda-\frac{s}{2}+1\right)\Gamma\left(\frac{s}{2}\right)\Gamma\left(n+\lambda+\frac{1}{2}\right)} P_n^{\left(\lambda-\frac{1}{2},\, \lambda-\frac{1}{2}\right)}(x). \tag{64}$$

Formula (65) from [11] implies

$$E_{R_{s,t}}(X_N) = \mathcal{O}\left(N^{1+\frac{s}{d}}\right). \tag{65}$$

Hence,

$$E_s^{(d)}(X_N) = E_{H_{s,t}}(X_N) + \mathcal{O}\left(N^{1+\frac{s}{d}}\right), \quad \lambda > s-1, \tag{66}$$

where we have used formulas (61), (62) and (65).

The polynomials $H_{s,t}(\langle \mathbf{x}, \mathbf{x}_j \rangle)$, $\mathbf{x} \in \mathbb{S}^d$, are spherical polynomials of degree t for each $j = 1, \dots, N$ with note set X_N. So, an equal weight integration rule with node set X_N integrates $H_{s,t}$ exactly, and

$$\begin{aligned} E_{H_{s,t}}(X_N) =& \sum_{j=1}^{N} \sum_{i=1}^{N} H_{s,t}(\langle \mathbf{x}_i, \mathbf{x}_j \rangle) - \sum_{j=1}^{N} \sum_{\substack{i=1, \\ \mathbf{x}_i \in S(\pm \mathbf{x}_j; \alpha_N)}}^{N} H_{s,t}(\langle \mathbf{x}_i, \mathbf{x}_j \rangle) + \mathcal{O}\big(N H_{s,t}(1)\big) \\ =& N^2 \int\limits_{\mathbb{S}^d} H_{s,t}(\langle \mathbf{x}, \mathbf{y} \rangle) d\sigma_d(\mathbf{x}) + \mathcal{O}\big(N H_{s,t}(1)\big), \quad \mathbf{y} \in \mathbb{S}^d. \end{aligned} \tag{67}$$

From relations (13), (14), (16) and (63) we obtain

$$\begin{aligned} &H_{s,t}(1) \\ &= 2^{1-s} \Gamma\left(\lambda - \frac{s}{2} + \frac{1}{2}\right) \sum_{n=0}^{t-1} \frac{(n+\lambda)\Gamma\big(n+\frac{s}{2}\big)\Gamma(n+2\lambda)}{\Gamma\big(n+2\lambda-\frac{s}{2}+1\big)\Gamma\big(\frac{s}{2}\big)\Gamma\big(n+\lambda+\frac{1}{2}\big)} P_n^{\left(\lambda-\frac{1}{2},\, \lambda-\frac{1}{2}\right)}(1) \\ &= 2^{1-s} \frac{\Gamma\big(\lambda - \frac{s}{2} + \frac{1}{2}\big)}{\Gamma\big(\frac{s}{2}\big)\Gamma\big(\lambda+\frac{1}{2}\big)} \sum_{n=0}^{t} \frac{(n+\lambda)\Gamma\big(n+\frac{s}{2}\big)\Gamma(n+2\lambda)}{\Gamma\big(n+2\lambda-\frac{s}{2}+1\big)\Gamma(n+1)} \\ &\ll \sum_{n=1}^{t} n(n+2\lambda)^{\frac{s}{2}-1} n^{\frac{s}{2}-1} \ll t^s. \end{aligned} \tag{68}$$

Let now estimate the integral from (67). Substituting $\gamma = \lambda - \frac{1}{2}$, $\alpha = \frac{d}{2} - 1$ in formula (18), we have

$$\begin{aligned} &P_n^{(\lambda-\frac{1}{2},\lambda-\frac{1}{2})}(x) \\ &= \frac{\left(\lambda+\frac{1}{2}\right)_n}{(2\lambda)_n} \sum_{k=0}^{[\frac{n}{2}]} \frac{(d-1)_{n-2k}}{\left(\frac{d}{2}\right)_{n-2k}} \frac{(\lambda)_{n-k}\big(\frac{d}{2}+\frac{1}{2}\big)_{n-2k}\big(\lambda-\frac{d}{2}+\frac{1}{2}\big)_k}{\big(\frac{d}{2}+\frac{1}{2}\big)_{n-k}\big(\frac{d}{2}-\frac{1}{2}\big)_{n-2k} k!} P_{n-2k}^{\left(\frac{d}{2}-1,\frac{d}{2}-1\right)}(x). \end{aligned} \tag{69}$$

Since

$$\int\limits_{\mathbb{S}^d} P_n^{\left(\frac{d}{2}-1,\frac{d}{2}-1\right)}(\mathbf{x}) d\sigma_d(\mathbf{x}) = \gamma_d \int\limits_{-1}^{1} P_n^{\left(\frac{d}{2}-1,\frac{d}{2}-1\right)}(t)(1-t^2)^{\frac{d}{2}-1} = 0, \quad n \geq 1, \tag{70}$$

where we have used Func–Hecke formula and orthogonality relation for Jacobi polynomials, then (69) yields

$$\int_{\mathbb{S}^d} P_n^{\left(\lambda-\frac{1}{2},\lambda-\frac{1}{2}\right)}(\mathbf{x})d\sigma_d(\mathbf{x}) = \begin{cases} 0 & \text{if } n = 2m+1, \\ \frac{\left(\lambda+\frac{1}{2}\right)_n}{(2\lambda)_n} \frac{(\lambda)_{\frac{n}{2}}\left(\lambda-\frac{d}{2}+\frac{1}{2}\right)_{\frac{n}{2}}}{\left(\frac{d}{2}+\frac{1}{2}\right)_{\frac{n}{2}}\left(\frac{n}{2}\right)!} & \text{if } n = 2m. \end{cases} \tag{71}$$

So,

$$\int_{\mathbb{S}^d} H_{s,t}(\langle \mathbf{x}, \mathbf{y}\rangle)d\sigma_d(\mathbf{x})$$

$$= 2^{1-s}\frac{\Gamma\left(\lambda-\frac{s}{2}+\frac{1}{2}\right)}{\Gamma\left(\frac{s}{2}\right)}\sum_{n=0}^{\left[\frac{t-1}{2}\right]}\frac{(2n+\lambda)\Gamma\left(2n+\frac{s}{2}\right)\Gamma(2n+2\lambda)}{\Gamma\left(2n+2\lambda-\frac{s}{2}+1\right)\Gamma\left(2n+\lambda+\frac{1}{2}\right)}\frac{\left(\lambda+\frac{1}{2}\right)_{2n}}{(2\lambda)_{2n}}\frac{(\lambda)_n\left(\lambda-\frac{d}{2}+\frac{1}{2}\right)_n}{\left(\frac{d}{2}+\frac{1}{2}\right)_n n!}, \tag{72}$$

where we have used (39) and (71).

Thus, if $s > d$, then

$$\int_{\mathbb{S}^d} H_{s,t}(\langle \mathbf{x}, \mathbf{y}\rangle)d\sigma_d(\mathbf{x}) \ll t^{s-d} \ll N^{-1+\frac{s}{d}} \tag{73}$$

and the relations (67), (68) and (73) imply

$$E_s^{(d)}(X_N) \ll N^{1+s}. \tag{74}$$

This implies (27).

If $s = d$, then using (13) and (14) in (72) we have

$$\int_{\mathbb{S}^d} H_{d,t}(\langle \mathbf{x}, \mathbf{y}\rangle)d\sigma_d(\mathbf{x})$$

$$= 2^{2\lambda-d}\gamma_d\sum_{n=0}^{\left[\frac{t-1}{2}\right]}\frac{(2n+\lambda)\Gamma\left(2n+\frac{d}{2}\right)}{\Gamma\left(2n+2\lambda-\frac{d}{2}+1\right)}\frac{\Gamma(n+\lambda)\Gamma\left(n+\lambda-\frac{d}{2}+\frac{1}{2}\right)}{\Gamma\left(n+\frac{d}{2}+\frac{1}{2}\right)\Gamma(n+1)}$$

$$= 2^{2\lambda-d}\gamma_d\sum_{n=1}^{\left[\frac{t-1}{2}\right]}\frac{(2n+\lambda)\left(n^{2\lambda-d-1}+Q_{2\lambda-d-1}(n)\right)}{(2n)^{2\lambda-d+1}+Q_{2\lambda-d}(n)}$$

$$= \sum_{n=1}^{\left[\frac{t-1}{2}\right]} n^{-1} + \mathcal{O}(1) = \gamma_d \log t + \mathcal{O}(1), \tag{75}$$

where $Q_m(n)$ denotes an algebraic polynomial of order m.

Formulas (67), (68) and (73) imply (27).

As there exist $C_1, C_2 > 0$, such that

$$\log t \leq \log(C_1 N^{\frac{1}{d}}) = \log N + \mathcal{O}(1)$$

and

$$\log t \geq \log(C_2 N^{\frac{1}{d}}) = \log N + \mathcal{O}(1),$$

we have that (29) holds.

Theorem 2 is proved. □

Acknowledgements The author is supported by the Austrian Science Fund FWF project F5503 part of the Special Research Program (SFB) "Quasi-Monte Carlo Methods: Theory and Applications" and partially is supported by grant of NAS of Ukraine for groups of young scientists (project No16-10/2018).

References

1. Andrews, G.E., Askey, R., Roy, R.: Special Functions. Cambridge University Press (1999)
2. Betermin, L., Sandier, E.: Renormalized energy and asymptotic expansion of optimal logarithmic energy on the sphere. Constr. Approx. **47**(1), 39–44 (2018)
3. Bondarenko, A., Radchenko, D., Viazovska, M.: Optimal asymptotic bounds for spherical designs. Ann. Math. **178**(2), 443–452 (2013)
4. Bondarenko, A., Radchenko, D., Viazovska, M.: Well-separated spherical designs. Constr. Approx. **41**(1), 93–112 (2015)
5. Boyvalenkov, P.G., Dragnev, P.D., Hardin, D.P., Saff, E.B., Stoyanova, M.M.: Universal upper and lower bounds on energy of spherical designs. Dolomites Res. Notes Approx. **8**(Special Issue), 51–65 (2015)
6. Brauchart, J.S.: Optimal logarithmic energy points on the unit sphere. Math. Comput. **77**(263), 1599–1613 (2008)
7. Brauchart, J.S., Grabner, P.J.: Distributing many points on spheres: minimal energy and designs. J. Complex. **31**(3), 293–326 (2015)
8. Brauchart, J.S., Hardin,D.P., Saff, E.B.: The next-order term for optimal Riesz and logarithmic energy asymptotics on the sphere. Contemp. Math. **578**, 31–61 (2012)
9. Brauchart, J.S., Hesse, K.: Numerical integration over spheres of arbitrary dimension. Constr. Approx. **25**(1), 41–71 (2007)
10. Delsarte, P., Goethals, J.M., Seidel, J.J.: Spherical codes and designs. Geom. Dedicata **6**(3), 363–388 (1977)
11. Grabner, P.J., Stepanyuk, T.A.: Comparison of probabilistic and deterministic point sets on the sphere. J. Approx. Theory **239**, 128–143 (2019)
12. Hesse, K.: The s-energy of spherical designs on S^2. Adv. Comput. Math. **30**(1), 37–59 (2009)
13. Hesse, K., Leopardi, P.: The coulomb energy of spherical designs on S^2. Adv. Comput. Math. **28**(4), 331–354 (2008)
14. Kuijlaars, A.B.J., Saff, E.B.: Asymptotics for minimal discrete energy on the sphere. Trans. Am. Math. Soc. **350**(2), 523–538 (1998)
15. Magnus, W., Oberhettinger, F., Soni, R.P.: Formulas and theorems for the special functions of mathematical physics, 3rd enlarged edn. Die Grundlehren der mathematischen Wissenschaften, vol. 52. Springer New York, Inc., New York (1966)

16. Müller, C.: Spherical harmonics. Lecture Notes in Mathematics, vol. 17. Springer, Berlin-New York (1966)
17. Szegő, G.: Orthogonal polynomials, 4th edn. American Mathematical Society, Providence, R.I., American Mathematical Society, Colloquium Publications, Vol. XXIII (1975)
18. Wagner, G.: On the means of distances on the surface of a sphere II (upper bounds). Pacific J. Math. **154**(2), 381–396 (1992)

Rank-1 Lattices and Higher-Order Exponential Splitting for the Time-Dependent Schrödinger Equation

Yuya Suzuki and Dirk Nuyens

Abstract In this paper, we propose a numerical method to approximate the solution of the time-dependent Schrödinger equation with periodic boundary condition in a high-dimensional setting. We discretize space by using the Fourier pseudo-spectral method on rank-1 lattice points, and then discretize time by using a higher-order exponential operator splitting method. In this scheme the convergence rate of the time discretization depends on properties of the spatial discretization. We prove that the proposed method, using rank-1 lattice points in space, allows to obtain higher-order time convergence, and, additionally, that the necessary condition on the space discretization can be independent of the problem dimension d. We illustrate our method by numerical results from 2 to 8 dimensions which show that such higher-order convergence can really be obtained in practice.

Keywords Time-dependent schrödinger equation · Quasi-Monte Carlo · Pseudo-spectral method · Higher-order operator splitting

1 Introduction

Rank-1 lattice points have been widely used in the context of high-dimensional problems. Their traditional usage is in numerical integration, see, e.g., [5, 19] and references therein. In this work, we use rank-1 lattice points for function approximation, to approximate the solution of the time-dependent Schrödinger equation (TDSE). Function approximation using rank-1 lattice points has recently received more attention, see, e.g., [3, 11–14, 21]. In [13], Li and Hickernell introduced the pseudo-spectral Fourier collocation method using rank-1 lattice rules. Due to the rank-1 lattice struc-

Y. Suzuki (✉) · D. Nuyens
Computer Science Department, Numerical Analysis and Applied Mathematics Section, KU Leuven, Celestijnenlaan 200a, 3001 Leuven, Belgium
e-mail: yuya.suzuki@cs.kuleuven.be

D. Nuyens
e-mail: dirk.nuyens@cs.kuleuven.be

B. Tuffin and P. L'Ecuyer (eds.), *Monte Carlo and Quasi-Monte Carlo Methods*, Springer Proceedings in Mathematics & Statistics 324,
https://doi.org/10.1007/978-3-030-43465-6_24

ture, Fourier pseudo-spectral methods can be efficiently implemented using one-dimensional Fast Fourier transformations (FFTs). This is well known, and we state the exact form in Theorem 1 together with other useful properties of approximations on rank-1 lattice points.

To simulate many particles in the quantum world is a computationally challenging problem. For the TDSE, the dimensionality of the problem increases with the number of particles of the system. In the present paper, the following form is considered:

$$\begin{aligned} \mathrm{i}\,\gamma\,\frac{\partial u}{\partial t}(\boldsymbol{x},t) &= -\frac{\gamma^2}{2}\,\Delta u(\boldsymbol{x},t) + v(\boldsymbol{x})\,u(\boldsymbol{x},t), \\ u(\boldsymbol{x},0) &= g(\boldsymbol{x}), \end{aligned} \tag{1}$$

where i represents the imaginary unit, $\boldsymbol{x}$ is the spatial position in the d-dimensional torus $\mathbb{T}^d = \mathbb{T}([0,1]^d) \simeq [0,1]^d$, the time t is positive valued, and γ is a small positive parameter. The function $u(\boldsymbol{x},t)$ is the sought solution, while $v(\boldsymbol{x})$ and $g(\boldsymbol{x})$ are the potential and initial conditions respectively. The Laplacian can be interpreted as $\Delta = \sum_{i=1}^{M}\sum_{j=1}^{D} \partial^2/\partial x_{i,j}^2 = \sum_{i=1}^{d} \partial^2/\partial x_i^2$ where M is the number of particles and D is the physical dimensionality with $MD = d$. We note that the above form of the TDSE becomes equivalent after substitution to the following form which is common in the context of physics:

$$\mathrm{i}\,\hbar\,\frac{\partial \psi}{\partial t}(\boldsymbol{x},t) = -\frac{\hbar^2}{2m}\,\Delta \psi(\boldsymbol{x},t) + v(\boldsymbol{x})\,\psi(\boldsymbol{x},t),$$

where $\hbar$ is the reduced Planck constant and m is the mass.

The form (1) of the TDSE has been studied from various perspectives of numerical analysis [7, 9, 15, 22]. In the present paper, we focus on two perspectives; high-dimensionality and higher-order convergence in time stepping. For the first point, Gradinaru [7] proposed to use sparse grids for the physical space. In [21], the current authors used rank-1 lattice points to prove second order convergence for the time discretization using Strang splitting and numerically compared results with the sparse grid approach from [7]. The numerical result using rank-1 lattices showed the expected second order convergence even up to 12 dimensions. Hence rank-1 lattice points perform thereby much better than the sparse grid approach. The second point, higher-order convergence in time stepping, is successfully achieved by Thalhammer [22] using higher-order exponential operator splitting. In that paper, the spatial discretization was done by a full grid and therefore was limited to lower dimensional cases ($d \leq 3$).

The rest of this paper is organized as follows: Sect. 2 describes the proposed method consisting of the higher-order exponential splitting method and Fourier pseudo-spectral method using rank-1 lattices. Section 3 shows numerical results with various settings. The main aim here is to show higher-order time stepping convergence in higher-dimensional cases. Finally, Sect. 4 concludes the present paper with a short summary. Throughout the present paper, we denote the set of integer numbers by $\mathbb{Z}$ and the ring of integers modulo n by $\mathbb{Z}_n := \{0, 1, \ldots, n-1\}$. We distinguish between the normal equivalence in congruence modulo n as $a \equiv b \pmod{n}$ and the binary operation modulo n denoted by $\bmod n$ which returns the corresponding value in $\mathbb{Z}_n$ for $\bmod n$ and in $\mathbb{T}$ for $\bmod 1$.

2 The Numerical Method

In this section, we describe necessary ingredients of our method. For the conciseness, we restrict ourselves to the rank-1 lattice points instead of general rank-r lattice points. However, our method is indeed possible to generalize to rank-r lattice points, similar as in [21].

We use a rank-1 lattice point set and an associated anti-aliasing set for the Fourier pseudo-spectral method. For using the Fourier pseudo-spectral method, one obvious choice is regular grids [22], but the number of points increases too quickly in terms of the number of dimensions. To mitigate this problem, Gradinaru [6, 7] proposed to use sparse grids. For the same reason we introduced lattice points in [21] to get first and second order time convergence, and obtained much better results compared to [6, 7].

2.1 *Rank-1 Lattice Point Sets and the Associated Anti-aliasing Sets*

A rank-1 lattice point set $\Lambda(z, n)$ is fully determined by the modulus n and a generating vector $z \in \mathbb{Z}_n^d$:

$$\Lambda(z, n) := \left\{ \frac{zk}{n} \bmod 1 : k \in \mathbb{Z} \right\}.$$

Usually, all components of the generating vector are chosen to be relatively prime to n which means all points have different values in each coordinate and the number of points is exactly n. The generating vector determines the *quality* of the rank-1 lattice points. Of course, the quality criterion needs to take into account what the lattice points will be used for. A well studied setting is numerical integration, e.g., [17, 19] and [16, Chap. 5]. Function approximation using lattice points is relatively new. In that context, we refer to [3, 11, 12]. We call $\mathscr{A}(z, n) \subset \mathbb{Z}^d$ an anti-aliasing set for the lattice point set $\Lambda(z, n)$ if

$$z \cdot \boldsymbol{h} \not\equiv z \cdot \boldsymbol{h}' \pmod{n} \quad \text{for all } \boldsymbol{h}, \boldsymbol{h}' \in \mathscr{A}(z, n), \quad \boldsymbol{h} \neq \boldsymbol{h}'.$$

We remark that the anti-aliasing set is not uniquely determined and the cardinality $|\mathscr{A}(z,n)| \le n$. By using the *dual lattice* $\Lambda^\perp(z,n) := \{\boldsymbol{h} \in \mathbb{Z}^d : \boldsymbol{z}\cdot\boldsymbol{h} \equiv 0 \pmod{n}\}$, we can rewrite the condition as $\boldsymbol{h} - \boldsymbol{h}' \notin \Lambda^\perp(z,n)$ for $\boldsymbol{h} \neq \boldsymbol{h}'$. If we have the full cardinality $|\mathscr{A}(z,n)| = n$, we can divide $\mathbb{Z}^d$ into conjugacy classes:

$$\begin{aligned}\mathbb{Z}^d &= \biguplus_{\boldsymbol{h}\in\Lambda^\perp(z,n)} (\boldsymbol{h} + \mathscr{A}(z,n)) \\ &= \biguplus_{\boldsymbol{h}\in\mathscr{A}(z,n)} \{\boldsymbol{h}' \in \mathbb{Z}^d : \boldsymbol{z}\cdot\boldsymbol{h}' \equiv \boldsymbol{z}\cdot\boldsymbol{h} \pmod{n}\} \\ &= \biguplus_{j\in\mathbb{Z}_n} \{\boldsymbol{h} \in \mathbb{Z}^d : \boldsymbol{z}\cdot\boldsymbol{h} \equiv j \pmod{n}\},\end{aligned} \tag{2}$$

where $\uplus$ is the union of conjugacy classes.

2.2 Korobov Spaces

Rank-1 lattices are closely related to *Korobov spaces* which are reproducing kernel Hilbert spaces of Fourier series. The Korobov space $E_\alpha(\mathbb{T}^d)$ is given by

$$E_\alpha(\mathbb{T}^d) := \left\{ f \in L_2(\mathbb{T}^d) : \|f\|^2_{E_\alpha(\mathbb{T}^d)} := \sum_{\boldsymbol{h}\in\mathbb{Z}^d} |\widehat{f}(\boldsymbol{h})|^2\, r_\alpha^2(\boldsymbol{h}) < \infty \right\},$$

where

$$r_\alpha^2(\boldsymbol{h}) := \prod_{j=1}^{d} \max(|h_j|^{2\alpha}, 1). \tag{3}$$

The parameter $\alpha \ge 1/2$ is called the smoothness parameter which determines the rate of decay of the Fourier coefficients. To ensure regularity of the solution of the TDSE (1) and to prove that our method gives higher-order convergence for the temporal discretization, we will assume that the initial condition $g(\boldsymbol{x})$ and the potential function $v(\boldsymbol{x})$ are in the Korobov space with given smoothness, see Lemma 1 and Theorem 2.

2.3 Fourier Pseudo-Spectral Methods Using Rank-1 Lattices

We approximate the solution of the TDSE (1) by the truncated Fourier series. To ensure the solution to be regular enough so that the Fourier expansion makes sense (e.g., uniqueness, continuity, point-wise convergence), we require all functions to be in *Wiener algebra* $A(\mathbb{T}^d)$:

$$A(\mathbb{T}^d) := \{f \in L_2(\mathbb{T}^d) : \|f\|_{A(\mathbb{T}^d)} := \sum_{\boldsymbol{h}\in\mathbb{Z}^d} |\widehat{f}(\boldsymbol{h})| < \infty\}.$$

For $\alpha > 1/2$, we have $E_\alpha(\mathbb{T}^d) \subset A(\mathbb{T}^d)$. The following lemma shows the regularity of the solution, and the TDSE (1) in terms of Fourier coefficients and was already stated and proven in [21].

Lemma 1 (Regularity of solution and Fourier expansion) *Given the TDSE* (1) *with* $v, g \in E_\alpha(\mathbb{T}^d)$ *and* $\alpha \geq 2$*, then the solution* $u(\boldsymbol{x}, t) \in E_\alpha(\mathbb{T}^d)$ *for all finite* $t \geq 0$ *and therefore*

$$u(\boldsymbol{x}, t) = \sum_{\boldsymbol{h}\in\mathbb{Z}^d} \widehat{u}(\boldsymbol{h}, t) \exp(2\pi \mathrm{i}\, \boldsymbol{h}\cdot\boldsymbol{x}), \tag{4}$$

with

$$\mathrm{i}\,\gamma\, \widehat{u}'(\boldsymbol{h}, t) = 2\pi^2\gamma^2\, \|\boldsymbol{h}\|_2^2\, \widehat{u}(\boldsymbol{h}, t) + \widehat{f}(\boldsymbol{h}, t), \tag{5}$$

for all $\boldsymbol{h} \in \mathbb{Z}^d$*, with* $\widehat{u}'(\boldsymbol{h}, t) = (\partial/\partial t)\,\widehat{u}(\boldsymbol{h}, t)$ *and* $\widehat{f}(\boldsymbol{h}, t)$ *the Fourier coefficients of* $f(\boldsymbol{x}, t) := u(\boldsymbol{x}, t)\, v(\boldsymbol{x})$.

We then truncate the Fourier series (4) to a finite sum on an anti-aliasing set $\mathscr{A}(z, n)$ associated to a rank-1 lattice $\Lambda(z, n)$ to get the approximation

$$u_a(\boldsymbol{x}, t) := \sum_{\boldsymbol{h}\in\mathscr{A}(z,n)} \widehat{u}_a(\boldsymbol{h}, t) \exp(2\pi \mathrm{i}\, \boldsymbol{h}\cdot\boldsymbol{x}), \tag{6}$$

with the approximated coefficients calculated by the rank-1 lattice rule

$$\widehat{u}_a(\boldsymbol{h}, t) := \frac{1}{n} \sum_{\boldsymbol{p}\in\Lambda(z,n)} u(\boldsymbol{p}, t) \exp(-2\pi \mathrm{i}\, \boldsymbol{h}\cdot\boldsymbol{p}). \tag{7}$$

The subscript a of $u_a(\boldsymbol{x}, t)$ and $\widehat{u}_a(\boldsymbol{h}, t)$ indicates that these are approximations of $u(\boldsymbol{x}, t)$ and $\widehat{u}(\boldsymbol{h}, t)$ respectively. For simplicity of notation, we omit the time t in the rest of this section. Due to the rank-1 lattice structure and by choosing the anti-aliasing set to be of full size, we have the following properties:

Theorem 1 *Given a rank-*1 *lattice point set* $\Lambda(z, n)$ *and a corresponding anti-aliasing set* $\mathscr{A}(z, n)$ *with* $|\mathscr{A}(z, n)| = n$*, the following properties hold.*

(i) (Character property and dual character property) For any two vectors $\boldsymbol{h}$, $\boldsymbol{h}' \in \mathscr{A}(z, n)$

$$\frac{1}{n} \sum_{\boldsymbol{p}\in\Lambda(z,n)} \exp(2\pi \mathrm{i}\,(\boldsymbol{h} - \boldsymbol{h}')\cdot\boldsymbol{p}) = \delta_{\boldsymbol{h},\boldsymbol{h}'}, \tag{8}$$

where $\delta_{\boldsymbol{p},\boldsymbol{p}'}$ is the Kronecker delta function that is 1 if $\boldsymbol{p} = \boldsymbol{p}'$ and 0 otherwise. Also, for any two lattice points $\boldsymbol{p}, \boldsymbol{p}' \in \Lambda(\boldsymbol{z}, n)$

$$\frac{1}{n} \sum_{\boldsymbol{h} \in \mathscr{A}(\boldsymbol{z},n)} \exp(2\pi \mathrm{i}\, \boldsymbol{h} \cdot (\boldsymbol{p} - \boldsymbol{p}')) = \delta_{\boldsymbol{p},\boldsymbol{p}'}. \tag{9}$$

(ii) (Interpolation condition) If u_a is the approximation of a function $u \in A(\mathbb{T}^d)$ by truncating its Fourier series expansion to the anti-aliasing set $\mathscr{A}(\boldsymbol{z}, n)$ and by calculating the coefficients by the rank-1 lattice rule, cfr. (6) *and* (7)*, then for any $\boldsymbol{p} \in \Lambda(\boldsymbol{z}, n)$*

$$u_a(\boldsymbol{p}) = u(\boldsymbol{p}). \tag{10}$$

(iii) (Mapping through FFT) Define the following vectors:

$$\begin{aligned}
\boldsymbol{u} &:= \big(u(\boldsymbol{p}_k)\big)_{k=0,\ldots,n-1}, \\
\boldsymbol{u}_a &:= \big(u_a(\boldsymbol{p}_k)\big)_{k=0,\ldots,n-1}, \\
\widehat{\boldsymbol{u}}_a &:= \big(\widehat{u}_a(\boldsymbol{h}_\xi)\big)_{\xi=0,\ldots,n-1},
\end{aligned}$$

with $\boldsymbol{p}_k = \boldsymbol{z}k/n \bmod 1 \in \Lambda(\boldsymbol{z}, n)$, and where $\boldsymbol{h}_\xi \in \mathscr{A}(\boldsymbol{z}, n)$ is chosen such that $\boldsymbol{h} \cdot \boldsymbol{z}_\xi \equiv \xi \pmod{n}$. Then $\boldsymbol{u} = \boldsymbol{u}_a$ (by (ii)) is the collection of function values $u(\boldsymbol{p})$ on the lattice points $\boldsymbol{p} \in \Lambda(\boldsymbol{z}, n)$ and $\widehat{\boldsymbol{u}}_a$ is the collection of Fourier coefficients $\widehat{u}_a(\boldsymbol{h})$ (by using the lattice rule, cfr. (6) *and* (7)*) on the anti-aliasing indices $\boldsymbol{h} \in \mathscr{A}(\boldsymbol{z}, n)$. The 1-dimensional discrete Fourier transform and its inverse now maps $\boldsymbol{u}_a \in \mathbb{C}^n$ to $\widehat{\boldsymbol{u}}_a \in \mathbb{C}^n$ and back.*

(iv) (Aliasing) The approximated Fourier coefficients (7) *through the lattice rule $\Lambda(\boldsymbol{z}, n)$ alias the true Fourier coefficients in the following way*

$$\widehat{u}_a(\boldsymbol{h}) = \sum_{\boldsymbol{h}' \in \Lambda^\perp(\boldsymbol{z},n)} \widehat{u}(\boldsymbol{h} + \boldsymbol{h}') = \widehat{u}(\boldsymbol{h}) + \sum_{\boldsymbol{0} \neq \boldsymbol{h}' \in \Lambda^\perp(\boldsymbol{z},n)} \widehat{u}(\boldsymbol{h} + \boldsymbol{h}').$$

Proof We refer to [21, Theorem 2 and Lemma 3] where more general statement for rank-r lattices can be found. □

We remark that the above theorem can also be understood in terms of Fourier analysis on a finite Abelian group where the group, normally denoted as G, is the rank-1 lattice point set $\Lambda(\boldsymbol{z}, n)$ and the associated character group (Pontryagin dual) $\widehat{G} := \{\exp(2\pi \mathrm{i}\, \boldsymbol{h} \cdot \circ) : \boldsymbol{h} \in \mathscr{A}(\boldsymbol{z}, n)\}$ with $|\mathscr{A}(\boldsymbol{z}, n)| = n$. The (dual) character property is then to be understood as orthonormality of $\widehat{G}$ on $L_2(G)$. The interpolation condition can be seen as the representability of functions by using Fourier series. Due to this structure, the Plancherel theorem also holds:

$$\sum_{\boldsymbol{p} \in \Lambda(\boldsymbol{z},n)} f(\boldsymbol{p})\, \overline{g(\boldsymbol{p})} = \sum_{\boldsymbol{h} \in \mathscr{A}(\boldsymbol{z},n)} \widehat{f_a}(\boldsymbol{h})\, \overline{\widehat{g_a}(\boldsymbol{h})}$$

for $f, g \in L_2(G)$.

For readers who are not familiar with Fourier transforms on a rank-1 lattice, one intuitive way of seeing why one-dimensional FFTs are available is the following. The usual one-dimensional Fourier transform for equidistant points which is a scalar multiple of a unitary Fourier transform, for a function $f : \mathbb{T} \to \mathbb{C}$, can be written as

$$\widehat{f}(h) = \frac{1}{n} \sum_{k=0}^{n-1} f(k/n) \exp(-2\pi \mathrm{i}\, hk/n),$$

and the inverse

$$f(k/n) = \sum_{h=0}^{n-1} \widehat{f}(h) \exp(2\pi \mathrm{i}\, hk/n).$$

Now we see that the Fourier transform on a rank-1 lattice has the exact same structure for a function $f : \mathbb{T}^d \to \mathbb{C}$,

$$\widehat{f}(\boldsymbol{h}_\xi) = \frac{1}{n} \sum_{k=0}^{n-1} f(\boldsymbol{p}_k) \exp(-2\pi \mathrm{i}\, \boldsymbol{h}_\xi \cdot \boldsymbol{p}_k) = \frac{1}{n} \sum_{k=0}^{n-1} f(\boldsymbol{p}_k) \exp(-2\pi \mathrm{i}\, \xi k/n),$$

and

$$f(\boldsymbol{p}_k) = \sum_{\xi=0}^{n-1} \widehat{f}(\boldsymbol{h}_\xi) \exp(2\pi \mathrm{i}\, \boldsymbol{h}_\xi \cdot \boldsymbol{p}_k) = \sum_{k=0}^{n-1} \widehat{f}(\boldsymbol{h}_\xi) \exp(2\pi \mathrm{i}\, \xi k/n),$$

where we note $\boldsymbol{p}_k = \boldsymbol{z}k/n \bmod 1$ and $\boldsymbol{h}_\xi \cdot \boldsymbol{z} \equiv \xi \pmod{n}$. Hence we only need one-dimensional FFTs to transform functions on $\mathbb{T}^d$.

2.4 *Higher-Order Exponential Splitting*

For the temporal discretization, we employ a higher-order exponential splitting scheme (also called an exponential propagator), see, e.g., [1, 20, 22]. To describe the higher-order exponential splitting, let us consider the following ordinary differential equation:

$$y'(t) = (A + B)\, y(t), \qquad y(0) = y_0, \tag{11}$$

where A and B are differential operators. The solution for the Eq. (11) is $y(t) = \mathrm{e}^{(A+B)t} y_0$. However, often it is not possible to compute this exactly, and one needs to approximate the quantity with cheap computational cost. When both e^{At} and e^{Bt} can be computed easily, the higher-order exponential splitting is a powerful tool to approximate the solution $\mathrm{e}^{(A+B)t} y_0$. The approximated solution for this case is given by:

$$y(t + \Delta t) \approx \mathrm{e}^{b_1 B \, \Delta t} \, \mathrm{e}^{a_1 A \, \Delta t} \cdots \mathrm{e}^{b_s B \, \Delta t} \mathrm{e}^{a_s A \, \Delta t} \, y(t), \tag{12}$$

where a_i and b_i, $i = 1, \ldots, s$, are coefficients determined by the desired order of convergence p. In other words, if the splitting (12) satisfies

$$\| \mathrm{e}^{b_1 B \, \Delta t} \, \mathrm{e}^{a_1 A \, \Delta t} \cdots \mathrm{e}^{b_s B \, \Delta t} \mathrm{e}^{a_s A \, \Delta t} \, y(t) - \mathrm{e}^{(A+B)\Delta t} \, y(t) \|_X \le C (\Delta t)^{p+1}, \tag{13}$$

for some normed space X, where the constant C is independent of Δt, then the splitting is said to have p-th order. The number of steps s and the coefficients a_i, b_i can be determined according to the order p, see [8] for details. We evolve the time using this discretization from time 0, i.e.,

$$y_{k+1} = \mathrm{e}^{b_1 B \, \Delta t} \, \mathrm{e}^{a_1 A \, \Delta t} \cdots \mathrm{e}^{b_s B \, \Delta t} \mathrm{e}^{a_s A \, \Delta t} \, y_k, \qquad y_{\{k=0\}} = y_0.$$

By summing up the local errors (13) of each step $k = 1, \ldots, m$, where $t = m\Delta t$, gives the total error:

$$\| y_m - y(t) \|_X \le C \; m \Delta t \; (\Delta t)^p = C t (\Delta t)^p.$$

We call this quantity the total error in the L_2 sense, and this is the reason why the splitting is called to be of p-th order. The error coming from the exponential splitting can be related to commutators of two operators A and B, namely $[A, B] := AB - BA, [A, [A, B]] := A^2 B - 2ABA + BA^2$, etc. We introduce the notation for the p-th commutator by following [22]:

$$\mathrm{ad}_A^p(B) = [A, \mathrm{ad}_A^{p-1}(B)], \;\; \mathrm{ad}_A^0(B) = B,$$

where $p \ge 1$. When the p-th commutator is bounded, it is known that the p-th order exponential splitting gives the desired order, see [22, Lemma 1 and Theorem 1]. We also refer to [9, Theorem 2.1] for the second-order splitting (namely, Strang splitting) in a more abstract setting.

2.5 Higher-Order Exponential Splitting on Rank-1 Lattices

We apply the higher-order exponential splitting to the space discretized TDSE in this section. For solving the TDSE (1) in the dual space with finite number of Fourier basis functions, we will rewrite the problem in vector form. We let $\widehat{\boldsymbol{u}}_t := \big(\widehat{u}_a(\boldsymbol{h}_0, t), \ldots, \widehat{u}_a(\boldsymbol{h}_{n-1}, t)\big)$ the approximated solution at time t. Throughout time evolution, we use a fixed anti-aliasing set $\mathscr{A}(\boldsymbol{z}, n) = \{\boldsymbol{h}_\xi : \xi = 0, \ldots, n-1\}$ of full size $|\mathscr{A}(\boldsymbol{z}, n)| = n$, where we denote $\boldsymbol{h}_\xi \in \mathscr{A}(\boldsymbol{z}, n)$ as such a vector that $\boldsymbol{h}_\xi \cdot \boldsymbol{z} \equiv \xi \pmod{n}$. We obtain the following relation by imposing that (5) holds for all $\boldsymbol{h} \in \mathscr{A}(\boldsymbol{z}, n)$,

$$\mathrm{i}\,\gamma\,\widehat{\boldsymbol{u}}_t' = \frac{1}{2}\gamma^2 D_n \widehat{\boldsymbol{u}}_t + W_n \widehat{\boldsymbol{u}}_t, \tag{14}$$

with the initial condition $\widehat{\boldsymbol{u}}_0 = \widehat{\boldsymbol{g}}_a := (\widehat{g}_a(\boldsymbol{h}_0), \ldots, \widehat{g}_a(\boldsymbol{h}_{n-1}))$,

$$D_n := \operatorname{diag}\left((4\pi^2 \|\boldsymbol{h}_\xi\|_2^2)_{\xi=0,\ldots,n-1}\right), \tag{15}$$

and the potential multiplication operator $W_n := F_n V_n F_n^{-1}$ with

$$V_n := \operatorname{diag}\left(\left(v(\boldsymbol{p}_k)\right)_{k=0,\ldots,n-1}\right), \tag{16}$$

where F_n is the unitary Fourier matrix

$$F_n = \left(\frac{1}{\sqrt{n}} \exp(-2\pi\mathrm{i}\,\xi\xi'/n)\right)_{\xi,\xi'=0,\ldots,n-1}.$$

The approximation of the multiplication operator, W_n, is justified by the following lemma which is taken from [21].

Lemma 2 (Multiplication operator on rank-1 lattices) *Given a rank-1 lattice point set $\Lambda(\boldsymbol{z}, n)$ and corresponding anti-aliasing set $\mathscr{A}(\boldsymbol{z}, n)$ of full size, a potential function $v \in E_\alpha(\mathbb{T}^d)$ with $\alpha \geq 2$ and a function $u_a \in E_\beta(\mathbb{T}^d)$ with $\beta \geq 2$ with Fourier coefficients only supported on $\mathscr{A}(\boldsymbol{z}, n)$. Then the action in the Fourier domain restricted to $\mathscr{A}(\boldsymbol{z}, n)$ of multiplying with v, that is $f_a(\boldsymbol{x}) = v(\boldsymbol{x})\,u_a(\boldsymbol{x})$, on the nodes of the rank-1 lattice, and with f_a having Fourier coefficients restricted to the set $\mathscr{A}(\boldsymbol{z}, n)$, can be described by a circulant matrix $W_n \in \mathbb{C}^{n\times n}$ with $W_n = F_n\,V_n\,F_n^{-1}$, with V_n given by (16) and F_n the unitary Fourier matrix, where the element at position (ξ, ξ') of W_n is given by*

$$w_{\xi,\xi'} = w_{(\xi-\xi') \bmod n} = \sum_{\substack{\boldsymbol{h}\in\mathbb{Z}^d \\ \boldsymbol{h}\cdot\boldsymbol{z}\equiv\xi-\xi' \pmod{n}}} \widehat{v}(\boldsymbol{h}). \tag{17}$$

Proof We refer to [21, Lemma 5]. □

We approximate the solution of the ordinary differential equation (14)

$$\widehat{\boldsymbol{u}}_t = \mathrm{e}^{-\frac{\mathrm{i}}{\gamma} W_n t - \frac{\mathrm{i}\gamma}{2} D_n t}\,\widehat{\boldsymbol{u}}_0,$$

by applying the higher-order exponential splitting method (12):

$$\widehat{\boldsymbol{u}}_a^{k+1} = \mathrm{e}^{-b_1 \frac{\mathrm{i}}{\gamma} W_n \Delta t}\,\mathrm{e}^{-a_1 \frac{\mathrm{i}\gamma}{2} D_n \Delta t} \cdots \mathrm{e}^{-b_s \frac{\mathrm{i}}{\gamma} W_n \Delta t}\,\mathrm{e}^{-a_s \frac{\mathrm{i}\gamma}{2} D_n \Delta t}\,\widehat{\boldsymbol{u}}_a^k \quad \text{for } k = 0, 1, \ldots, m-1, \tag{18}$$

where again the coefficients a_i, b_i are determined according to the desired order of convergence, and

$$\mathrm{e}^{-\frac{\mathrm{i}}{2} W_n \Delta t} = F_n \operatorname{diag}\left((\mathrm{e}^{-\frac{\mathrm{i}}{2} v(\boldsymbol{p}_k) \Delta t})_{k=0,\dots,n-1} \right) F_n^{-1}.$$

The approximated solution at the time $t = k\Delta t$ is then obtained by stepping time Δt iteratively by (18). In the following we show the *commutator bounds* which correspond to [22, Hypothesis 3] and lead us to the total bound as in [22, Theorem 1].

Theorem 2 (p-th commutator bound and total error bound) *Given a rank-*1 *lattice with generating vector $z \in \mathbb{Z}^d$ and modulus n and a TDSE with a potential function $v \in E_\alpha(\mathbb{T}^d)$ with $\alpha > 2p + 1/2$ and an initial condition $g \in E_\beta(\mathbb{T}^d)$ with $\beta \geq 2$. Let $D = \frac{\gamma}{2} D_n$ and $W = \frac{1}{\gamma} W_n$ with D_n and $W_n = F_n V_n F_n^{-1}$ as defined in* (15) *and* (17), *and with V_n as defined in* (16) *using the potential function v.*

If the anti-aliasing set $\mathscr{A}(z, n) = \{\boldsymbol{h}_\xi \in \mathbb{Z}^d : \boldsymbol{h}_\xi \cdot z \equiv \xi \pmod{n}$ for $\xi = 0, \dots, n-1\}$, with full cardinality, is chosen such that its elements $\boldsymbol{h}_\xi$ have minimal ℓ_2 norm in the sense that,

$$\|\boldsymbol{h}_\xi\|_2 = \min_{\boldsymbol{h}' \in A(z,n,\xi)} \|\boldsymbol{h}'\|_2, \tag{19}$$

with

$$A(z, n, \xi) := \left\{ \boldsymbol{h} \in \mathbb{Z}^d : \boldsymbol{h} \cdot z \equiv \xi \pmod{n} \right\},$$

then for all $\boldsymbol{y} \in \mathbb{R}^n$ we have the following bound for the p-th commutator:

$$\|\mathrm{ad}_D^p(W)\, \boldsymbol{y}\|_2 \leq c\, \|(D + I)^p\, \boldsymbol{y}\|_2,$$

where c is a constant independent of n and $\boldsymbol{y}$.
This commutator condition and [22, Theorem 1] directly give us the total error bound for (14):

$$\|\widehat{\boldsymbol{u}}_t - \widehat{\boldsymbol{u}}_a^m\|_2 \leq C \|\widehat{\boldsymbol{u}}_0 - \widehat{\boldsymbol{u}}_a^0\|_2 + C'(\Delta t)^p \|(D + I)^p \widehat{\boldsymbol{u}}_0\|_2,$$

where $m\Delta t = t$ and the constants depend on t but not on m or Δt.

Proof Let $M := \mathrm{ad}_D^p(W)\, (D + I)^{-p}$. Since $(D + I)^p$ is is non-singular, the claim of the theorem is equivalent to the assertion that the induced ℓ_2 norm of the matrix $\|M\|_2 := \sup_{\boldsymbol{0} \neq \boldsymbol{y} \in \mathbb{R}^n} \|M\, \boldsymbol{y}\|_2 / \|\boldsymbol{y}\|_2$ is bounded independent of n. Each element of the matrix M is given by,

$$M = \left(\frac{(\|\boldsymbol{h}_\xi\|_2^2 - \|\boldsymbol{h}_{\xi'}\|_2^2)^p}{\gamma (\|\boldsymbol{h}_{\xi'}\|_2^2 + c_1)^p} w_{\xi,\xi'} \right)_{\xi,\xi'=0,\dots,n-1},$$

where the constant $c_1 = 1/(2\pi\gamma)^p > 0$. Now we bound $\|M\|_2$ by using $\|M\|_2 \leq \sqrt{\|M\|_1 \|M\|_\infty}$. First we bound $\|M\|_1$:

$$\|M\|_1 = \frac{1}{\gamma} \max_{\xi' \in \mathbb{Z}_n} \sum_{\substack{\xi=0 \\ \xi \neq \xi'}}^{n-1} \left| \frac{(\|\boldsymbol{h}_\xi\|_2^2 - \|\boldsymbol{h}_{\xi'}\|_2^2)^p}{(\|\boldsymbol{h}_{\xi'}\|_2^2 + c_1)^p} w_{\xi,\xi'} \right|$$

$$\leq \frac{1}{\gamma} \max_{\xi' \in \mathbb{Z}_n} \sum_{\substack{\xi=0 \\ \xi \neq \xi'}}^{n-1} \left| \frac{(\max(\|\boldsymbol{h}_\xi\|_2^{2p}, \|\boldsymbol{h}_{\xi'}\|_2^{2p})}{(\|\boldsymbol{h}_{\xi'}\|_2^2 + c_1)^p} w_{\xi,\xi'} \right|.$$

We notice that the diagonal components of M ($\xi = \xi'$) is always 0, hence we exclude such cases in the following argument. Because we collect the anti-aliasing set by minimizing the ℓ_2 norm (19), we have $\|\boldsymbol{h}_\xi\|_2 \leq \|\boldsymbol{h}'_\xi\|_2$ for any $\boldsymbol{h}'_\xi \in A(\boldsymbol{z}, n, \xi)$. In particular, this holds for $\boldsymbol{h}'_\xi = \boldsymbol{h}_{\xi-\xi'} + \boldsymbol{h}_{\xi'}$ since $(\boldsymbol{h}_{\xi-\xi'} + \boldsymbol{h}_{\xi'}) \cdot \boldsymbol{z} \equiv \xi \pmod{n}$ for any choice of $\xi' = 0, \ldots, n-1$. This gives us the connection between $\|\boldsymbol{h}_\xi\|_2$ and $\|\boldsymbol{h}_{\xi'}\|_2$ using $\|\boldsymbol{h}_{\xi-\xi'}\|_2$:

$$\frac{\|\boldsymbol{h}_\xi\|_2^2}{\|\boldsymbol{h}_{\xi'}\|_2^2 + c_1} \leq \frac{\|\boldsymbol{h}_{\xi'} + \boldsymbol{h}_{\xi-\xi'}\|_2^2}{\|\boldsymbol{h}_{\xi'}\|_2^2 + c_1} \leq 4\|\boldsymbol{h}_{\xi-\xi'}\|_2^2,$$

for $\xi \neq \xi'$. We continue from the above bound of $\|M\|_1$,

$$\|M\|_1 \leq \frac{1}{\gamma} \max_{\xi' \in \mathbb{Z}_n} \sum_{\substack{\xi=0 \\ \xi \neq \xi'}}^{n-1} \left| \frac{(\max(\|\boldsymbol{h}_\xi\|_2^{2p}, \|\boldsymbol{h}_{\xi'}\|_2^{2p})}{(\|\boldsymbol{h}_{\xi'}\|_2^2 + c_1)^p} w_{\xi,\xi'} \right|$$

$$\leq \frac{1}{\gamma} \max_{\xi' \in \mathbb{Z}_n} \sum_{\substack{\xi=0 \\ \xi \neq \xi'}}^{n-1} \left| \max\left(\frac{\|\boldsymbol{h}_\xi\|_2^2}{\|\boldsymbol{h}_{\xi'}\|_2^2 + c_1}, 1 \right)^p w_{\xi,\xi'} \right|$$

$$\leq \frac{1}{\gamma} \max_{\xi' \in \mathbb{Z}_n} \sum_{\substack{\zeta=0 \\ \xi \neq \xi'}}^{n-1} \left| \max\left(4^p \|\boldsymbol{h}_{\xi-\xi'}\|_2^{2p}, 1\right) w_{\xi,\xi'} \right|$$

$$= \frac{1}{\gamma} \max_{\xi' \in \mathbb{Z}_n} \sum_{\substack{\xi=0 \\ \xi \neq \xi'}}^{n-1} \left| \left(4^p \|\boldsymbol{h}_{\xi-\xi'}\|_2^{2p}\right) w_{\xi,\xi'} \right|$$

$$\leq \frac{4^p}{\gamma} \max_{\xi' \in \mathbb{Z}_n} \sum_{\substack{\xi=0 \\ \xi \neq \xi'}}^{n-1} \|\boldsymbol{h}_{\xi-\xi'}\|_2^{2p} \left| \sum_{\boldsymbol{h} \in A(\boldsymbol{z}, n, \xi-\xi')} \widehat{v}(\boldsymbol{h}) \right|$$

$$\leq \frac{4^p}{\gamma} \max_{\xi' \in \mathbb{Z}_n} \sum_{\substack{\xi=0 \\ \xi \neq \xi'}}^{n-1} \sum_{\boldsymbol{h} \in A(\boldsymbol{z}, n, \xi-\xi')} \|\boldsymbol{h}\|_2^{2p} \, |\widehat{v}(\boldsymbol{h})|$$

$$\leq \frac{4^p}{\gamma} \sum_{\boldsymbol{h} \in \mathbb{Z}^d} \|\boldsymbol{h}\|_2^{2p} \, |\widehat{v}(\boldsymbol{h})|.$$

For the last inequality, we used the conjugacy decomposition (2). By using Cauchy–Schwarz inequality and multiplying and dividing by r_α, we have

$$\begin{aligned}\sum_{\boldsymbol{h}\in\mathbb{Z}^d} \|\boldsymbol{h}\|_2^{2p}\,|\widehat{v}(\boldsymbol{h})| &\le \left(\sum_{\boldsymbol{h}\in\mathbb{Z}^d} r_\alpha^2(\boldsymbol{h})\,|\widehat{v}(\boldsymbol{h})|^2\right)^{1/2} \left(\sum_{\boldsymbol{h}\in\mathbb{Z}^d} \frac{\|\boldsymbol{h}\|_2^{4p}}{r_\alpha^2(\boldsymbol{h})}\right)^{1/2} \\ &\le \|v\|_{E_\alpha(\mathbb{T}^d)} \left(\sum_{\boldsymbol{h}\in\mathbb{Z}^d} \frac{(\sqrt{d}\,\|\boldsymbol{h}\|_\infty)^{4p}}{r_\alpha^2(\boldsymbol{h})}\right)^{1/2} \\ &\le \|v\|_{E_\alpha(\mathbb{T}^d)} \left(\sum_{\boldsymbol{h}\in\mathbb{Z}^d} \frac{d^{2p}}{r_{\alpha-2p}^2(\boldsymbol{h})}\right)^{1/2} \\ &\le \|v\|_{E_\alpha(\mathbb{T}^d)} \left(d^{2p}\,(1+2\,\zeta(2\alpha-4p))^d\right)^{1/2} < \infty.\end{aligned}$$

This means we have bounded $\|M\|_1$ independent of n. For $\|M\|_\infty$ we can proceed in a similar way to obtain

$$\begin{aligned}\|M\|_\infty &= \max_{\xi\in\mathbb{Z}_n} \sum_{\substack{\xi'=0\\ \xi'\neq\xi}}^{n-1} \left| \frac{\left(\|\boldsymbol{h}_\xi\|_2^2 - \|\boldsymbol{h}_{\xi'}\|_2^2\right)^p}{\left(\|\boldsymbol{h}_{\xi'}\|_2^2 + c_1\right)^p}\, w_{\xi,\xi'} \right| \\ &\le \frac{4^p}{\gamma} \|v\|_{E_\alpha(\mathbb{T}^d)} \left(d^{2p}\,(1+2\,\zeta(2\alpha-4p))^d\right)^{1/2} < \infty.\end{aligned}$$

Therefore, we have $\|M\|_2 < \infty$ independent of n. The total error bound directly follows from this commutator bound and [22, Theorem 1]. □

3 Numerical Results

We demonstrate our method by showing some numerical results in this section. We construct rank-1 lattices by using the component-by-component (CBC) construction [4, 17]. The code for producing the rank-1 lattice is available online [18], `fastrank1expt.m`. With the script, we choose n being a power of 2 and generate the vector z which is optimized for integration in (unweighted) Korobov space with first order mixed derivatives, i.e., $\alpha = 1$. In Table 1 we display the generating vector z and the number of points n for the following numerical results. Using given n and z, we construct the anti-aliasing set in accordance with Theorem 2 in the following manner: (i) first we generate all integer vector $\boldsymbol{h} \in \mathbb{Z}^d$ in a bounded region $\|\boldsymbol{h}\| \le R$ for a well chosen R; (ii) then we sort the obtained set according to the ℓ_2 distance in ascending order; (iii) we calculate the value $m_{\boldsymbol{h}} := \boldsymbol{h} \cdot z \bmod n$ in the sorted order and store $\boldsymbol{h}$ in $\mathscr{A}(z, n)$ if the value $m_{\boldsymbol{h}}$ has not appeared before. We repeat this step (iii) until we have the full cardinality $|\mathscr{A}(z, n)| = n$.

Table 1 Parameters of the rank-1 lattice points for our numerical results

d	n	$z^\top$
2	2^{16}	$(1, 100135)$
4	2^{20}	$(1, 443165, 95693, 34519)$
6	2^{24}	$(1, 6422017, 7370323, 2765761, 8055041, 2959639)$
8	2^{24}	$(1, 6422017, 7370323, 2765761, 8055041, 2959639, 7161203, 4074015)$

3.1 Convergence with Respect to Time Step Size

We consider a common numerical setting as it is considered in [7, 9, 22] where Fourier pseudo-spectral methods are used. We calculate the error with different value of time steps against a reference solution. For the initial condition $g(\boldsymbol{x})$, we choose the *Gaussian wave packet* given by:

$$g(\boldsymbol{x}) := \left(\frac{2}{\pi\gamma}\right)^{d/4} \exp\left(-\frac{\sum_{j=1}^{d}\left(2\pi x_j - \pi\right)^2}{\gamma}\right)\frac{1}{c},$$

where the constant c is a normalizing constant to make $\|g\|_{L_2} = 1$. For the potential function v, we consider a *smooth potential* function

$$v_1(\boldsymbol{x}) = \prod_{j=1}^{d}(1 - \cos(2\pi x_j)),$$

and a *harmonic potential* function

$$v_2(\boldsymbol{x}) = \frac{1}{2}\sum_{j=1}^{d}(2\pi x_j - \pi)^2.$$

Our aim is to show the temporal discretization error $\|u_a(\boldsymbol{x}, t) - u_a^m(\boldsymbol{x})\|_{L_2}$ at fixed time $t = m\,\Delta t = 1$, for that sake we calculate a reference solution $u_a^M(\boldsymbol{x})$ with the finest time step size $\Delta t = 1/M = 1/10000$, as an approximation of $u_a(\boldsymbol{x}, t)$. We then vary the time step size $\Delta t = 1/m = 1/5, \ldots, 1/1000$ and calculate $u_a^m(\boldsymbol{x})$ to see the convergence plot of $\|u_a^M(\boldsymbol{x}) - u_a^m(\boldsymbol{x})\|_{L_2}$.

Table 2 Coefficients for the sixth-order method, calculated based on [10]

	a_j		b_j
$j = 1, 9$	0.392161444007314	$j = 1, 10$	0.196080722003657
$j = 2, 8$	0.332599136789359	$j = 2, 9$	0.362380290398337
$j = 3, 7$	−0.706246172557639	$j = 3, 8$	−0.186823517884140
$j = 4, 6$	0.0822135962935508	$j = 4, 7$	−0.312016288132044
$j = 5$	0.798543990934830	$j = 5, 6$	0.440378793614190
$j = 10$	0		

3.2 Sixth-Order Splitting

We recall that the higher-order exponential splitting is written as

$$y_{k+1} = \mathrm{e}^{b_1 B\, \Delta t}\, \mathrm{e}^{a_1 A\, \Delta t} \cdots \mathrm{e}^{b_s B\, \Delta t} \mathrm{e}^{a_s A\, \Delta t}\, y_k.$$

For the sixth-order method, we employ the coefficients a_j and b_j from [10] denoted as "s9odr6a" therein. We exhibit the coefficients in Table 2. We plot the results for dimension 2 to 8 in Fig. 1. The potential v_1 is not smooth enough on the boundary of $[0, 1]^d$ so it does not satisfy the required condition in the strict sense. The initial condition g and the potential v_2 meet all the required conditions. The expected sixth-order convergence is consistent in every plot. When the error reaches to the machine precision, the plot becomes flat. For the 2-dimensional case with the potential v_2, we see the convergence happening when the time step size is very small. This can be explained by a phenomenon, called instability of exponential splitting; this is caused by negative coefficients of the exponential splitting a_j and b_j, and is discussed in e.g., [2]. Especially in [2], commutator-free quasi-Magnus exponential integrators are proposed to avoid the issue, however, this is out of the scope of the present paper. The instability issue does not happen in a higher-dimensional settings.

3.3 Eighth-Order Splitting

For the eighth-order method, we employ the coefficients again from [10] denoted as "s17odr8a". The coefficients are shown in Table 3. The results are shown in Fig. 2 and we again see that the convergence rate is consistently eighth order in each plot. Most of the plot seems to be similar to Fig. 1 but with faster convergence, therefore they reach to the machine precision more quickly.

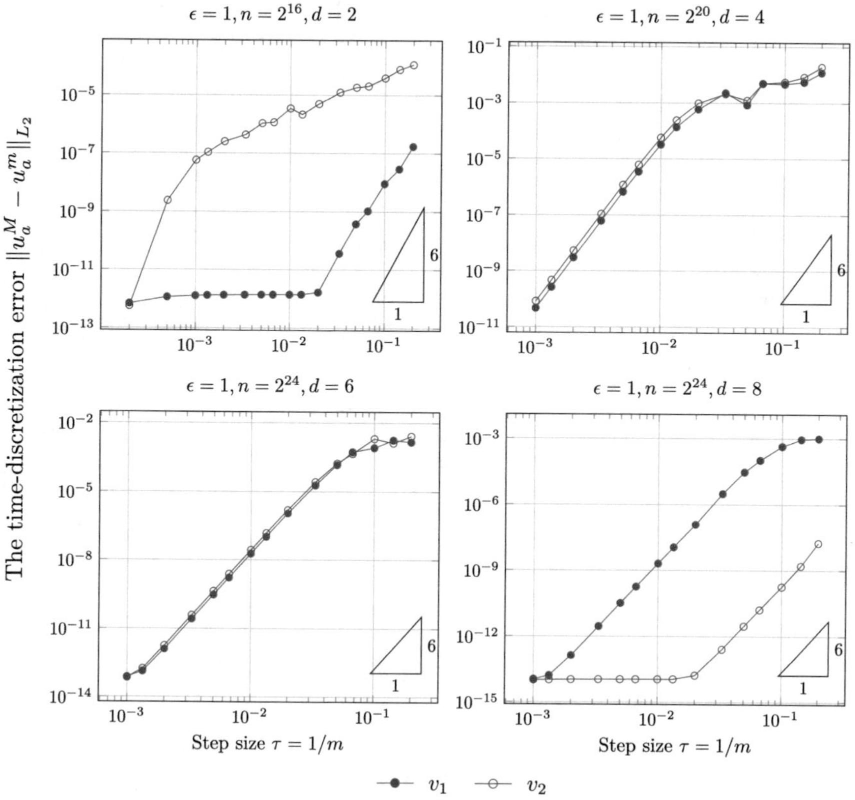

Fig. 1 The time-discretization error with the sixth-order method

Table 3 Coefficients for the eighth-order method, calculated based on [10]

	a_j		b_j
$j = 1, 17$	0.130202483088890	$j = 1, 18$	0.0651012415444150
$j = 2, 16$	0.561162981775108	$j = 2, 17$	0.345682732431999
$j = 3, 15$	−0.389474962644847	$j = 3, 16$	0.0858440095651306
$j = 4, 14$	0.158841906555156	$j = 4, 15$	−0.115316528044846
$j = 5, 13$	−0.395903894133238	$j = 5, 14$	−0.118530993789041
$j = 6, 12$	0.184539640978316	$j = 6, 13$	−0.105682126577461
$j = 7, 11$	0.258374387686322	$j = 7, 12$	0.221457014332319
$j = 8, 10$	0.295011723609310	$j = 8, 11$	0.276693055647816
$j = 9$	−0.605508533830035	$j = 9, 10$	−0.155248405110362
$j = 18$	0		

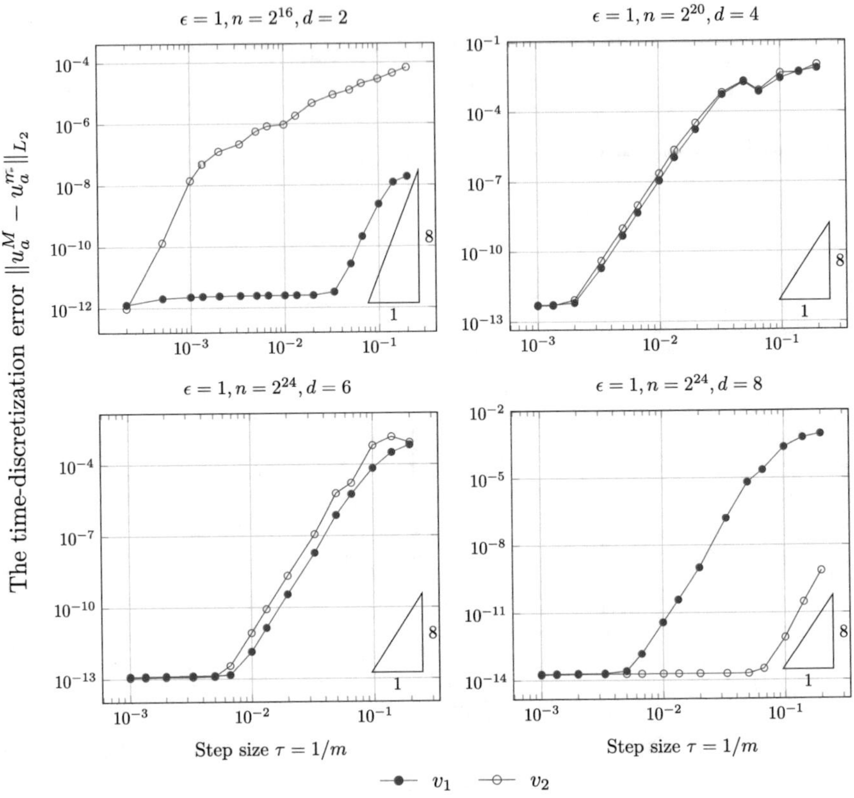

Fig. 2 The time-discretization error with the eighth-order method

4 Conclusion

We proposed a numerical method to solve the TDSE. With our method using the time step size Δt, the temporal discretization error converges like $\mathcal{O}((\Delta t)^p)$ given that the potential function is in Korobov space of smoothness greater than $2p + 1/2$. The numerical results (which are performed from 2 up to 8 dimensions) confirmed the theory and the rate of error convergence is consistent. By using rank-1 lattices, calculations of the time stepping operator and multiplications are efficiently done by only using one-dimensional FFTs.

Pseudo-spectral methods are widely used technique for solving partial differential equations. It is a common choice to use regular grids, but the number of nodes increases exponential with d. We have shown an alternative, rank-1 lattice pseudo-spectral methods where the number of points can be chosen freely by the user. In combination with higher-order splitting methods, the proposed method solves the TDSE with higher-order convergence in time.

Acknowledgements We thank three anonymous referees for their valuable comments which helped to improve readability of the manuscript. We also acknowledge financial support from the KU Leuven research fund (OT:3E130287).

References

1. Bandrauk, A.D., Shen, H.: Higher order exponential split operator method for solving time-dependent schrödinger equations **70**(2), 555–559 (1992)
2. Blanes, S., Casas, F., Thalhammer, M.: High-order commutator-free quasi-Magnus exponential integrators for non-autonomous linear evolution equations. Comput. Phys. Commun. **220**, 243–262 (2017)
3. Byrenheid, G., Kämmerer, L., Ullrich, T., Volkmer, T.: Tight error bounds for rank-1 lattice sampling in spaces of hybrid mixed smoothness. Numer. Math. **136**(4), 993–1034 (2017)
4. Cools, R., Kuo, F.Y., Nuyens, D.: Constructing embedded lattice rules for multivariable integration. SIAM J. Sci. Comput. **28**(6), 2162–2188 (2006)
5. Cools, R., Nuyens, D.: A Belgian view on lattice rules. In: Monte Carlo and Quasi-Monte Carlo Methods 2006, pp. 3–21. Springer, Berlin (2008)
6. Gradinaru, V.: Fourier transform on sparse grids: code design and the time dependent Schrödinger equation. Comput. Arch. Sci. Comput. **80**(1), 1–22 (2007)
7. Gradinaru, V.: Strang splitting for the time-dependent Schrödinger equation on sparse grids. SIAM J. Numer. Anal. **46**(1), 103–123 (2007/08)
8. Hairer, E., Lubich, C., Wanner, G.: Geometric numerical integration. Structure-Preserving Algorithms for Ordinary Differential Equations, 2nd edn. Springer Series in Computational Mathematics, vol. 31. Springer, Berlin (2006).
9. Jahnke, T., Lubich, C.: Error bounds for exponential operator splittings. BIT. Numer. Math. **40**(4), 735–744 (2000)
10. Kahan, W., Li, R.C.: Composition constants for raising the orders of unconventional schemes for ordinary differential equations. Math. Comput. **66**(219), 1089–1099 (1997)
11. Kuo, F.Y., Sloan, I.H., Woźniakowski, H.: Lattice rule algorithms for multivariate approximation in the average case setting. J. Complex. **24**(2), 283–323 (2008)
12. Kuo, F.Y., Wasilkowski, G.W., Woźniakowski, H.: Lattice algorithms for multivariate L_∞ approximation in the worst-case setting. **30**(3), 475–493 (2009)
13. Li, D., Hickernell, F.J.: Trigonometric spectral collocation methods on lattices. In: Recent Advances in Scientific Computing and Partial Differential Equations (Hong Kong, 2002). Contemporary Mathematics, vol. 330, pp. 121–132. American Mathematical Society, Providence, RI (2003)
14. Munthe-Kaas, H., Sørevik, T.: Multidimensional pseudo-spectral methods on lattice grids. Appl. Numer. Math. IMACS J. **62**(3), 155–165 (2012)
15. Neuhauser, C., Thalhammer, M.: On the convergence of splitting methods for linear evolutionary Schrödinger equations involving an unbounded potential. BIT. Numeri. Math. **49**(1), 199–215 (2009)
16. Niederreiter, H.: Random number generation and quasi-Monte Carlo methods, CBMS-NSF Regional Conference Series in Applied Mathematics, vol. 63. Society for Industrial and Applied Mathematics (SIAM), Philadelphia, PA (1992)
17. Nuyens, D.: The construction of good lattice rules and polynomial lattice rules. In: Kritzer, P., Niederreiter, H., Pillichshammer, F., Winterhof, A. (eds.) Uniform Distribution and Quasi-Monte Carlo Methods: Discrepancy, Integration and Applications, Radon Series on Computational and Applied Mathematics, vol. 15, pp. 223–256. De Gruyter, Berlin, Boston (2014)
18. Nuyens, D., Cools, R.: Fast component-by-component construction of rank-1 lattice rules with a non-prime number of points. J. Complex. **22**(1), 4–28 (2006)

19. Sloan, I.H., Joe, S.: Lattice Methods for Multiple Integration. Oxford Science Publications. The Clarendon Press, Oxford University Press, New York (1994)
20. Suzuki, M.: Fractal decomposition of exponential operators with applications to many-body theories and Monte Carlo simulations. Phys. Lett. A **146**(6), 319–323 (1990)
21. Suzuki, Y., Suryanarayana, G., Nuyens, D.: Strang splitting in combination with rank-1 and rank-*r* lattices for the time-dependent Schrödinger equation. SIAM J. Sci. Comput. to appear
22. Thalhammer, M.: High-order exponential operator splitting methods for time-dependent Schrödinger equations. SIAM J. Numer. Anal. **46**(4), 2022–2038 (2008)

An Analysis of the Milstein Scheme for SPDEs Without a Commutative Noise Condition

Claudine von Hallern and Andreas Rößler

Abstract In order to approximate solutions of stochastic partial differential equations (SPDEs) that do not possess commutative noise, one has to simulate the involved iterated stochastic integrals. Recently, two approximation methods for iterated stochastic integrals in infinite dimensions were introduced in [8]. As a result of this, it is now possible to apply the Milstein scheme by Jentzen and Röckner [2] to equations that need not fulfill the commutativity condition. We prove that the order of convergence of the Milstein scheme can be maintained when combined with one of the two approximation methods for iterated stochastic integrals. However, we also have to consider the computational cost and the corresponding effective order of convergence for a meaningful comparison with other schemes. An analysis of the computational cost shows that, in dependence on the equation, a combination of the Milstein scheme with any of the two methods may be the preferred choice. Further, the Milstein scheme is compared to the exponential Euler scheme and we show for different SPDEs depending on the parameters describing, e.g., the regularity of the equation, which of the schemes achieves the highest effective order of convergence.

Keywords SPDE · Milstein scheme · Euler scheme · Non-commutative noise · Iterated stochastic integral

C. von Hallern
Department of Mathematics, Christian-Albrechts-Universität zu Kiel, Christian-Albrechts-Platz 4, 24118 Kiel, Germany
e-mail: vonhallern@math.uni-kiel.de; leonhard@math.uni-kiel.de

A. Rößler (✉)
Institute of Mathematics, Universität zu Lübeck, Ratzeburger Allee 160, 23562 Lübeck, Germany
e-mail: roessler@math.uni-luebeck.de

B. Tuffin and P. L'Ecuyer (eds.), *Monte Carlo and Quasi-Monte Carlo Methods*, Springer Proceedings in Mathematics & Statistics 324,
https://doi.org/10.1007/978-3-030-43465-6_25

1 Introduction

It is well known that for a commutative stochastic differential equation the Milstein scheme can be easily implemented as no iterated stochastic integrals have to be simulated. However, if we deal with an SPDE which does not fulfill the commutativity condition, it is, in general, not possible to rewrite the expression in such a way that implementation becomes straightforward. In the following, we consider SPDEs of type

$$\mathrm{d}X_t = \big(AX_t + F(X_t)\big)\,\mathrm{d}t + B(X_t)\,\mathrm{d}W_t, \quad t \in (0,T], \quad X_0 = \xi. \tag{1}$$

In this work, we are concerned with the efficient approximation of the mild solution of equation (1) which does not need to have commutative noise by a higher order scheme, that is, we deal with equations where the commutativity condition

$$\big(B'(v)(B(v)u)\big)\tilde{u} = \big(B'(v)(B(v)\tilde{u})\big)u \tag{2}$$

for all $v \in H_\beta$, $u, \tilde{u} \in U_0$ does *not* have to be fulfilled. We consider the Milstein scheme for SPDEs recently proposed in [2] which reads as $Y_0^{N,K,M} = P_N\xi$ and

$$\begin{aligned} Y_{m+1}^{N,K,M} = P_N e^{Ah}\Bigg(& Y_m^{N,K,M} + hF(Y_m^{N,K,M}) + B(Y_m^{N,K,M})\Delta W_m^{K,M} \\ & + \int_{t_m}^{t_{m+1}} B'(Y_m^{N,K,M})\Big(\int_{t_m}^{s} B(Y_m^{N,K,M})\,\mathrm{d}W_r^K\Big)\,\mathrm{d}W_s^K\Bigg) \end{aligned} \tag{3}$$

for some $N, M, K \in \mathbb{N}$, $h = \frac{T}{M}$ and $m \in \{0, \ldots, M-1\}$. For details on the notation, we refer to Sects. 2.1 and 2.2. The main difficulty in the approximation of equations with non-commutative noise is the simulation of the iterated stochastic integrals, since it is not possible to rewrite integrals such as

$$\int_t^{t+h} B'(X_t)\Big(\int_t^s B(X_t)\,\mathrm{d}W_r^K\Big)\,\mathrm{d}W_s^K$$

for $h > 0$, $t, t+h \in [0,T]$ and $K \in \mathbb{N}$ in terms of increments of the approximation $(W_t^K)_{t\in[0,T]}$ of the Q-Wiener process $(W_t)_{t\in[0,T]}$ like in the commutative case, see [2]. Since the iterated stochastic integrals can, in general, not be computed explicitly, we need to approximate these terms. In [8], the authors recently proposed two algorithms to approximate integrals of type

$$\int_t^{t+h} \Psi\left(\Phi \int_t^s \mathrm{d}W_r\right)\mathrm{d}W_s \tag{4}$$

with $t \geq 0$, $h > 0$ for some operators $\Psi \in L(H, L(U,H)_{U_0})$, $\Phi \in L(U,H)_{U_0}$ and a Q-Wiener process $(W_t)_{t\in[0,T]}$. Applying these algorithms, it is possible to implement the Milstein scheme stated in (3) if we choose $\Psi = B'(Y_t)$ and $\Phi = B(Y_t)$ for some

$B : H_\beta \to L(U, H)_{U_0}$ and an approximation $Y_t \in H_\beta$ with $t \geq 0$ and $\beta \in [0, 1)$. For more details on the operators, we refer to [8] and Sect. 2.1. In this work, we combine the Milstein scheme with the approximation of the iterated stochastic integrals.

For finite dimensional stochastic differential equations, the issue of how to simulate iterated stochastic integrals was answered, e.g., by [3, 12]. In this setting, the Milstein scheme combined with the approximation as specified by [12] outperforms the method that was introduced in [3] in terms of the computational cost when the temporal order of convergence of the Milstein scheme is to be preserved. The results in [8] suggest that in the infinite dimensional setting of SPDEs, it is not obvious which of the two methods requires less computational effort. Therefore, in this work, we analyze the cost involved in the simulation for each of the two methods in detail and also compare the Milstein scheme combined with each method to the exponential Euler scheme.

2 Analysis of the Numerical Scheme

We present two versions of the Milstein scheme for non-commutative SPDEs in this section. To be precise, we analyze two schemes which differ by the method that is used to approximate the iterated stochastic integrals that are involved. We prove in Sect. 2.2 that the order of convergence that the Milstein scheme obtains for commutative equations, see [2], can be maintained if the iterated integrals are approximated by the methods introduced in [8]. In Sect. 2.3, these two versions of the Milstein scheme are compared to each other and to the exponential Euler scheme when the computational cost is also taken into account.

2.1 Setting and Assumptions

The setting that we work in is similar to the one considered for the Milstein scheme in [2] except that the commutativity condition (24) in their paper (see also Eq. (2)) does not have to be fulfilled, that we replace the space $L_{HS}(U_0, H)$ by $L(U, H)_{U_0} \subset L_{HS}(U_0, H)$ in assumption (A3) and that we introduce a projection operator in (A3).

Let $T \in (0, \infty)$ be fixed, let $(H, \langle\cdot,\cdot\rangle_H)$ and $(U, \langle\cdot,\cdot\rangle_U)$ denote some separable real-valued Hilbert spaces. We fix some probability space $(\Omega, \mathscr{F}, P)$ and denote by $(W_t)_{t\in[0,T]}$ a U-valued Q-Wiener process with respect to the filtration $(\mathscr{F}_t)_{t\in[0,T]}$ which fulfills the usual conditions. The operator $Q \in L(U)$ is assumed to be non-negative, symmetric and to have finite trace. We denote its eigenvalues by η_j with corresponding eigenvectors $\tilde{e}_j$ for $j \in \mathscr{J}$ with some countable index set $\mathscr{J}$ forming an orthonormal basis of U [10]. We employ the following series representation of the Q-Wiener process, see [10],

$$W_t = \sum_{\substack{j \in \mathcal{J} \\ \eta_j \neq 0}} \sqrt{\eta_j}\, \tilde{e}_j\, \beta_t^j, \quad t \in [0, T].$$

Here, $(\beta_t^j)_{t\in[0,T]}$ denote independent real-valued Brownian motions for all $j \in \mathcal{J}$ with $\eta_j \neq 0$. By means of the operator Q, we define the subspace $U_0 \subset U$ as $U_0 = Q^{\frac{1}{2}} U$. The set of Hilbert–Schmidt operators mapping from U to H is denoted by $L_{\mathrm{HS}}(U, H)$ and the space of linear bounded operators on U restricted to U_0 by $(L(U, H)_{U_0}, \|\cdot\|_{L(U,H)})$ with $L(U, H)_{U_0} := \{T : U_0 \to H \mid T \in L(U, H)\}$. Moreover, we designate $L^{(2)}(U, H) = L(U, L(U, H))$ and $L_{\mathrm{HS}}^{(2)}(U, H) = L_{\mathrm{HS}}(U, L_{\mathrm{HS}}(U, H))$.

Our aim is to approximate the mild solution of SPDE (1) and, therefore, we impose the following assumptions.

(A1) The linear operator $A\colon \mathcal{D}(A) \subset H \to H$ generates an analytic semigroup $S(t) = e^{At}$ for all $t \geq 0$. Let $\lambda_i \in (0, \infty)$ denote the eigenvalues of $-A$ with eigenvectors e_i for $i \in \mathcal{I}$ and some countable index set $\mathcal{I}$, i.e., it holds $-Ae_i = \lambda_i e_i$ for all $i \in \mathcal{I}$. Moreover, assume that $\inf_{i\in\mathcal{I}} \lambda_i > 0$ and that the eigenfunctions $\{e_i : i \in \mathcal{I}\}$ of $-A$ form an orthonormal basis of H, see [11]. Furthermore,

$$Av = \sum_{i\in\mathcal{I}} -\lambda_i \langle v, e_i\rangle_H e_i$$

for all $v \in \mathcal{D}(A)$. By means of A, we define the real Hilbert spaces $H_r := \mathcal{D}((-A)^r)$ for $r \in [0, \infty)$ with norm $\|x\|_{H_r} = \|(-A)^r x\|_H$ for $x \in H_r$.

(A2) For some $\beta \in [0, 1)$, assume that $F\colon H_\beta \to H$ is twice continuously Fréchet differentiable with $\sup_{v\in H_\beta} \|F'(v)\|_{L(H)} < \infty$ and $\sup_{v\in H_\beta} \|F''(v)\|_{L^2(H_\beta,H)} < \infty$.

(A3) The operator $B\colon H_\beta \to L(U, H)_{U_0}$ is twice continuously Fréchet differentiable with $\sup_{v\in H_\beta} \|B'(v)\|_{L(H,L(U,H))} < \infty$, $\sup_{v\in H_\beta} \|B''(v)\|_{L^{(2)}(H_\beta,L_{\mathrm{HS}}(U_0,H))} < \infty$. Assume that $B(H_\delta) \subset L_{\mathrm{HS}}(U_0, H_\delta)$ for some $\delta \in (0, \frac{1}{2})$ and that

$$\begin{aligned} \|B(u)\|_{L_{\mathrm{HS}}(U_0,H_\delta)} &\leq C(1 + \|u\|_{H_\delta}), \\ \|B'(v)PB(v) - B'(w)PB(w)\|_{L_{\mathrm{HS}}^{(2)}(U_0,H)} &\leq C\|v - w\|_H, \\ \|(-A)^{-\vartheta} B(v) Q^{-\alpha}\|_{L_{\mathrm{HS}}(U_0,H)} &\leq C(1 + \|v\|_{H_\gamma}) \end{aligned}$$

for some constant $C > 0$, all $u \in H_\delta$, $v, w \in H_\gamma$, where $\gamma \in \left[\max(\beta, \delta), \delta + \frac{1}{2}\right)$, $\alpha \in (0, \infty)$, $\vartheta \in \left(0, \frac{1}{2}\right)$, any projection operator $P\colon H \to \operatorname{span}\{e_i : i \in \tilde{\mathcal{I}}\} \subset H$ with finite index set $\tilde{\mathcal{I}} \subset \mathcal{I}$ and the case that P is the identity.

(A4) Assume that the initial value $\xi\colon \Omega \to H_\gamma$ fulfills $\mathrm{E}\left[\|\xi\|_{H_\gamma}^4\right] < \infty$ and that it is $\mathcal{F}_0$-$\mathcal{B}(H_\gamma)$-measurable.

In the following, we do not distinguish between the operator B and its extension $\tilde{B}\colon H \to L(U, H)_{U_0}$ which is globally Lipschitz continuous; this holds as $H_\beta \subset H$ is dense. With F, we proceed analogously. Conditions (A1)–(A4) imply the existence of a unique mild solution $X\colon [0, T] \times \Omega \to H_\gamma$ for SPDE (1), see [1, 2].

2.2 The Milstein Scheme for Non-commutative SPDEs

We define the numerical scheme under consideration and introduce the corresponding discretizations of the infinite dimensional spaces. To be precise, we need to discretize the time interval $[0, T]$, project the Hilbert space H to some finite dimensional subspace and we need an approximation of the infinite dimensional stochastic process $(W_t)_{t\in[0,T]}$. For the discretization of the solution space H, we define a projection operator $P_N : H \to H_N$ that maps H to the finite dimensional subspace $H_N := \operatorname{span}\{e_i : i \in \mathscr{I}_N\} \subset H$ for some fixed $N \in \mathbb{N}$. This projection is expressed by the index set $\mathscr{I}_N \subset \mathscr{I}$ with $|\mathscr{I}_N| = N$ that picks N basis functions. We specify this operator as

$$P_N x = \sum_{i\in\mathscr{I}_N} \langle x, e_i\rangle_H e_i, \quad x \in H.$$

Similarly, we approximate the Q-Wiener process. For $K \in \mathbb{N}$, we define the projected Q-Wiener process $(W_t^K)_{t\in[0,T]}$ taking values in $U_K := \operatorname{span}\{\tilde{e}_j : j \in \mathscr{J}_K\} \subset U$ by

$$W_t^K := \sum_{j\in\mathscr{J}_K} \sqrt{\eta_j}\tilde{e}_j \beta_t^j, \quad t \in [0, T],$$

for some index set $\mathscr{J}_K \subset \mathscr{J}$ with $|\mathscr{J}_K| = K$ and $\eta_j \neq 0$ for $j \in \mathscr{J}_K$. For the temporal discretization, we choose an equidistant time step for legibility of the representation. Let $h = \frac{T}{M}$ for some $M \in \mathbb{N}$ and denote $t_m = m \cdot h$ for $m \in \{0, \ldots, M\}$. On this grid, we define the increments of the projected Q-Wiener process

$$\Delta W_m^{K,M} := W_{t_{m+1}}^K - W_{t_m}^K = \sum_{j\in\mathscr{J}_K} \sqrt{\eta_j}\, \Delta\beta_m^j\, \tilde{e}_j$$

where the increments of the real-valued Brownian motions are given by $\Delta\beta_m^j = \beta_{t_{m+1}}^j - \beta_{t_m}^j$ for $m \in \{0, \ldots, M-1\}$, $j \in \mathscr{J}_K$. We apply these discretizations to the setting described above. Then, the Milstein scheme yields a discrete-time stochastic process which we denote by $(\bar{Y}_m^{N,K,M})_{m\in\{0,\ldots,M\}}$ such that $\bar{Y}_m^{N,K,M}$ is $\mathscr{F}_{t_m}$-$\mathscr{B}(H)$-measurable for all $m \in \{0, \ldots, M\}$, $M \in \mathbb{N}$. We define the Milstein scheme (MIL) for non-commutative SPDEs based on [2] as $\bar{Y}_0^{N,K,M} = P_N\xi$ and

$$\begin{aligned}\bar{Y}_{m+1}^{N,K,M} &= P_N e^{Ah}\Big(\bar{Y}_m^{N,K,M} + hF(\bar{Y}_m^{N,K,M}) + B(\bar{Y}_m^{N,K,M})\Delta W_m^{K,M} \\ &\quad + \sum_{i,j\in\mathscr{J}_K} B'(\bar{Y}_m^{N,K,M})\big(P_N B(\bar{Y}_m^{N,K,M})\tilde{e}_i, \tilde{e}_j\big)\bar{I}_{(i,j),m}^Q\Big)\end{aligned} \tag{5}$$

for all $m \in \{0, \ldots, M-1\}$. Compared to the Milstein scheme (3) proposed in [2], we added an additional projector and replaced the iterated stochastic integrals

$$I^Q_{(i,j),m} := \int_{t_m}^{t_{m+1}} \int_{t_m}^{s} \langle \mathrm{d}W_r, \tilde{e}_i \rangle_U \langle \mathrm{d}W_s, \tilde{e}_j \rangle_U$$

by an approximation $\bar{I}^Q_{(i,j),m}$ for all $i, j \in \mathscr{J}_K$ and $m \in \{0, \dots, M-1\}$. We can show that the error estimate for the Milstein approximation that is obtained in the commutative case remains valid for the scheme MIL in (5) if $\bar{I}^Q_{(i,j),m}$ represents an approximation obtained by one of the methods introduced in [8] provided the accuracy for these approximations is chosen appropriately. If Algorithm 1 in [8] is employed to approximate the iterated integrals, we denote the numerical scheme (5) by MIL1 and the approximation $\bar{I}^Q_{(i,j),m}$ of $I^Q_{(i,j),m}$ is denoted by $\bar{I}^{Q,(D),(1)}_{(i,j),m}$. This algorithm is based on a series representation of the iterated stochastic integral which is truncated after D summands for some $D \in \mathbb{N}$, see [3, 8]. If we employ Algorithm 2 instead, the scheme (5) is called MIL2 and we denote the approximation $\bar{I}^Q_{(i,j),m}$ of $I^Q_{(i,j),m}$ by $\bar{I}^{Q,(D),(2)}_{(i,j),m}$. The main difference compared to Algorithm 1 is that the series is not only truncated but the remainder is approximated by a multivariate normally distributed random vector additionally, see [8, 12] for details. Let

$$\begin{aligned}\mathscr{E}(M, K) = \bigg(\mathrm{E}\bigg[\bigg\| \int_{t_l}^{t_{l+1}} B'(\bar{Y}_l)\Big(\int_{t_l}^{s} P_N B(\bar{Y}_l)\,\mathrm{d}W_r^K\Big)\,\mathrm{d}W_s^K \\ - \sum_{i,j\in\mathscr{J}_K} \bar{I}^Q_{(i,j),l} B'(\bar{Y}_l)(P_N B(\bar{Y}_l)\tilde{e}_i, \tilde{e}_j)\bigg\|_H^2\bigg]\bigg)^{\frac{1}{2}}\end{aligned}$$

for all $l \in \{0, \dots, m-1\}$, $m \in \{1, \dots, M\}$ and $M, K \in \mathbb{N}$ denote the approximation error of the iterated integral term. Then, we obtain the following error estimate.

Theorem 1 (Convergence of Milstein scheme) *Let assumptions (A1)–(A4) hold. Then, there exists a constant $C_{Q,T} \in (0, \infty)$, independent of N, K and M, such that for $(\bar{Y}_m^{N,K,M})_{0\le m\le M}$, defined by the Milstein scheme in (5), it holds*

$$\begin{aligned}&\Big(\mathrm{E}\Big[\big\|X_{t_m} - \bar{Y}_m^{N,K,M}\big\|_H^2\Big]\Big)^{\frac{1}{2}} \\ &\le C_{Q,T}\Big(\Big(\inf_{i\in\mathscr{I}\setminus\mathscr{I}_N} \lambda_i\Big)^{-\gamma} + \Big(\sup_{j\in\mathscr{J}\setminus\mathscr{J}_K} \eta_j\Big)^{\alpha} + M^{-q_{MIL}} + \mathscr{E}(M,K)M^{\frac{1}{2}}\Big)\end{aligned}$$

with $q_{MIL} = \min(2(\gamma - \beta), \gamma)$ and for all $m \in \{0, \dots, M\}$ and all $N, K, M \in \mathbb{N}$. The parameters are determined by assumptions (A1)–(A4).

The proof of this statement is given at the end of this section.

Depending on the choice of the algorithm, we get a different error bound for $\mathscr{E}(M, K)$. We set $\Psi = B'(\bar{Y}_l)$ and $\Phi = P_N B(\bar{Y}_l)$ in (4). Then, we can transfer the error estimates given in [8, Corollaries 1, 2, Theorem 4] to our setting. Thus, for Algorithm 1 there exists some constant $C_{Q,T} > 0$ such that

$$\mathscr{E}(M, K) = \mathscr{E}^{(D),(1)}(M, K) \leq C_{Q,T} \frac{1}{M\sqrt{D}} \tag{6}$$

for all $D, K, M \in \mathbb{N}$. In contrast, for Algorithm 2, we get an estimate that converges in D with order 1. It is, however, also dependent on the number K which controls the approximation of the Q-Wiener process as well as on the eigenvalues η_j, $j \in \mathscr{J}_K$, of the operator Q. There exists some constant $C_{Q,T} > 0$ such that

$$\mathscr{E}(M, K) = \mathscr{E}^{(D),(2)}(M, K) \leq C_{Q,T} \frac{\min\left(K\sqrt{K-1}, (\min_{j\in\mathscr{J}_K} \eta_j)^{-1}\right)}{M\,D} \tag{7}$$

for all $D, K, M \in \mathbb{N}$. For example, if we assume $\eta_j \asymp j^{-\rho_Q}$, $\rho_Q > 1$ and all $j \in \mathscr{J} = \mathbb{N}$, then in the case $\rho_Q < \frac{3}{2}$ it holds $\mathscr{E}^{(D),(2)}(M, K) \leq C_{Q,T} (\min_{j\in\mathscr{J}_K} \eta_j)^{-1} M^{-1} D^{-1}$ and $\mathscr{E}^{(D),(2)}(M, K) \leq C_{Q,T}\, K(K-1)^{\frac{1}{2}} M^{-1} D^{-1}$ for $\rho_Q \geq \frac{3}{2}$. The proofs of these error estimates can be found in [8]. It is not immediately obvious which of the two algorithms is superior, see also [8] for a discussion of this issue. Here, we repeat the considerations in short. For the two algorithms stated above, we want to select the integer D such that the order of convergence stated in Theorem 1 is not reduced. Therefore, we need to choose $D \geq M^{2\min(2(\gamma-\beta),\gamma)-1}$ for Algorithm 1. In contrast, for Algorithm 2, we require $D \geq M^{\min(2(\gamma-\beta),\gamma)-\frac{1}{2}} (\min_{j\in\mathscr{J}_K} \eta_j)^{-1}$ or $D \geq M^{\min(2(\gamma-\beta),\gamma)-\frac{1}{2}} K\sqrt{K-1}$. Alternatively, one can choose $D \geq M^{-1} (\sup_{j\in\mathscr{J}\setminus\mathscr{J}_K} \eta_j)^{-2\alpha}$ for Algorithm 1 and $D \geq M^{-\frac{1}{2}} (\min_{j\in\mathscr{J}_K} \eta_j)^{-1} (\sup_{j\in\mathscr{J}\setminus\mathscr{J}_K} \eta_j)^{-\alpha}$ or $D \geq M^{-\frac{1}{2}} K\sqrt{K-1} (\sup_{j\in\mathscr{J}\setminus\mathscr{J}_K} \eta_j)^{-\alpha}$ for Algorithm 2. This shows that the choice of D depends on γ, β, K, $(\eta_j)_{j\in\mathscr{J}_K}$ and on α additionally. Therefore, the choice of D, and with this the computational effort for the simulation of the iterated stochastic integrals is dependent on the equation to be solved. We cannot identify one scheme to be superior in general and refer to Sect. 2.3 for details. Now, we prove the statement on the convergence of the schemes MIL1 and MIL2.

Proof (Proof of Theorem 1) The proof of convergence of the Milstein scheme in [2] does not use the commutativity assumption, therefore, it remains valid also in our setting. To ease the notations, we denote by $(Y_m)_{m\in\{0,\ldots,M-1\}}$ the Milstein approximation which does not involve an approximation of the iterated stochastic integrals

$$\begin{aligned} Y_{m+1} = P_N e^{Ah} \bigg(& Y_m + hF(Y_m) + B(Y_m)\Delta W_m^{K,M} \\ & + \int_{t_m}^{t_{m+1}} B'(Y_m)\Big(\int_{t_m}^{s} P_N B(Y_m)\,\mathrm{d}W_r^K\Big)\,\mathrm{d}W_s^K\bigg). \end{aligned} \tag{8}$$

Analogously to the proof for Theorem 1 in [2], we get an estimate for (8) of the form

$$\Big(\mathrm{E}\big[\|X_{t_m} - Y_m\|_H^2\big]\Big)^{\frac{1}{2}}$$
$$\leq C_{Q,T}\Big(\Big(\inf_{i\in\mathscr{I}\setminus\mathscr{I}_N} \lambda_i\Big)^{-\gamma} + \Big(\sup_{j\in\mathscr{J}\setminus\mathscr{J}_K} \eta_j\Big)^{\alpha} + M^{-\min(2(\gamma-\beta),\gamma)}\Big).$$

The proof for the scheme given in (8) can be conducted in the same way as for the scheme in (3) except that the projection operator P_N in (8) has to be taken into account, see also the comments in [7] and the detailed proof in [6]. It remains to prove the expression for the error caused by the approximation of the iterated stochastic integrals. We denote $\bar{Y}_m := \bar{Y}_m^{N,K,M}$ for all $m \in \{0, \ldots, M\}$ and compute the following two terms

$$\Big(\mathrm{E}\big[\|Y_m - \bar{Y}_m\|_H^2\big]\Big)^{\frac{1}{2}} \leq \Big(\mathrm{E}\big[\|Y_m - Y_{m,\bar{Y}}\|_H^2\big]\Big)^{\frac{1}{2}} + \Big(\mathrm{E}\big[\|Y_{m,\bar{Y}} - \bar{Y}_m\|_H^2\big]\Big)^{\frac{1}{2}} \tag{9}$$

where

$$Y_{m,\bar{Y}} = P_N\Big(e^{At_m}X_0 + \sum_{l=0}^{m-1}\int_{t_l}^{t_{l+1}} e^{A(t_m-t_l)}F(\bar{Y}_l)\,\mathrm{d}s + \sum_{l=0}^{m-1}\int_{t_l}^{t_{l+1}} e^{A(t_m-t_l)}B(\bar{Y}_l)\,\mathrm{d}W_s^K$$
$$+\sum_{l=0}^{m-1}\int_{t_l}^{t_{l+1}} e^{A(t_m-t_l)}B'(\bar{Y}_l)\Big(P_N\int_{t_l}^{s} B(\bar{Y}_l)\,\mathrm{d}W_r^K\Big)\,\mathrm{d}W_s^K\Big).$$

We insert this expression and obtain

$$\mathrm{E}\big[\|Y_m - Y_{m,\bar{Y}}\|_H^2\big] = \mathrm{E}\Big[\Big\|P_N\Big(\sum_{l=0}^{m-1}\int_{t_l}^{t_{l+1}} e^{A(t_m-t_l)}\big(F(Y_l) - F(\bar{Y}_l)\big)\,\mathrm{d}s$$
$$+\sum_{l=0}^{m-1}\int_{t_l}^{t_{l+1}} e^{A(t_m-t_l)}\big(B(Y_l) - B(\bar{Y}_l)\big)\,\mathrm{d}W_s^K$$
$$+\sum_{l=0}^{m-1}\Big(\int_{t_l}^{t_{l+1}} e^{A(t_m-t_l)}B'(Y_l)\Big(P_N\int_{t_l}^{s} B(Y_l)\,\mathrm{d}W_r^K\Big)\,\mathrm{d}W_s^K$$
$$-\int_{t_l}^{t_{l+1}} e^{A(t_m-t_l)}B'(\bar{Y}_l)\Big(P_N\int_{t_l}^{s} B(\bar{Y}_l)\,\mathrm{d}W_r^K\Big)\,\mathrm{d}W_s^K\Big)\Big)\Big\|_H^2\Big]$$
$$\leq CMh\sum_{l=0}^{m-1}\int_{t_l}^{t_{l+1}} \mathrm{E}\Big[\Big\|F(Y_l) - F(\bar{Y}_l)\Big\|_H^2\Big]\,\mathrm{d}s$$
$$+C\sum_{l=0}^{m-1}\int_{t_l}^{t_{l+1}} \mathrm{E}\Big[\Big\|B(Y_l) - B(\bar{Y}_l)\Big\|_{L_{HS}(U_0,H)}^2\Big]\,\mathrm{d}s$$
$$+C\sum_{l=0}^{m-1}\int_{t_l}^{t_{l+1}} \mathrm{E}\Big[\Big\|B'(Y_l)\Big(P_N\int_{t_l}^{s} B(Y_l)\,\mathrm{d}W_r^K\Big)$$

$$- B'(\bar{Y}_l)\Big(P_N \int_{t_l}^{s} B(\bar{Y}_l)\,\mathrm{d}W_r^K\Big)\Big\|_{L_{\mathrm{HS}}(U_0,H)}^2\Big]\,\mathrm{d}s$$

$$\leq C_T h \sum_{l=0}^{m-1} \mathrm{E}\Big[\big\|Y_l - \bar{Y}_l\big\|_H^2\Big]$$

where the computations are the same as in [2, Sect. 6.3]. This estimate mainly employs the Lipschitz continuity of the involved operators.

Next, we analyze the second term in (9). By the stochastic independence of $I^Q_{(i,j),l}$ and $\bar{I}^Q_{(i,j),l}$ from $I^Q_{(i,j),k}$ and $\bar{I}^Q_{(i,j),k}$ for $l \neq k$, we obtain

$$\mathrm{E}\big[\|Y_{m,\bar{Y}} - \bar{Y}_m\|_H^2\big] = \mathrm{E}\Big[\Big\|P_N\Big(\sum_{l=0}^{m-1}\sum_{j\in\mathcal{J}_K} e^{A(t_m-t_l)}\Big(B'(\bar{Y}_l)\Big(\sum_{i\in\mathcal{J}_K} P_N B(\bar{Y}_l)\tilde{e}_i I^Q_{(i,j),l}, \tilde{e}_j\Big)$$

$$- B'(\bar{Y}_l)\Big(\sum_{i\in\mathcal{J}_K} P_N B(\bar{Y}_l)\tilde{e}_i \bar{I}^Q_{(i,j),l}, \tilde{e}_j\Big)\Big)\Big)\Big\|_H^2\Big]$$

$$\leq C\sum_{l=0}^{m-1}\mathrm{E}\Big[\Big\|\int_{t_l}^{t_{l+1}} B'(\bar{Y}_l)\Big(\int_{t_l}^{s} P_N B(\bar{Y}_l)\,\mathrm{d}W_r^K\Big)\,\mathrm{d}W_s^K$$

$$- \sum_{i,j\in\mathcal{J}_K} \bar{I}^Q_{(i,j),l} B'(\bar{Y}_l)(P_N B(\bar{Y}_l)\tilde{e}_i, \tilde{e}_j)\Big\|_H^2\Big]$$

$$= C\sum_{l=0}^{m-1}\mathcal{E}(M,K)^2.$$

In total, we get with Gronwall's lemma

$$\mathrm{E}\big[\|Y_m - \bar{Y}_m\|_H^2\big] \leq C_T h \sum_{l=0}^{m-1}\mathrm{E}\Big[\big\|Y_l - \bar{Y}_l\big\|_H^2\Big] + C\sum_{l=0}^{m-1}\mathcal{E}(M,K)^2$$

$$\leq CM\mathcal{E}(M,K)^2,$$

which completes the proof. □

2.3 Comparison of Computational Cost

In order to compare the numerical methods introduced in this work, we consider the effective order of convergence based on a cost model introduced in [7]. This number combines the theoretical order of convergence, as stated for example in Theorem 1, with the computational cost involved in the calculation of an approximation by a particular scheme. For the computational cost model, we assume that each evaluation of a real valued functional and each generation of a standard normally distributed

Table 1 Computational cost determined by the number of necessary evaluations of real-valued functionals and independent $N(0,1)$-distributed random variables for each time step. The choice of D differs for MIL1 and MIL2

	# of evaluations of functionals			# of $N(0,1)$ r. v.
Scheme	$P_N F(\cdot)\|_{H_N}$	$P_N B(\cdot)\|_{U_K}$	$P_N B'(\cdot)\|_{H_N,U_K}$	
EES	N	KN	–	K
MIL1	N	KN	KN^2	$K(1+2D)$
MIL2	N	KN	KN^2	$K(1+2D)+\frac{1}{2}K(K-1)$

random number is of some cost $c \geq 1$ whereas each elementary arithmetic operation is of unit cost 1, see [7] for details. Then, the computational cost for one time step and each scheme under consideration can be determined by the corresponding values listed in Table 1. We compare the two Milstein schemes MIL1 and MIL2 to the exponential Euler scheme (EES). For the EES, we employ the version introduced in [9] combined with a Galerkin approximation. The convergence results for the exponential Euler scheme in this setting can be obtained similarly as in the proof of the Milstein scheme in [2], see also [5, Theorem 3.2]. We state the result without giving a proof.

Proposition 1 (Convergence of EES) *Assume that (A1)–(A4) hold. Then, there exists a constant $C_T \in (0,\infty)$, independent of N, K and M, such that for the approximation process $(Y_m^{EES})_{0\leq m\leq M}$, defined by the EES, it holds*

$$\Big(\mathrm{E}\Big[\big\|X_{t_m} - Y_m^{EES}\big\|_H^2\Big]\Big)^{\frac{1}{2}} \leq C_T\bigg(\Big(\inf_{i\in\mathcal{I}\setminus\mathcal{I}_N} \lambda_i\Big)^{-\gamma} + \Big(\sup_{j\in\mathcal{J}\setminus\mathcal{J}_K} \eta_j\Big)^{\alpha} + M^{-q_{EES}}\bigg)$$

with $q_{EES} = \min(\frac{1}{2}, 2(\gamma-\beta), \gamma)$ and for all $m \in \{0,\ldots,M\}$ and all $N, K, M \in \mathbb{N}$. The parameters are determined by assumptions (A1)–(A4).

Note that for the EES we can dispense with some of the conditions specified in (A3), e.g., no assumptions are needed for the second derivative of B and the estimate for $B'(v)PB(v) - B'(w)PB(w)$ can be suspended. In the following, let q denote the order of convergence w.r.t. the step size $h = \frac{T}{M}$. Obviously, it holds $q_{\text{MIL}} = \min(2(\gamma-\beta),\gamma) \geq \min(\frac{1}{2}, 2(\gamma-\beta),\gamma) = q_{\text{EES}}$. However, we need to take into account the computational cost in order to determine the scheme that is superior as we do not need to simulate the iterated integrals in the Euler scheme after all. Therefore, we derive the effective order of convergence for each of the schemes MIL1, MIL2 and EES, see [7] for details.

For each approximation $(Y_m)_{m\in\{0,\ldots,M\}}$ under consideration, we minimize the error term

$$\sup_{m\in\{0,\ldots,M\}} \big(\mathrm{E}\big[\|X_{t_m} - Y_m\|_H^2\big]\big)^{\frac{1}{2}}$$

over all $N, M, K \in \mathbb{N}$ under the constraint that the computational cost does not exceed some specified value $\bar{c} > 0$. If we assume that $\sup_{j \in \mathscr{J} \setminus \mathscr{J}_K} \eta_j = \mathscr{O}(K^{-\rho_Q})$ and $(\inf_{i \in \mathscr{I} \setminus \mathscr{I}_N} \lambda_i)^{-1} = \mathscr{O}(N^{-\rho_A})$ for some $\rho_A > 0$ and $\rho_Q > 1$, we obtain the following expression for all $N, M, K \in \mathbb{N}$ and some $C > 0$, see also [7],

$$\mathrm{err}(\mathrm{SCHEME}) = \sup_{m \in \{0,\ldots,M\}} \Big(\mathrm{E}\Big[\big\| X_{t_m} - Y_m \big\|_H^2 \Big] \Big)^{\frac{1}{2}} \leq C \big(N^{-\gamma\rho_A} + K^{-\alpha\rho_Q} + M^{-q} \big).$$

The parameter $q > 0$ is determined by the scheme that is considered. Then, optimization yields the effective order of convergence, denoted by EOC(SCHEME), which is given as

$$\mathrm{err}(\mathrm{SCHEME}) = \mathscr{O}\big(\bar{c}^{\,-\mathrm{EOC(SCHEME)}}\big).$$

First, we consider Algorithm 1. For the scheme MIL1, the computational cost amounts to $\bar{c} = \mathscr{O}(MKN^2) + \mathscr{O}(KM^{2q_{\mathrm{MIL}}})$, see Table 1 and the discussion in the previous section. We solve the optimization problem and obtain

$$\begin{aligned} M &= \mathscr{O}\Big(\bar{c}^{\frac{\gamma\rho_A\alpha\rho_Q}{(2\alpha\rho_Q+\gamma\rho_A)q_{\mathrm{MIL}}+\alpha\rho_Q\gamma\rho_A}}\Big), \quad N = \mathscr{O}\Big(\bar{c}^{\frac{\alpha\rho_Q q_{\mathrm{MIL}}}{(2\alpha\rho_Q+\gamma\rho_A)q_{\mathrm{MIL}}+\alpha\rho_Q\gamma\rho_A}}\Big), \\ K &= \mathscr{O}\Big(\bar{c}^{\frac{\gamma\rho_A q_{\mathrm{MIL}}}{(2\alpha\rho_Q+\gamma\rho_A)q_{\mathrm{MIL}}+\alpha\rho_Q\gamma\rho_A}}\Big) \end{aligned} \tag{10}$$

in the case of $\gamma\rho_A(2q_{\mathrm{MIL}} - 1) \leq 2q_{\mathrm{MIL}}$, denoted as condition M1C2. This condition makes sure that the computational cost is of order $\bar{c} = \mathscr{O}(MKN^2)$. Therefore, we obtain the effective order of convergence from

$$\mathrm{err}(\mathrm{MIL1}) = \mathscr{O}\Big(\bar{c}^{\,-\frac{\gamma\rho_A\alpha\rho_Q q_{\mathrm{MIL}}}{(2\alpha\rho_Q+\gamma\rho_A)q_{\mathrm{MIL}}+\alpha\rho_Q\gamma\rho_A}}\Big), \tag{11}$$

which is the same result as for the Milstein scheme in the case of SPDEs with commutative noise, see the computations in [7].

On the other hand, in the case of $\gamma\rho_A(2q_{\mathrm{MIL}} - 1) \geq 2q_{\mathrm{MIL}}$, denoted as condition M1C1, it holds $\bar{c} = \mathscr{O}(KM^{2q_{\mathrm{MIL}}})$ and optimization yields

$$M = \mathscr{O}\Big(\bar{c}^{\frac{\alpha\rho_Q}{(2\alpha\rho_Q+1)q_{\mathrm{MIL}}}}\Big), \quad N = \mathscr{O}\Big(\bar{c}^{\frac{\alpha\rho_Q}{(2\alpha\rho_Q+1)\gamma\rho_A}}\Big), \quad K = \mathscr{O}\Big(\bar{c}^{\frac{1}{2\alpha\rho_Q+1}}\Big) \tag{12}$$

and the effective order of convergence equals

$$\mathrm{err}(\mathrm{MIL1}) = \mathscr{O}\Big(\bar{c}^{\,-\frac{\alpha\rho_Q}{2\alpha\rho_Q+1}}\Big). \tag{13}$$

In order to facilitate computation, we distinguish the case $(\min_{j \in \mathscr{J}_K} \eta_j)^{-1} = o(K^{\frac{3}{2}})$ which results in $\rho_Q < \frac{3}{2}$ and the case that $\min_{j \in \mathscr{J}_K} \eta_j = \mathscr{O}(K^{-\frac{3}{2}})$ where we choose $\rho_Q \geq \frac{3}{2}$ maximal admissible. In the following, we always assume that ρ_Q

is chosen maximal such that $\sup_{j \in \mathscr{J} \setminus \mathscr{J}_K} \eta_j = \mathscr{O}(K^{-\rho_Q})$ is fulfilled and we refer to these two cases by simply writing case $\rho_Q < \frac{3}{2}$ and case $\rho_Q \geq \frac{3}{2}$, respectively.

For Algorithm 2, we have to take $\bar{c} = \mathscr{O}(MKN^2) + \mathscr{O}(K \min(K^{\frac{3}{2}}, K^{\rho_Q}) M^{q_{\mathrm{MIL}}+\frac{1}{2}}) + \mathscr{O}(MK^2)$ into account. As above, we need to treat several cases. We detail the case $\min(K^{\frac{3}{2}}, K^{\rho_Q}) = K^{\frac{3}{2}}$, that is, $\rho_Q \geq \frac{3}{2}$; the results for $\rho_Q < \frac{3}{2}$ can be obtained analogously and are stated in Table 2. The first case corresponds to $\bar{c} = \mathscr{O}(MKN^2)$. For $\rho_Q \geq \frac{3}{2}$, $\gamma\rho_A \leq 2\alpha\rho_Q$ and $\frac{3}{2}\gamma\rho_A q_{\mathrm{MIL}} + (q_{\mathrm{MIL}} - \frac{1}{2})\gamma\rho_A\alpha\rho_Q \leq 2\alpha\rho_Q q_{\mathrm{MIL}}$, denoted as condition M2C1a, we get the same choice for M, N, K and the same effective order as for the scheme for SPDEs with commutative noise given in (10) and (11). In case of $\bar{c} = \mathscr{O}(MK^2)$, that is, if $\gamma\rho_A \geq 2\alpha\rho_Q$ and $q_{\mathrm{MIL}} \leq \frac{\alpha\rho_Q}{1+2\alpha\rho_Q}$, denoted as condition M2C2a, we obtain

$$M = \mathscr{O}\Big(\bar{c}^{\frac{\alpha\rho_Q}{\alpha\rho_Q + 2q_{\mathrm{MIL}}}}\Big), \quad N = \mathscr{O}\Big(\bar{c}^{\frac{\alpha\rho_Q q_{\mathrm{MIL}}}{\gamma\rho_A\alpha\rho_Q + 2\gamma\rho_A q_{\mathrm{MIL}}}}\Big), \quad K = \mathscr{O}\Big(\bar{c}^{\frac{q_{\mathrm{MIL}}}{\alpha\rho_Q + 2q_{\mathrm{MIL}}}}\Big) \tag{14}$$

with effective order of convergence given by

$$\mathrm{err}(\mathrm{MIL2}) = \mathscr{O}\Big(\bar{c}^{-\frac{\alpha\rho_Q q_{\mathrm{MIL}}}{\alpha\rho_Q + 2q_{\mathrm{MIL}}}}\Big). \tag{15}$$

Note that in this case, it follows $q_{\mathrm{MIL}} < \frac{1}{2}$. Next, we consider the case of $2\alpha\rho_Q q_{\mathrm{MIL}} \leq \frac{3}{2}\gamma\rho_A q_{\mathrm{MIL}} + (q_{\mathrm{MIL}} - \frac{1}{2})\gamma\rho_A\alpha\rho_Q$ and $q_{\mathrm{MIL}} \geq \frac{\alpha\rho_Q}{1+2\alpha\rho_Q}$, denoted as condition M2C3a, i.e., where $\bar{c} = \mathscr{O}(M^{q_{\mathrm{MIL}}+\frac{1}{2}} K^{\frac{5}{2}})$. Then, we get

$$\begin{aligned} M &= \mathscr{O}\Big(\bar{c}^{\frac{\alpha\rho_Q}{\alpha\rho_Q(q_{\mathrm{MIL}}+\frac{1}{2}) + \frac{5}{2}q_{\mathrm{MIL}}}}\Big), \qquad N = \mathscr{O}\Big(\bar{c}^{\frac{\alpha\rho_Q q_{\mathrm{MIL}}}{\gamma\rho_A(\alpha\rho_Q(q_{\mathrm{MIL}}+\frac{1}{2}) + \frac{5}{2}q_{\mathrm{MIL}})}}\Big), \\ K &= \mathscr{O}\Big(\bar{c}^{\frac{q_{\mathrm{MIL}}}{\alpha\rho_Q(q_{\mathrm{MIL}}+\frac{1}{2}) + \frac{5}{2}q_{\mathrm{MIL}}}}\Big) \end{aligned} \tag{16}$$

with

$$\mathrm{err}(\mathrm{MIL2}) = \mathscr{O}\Big(\bar{c}^{-\frac{\alpha\rho_Q q_{\mathrm{MIL}}}{\alpha\rho_Q(q_{\mathrm{MIL}}+\frac{1}{2}) + \frac{5}{2}q_{\mathrm{MIL}}}}\Big). \tag{17}$$

Finally, we want to mention one case for $\rho_Q < \frac{3}{2}$ explicitly where we assume $2\alpha q_{\mathrm{MIL}} \leq \gamma\rho_A q_{\mathrm{MIL}} + (q_{\mathrm{MIL}} - \frac{1}{2})\alpha\gamma\rho_A$ and $q_{\mathrm{MIL}} + \frac{1}{2}\alpha\rho_Q \leq (1+\alpha)\rho_Q q_{\mathrm{MIL}}$, which are the conditions denoted as M2C3b. In this case, it holds that $\bar{c} = \mathscr{O}(M^{q_{\mathrm{MIL}}+\frac{1}{2}} K^{\rho_Q+1})$ which is the only case where the dominating term for $\bar{c}$ depends on ρ_Q explicitly. Here we get

$$\begin{aligned} M &= \mathscr{O}\Big(\bar{c}^{\frac{\alpha\rho_Q}{\alpha\rho_Q(q_{\mathrm{MIL}}+\frac{1}{2}) + q_{\mathrm{MIL}}(\rho_Q+1)}}\Big), \qquad N = \mathscr{O}\Big(\bar{c}^{\frac{\alpha\rho_Q q_{\mathrm{MIL}}}{\gamma\rho_A(\alpha\rho_Q(q_{\mathrm{MIL}}+\frac{1}{2}) + q_{\mathrm{MIL}}(\rho_Q+1))}}\Big), \\ K &= \mathscr{O}\Big(\bar{c}^{\frac{q_{\mathrm{MIL}}}{\alpha\rho_Q(q_{\mathrm{MIL}}+\frac{1}{2}) + q_{\mathrm{MIL}}(\rho_Q+1)}}\Big) \end{aligned} \tag{18}$$

with

Table 2 Conditions M1C1 and M1C2 are the ones that have to be considered for MIL1, whereas the remaining conditions belong to MIL2. Under each given condition, the corresponding scheme MIL1 or MIL2 possesses computational cost $\bar{c}$, respectively. Note that $\rho_Q > 1$ and we denote $q = q_{\text{MIL}}$

Abbrev.	Condition	$\bar{c}$	EOC
M1C1	$\gamma\rho_A(2q-1) \geq 2q$	$\mathcal{O}(M^{2q}K)$	(13)
M1C2	$\gamma\rho_A(2q-1) \leq 2q$	$\mathcal{O}(MKN^2)$	(11)
M2C1a	$\rho_Q \geq \frac{3}{2} \wedge \gamma\rho_A \leq 2\alpha\rho_Q \wedge$ $\frac{3}{2}\gamma\rho_A q + (q-\frac{1}{2})\alpha\rho_Q\gamma\rho_A \leq 2\alpha\rho_Q q$	$\mathcal{O}(MKN^2)$	(11)
M2C1b	$\rho_Q < \frac{3}{2} \wedge \gamma\rho_A \leq 2\alpha\rho_Q \wedge$ $\gamma\rho_A q + (q-\frac{1}{2})\alpha\gamma\rho_A \leq 2\alpha q$	$\mathcal{O}(MKN^2)$	(11)
M2C2a	$\rho_Q \geq \frac{3}{2} \wedge 2\alpha\rho_Q \leq \gamma\rho_A \wedge q \leq \frac{\alpha\rho_Q}{2\alpha\rho_Q+1}$	$\mathcal{O}(MK^2)$	(15)
M2C2b	$\rho_Q < \frac{3}{2} \wedge 2\alpha\rho_Q \leq \gamma\rho_A \wedge q < \frac{\alpha\rho_Q}{2\alpha\rho_Q+2(\rho_Q-1)}$	$\mathcal{O}(MK^2)$	(15)
M2C3a	$\rho_Q \geq \frac{3}{2} \wedge 2\alpha\rho_Q q \leq \frac{3}{2}\gamma\rho_A q + (q-\frac{1}{2})\alpha\rho_Q\gamma\rho_A \wedge$ $q \geq \frac{\alpha\rho_Q}{2\alpha\rho_Q+1}$	$\mathcal{O}(M^{q+\frac{1}{2}}K^{\frac{5}{2}})$	(17)
M2C3b	$\rho_Q < \frac{3}{2} \wedge 2\alpha q \leq \gamma\rho_A q + (q-\frac{1}{2})\alpha\gamma\rho_A \wedge$ $q \geq \frac{\alpha\rho_Q}{2\alpha\rho_Q+2(\rho_Q-1)}$	$\mathcal{O}(M^{q+\frac{1}{2}}K^{\rho_Q+1})$	(19)

$$\text{err(MIL2)} = \mathcal{O}\Big(\bar{c}^{\,-\frac{\alpha\rho_Q q_{\text{MIL}}}{\alpha\rho_Q(q_{\text{MIL}}+\frac{1}{2})+q_{\text{MIL}}(\rho_Q+1)}}\Big). \tag{19}$$

All possible cases M1C1 and M1C2 for MIL1 as well as M2C1a, M2C1b, M2C2a, M2C2b, M2C3a and M2C3b for MIL2 together with their effective orders of convergence are summarized in Table 2. Further, the optimal choice for M, N and K for the cases not detailed is given by the case with the same effective order of convergence listed above.

In order to determine the scheme with the highest effective order of convergence, we compare the schemes MIL1 and MIL2 to each other and to the exponential Euler scheme. For the EES, the optimal choice for M, N and K is given by

$$\begin{aligned} M &= \mathcal{O}\Big(\bar{c}^{\frac{\gamma\rho_A\alpha\rho_Q}{(\alpha\rho_Q+\gamma\rho_A)q_{\text{EES}}+\alpha\rho_Q\gamma\rho_A}}\Big), \quad N = \mathcal{O}\Big(\bar{c}^{\frac{\alpha\rho_Q q_{\text{EES}}}{(\alpha\rho_Q+\gamma\rho_A)q_{\text{EES}}+\alpha\rho_Q\gamma\rho_A}}\Big), \\ K &= \mathcal{O}\Big(\bar{c}^{\frac{\gamma\rho_A q_{\text{EES}}}{(\alpha\rho_Q+\gamma\rho_A)q_{\text{EES}}+\alpha\rho_Q\gamma\rho_A}}\Big) \end{aligned} \tag{20}$$

with the effective order of convergence

$$\text{err(EES)} = \mathcal{O}\Big(\bar{c}^{\,-\frac{q_{\text{EES}}\gamma\rho_A\alpha\rho_Q}{(\alpha\rho_Q+\gamma\rho_A)q_{\text{EES}}+\gamma\rho_A\alpha\rho_Q}}\Big) \tag{21}$$

where $q_{\text{EES}} = \min(\frac{1}{2}, 2(\gamma-\beta), \gamma)$, see [7].

Obviously, our main interest is in parameter constellations such that $q_{\text{MIL}} > q_{\text{EES}}$ which implies that $q_{\text{EES}} = \frac{1}{2}$. In case of $q_{\text{MIL}} = q_{\text{EES}} \leq \frac{1}{2}$ the EES is always the optimal choice compared to MIL1 and MIL2. Therefore, we assume $q_{\text{MIL}} > q_{\text{EES}} = \frac{1}{2}$

in the following. Then, by comparing the different effective orders of convergence across parameter sets, one can show that except for one case the Milstein scheme always has a higher effective order of convergence than the exponential Euler scheme. We refer to Table 3 for an overview; this shows that for larger q_{MIL} the Milstein scheme is favoured over the exponential Euler scheme. Here, we only elaborate one case. Assume that the parameters take values such that either the scheme MIL1 or the scheme MIL2 obtains the same effective order of convergence as the scheme for SPDEs with commutative noise (11). Note that (11) is the highest effective order that can be attained by MIL1 and MIL2 for $q_{\mathrm{MIL}} > \frac{1}{2}$ anyway. We compare the effective order (11) with that of the exponential Euler scheme in (21)

$$\frac{q_{\mathrm{MIL}}\gamma\rho_A\alpha\rho_Q}{(2\alpha\rho_Q+\gamma\rho_A)q_{\mathrm{MIL}}+\gamma\rho_A\alpha\rho_Q} \lesseqgtr \frac{q_{\mathrm{EES}}\gamma\rho_A\alpha\rho_Q}{(\alpha\rho_Q+\gamma\rho_A)q_{\mathrm{EES}}+\gamma\rho_A\alpha\rho_Q}.$$

This can be rewritten such that we obtain

$$q_{\mathrm{MIL}}(\gamma\rho_A - q_{\mathrm{EES}}) \lesseqgtr q_{\mathrm{EES}}\gamma\rho_A.$$

For $q_{\mathrm{MIL}} > q_{\mathrm{EES}} = \frac{1}{2}$, this results in

$$\gamma\rho_A \lesseqgtr \frac{q_{\mathrm{MIL}}}{2q_{\mathrm{MIL}}-1}.$$

The condition $\gamma\rho_A > \frac{q_{\mathrm{MIL}}}{2q_{\mathrm{MIL}}-1}$ is required for a higher effective order of the Milstein scheme whereas $\gamma\rho_A \leq \frac{q_{\mathrm{MIL}}}{2q_{\mathrm{MIL}}-1}$ results in a higher order for the exponential Euler scheme. Clearly, either condition M1C1 or condition M1C2 has to be fulfilled and in case of M1C1 the effective order of convergence for MIL1 in (13) is greater than that in (21) for the EES scheme if $q_{\mathrm{MIL}} > q_{\mathrm{EES}}$. Thus, in the case that M1C1 is fulfilled it only remains to check whether MIL2 attains an even higher effective order of convergence than (13). These calculations can be conducted in a similar way as above.

Based on the effective order of convergence, it is not possible to identify one scheme that dominates the others across all parameter constellations. The results of a comparison are summarized in Table 3; this overview clearly illustrates the dependence on the parameters $q_{\mathrm{MIL}}, \alpha, \gamma, \rho_A$ and ρ_Q. For completeness, we want to note that parts of (A3) do not have to be fulfilled for the exponential Euler scheme. Therefore, there exist equations where this scheme might indeed be beneficial for parameter sets other than the combinations stated in Table 3. The effective order for the Milstein scheme indicates that, compared to the Euler schemes, the increase in the computational cost that results from the approximation of the iterated stochastic integrals is, in most cases, significantly compensated by the higher theoretical order of convergence q_{MIL} w.r.t. the time steps that the Milstein scheme attains.

Table 3 For a given parameter set, the conditions in this table have to be checked in order to determine the optimal scheme among the schemes EES, MIL1 and MIL2 for the case of $q = q_{\text{MIL}} > q_{\text{EES}} = \frac{1}{2}$. In case of $q_{\text{MIL}} = q_{\text{EES}} \leq \frac{1}{2}$, the exponential Euler scheme is always the optimal choice

Conditions	Optimal scheme	Optimal M, N, K	EOC
M1C1 $\wedge$ M2C1a	MIL2	(10)	(11)
M1C1 $\wedge$ M2C1b	MIL2	(10)	(11)
M1C1 $\wedge$ M2C3a $\wedge$ $(2\alpha\rho_Q - 3)q < \alpha\rho_Q$	MIL1	(12)	(13)
M1C1 $\wedge$ M2C3a $\wedge$ $(2\alpha\rho_Q - 3)q \geq \alpha\rho_Q$	MIL2	(16)	(17)
M1C1 $\wedge$ M2C3b $\wedge$ $\alpha(2q-1) < 2q$	MIL1	(12)	(13)
M1C1 $\wedge$ M2C3b $\wedge$ $\alpha(2q-1) \geq 2q$	MIL2	(18)	(19)
M1C2 $\wedge$ $\gamma\rho_A(2q-1) \leq q$	EES	(20)	(21)
M1C2 $\wedge$ M2C1a $\wedge$ $\gamma\rho_A(2q-1) > q$	MIL1=MIL2	(10)	(11)
M1C2 $\wedge$ M2C1b $\wedge$ $\gamma\rho_A(2q-1) > q$	MIL1=MIL2	(10)	(11)
M1C2 $\wedge$ M2C3a $\wedge$ $\gamma\rho_A(2q-1) > q$	MIL1	(10)	(11)
M1C2 $\wedge$ M2C3b $\wedge$ $\gamma\rho_A(2q-1) > q$	MIL1	(10)	(11)

2.4 Example

Finally, we illustrate the theoretical results on the effective order of convergence and the consequences for the choice of a particular scheme, summarized in Table 3, with an example.

Throughout this section, we fix the following setting. Let $H = U = L^2((0,1), \mathbb{R})$, set $T = 1$, $\beta = 0$ and $\mathcal{I} = \mathcal{J} = \mathbb{N}$. We choose A to be the Laplacian with Dirichlet boundary conditions; to be precise, $A = \frac{1}{100}\Delta$. Thus, it holds for the eigenvalues $\lambda_i = \frac{\pi^2 i^2}{100}$, for the eigenvectors $e_i = \sqrt{2}\sin(i\pi x)$ for $i \in \mathbb{N}$, $x \in (0,1)$ and on the boundary, we have $X_t(0) = X_t(1) = 0$ for all $t \in (0, T]$. The operator Q is defined by $\eta_j = j^{-3}$ and $\tilde{e}_j = \sqrt{2}\sin(j\pi x)$ for $j \in \mathbb{N}$, $x \in (0,1)$. As a result of this, it holds $\rho_A = 2$ and $\rho_Q = 3$. Moreover, we choose $F(y) = 1 - y$, $y \in H$ and $\xi(x) = X_0(x) = 0$ for all $x \in (0,1)$. The operator B is defined in the following. It fits into the general setting introduced for the numerical analysis in [7, Sect. 5.3], which we repeat here in short only. Let some functionals $\mu_{ij} : H_\beta \to \mathbb{R}$, $\phi_{ij}^k : H_\beta \to \mathbb{R}$ be given for $i, k \in \mathcal{I}$, $j \in \mathcal{J}$ such that ϕ_{ij}^k is the Fréchet derivative of μ_{ij} in direction e_k. Then, we define

$$B(y)u = \sum_{i \in \mathcal{I}} \sum_{j \in \mathcal{J}} \mu_{ij}(y) \langle u, \tilde{e}_j \rangle_U e_i$$

and it holds that

$$\big(B'(y)(B(y)v)\big)u = \sum_{i,k \in \mathcal{I}} \sum_{j,r \in \mathcal{J}} \phi_{ij}^k(y) \mu_{kr}(y) \langle v, \tilde{e}_r \rangle_U \langle u, \tilde{e}_j \rangle_U e_i$$

for $y \in H_\beta$ and $u, v \in U_0$. For details, we refer to [7, Sect. 5.3].

Here, we choose $\mu_{ij}(y) = \frac{\langle y, e_j \rangle_H}{i^4+j^4}$ for all $i \in \mathcal{I}$, $j \in \mathcal{J}$ and $y \in H$. With this choice, we get $\phi_{ij}^k(y) = \begin{cases} 0, & k \neq j \\ \frac{1}{i^4+j^4}, & k = j \end{cases}$ for all $i, k \in \mathcal{I}$, $j \in \mathcal{J}$, $y \in H$. We show that assumptions (A1)–(A4) are fulfilled in this setting. For conditions (A1), (A2) and (A4) this is obvious. It remains to examine (A3). We use the expressions that have been computed in [7, Sect. 5.3], that is,

$$\|B(y)\|_{L(U,H_\delta)} \leq \sum_{i \in \mathcal{I}} \sum_{j \in \mathcal{J}} \lambda_i^\delta |\mu_{ij}(y)| \leq \sum_{i \in \mathcal{I}} \sum_{j \in \mathcal{J}} \frac{1}{j^{2-2\delta}} \frac{\pi^{2\delta}}{100^\delta} \frac{1}{i^{2-2\delta}} \|y\|_{H_\delta}$$

for all $y \in H_\delta$. Thus, we get $\|B(y)\|_{L(U,H_\delta)} \leq C(1 + \|y\|_{H_\delta})$ for all $y \in H_\delta$ if $\delta < \frac{1}{2}$, where we select the maximal value for δ. Moreover, we check

$$\begin{aligned} \|(-A)^{-\vartheta} B(z) Q^{-\alpha}\|_{L_{HS}(U_0,H)} &= \Big(\sum_{j \in \mathcal{J}} \eta_j^{1-2\alpha} \sum_{i \in \mathcal{I}} \lambda_i^{-2\vartheta} \mu_{ij}^2(z) \Big)^{\frac{1}{2}} \\ &\leq C \Big(\sum_{j \in \mathcal{J}} \frac{1}{j^{3(1-2\alpha)+8+4\gamma}} \sum_{i \in \mathcal{I}} \frac{1}{i^{4\vartheta}} \|z\|_{H_\gamma}^2 \Big)^{\frac{1}{2}} \end{aligned}$$

for all $z \in H_\gamma$. This shows that $\|(-A)^{-\vartheta} B(z) Q^{-\alpha}\|_{L_{HS}(U_0,H)} \leq C(1 + \|z\|_{H_\gamma})$ is fulfilled for all $z \in H_\gamma$ if $\alpha < \frac{7}{3}$. The remaining conditions in (A3) hold as well. These are not stated here as they do not restrict the parameters. Finally, we show that the commutativity condition (2), expressed in the notation presented above, is actually not fulfilled. On the one hand, we get

$$\sum_{k \in \mathcal{I}} \phi_{im}^k(y) \mu_{kn}(y) = \frac{1}{i^4 + m^4} \frac{\langle y, e_n \rangle_H}{m^4 + n^4}$$

but

$$\sum_{k \in \mathcal{I}} \phi_{in}^k(y) \mu_{km}(y) = \frac{1}{i^4 + n^4} \frac{\langle y, e_m \rangle_H}{n^4 + m^4}$$

holds for $y \in H$ and $i \in \mathcal{I}$, $n, m \in \mathcal{J}$. Obviously, these two terms are not equal for all $n, m \in \mathcal{J}$.

From the parameter values stated above, we compute $\gamma \in [\frac{1}{2}, 1)$. With this information, we can identify the scheme that is superior according to Table 3. Let $\varepsilon \in (0, \frac{1}{2})$ be arbitrarily small and choose $q_{\text{MIL}} = \gamma = 1 - \varepsilon > q_{\text{EES}}$ and $\alpha = \frac{7}{3} - \varepsilon$. First, we check condition M1C2, see Table 2, which holds as

$$\gamma \rho_{\text{A}}(2q_{\text{MIL}} - 1) \leq 2q_{\text{MIL}} \quad \Leftrightarrow \quad 2(1-\varepsilon)(1-2\varepsilon) \leq 2(1-\varepsilon).$$

Table 4 Error and standard deviation obtained from 200 paths. The computational cost $\bar{c}$ is computed as $\bar{c}(\text{MIL1}) = MKN^2 + KM^{2q_{\text{MIL}}} + M(K+N+KN)$, $\bar{c}(\text{MIL2}) = MKN^2 + M^{q_{\text{MIL}}+\frac{1}{2}}K^{\frac{5}{2}} + MK^2 + M(K+N+KN)$ and $\bar{c}(\text{EES}) = MKN + MN + MK$

			MIL1			MIL2		
N	M	K	$\bar{c}$	Error	Std	$\bar{c}$	Error	Std
2	4	$\lceil 2^{\frac{2}{7}} \rceil$	64	$2.9 \cdot 10^{-2}$	$7.2 \cdot 10^{-3}$	71	$2.9 \cdot 10^{-2}$	$7.2 \cdot 10^{-3}$
4	2^4	$\lceil 2^{\frac{4}{7}} \rceil$	1024	$2.5 \cdot 10^{-2}$	$5.4 \cdot 10^{-4}$	758	$2.5 \cdot 10^{-2}$	$5.4 \cdot 10^{-4}$
8	2^6	$\lceil 2^{\frac{6}{7}} \rceil$	16384	$1.7 \cdot 10^{-2}$	$1.1 \cdot 10^{-4}$	9897	$1.7 \cdot 10^{-2}$	$1.1 \cdot 10^{-4}$
16	2^8	$\lceil 2^{\frac{8}{7}} \rceil$	393216	$6.3 \cdot 10^{-3}$	$2.8 \cdot 10^{-5}$	220196	$6.3 \cdot 10^{-3}$	$2.8 \cdot 10^{-5}$
32	2^{10}	$\lceil 2^{\frac{10}{7}} \rceil$	6291456	$1.6 \cdot 10^{-3}$	$2.6 \cdot 10^{-5}$	3325212	$1.6 \cdot 10^{-3}$	$2.6 \cdot 10^{-5}$

			Exponential Euler		
N	M	K	$\bar{c}$	Error	Std
2	2^4	$\lceil 2^{\frac{2}{7}} \rceil$	64	$2.1 \cdot 10^{-2}$	$6.0 \cdot 10^{-3}$
4	2^8	$\lceil 2^{\frac{4}{7}} \rceil$	2048	$2.7 \cdot 10^{-2}$	$7.2 \cdot 10^{-4}$
8	2^{12}	$\lceil 2^{\frac{6}{7}} \rceil$	65536	$1.7 \cdot 10^{-2}$	$2.1 \cdot 10^{-4}$
16	2^{16}	$\lceil 2^{\frac{8}{7}} \rceil$	3145728	$6.1 \cdot 10^{-3}$	$4.4 \cdot 10^{-5}$
32	2^{20}	$\lceil 2^{\frac{10}{7}} \rceil$	100663296	$1.5 \cdot 10^{-3}$	$6.6 \cdot 10^{-6}$

Moreover, condition M2C1a in Table 2 is fulfilled as well because it holds $\rho_Q = 3$, $\gamma\rho_A = 2(1-\varepsilon) \leq 6(\frac{7}{3} - \varepsilon) = 2\alpha\rho_Q$ and it is easy to check that

$$\frac{3}{2}\gamma\rho_A q_{\text{MIL}} + \Big(q_{\text{MIL}} - \frac{1}{2}\Big)\gamma\rho_A\alpha\rho_Q \leq 2\alpha\rho_Q q_{\text{MIL}}$$
$$\Leftrightarrow \quad 3(1-\varepsilon) + 6\Big(\frac{1}{2} - \varepsilon\Big)\Big(\frac{7}{3} - \varepsilon\Big) \leq 6\Big(\frac{7}{3} - \varepsilon\Big)$$

is fulfilled due to $\gamma = q_{\text{MIL}}$, which proves condition M2C1a. From Table 3, we expect that both schemes MIL1 and MIL2 obtain the same effective order of convergence (11) which exceeds the order of the exponential Euler scheme in this case. For some fixed $N \in \mathbb{N}$, we compute the relation of N, M, K from (10). This yields $M = N^2$ and $K = \lceil N^{\frac{2}{7}} \rceil$ for the Milstein schemes. Moreover, we calculate the effective order of convergence as $\text{error(MIL1)} = \text{error(MIL2)} = \mathcal{O}(\bar{c}^{-\frac{7}{15}+\varepsilon})$ for some arbitrarily small $\varepsilon > 0$. For the EES, on the other hand, we obtain $M = N^4$, $K = \lceil N^{\frac{2}{7}} \rceil$ and $\text{error(EES)} = \mathcal{O}(\bar{c}^{-\frac{14}{37}+\varepsilon})$.

In the numerical analysis, we simulate 200 paths with the schemes MIL1, MIL2 and EES. The results are compared to a substitute for the exact solution—an approximation computed with the linear implicit Euler scheme [4] with $N = 2^5$, $K = \lceil 2^{\frac{10}{7}} \rceil$ and $M = 2^{16}$. Our findings are summarized in Table 4 and Fig. 1. In Fig. 1, we plot the errors versus the computational cost based on the cost model that is used for

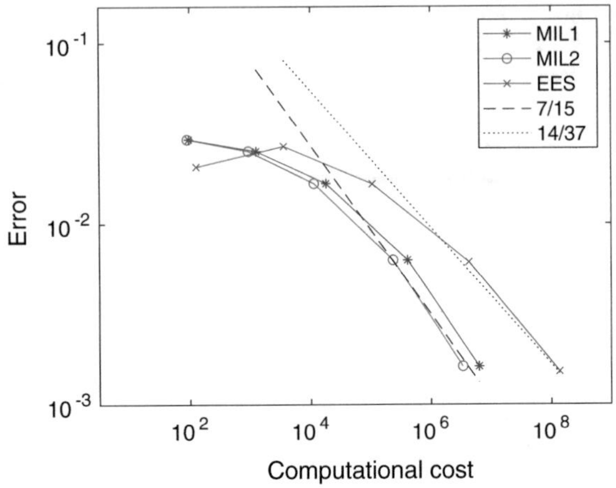

Fig. 1 Error against computational cost computed from 200 paths for $N \in \{2, 4, 8, 16, 32\}$ in log-log scale

the analysis. Here, one observes that the Milstein schemes obtain a higher effective order of convergence than the Euler scheme. Moreover, Table 4 illustrates the difference in the computational costs of these schemes. The Euler scheme involves costs which are significantly higher. A comparison of MIL1 and MIL2 shows for this example that the Milstein scheme in combination with Algorithm 2 involves a lower computational cost than the Milstein scheme combined with Algorithm 1.

Acknowledgements This project was funded by the Cluster of Excellence "The Future Ocean". The "Future Ocean" is funded within the framework of the Excellence Initiative by the Deutsche Forschungsgemeinschaft (DFG) on behalf of the German federal and state governments.

References

1. Jentzen, A., Röckner, M.: Regularity analysis for stochastic partial differential equations with nonlinear multiplicative trace class noise. J. Differ. Equ. **252**(1), 114–136 (2012)
2. Jentzen, A., Röckner, M.: A Milstein scheme for SPDEs. Found. Comput. Math. **15**(2), 313–362 (2015)
3. Kloeden, P.E., Platen, E., Wright, I.W.: The approximation of multiple stochastic integrals. Stoch. Anal. Appl. **10**(4), 431–441 (1992)
4. Kloeden, P.E., Shott, S.: Linear-implicit strong schemes for Itô-Galerkin approximations of stochastic PDEs. J. Appl. Math. Stoch. Anal. **14**(1), 47–53 (2001)
5. Leonhard, C.: Derivative-free numerical schemes for stochastic partial differential equations. Ph.D. thesis, Institute of Mathematics, Universität zu Lübeck (2016)
6. Leonhard, C., Rößler, A.: Enhancing the order of the Milstein scheme for stochastic partial differential equations with commutative noise. ArXiv e-prints, v2 (2018)
7. Leonhard, C., Rößler, A.: Enhancing the order of the Milstein scheme for stochastic partial differential equations with commutative noise. SIAM J. Numer. Anal. **56**(4), 2585–2622 (2018)
8. Leonhard, C., Rößler, A.: Iterated stochastic integrals in infinite dimensions: approximation and error estimates. Stoch. Partial Differ. Equ. Anal. Comput. **7**(2), 209–239 (2019)
9. Lord, G.J., Tambue, A.: Stochastic exponential integrators for the finite element discretization of SPDEs for multiplicative and additive noise. IMA J. Numer. Anal. **33**(2), 515–543 (2013)

10. Prévôt, C., Röckner, M.: A Concise Course on Stochastic Partial Differential Equations. Lecture Notes in Mathematics, vol. 1905. Springer, Berlin (2007)
11. Sell, G.R., You, Y.: Dynamics of Evolutionary Equations. Applied Mathematical Sciences, vol. 143. Springer, New York (2002)
12. Wiktorsson, M.: Joint characteristic function and simultaneous simulation of iterated Itô integrals for multiple independent Brownian motions. Ann. Appl. Probab. **11**(2), 470–487 (2001)

QMC Sampling from Empirical Datasets

Fei Xie, Michael B. Giles and Zhijian He

Abstract This paper presents a simple idea for the use of quasi-Monte Carlo sampling with empirical datasets, such as those generated by MCMC methods. It also presents and analyses a related idea of taking advantage of the Hilbert space-filling curve. Theoretical and numerical analyses are provided for both. We find that when applying the proposed QMC sampling methods to datasets coming from a known distribution, they give similar performance as the standard QMC method directly sampling from this known distribution.

Keywords QMC · Empirical datasets · Recursive bisection · The Hilbert space-filling curve

1 Introduction

If U is a random variable uniformly distributed over the d-dimensional cube $[0, 1]^d$, the standard MC (Monte Carlo) estimate for $\mathbb{E}[f(U)]$ using n i.i.d. sample points is

$$\bar{f}_n = \frac{1}{n}\sum_{j=1}^{n} f(U_j)$$

F. Xie (✉)
Tsinghua University, Beijing, China
e-mail: xiefei901108@126.com

M. B. Giles
Mathematical Institute, University of Oxford, Oxford, UK
e-mail: mike.giles@maths.ox.ac.uk

Z. He
South China University of Technology, Guangzhou, China
e-mail: hezhijian@scut.edu.cn

B. Tuffin and P. L'Ecuyer (eds.), *Monte Carlo and Quasi-Monte Carlo Methods*, Springer Proceedings in Mathematics & Statistics 324,
https://doi.org/10.1007/978-3-030-43465-6_26

If the variance $\mathbb{V}[f]$ is finite then the error is $O(n^{-1/2})$. The QMC (Quasi-Monte Carlo) estimate is similar, except that the n points are chosen carefully [3] so that the error is $O((\log n)^d/n)$ for integrands satisfying appropriate conditions [8]. This extends naturally to more general distributions, with

$$\mathbb{E}[f(X)] \approx \bar{f}_n = \frac{1}{n}\sum_{j=1}^{n} f(X_j)$$

where for the MC estimate the X_j are i.i.d. samples from the distribution of the random variable X. For the corresponding QMC estimate $X_j = \psi(U_j)$ with the U_j being the same set of QMC points in the unit cube as before, and $\psi(U)$ is a mapping which gives the desired probability density function for X, i.e., if X has probability density function $\rho(x)$ then it is related to the determinant of the Jacobian of the mapping through

$$\rho(\psi(u)) = \left|\det\left(\frac{\partial\psi}{\partial u}\right)\right|^{-1}.$$

The question we address in this paper is what should we do when we do not know $\rho(x)$, and instead all we have is a large set of samples $X_j, j = 1, \ldots, N$ which constitutes an empirical dataset?

The medical decision application [1] which motivated this research has inputs which are multiple independent multi-dimensional random variables. Some of these come from known distributions, but some correspond to Bayesian posterior distributions defined solely by a set of values generated by an MCMC process. In order to obtain a given number of samples from this posterior distribution, currently these MCMC datasets are approximated by multi-variate Gaussian distributions, and then QMC can be used very effectively. However, it might be better to work directly with these empirical datasets without introducing any approximation error, and the question is then how to use QMC. Although the overall application is fairly high-dimensional (15–40), the individual MCMC datasets are low-dimensional (2 or 3).

For the purpose of sampling from an empirical dataset using QMC, we propose a novel QMC sampling method based on recursive bisection. The proposed QMC sampling method is described in Sect. 2, and an error analysis for the case $d = 1$ is given. Section 3 presents an alternative method based on a Hilbert space-willing curve, and a corresponding numerical analysis. Section 4 presents numerical results which compare the two QMC sampling methods with the standard QMC method. Conclusions are given in Sect. 5.

2 QMC Sampling

Suppose we have an empirical dataset of $N = 2^M$ d-dimensional samples $\{X_j, j = 1, \ldots, N\}$, where M is a positive integer. With QMC sampling, the challenge is to

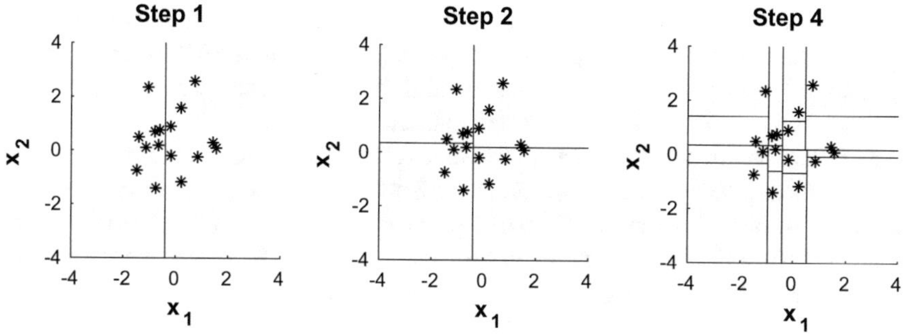

Fig. 1 Steps 1, 2, and 4 in the recursive bisection of a 2-dimensional dataset with 16 points

construct a mapping from QMC points in the hypercube $(0, 1)^d$ to points X_i in the dataset.

2.1 *Recursive Bisection-Based Sorting*

Borrowing ideas from recursive graph bisection for distributed-memory parallel computing [2, 6] the central idea is to repeatedly bisect the dataset in alternating directions.

For example, for a 2-dimensional dataset $\{(x_1, x_2)_j, j = 1, \ldots, N\}$, firstly we find the median value $x^{(1)}_{\mathrm{mid}}$ of their x_1 coordinates by sorting the points based on x_1, and then re-arrange the set into two halves, the first half with $x_1 < x^{(1)}_{\mathrm{mid}}$ and the second half with $x_1 > x^{(1)}_{\mathrm{mid}}$; if there are two or more points for which $x_1 = x^{(1)}_{mid}$ they can be assigned to the two halves based on their indices to achieve a bisection. Each of the two halves is then sorted within each half based on x_2, and bisected and re-arranged using the median value of their x_2 coordinates (with the median being different for each half). In the next step, the points in each quarter are sorted again based on x_1, and each quarter is further bisected. This is repeated, alternating between the x_1 and x_2 coordinates, until each subset has just one point. The empirical dataset is thus sorted by the recursive bisection procedure. The process is illustrated in Fig. 1 for a small dataset with just 16 sample points.

2.2 *Mapping of QMC Points*

For application of QMC on selecting samples from a dataset of size N, firstly the dataset is recursively bisected into N subsets. Then the location of the subsets to be selected is characterized by the relative position of the QMC points in the hypercube $(0, 1)^d$. In this way, we pass the uniformity of QMC points into an integer index between 0 and $N - 1$ representing the particular position in a data sequence. In this subsection, we construct a mapping from points in $(0, 1)^d$ to the dataset sorted by the recursive bisection procedure. It is then straightforward to use QMC (or randomized QMC) points to sample from the sorted dataset.

Let us start with the simple case $M = d \times k$, where k is a positive integer. Suppose the low discrepancy sequence we use is the Sobol' sequence [10], which is a (t, d)-sequence in base 2 for all d. To map a point $Q = (q_1, \ldots, q_d) \in (0, 1)^d$ from the hypercube $(0, 1)^d$ to one of the indices $\{0, 1, \ldots, 2^{dk} - 1\}$, there are two steps. In the first step, the point Q is mapped to the d-dimensional integer space $\{0, \ldots, 2^k - 1\}^d$ by

$$(q_1, \ldots, q_d) \mapsto (\lfloor 2^k q_1 \rfloor, \ldots, \lfloor 2^k q_d \rfloor),$$

where $\lfloor x \rfloor$ denotes x rounded down to the nearest integer.

In the second step, the d-dimensional integer index $(i_1, \ldots, i_d) \in \{0, \ldots, 2^k - 1\}^d$ needs to be bijectively mapped onto an integer index $i \in \{0, \ldots, N - 1\}$ as the data points are stored in a sequence of length N. To this end, we write each integer coordinate of the index $(i_1, \ldots, i_d)$ in its k-bit binary representation $(i_j)_2 = b_{j,k-1} \ldots b_{j,0},\ j = 1, \ldots, d$, where

$$i_j = \sum_{\ell=0}^{k-1} b_{j,\ell} 2^\ell.$$

Each bit of i_j's binary representation can be viewed as the indication of which half the desired point is located in each time the dataset is bisected according to the j-th coordinate. i.e., $b_{j,k-1}$ indicates whether the desired point is in the first half or the second when the dataset is bisected according to the j-th coordinate for the first time, $b_{j,k-2}$ indicates when the dataset is bisected in this direction for the second time, and so on. As we are doing the bisection in alternating directions, interleaving the bits of each integer coordinate gives the complete information of selection in each successive bisection step

$$(i)_2 = b_{1,k-1} \ldots b_{d,k-1} b_{1,k-2} \ldots b_{d,k-2} \ldots b_{1,0} \ldots b_{d,0},$$

which can then be easily transformed into the corresponding index

$$i = \sum_{\ell=0}^{k-1} 2^{d\ell} \left(\sum_{j=1}^{d} b_{j,\ell} 2^{d-j} \right) \in \{0, \ldots, N - 1\}.$$

To better understand the above formula, let us consider the 2-dimensional index $(i_1, i_2) = (5, 2)$. Writing it in 3-bit binary representation and interleaving their bits results in the 6-bit binary representation of the index i.

$$\begin{aligned} i_1 &= 1\ \ 0\ \ 1 \\ i_2 &= \ \ 0\ \ 1\ \ 0 \\ \Rightarrow\ i &= 1\,0\,0\,1\,1\,0 \end{aligned}$$

Thus, the index i corresponding to $(i_1, i_2) = (5, 2)$ is 38 in a dataset with $N = 2^{2\times 3} = 64$ points.

For computational efficiency, a final step is to re-arrange the dataset once more so that the $(i_1, \ldots, i_d)$ point is stored at linear memory location

$$\sum_{\ell=1}^{d} 2^{(\ell-1)k} i_\ell.$$

This simplifies the subsequent implementation of the QMC algorithm.

Now we generalize to the case $M = d \times k - c$ where $0 < c < k$. For this case, the point Q is firstly mapped to the d-dimensional integer space $\Omega(d, k, c) = \{0, \ldots, 2^k - 1\}^{d-c} \times \{0, \ldots, 2^{k-1} - 1\}^c$, where the last c coordinates are in $\{0, \ldots, 2^{k-1} - 1\}$, and the remaining coordinates are in $\{0, \ldots, 2^k - 1\}$. This is achieved by

$$(q_1, \ldots, q_d) \mapsto (\lfloor 2^k q_1 \rfloor, \ldots, \lfloor 2^k q_{d-c} \rfloor, \lfloor 2^{k-1} q_{d-c+1} \rfloor, \ldots, \lfloor 2^{k-1} q_d \rfloor).$$

Similar to the case where $c = 0$, each integer coordinate of the index $(i_1, \ldots, i_d) \in \Omega(d, k, c)$ is written in its binary representation, but with different bits:

$$(i_j)_2 = \begin{cases} b_{j,k-1} \ldots b_{j,0}, & j = 1, \ldots, d - c \\ b_{j,k-1} \ldots b_{j,1}, & j = d - c + 1, \ldots, d \end{cases}.$$

Note that here $b_{j,k-l}$ still indicates whether the desired point is in the first half or the second when the dataset is bisected according to the j-th coordinate for the lth time. The difference is that the dataset is bisected for k times in total according to each of the first $d - c$ coordinates, while it is only bisected for $k - 1$ times according to each of the last c coordinates. Again, by interleaving the bits of each integer coordinate we get

$$(i)_2 = b_{1,k-1} \ldots b_{d,k-1} b_{1,k-2} \ldots b_{d,k-2} \ldots b_{1,1} \ldots b_{d,1} b_{1,0} \ldots b_{d-c,0}.$$

The corresponding index can be expressed as

$$i = \sum_{\ell=1}^{k-1} 2^{d\ell - c} \left(\sum_{j=1}^{d} b_{j,\ell} 2^{d-j} \right) + \left(\sum_{j=1}^{d-c} b_{j,0} 2^{d-c-j} \right) \in \{0, \ldots, N-1\}.$$

Again consider the example of the 2-dimensional index $(i_1, i_2) = (5, 2)$. We now calculate its corresponding index i in a dataset with $N = 2^{2\times 3-1} = 32$ points. Writing

i_1 and i_2 in 3-bit and 2-bit binary representation respectively, and interleaving their bits results in the 5-bit binary representation of the index i.

$$\begin{aligned} i_1 &= 1\quad 0\quad 1 \\ i_2 &= \quad 1\quad 0 \\ \Rightarrow\ i &= 1\,1\,0\,0\,1 \end{aligned}$$

Thus, the index i corresponding to $(i_1, i_2) = (5, 2)$ is 25 in a dataset with $N = 32$ points. We see that for different numbers of points N the same 2-dimensional index (i_1, i_2) is mapped to different indices i.

2.3 Error Analysis

Let $\tau : [0, 1]^d \to \mathscr{X}=\{X_1, \ldots, X_N\}$ be the mapping described above, where $N=2^M$ for some integer $M \geq 1$. Based on the empirical dataset $\mathscr{X}$, one may use the following sample average

$$\hat{\mu}_N = \frac{1}{N}\sum_{i=1}^{N} f(X_i) \tag{1}$$

as an estimate of $\mu = \mathbb{E}[f(X)]$. The QMC sampling for the empirical dataset gives an estimate

$$\hat{\mu}_n^{\mathrm{I}} = \frac{1}{n}\sum_{i=1}^{n} f(\tau(U_i)), \tag{2}$$

where $U_i \in (0, 1)^d$ are RQMC (randomised QMC) points with the sample size $n \leq N$. We define the difference between these two estimates as the selection error from QMC sampling. We next study the selection error with QMC sampling when $d = 1$.

When $d = 1$, the mapping τ reduces to the inverse function of the empirical CDF of the dataset. More specifically, let the empirical CDF of the dataset be

$$F_N(x) = \frac{1}{N}\sum_{i=1}^{N} \mathbf{1}\{X_i \leq x\}.$$

Then we have

$$\tau(u) = F_N^{-1}(u) := \inf\{x : F_N(x) \geq u\}.$$

Now consider the case of using a scrambled $(0, m, 1)$-net in base $b = 2$ as inputs in (2), where $n = 2^m$ and $m \leq M$. Let $U_{(i)}$ be the order statistic of $U_i,\ i = 1, \ldots, n$, and similarly for $X_{(i)}$. By the property of a scrambled $(0, m, 1)$-net, we have $U_{(i)} \sim \mathbb{U}([(i-1)/n, i/n))$ independently with probability one. Note that one-dimensional stratified sampling has the same effect. Our results in this section also apply for stratified sampling.

Let $R = N/n = 2^{M-m}$, and $Y_j := \tau(U_{(i)})$. The estimate becomes

$$\hat{\mu}_n^{\mathrm{I}} = \frac{1}{n}\sum_{i=1}^{n} f(Y_i), \tag{3}$$

where Y_j is randomly and independently selected from the block $\{X_{(iR-R+1)}, \ldots, X_{(iR)}\}$ for $i = 1, \ldots, n$.

To fix our idea, let's consider the simple case $X_i \overset{\text{iid}}{\sim} \mathbb{U}((0, 1))$ for $i = 1, \ldots, N$. For this case, the kth order statistic of the uniform distribution is a Beta random variable, i.e., $X_{(k)} \sim \text{Beta}(k, N + 1 - k)$. We next bound the selection error $\mathbb{E}[(\hat{\mu}_n^{\mathrm{I}} - \hat{\mu}_N)^2]$. If f is Lipschitz continuous with constant $C > 0$, we find that

$$\mathbb{E}[(\hat{\mu}_n^{\mathrm{I}} - \hat{\mu}_N)^2 | \mathscr{X}] = \frac{1}{n^2}\sum_{i=1}^{n} \mathbb{E}\left[\left(f(Y_i) - \frac{1}{R}\sum_{j=1}^{R} f(X_{(iR-R+j)})\right)^2 \middle| \mathscr{X}\right] \tag{4}$$

$$= \frac{1}{n^2}\sum_{i=1}^{n} \mathbb{E}\left[\left(\frac{1}{R}\sum_{j=1}^{R}(f(Y_i) - f(X_{(iR-R+j)}))\right)^2 \middle| \mathscr{X}\right] \tag{5}$$

$$\leq \frac{C^2}{n^2}\sum_{i=1}^{n} \mathbb{E}\left[\left(X_{(iR)} - X_{(iR-R+1)}\right)^2 \middle| \mathscr{X}\right]. \tag{6}$$

The first equality (4) is due to the fact that given $\mathscr{X}$, Y_i is randomly and independently selected from the block $\{X_{(iR-R+1)}, \ldots, X_{(iR)}\}$. This gives

$$\begin{aligned}
\mathbb{E}[(\hat{\mu}_n^{\mathrm{I}} - \hat{\mu}_N)^2] &\leq \frac{C^2}{n^2}\sum_{i-1}^{n} \mathbb{E}\left[\left(X_{(iR)} - X_{(iR-R+1)}\right)^2\right] \\
&= \frac{C^2}{n^2}\sum_{i=1}^{n} \frac{(R-1)^2 + R - 1}{(N+1)(N+2)} \qquad (7)\\
&= \frac{C^2(R^2 - R)}{(N+1)(N+2)n} \\
&\leq \frac{C^2R^2}{N^2n} = C^2n^{-3},
\end{aligned}$$

where we use Lemma 1 from the Appendix in establishing (7). More generally, we have the following theorem.

Theorem 1 *Suppose that X_i are iid with a density $q(x)$ on a bounded support $\mathscr{D} = [a, b]$. Assume that $q_{\min} = \inf_{x\in\mathscr{D}} q(x) > 0$. If f is Lipschitz continuous with constant $C > 0$, then*

$$\mathbb{E}[(\hat{\mu}_n^{\mathrm{I}} - \hat{\mu}_N)^2] \le \frac{C^2}{q_{\min}^2 n^3}. \tag{8}$$

If f is piecewise Lipschitz continuous with jumps at position a_j, $j = 1, \ldots, k$, then

$$\mathbb{E}[(\hat{\mu}_n^{\mathrm{I}} - \hat{\mu}_N)^2] \le \frac{C_1^2}{q_{\min}^2 n^3} + \frac{4kC_2^2}{n^2}, \tag{9}$$

where C_1 is the Lipschitz constant and $C_2 = \sup_{x\in\mathscr{D}} |f(x)|$.

Proof Note that for any $1 \le i, j \le N$,

$$\left|X_i - X_j\right| = \left|F(X_i) - F(X_j)\right| / q(\xi) \le \left|F(X_i) - F(X_j)\right| / q_{\min}, \tag{10}$$

where ξ is a constant between X_i and X_j, and F is the CDF of the X_i. Let $\tilde{Y}_i = F(X_{(i)})$. Then $\tilde{Y}_i$ are the order statistic of the uniform distribution. If f is Lipschitz continuous, by (6) and (10), we have

$$\mathbb{E}\left[(\hat{\mu}_n^{\mathrm{I}} - \hat{\mu}_N)^2\right] \le \frac{C^2}{n^2 q_{\min}^2} \sum_{i=1}^{n} \mathbb{E}\left[\left(\tilde{Y}_{iR} - \tilde{Y}_{iR-R+1}\right)^2\right] \le \frac{C^2}{q_{\min}^2} n^{-3}.$$

Now assume that f is piecewise Lipschitz continuous with jumps at position a_j, $j = 1, \ldots, k$. There are at most k intervals of the form $[X_{(iR-R+1)}, X_{(iR)})$ in which f is not Lipschitz continuous. For each of them,

$$\mathbb{E}\left[\left(\frac{1}{R}\sum_{j=1}^{R}(f(Y_i) - f(X_{(iR-R+j)}))\right)^2 \middle| \mathscr{X}\right] \le 4C_2^2.$$

We thus establish (9). □

The mean squared error (MSE) of the QMC estimate (2) can be decomposed into two terms, that is

$$\begin{aligned}
\mathrm{MSE}(\hat{\mu}_n^{\mathrm{I}}) &= \mathbb{E}[(\hat{\mu}_n^{\mathrm{I}} - \mu)^2] = \mathbb{E}[(\hat{\mu}_n^{\mathrm{I}} - \hat{\mu}_N + \hat{\mu}_N - \mu)^2] \\
&= \mathbb{E}[(\hat{\mu}_n^{\mathrm{I}} - \hat{\mu}_N)^2] + \mathbb{E}[(\hat{\mu}_N - \mu)^2] + 2\mathbb{E}[(\hat{\mu}_n^{\mathrm{I}} - \hat{\mu}_N)(\hat{\mu}_N - \mu)] \\
&= \mathbb{E}[(\hat{\mu}_n^{\mathrm{I}} - \hat{\mu}_N)^2] + \mathbb{E}[(\hat{\mu}_N - \mu)^2],
\end{aligned} \tag{11}$$

where we use the fact that $\mathbb{E}[\hat{\mu}_n^{\mathrm{I}} - \hat{\mu}_N | \mathscr{X}] = 0$ and thus

$$\begin{aligned}
\mathbb{E}[(\hat{\mu}_n^{\mathrm{I}} - \hat{\mu}_N)(\hat{\mu}_N - \mu)] &= \mathbb{E}[\mathbb{E}[(\hat{\mu}_n^{\mathrm{I}} - \hat{\mu}_N)(\hat{\mu}_N - \mu)|\mathscr{X}]] \\
&= \mathbb{E}[(\hat{\mu}_N - \mu)\mathbb{E}[\hat{\mu}_n^{\mathrm{I}} - \hat{\mu}_N|\mathscr{X}]] = 0.
\end{aligned}$$

The first term in (11) is the error due to selecting a few samples from the original datasets, and the second term is due to the original sampling error. The mean square selection error $\mathbb{E}[(\hat{\mu}_n^{\mathrm{I}} - \hat{\mu}_N)^2]$ is related to n, the size of the thinned sample set, and the dimension d, and intuitively this quantity should decrease with increasing n. What we are interested in determining is the minimum size of n to ensure the selection error is no bigger than the sampling error, so that at this level of n, the accuracy of the estimate $\hat{\mu}_n^{\mathrm{I}}$ is comparable to $\hat{\mu}_N$. For an iid dataset, the mean square error $\mathbb{E}[(\hat{\mu}_N - \mu)^2] = \sigma^2/N$, where σ^2 is the variance of X_i. To make the two terms in the right-hand side of (11) comparable, the optimal n is

$$n^* = \begin{cases} O(N^{1/3}), \text{ if } & f \text{ is Lipschitz,} \\ O(N^{1/2}), \text{ if } & f \text{ is piecewise Lipschitz.,} \end{cases} \tag{12}$$

for $d = 1$. As a result, the MSE of the QMC estimate is

$$\mathrm{MSE}(\hat{\mu}_n^{\mathrm{I}}) = \begin{cases} O(n^{-3}), \text{ if } & f \text{ is Lipschitz,} \\ O(n^{-2}), \text{ if } & f \text{ is piecewise Lipschitz,} \end{cases} \tag{13}$$

for $n \le n^*$ and as $N \to \infty$.

Things become more complicated for $d > 1$. We leave these cases for future research. Instead, we give numerical analysis for an alternative way of QMC sampling by making use of the Hilbert space-filling curve.

3 Sorting the Dataset via the Hilbert Space-Filling Curve

Instead of using the recursive bisection mapping in Sect. 2, we map the dataset in dimension $d > 2$ to the one-dimensional unit interval via a Hilbert space-filling curve. The data then are simply sorted in natural order. Formally, let $H : [0, 1] \to [0, 1]^d$ be the Hilbert curve with $H(0) = (0, \dots, 0)$. It is important to noting that the mapping is not a bijection because certain points in $[0, 1]^d$ have more than one preimage through H. However, the set of such points is of Lebesgue measure 0. There exists a one-to-one Borel measurable pseudo-inverse $h : [0, 1]^d \to [0, 1]$ such that $H(h(x)) = x$ for all $x \in [0, 1]^d$. There are some properties of the Hilbert curve and its inverse h:

- For any measurable set $I \subset [0, 1]$, $\lambda_d(H(I))) = \lambda_1(I)$;
- For any $x, y \in [0, 1]$, then

$$\|H(y) - H(x)\| \le 2\sqrt{d+3}\,|y - x|^{1/d}\,; \tag{14}$$

- If $u \sim \mathbb{U}([0, 1])$, then $H(u) \sim \mathbb{U}([0, 1]^d)$;
- If $U \sim \mathbb{U}([0, 1]^d)$, then $h(U) \sim \mathbb{U}([0, 1])$.

See [5] for the properties of the Hilbert curve. Gerber and Chopin [4] uses the inverse function h to sort the particles in the particle filtering algorithm to combine with QMC. Let $\mathscr{D}$ be the support of X_i, and let

$$x_i = h \circ \psi(X_i) \in [0, 1], \tag{15}$$

where $\psi : \mathscr{D} \to [0, 1]^d$ is some user-chosen bijection between $\mathscr{D}$ and $\psi(\mathscr{D}) \subset [0, 1]^d$, and X_is are still the points of an empirical dataset as in Sect. 2. We may simply take $\psi(X_i) = (\psi_1(x_{i1}), \ldots, \psi_d(x_{id}))$, where x_{ij} is the jth component of X_i, and the ψ_is are continuous and strictly monotone. Particularly, if the X_i are in a bounded cube $\prod_{i=1}^d [a_i, b_i]$, then we can simply take

$$\psi_i(t) = \frac{t - a_i}{b_i - a_i}, \quad i = 1, \ldots, d.$$

When $\mathscr{D} = \mathbb{R}^d$, [4] suggested to use the logistic transformation componentwise. Let $\tilde{F}_N(x)$ be the empirical CDF of the transformed dataset $\{x_1, \ldots, x_N\}$:

$$\tilde{F}_N(x) = \frac{1}{N} \sum_{i=1}^{N} \mathbf{1}\{x_i \le x\}. \tag{16}$$

Let $\tilde{\tau}$ be a mapping from [0, 1] to $\mathscr{D}$, defined by

$$\tilde{\tau}(u) = \psi^{-1}(H(\tilde{F}_N^{-1}(u))).$$

This gives an estimate

$$\hat{\mu}_n^{\mathrm{II}} = \frac{1}{n} \sum_{i=1}^{n} f(\tilde{\tau}(u_i)), \tag{17}$$

where $u_i \in [0, 1]$ are one-dimensional RQMC points. We focus on the case of using a scrambled $(0, m, 1)$-net in base $b = 2$ as inputs.

For simplicity, let's consider the case $X_j \overset{\text{iid}}{\sim} \mathbb{U}([0, 1]^d)$ for $j = 1, \ldots, N$. Now ψ is set to the identity function. By the properties of h listed above, we have $x_j \overset{\text{iid}}{\sim} \mathbb{U}([0, 1])$, $j = 1, \ldots, N$. Again, by the properties of a scrambled $(0, m, 1)$-net, we have the order statistic $u_{(i)} \sim \mathbb{U}([(i-1)/n, i/n))$ independently with probability one. Let $Y_i := \tilde{\tau}(u_{(i)})\ i = 1, \ldots, n$, and let $\tilde{Y}_j = \psi^{-1} \circ H(x_{(j)}),\ j = 1, \ldots, N$. The Hilbert curve based estimate becomes

$$\hat{\mu}_n^{\mathrm{II}} = \frac{1}{n} \sum_{i=1}^{n} f(Y_i), \tag{18}$$

where Y_i is randomly and independently selected from the block $\{\tilde{Y}_{iR-R+1}, \ldots, \tilde{Y}_{iR}\}$ for $i = 1, \ldots, n$. Thus if f is Lipschitz continuous with constant $C > 0$, by (14), we have

$$
\begin{aligned}
\left|f(Y_i) - f(\tilde{Y}_{iR-R+j})\right| &\le C\left\|Y_i - \tilde{Y}_{iR-R+j}\right\| \\
&\le C \sup_{k,\ell=1,\dots,R} \left\|\tilde{Y}_{iR-R+k} - \tilde{Y}_{iR-R+\ell}\right\| \\
&= C \sup_{k,\ell=1,\dots,R} \left\|\psi^{-1} \circ H(x_{(iR-R+k)}) - \psi^{-1} \circ H(x_{(iR-R+\ell)})\right\| \\
&= C \sup_{k,\ell=1,\dots,R} \left\|H(x_{(iR-R+k)}) - H(x_{(iR-R+\ell)})\right\| \\
&\le 2C\sqrt{d+3}(x_{(iR)} - x_{(iR-R+1)})^{1/d}
\end{aligned}
\tag{19}
$$

for $j = 1, \dots, R$. Similarly to (5), we have

$$
\mathbb{E}[(\hat{\mu}_n^{\text{II}} - \hat{\mu}_N)^2|\mathscr{X}] \le \frac{4C^2(d+3)}{n^2}\sum_{i=1}^{n}\mathbb{E}\left[(x_{(iR)} - x_{(iR-R+1)})^{2/d}\middle|\mathscr{X}\right]. \tag{20}
$$

As a result, for any $d \ge 2$,

$$
\begin{aligned}
\mathbb{E}[(\hat{\mu}_n^{\text{II}} - \hat{\mu}_N)^2] &\le \frac{4C^2(d+3)}{n^2}\sum_{i=1}^{n}\mathbb{E}\left[(x_{(iR)} - x_{(iR-R+1)})^{2/d}\right] \\
&\le \frac{4C^2(d+3)}{n^2}\sum_{i=1}^{n}\left(\mathbb{E}[x_{(iR)} - x_{(iR-R+1)}]\right)^{2/d} &(21) \\
&= \frac{4C^2(d+3)}{n^2}\sum_{i=1}^{n}\left(\frac{R-1}{N+1}\right)^{2/d} &(22) \\
&\le \frac{4C^2(d+3)}{n}\left(\frac{R-1}{N+1}\right)^{2/d} \\
&\le 4C^2(d+3)n^{-1-2/d}, &(23)
\end{aligned}
$$

where in (21) we use Jensen's inequality (because $2/d \le 1$), and we use Lemma 1 in establishing (22).

Theorem 2 *Suppose that X_i are iid with a density $q(X)$ on a bounded support $\mathscr{D} = \prod_{i=1}^{d}[a_i, b_i]$. Assume that $q_{\min} = \inf_{X\in\mathscr{D}} q(X) > 0$. Let*

$$
\psi_i(t) = \frac{t - a_i}{b_i - a_i},\ i = 1, \dots, d.
$$

If f is Lipschitz continuous with constant $C > 0$, then

$$\mathbb{E}[(\hat{\mu}_n^{\mathrm{II}} - \hat{\mu}_N)^2] \leq \frac{4C^2L_1^2(d+3)}{q_{\min}^{2/d} L_2^2 n^{1+2/d}}, \tag{24}$$

where $L_1 = \max_{i=1,\ldots,d}\{b_i - a_i\}$ *and* $L_2 = \min_{i=1,\ldots,d}\{b_i - a_i\}$.

Proof Let $F_1(t)$ be the CDF of $x = h \circ \psi(X)$. For any $t_1 < t_2$, we have

$$\begin{aligned} F_1(t_2) - F_1(t_1) &= P[t_1 < h \circ \psi(X) \leq t_2] = P[\psi(X) \in H((t_1, t_2])] \\ &= \prod_{i=1}^{d}(b_i - a_i) \int_{Y \in H((t_1,t_2])} q(\psi^{-1}(Y)) \mathrm{d}Y \\ &\geq q_{\min} \prod_{i=1}^{d}(b_i - a_i)\lambda_d(H((t_1, t_2])) \\ &= q_{\min}(t_2 - t_1)\prod_{i=1}^{d}(b_i - a_i). \end{aligned}$$

Let $\tilde{x}_i = F_1(x_i),\ i = 1, \ldots, N$. Then for any $1 \leq i, j \leq N$,

$$|x_i - x_j| \leq \frac{|\tilde{x}_i - \tilde{x}_j|}{q_{\min}\prod_{i=1}^{d}(b_i - a_i)}. \tag{25}$$

Then $\tilde{x}_{(i)}$s are the order statistics of the uniform distribution. By (19) and (25), we have

$$\begin{aligned} \left|f(Y_i) - f(\tilde{Y}_{iR-R+j})\right| &\leq C \sup_{k,\ell=1,\ldots,R} \left\|\psi^{-1} \circ H(x_{(iR-R+k)}) - \psi^{-1} \circ H(x_{(iR-R+\ell)})\right\| \\ &\leq CL_1 \sup_{k,\ell=1,\ldots,R} \left\|H(x_{(iR-R+k)}) - H(x_{(iR-R+\ell)})\right\| \\ &\leq 2CL_1\sqrt{d+3}(x_{(iR)} - x_{(iR-R+1)})^{1/d} \\ &\leq \frac{2CL_1\sqrt{d+3}}{q_{\min}^{1/d}\prod_{i=1}^{d}(b_i - a_i)^{1/d}}(\tilde{x}_{(iR)} - \tilde{x}_{(iR-R+1)})^{1/d} \\ &\leq \frac{2CL_1\sqrt{d+3}}{q_{\min}^{1/d}L_2}(\tilde{x}_{(iR)} - \tilde{x}_{(iR-R+1)})^{1/d}, \end{aligned}$$

where $L_1 = \max_{i=1,\ldots,d}\{b_i - a_i\},\ L_2 = \min_{i=1,\ldots,d}\{b_i - a_i\}$. If f is Lipschitz continuous, similarly to (23), we have

$$\mathbb{E}[(\hat{\mu}_n^{\mathrm{II}} - \hat{\mu}_N)^2] \leq \frac{4C^2L_1^2(d+3)}{n^2 q_{\min}^{2/d} L_2^2} \sum_{i=1}^{n} \mathbb{E}(\tilde{x}_{(iR)} - \tilde{x}_{(iR-R+1)})^{2/d}] \leq \frac{4C^2L_1^2(d+3)}{q_{\min}^{2/d} L_2^2 n^{1+2/d}}.$$

□

As a result, for Lipschitz functions in dimensions $d \geq 2$, we may take $n = O(N^{d/(d+2)})$ so that $\mathrm{MSE}(\hat{\mu}_n^{\mathrm{II}}) = O(n^{-1-2/d})$. This error rate is in line with the one reported in [5], where for an RQMC quadrature based on the Hilbert curve an MSE of size $O(n^{-1-2/d})$ is found for a class of Lipschitz functions.

4 Numerical Tests

To test the effectiveness of the QMC sampling methods, we apply it to datasets generated from four different distributions to estimate $\mu = \mathbb{E}[f(X_1, \ldots, X_d)]$ with two different integrand functions, and compare the accuracy to the standard MC and QMC methods applied to the same problems.

The integrands include a continuous one

$$f(x_1, \ldots, x_d) = \prod_{i=1}^{d} (x_i^2 + 1), \tag{26}$$

and a discontinuous one

$$f(x_1, \ldots, x_d) = 1\left\{\sum_{i=1}^{d} x_i > d/2\right\}. \tag{27}$$

The four distributions, based on independent Normals $Z_1, Z_2, Z_3 \stackrel{\text{iid}}{\sim} N(0, 1)$ and uniformly distributed $U \sim U(0, 1)$ are:

- Correlated Normals: $X_1 = Z_1 + 0.5Z_2$, $X_2 = Z_1 - 0.5Z_2$,
- Distorted Gaussian: $X_1 = Z_1$ and $X_2 = X_1^2 + Z_2$;
- Double well: $X_1 = \mathrm{sign}(U - 0.5)\left(2 + \Phi^{-1}(2|U - 0.5|)\right)$ and $X_2 = 2Z_2$
- Unit ball surface: $X_i = Z_i / \sqrt{\sum_{j=1}^{3} Z_j^2}$, $i = 1, 2, 3$

The empirical dataset is generated via crude MC sampling. We focus on four schemes of sampling:

- Standard MC (labelled as "std MC"), which estimates it by averaging function values based on n random samples;
- Standard QMC (labelled as "std QMC"), which generates samples from the two or three dimensional Sobol' points u using the mapping $z = \Phi^{-1}(u)$ to generate the quasi-Normal random vectors z;
- QMC sampling based on recursive bisection (labelled as "bQMC"), which selects n samples from a dataset of size $N = 2^{20}$ using two or three dimensional Sobol' points according to the procedure described in Sect. 2.
- QMC sampling based on the Hilbert curve (labelled as "hQMC"), which also selects n samples from the Hilbert sorted dataset of size $N = 2^{20}$.

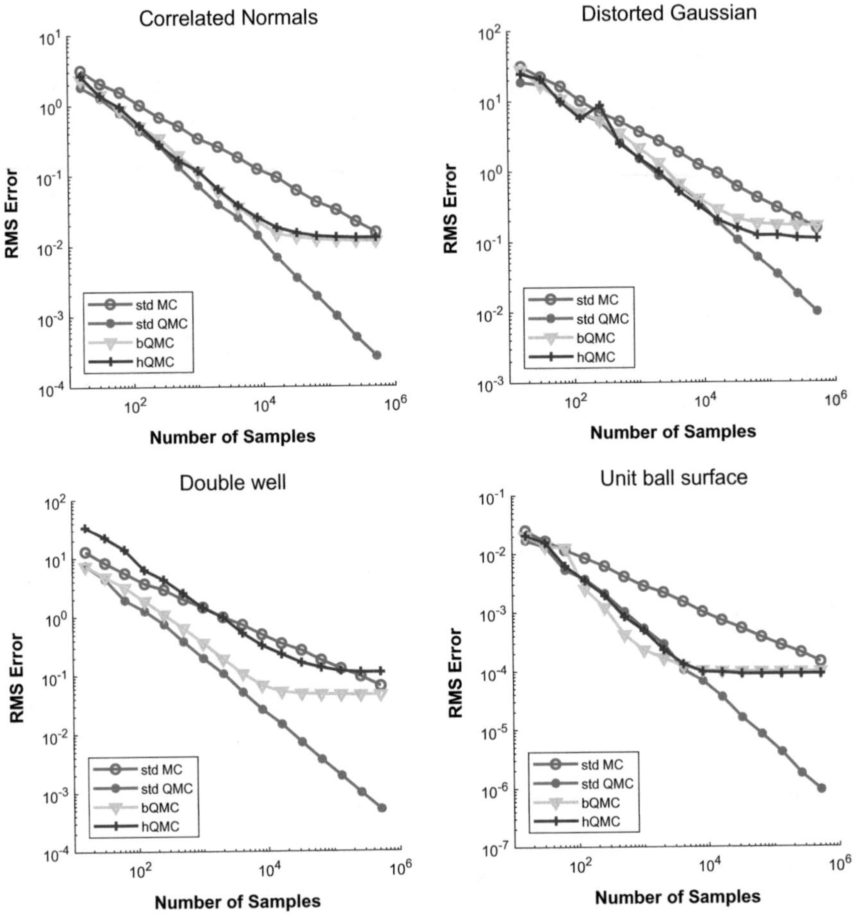

Fig. 2 Results for continuous test function

Under each scheme, the RMS (root mean squared) error is calculated for different numbers of samples $n = 2^m$, $m = 1, \ldots, 19$ and plotted for comparison. To obtain these values, all experiments are replicated $r = 256$ times, resulting in estimates $\hat{\mu}_{N,j}$ (dataset), $\hat{\mu}^{\text{I}}_{n,j}$ (bQMC) and $\hat{\mu}^{\text{II}}_{n,j}$ (hQMC), respectively, for $j = 1, \ldots, r$. The Sobol' points are randomized by the digital scrambling [9]. The "true" value μ is taken to be the average over the $r = 256$ estimates based on the datasets when estimating RMS errors for "bQMC" and "hQMC", while for "std MC" and "std QMC", we use the standard deviation as an estimate of the RMS error since they are unbiased.

Figure 2 presents the results for estimating the expectation of the continuous integrand (26) with datasets from the four distributions. Generally speaking, the proposed QMC sampling method based on recursive bisection has an accuracy which is comparable to the standard QMC method until it reaches a plateau which corresponds

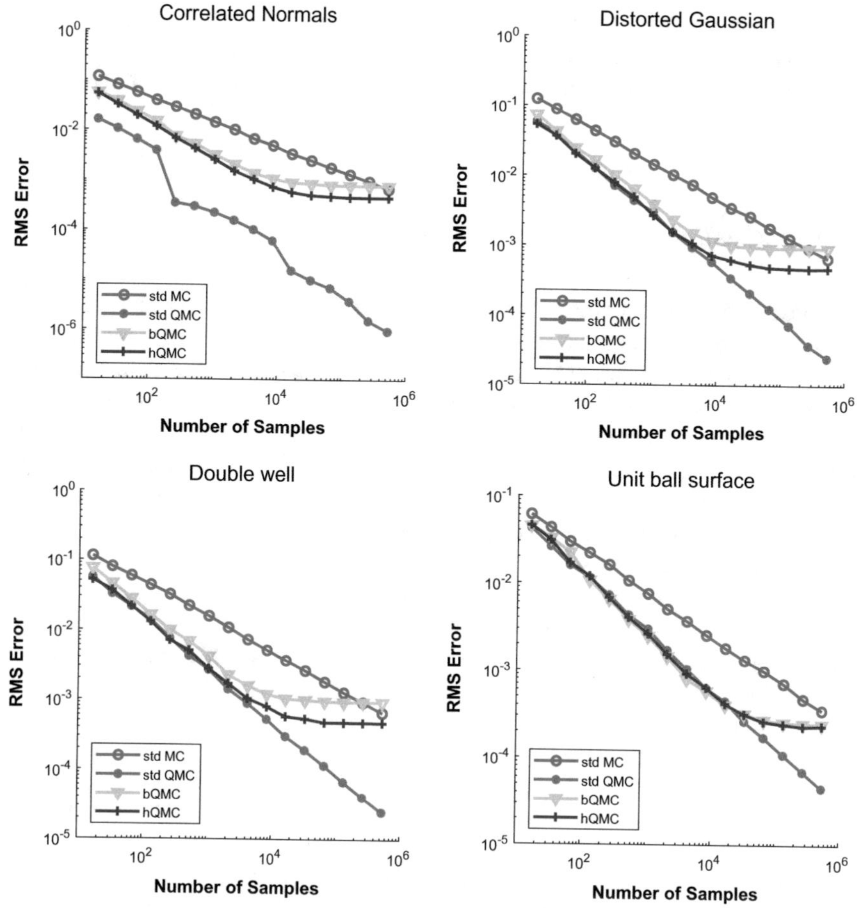

Fig. 3 Results for discontinuous test function

to the inherent error, i.e., the original sampling error. In particular, with the unit ball surface distribution, the QMC sampling method converges faster than the standard QMC method before it reaches the plateau. The mapping via a Hilbert curve gives similar results.

Figure 3 presents the results for the discontinuous integrand (27). Again the two QMC sampling methods result in a comparable accuracy to the standard QMC, with the Hilbert curve mapping scheme showing small advantage over the recursive bisection scheme.

5 Conclusions

The proposed QMC sampling technique provides a novel way in which to improve the accuracy of estimates obtained using empirical datasets, such as those produced by MCMC methods. Error analysis for the $d = 1$ case is provided, and the numerical results for the 2-dimensional and 3-dimensional cases are presented. These show that when applied to datasets coming from known distributions the effectiveness is very similar to the standard QMC method applied to those distributions. Furthermore, the technique is very easy to apply. In addition, the Hilbert space-filling curve can also be made use of for mapping QMC points to the dataset. Theoretical and empirical error analyses are also provided for this QMC sampling method based on the Hilbert curve. Numerical comparison results show that these two QMC sampling methods demonstrate similar performance, both are comparable to the standard QMC method. Supporting theory for the proposed recursive bisection based QMC sampling method will need to be the subject of future research.

6 Appendix

Lemma 1 *If Y_k, $k = 1, \ldots, N$, are the order statistics of N iid uniformly distributed random variables over* $[0, 1]$, *then for any $1 \le i < j \le N$, we have*

$$\mathbb{E}[Y_j - Y_i] = \frac{j - i}{N + 1},$$

and

$$\mathbb{E}[(Y_j - Y_i)^2] = \frac{(j - i)^2 + (j - i)}{(N + 1)(N + 2)}.$$

Proof Since $Y_k \sim Beta(k, N + 1 - k)$ for any $1 \le k \le N$, we have

$$\mathbb{E}[Y_k] = \frac{k}{N + 1}, \quad \mathbb{E}[Y_k^2] = \frac{k(k + 1)}{(N + 1)(N + 2)}.$$

By [7], we have

$$\mathbb{E}[Y_i Y_j] = \frac{i(j + 1)}{(N + 1)(N + 2)}.$$

We therefore obtain

$$
\begin{aligned}
\mathbb{E}[(Y_j - Y_i)^2] &= \mathbb{E}[Y_i^2] + \mathbb{E}[Y_j^2] - 2\mathbb{E}[Y_i Y_j] \\
&= \frac{i(i+1) + j(j+1)}{(N+1)(N+2)} - \frac{2i(j+1)}{(N+1)(N+2)} \\
&= \frac{(j-i)^2 + (j-i)}{(N+1)(N+2)}.
\end{aligned}
$$

□

Acknowledgements FX would like to acknowledge support from China Scholarship Council for her visit to Oxford University which led to this research. FX was also supported by the National Science Foundation of China (No. 71471100) and the National Key R&D Program of China (No. 2016QY02D0301). MBG would like to acknowledge the support of SAMSI (the Statistical and Applied Mathematical Sciences Institute under National Science Foundation Grant DMS-1127914) which hosted the workshop at which he first presented this work, and to thank various participants for their very helpful feedback. ZH was supported by the National Science Foundation of China (No. 71601189), the Fundamental Research Funds for the Central Universities (No. 2019MS106), and the Research Funds of South China University of Technology (No. D6191160).

References

1. Ades, A., Lu, G., Claxton, K.: Expected value of sample information calculations in medical decision modeling. Med. Decis. Mak. **24**(2), 207–227 (2004)
2. Barnard, S., Simon, H.: Fast multilevel implementation of recursive spectral bisection for partitioning unstructured problems. Concurr. Comput.: Pract. Exp. **6**(2), 101–117 (1994)
3. Dick, J., Kuo, F.Y., Sloan, I.H.: High-dimensional integration: the quasi-Monte Carlo way. Acta Numer. **22**, 133–288 (2013)
4. Gerber, M., Chopin, N.: Sequential quasi Monte Carlo. J. R. Stat. Soc.: Ser. B (Statistical Methodology) **77**(3), 509–579 (2015)
5. He, Z., Owen, A.B.: Extensible grids: uniform sampling on a space filling curve. J. R. Stat. Soc.: Ser. B (Statistical Methodology) **78**(4), 917–931 (2016)
6. Hendrickson, B., Leland, R.: A multi-level algorithm for partitioning graphs. Supercomputing **95**(28) (1995)
7. Moghadam, S., Pazira, H.: The relations among the order statistics of uniform distribution. Trends Appl. Sci. Res. **6**(7), 719–723 (2011)
8. Niederreiter, H.: Random Number Generation and Quasi-Monte Carlo Methods. SIAM, Philadelphia (1992)
9. Owen, A.B.: Scrambling Sobol and Niederreiter-Xing points. J. Complex. **14**(4), 466–489 (1998)
10. Sobol', I.M.: On the distribution of points in a cube and the approximate evaluation of integrals. Comput. Math. Math. Phys. **7**(4), 86–112 (1967)

Printed in the United States
by Baker & Taylor Publisher Services